JOURNAL OF THE TRANSPORTATION RESEARCH FORUM

Volume XXVIII Number 1 1987

TRANSPORTATION RESEARCH FORUM
In conjunction with

CANADIAN TRANSPORTATION RESEARCH FORUM

Published and Distributed by
Transportation Research Forum
1133 Fifteenth Street, N.W., Suite 1000
Washington, D.C. 20005
(202) 293-5913 TLX: 292046IMGUR

FROM THE EDITORS

Since the late 1970's, transportation has undergone substantial change. This includes the implementation of regulatory reform in the airline, motor carrier and railroad industries, fluctuations in energy costs, growing interest in private sector provision of public transportation services, a variety of technological innovations and advances in the ways that shippers purchase and use transportation services.

Along with these changes in transportation, have come changes in professional organizations, and publications oriented towards transportation researchers and practitioners. This first issue of the *Journal of the Transportation Research Forum* represents such a change.

The dynamic nature of transportation research and practice demands that information be provided in a timely fashion in a format that can be readily useable and understandable. Transportation researchers, managers, planners and policy makers require access to major contributions to the literature which may significantly affect research and practice in the future. They also need publications of consistent high quality. The goals of this journal are to provide information to the transportation community that meets these needs both in the present and in the future as transportation research and practice continues to change.

This first issue of the JTRF carries forward the volume numbers that had been used by its predecessor, the *Proceedings of the Transportation Research Forum*. This signals its link to the rich traditions of the Transportation Research Forum. The journal is dedicated to publishing articles which apply research and theory to practical problems. It strives to be more than just a publication outlet, but an important source of information that can be used by researchers and practitioners.

Papers appearing in this publication have all undergone a full peer review process. The Editorial Review Board has been drawn from academia, industry, and government. The use of this process signals TRF's determination to improve the quality of the publication, while preserving its applied flavor and emphasis on management, economics and policy.

Over the next few years, we will instituting a number of changes intended to make the JTRF a more useful medium for communicating ideas. Among these will be the publication of more than one issue each year, and the addition of comments, replies and book reviews. Other changes may be instituted as the times and needs dictate. We welcome your suggestions and comments.

Finally, we would be remiss if the Associate Editors, Editorial Review Board and the support of the TRF National Council were not acknowledged. Without their efforts, the JTRF simply would not happen.

Richard Beilock
Anthony M. Pagano
Co-General Editors

Editorial Review Board

Newton A. Graves
Yellow Freight Systems, Inc.

Curtis Grimm
University of Maryland

John F. Haifley
American Automobile Association

Frank E. Hawley
Western Highway Institute

Kevin H. Horn
University of North Florida

Mark Hornung
ALK Associates, Inc.

Jeffery F. Hudson
Transamerica Transportation Services

William Huneke
Association of American Railroads

Arthur C. Jacoby
Pennsylvania State University

Jeffrey Jordan
Georgia Experiment Station

Thomas G. Kenefick
American Trucking Associations, Inc.

Rajendra P. Khandekar
University of Colorado at Denver

Wesley R. Kriebel
US Department of Agriculture

Paul N. Kromberg
Association of American Railroads

Richard Lill
American Trucking Associations, Inc.

George List
Renneslear Polytechnic Institute

James B. Lloyd
The Barlett Co.

James MacDonald
US Department of Agriculture

Carl D. Martland
Massachusetts Institute of Technology

James R. Matthews
Southern Pacific Transportation Company

Douglas J. McKelvey
Federal Highway Administration

Claire E. McKnight
City College of New York

Tom Mentzer
Virginia Polytechnic Institute and State University

David F. Miller
Association of American Railroads

Joseph R. Morris
Transportation Research Board

Raymond Mundy
University of Tennessee

Bengt Muten
DNS Associates

H. Chris Nuckols, Jr.
Norfolk Southern Corp.

William C. Osswald
Challenge Consulting Company

Clinton V. Oster, Jr.
Indiana University

Tom O'Connor
DNS Associates

Robert Piper
Athearn Transportation Consultants

Terry Priest
Adolph Coors Company

James P. Rakowski
Memphis State University

John L. Reith
American Trucking Associations, Inc.

Philip B. Schary
Oregon State University

Joseph P. Schwieterman
United Airlines

Peter Shaw
California State University-Long Beach

Al Sloup
Union Pacific Railroad Corp.

Myron B. Slovin
Louisiana State University

Merrill A. Smith
Canadian Department of Agriculture

Thomas Stalberger
Emery Worldwide

Edward C. Sullivan
University of California-Berkeley

Kristofer Sweger
Manalytics, Inc.

Mark E. Thompson
Union Pacific Railroad Corp.

William B. Tye
Putnam, Hayes & Bartlett

Theodore Wallin
Syracuse University

Clyde K. Walter, Jr.
Iowa State University

Robert C. Waters
George Washington University

Dabney T. Waring, Jr.
Motor Common Carrier Associations

Porter K. Wheeler
Management Analysis Center, Inc.

Kenneth Williamson
James Madison University

Don Wood
San Francisco State University

Daniel L. Zink
North Dakota State University

TABLE OF CONTENTS

FROM THE EDITORS i

EDITORS ii

EDITORIAL REVIEW BOARD iii

NOVEMBER 1987
TRANSPORTATION RESEARCH FORUM

AIRLINE PERFORMANCE AND SAFETY

DEREGULATION'S IMPACT ON AIRLINE SAFETY
Clinton V. Oster, Jr. and C. Kurt Zorn 3

THE UNIT COSTS OF FAA AIR TRAFFIC CONTROL SERVICES
Richard Golaszewski 13

A PRELIMINARY ANALYSIS OF THE IMPACT OF THE PATCO STRIKE ON AIRLINE SAFETY
William A. Cunningham and Grant M. Davis 21

AIRLINE PERFORMANCE UNDER DEREGULATION: THE ROLE OF ON-TIME ARRIVALS
Ray M. Haynes and Martin T. Farris 25

AIRLINES' STRATEGIES IN HUB DOMINANCE: A STATISTICAL ANALYSIS
Atef Ghobrial, Jorge Sousa and Colleen Callahan 33

AN ANALYSIS OF ESSENTIAL AIR SERVICE TO THE SOUTHEASTERN U.S. SINCE THE AIRLINE DEREGULATION ACT OF 1978
David B. Vellenga and Daniel R. Vellenga 40

PUBLIC TRANSPORTATION

VALUE CAPTURE IN TRANSIT
W. Bruce Allen 50

SCALE ECONOMIES AMONG UNITED STATES PUBLIC BUS SYSTEMS
Michael Keough 58

LABOR-MANAGEMENT COOPERATION IN MASS TRANSIT: CONCEPTUAL AND APPLIED CONSIDERATIONS
Ken Jennings, Jay A. Smith and Earle C. Traynham 65

PRIVATIZING BUS TRANSIT: COST SAVINGS FROM COMPETITIVE CONTRACTING
Edward K. Morlok 72

THE DYNAMICS OF WEST COAST CONTAINER PORT COMPETITION
Robert L. Hanelt and Daniel S. Smith 82

ASSESSMENT OF CONTAINER-ON-BARGE SERVICE ON THE MISSISSIPPI RIVER
James Crew and Kevin Horn 92

THE INTERMODAL EQUIPMENT DISTRIBUTION MODEL
Kenneth C. K. Chih and Carl D. Van Dyke 97

AGRICULTURAL TRANSPORTATION

CARGO PREFERENCE: ITS IMPACT ON THE U.S. WHEAT SECTOR
Mechel Paggi, Stephen Fuller and Warren R. Grant 104

A STATEWIDE NETWORK MODEL FOR RATIONALIZING THE GRAIN AND LIVESTOCK TRANSPORTATION SYSTEM IN RURAL NEBRASKA
Azzeddine Azzam and Dean Linsenmeyer 112

THE EFFECT OF ELEVATOR COMPETITION UPON MARKET AREAS
Frank J. Dooley and Wesley W. Wilson 120

OPTIMAL CROSS-COUNTRY SHIPMENT OF PERISHABLE COMMODITIES: AN ECONOMIC APPROACH
Jeffery L. Jordan, S. E. Prussia, R. L. Shewfelt and W. A. Flanders 126

A ROLE OF TRANSPORTATION IN AGRICULTURAL PRODUCT TRADE
Won W. Koo 132

QUANTIFYING THE COST & UNCERTAINTIES OF TRANSPORTING GRAINS FROM FARMS TO ASSEMBLY POINTS
Neil Meyer and G. Sparks 140

IMPACT OF RAIL MERGERS ON KANSAS EXPORT WHEAT RAIL RATES AND COSTS
Kenneth Lemke and Michael W. Babcock 151

EFFECT OF DEREGULATION IN EXPORT-GRAIN RAIL RATES IN THE PLAINS AND CORN BELT
Stephen Fuller, David Bessler, James MacDonald and Michael Wohlgenant 160

USE OF RAIL RATE CONTRACTING ON VARYING TRANSPORTATION COMPETITIVE ENVIRONMENTS AND ITS IMPACT ON THE COMPETITIVE STRUCTURE OF GRAIN MERCHANDISING
Henry M. Bahn, Kenneth L. Casavant, Gene C. Griffin and Daniel L. Zink 168

TRANSPORTATION SAFETY

ECONOMIC PRESSURE, LONG DISTANCE TRUCKING, AND SAFETY
Richard Beilock and Russell B. Capelle, Jr. 177

WHAT IF EVERYTHING WE KNOW ABOUT SAFETY IS WRONG?
John Semmens and Dianne Kresich 187

MOTOR CARRIER COSTS

MOTOR CARRIER COST ANALYSIS
Russell S. Elliott 194

A MODEL OF FIRM BEHAVIOR AND COSTS IN EXEMPT MOTOR CARRIAGE
Wesley W. Wilson 200

LOGISTICAL MANAGEMENT

IDENTIFICATION OF PLANT RATIONALIZATION STRATEGIES USING OPTIMIZATION TECHNIQUES

George List 209

ANALYSIS OF THE TRANSPORTATION ELEMENTS OF ALTERNATIVE LOGISTICS CONCEPTS FOR DISPOSAL OF SPENT NUCLEAR FUEL

Michael S. Bronzini, David P. Middendorf and Robert E. Stammer, Jr. 221

ASSESSING THE IMPACT OF CONSOLIDATION ON INBOUND VENDOR TRAFFIC

Charles F. Myers, Joseph L. Fanelli and Dan C. Boger 230

POST STAGGERS RAILROAD COSTS

RAILROAD COST STRUCTURE—REVISITED

Anthony Barbera, Curtis M. Grimm, Kent A. Phillips and Leslie J. Selzer 237

THE COST STRUCTURE OF THE U.S. RAILROAD INDUSTRY UNDER DEREGULATION

Tenpao Lee and C. Phillip Baumel 245

RAIL OPERATIONS AND ANALYSIS

THE UP-MKT MERGER: OPERATING PLAN AND OPERATING SAVINGS

Michael B. Godfrey 254

A BRANCH AND BOUND PROCEDURE FOR OPTIMAL TRAIN DISPATCHING

Edwin R. Kraft 263

SHORT LINE OPERATION OF LIGHT DENSITY RAIL NETWORKS: ECONOMICS AND PUBLIC POLICY

Denver D. Tolliver, Frank J. Dooley and Daniel L. Zink 277

MANDATORY TRACKAGE RIGHTS: ECONOMIC CONSEQUENCES

Allen R. Ferguson 283

AN ASSESSMENT OF THE RAIL COMPETITIVE MOTOR CARRIER INDUSTRY (1977-1986)

K. Eric Wolfe, Vijaya Krishna and William F. Huneke 289

MOTOR CARRIER MANAGEMENT ISSUES IN A DEREGULATED ENVIRONMENT

MOBILITY BARRIERS IN THE MOTOR CARRIER INDUSTRY

Thomas M. Corsi and Curtis M. Grimm 302

SALES FORCE MANAGEMENT IN A DEREGULATED ENVIRONMENT: GENERAL FREIGHT CARRIERS

Paul R. Murphy, Jr. and Thomas M. Corsi 310

THE DEFINITION AND MEASUREMENT OF FINANCIAL FITNESS IN THE TRUCKING INDUSTRY

Garland Chow, Mark Vasina, Bahram Adrangi and Richard D. Gritta 319

TRANSPORT MARKETS AND FIRM BEHAVIOR: THE BACKHAUL PROBLEM

Wesley W. Wilson 325

HIGHWAY USER TAXES AND INFRASTRUCTURE IMPROVEMENTS: THE QUESTION OF BENEFITS
Edward A. Morash and Charles R. Enis 334

SEPARATE FREIGHT RATES FOR WEIGHT AND CUBE
Miles H. Sonstegaard 342

PIPELINE TRANSPORTATION

PETROLEUM PIPELINE DEREGULATION: A CRITICAL ANALYSIS OF THE PRIMARY ANALYTIC JUSTIFICATIONS FOR DEREGULATION
Leonard L. Coburn 346

STUDENT PAPER AWARDS

IMPROVING TRUCK TRANSPORTATION OF AGRICULTURAL COMMODITIES: A CASE STUDY OF FRESH PRODUCE SHIPPED FROM FLORIDA FARMERS' MARKETS
N. Devadoss 355

TOWARDS A LEVEL-OF-SERVICE MEASURE FOR MASS TRANSIT
Jonathan C. Levine 362

TL-PIGGY, THE INTERMODAL DECISION ASSISTANT
Paul C. Heymann 368

JUNE 1987 MEETING CANADIAN TRANSPORTATION RESEARCH FORUM

AN ALTERNATIVE MECHANISM FOR PAYMENT OF THE CROW SUBSIDY: THE WESTERN GRAIN STABILIZATION PROGRAM
Paul D. Earl and Edward W. Tyrchniewicz 383

PRODUCTIVITY IMPLICATIONS OF VEHICLE WEIGHT AND DIMENSION REGULATIONS
Gordon A. Sparks and Russell D. Neudorf 392

INSTITUTIONAL CONSTRAINTS ON THE MOVEMENT OF CANADIAN GRAIN TO EXPORT BY ALTERNATIVE ROUTES
A. G. Wilson and P. Miller 402

THE ECONOMIC COST OF TRANSPORTING GRAINS FROM FARM TO MARKET
N. Meyer and G. Sparks 408

TRANSIT SERVICE DISTRIBUTION—EVALUATING LEVEL OF SERVICE
M. A. Gallagher and M. C. Ircha 419

PORT STATE CONTROL: DEVELOPING A CANADIAN POSITION
Mary R. Brooks and Wade Eliott 428

BENEFIT COST ANALYSIS OF A TOLL HIGHWAY—BRITISH COLUMBIA'S COQUIHALLA
W. G. Waters II and Shane J. Meyers 434

FORECASTING FOR SUBSIDY PURPOSES—IDIOSYNCRACIES IN THE WESTERN GRAIN TRANSPORTATION ACT
E. Tyrchniewicz and E. R. Wansbutter 444

MODELLING THE EFFECT OF PASSENGER/FREIGHT INTERFERENCE ON RAILWAY COSTS
Vivian Wei 451

THE FEASIBILITY OF SUBMARINE TRANSPORTATION OF ARCTIC RESOURCES
D. Molyneux 460

APPENDICES: STATEMENT OF PURPOSE 473
HISTORY AND ORGANIZATION OF THE TRF 473
TRF ANNUAL MEETINGS 473
TRF COUNCIL ROSTER—1987 473
TRF CHAPTER PRESIDENTS—1987 474
1987 TRF ANNUAL MEETING PROGRAM COMMITTEE 475
TRF PAST PRESIDENTS 475
CTRF OFFICERS—1987 475

Journal of the
Transportation Research Forum

DAJ9760

Deregulation's Impact On Airline Safety

by Clinton V. Oster, Jr. and C. Kurt Zorn***

ABSTRACT

Concomitant with changes in the structure and operations of the domestic airline industry since deregulation have come three major concerns about airline deregulation's potential effects on safety. First, heightened competition accompanying deregulation was feared to result in less safe operations and maintenance by the established jet carriers themselves. Critics point to dramatic fines recently imposed on several carriers for safety violations, well-publicized crashes of several aircraft from all segments of the industry in 1985 and 1986, and the seemingly increasing number of "near misses" as evidence of deteriorating safety. Second, deregulation permitted widespread entry into the industry by entirely new jet airline companies some of whom were feared to lack the necessary operational experience of the established trunk and local service airlines. Third, as entry and exit requirements were eased, jet carriers withdrew service to many small communities and were replaced by commuter airlines leading to concern that safety would degrade because of the historically poorer safety record of the commuter airline industry.

This paper analyzes safety in the airline industry from both a cross-sectional and time-series perspective. Systematic differences in safety performance among various segments of the airline industry are examined in both the pre- and post-deregulation periods to determine whether the safety the industry or any of its subsets has deteriorated since deregulation. The analyses go beyond simple comparisons of aggregate fatality rates and examine both injuries and non-fatal accidents as well as the frequency of various types of accidents.

I. INTRODUCTION

The U.S. domestic airline industry has changed dramatically since passage of the Airline Deregulation Act of 1978 and the accompanying increase in service and fare competition. While deregulation's impact on safety raised some concern prior to passage of the act, such concerns have grown, fueled by a series of well-publicized accidents in 1985, recent fines assessed against large well-established carriers for maintenance and safety violations, and reports of growing numbers of "near misses" in congested airspace. While the most highly publicized accidents of 1985 and 1986 involved foreign or charter carriers or seemed clearly related to weather and were thus unrelated to deregulation, the heightened concern about aviation safety has nevertheless focused attention on the airline industry's adaptation to deregulation and the possible effect on safety. Specifically, four questions have arisen:

Have the economic pressures for cost reductions accompanying deregulation led to an increase in accidents that might likely result from shortcomings in maintenance procedures?

Has increased pressure on the Air Traffic Control System from the combination of deregulation-induced changes in routes and schedules coupled with the 1981 dismissal of the striking PATCO controllers reduced safety?

Has the growth of new entrant and former intrastate jet carriers degraded the safety of the industry?

Has the substitution of commuter airlines for the former trunk and local service airlines in small community service increased risk for travelers to and from these communities?

A. Airline Industry Changes Since Deregulation

Deregulation has, of course, brought forth a wide array of changes in the domestic airline industry (Meyer and Oster,, 1981; Bailey, Graham, and Kaplan, 1985). From the standpoint of airline safety, however, the operational changes induced by route and scheduling strategies as well as the impact on finances would seem of greatest importance. A dominant feature of post-deregulation route strategies has been the expansion and strengthening of hub and spoke route networks by all of the established jet carriers (Meyer and Oster, 1987; Morrison and Winston, 1986). Hub and spoke route networks, of course, rely on connections with more spokes providing more connecting opportunities so that large hub and spoke systems have service advantages over small systems, at least up to the point where the hub airport's landing and takeoff capacity becomes saturated with the accompanying rise in delays. The concentration of operations during hub and spoke operations can put intense pressure on the air traffic control system. Moreover, the reliance on connections increases difficulties for passengers and airlines caused by late flights and thus may increase pressure on airlines to maintain schedules.

A second important feature of post-deregulation route strategies has been the transfer of low-density service to and from small communities from local service airlines (and to a lesser extent trunks) to commuter carriers (Meyer and Oster, 1984). With deregulation's route entry and exit freedoms, the local service and trunk airlines' jets could be deployed far more profitably in denser and longer haul markets than in those serving small communities. This aircraft reassignment opened up replacement service opportunities for commuters and helps explain the commuters' rapid post-deregulation growth. The commuter airlines' market niche even before deregulation had been to provide service to smaller communities whenever and wherever the larger jet carriers pulled out in search of better opportunities elsewhere. Deregulation simply enhanced those opportunities and hastened this "rationalization" of small community airline service.

The accelerated transfer of small community service from jet airlines to commuters gave rise to the concern that travelers to and from small communities might face dramatically increased risks from the poorer safety performance of commuters. Widely cited statistics based on fatalities per passenger mile gave credence to the perception that commuters were less safe. Commuters were often portrayed as from 10 to 30 times more dangerous than the jet carriers they replaced. Although the differences between the risk faced by passengers on commuters and passengers on jet carriers have been shown to be not nearly as great as distance-based measures would imply, the safety of commuter airlines remains a concern of many travelers (Oster and Zorn, 1983a).

Another important post-deregulation development was the rapid growth of "new entrant" jet carriers—either entirely new companies, such as Midway or People Express, or those that emerged from the ranks of intrastate carriers previously confined by CAB regulation to operations within a single state or from the ranks of supplemental carriers confined by CAB regulation to nonscheduled charter operations (Meyer and Oster, 1984). With deregulation entry and fare freedoms, such carriers expanded rapidly, raising concerns whether maintenance and training procedures could keep up with such growth.

Finally, intense post-deregulation competition increased pressure for cost-cutting, raising fears that some carriers, particularly those in financial difficulty, may be devoting inadequate resources to maintenance. As evidence, it is sometimes pointed out that two of the carriers receiving large fines for maintenance violations were financially troubled Eastern and Pan Am. Although deregulation has probably helped rather than hurt carrier finances, the concerns about cost-cutting pressure and maintenance remain (Gomez-Ibanez, Oster, and Pickrell; Morrison and Winston; Meyer and Oster, 1987).

II. CHARACTERISTICS OF THE ANALYSIS

Because of the domestic airline industry's diversity, highly aggregated safety statistics can provide misleading comparisons of air travel safety with other modes or within the airline industry. Measures of air travel safety must adjust for varying stage lengths, different mixes of equipment, the variety of airports served, varied maintenance requirements and practices, and differing training and operational procedures. This analysis separates the industry into subsets of similar operating characteristics and examines the variation in safety rates for these segments over time. Some of the factors thought to affect safety within the industry can thus be isolated and examined.

The airline industry falls easily into two distinct groups—jet carriers and commuters. Jet carriers are defined as those airlines with aircraft fleets predominately powered by jet engines, whose operations are subject to Federal Aviation Administration (FAA) Part 121 operating regulations. Jet carriers are divided into the established trunk and local service airlines and the other jet carriers newly permitted under deregulation's freedoms to offer scheduled interstate passenger service. Trunk carriers are those that evolved from the 16 carriers certificated when economic regulation of the industry began in 1938. Local service carriers were certificated on an experimental basis in 1944 to extend air service to small communities and became permanently certificated in 1955. Over the years, some local service airlines have evolved to resemble small trunk operations but throughout most of the 1970–1985 period their route structures and operating policies still differed from the trunks in important ways (Meyer and Oster 1981). While the local service airlines have operated some propellor equipment, they are nevertheless considered as jet carriers for this analysis.

Commuter airlines, as defined in this study, operate predominately propellor-driven aircraft seating 60 or fewer passengers, are covered by FAA Part 135 regulations, and conduct relatively short-haul flights on low-density routes (Oster and Zorn, 1983b). Carriers operating fleets with mostly propellor-driven aircraft are considered commuters even if they acquired some jet aircraft after deregulation or operate under Part 121 regulations. This definition of commuters parallels the definition of regional airlines used by the Regional Airline Association.

A. Measures of Safety

This study focuses on the likelihood that a passenger will be killed or seriously injured while taking an airline flight, how and why that likelihood varies across segments of the industry, and how and why that likelihood may have changed over time. Thus, passenger fatalities, serious passenger injuries, accidents resulting in passenger fatalities, and accidents resulting in serious injuries to passengers are used as measures of risk in the numerator of safety rates. Accidents resulting in minor injuries or no injuries to passengers are also included. While fatal and serious injury accidents are obvious cause for immediate concern, frequent non-injury accidents may be a portent of problems that could lead to more severe accidents in the future. Moreover, the difference between an accident that kills many passengers and one in which passengers escape unharmed is often very small.

In transportation exposure to risk is often measured by either the distance traveled or the number of trips. For airline travel, most accidents occur during the takeoff and landing phases of flight, so the risk of being killed, injured, or involved in an accident is much more strongly related to the number of takeoffs and landings than to the distance traveled (Oster and Zorn, 1982). Long flights amass many passenger miles yet expose passengers to the same takeoff and landing risk as short flights. Thus, passenger fatality rates based on passenger departures or aircraft accident rates based on aircraft departures are far more useful for comparing risk in different segments of the industry than mileage-based measures.

National Transportation Safety Board (NTSB) Accident Briefs for Part 121 and Part 135 operations provide a rich source of accident data for both jet carriers and commuters including the number of fatalities and injuries for both passengers and crew, a description of the aircraft involved, details concerning the circumstances surrounding the accident, and probable cause and contributing factors. Data regarding passenger fatalities, serious injuries, fatal accidents, serious injury accidents, and other accidents were derived from this source.

Operations data for jet carriers, including aircraft departures and passenger enplanements, were obtained from Civil Aeronautics Board (CAB) Form 41. The U.S. Department of Transportation has continued collection of these data after the sunset of the CAB. Commuter operations data are more limited than jet carrier operations data. Passenger enplanements were obtained from CAB Form 298 for the 1970 through 1980 time period. Data for the 1981 through 1985 period were obtained from the annual reports of the Regional Airline Association. For those carriers operating under Part 121 but still classified as commuter carriers for purposes of this study, enplanement data were obtained from CAB Form 41 and Regional Airline Association annual reports.

Limitations on operations data require some compromises in constructing safety measures. For example, aircraft departure data are not available for many commuters. As a result, aircraft accident rates based on aircraft departures could not be constructed for commuters as they were for jet carriers. Similarly, there are no data on passenger departures for either jet carriers or commuters so enplanement data must act as a reasonable proxy. A passenger is counted as a single enplanement for each flight boarded but if the flight involves multiple stops a passenger is still counted as one enplanement even though that passenger is subjected to additional risk with each takeoff and landing. For flights with intermediate stops, enplanements are less than passenger departures so that passenger fatality rates based on passenger departures would be lower than those based on enplanements. The differences are not large, however. For example, 88.9 percent of commuter flights were either nonstop or one-stop in 1978 with the percentage increasing to 89.1 percent in 1983. Similarly, for trunk and local service airlines, 71.3 percent of flights were either nonstop or one-stop in 1978 with the percentage increasing to 81.2 percent in 1983. Thus, any bias introduced by using enplanements is small and, more importantly, constant over time for commuters where the enplanement-based measures are relied upon most heavily. For jet carriers, the bias may be somewhat larger but has declined since deregulation. In any event, the bias is less of an issue for jet carriers because aircraft accident measures can be used.

To focus on the effects of deregulation, the analysis was further limited to domestic scheduled passenger service with international, caribbean, and charter operations and accidents excluded. Operations in the non-continental states (Alaska and Hawaii) were also excluded because commuters, in particular, serve a much different role in these areas and frequently face a much harsher operating environment making comparisons across segments of the industry problematic (Oster and Zorn, 1982).

III. AIRLINE SAFETY SINCE DEREGULATION

In the wake of deregulation, competition among carriers, enhanced flexibility in setting route strategies, and shifts in service responsibilities among segments of the industry have kindled rising public concern over safety within the industry. Table 1 suggests, however, that this concern about worsening safety as a result of deregulation may well be unwarranted no matter what measures of airline safety are used. During the post-regulation period (1979-1985), passenger fatality, passenger injury, and aircraft accident rates were consistently lower for both jet carriers and commuters than had been the case in the pre-deregulation period (1970-1978). Moreover, as the table indicates, these differences were generally statistically significant. Only the passenger fatality rate for trunk carriers did not drop and it was essentially unchanged.

A comparison of commuter and jet carrier rates in the table also suggests that commuters are, indeed, less safe in the aggregate than jet carriers. As will be discussed later, however, several other factors must also be considered in assessing the impact of commuter substitution on airline safety.

A. Economic Pressures and Airline Maintenance

The record high fines levied against American, Pan Am, and Eastern coupled with several post-deregulation commuter accidents in which equipment failure was a clear contributing factor have combined to raise the question of whether the increased competitive pressures accompanying deregulation have caused the airlines to cut corners in maintenance practices in an attempt to lower costs (Air Transport World, 1987). One approach to assessing the impact of deregulation on maintenance practices is to examine the frequency of accidents in which inadequate maintenance might have been a contributing factor. Thus for the purposes of this study, accidents for jet carriers (and commuter airlines) were put into one of eight categories according to their "primary contributing factor":

1) Equipment Failure
2) Seatbelt Not Fastened
3) Weather
4) Pilot Error
5) Air Traffic Control
6) Ground Crew Error
7) General Aviation
8) Other

Many, indeed most, accidents result from several factors coming together at the same time. Determining a primary contributing factor for multiple-cause accidents is a difficult task requiring admittedly subjective judgements. The goal, however, was not to determine the "main" cause, but rather to apply a consistent set of criteria that would allow a longitudinal analysis of the relative frequency of types of accidents. Thus before presenting the results of the analyses, it is important to be clear about what sorts of accidents were included in each category.

1. Equipment Failure. If the events that culminated in the accident were precipitated by mechanical or electrical malfunction in the aircraft, the accident was considered as equipment failure even if there were other important factors. In some cases, such as improper installation of a part or failure to detect cracks or corrosion, more meticulous maintenance might have prevented the accident. In other cases, such as failure of a tire that did not show

TABLE 1
AIRLINE SAFETY RATES
DOMESTIC SCHEDULED SERVICE
1970-1985

Industry Segment	1970-1985	1970-1978	1979-1985
Passenger fatalities per one million enplanements:			
Trunks	0.37	0.36	0.38
Local Service	0.29	0.65	0.01**
Other Jet Carriers	n.a.	n.a.	0.42
Commuters	1.66	2.65	1.27*
Serious passenger injuries per one million enplanements:			
Trunks	0.11	0.19	0.03**
Local Service	0.25	0.52	0.04*
Other Jet Carriers	n.a.	n.a.	0.05
Commuters	0.93	1.62	0.66*
Fatal aircraft accidents per one million aircraft departures:			
Trunks	0.31	0.38	0.22
Local Service	0.45	0.62	0.22*
Other Jet Carriers	n.a.	n.a.	0.90
Commuters	n.a.	n.a.	n.a.
Serious aircraft accidents per one million aircraft departures:			
Trunks	1.78	2.42	0.86**
Local Service	0.86	0.92	0.77
Other Jet Carriers	n.a.	n.a.	0.90
Commuters	n.a.	n.a.	n.a.
Minor aircraft accidents per one million aircraft departures:			
Trunks	2.34	3.08	1.30**
Local Service	2.13	2.54	1.53*
Other Jet Carriers	n.a.	n.a.	1.51
Commuters	n.a.	n.a.	n.a.

n.a. - not available due to data limitations

* - lower than the 1970-78 rate at the 90 percent confidence level
** - lower than the 1970-78 rate at the 95 percent confidence level

Source: Derived from computer printout of Part 121 and 135 operation accident briefs provided by the National Transportation Safety Board; U.S. Civil Aeronautics Board, Forms 41 and 298; and Regional Airline Association, Annual Reports, various years.

excessive wear, maintenance practices could not reasonably be blamed. No attempt wwas made to distinguish whether inadequate maintenance was at fault. Rather all equipment failure accidents were treated the same on the assumption that equipment failures where maintenance was not a factor would be randomly distributed over time so that any time trend in equipment failures resulted from changing maintenance practices or improved aircraft design.

2. Seatbelt Not Fastened. A surprisingly common source of serious passenger injury is passengers not having their seatbelts fastened when turbulence is encountered despite the seatbelt sign in the cabin being illuminated and passengers being asked to fasten their seatbelts. If the seatbelt sign was illuminated in sufficient time for passengers to return to their seats and fasten their seatbelts prior to the injury, the accidents were placed in this category, If, however, the turbulence was unexpected and the seatbelt sign was not on, the accident was placed in the "weather" category discussed below.

3. Weather. Weather is a factor in many airline accidents but is frequently not regarded by the National Transportation Safety Board as the cause of the accident. Rather, the NTSB often determines the cause to be the cockpit crew not responding properly

to weather conditions. This approach taken for this study differs from the NTSB approach in that accidents were rarely classified as "pilot error" if weather was a significant contributing factor. Thus an accident was usually classified as weather even if the pilot taking precisely the right action at precisely the right time could have prevented the accident. There were, however, some exceptions to this approach. If an aircraft took off under weather conditions that led to a takeoff accident, the accident was considered pilot error. If an aircraft was unable to stop after landing on a slick runway, it was considered a weather accident, unless the pilot landed excessively long on the runway, in which case it was considered pilot error. Similarly, if a pilot attempted to land when the airport was below minimums and an alternate airport was available, it was judged pilot error.

4. Pilot Error. In addition to the weather-related examples cited above, an accident was classified as pilot error only in those cases where the error appeared undeniable. Examples of pilot error include attempting a landing without lowering the landing gear, taxiing into a stationary object, or running out of fuel because of failure to switch fuel tanks during flight.

5. Air Traffic Control. An accident was classified as air traffic control when normal action by a controller could have prevented the accident such as ATC failing to advise a flight of unsafe weather.

6. Ground Crew. Accidents attributed to ground crew error included such cases as a service truck colliding with a parked aircraft and a ground crew member walking into a propellor while delivering a message to the pilot of an aircraft about to depart.

7. General Aviation. General aviation accidents included those where the accident would not have occurred without general aviation aircraft operating in the area. It was not necessary that the general aviation aircraft was judged at fault, just that its presence contributed to the accident.

8. Other. Accidents not falling into one of the above categories were classified as "other." Such accidents included a wide array of causes ranging from a passenger tripping over a baby bottle in the aisle to injuries sustained during an evacuation due to a bomb threat. There were also two accidents in late 1984 and seven accidents in 1985 that were classified as other because NTSB accident investigations had not yet released findings of the causes of the accidents.

B. Jet Carriers Before and After Deregulation

Table 2 shows the total accident rates for jet carriers broken down among the eight principal contributing factors. The two columns on the left show the

TABLE 2

JET CARRIERS
TOTAL ACCIDENT RATE BY PRINCIPAL CONTRIBUTING FACTOR
DOMESTIC SCHEDULED SERVICE

Aircraft Accidents per One Million Aircraft Departures

Contributing Factor	Combined Trunk and Local Service Carriers 1970-78	Combined Trunk and Local Service Carriers 1979-85	Trunk 1979-85	Local Service 1979-85	Other Jet 1979-85
Equipment Failure	1.49	0.43**	0.59	0.11++	0.30
Seatbelt Not Fastened	1.49	0.68**	0.70	0.66	0.60
Weather	0.82	0.33**	0.32	0.33	0.60
Pilot Error	0.54	0.21**	0.16	0.33	0.60+
Air Traffic Control	0.26	0.11*	0.05	0.22	0.60++
Ground Crew Error	0.23	0.11	0.10	0.11	0.30
General Aviation	0.10	0.04	0.00	0.11+	0.00
Other	0.39	0.50	0.43	0.66	0.30
Total	5.28	2.42**	2.38	2.52	3.31

* - lower than the 1970-78 rate at the 90 percent confidence level
** - lower than the 1970-78 rate at the 95 percent confidence level

\+ - different from the Trunk rate at the 90 percent confidence level
++ - different from the Trunk rate at the 95 percent confidence level

Source: Derived from computer printout of Part 121 operation accident briefs provided by the National Transportation Safety Board; U.S. Civil Aeronautics Board, Form 41.

combined accident rates for the trunk and local service carriers for both the pre-deregulation and post-deregulation periods. As can be seen in these columns, the rates declined for seven of the eight types of accidents; the only exception was the "other" category.

Perhaps the most striking observation is that the rate for equipment failure related accidents in the post-deregulation period is seen to be less than one third of the pre-deregulation rate. Had deregulation induced shortcuts in aircraft maintenance, the rate of equipment failures would be expected to have increased. The sharp decrease suggests that, at least through 1985, deregulation has not caused widespread maintenance deficiencies.

The accident rates are also down for both air traffic control and for general aviation although the rates are low both before and after deregulation. Part of the post-deregulation period was prior to the PATCO strike (in August 1981), but much of the period was after the striking controllers had been dismissed. In terms of frequency of accidents, there is no evidence that the air traffic control system has functioned less safely after deregulation than before. Similarly the rate of general aviation related accidents has not increased since deregulation. Thus despite the added pressure on the air traffic control system and on the airspace surrounding large airports because of the periodic concentration of flights from intensified hub and spoke operations, the rates for the types of accidents that might have been expected to increase have instead gone down.

Deregulation has also increased pressure on airline labor for less restrictive work rules. Pilots and cabin attendants fly more hours per month and ground crews perform a wider variety of tasks than prior to deregulation (Meyer and Oster, 1987). Despite these pressures, the rates for pilot error and grow crew error have both declined. The rate for accidents where the seatbelt was not fastened has also declined, suggesting that if anything cabin attendants have been more rather than less effective in making sure that passengers fasten their seatbelts when turbulence is expected. Again the types of accidents that might have been expected to have increased because of labor practice changes induced by deregulation have been less frequent following deregulation than they were before.

The rapid growth and route expansion of the new entrant jet carriers has been a major development following deregulation and has raised concerns about these carriers' impacts on safety. One fear was that it would be difficult to achieve adequate training and maintenance practices in the face of rapid growth and a second fear, raised by the Air Florida accident in Washington in 1982, was that expansion might confront these carriers with weather and operating conditions for which they were not fully prepared. As table 1 indicated, however, these other jet carriers have a safety record very similar to other segments of the jet carrier industry. Of the five measures in table 1, only the rate of fatal accidents per one million departures is higher than comparable rates for trunk airlines. In essence, the other jet carriers experienced three such accidents rather than the one that might have been expected based on the trunk and local service airlines' experience.

The three columns on the right in table 2 break down the accident rates by primary contributing factor for all three segments of the jet industry. The other jet carriers' experiences are seen to be quite similar to the trunks and local service airlines. The accidents are evenly distributed among contributing factors with no factors appearing disproportionately represented. While tests for statistical significance indicate that the rates for two of the categories are higher than for the trunks, these same tests indicate that the total accident rate is not higher than the trunk rate. In sum, while these other jet carriers may have experienced a somewhat higher rate of fatality-causing accidents, their safety experience does not appear to differ markedly from the other segments of the jet industry.

This "other jet carrier" category also includes carriers that might not be considered new entrant airlines in the manner of carriers such as Air Florida, Midway or People Express. World Airways, for example, was new to scheduled passenger service but had considerable experience in charter service with the type of aircraft (DC-10) involved in their post-deregulation accident in Boston. Similarly, carriers such as Southwest and PSA had many years of experience flying jet aircraft in intrastate service prior to deregulation. Thus, the experience of these other jet carriers reported in tables 1 and 2 does not support the contention that the new entrant jet carriers have degraded the safety of the airline industry.

C. Commuter Substitution in Small Community Service

As commuter carriers replace withdrawing jet carriers, small community residents have been concerned about degradation in their air transportation safety. Table 1 suggested that such substitutions may indeed increase risk to travelers to and from small communities. During the entire 1970-1985 period passengers were more than four times as likely to be killed when boarding a commuter flight than one operated by trunk or local service carriers. More importantly, in the post-deregulation period commuter air travel remained more risky than jet air travel, by more than a factor of three.

However, table 3 demonstrates that safety rates within the commuter industry decline with increasing carrier size. In the 1970-1978 period, the top twenty carriers, in terms of passenger enplanements, were more than 4.5 times safer than the rest of the top fifty carriers (carriers 21 through 50 based on size) in terms of passenger fatalities per one million enplanements, and more than 19 times safer than the rest of the industry, the relatively small carriers who make up the non-top fifty category. In terms of passenger serious injuries per one million enplanements, the top twenty carriers were almost 2.5 times safer than the rest of the industry. The top twenty carriers provide most of the replacement service for withdrawing jet carriers.

The implications for commuter replacement of different safety records for commuters and jet carriers must also consider that service patterns typically change when commuters replace jet carriers. For example, between 1978 and 1986 there were 55 communities which lost jet carrier service and had commuter carriers provide replacement service. Commuter replacement service in the 60 city-pair markets serving these cities resulted in an increase in the average frequency of weekday flights of from 2.88 in 1978 to 6.29 in 1986.[1] In addition, the

TABLE 3
COMMUTER CARRIERS
PASSENGER FATALITY AND SERIOUS INJURY RATES
DOMESTIC SCHEDULED SERVICE
1970-78 VERSUS 1979-85

	1970-78	1979-85
Total Industry:		
Passenger Fatalities per One Million Enplanements	2.65	1.27*
Serious Passenger Injuries per One Million Enplanements	1.62	0.66*
Top Twenty Carriers:		
Passenger Fatalities per One Million Enplanements	0.69	0.67
Serious Passenger Injuries per One Million Enplanements	0.93	0.21
Rest of the Top Fifty:		
Passenger Fatalities per One Million Enplanements	3.27	1.21+
Serious Passenger Injuries per One Million Enplanements	2.27	0.48
Rest of the Industry:		
Passenger Fatalities per One Million Enplanements	13.32	4.08**++
Serious Passenger Injuries per One Million Enplanements	4.30	3.00++

* - lower than the 1970-78 rate at the 90 percent confidence level
** - lower than the 1970-78 rate at the 95 percent confidence level

\+ - higher than the Top Twenty rate at the 90 percent confidence level
++ - higher than the Top Twenty rate at the 95 percent confidence level

Source: Derived from computer printout of Part 121 and 135 operation accident briefs provided by the National Transportation Safety Board; U.S. Civil Aeronautics Board, Forms 41 and 298; and Regional Airline Association, Annual Reports, various years.

average number of intermediate stops on flights between the small community and its nearest major hub decreased from 0.59 in 1978 to 0.30 in 1986. Thus on average, service has improved in these communities where commuters have provided substitute service for jets because passengers have more frequent connections with fewer intermediate stops to hubs where they can link up with jet carrier service.

Because the replacement service includes fewer intermediate stops, the relative safety rates of commuters and jet carriers should be adjusted to reflect the additional risk from extra takeoffs and landings on the jet flights. The average number of takeoffs and landings associated with jet service to these small communities was 1.59 (the original takeoff and subsequent landing and 0.59 takeoffs and landings at intermediate stops), but only 1.30 for commuters serving the same cities. Thus, to get comparable safety measures that take into account this improvement in service and its impact on safety, the jet carrier rate should be inflated by 22.3 percent and compared with the rates for the larger commuter carriers.

Even with this adjustment a simple comparison between jet carriers and aggregate commuter safety measures is not appropriate. In 55 of the 60 city-pair markets, the replacement commuter carrier was either a top twenty carrier or a top fifty carrier in a code-sharing marketing alliance with a major jet carrier (Oster and Pickrell, 1986). As can be seen by comparing tables 3 and 1, the safety record of the top twenty commuters, even without adjusting for fewer intermediate steps, is about the same as the pre-deregulation record of the local service airlines that had served 45 of the 60 markets prior to 1978.

A final consideration is that the typical result of increased commuter service frequency in small communities in the wake of jet carrier exit was increased ridership (Meyer and Oster 1981). The added convenience offered by commuters attracts passengers who had previously opted for the automobile. Assuming that the average commuter flight length of between 120 and 130 miles substitutes for a more circuitous auto trip to the same hub airport of about 150 miles and further assuming that average auto occupancy for such trip is about 1.5 (slightly below the average for all intercity auto trips), the comparable fatality rate associated with motor vehicle trips is somewhere between 1.9 and 2.3 passenger fatalities per one million passenger trips. Comparing this motor vehicle fatality rate to that of the top fifty commuters indicates that travel is safer for those passengers switching from auto to commuter as a result of improved commuter service. Thus commuter service to small communities may actually improve the transportation safety record once these appropriate adjustments have been made.

D. Commuters Before and After Deregulation

Part of the improvement in the safety record of commuter carriers that is evident in both tables 1 and 3 can be attributed, at least in part, to revisions in commuter safety regulations that occurred in 1978. In response to concerns about commuter airline safety the FAA tightened regulations in three areas. First, pilot qualifications were upgraded to parallel pilot qualifications for established jet carriers. Second, maintenance requirements were improved to include more detailed and extensive procedures for all types of classes of aircraft used by Part 135 operators. Third, training programs of Part 135 operators were enhanced (Oster and Zorn 1983b). The safest portion of the industry, the top twenty commuters, experienced the smallest improvement in their safety records, most likely because they had previously instituted many of the procedures that the FAA mandated in 1978 and thus the safety revisions had little additional effect. On the other hand, the dramatic improvement in safety for the rest of the top fifty and the rest of the industry are most likely the direct result of the 1978 safety revisions.

Table 4 breaks down the passenger fatality rate for the entire commuter industry into the eight principal contributing factor categories for both the 1970-1978 pre-deregulation period and the

TABLE 4

COMMUTER CARRIERS
PASSENGER FATALITY RATE BY PRINCIPAL CONTRIBUTING FACTOR
DOMESTIC SCHEDULED SERVICE
1970-78 VERSUS 1979-85

Passenger Fatalities per One Million Enplanements

Contributing Factor	All Commuters 1970-78	All Commuters 1979-85	Top Twenty 1979-85	Rest of Top Fifty 1979-85	Rest of Industry 1979-85
Equipment Failure	1.07	0.35*	0.11	0.52++	1.15++
Seatbelt Not Fastened	0.00	0.00	0.00	0.00	0.00
Weather	0.61	0.27	0.15	0.03	1.21++
Pilot Error	0.46	0.05	0.07	0.00	0.06
Air Traffic Control	0.04	0.00	0.00	0.00	0.00
Ground Crew Error	0.00	0.01	0.01	0.00	0.00
General Aviation	0.22	0.20	0.00	0.45++	0.64++
Other	0.24	0.39	0.32	0.21	1.02++
Total	2.65	1.27*	0.67	1.21+	4.08++

* - lower than the 1970-78 rate at the 90 percent confidence level

\+ - higher than the Top Twenty rate at the 90 percent confidence level
++ - higher than the Top Twenty rate at the 95 percent confidence level

Source: Derived from computer printout of Part 121 and 135 operation accident briefs provided by the National Transportation Safety Board; U.S. Civil Aeronautics Board, Forms 41 and 298; and Regional Airline Association, Annual Reports, various years.

1979-1985 post-deregulation period. As the two columns on the left indicate, most of the improvement in the commuter airline passenger fatality rate between the two periods comes from a reduction in fatalities due to three reasons: equipment failure, pilot error, and weather. Given the thrust of many of the 1978 FAA safety regulation revisions, this pattern of fatality rate reduction might have been expected. The tightened maintenance procedures have probably contributed to the reduction in accidents involving equipment failure. The added pilot certification requirements coupled with added recurrent training have likely contributed to both reduced pilot error and reduced weather-related fatalities. Weather-related fatalities may have also been reduced by tightened regulations regarding navigational equipment for IFR flights.

The three columns on the right of table 4 show the 1979-1985 passenger fatality rate by type of accident for three segments of the commuter industry: the top 20 carriers, the rest of the top 50 (those carriers ranked in size from 21 through 50), and the rest of the commuter industry (those carriers ranked 51 and below). The most striking pattern in the table is the increasing rate of equipment failure related fatalities as the carrier size decreases. Even with the tightened FAA regulations, the larger carriers may still have more effective maintenance practices than the smallest carriers.

One hypothesis as to why carrier size might contribute to lower rates of equipment failure is that a larger carrier can afford greater specialization in maintenance and a more extensive inventory of parts. Another possible reason for the higher rate of equipment failure among the smallest carriers is that they are least likely to operate a fleet of all turbine-engine aircraft and more likely to operate small piston-engine aircraft seating nine or fewer passengers. Piston engines have far more moving parts and are more complex to maintain, particularly it appears for small carriers. A mixed fleet of both piston and turbine engines aggravates the maintenance specialization and spare parts inventory, again particularly for smaller carriers. Moreover, many of the 1978 FAA revisions did not apply to operators of aircraft seating nine or fewer passengers so that these aircraft may not be subject to some of the same tighter maintenance requirements (General Accounting Office, 1984). Indeed, even aircraft seating 10 to 19 passengers had somewhat less extensive requirements than for larger aircraft, a difference that may have contributed to some accidents (NTSB, 1986; Aviation Week and Space Technology, 1986).

IV. SUMMARY AND CONCLUSIONS

For both commuters and jet carriers, the rate of accidents related to equipment failure are substantially lower in the post-deregulation period suggesting that at least through 1985 there is no evidence of increased accidents from worsening maintenance practices. Despite increased pressure on the air traffic control system because of increased emphasis on hub and spoke operations and the reduction in the number of air traffic controllers, the rates of air traffic control related accidents are lower in the post-deregulation period than prior to deregulation. While the accident rates, based on limited operations, appear higher for the new entrant and former charter and intrastate jet carriers than for the established trunk and local service carriers, the differences are small and do not suggest a serious degradation of safety.

While the commuter industry as a whole has a safety record clearly inferior to that of the established jet carriers, three factors suggest that there has not been a substantial reduction in safety for travelers to and from small communities as a result of the transition to commuter service. First, the larger commuter carriers have a safety record nearly comparable to that of the local service jet carriers before deregulation and these larger carriers have provided most of the replacement service. Second, commuter flights between small communities and large hubs typically operate with fewer intermediate stops than the jet flights they replace thus exposing passengers to fewer takeoffs and landings. Third, many of the commuter passengers had previously used auto for their trips because of the relatively low frequency of the prior jet flights thus shifting those passengers to a safer mode of travel. In sum, airline safety has been improving for decades because of such factors as improved aircraft design and construction, improved pilot training particularly with the use of cockpit simulators, and better air traffic control procedures. Airline deregulation does not appear to have interfered witth this continuing trend of improvement.

ENDNOTES

*Transportation Research Center, Indiana University, Bloomington, IN.

**School of Public and Environmental Affairs, Indiana University, Bloomington, IN.

1. These 55 communities were identified in U.S. Civil Aeronautics Board Staff Study, *Report on Airline Service, Fares, Traffic, Load Factors, and Market Shares, Service Status on September 1, 1984* (Washington: Government Printing Office, December 1984). The service levels for 1978 and 1986 were derived from the July 1 editions of the *Official Airline Guide*, North American Edition, for 1978 and 1986.

REFERENCES

Air Transport World (1987), "FAA points finger at airline maintenance systems," February.

Aviation Week and Space Technology (1986), "NTSB Recommends Improvements in Commuter Safety Standards," October 6.

Elizabeth E. Bailey, David R. Graham, and Daniel P. Kaplan (1985), *Deregulating the Airlines*, The MIT Press, Cambridge.

General Accounting Office (1984), "Safety Standards On Small Passenger Aircraft—With Nine Or Fewer Seats—Are Significantly Less Stringent Than On Larger Aircraft," GAO/RCED-84-2, Washington.

Jose A. Gomez-Ibanez, Clinton V. Oster, Jr., and Don H. Pickrell (1983), "Airline Deregulation: What's Behind the Recent Losses?" *Journal of Policy Analysis and Management*, Volume 3, Number 1.

John R. Meyer and Clinton V. Oster, Jr. (eds.) (1981), *Airline Deregulation: The Early Experience,* Auburn House Publishing Company, Boston.

John R. Meyer and Clinton V. Oster, Jr. (1984) *Deregulation and the New Airline Entrepreneurs,* The MIT Press, Cambridge.

John R. Meyer and Clinton V. Oster, Jr. (1987), *Deregulation and the Future of Intercity Passenger Travel,* The MIT Press, Cambridge.

Steven Morrison and Clifford Winston (1986), *The Economic Effects of Airline Deregulation,* The Brookings Institution, Washington.

National Transportation Safety Board (1986), "Safety Recommendation A-86-98 through -118," Washington.

Clinton V. Oster, Jr. and Don H. Pickrell (1986), "Marketing Alliances and Competitive Strategy in the Airline Industry," *The Logistics and Transportation Review,* Volume 22, Number 4.

Clinton V. Oster, Jr. and C. Kurt Zorn (1982), "Commuter Airline Safety," U.S. Department of Transportation, Washington.

Clinton V. Oster, Jr. and C. Kurt Zorn (1983a), "Airline Deregulation, Commuter Safety, and Regional Air Transportation," *Growth and Change,* Volume 14, Number 3.

Clinton V. Oster, Jr. and C. Kurt Zorn (1983b), "Deregulation and Commuter Airline Safety," *Journal of Air Law and Commerce,* Volume 49, Number 2.

The Unit Costs of FAA Air Traffic Control Services

*by Richard Golaszewski**

I. INTRODUCTION AND SUMMARY

This paper presents estimates of the unit costs of providing air traffic control (ATC) services by the Federal Aviation Administration (FAA). It is based on work conducted for the (FAA) Cost Allocation Study (CAS) by Gellman Research Associates and its subcontractor, Ernst and Whinney. However, the opinions expressed in this paper are those of author.

The costs of ATC services are very relevant to the current discussions surrounding alternative organizational forms for FAA. The Air Transport Association has proposed that some FAA functions be organized into a federal corporation. The Reason Foundation has called for a shift of ATC functions to the private sector. Both proposals assert that such changes will promote efficiency and increased responsiveness to users. A federal or private corporation may choose to price its services so that users receive the type and quantity of services they are willing to pay for. Also there is an apparent scarcity of ATC resources (at least at some times) which is manifest by the increasing level of delays. Peak-level pricing is one means of alleviating such delays.

The cost and price data developed in this paper provide an indication of the likely level of average fees for ATC services necessary to operate an organization such as FAA. The data which underlie this paper provide a basis for the development of alternative pricing structures for ATC services.

FAA provides ATC services to various classes of aviation users, such as air carriers, commuter air carriers, general aviation and the military. It provides these at four types of facilities:

- Air route traffic control centers (ARTCCs);
- Terminal radar control areas (TRACONs);
- Air traffic control towers (ATCTs);
- Flight service stations (FSSs).

The CAS used statistical cost functions which measured the cost of providing services to specific classes of users at each of the above facility types. All costs and outputs were measured on an site-specific basis. As noted by Kanafani and Chaudhuri (page 12), cross-sectional analyses of cost-output relationships across facilities of different sizes provide an estimate of long-run costs. Site costs included ATC labor (including fringe benefits, supervision, etc.), ATC maintenance and communications. Available data did not permit the inclusion of capital costs in the cost functions.

A key step in estimating any economic function is testing for alternative functional forms. In many cases, economic theory provides a priori expectations about which functional form is consistent with market expectations. However, in the case at hand, there were no theoretic reasons for selecting a particular form. Linear, log-linear and quadratic forms were tested. In all cases, the linear form provided produced the most robust results.

The summary results of the marginal costs of ATC services are shown in Table A. It can be seen that ARTCC marginal costs for all user types, except military, are of the same magnitude. (When identical marginal costs are reported across user types, the analysis did not support separate estimation of costs for these user types.) At TRACONs, all user types impose identical marginal costs except for general aviation. The inclusion of seconds and overs in the TRACON activity measure reflects provision of ATC services to users who did not land at the primary airport. At ATCTs, air carrier marginal costs are almost twice as high for those of the military and four to five times higher than those for commuter air carriers and general aviation.

FSSs provides services primarily to general aviation. In this case, marginal costs were estimated for the various service types. They range from \$3.87 for an air contact to \$13.68 for opening and closing a VFR flight plan. These estimates compare favorably with a time and motion analysis of the labor effort included in producing services at this facility type. *(Automation Evaluation Study).*

The estimated marginal costs by user class reflect, in part, the facility establishment criteria (FAA Order 7031.2C) used by FAA for determining when a specific type of ATC facility or equipment should be installed. These criteria as applied in FAA benefit-cost studies give more weight to air carrier, commuter and military activity in facility establishment. Also, the estimates are not inconsistent with the identified effort in providing services to the various classes of users.

It is clear that FAA's costs of providing air traffic control services also may differ based on capacity utilization or other site-specific characteristics. The marginal costs estimated in this paper represent an average across each facility type. However, if pricing ATC services is contemplated, economic efficiency would dictate that optimal prices reflect variations in cost and variations in users' demand elasticities.

In order to provide order of magnitude estimates of prices for ATC services, other ATC system-related costs were identified to determine the proportion of total costs accounted for by marginal costs given the service units provided in 1985. These other costs include: site joint costs, ATC maintenance costs not allocated to sites, research and development, facilities and equipment and agency overhead. Marginal costs represent from 20 percent to 40 percent of total costs, which means that total costs are from two and one-half to five times as large as the estimated marginal costs in Table A. If the FAA ATC functions were placed in a federal or private corporation and it

TABLE A
SUMMARY RESULTS
MARGINAL COSTS OF ATC SERVICES

Facility Type	Activity Measure	Air Carrier	Commuter	General Aviation	Military
		Estimated Marginal Costs			
ARTCC	Total Handles[1]	$13.93	$13.93	$12.63	$21.30
TRACON	Operations, Seconds and Overs	$12.80	$12.80	$3.44	$12.80
ATCT	Operations[2]	$7.91	$1.86	$1.44	$4.45
FSS	Various (costs same for all users)	Pilot Briefs	IFRFP[3]	VFRFP[3]	Air Contacts
		$6.86	$6.86	$13.68	$3.87

[1]Handles - Two times IFR departures plus overs

[2]Equation has dummy variable of - $85,133 for Level I VFR towers.

[3]IFRFP = Instrument flight rules flight plan.
VFRFP = Visual flight rules flight plan.

charged for individual services, these data provides a basis to derive order of magnitude estimates of the average user fees necessary to operate the entity.

II. METHODOLOGY

The first step in developing cost functions for FAA ATC facilities was to establish site-level databases of input costs and services provided. (See Volume 5: *Econometric Cost Functions,* FAA CAS.) Conceptually, input costs should include all costs—such as labor, equipment, outside services and buildings—used to produce services. However, available data limited the study to the following cost categories:

Site ATC Labor—includes wages, fringe benefits and other miscellaneous costs contained in the FAA Consolidated Report of Obligations. Total staffing by GS grade at each site was obtained as of the end of FY1985 from FAA's Personnel Management Information System and was costed out at Step 5 for each GS grade. Staffing costs were summed across all sites and divided into total obligations for each facility type to produce an average mark-up on direct labor to allow for fringe benefits and other labor-related costs. The direct mark-up was applied to the direct labor costs at each ATC site. (This mark-up represents largely labor related costs such as fringe benefits, overtime and shift differential pay. As such, it is applied to GS labor rate to measure the total cost of a labor hour to FAA.)

Site Maintenance Labor—includes wages and a mark-up for fringe benefits and supervision. These costs were developed from the standard labor-hours to maintain equipment as contained in the FAA Facility Master File (FMF). These hours were divided into the total personnel costs for ATC systems maintenance to yield a fully-allocated cost per hour. This was applied to standard hours of maintenance for the equipment used at each site as identified in the FMF.

Site Communication Costs—includes leased circuit costs for each ATC site. Circuit costs were assigned to a facility based on the proportion of use of that circuit by the facility. Data were obtained from the DOT Transportation Systems Center FAACIS database.

The most significant cost element which could not be included in the site data was the allocation of capital costs of air traffic control equipment. Although the FMF provides information on the types of ATC equipment installed at each facility, it was not possible to determine the historical or replacement cost of such capital equipment. FAA expenses capital equipment in the year purchased. Depreciation-based historical cost records are not maintained.

A major issue is what type of cost function is estimated from cross-sectional analyses of FAA facilities. A prior study (Kanafani and Chaudhuri) notes that cross-sectional analyses of different sizes of facilities produces a long-run marginal cost curve. The rationale for this is that this represents a path of changing capacity for a specific facility type.

A related assumption is that the costs of existing ATC equipment is sunk. That is, there is no value of this equipment in alternative uses. Therefore, no return is assumed to this past investment. Moreover, it is assumed that the costs of existing equipment are not avoidable in either the long- or short-run.

Another concept which is relevant to the cost of producing ATC services is that of joint versus separable costs. A cost is separable if it can be traced directly to the provision of a specific service for a specific user. Separable costs are avoidable in that they vanish if the service to which they are attributed is no longer produced.

Joint costs are those costs that are incurred as a result of the production of more than one service. In general, joint costs will not disappear if an operating site is relieved of the responsibility of producing any given single service or of providing service to a specific user group. Joint (or non separable) costs are avoidable only if no services are produced by a facility.

A joint cost function expresses a relationship between costs and some measure of joint output or productivity. For example, let Ci and Qi represent, respectively, total direct costs and a measure of joint output for the ith facility. The cost function can be written:

$$C_i = f(Q_i)$$

If the cost function is known, the marginal cost of the jth service at the ith facility, MC(Qij) can be computed as:

$$MC(Q_{ij}) = \partial C_i / \partial Q_{ij}$$

were Qij is the output of the jth service at the ith facility.

Given MC(Qij), the avoidable costs of producing the jth service at the ith facility, V(Qij), can be computed as:

$$V(Q_{ij}) = \int_0^{Q_{ij}} MC(Q_{ij})\,d\,Q_{ij}$$

In reality, where costs can change by user groups, the relevant marginal cost is MC(Qijk) where k represent the kth user group. This is deleted from the derivation above for ease of exposition. The avoidable costs of producing the jth service for the entire system can be computed by summing V(Qij) over all facilities of a particular type.

Several tests of the properties of the cost functions for each facility type were conducted. They examined the following issues:

- Appropriate functional form (linear, non-linear or quadratic);
- Differences in the costs of specific service among user groups;
- Differences in costs among services produced;
- Multicollinearity.

The results of tests of the cost functions are reported below in the discussion of each type of site cost.

FAA maintains databases of site-specific activity measures for each facility type. These output measures have been used in the econometric cost functions described below. Data are available in an annual publication *FAA Air Traffic Activity.* The data measure activity related to air carrier, air taxi, general aviation and military users. Data on commuter air carrier operations at ATCT and TRACON facilities were developed through special analyses of the Official Airline Guide (see: Appendix A of Volume 5, *Cost Allocation Study* for a description of the method used). In this analysis, air taxi use of ATCT and TRACON facilities was divided between commuter and general aviation users. Adjustments to ARTCC activity counts to calculate commuter air carrier use of this facility type also were made (ibid).

Marginal costs were converted to total unit costs of services by examining the other ATC-related costs allocated to the four user types in the FAA CAS (Volume 1, Chapter 2).

III. THE MARGINAL COSTS OF ATC SERVICES

The sections below report the estimated marginal costs of ATC services for each facility type. The discussions report the average annual cost of each facility type along with average annual activity for each user group. They also show the specifications of cost function and the resultant estimates. The explanatory power of the equations and significance of coefficients are reported along with a discussion of the properties of the cost functions.

1. *Air Route Traffic Control Centers*

Air route traffic control centers (ARTCCs) provide separation services for aircraft flying under instrument flight procedures. In addition, they provide approach control service at some airports where terminal radar control (TRACON) facilities do not exist. ARTCC controllers provide separation among aircraft using computer-enhanced secondary radar returns. Extensive air-ground and ground-ground coordination communications are used. Work load is measured in IFR aircraft departures and aircraft overs. These are measured for four specific user types—air carrier, air taxi, general aviation and military. Total workload is counted as aircraft handled which is defined as two times IFR departures plus overs.

There are twenty-two ARTCCs in the FAA system. The mean values for ARTCC facilities are as follows:

• Total cost	$22.7 million
• Annual air carrier handles	682,507
• Annual commuter handles	115,816
• Annual GA handles	416,190
• Annual military handles	220,354

Statistical tests rejected the log-linear and quadratic forms of cost functions. Analysis of variance indicated that the cost coefficients for air carrier and commuter handles did not differ.

The following linear regression equation was estimated

$$\text{Cost} = \text{constant} + A_1\,[ACH + CH] + A2\ GAH + A3\ MH$$

where ACH = Air Carrier Handle
CH = Commuter Handle
GAH = General Aviation Handle
MH = Military Handle

It produced the following estimates:

$$\text{Cost} = \$4.1 \text{ million} + \underset{(3.83)}{\$13.93} [\text{ACH} + \text{CH}] + \underset{(4.31)}{\$12.63} \text{ GAH} + \underset{(4.55)}{\$21.30} \text{ MH}$$

$R2 = 0.896$

The t-values are shown in parentheses underneath each coefficient.

Visual inspection of the data indicated a linear relation between costs and output. No indications of multicollinearity were found.

The estimated costs for ARTCCs are a function of the variation in costs and output at each facility. The estimated marginal costs for three of the four groups are equivalent. It is hypothesized that the relatively high military cost coefficient may result from additional staff, equipment and communications required to handle military operations and to separate civil aircraft from military Special Use Airspace.

2. *Flight Service Stations*

The 306 flight service stations (FSSs) in the FAA system* provide weather briefings, IFR and VFR flight plan filing, and inflight flight advisory service to aircraft as the principal measured outputs. Myriad other services are provided but not measured (see Chapter 2, *Flight Service Station Privatization Evaluation Report* for a description of these other services). The mean values of cost and activity for 1985 used to estimate the marginal cost of FSS services are as follows:

- Annual total cost—$769,899
- Pilot briefs—50,043
- IFR flight plans—21,167
- VFR flight plans—6,312
- Air contacts—27,919

Analysis of variance indicated that FSS marginal costs did not differ among user groups. (These facilities serve principally general aviation users; most air carrier and some commuter IFR flight plans are filed automatically into ARTCC computers.) A time and motion study conducted by FAA indicated that IFR flight plans and pilot briefs required equivalent effort (Automation Evaluation Study). These outputs were combined for estimation of marginal costs. Tests for non-linearity proved to be negative, with both the log-linear and quadratic forms having less desirable properties than the linear form.

An equation of the following form was estimated

$$\text{Cost} = \text{constant} + \text{A1 [IFRFP + PB]} + \text{A2 VFRFP} + \text{A3 QAC}$$

where:

PB = Pilot briefing
IFRFP = IFR flight plan
VFRFP = VFR flight plan
QAC = Aircraft contacted

It produced the following results:

$$\text{Cost} = \$89{,}919 + \underset{(44.96)}{\$6.86} [\text{PB} + \text{IFRFP}] + \underset{(6.82)}{\$13.68} \text{ VFRFP} + \underset{(4.62)}{\$3.87} \text{ QAC}$$

$R2 = 0.929$

The t-values are shown in parentheses below the marginal cost coefficients. It should be noted that the relative costs of service are consistent with the FAA time and motion study which found that VFR flight plans entail twice as much and air contacts one-half as much staff time as do IFR flight plans or pilot briefs.

Multicollinearity existed between IFR flight plans and pilot briefs; however, this was overcome by specifying them together. Multicollinearity also existed between VFR flight plans and air contacts; however, alternative specifications were inconsistent with the FAA time and motion study.

3. *Terminal Radar Control Facilities*

Terminal radar approach control (TRACON) facilities provide radar separation to aircraft arriving in and departing from terminal areas. Most TRACONs include an air traffic control tower (ATCT) at the primary airport and typically provide radar services to aircraft at surrounding airports which may or may not have an ATCT. Some independent TRACONs serve more than one primary airport. In these cases TRACON costs were allocated among multiple primary airports based on activity to create pseudo-facilities that produced services equivalent to the majority of TRACONs. The estimated costs include both the radar approach control and ATCT outputs.

TRACON activity is measured as total operations (itinerant and local) at the primary airport, secondary operations (approach control at outlying airports) and overs (radar services to overflights through the terminal airspace). Analysis of variance indicated that the costs of producing operations, seconds and overs closely corresponded with one another. Also analyses of variance indicated that there were two groups—general aviation and all other—for which the cost of producing services differed.

The 1985 average cost and activity data for the 185 TRACONs are as follows:

Total cost—$2.75 million
- Air carrier TSOs—60,432
- Commuter TSOs—31,967
- GA TSOs—126,951
- Military TSOs—21,494

The marginal cost function was specified as:

$$\text{Cost} = \text{constant} + A_1 [\text{AC} + \text{COM} + \text{MIL}] \text{ TSO} + A_2\text{GATSO}$$

where

ACTSO = Air carrier operations, seconds and overs
COMTSO = Commuter operations, seconds and overs
MILTSO = Military operations, seconds and overs
GATSO = General aviation operations, seconds and overs

The results of the estimation produced the following:

$$\text{Cost} = \$850{,}312 + \underset{(28.38)}{\$12.80} [\text{AC} + \text{COM} + \text{MIL}] \text{ TSO} + \underset{(4.18)}{\$3.44} \text{ GATSO}$$

$R2 = 0.867$

The t-statistics are shown in parentheses below the marginal cost coefficients. Visual inspection of the data showed a linear relationship between activity and costs. There were no indications of multicollinearity.

General aviation activity exhibits a marginal cost that is less than one-third of that of other user groups. This may be accounted for by the fact that general aviation activity is given a lower weight in FAA benefit-cost criteria for facility establishment. Also, the operational characteristics of general aviation activity at TRACONs may make it less costly for FAA to provide services to these users.

4. *Air Traffic Control Towers*

Air traffic control towers (ATCTs) provide control during landing, take-off and taxi of aircraft at airports. The ATCTs in this section are those without primary radar approach control services. ATCTs without radar exist at airports that serve mostly general aviation traffic. There are five levels of FAA ATCTs based on different activity levels. In this analysis, a variable was used to differentiate Level 1 VFR ATCTs from all others. Operations (itinerant and local) are the FAA measure of output at ATCTs.

The 1985 data for average costs and activity at the 203 ATCTs are as follows:

- Annual total cost—$551,260
- Air carrier operations—1,281
- Commuter operations—5,733
- Military operations—4,106
- General aviation operations—110,869
- Percent Level 1 facilities—43%

Alternative functional forms were tested, including log-linear and quadratic forms; the results indicated that the linear cost function had superior results. Analysis of variance did not provide a basis for combining any user types.

The cost equation was specified as follows:

$$\text{Cost} = \text{constant} + A_1\,AC + A_2\,COM + A_3\,MIL + A_4\,GA + \text{Level 1}$$

where:

AC = Air carrier operations
COM = Commuter operations
MIL = Military operations
GA = General aviation operations
Level 1 = 1 if Level 1 ATCT; 0 if otherwise

The results are as follows:

$$\text{Cost} = \$386{,}623 + \underset{(2.53)}{\$7.91\,AC} + \underset{(1.76)}{\$1.86\,COM} + \underset{(3.83)}{\$4.45\,MIL} + \underset{(9.60)}{\$1.44\,GA}$$

$$\text{Level 1} = \underset{(3.49)}{-\$85{,}133} \qquad R2 = 0.555$$

The t-values are shown in parentheses below the marginal cost and Level 1 coefficients. There were no indications of multi collinearity under alternative tests of the independent variables. The R2 is lower in the ATCT regression than for the other facility types. Also, the coefficient for commuter operations is only marginally significant. The size of the constant term relative to average total facility costs indicates that costs do not vary greatly with activity. These are FAA's smallest facilities in terms of staffing and other input costs. Fixed costs may be a large part of total costs.

IV. FROM MARGINAL COSTS TO TOTAL COSTS

The marginal costs estimated above do not cover the entire cost of providing ATC services. That is, if these marginal costs were multiplied by average output during 1985, fixed costs (represented by the constant terms) in the econometric cost functions would not be covered. In addition, ATC system research and development, ATC system management and some portion of ATC system maintenance and communications could not be assigned to sites. As noted above, capital costs (buildings and ATC equipment) were not included in site costs and are not represented in the marginal cost coefficients. All these items and general agency overhead must be added to the marginal cost of service to produce total costs. This section makes these adjustments to produce estimates of 1985 ATC service costs. While any allocation of these fixed cost categories will be somewhat arbitrary (Thomas 1969 and Thomas 1974), the FAA CAS used three general methods to allocate joint and fixed costs.

The proportion of site operations and maintenance costs represented by marginal costs, using the 1985 cost and activity measures, is shown in Table 1. These result from the application of marginal costs for each facility type to the number of service units consumed by each user group (FAA CAS, Volume 1). The site joint costs represent the constant terms in the econometric cost functions multiplied by the number of facilities of each type. Joint costs were allocated to users in the CAS using Ramsey Pricing which minimizes the distortion is the quantity of services demanded by user groups. One subset of the general aviation population—piston airplane users—was assumed to have a demand elasticity twice that of all other users.

There are a number of other costs incurred by FAA in operating the ATC system beyond those which could be allocated to sites in the CAS. The cost categories and general bases for allocation are as follows:

- ATC Equipment Maintenance not allocated to sites- allocated using Ramsey Pricing.
- Facilities and Equipment (F&E) expenditures are largely for ATC system equipment. They are allocated first to user groups based on the avoidable cost concept. Joint costs allocated using Ramsey Pricing. Allocation factors were developed using five years of F&E expenditure data.
- Research and Development expenditures are largely for ATC system equipment. They are allocated first to users based on the avoidable cost concept. Joint costs allocated using Ramsey Pricing.
- General Overhead—assigned to users with cross allocation or multiple reporting method. The avoidable cost concept assigns costs to a group or groups of users by examining whether they could be avoided if that user group (or groups) did not exist. The cross allocation or multiple reporting method recognizes the interdependence of support units in a complex organization such as FAA. This method is explained in detail in Volume 1 of the FAA CAS.

TABLE 1
MARGINAL AND JOINT COSTS OF ATC SERVICES BY FACILITY AND USER GROUP
($ MILLIONS—1985)

Facility Type	Air Carrier	Commuter	General Aviation	Military
ARTCC Marginal Costs	$216.1	$38.1	$119.4	$103.8
ARTCC Joint Costs	$41.6	$7.3	$19.6	$20.0
ARTCC Total Costs	$257.7	$45.4	$139.0	$123.8
FSS Marginal Costs	$9.9	$12.0	$167.6	$22.9
FSS Joint Costs	$1.9	$2.3	$18.6	$4.3
FSS Total Costs	$11.8	$14.2	$186.2	$27.3
TRACON Marginal Costs	$149.8	$69.6	$87.5	$55.1
TRACON Joint Costs	$76.2	$35.4	$24.3	$28.0
TRACON Total Costs	$226.0	$105.0	$111.8	$83.1
ATCT Marginal Costs	$2.1	$2.3	$33.7	$3.7
ATCT Joint Costs	$7.3	$7.8	$43.0	$12.8
ATCT Total Costs	$9.4	$10.1	$76.7	$16.6
Marginal Cost Proportion	74.9%	69.8%	79.5%	74.0%

Details may not add because of rounding.

Source: FAA CAS Volume 1 Tables 2.3.3.2, 2.3.4.1, 2.3.5.1 and 2.3.6.1.

Table 2 shows the costs allocated to each user group employing the above methods. ATC system facilities and equipment expenditures represent the single largest category. Ideally, these should be an input to the site econometric models. The data in Table 2 indicate that marginal costs represent from 20.7 percent (air carriers) to 38.8 percent (general aviation) of total ATC system costs. Thus, marginal costs would have to be increased by from 2.6 to 4.8 times to equal total costs.

There have been a number of proposals to turn over operation of all or part of the FAA ATC system to a federal corporation or to the private sector. One purpose of such an action would be to promote efficiency. If this were the case, such an entity may price its services so that users consider the resources required to provide them in making consumption decisions. In order to be self-sustaining and replenish its capital equipment, this entity would have price its services at a level to recover total costs. Table 3 shows order of magnitude estimates of the total unit costs of providing services to the four user groups.

It must be recognized that some users may consume a bundle of FAA services on each flight. For example, a single air carrier flight may involve an ARTCC IFR departure, several ARTCC overs and two TRACON operations. (Air carriers typically do not use FSS services). Some general aviation flights may not use any FAA services—a local flight at a non-towered airport or a cross country flight between two non-towered airports where the pilot obtains its required weather briefing from a private sector provider.

Imposition of fees based on total costs is likely to alter the quantity of services demanded. Those user groups with the highest demand elasticities are likely to reduce their consumption the most. Some

TABLE 2
DEVELOPMENT OF ATC SYSTEM COSTS
($ MILLIONS—1985)

Cost Category	Air Carrier	Commuter	General Aviation	Military
Site Marginal Costs	$377.9	$122.0	$408.2	$185.5
Site Joint Costs	$127.0	$52.8	$105.5	$65.1
ATC Equipment Maintenance not allocated to sites	$155.0	$50.0	$107.1	$76.1
Facilities and Equipment	$753.6	$215.3	$226.8	$150.8
Research and Development	$177.3	$56.1	$20.3	$11.8
General Overhead	$239.0	$82.0	$185.2	$121.3
Estimated ATC System Costs	$1829.8	$578.2	$1053.1	$610.1
Marginal Costs as a proportion of total costs	20.7%	21.1%	38.8%	30.4%
Total Cost Factor	4.83	4.74	2.58	3.29

Sources 1. Table 1

2. Volume 1 FAA CAS Tables 2.7.2, 2.4.1, 2.5.1 and 2.8.3.1

TABLE 3
APPROXIMATE UNIT TOTAL
COSTS OF ATC SERVICES
($ 1985)

		Unit Total Costs			
Facility Type	Output Type	Air Carrier	Commuter	General Aviation	Military
ARTCC	IFR Departure	$134.56	$132.06	$65.17	$140.15
	Over	$67.28	$66.02	$32.59	$70.08
TRACON	Operation, Second or Over	$61.82	$60.67	$8.88	$42.11
ATCT	Operation	$38.21	$8.82	$3.72	$14.64
FSS	Pilot Brief	$33.13	$32.52	$17.70	$22.57
	IFR Flight Plan	$33.13	$32.52	$17.70	$22.57
	VFR Flight Plan	$66.07	$64.84	$35.29	$45.00
	Air Contact	$18.69	$18.34	$9.98	$12.73

Source: Marginal cost of service times total cost factors reported in Table 2.

pilots may elect to forego important safety services with adverse consequences (Flight Service Station Privatization Evaluation Report, Chapter 7). Some users are required to consume FAA safety services by regulation and could not legally fly without certain services.

Optimal prices may require consideration of whether the true cost of service varies by the intensity of use or, at specific FAA facility sites. The total costs reported in this paper should be viewed as order of magnitude estimates of ATC service costs. The ability to produce more accurate estimates is limited by available data.

The prices shown in Table 3 are relevant to the ongoing discussions of alternative organizational forms to provide FAA services. As noted in the FAA CAS, the revenues from air carrier passenger ticket tax approximately recover the resources consumed in providing services to this user group. Thus, moving to system of fee for service would not alter the aggregate costs imposed on this industry. However, to the extent the ticket tax is not a good surrogate for resource consumption, user fees may promote a more efficient use of resources.

In the case of general aviation the situation is quite different. The level of revenues from the fuel tax on general aviation fuel accounts for less than 20 percent of resources consumed in providing services to this sector (FAA CAS). In position of fees of the magnitude shown in Table 3 would represent a significant change in the operating economics of this sector. In fact, fees of this magnitude would alter considerably the level of general aviation activity at FAA facilities. Such fees may be well beyond the ability to pay of some general aviation users.

Thus, a move to alternative organizational forms for ATC services portends an important decision for policymakers. In order to achieve the suggested improvements in efficiency, user fees should be imposed. However, user fees of a magnitude sufficient to assure revenue adequacy for the organization, would represent a pronounced increase in the current costs of general aviation flying. Therefore, either other users (air carriers) or some other group (the taxpayer) would have to subsidize general aviation in order for it to operate in a fee for service ATC system. Otherwise, general aviation would have to significantly reduce its use of ATC services. To the extent general aviation requires such services for safety or mobility of the population, there may be important public policy considerations in decisions regarding fee levels for ATC services.

ENDNOTES

* Gellman Research Associates, Inc.

1 The number of FSSs is being reduced under an automation program. When fully implemented, only 61 automated FSSs will remain.

A Preliminary Analysis of the Impact of the PATCO Strike on Airline Safety

by William A. Cunningham and Grant M. Davis***

ABSTRACT

Contentious debate has evolved recently over the issue of airline safety. A recent survey conducted by the DALLAS NEWS indicated that, in the opinion of airline pilots, safety has deteriorated.

One of the principal reasons stated in the Dallas survey, and offered by others who perceive a decline in safety, is the air traffic controllers' strike of 1981. Another related factor contributing to decreasing safety is the increasing number of flights offered since deregulation, especially in the commuter airline segment.

This brief study investigates the issue of safety from two perspectives. The first is the effect, if any, of the controllers' strike and concurrent relation in personnel. The second is concerned with the data on safety by type of service (for instance, general; computer; air taxi; and trunk). To analyze the data ANOVA is performed on data before and after the strike to attempt to discern any effects on accidents and fatalities that can be attributed to the strike. The analysis, moreover, delineates any categorical variations or changes in safety due to increasing frequency of service.

While there is a plethora of literature relating to the issue of safety, this study provides some preliminary statistical insight into this highly controversial issue.

I. INTRODUCTION

On October 22, 1981, the Federal Labor Relations Authority removed the Professional Air Traffic Controllers Organization's (PATCO) right to represent the controllers.[1] This action was taken in direct response to the August 3 illegal strike called by PATCO President, Robert E. Poli (BUSINESS WEEK, August, 1981).

Of the approximately 15,000 PATCO members, employed in 1981, more than 12,000 ignored the return-to-work order issued by the Reagan Administration (WALL STREET JOURNAL, August, 1981). In November of 1981, there remained only 10,000 functioning controllers (supervisors, military controllers, nonstrikers, and new trainees) compared with a prestrike number of 16,000 (BUSINESS WEEK, November, 1981).

Irrespective of this dramatic change in the controller employment situation, in 1982 the FAA began plans to substantially alter their safety regulations in order to provide airlines additional operational flexibility in complying with these regulations. Designated by the FAA as "regulation by objective", the new regulations would offer broad rules as to how airlines should operate replacing the "how to" rules then in existence (FLIGHT, 1982). The actual method or technique of complying with the new rules would be up to the individual airlines who in turn would be required to submit to the FAA a detailed statement of how compliance with the new regulations would be achieved. Putative benefits would be the fostering of the development of new safety methods and a simultaneous reduction in the volume of regulations (JOURNAL OF COMMERCE, 1983). This new approach was not universally accepted by aviation interests as leading to improvements in safety.

While there are almost infinite number of aviation safety issues; these are but two, albeit major, changes that occurred that produced substantial debate regarding the airline safety issue. The first impact, that of the PATCO strike, generated the most controversy. Numerous aviation practitioners in the industry (pilots, new and old controllers, bureaucrats, among others) and outside experts as well who question whether or not the level of airline safety is adequate. The quintessential question regarding the effect of deregulation on safety has, and continues to be explored and is contentious.

Although a myriad of opinions exists concerning the level of safety and the impact of low rates and a declining number of controllers, there is a notable absence of empirical verification of the effect, if any, there has been on airline safety. The balance of this paper attempts to provide rudimentary evidence regarding safety experience in aviation and the diminished number of air controllers.

To empirically ascertain whether or not the PATCO strike and subsequent reduction in the number of air traffic controllers has contributed to a decline in airline safety, the data on safety will be examined dichotomously and compared over the pre-strike period of 1975–1980. The second time period will be the 1981–1984 post-strike years. Although the strike did not occur until the latter part of 1981, there had been several incidents of work slow-downs are growing dissention prior to the strike; ergo, 1981 should be included in the post-strike time period. This rather straight-forward methodology concentrates on aviation safety with a full compliment of controllers, and an environment with a reduced number.

II. COLLECTED DATA

The data examined in the analysis was collected from the FAA statistical handbook for the years 1975–1980, a publication containing data compiled from the National Transportation Safety Board (accidents, and miles flown for air carriers), and the FAA itself for data on general aviation. The FAA Safety

Board basically categorizes air carrier accidents into one of four groups. These groupings depend on which Federal Air Regulation were in use when the accident flights occurred. The groupings are (1) large airlines in scheduled service under Part 121 of the regulations; (2) commuter carriers in scheduled service under Part 135; (3) "on-demand" air taxis in unscheduled operations under Part 135; and (4) general aviation, a category for all other civil flying. These data do not include flights made by military aircraft.

As previously indicated, the analysis is essentially dichotomous. First, the data is examined to ascertain if the level of safety has changed over the two select time periods. Data on total accidents, fatal accidents, and total fatalities is employed. Secondly, an examination of these data is conducted for each of the four groupings. This approach permits isolating any changes occurring within categories, as well as identifying the direction of those changes. Even though the aggregate data alone would not necessarily indicate any change, significant differences do exist when examining by grouping. In addition to the above data classes (total accidents, fatal accidents, and total fatalities), analysis is conducted relative to accidents on a per 1,000,000 hours flown to correct for differences between groups in usage or exposure.

The ANOVA statistical technique is employed in this analysis of aviation safety data. Specifically, the data is examined with a view to discerning if significant differences in the mean value of the various measures exist in the pre-and-post strike periods. In this regard, the ANOVA results within the four groups is of particular interest. Due to the proliferation of small regional airlines and computer carriers caused in part by deregulation and in part by the adoption of hub-and-spike systems utilized by the major carriers. A general but pervasive contention has evolved recently that many of these airlines are under inordinate pressures associated with increased competition, and one which cuts have been made, at the expense of safety, has been maintenance. Another area relating to safety has been the utilization of less experienced pilots by the smaller commuter and regional airlines. Examining accident data on this breakdown by categories will assist in testing these hypotheses. Table 1 illustrates the results of the ANOVA's for the aggregate data. TA is total accidents, FA represents fatal accidents, FT is fatalities, and TAPHR is total accidents per 100,000 hours flown. Tables 2 through 5 depicts the same results for each category employed: large airlines; commuter; "on-demand" air taxis; and general aviation, respectively. Premean and postmean are the means of the data for the two time periods.

An examination of Table 1 reveals that the hypotheses regarding the means of the two different time periods are different would be rejected at the .90 confidence level (.10 level of significance) regardless of which variable is considered. This result suggests that commercial aviation is no less or no more safe after the PATCO strike than before. However, analysis contained in Tables 2–5 reveals this observation is not carried across all groupings of air carriers.

TABLE 1
Anova Results For Total Aggregated Data

	Premean	Postmean	F	Significance Level
TA	10499.65	7624.82	.306	.584
FA	1352.43	1380.94	.002	.968
FT	2408.17	2117.82	.092	.7653
TAPHR	.53	.09	1.456	.235

TABLE 2
Anova Results For Large Scheduled Airlines

	Premean	Postmean	F	Significance Level
TA	22.33	18.25	1.345	.280
FA	2.67	3.00	0.100	.760
FT	41.50	14.00	3.269	.108
TAPHR	.05	.04	.076	.790

TABLE 3
Anova Results For Commuter Airlines

	Premean	Postmean	F	Significance Level
TA	463.33	154.50	15.314	.004
FA	1116.67	352.25	11.309	.010
FT	3966.67	1214.00	7.705	.024
TAPHR	1.00	.04	18.172	.003

Table 2 indicates no significant difference between the pre and post periods (at the .10 level of significance), yet, safety is measured, with the exception of fatalities. The variable FT only marginally misses the .10 level.

In Table 3 the results for commuter airlines indicated a marked difference. In all variables and measures of safety there is significant difference in the means of the two periods, exceeding the .95 level of confidence in all cases. Examination of the means suggests that in all measures employed the post-strike period is safer than before the strike.

Table 4 shows the results for "on-demand", non-scheduled air taxis. In only one case, total accidents, is there a significant difference in safety, the post period being safer than the pre.

Table 5 provides somewhat mixed results for general aviation. Both TA and FT are significant at the .10 level, with the mean of total accidents declining, and that of fatalities declining. Interestingly, in four of the five cases, commuter airlines being the exception, the mean number of fatal accidents is greater in the period after the strike than before; none of the differences being statistically significant, with the exception of commuter airlines, at any acceptable level.

An examination of the relationship between safety measures and departures should prove enlightening regarding controller impact. An argument could be proffered that departure and landings are the most dangerous segment of flight times, safety wide, for aircraft, and hence are crucial times when the role of the air traffic controller comes into play. Unfortunately, data was not available for all groups for any safety measures related to departures, but, data was available for the commuter airline on total accidents per departure. This figure would also implicitly include landings, since each departure would require a subsequent landing. These data are presented in Table 6. These results again indicate that the commuter airline sector is safer post than pre period. The results are significant at the .006 level and the mean values have fallen.

III. CONCLUSIONS

The results of the brief analysis suggests several conclusions. First, our analysis of the FAA data indicates that the PATCO strike and actions taken by the Reagan Administration following the walkout had no significantly adverse effect on airline safety. Those individuals who contend that the general level of safety has deteriorated, and deteriorating, would appear to be erroneous in light of the data analyzed. One factor that conceivable may have contributed to this situation, or at least contributed to offsetting any potential adverse effects, is technological change in air safety. Newer types of radar have been installed, as well as new electronic aids to assist the air traffic controllers in their work. No attempt has been made to quantify this effect.

The second conclusion, and one that runs contrary to what has appeared in the media of late, is that the commuter air segment of the airline industry is safer in the post-strike period than before.[2] In fact, it is the only category that showed a significant improvement in *all* measures of safety. This is in spite of the increasing number of firms and aircraft now in existence, and is perhaps the most surprising result of the study.

One conclusion that can be reached as a result of this brief examination of aviation data is simply that analysis should be conducted before conclusions are reached. Safety, in particular, is an area in which opinions by professed "experts" are of miniscule use in determining the true state of events. In such an emotional area where the potential risks are great, the need for adequate research prior to conclusions and sweeping generalizations is acute.

In short, available data tends to suggest that reduction in air controllers levied no material impact on accident occurrences in aviation. Air commuter operations, moreover, are statistically more "safe" than generally perceived.But, before any meaningful conclusions can be reached, more exposure, metrological, and "incident" data needs to be gathered and analyzed. In terms of the impact of the PATCO strike, however, the analysis suggests no impact; yet the number of accidents such as "near misses" and safety citations will be more meaningful in analyzing the overall aviation safety environment. In short, the discomforting results of this manuscript remain preliminary in nature.

ENDNOTES

* Department of Marketing, Memphis State University, Memphis, TN.

** Center for Transportation Research, University of Arkansas, Fayatteville, AR.

[1] "Who Controls the Air," *Newsweek*, August 17, 1981, pp. 18-24, and "Turbudence in the Tower," *Time*, August 17, 1981, pp. 14-24.

[2] J. Richard Jones and Sheila I. Cocke, "A Performance Evaluation of Commuter Airlines" The Passenger's View," *Proceedings*, TRF, Vol. 22, #1 (Oxford, IN), Richard B. Cross Company, pp. 248-249.

** *Tables 4-6 were not included.*

Airline Performance Under Deregulation: The Role of On-Time Arrivals

by Ray M. Haynes and Martin T. Farris***

ABSTRACT

The airline industry has undergone considerable change in the past decade due to environmental dynamics, mainly passenger airline deregulation initiated in 1978. One significant effect of deregulation has been the level of customer service. This paper will examine one aspect of the customer service level, specifically, on-time arrival performance, from both the perspective of the traveler and the airline.

Since deregulation caused the demise of the CAB, where service level statistics had been collected, the nature of data collection was by questionnaire. The primary purpose of this paper is to design and test a methodological construct that may prove to be a valid mechanism for future research in this area.

There are many variables in airline customer service level. However, by comparing the performance of two similar regional carriers serving the same points over a given time frame, the effect of on-time arrival can be measured. The results of this study have confirmed the proposed paradigm in that carrier on-time performance can be directly linked to customer preference as represented by ticket sales. Non-parametric statistical procedures have been utilized in this study to facilitate the data analysis with the added benefit of retaining the maximum flexibility for future analysis due to the inherent simplicity of model construct.

I. INTRODUCTION

The airlines are the youngest segment in the transportation industry, having been established for approximately 60 years. The latter half of this time has evidenced the most growth and widespread use of aviation by both the business and pleasure traveler. This increase may be attributed to the emergence of commercial jet service coinciding with the general increase in the pace and scope of both business and pleasure travel.

Airlines typify the classical service operations organizaiton; delivering a "product" that is intangible, it lacks inventory potential, requires simultaneous production and consumption, and offers heterogeneous delivery of services based on the human involvement of flight crew, ground personnel, and reservations agents.

The service output considers customer satisfaction as a determinant of service quality with problems of standardized measurement. At the other extreme, "dis-satisfaction", the FAA maintains records on consumer complaints. Approximately one third of these are due to schedule delays and cancellations. However, complaints may not reflect the entire traveler population. For instance, in the first quarter of 1987, only one-half of one percent of airline passengers documented complaints. This does not necessarily imply that 99.5% are "satisfied" with schedule performance [3].

After 40 years of close Federal regulation, this industry was deregulated in 1978. The plan was to give the consumer better service while giving the carriers more flexibility to deliver this service and allowing more competition between carriers. The results are mixed and highly subject to individual consumer perceptions of the service delivery environment. The airline business has prospered in one sense; certificated carriers have grown from 37 to over 100 in the past eight years.

II. LITERATURE REVIEW

The general impact of airline deregulation has been well documented by Bailey, et al, [1] and Meyer and Oster [11] in the MIT Series. The latter evolved from a prior Meyer and Oster [10] evaluation of the early deregulation experience. Brenner, et al, [2] have focused on the structural components while Morrison and Winston [12] looked at the economic effects. While deregulation must consider the external environment like all business, there are several unique factors that should be considered.

1. Fare wars have been prevalent with start-up carriers seeking to establish a market niche. The entrenched majors can generally match these on a selective basis resulting in eventual economic stabilization at a level allowing for profitable operation.
2. Fare restrictions have allowed both the consumer and carrier to benefit. The first by lower fares and the second by forecasted demand planning enhancement due to advanced ticket purchase.
3. Computer reservation systems have provided several carriers with a competitive tool but have been subject to periodic abuse due to display bias favoring particular carriers. Another bias is in the area of schedules, both departure times and in-transit times can "bubble" a given flight to the top of a display sequence. Recent DOT-recommened changes have alleviated many of these problems.
4. Increased hub-and-spoke operations have been effective for some carriers, such as America West Airlines in Phoenix, Southwest at Dallas-Love Field, and American at DFW. Others have been beset with problems, such as Delta and Eastern at Hartsfield Atlanta, United at Chicago-O'Hare, and People Express at Newark. While theoretically sound, once problems

start happening there is a domino effect that can propagate delays throughout the system.

5. The emergence of mega-carriers has placed an umbrella of uncertainty on the industry. The mega-carrier can be exemplified by recent mergers such as Delta-Western, Northwest Republic, Continental-Frontier-Eastern-People Express-New York Air, TWA-Ozark, American-AirCal, and PSA-USAir and others. The apparent rationale for this "merger-mania" is achievement of economies of scale and competitive advantage that will yield increased profitability, however in many instances, this consolidation results in more problems in service delivery [6,14,15].

While the traveler appears to have benefited from the deregulation of the industry (Fares are lower than ten years previously and schedules are plentiful to most market areas) under this glossy surface, there appears to be an escalating concern about a deterioration of service level, particularly on-time performance.

During the Spring of 1987, the issue of on-time performance seemed to come to a head [3]. In April, the Department of Transportation (DOT) reported consumer complaints in the January to March quarter rose 43 percent from the same quarter in 1986. Significantly, over a third of these complaints concerned flight delays and cancellations. User complaints against Continental tripled from 8.6 complaints per 100,000 passengers in January, to 25.5 complaints per 100,000 passengers in March. In March, DOT allowed the airlines to meet collectively (by granting immunity from anti-trust laws) to adjust schedules at Chicago-O'Hare, Dallas-Fort Worth, Newark, Boston, and Atlanta-Hartsfield. Schedule adjustments of various degrees were accomplished at all but Atlanta with a view to increasing on-time arrivals by avoiding too many flights scheduled in the same 15-minute segment [9].

Various studies of the effects of airline deregulation have appeared in the transportation literature. Jones and Cocke [8] conducted a study focusing on commuter airlines alone. One finding of interest was that nearly 85% of their sample were repetitive airline users. Daley and LaForge [4] found in intercity commuter travelers that 70% were frequent airline passengers. Furthermore, their study of 12 service factors indicated that on-time performance was ranked first in importance. Davis and Dillard [5] investigated the effects of airline deregulation on the business traveler; results showed that 52% felt that service had deteriorated in some manner. Hu and Bruning [7] compared the characteristics of commuter and major (trunk) carriers using a 12-factor analysis with on-time performance prominently featured.

Finally, Rose reported, in 1981 [13], that system load factors had remained relatively constant following deregulation. No conclusions were reached other than that the pleasure traveler has benefited more than his business counterpart, primarily due to the fare re-structuring.

The basic goal of this paper is to build upon prior research and to consider on-time performance in detail. Additionally, there is an attempt to establish a methodological process that may prove beneficial to similar studies in the future on airline service and performance.

III. RESEARCH QUESTION

One of the key concerns since deregulation has been the level of service positioning. As noted, various studies have examined this issue using multiple methodologies. Customer surveys are the most popular mechanism but there is some suspicion that the micro-orientation of most does not lend itself to any degree of broad generalization based on the results. Many variables are involved in the attempts to measure airline performance in prior studies. However, one factor that consistently emerges prominent across the literature is the concern with on-time performance.

Without question airline traffic has grown markedly since deregulation. In 1978, some 275 million passengers flew on scheduled airlines. . . in 1986, the number jumped 66% to 415 million in spite of an apparent deterioration of service levels. But the numerous articles in the public press plus the actions of the FAA, in Spring of 1987, poses one significant question: just how important is on-time performance in measuring airline service?

IV. THE PROJECT

Basically, consumers have a choice of airlines to fly in most markets. Decisions are based on fares, schedules, routes, and other intangible factors. It is hypothesized that one of the major factors that contribute to this choice is on-time performance, or at least the perception of this by the passenger. In theory, it should be possible to establish a linkage between this characteristic and the economic success of an airline. As shown in Figure 1, on-time performance would be predictive of traffic and result in more ticket sales and subsequent profits.

Unfortunately, this premise is based on the principle of cetaris paribus, or all other things being equal. This is not the case for most carriers. As previously noted, fares, schedules, aircraft, and routing may vary considerably within a given market being served by multiple carriers. These would be expected to exhibit a mediating influence on the proposed model based on passenger perceptions and

Figure 1

ON-TIME PERFORMANCE ⟶ TICKET SALES ⟶ PROFITS

(observed) (result) (potential)

⟵ PASSENGER ⟶

observed behaviors. It is felt that an analysis of this magnitude is beyond the scope of this paper.

However, a survey has been done on a micro-level basis within a highly segmented market in the Southwestern United States which may achieve some degree of cetaris paribus. There are several key points that must be considered in this study.

1. Only two carriers were examined.
2. Both operate identical aircraft types, the Boeing 737 in various configurations.
3. Both serve the same markets for the most part with similar schedules available on a daily basis.
4. Fare structures are comparable and can be matched exactly for any designated city-pair service route.

The formal hypotheses are relatively simple and are as follows:

H1: On-time arrival performance is important.

H2: Carrier segmentation by on-time performance measurement is possible.

H3: Passengers will select an airline with better performance characteristics.

H4: The carrier with better on-time arrival performance will have more sales in a given market being served.

The first hypothesis was evaluated using a user survey designed to elicit specific information.

The second hypothesis was tested by an evaluation of a series of randomly sampled flight arrival times of each carrier in comparison to their scheduled arrival times.

The third and fourth hypotheses were examined in conjunction with a specific question on the user survey and reported financial information by each carrier. No consideration of actual operations expenses was made so that productivity elements are not a factor being tested.

V. PROCEDURES

The procedures utilized were straightforward and easy to implement in the research environment.

A. USER SURVEY: A one-page questionnaire was prepared based on examples of those found in the literature review. It was administered to 90 senior-level business students at a major university. Two classes in Business Policy (n = 33, n = 41) and one in Service Operations Management (n = 16) were sampled. The students were told that this was research for a study in the Transportation Department of the University and that their help was needed to establish a database. No incentives were offered, however, there was encouragement for participation based on the context of the courses. First, a pretest of a small group of students established the time allotted for the survey to be completed. There was minor explanation of the questions on the survey and there was no apparent confusion as to the nature or detail of questions being asked of the respondents. Each class was made aware that the results would be available by the end of the semester and that the information would be presented to them as a courtesy and general information. The survey instrument is presented as Table 1.

B. AIRLINE INFORMATION: Each airline was contacted and asked to provide a documented log of 50 consecutive flights with scheduled and actual arrival times listed. Data was requested to be taken from a "standard" day having no significant delays due to either equipment failures or weather conditions. The location of the airport contributed to the high expectancy of being able to establish this "standard" day. The airlines were briefed on the nature of the study and that it was being done at the University within the Transportation Department. Formal request letters on Department letterhead were sent to the individual identified as being the primary contact at each carrier. Again, there was an offer to share the results following the completion of the study and in each case, the carrier expressed an interest in receiving this information. The carrier logs are not presented due to the proprietary nature of the information.

C. FINANCIAL PERFORMANCE: This information was taken from published annual and quarterly report data. Certain corrections were considered to normalize the information for indicated market equivalences such as the total number of flights serving a designated airport, load factors, schedules, and other elements that may serve to make the passenger decisions more equitable. These corrections were not made due to the macro level of the financial information available from the indicated sources. A summary of the carrier information is presented in Table 2.

All information was collected over a period of approximately eight weeks during Fall Semester 1986. There were no apparent problems with either the data collection mechanism or the information content itself in this study.

VI. THE ANALYSIS

The statistical analysis was done using various non-parametric tests on the data collected. The use of these techniques can be justified based on the following rationale:

1. All data collected is from samples of convenience and there has been no randomization of data attempted.
2. There is no expectation of normal distributions or equal variances in either data sample.
3. The measurement scales utilized are not particularly robust.
4. There have not been any assumptions made which would suggest the use of parametric procedures being preferable to nonparametrics in this study.
5. The study is of the nature of a design exercise and the use of "quick-and-dirty" statistical analyses is warranted to establish a baseline for further investigation.

The primary elements of interest were population parameters and potential trends during the reporting periods. Standard nonparametric tests were used including the Tukey Quick Test, CoxStuart, Mann-Whitney-Wilcoxon and Extension of Median Test. The summary information from these tests is presented in Table 3.

VII. THE RESULTS

The results of the experiment confirmed the stated hypotheses presented as H1, H2, and H3. There was less confirmation of H4, primarily due to a lack of measurement sensitivity. A summary of the survey results is presented in Table 4.

H1: On-time arrival performance is rated as being important by 81% of the total respondents. The airlines reported average late times ranging from 8-10.4 minutes which nearly matched that of 9 minutes reported by the student sample. The students surveyed indicated that 80% had taken flights during the preceding twelve months with an average frequency being 4.57 flights per student. Of those flying, 61% reported late arrivals on flights taken. These numbers compare favorably to those presented earlier based on national statistics. The survey reported a maximum acceptable late arrival time of 18.09 minutes averaged across the sample. This implies that "on-time" in the mind of the traveler is the same as "acceptable late arrival time." It should be noted that in the strict definitional sense, the FAA allows a 15-minute buffer against schedule time, but also only logs delays due to air traffic controller delays once the aircraft has departed the gate. This precludes all carrier-originated gate-holds which account for the majority of delays. This is a sensitive subject and information does not flow freely from the carriers.

H2: Carrier segmentation by on-time performance was established based on the data provided.

TABLE 1
AIRLINE PASSENGER PERCEPTION SURVEY

PLEASE SELECT THE BEST ANSWER FROM THE CHOICES PRESENTED BELOW.
****

1. NUMBER OF FLIGHTS FLOWN DURING THE PAST 12 MONTHS ON A COMMERCIAL AIRLINE?

   1 OR LESS 2 3 4 5 6 7 OR MORE

2. HOW MANY OF THE FLIGHTS INDICATED ABOVE HAD LATE ARRIVAL TIME?

   1 OR LESS 2 3 4 5 6 7 OR MORE

3. WOULD YOU ESTIMATE THE AVERAGE LATE TIME IN MINUTES?

   2 OR LESS 2-4 4-6 6-8 8-10 10-12 12 OR MORE

4. RATE THE IMPORTANCE OF ON-TIME ARRIVAL TO YOU AS A PASSENGER.

   VERY 1 2 3 4 5 6 7 LITTLE

5. RATE THE IMPORTANCE OF ON-TIME ARRIVAL TO YOU AS A PERSON MEETING AN ARRIVING PASSENGER.

   VERY 1 2 3 4 5 6 7 LITTLE

6. WHAT IS THE MAXIMUM LATE ARRIVAL TIME IN MINUTES THAT IS ACCEPTABLE TO YOU.

   AS A PASSENGER [ ]

   MEETING A PASSENGER [ ]

7. WOULD YOU CHOOSE AN AIRLINE BASED ON THE PERCENTAGE (%) OF DOCUMENTED ON-LINE ARRIVALS IF ALL OTHER COMPETITIVE FACTORS (PRICE, SERVICE, AVAILABILITY, ETC.) WERE EQUAL?

   YES [ ] NO [ ]

THANK YOU FOR COMPLETING THIS QUESTIONNAIRE, THE RESULTS WILL BE AVAILABLE IN EARLY JANUARY AND YOU ARE WELCOME TO CONTACT THE AIRLINE SURVEY PROGRAM ADMINISTRATOR IN THE PTO DEPARTMENT.

Carrier A averaged a late arrival time of 8.0 minutes with 72% of the flights being late, 14% being early, and 14% on-time. Carrier B averaged a late arrival time of 10.4 minutes with 58% of the flights being late, 40% early, and only 2% exactly on-time. Since this data was based on self-report, there was some possibility of "loading". However, it is interesting to note that while numerically different, the order of magnitudes were equivalent. The experiment did not allow for any buffer but used actual arrival time compared to scheduled time of arrival. In fact, one carrier volunteered that their schedule was adjusted for such delays which may account for the high number of "early" arrivals reported.

H3: The student survey indicated that passengers will select an airline with demonstrated/documented on-time performance 85% of the time. It is not clear exactly how this information is garnered, however there seems to be a strong "grapevine" effect relative to frequent travelers. More likely, if one were given a choice between carriers where the only difference was on-time arrival, the survey indicates that

TABLE 2
NUMBERS OF INTEREST

1986		
ITEM	CARRIER B	CARRIER A
PASSENGERS	6.2 Million Down 6% Ranked #2	3.2 Million Up 41% Ranked #5
SCHEDULES		
PHX-LAX	16 Flights Daily	23 Flights Daily
PHX-LAS	7 Flights Daily	12 Flights Daily
LOAD FACTOR	55.9%	62.9%
BREAKEVEN	52% Load*	57.7% Load
UTILIZATION	11-12 Hours	10 Hours or less
AIRCRAFT	54 (B-737)	35 (B-737)
PROFIT	3.9% Revenues	5.9% Revenues
MARKET AREA	24 Cities/12 States	29 Cities/12 States + Canada
PHOENIX MARKET SHARE	16%	34%
PRIMARY HUB	DALLAS	PHOENIX
FARE STRUCTURE	Two-tier	Multiple

*Possibly due to aircraft configuration with higher density seating.

NOTE: Fares tend to be equivalent on same routes in and out of Phoenix except for certain advance reservation/purchase seating.

FINANCIAL TRENDS (thousands)

(Passenger Revenues Only)

1984	$535,900	$121,062
1985	679,700	237,805
1986	768,800	320,179

TABLE 3
NON-PARAMETRIC TESTS

AIRLINE ARRIVALS:

1. HO: SAMPLES FROM SAME POPULATION
   H1: DIFFERENT CARRIERS REPRESENTED

* TUKEY QUICK TEST NOT POSSIBLE DUE TO EXTREMES BEING IN ONE SAMPLE

* MANN-WHITNEY-WILCOXON UTILIZED — Test Statistic = 93 > 28
  REJECT HO — P-Value < .05

2. HO: ARRIVAL TIMES REFLECT NO TREND
   H1: THERE IS EITHER AN UPWARD OR DOWNWARD TREND

* COX-STUART UTILIZED
  CARRIER A — Test Statistic = .0577 > .025
  REJECT HO — P-Value = .1154
  CARRIER B — Test Statistic = .0059 < .025
  FAIL TO REJECT HO — P-Value = .0118

STUDENT SURVEY:

1. HO: ANY TWO SAMPLES ARE FROM IDENTICAL POPULATIONS
   H1: POPULATIONS ARE DIFFERENT

* TUKEY QUICK TEST UTILIZED ON 2 X 2 BASIS
  A. POLICY: POLICY — Test Statistic = 3.5 < 7
     FAIL TO REJECT HO — P-Value = .05
  B. POLICY 1:OPM — Test Statistic = 7.5 > 7
     REJECT HO — P-Value < .05
  C. POLICY 2:OPM — Test Statistic = 8.0 > 7
     REJECT HO — P-Value < .05

2. HO: ALL THREE POPULATIONS HAVE SAME MEDIAN
   H1: ONE POPULATION MEDIAN IS DIFFERENT

* EXTENSION OF MEDIAN TEST UTILIZED — CHI-SQUARE = 14.19 > 5.991
  REJECT HO — P-Value < .005

the choice would be made to the better performing airline.

H4: The actual airline financials do not break down into the specific information necessary to fully confirm the model in Figure 1. Passenger revenues are not segmented either by market or by ticket sales, however, both carriers showed a general profit trend which may be some degree of confirmation of the hypothesized relationship. The small differentials in performance (on-time arrivals) also may cloak basic problems in measurement sensitivity. It should be noted that the absolute profit margin reflects the carrier with the better arrival performance, but this may not necessarily reflect the margin achieved in the specific market tested.

VIII. DISCUSSION

The study as presented offers an approach that may be useful in designing a more thorough and rigorous project. The stated hypotheses are considered to be valid research questions and of significant interest based on the existing literature. The results confirm the stated propositions in general terms, however any generalization would be premature given six possible methodological problems.

First, the college student sample may not truly represent the actual airline passenger population parameters. Additional research must be done to ascertain that this is a valid sample for the population, although it appears to be acceptable.

TABLE 4
SUMMARY OF RESULTS

AIRLINE SAMPLE

CARRIER A: SAMPLE SIZE = 50 CONTINUOUS FLIGHTS IN ONE DAY

EARLY ARRIVALS	=	14%
AVERAGE EARLY TIME	=	3.43 MINUTES
ON-TIME ARRIVALS	=	14%
LATE ARRIVALS	=	72%
AVERAGE LATE TIME	=	8.0 MINUTES

CARRIER B: SAMPLE SIZE = 50 CONTINUOUS FLIGHTS IN ONE DAY

EARLY ARRIVALS	=	40%
AVERAGE EARLY TIME	=	7.15 MINUTES
ONE-TIME ARRIVALS	=	2%
LATE ARRIVALS	=	58%
AVERAGE LATE TIME	=	10.4 MINUTES

STUDENT SAMPLES

AMOUNT FLYING MORE THAN ONE FLIGHT IN LAST 12 MONTHS	=	80%
AVERAGE NUMBER OF FLIGHTS FLOWN BY THE ABOVE GROUP	=	4.57
AMOUNT OF FLIGHTS FLOWN THAT WERE LATE	=	61%
AVERAGE LATE TIME OF THESE FLIGHTS (MINUTES)	=	9.07
AMOUNT RATING ON-TIME PERFORMANCE IMPORTANT (AS A PASSENGER)	=	81%
AMOUNT RATING ON-TIME PERFORMANCE IMPORTANT (MEETING AN ARRIVING PASSENGER)	=	79%
MAXIMUM ACCEPTABLE LATE ARRIVAL TIME (RANGE 5-90)	=	18.09
AMOUNT THAT WOULD CHOOSE AN AIRLINE BASED ON PERFORMANCE CRITERIA	=	85%

NOTE: THE DOT/FAA/CAB HAD ORIGINALLY DEFINED LATE ARRIVALS AS BEING WITHIN 15 MINUTES OF SCHEDULE ARRIVAL TIMES BUT THIS SAMPLE ASSUMED NO BUFFER ZONE IN CALCULATIONS

Second, the survey questions, while apparently straightforward, do not consider any semantic differential and have not been validated for reliability. However, the survey was similar to others being done in this area of investigation.

Third, the airline logs are self-report over a fairly brief time sequence. Better information may be obtained by a longitudinal effort using third-party data collection. Ideally, this time period would match the financial performance period closely. In addition, the period should be long enough to effectively "smooth" the data being collected rather than using the concept of a "standard" day which may possibly introduce sampling biases.

Fourth, the actual financial performance must be isolated to a greater degree and focus on some measure of actual consumer preference such as ticket sales within a given market area that coincides with an arrival-time compilation for both the same time and period under consideration by the experiment.

Fifth, the use of non-parametric statistics was justified in this study as being a mechanism for quick evaluation of collected data. The transportation and operations management literature has not shown any strong predilection in this area, but parametric statistics should be considered as a matter of course.

Finally, the examination of this type of information on a micro, "snapshot" basis does not necessarily lend itself to any degree of validated generalization. The results are tidy and confirm the stated hypotheses, but a large-scale study would definitely be more suggestive for any predictive position. In fact, the dynamic nature of this industry is such that any study may lack any true predictive power over a longer period.

The system participants tend to point fingers in different directions. The traveler blames the airline. The DOT says that carriers "fudge" on their schedules. The carriers maintain that the ATC operation is obsolete and under staffed. The ATC points out that traffic exceeds physical capacity. And everyone blames the weather, which appears to account for 70% of all delays.

The answer to the original question posed is that deregulation, according to the results of this study, has not resulted in service deterioration as measured by on-time performance. This appears to be somewhat contradictory to actual facts of the situation as presented in the background information. The best explanation is that the homogeneity of the two populations sampled and the unique setting of the survey has fogged the primary issue of on-time performance. The paradigm remains appropriate but must be used in a macro study to gain the robustness necessary for proper analysis. Secondarily, it may be observed that the advantages of air travel offset the disadvantages of decreased service level. This would be consistent with the literature in service operations, in that the expectations of the traveler have been reduced and he accepts a degree of service degradation as unavoidable. In any event, this is a topic requiring more attention in the future in order to fully understand how the system can be made more productive for all participants . . . the airlines and the traveler. The experimental construct presented has been tested and serves to present an approach to the evaluation of airline performance under deregulation focusing on the specific issue of on-time arrivals and its effect on consumer preference.

REFERENCES

1. Bailey, E.E., D.R. Graham, and D.P. Kaplan, *Deregulating the Airlines,* Cambridge, MA: MIT Press 1985.
2. Brenner, M.A., J.O. Leet, and E. Schott, *Airline Deregulation,* Westport, CT: Eno Foundation for Transportation, Inc. 1985.
3. Dahl, J. "Consumer Complaints Against Airlines Rose Sharply in Quarter, U.S. Data Show". *Wall Street Journal*. April 13, 1987.
4. Daley, J.M. and R.W. LaForge. "An Analysis of Intercity Commuter Travel Behavior". *Transportation Research Forum*. 23:1982, pages 412-418.
5. Davis, G.M. and J.E. Dillard. "The Professional Traveler and The Airline Deregulation Act of 1978: An Appraisal". *Transportation Research Forum*. 23:1982, pages 419-426.
6. "Delays, Bumpings, and Lost Bags: Travelers Gripe About Air Travel". *Wall Street Journal*. November 10, 1986.
7. Hu, M. and E. Bruning. "Prior Experience and Misperceptions of Airline Service". *The Logistics and Transportation Journal*. 20,3:1986, pages 213-222.
8. Jones, R.J. and S.I. Cocke. "A Performance Evaluation of Commuter Airlines: The Passenger View". *Transportation Research Forum*. 23:1982, pages 419-426.
9. Karr, A.R. "Airlines Agree on New Schedules at More Airports". *Wall Street Journal*. April 13, 1987.
10. Meyer, J.R. and C.V. Oster, Jr., *Airline Deregulation: The Early Experience,* Boston: Auburn House 1981.
11. Meyer, J.R. and C.V. Oster, Jr., *Deregulation and The New Airline Entrepreneurs,* Cambridge, MA: MIT Press 1985.
12. Morrison, S.A. and C.W. Winston, *The Economic Effects of Airline Deregulation,* Washington, D.C.: Brooking Institution 1986.
13. Rose, W. "Three Years After Airline Passenger Deregulation in the United States: A Report Card on Trunkline Carriers". *Transportation Journal*. Winter 1981, pages 51-57.
14. "The Bumpy Takeoff for Northwest and Republic". *Business Week*. November 24, 1986.
15. "The Debate: Airline Service". *USA Today*. October 24, 1986.

ENDNOTES

*Department of Purchasing, Transportation and Operations, Arizona State University, Temple, AZ.

The following texts have been used for general information and have not been specifically cited in the text of the paper.

1. Daniel, W.W., *Applied Non-Parametric Statistics,* Boston: Houghton-Mifflin Company 1978.
2. Sampson, R.J., M.T. Farris, and D.L. Shrock, *Domestic Transportation: Practice, Theory and Policy,* Boston: Houghton-Mifflin Company 1985 (Fifth Edition).
3. Annual Reports for 1986 for both carriers in the study.

Airlines' Strategies in Hub Dominance: A Statistical Analysis

by Atef Ghobrial, Jorge Sousa,** and Colleen Callahan***

I. INTRODUCTION

Deregulation has led to radical changes in airline network planning. A main focal point for these changes has been the greatly increased emphasis on "hubbing." All carriers have sought to build up their previous hubs, and in many cases have created new ones. An examination of frequency concentration patterns at different airports suggests that most hubs are dominated by a few airlines which compete in order to capture a large share of airport traffic. Competition for hub dominance seems to take the form of increased frequency concentration and employing a relatively large aircraft.

This paper attempts to assess the effects of airlines' dominance strategies and to address their marketing and policy implications. The balance of this paper includes three sections. In section II, we review trends in airline dominance strategies and discuss theoretical bases for relating these trends to airline marketing. Our methodology, results and analysis are discussed in section III. Section IV assesses the implications of our findings for airline marketing and some policies regarding airline mergers and airport congestion.

II. HUB DOMINANCE STRATEGIES

The most direct way of quantifying airline service convenience is by measuring the waiting time (delay) required to gain service. That is, holding fare, speed, and other demand decision variables constant, passengers tend to favor airlines with the presumed greatest likelihood of offering convenient flights. Two delay components are usually referred to in evaluating service convenience. The first is frequency delay which is defined as the expected time differential between the most desired departure time and that of the closest departure. If these time differentials increase significantly, we would expect the number of air travelers to diminish, as passengers choose other modes, other destinations, or even cancel their trips. This aspect of convenience can be quantified crudely by the inverse of airline scheduled frequency.

The second determinant of service convenience is known as stochastic delay which is the expected length of delay a potential passenger faces because of the chance that his most preferred scheduled departure is booked up and he will have to select a later flight, and possibly even cancel his trip. The length of stochastic delay is related to the degree of dispersion of passenger demand for a flight and the aircraft seating capacity. As the average number of passengers on each flight approaches the average capacity of the aircraft, the probability of not obtaining a seat grows large, and the expected length of delay grows long. The summation of frequency and stochastic delays constitutes the expected difference between a potential traveler's preferred departure and his actual departure times.

In light of the above, one would expect airline management to pursue competitive policies that are aimed at minimizing passenger waiting time, thus capturing high market share. Two hub dominance strategies seem to blend in the context of the above discussion: increased frequency competition and employing a relatively large aircraft.

In order to illuminate further, we examine the strategies of American airlines in dominating the Dallas-Fort Worth (DFW) airport between 1978 and 1985. Figure 1 depicts the trends in American's share of total aircraft departures at DFW along with its share of passenger enplanements. The upward trend in hub dominance through frequency concentration is quite evident; American airlines increased its aircraft departure share at DFW by about 93.4 percent since 1978. More interesting is the fact that in every year in the period under review, American airlines' share of passenger enplanements is higher than its aircraft departure share. The disproportionality phenomenon between market share and frequency share deserves further discussion. Local travelers originating at the DFW area learn which carrier offers the greatest flight frequency (American in this case,) and given the existence of information costs, they tend to contact that carrier first. Passengers flying American via DFW will enhance their chances of enjoying single-plane service. Furthermore, through increased frequency concentration at DFW and a proper timing of flight arrivals and departures, American airlines can offer connecting passengers single-airline service (i.e., on-line connection,) thus reducing their chances of missing a connection or losing baggage, as well as eliminating the need to rush form one terminal to another at DFW airport. The phenomenon of disproportionality between airline market share and its frequency share seems to encourage carriers to pursue, in general, a policy that maximizes their frequency concentration.

Figure 2 depicts the trends in the average aircraft size (number of seats) for American airlines at DFW. Again, American increased its average aircraft seating capacity between 1978 and 1985 by 21 percent. As mentioned above, an increase in aircraft size will reduce passenger stochastic delays and add to airline service convenience.

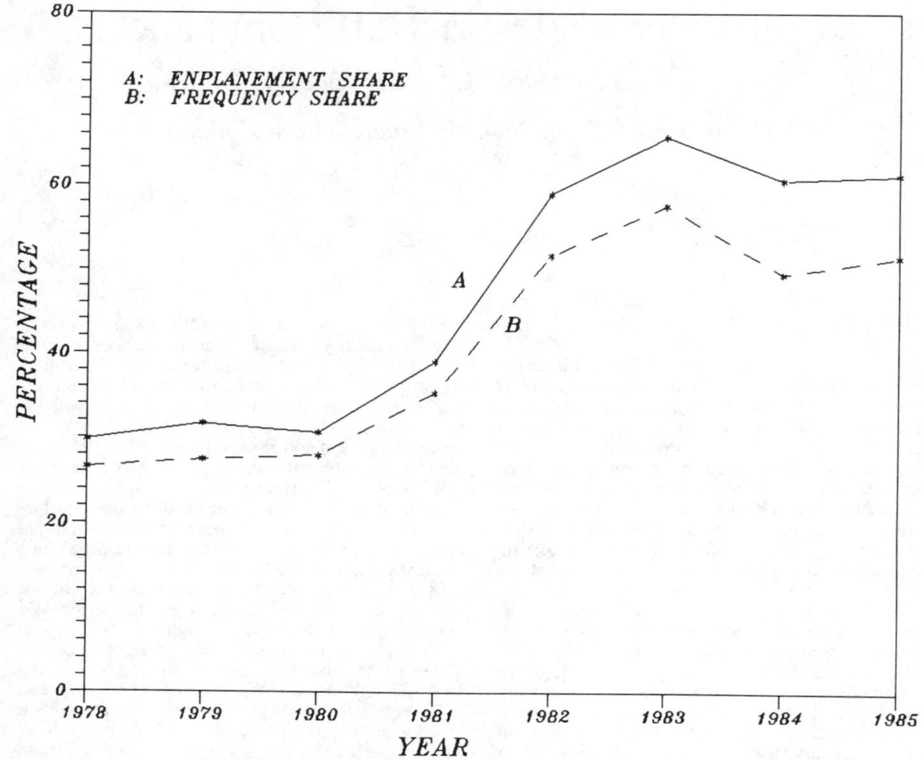

Figure 1: Frequency and enplanement share for American Airlines at Dallas/Fort Worth.

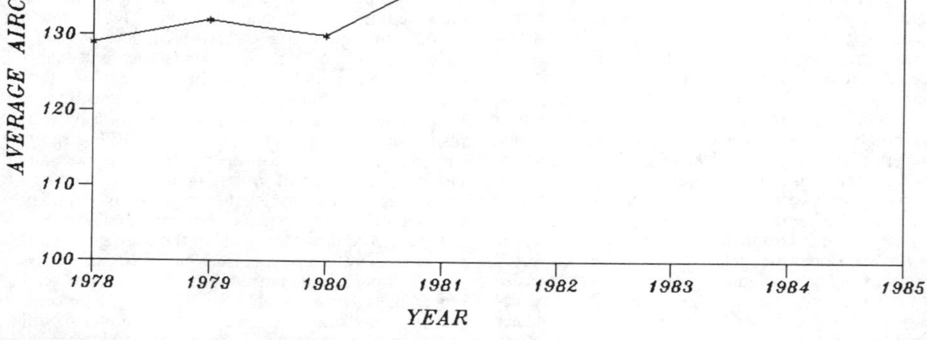

Figure 2: Trends in the Average Aircraft Size for American Airlines at DFW Airport

III. ANALYSIS

In order to quantify the effects of hub dominance strategies on airline marketing, an econometric model is employed. The model serves as a tool that simulates passenger perception of the above service convenience determinants. The dependent variable is the airline share of the total passenger enplanements for the airport. For the purpose of this study, two explanatory variables are considered: airline share of aircraft departures at an airport and its average aircraft size relative to that of other airlines at the airport. The multiplicative traffic share model is formulated as:

$$\text{LOG} \frac{ENP_i^m}{1 - ENP_i^m} = \alpha + \beta \text{ LOG} \frac{FREQ_i^m}{1 - FREQ_i^m} + \theta \text{ LOG} \frac{SIZE_i^m}{SIZE_c^m} + \varepsilon \quad \ldots(1)$$

Where ENP_i^m is passenger enplanements share for carrier i at airport m, $FREQ_i^m$ is aircraft departure share for carrier i of the total domestic flight departures at airport m, $SIZE_i^m$ is the average aircraft seating capacity for other competing carriers c at airport m. α, β and Θ are the regression coefficients to be estimated, and ϵ is the error term. Both ENP and FREQ variables have been transformed as is usually done in the logit analysis, Theil (1971). Untransformed, ENP and FREQ would always be a fraction between zero and one. A priori signs of both β and Θ are expected to be positive since an increase in airline frequency concentration and/or in its average aircraft size would yield better service convenience.

The above model was estimated using data from twenty airports ranging from small to large hubs. Data for aircraft departures and passenger enplanements were compiled from the Airport Activity Statistics for 1984. Aircraft seating capacity data were obtained from the Aircraft Performance Statistics published in the Official Airline Guide. Table 1 contains a summary of activity statistics at each airport. The pattern of hub dominance by a few airlines is quite evident. The three-airline concentration ratio (the sum of passenger enplanement shares of the top three airlines) ranges from 48.2 to 92.7 percent. As such, data for the top three airlines at

TABLE 1
Concentration Patterns At Selected Airports

AIRPORT	TOTAL AIRCRAFT DEPARTURES	NUMBER OF AIRLINES	THREE-AIRLINE CONCENTRATION RATIO
SAN FRANCISCO, CALIF	126969	22	62.8
SACRAMENTO, CALIF	19665	7	59.1
BURBANK, CALIF	22977	9	77.5
LOS ANGELES, CALIF	147769	28	52.5
ORANGE COUNTY, CALIF	14519	6	66.8
SAN DIEGO, CALIF	45197	16	51.7
MIAMI, FLA	72305	24	67.0
TAMPA, FLA	57869	19	55.4
BOSTON, MASS	94419	24	49.8
NEWARK, N.J	123481	22	63.3
JOHN F. KENNEDY, N.Y	68748	17	51.9
PHILADELPHIA, PA	66630	22	58.2
DETROIT, MICH	92085	24	55.7
MINNEAPOLIS, MINN	96941	18	80.7
ST. LOUIS, MISS	131127	21	80.4
DENVER, COLO	170751	21	81.0
ATLANTA, GA	268661	14	92.7
O'HARE, ILL	272286	25	67.1
DALLAS/FORT WORTH, TEX	197622	25	84.1
SEATTLE, WASH	68030	19	48.2

SOURCE: AIRPORT ACTIVITY STATISTICS, 1984

each airport, comprising 60 data points, were used to estimate the model. The results of the cross-section regression analysis together with t-statistics (shown in parentheses) are presented in Table 2. The signs of β and Θ agree with their a priori signs, and both explanatory variables are highly significant. Also, the hypothesis that the coefficient β is greater than one seems to be statistically significant, meaning that high frequency shares are associated with even higher market share and conversely. From Equation 1, the elasticity of an airline's market share with respect to its frequency share is derived as:

$$\epsilon_f = \frac{\beta(1-ENP)}{(1-FREQ)} \quad ...(2)$$

Where ϵ_f is the frequency elasticity, ENP and FREQ were defined earlier. From Equation 2, one can see that the coefficient β is the frequency share elasticity at the sample mean point (i.e., ENP = FREQ = ⅓) meaning that a ten percent increase in an airline's frequency share would likely be associated with 11.53 percent increase in its share of passenger enplanements.

The aircraft size variable appears to have a positive impact on airline market share but to a lesser extent than that of frequency dominance. This is in fact not surprising. Airlines tend to operate with load factors in the range of 55-70 percent, and stochastic delay is more likely encountered at peak hour flights. On the other hand, the phenomenon of increased emphasis on airline hubbing through routing passengers via a connecting hub will likely induce load factors to grow, further increasing stochastic delays, and the need for flying a relatively large aircraft will then exist. From Equation 1, the elasticity of market share with respect to aircraft size is expressed as:

TABLE 2
Airline Service Characteristics At Selected Airports

AIRPORT	AIRLINE	NUMBER OF AIRCRAFT DEPARTURES	PASSENGER ENPLANEMENSTS	AVERAGE SEATING CAPACITY
SAN FRANCISCO	UA	45517	3846124	163
	PS	22104	1537829	145
	OC	12150	611565	133
SACRAMENTO	PS	3890	318536	146
	OC	4092	242813	131
	UA	3645	215213	155
BURBANK	PS	10649	700292	138
	OC	4052	245581	148
	RC	3095	150272	108
LOS ANGELES	UA	27792	5707636	205
	WA	25420	2025242	151
	PS	24345	1667382	144
ORANGE COUNTY	OC	4435	482434	148
	RC	3083	457380	148
	PS	2175	223506	150
SAN DIEGO	PS	10418	924422	148
	UA	5715	578497	187
	WN	7221	283106	125
MIAMI	EA	27815	2673558	202
	PA	14172	1036209	137
	DL	6437	496311	372
TAMPA	DL	11017	890540	200
	EA	15509	847521	146
	NW	5549	353931	181

TABLE 2: CONT'D

AIRPORT	AIRLINE	NUMBER OF AIRCRAFT DEPARTURES	PASSENGER ENPLANEMENSTS	AVERAGE SEATING CAPACITY
BOSTON	EA	16874	1797983	207
	DL	18353	1561474	185
	AL	11812	739491	123
NEWARK	PE	55525	5705398	134
	EA	11467	1163066	209
	PI	11148	790336	133
JOHN F. KENNEDY	TW	13390	1440910	215
	PA	12922	1373923	215
	EA	9335	991317	208
PHILADELPHIA	AL	19332	1027353	123
	EA	12873	873543	169
	DL	6554	455363	163
DETROIT	RC	33462	1559422	128
	NW	10613	587781	180
	DL	7221	502523	166
MINNEAPOLIS/ST. PAUL	NW	36013	2861648	179
	RC	39003	1924767	124
	UA	3176	227358	157
ST. LOUIS	TW	57951	4540866	167
	OZ	43036	2219819	119
	DL	4376	185130	127
DENVER	UA	60470	5178504	167
	FL	44828	3153455	127
	CO	33054	2522702	139
ATLANTA	DL	119724	9675180	174
	EA	109549	7502927	154
	RC	19816	694648	118
O'HARE	UA	102123	8709335	168
	AA	63081	4768013	156
	RC	17379	685087	126
DALLAS/FT. WORTH	AA	97834	9079779	155
	DL	49243	3297459	167
	BN	19218	1023035	145
SEATTLE	UA	13623	1323835	187
	AS	11818	744843	132
	NW	7360	658100	229

SOURCES: AIRPORT ACTIVITY STATISTICS, 1984
AIRCRAFT PERFORMANCE STATISTICS, OFFICIAL AIRLINE GUIDE

$$\varepsilon_s = \theta\ (1\text{-ENP}) \qquad \ldots.(3)$$

Where ϵ_s is the aircraft size elasticity. For any value of ENP, Equation 3 shows the relationship between airline market share and aircraft size to be inelastic. At the sample mean point (i.e., ENP = ⅓,) a ten percent increase in the average aircraft size of an airline will likely result in 5.8 percent increase in its market share. Realizing the fact that market share is more sensitive to frequency competition than to aircraft size, Pacific Southwest Airlines (PSA) has apparently replaced its B727-200 aircraft (145 seats) with BAC-146-1 (88 seats) and BAC-146-2 (100 seats), thus attempting to dominate the air shuttle service between airports on the West Coast.

Using the estimated Equation 1, Figure 3 depicts the relationship between frequency share, relative aircraft size, and enplanement share. For any value of relative aircraft size, the relationship between market share and frequency share seems to follow an S-curve as was suggested by the fact that the coefficient β is greater than one. Also, note that in the absence of aircraft size competition (curve B), the inflection point—where the slope of the curve changes—takes place when the three competing airlines share aircraft departures equally, and the corresponding value of each market share is roughly one third.

IV. SOME POLICY IMPLICATIONS

The above discussion shows that hub dominance through increased frequency concentration appears to be the prominent airline strategy. Several issues emerge from this tendency towards airline hubbing. First, one can view the movement in airline mergers as a means for resource consolidation at airports. In addition to achieving integrated route networks, mergers between airlines hubbing at the same airport will most likely create a dominance power that not only increases the market share of the merged airlines, but might influence their pricing policy as

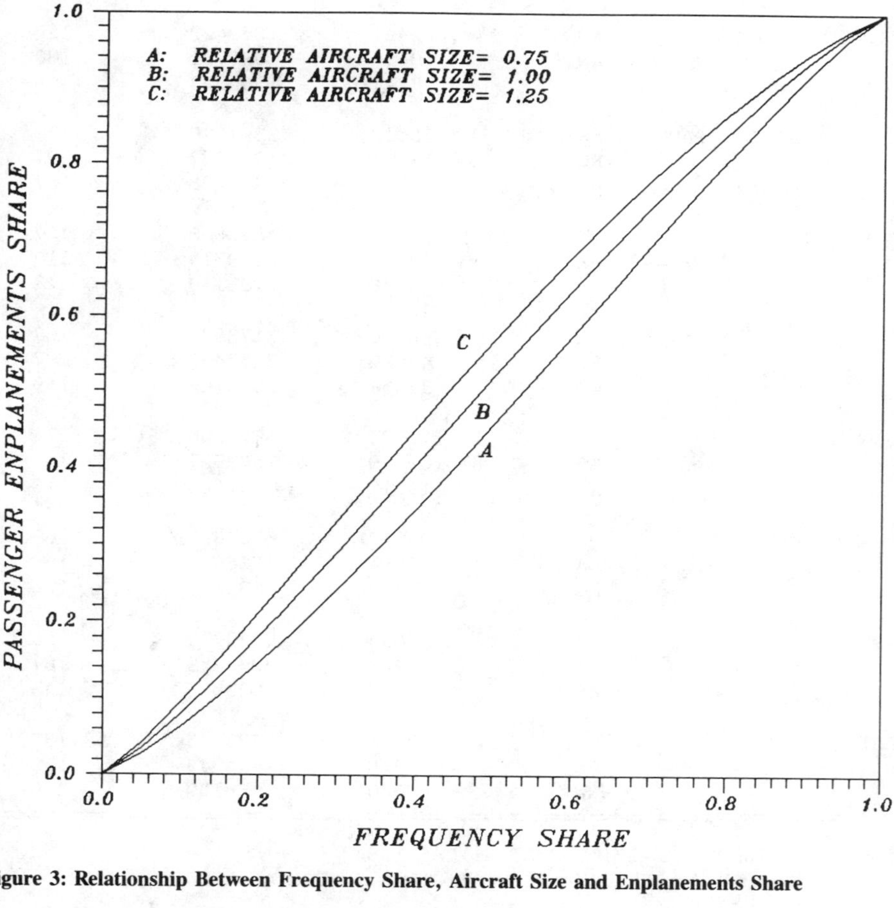

Figure 3: Relationship Between Frequency Share, Aircraft Size and Enplanements Share

TABLE 3
Results of Model Estimation*

CONSTANT α	FREQ β	SIZE Θ	R^2
0.0836 (1.79)	1.153 (29.90)	0.869 (6.635)	.938

* t-ratios in parentheses

well. A dominant airline with higher market share than its frequency share may charge its passengers higher fares as premiums for offering convenient flight schedule, single-plane service, and frequent on-line connections. From the perspective of policy making, this finding ought to be considered when approving mergers between airlines hubbing at the same airport.

Second, while hub dominance offers opportunities both for airlines and passengers, it may pose problems as well. Excessive frequency concentration at airports may result in some negative economic impacts such as congestion, pollution and noise. In an effort to alleviate congestion at selected airports, the Secretary of Transportation recently granted airlines some immunity from the antitrust laws by allowing them to engage in scheduling and capacity agreements. Under these arrangements, one would expect airline management to replace some of their flights with large-size planes in order to maintain the same market share.

Finally, this study focused on quantifying the effects of airlines' dominance strategies on their market shares. An extension of this study is to develop cost models for each strategy so as to enable airline management to conduct a cost-benefit analysis of different operating schemes.

REFERENCES

Aviation Weekly and Space Technology, March 23, 1987, pp. 33, McGraw Hill Publications.

Douglas, George and James Miller (1974), *Economic Regulation of Domestic Air Transport,* Washington: Brookings.

Ghobrial, Atef (1983), "Analysis of Airline Network Structure: The Hubbing Phenomenon," Dissertation, University of California, Berkeley, UCB-ITS-DS-83-2, 138 pp.

Kanafani, Adib and Atef Ghobrial (1985), "Airline Hubbing—Some Implications for Airport Economics," *Transportation Research,* 19A:1 (15-27).

Official Airline Guide, The Dun and Bradstreet Corporation, Illinois.

Theil, Henri (1971), *Principles of Econometerics,* New York: John Wiley.

US Department of Transportation, The Federal Aviation Administration, *Airport Activity Statistics of Certificated Route Air Carriers,* Twelve Months Ended December 1984

ENDNOTES

* Division of Business, Southern University, New Orleans, LA.
** Optima, Inc.

An Analysis of Essential Air Service to the Southeastern U.S. Since the Airline Deregulation Act of 1978

by David B. Vellenga and Daniel R. Vellenga***

ABSTRACT

This paper explores the impact of the 1978 Airline Deregulation Act upon air service to Essential Air Service (EAS) points in eight Southeastern U.S. states between 1978 and 1986. Factors such as service frequencies, seat availability, enplanement trends, interruptions or loss of service, fare levels, interline services, subsidy requirements, state promotion and aid, and the likelihood of post-1988 federal funding are evaluated. Conclusions based on this eight state sample are presented.

I. INTRODUCTION

More than eight years have elapsed since the U.S. Airline Deregulation Act of 1978 (hereafter termed ADA) was signed into law (Public Law 95-504) on October 24, 1978. Since that time numerous studies concerning the impact of ADA upon air service frequencies, profits and losses, adequacy of service, fare levels, load factors and cost structures have been presented in the transportation literature. Numerous writers have reported or demonstrated positive benefits (for example see Ellison, 1982; Graham et.al., 1982, Stephenson, et.al., 1981 and U.S. GAO reports 1981,1982,1985). Meanwhile other authors were less optimistic in their writings (for example see Langeland, 1982; Meyer, 1981; *Traffic World,* June 20, 1983; Wall Street Journal July 31, 1984 and Yoe, 1983.)

However one issue concerning airline deregulation, namely service levels to small communities, continues to be of mutual interest to both supporters and critics of the ADA. Therefore it was deemed appropriate to study the impact of deregulation on service to smaller cities in a particular region of the U.S. This study will analyze airline service to small communities in eight Southeastern states from 1978-1986. Specifically the states of Alabama, Florida, Georgia, Kentucky, Mississippi, North Carolina, Southern Carolina and Tennessee were chosen. Prior to the analysis a review of Essential Air Service (EAS) provisions of the ADA is necessary.

II. OVERVIEW OF ESSENTIAL AIR SERVICE PROVISIONS

The EAS provisions of the 1978 ADA were vital components in order to assure sufficient votes for its passage from representatives and senators from rural and less populous states and communities. These elected officials were primarily concerned over the possible loss of airline service to cities that they represented. Therefore the ADA guaranteed air service to designated EAS points for ten years (1978-1988). Other provisions of the EAS program are summarized below.

1) The first task was to define what constituted an adequate level of service at EAS points. An interim definition of EAS was two flights per week day to/from a hub (one flight per day on weekends) where a passenger could connect to the total U.S. airline system. Later specific calculations were made for each point, based on certain load factors and market demands, and specified a certain number of seats per day in a given market (See Addus, 1984; Nammeck, 1983; and the U.S. GAO, May 18, 1983 report especially pages 1-5 for more details on the EAS program).
2) A second goal of the EAS program was to minimize subsidy payments made to airlines serving small communities. Formerly these subsidies were paid to local service (feeder) carriers and were termed Section 406 (CAB Act) subsidies. The intent of the Section 406 program under regulation was to insure that local service carriers received adequate revenue to provide service in these high cost markets and to promote airline service to rural America. These payments peaked in 1981 at approximately ninety-five million dollars (CAB Form 270, 1984). Under EAS provisions a new source of subsidies, Section 419, was implemented. This subsidy program was primarily tied to the needs of small communities for air service rather than to the revenue needs of the carriers. It also was intended to promote the entry of commuter carriers (small regionals) into EAS markets. It was anticipated that total subsidy requirements would decline significantly as more efficient carriers would serve these routes at lesser (or no) subsidy rates. In actuality two Section 419 programs exit. Section 419(a) provides funding for ten years to EAS points. Section 419(b) was intended to be a funding source for potential new routes/markets that developed in the deregulated environment.
3) The program also mandated that continuous service be provided at EAS points. Therefore a major or national carrier could not exit a market unless a replacement carrier was in position to provide service. Potential and actual service interruptions could (did) exist (See Addus, 1984; Meyer, 1981; Oster et.al., 1984; U.S. GAO Report, May 18, 1983).
4) In the most extreme cases a replacement for a replacement carrier had to be found and autho-

rized by the CAB (Office of Essential Air Service, D.O.T. since 1985). Therefore the original carrier had to resume service (or be "held-in") in the market until an adequate replacement service was in place.

5) Proposals for "bumping rights" were also made (D.O.T. Replacement Carrier Procedures, 1984). In this case a more efficient carrier (defined as one requiring less or no (Section 419) subsidies could bump the current airline from, the market.

III. ADDITIONAL CONSIDERATIONS/ CONCERNS

An evaluation of essential air service should consider additional factors. These include flight frequencies/departures, the number of seats per market (See Ahmed, 1984; Stephenson et. al.; 1981, Warren, 1984), the safety record of commuter carriers vis-a-vis majors or nationals (See Oster et. al., 1984) and the level of fares at EAS points compared to fares at major cities (See Langland, 1982; Rose 1981). Other considerations deal with the partnerships between commuter/regional air lines and major/nationals and services such as reservations, interline fares, baggage checks, tickets and the availability of gate and ticket counter space at hubs. (See Junod, 1983; Kilman, 1984; *Wall Street Journal,* July 31, 1984).

Longer run considerations include the roles of local governments and state level departments of transportation in initiating, funding and promoting air service at EAS points. (Williamson et. al., 1982) Another issue is whether air service will continue at EAS points when Section 419 funds end in 1988. Additionally do these states support current efforts of the National Committee of Cities and States for Airline Service (NCCSAS) task force to develop legislation that will ensure the continuation of the EAS program (Regional Airline Association *Times, 1986).*

IV. METHODOLOGY

The topic of the level of air service to small communities since deregulation has been of considerable interest to many from industry, government, professional associations, academics and consulting firms. Therefore, an analysis of the deregulation experiences in eight Southeastern states from 1978-1986 was undertaken. These states are Alabama, Florida, Georgia, Kentucky, Mississippi, North Carolina and South Carolina. The data collection consisted of three distinct phases.

1) Initially a questionnaire concerning air service at EAS points was sent to the state director of aeronautics (or equivalent) at each of these eight states. In addition a personal interview was made with one director and phone interviews were made with directors or associates in most of the remaining states.
2) The second phase consisted of collecting EAS data from secondary sources. The primary sources were the annual reports of the Regional Airline Association from 1982-1986.
3) The third phase included data collection and an interview with an official from the Office of Essential Air Service in Washington, D.C. (Federal D.O.T.). He proved to be a valuable source of data concerning subsidy levels, air service levels, replacement carriers etc.

V. RESEARCH QUESTIONS/TOPICS

Specifically this study asked the following questions related to EAS from 1978-86:

1) How do levels of air service compare during this period as measured by the number of weekly departures and available seats to/from an EAS point?
2) Have there been trends (positive or negative) in passenger enplanements?
3) Has air service been interrupted at any of these cities due to carrier failures or lack of suitable replacements? Have any carrier been "held-in" a market?
4) Have carriers exercised "bumping rights" in any of these markets?
5) Have any new promotional markets been established under the Section 419(b) program?
6) What changes/trends exist in subsidy payments? A comparison between Section 406 and 419 programs is also made.
7) Have interline fares, ticketing, baggage transfers been available?
8) What will happen to EAS points post 1988? Do these states support new legislative efforts to extend the program?
9) What special steps have states taken to develop or promote air service at these communities?

The remaining sections of this paper will summarize the findings of this study. Initially an illustration of the data collection method and format is presented. Secondly the results are shown on a variable by variable basis across all eight states. Finally, generalizations are summarized in the conclusion section.

VI. DATA COLLECTION, ILLUSTRATION AND FORMAT

Table 1 shows that Alabama has five EAS points. The initial portion of Table 1 (and also Tables 2-8) shows the EAS city, its designated hub (or hubs), its guaranteed daily level of EAS (1978 CAB determination in most cases) and the actual levels of service between 1983 and 1986. As an illustration Anniston is the first EAS point listed for Alabama. Its designated hub is Altanta (ATL). EAS was defined as 2 daily (weekday) round trips (RT) and eighty seats to/ from Atlanta. Service levels have exceeded required minimums in numbers of RTs (4 or 5 versus 2 required) but the number of seats fluctuated between 90 and 250. The most recent flight frequency and seat data (November, 1986) summarized in Table 9 show that three points experienced increases in flight frequencies while two had declines. All five points show a decrease in the number of seats ranging from −8.3 to −73.2%. In no instances did service levels dip significantly below EAS guaranteed minimums. Enplanement trends are all negative. (Enplanement statistics for non-subsidized EAS points were not readily obtainable for 1985).

TABLE 1

STATE OF: ALABAMA (5) EAS CITIES	HUB	DAILY GUARANTED 1) EAS LEVELS ROUND TRIPS	SEATS	1983 (1) R.T./SEATS	ACTUAL DAILY SERVICE RATES 1984 (2) R.T./SEATS	1985 (3) R.T./SEATS	1986 (4) R.T./SEATS	SEC 406 SUBSIDIES 1981(5)	1982(1)	SEC 419 SUBSIDIES 1983(2)	1984(3)	1986(4)	PASSENGER ENPLANEMENTS 1978(6)	1985(6)	1986(7)
Anniston	ATL	2	80	5 250	5 94	4 144	4 90						30,408	16,808	---
Dothan	ATL	2	80	6/5 108/569	5/4 126/409	7/4 262/355	11 338	494,139					72,808	56,250	
Gadsden	ATL	2	39	3 54	3 55	3 54	2 36	407,749		295,929	232,986	111,409	6,508	2,712	2,191
Muscle	ATL	2	80	4/2 72/181	2/2 36/124	1 36	2 36						24,751	17,203	--
Shoals/	MEM			1/3 18/168	3 132	4 73	5 93	558,301							
Florence/															
Sheffield/															
Tuscumbia															
Tuscaloosa	ATL	2	80	2/2 36/142	3/2 54/134	2 36	3 90						34,569	18,386	
	MEM			3/2 54/112	- -	2 36		460,216							
							TOTALS $	1,920,405	N/A	295,929	232,986	111,409			

*Sources: (1) 1982 Regional/Commuter Airline Industry Annual Report (pp. 133 - 149)
(2) 1983 Regional Airline Industry Annual Report (pp. 97 - 111)
(3) 1985 Regional Airline Industry Annual Report (pp. 103 - 120)
(4) 1986 Regional Airline Industry Annual Report (pp. 86 - 103)
(5) National Association of State Aviation Officials, Airline Subsidy Payments, March 18, 1981.
(6) Cab Report to Congress, "Implementation of the Provisions of the Airline Deregulation Act of 1978," Jan. 31, 1984
(7) Internal DOT Report, Office of Essential Air Service, March 22, 1985. (Only available for subsidized points)

*Note: Sources apply to Tables 1-8.

Tables 1 and 10 also show that subsidy levels have declined significantly. Section 406 subsidies (prior to the ADA of 1978) were over $1.9 million in 1981. Subsidies under the Section 419 program fell to $111,409 in 1986. This is a very positive trend.

Unfortunately it is not possible to answer in detail some of the other questions addressed in this study since Alabama officials did not respond to the questionnaire. Based on other data, however it does not appear that Alabama is a supporter of proposed legislation to extend the EAS program and subsidies beyond 1988.

Space limitations preclude making a detailed state by state analysis as was made for the illustration State of Alabama. Instead the data are summarized on a variable by variable basis.

VII. VARIABLE BY VARIABLE SUMMARIES

A. Weekly Departures and Weekly Seats at EAS Points

Typically air service at EAS points is measured by the gain (or loss) in weekly flight departures and the gain (or loss) in the number of available seats per week. Furthermore, it was usually anticipated that flight frequencies (weekly departures) would increase during the post deregulation period as commuter carriers replaced larger airlines in small markets. It was predicted that the number of seats would decline since commuters use smaller aircraft. Table 9 presents these data for our eight state sample

TABLE 2

STATE OF: FLORIDA (6) EAS CITIES	HUB	DAILY GUARANTED 1) EAS LEVELS ROUND TRIPS	SEATS	1983 (1) R.T./SEATS	ACTUAL DAILY SERVICE RATES 1984 (2) R.T./SEATS	1985 (3) R.T./SEATS	1986 (4) R.T./SEATS	SEC 406 SUBSIDIES 1981(5)	1982(1)	SEC 419 SUBSIDIES 1983(2)	1984(3)	1986(4)	PASSENGER ENPLANEMENTS 1978(6)	1985(6)	1986(7)
Eglin	ATL	2	80	2/4 36/542	4/4 72/493	6/4 238/370	8 344	487,042					116,323	106,744	
A.F.B.	MCO			2/2 16/180	4/1 32/139	4/1 60/100	3 45								
Gaines-	ATL/	2	80	5 665	4 489	4 488	4/4 120/483	--					143,466	132,293	
ville	MCO			1 8	- --	- -	5 40								
	TPA/			3 24	3 31	5 50	4 32								
	MIA			1 125	2/1 30/139	4/1 62/99	3/5 24/464								
Keywest	MIA	2	80	15/1 612/122	28 674	31 510	16/5 388/367	--					54,179	106,537	
Panama	ATL	2	80	4 458	4 479	11/2 204/170	11 322						78,708	64,719	
City	MCO			2 215	1/2 8/215	8 92	3 38								
	TPA			3 34	4 32	11 113	8 92	373,102							
Lakeland*	TPA	-	--	- -	- -	- -	- -								
Melbourne	ATL	2	80	5 687	8 998	8 930	8 923						161,058	193,233	
	MCO			2/1 16/139	9/1 160/145	3/1 24/149	5/1 40/149								
	TAP			4/2 72/225	1 145	4 42	- -								
	MIA			6 58	4 32	4 32	8 64								
							TOTAL $	860,144	N/A	N/A	N/A				N/A

*Eligible for essential service, but served thru Tampa

TABLE 3

STATE OF: GEORGIA (9) EAS CITIES	HUB	DAILY GUARANTED 1) EAS LEVELS ROUND TRIPS	SEATS	1983 1) R.T./SEATS	ACTUAL DAILY SERVICE RATES 1984 2) R.T./SEATS	1985 3) R.T./SEATS	1986 4) R.T./SEATS	SEC 406 SUBSIDIES 1981 5)	1982 1)	SEC 419 SUBSIDIES 1983 2)	1984 3)	1986 4)	PASSENGER ENPLANEMENTS 1978 6)	1982 6)	1985 7)
Albàny	ATL	2	80	4/4 72/395	4/4 104/360	12 382	11 325	$619,012	--	--	--	--	67,811	60,984	50,066*
Athens	ATL	2	80	4 108	4 75	4 102	2 30	--	$264,920	$398,683	$398,683	$250,033	15,486	3,966	8,816*
Augusta	ATL	2	80	5 725	4/6 62/810	10/7 224/710	10/6 258/588	--	--	--	--	--	203,065	137,007	200,891*
Brunswick/ St. Simons Island	ATL	2	80	4 104	4 72	4 90	4 72	--	$184,000	--	--	--	15,856	10,544	11,335*
Columbus	ATL	2	80	5/4 95/520	4/4 200/520	5/4 176/412	9/9 256/442	--	--	--	--	--	170,732	94,112	105,873*
Macon	ATL	2	80	8/2 211/250	7/2 192/250	8/2 244/176	9/9 304/196	--	--	--	--	--	96,466	53,893	53,600*
Moultrie/ Thomasville	ATL	2	36	2 36	3 27	3 27	2 38	$427,867	$262,000	$478,899	$478,899	$404,240	4,302	845	1,565
Valdosta	ATL	2	80	3/2 54/180	3/2 54/180	4 154	4 108	$372,157	--	--	--	--	27,090	20,135	14,994*
Waycross	ATL	2	20	- -	2 16	no longer in service	no longer in service	--	--	$249,775	$249,775	--	468	--	--
							TOTALS	$1,419,036	$710,920	$1,127,357	$1,127,357	654,273			

Source: *GA-D.O.T.., 1985

TABLE 4

STATE OF: KENTUCKY (3) EAS CITIES	HUB	DAILY GUARANTED 1) EAS LEVELS ROUND TRIPS	SEATS	1983 (1) R.T./SEATS	ACTUAL DAILY SERVICE RATES 1984 (2) R.T./SEATS	1985 (3) R.T./SEATS	1986 (4) R.T./SEATS	SEC 406 SUBSIDIES 1981(5)	1982(1)	SEC 419 SUBSIDIES 1983(2)	1984(3)	1986(4)	PASSENGER ENPLANEMENTS 1978(6)	1985(6)	1986(7)
London/ Corbin	SDF	2	18	2 30	2 30	1 8	- -*	241,214	250,460	250,460	291,243		3,058	613	313
Owensboro	SDF	2	72	4 60	5 75	6 90	4 60						11,802	8,879	--
Paducah	SDF	2	80	4 60	5 75	2 45	3 45						55,316	29,571	
	STL			4 76	17 284	7 129	8 180								
							TOTALS $	241,214	250,460	250,460	291,243				

*EAS redefined as zero

TABLE 5

STATE OF: MISSISSIPPI (10) EAS CITIES	HUB	DAILY GUARANTED 1) EAS LEVELS ROUND TRIPS	SEATS	1983 (1) R.T./SEATS	ACTUAL DAILY SERVICE RATES 1984 (2) R.T./SEATS	1985 (3) R.T./SEATS	1986 (4) R.T./SEATS	SEC 406 SUBSIDIES 1981(5)	1982(1)	SEC 419 SUBSIDIES 1983(2)	1984(3)	1986(4)	PASSENGER ENPLANEMENTS 1978(6)	1985(6)	1986(7)
Columbus	ATL	2	80	2/3 36/236	3 224	4/3 72/255	3 54	694,961					46,307	38,148	
	MEM			3/4 54/292	4 268	3/4 54/355	8 208								
Greenville	MEM	2	80	4 293	3 178	2 170	9 182						34,999	21,484	
Greenwood	MEM	2	28	3 34	2 36	2 28	-* -	258,325	97,370[1]	430,901[1]	350,318		2,392	813	626
Gulfport/ Biloxi	ATL	2	80	4 465	4 465	3 270	3 108	727,300					91,443	84,614	
	MSY			4 395	2 250	- -	- -								
Jackson/ Vicksburg	ATL	2	80	7 1,234	8 1,120	8 988	8 996						429,086	302,781	
	DFW			8 1,184	8 1,160	5 656	6 882								
Laurel/ Haites- burg	MEM	2	80	2 112	1 90	2 100	2 36						20,899	19,478	
	ATL			2 180	2 88	3 55	3 38								
Meridian	MSY	2	80	2 112	2 180	3 150	3 90	464,083					43,376	30,649	
	ATL			2 180	1/2 18/88	4 72	9 167	371,011	359,052	195,108	195,108				
Natchez	MSY	2	24	2 36	2 36	1 18	1 18						3,685	1,347	1,252
Tupelo	ATL	2	80	1/1 18/56	1 44	- -	- -	612,677				114,681	15,635	12,564	
	MEM			2/3 36/168	3 132	3 81	7 148								
University/ Oxford	MEM	2	12	3 36	2 36	2 16	2 16	491,099	(1)	(1)	322,210		1,696	625	313
							TOTAL $	3,619,456	456,422	626,009	867,636	114,681			

(1)
Represents total subsidy at Greenwood & University/Oxford (MS).
*EAS is redefined as zero.

TABLE 6

STATE OF: No. CAROLINA (10) EAS CITIES	HUB	DAILY GUARANTED 1) EAS LEVELS ROUND TRIPS	SEATS	ACTUAL DAILY SERVICE RATES 1983 (1) R.T./SEATS		1984 (2) R.T./SEATS		1985 (3) R.T./SEATS		1986 (4) R.T./SEATS		SEC 406 SUBSIDIES 1981(5)	1982(1)	SEC 419 SUBSIDIES 1983(2)	1984(3)	1986(4)	PASSENGER ENPLANEMENTS 1978(6)	1985(6)	1986(7)
Asheville	ATL	2	80	3/3	86/366	3/3	150/366	3/2	136/130	9	235	-	-	-	-	-	191,292	149,245	-
	DCA			1	122	6/4	90/512	6/4	90/500	6/4	90/500								
Fayette-	ATL	2	80	4	488	3	366	?	?	?	?	-	-	-	-	-	164,893	119,606	-
ville	DCA			2	244	2/2	16/244												
Hickory	ATL	2	80	2	30	9	87	12	234	10	237	-	-	-	-	-	21,409	13,735	-
Jackson	RDU	2	80	1	122	1	122	-	-	-	-	-	-	-	-	-	48,380	51,575	-
ville/	ORF			1	122	-	-	-	-	-	-	-	-	-	-	-			
Camp Lejune	RIC			-	-	2	245	1	112										
	DCA			1	122	-	-	-	-	3	336								
	ATL			1	122														
	CLT			2	244														
Kinston/	RDU	2	80	-	-	-	-	4	39	-	-	-	-	-	-	-			
Goldsburg	ORF			-	-	1	122	-	-	-	-								
Greenville	RIC			1	122	1	122	1	145	1	164								
	DCA			1	122	1	122	1	65	-	-								
	ATL			1	122	-	-	-	-	1/3	15/388	783,450	-	-	-	-	58,419	46,580	-
	CLT			1	122	2	245	2/2	30/224										
New Bern/	RDU	2	80	3	62	1	18	5	95	-	-								
Morehead	ORF			-	-	3	57	-	-	4	144								
City/	RIC			2	36	-	-	-	-										
Beanford	DCA			2	36	-	-	-	-			-	-	-	-	-	32,609	12,176	-
Rocky Mt/ Wilson	RDU	2	58	1	18	1	18	2	38	3	57		270,766	270,766	131,820	147,222	9,049	2,005	626
Wilmington	ATL	2	80	3	366	2	245	2	177	1	112								
	RDU			3/1	31/122	3/1	31/122	2	16	3	35	1,046,254	-	-	-		88,097	77,625	
	ORF			1	122	1	122	1	112	1	112								
	RIC			2	244	1/1	15/122	2	130	-	-								
	DCA			1/1	15/122	1/2	15/245												
Winston-	.DCA	2	80	2	244	-	-	-	-	-	-	1,188,571	-	-	270,409	146,897	57,804	22,196	2817
Salem	ATL			2	244	-	-	-	-	-	-								
	CLT			-	-	5	62	4	60	5	90								
										TOTAL $		3,319,590	270,766	270,766	402,229	294,119			

*or any hub in NC or further north.

TABLE 7

STATE OF: So. CAROLINA (3) EAS CITIES	HUB	DAILY GUARANTED 1) EAS LEVELS ROUND TRIPS	SEATS	ACTUAL DAILY SERVICE RATES 1983 (1) R.T./SEATS		1984 (2) R.T./SEATS		1985 (3) R.T./SEATS		1986 (4) R.T./SEATS		SEC 406 SUBSIDIES 1981(5)	1982(1)	SEC 419 SUBSIDIES 1983(2)	1984(3)	1986(4)	PASSENGER ENPLANEMENTS 1978(6)	1985(6)	1986(7)
Florence	ATL	2	80	4	76	5	95	7	100	3	57	381,785	-	-	-	-	32,281	21,204	
	CLT			4	76	12	146	11	149	8	218								
	DCA				-	-	-	-	-										
Myrtle	ATL	2	80	2	244	2/2	38/245	2	244	4/2	76/130	704,618	-	-	-	-	85,220	91,916	
Beach	CLT			4/2	76/244	5/4	51/490	4	448	2/5	38/560								
	DCA			1	122	-	-												
Greenville/	ATL	2	80	-	-	2/2	38/245	6/7	192/801	5/14	194/2625		-	-	-	-	334,366	255,868	
Spartan- burg	CLT	(1984)	(1984)			5/4	51/490	1/7	30/809	2/8	60/894								
										TOTAL $		1,086,403	N/A	N/A	N/A	N/A			

TABLE 8

STATE OF: TENNESSEE (2) EAS CITIES	HUB	DAILY GUARANTED 1) EAS LEVELS ROUND TRIPS	SEATS	ACTUAL DAILY SERVICE RATES 1983 (1) R.T./SEATS		1984 (2) R.T./SEATS		1985 (3) R.T./SEATS		1986 (4) R.T./SEATS		SEC 406 SUBSIDIES 1981(5)	1982(1)	SEC 419 SUBSIDIES 1983(2)	1984(3)	1986(4)	PASSENGER ENPLANEMENTS 1978(6)	1985(6)	1986(7)
Clarksville/ Ft. Cambell/ Hopkinsville(KY)	BND	2	27	2	30	2	30	2	30	3	24	-	116,174	125,239	125,239	299,715	4,100	1,474	1545
Jackson	MEM	2	80	4	76	3	57	3	57	3	57	705,790	-	-	-	-	17,467	8,987	-
										TOTAL $		705,790	116,174	125,239	125,239	299,715			

between November, 1978 and November, 1986. An overall summary is shown here.

Number of EAS Points Showing Gains or Losses in:

	Weekly Departures		Weekly Seats	
	(November, 1978 to November, 1986)			
	Gains	Losses	Gains	Losses
Alabama	3	2	0	5
Florida	5	0	3	2
Georgia	6	3	1	8
Kentucky	2	1	1	2
Mississippi	7	3	2	8
North Carolina	6	4	2	8
South Carolina	2	1	2	1
Tennessee	1	1	0	2
Totals	32	15	11	36

As can be seen the results are as anticipated by better than a 2:1 ratio in increased departures per week and by more than a 3:1 ratio in a decline in weekly seats. These results indicate that commuter carriers have entered many markets and offer greater frequencies in smaller aircraft.

B. Levels of Service Above or Below EAS Guaranteed Minimums

As mentioned in the overview section EAS points are guaranteed minimum levels of service (usually two round trips per week day to a hub airport) by the D.O.T. until 1988. It is pos- sible to determine whether current levels of service exceed or fall below these standards. A summary of the data from Tables 1- 8 is presented below.

Levels of Air Service at EAS Points 1978 - 1986

	Above Minimums	Below Minimums
Alabama	4	1
Florida	5	0
Georgia	7	2
Kentucky	2	1
Mississippi	7	3
North Carolina	8	2
South Carolina	3	0
Tennessee	2	0
Totals	38	9

The vast majority of EAS cities now receive service above minimum guaranteed levels (greater than a 4:1 ratio). In many cases the levels of service greatly exceeds requirements (See Tables 1-8). This also includes cases where service is provided to multiple hubs. Conversely some points below minimums are barely below minimums or service has been terminated with little or no complaint from the community or state D.O.T.

C. Enplanement Trends

Still another indicator of the success (or lack thereof) of the EAS program is to analyze passenger boarding trends at these points. Increased boardings would indicated customer approval and acceptance of these services and demonstrate a viable market. Conversely a lack of growth would indicate less potential for a long-term market. An overview of enplanement data from Tables 18 is below.

Enplanement Trends 1978 - 1985 (or 1986 if available)

	Upward	Downward
Alabama	0	5
Florida	2	3
Georgia	0	9(some improvement 1985-86)
Kentucky	0	3
Mississippi	0	10
North Carolina	2	8
South Carolina	1	2
Tennessee	0	2
Totals	5	42

Overall enplanement trends from 1978 to 1985 (or 86) are quite negative with declines over gains by over an 8:1 ratio. In a few instances there have been improvements in 1986 over 1985 (for example see Georgia data Table 3). Overall this finding is discouraging and doesn't predict growth in many markets.

D. Service Interruptions or Inability to Find Adequate Replacement Carriers

A key provision of the EAS program is that carriers cannot exit a market unless an adequate replacement carrier is found. In some instances replacements for replacements might be necessary. Therefore, the questionnaire sent to the state D.O.T. Aeronautics Division representatives asked if this had been a problem.

Overall the responses were quite positive. Only in two or three instances did service cease for brief periods. Replacement carriers were usually found with little difficulty. Georgia experienced problems in Brunswick and Athens in finding replacement carriers in earlier years. (In the case of Athens three airlines have been in the market). South Carolina reported a similar experience in Florence.

These problems do not appear to be of much concern to state or local officials in this sample group of states.

E. Exercise of Bumping Rights

In 1984 the D.O.T. made a proposal termed "bumping rights". Under this program a more efficient airline (one defined as requiring less or no Section 419 subsidies than the incumbent) could "bump out" or take over a route upon approval of the D.O.T.

To date no carrier has exercised "bumping rights" in these eight states. This pattern is consistent with findings for other areas of the country. No carriers have yet been awarded routes under this program by the D.O.T.

F. Establishment of Section 419(b) Promotional Markets/Routes

Efforts were made to promote fundings (Section 419(b)) of new routes that offered high potentials for air service. (These cities were not receiving air service when the EAS guarantees were made in 1978). State aviation officials were asked if they had applied for this program.

Only Georgia had attempted to promote/develop markets under this program. One attempt was in Waycross where service was actually implemented for one year. However, due to the controversies with the CAB concerning the levels and timing of subsidy payments and lack of passengers the service was terminated in 1984. The application for service to Rome was not approved by the CAB.

G. Subsidy Requirements/Trends

As mentioned in the overview section one major goal of the EAS program was to reduce the amounts

TABLE 9

CHANGE IN AIR SERVICE LEVELS - NOVEMBER, 1978 TO NOVEMBER, 1986

ALABAMA	DEPARTURES/WK 11/78	11/86	%CHANGE	SEATS/WK 11/78	11/86	%CHANGE
Anniston	80	31	- 61.2	1,275	589	- 53.8
Dothan	53	83	56.6	4,240	2,409	- 43.1
Gadsden	20	18	- 10.0	360	330	- 8.3
Muscle Shoals	53	59	11.3	1,462	1,298	- 11.2
Tuscaloosa	40	45	12.5	3,200	855	- 73.2
FLORIDA						
Eglin A.F.B.	74	152	105.4	6,200	6,045	- 2.5
Gainesville	70	217	210.0	6,461	6,649	2.9
Keywest	71	411	478.8	8,206	5,574	211.7
Lakeland	--	--	--	--	--	--
Melborurne	91	192	110.9	10,906	9,240	- 15.2
Panama City	57	154	170.1	4,900	4,905	0.1
GEORGIA						
Albany	49	83	69.3	4,900	2,289	-53.2
Athens	34	61	79.4	612	939	53.4
Augusta	127	155	22.0	13,833	9,091	-34.2
Brunswick/ St. Simons Island	22	38	72.7	948	722	-23.8
Columbus	112	105	- 6.2	9,842	4,729	-51.9
Macon	29	73	151.7	2,569	2,429	- 5.4
Moultrie/ Thomasville	13	11	-15.3	234	209	-10.6
Valdosta	21	38	80.9	2,100	722	-65.6
Waycross	10	0	-100.0	340	0	-100.0
KENTUCKY						
London	21	0	-100.0	1,218	0	-100.0
Owensboro	50	74	48.0	750	1,167	55.6
Paducah	60	151	151.6	4,076	2,751	- 32.5
MISSISSIPPI						
Columbus	41	97	136.5	3,280	2,305	- 29.7
Greenville	28	45	60.7	2,240	1,095	- 51.1
Greenwood	14	0	-100.0	252	0	-100.0
Gulfport/Biloxi	42	70	66.6	3,360	2,931	- 12.7
Jackson/Vicksburg	288	305	5.9	26,323	27,186	3.2
Laurel/Haitiesburg	35	39	11.4	1,638	936	- 42.8
Meridian	56	90	60.7	3,570	2,107	- 40.9
Natchez	31	0	-100.0	527	0	-100.0
Tupelo	40	85	112.5	700	2,254	213.0
University	28	0	-100.0	504	0	-100.0
NORTH CAROLINA						
Asheville	143	186	30.0	12,321	8,922	- 27.5
Fayetteville	117	130	11.1	12,368	11,628	- 5.9
Greenville	20	62	210.0	255	1,484	481.9
Hickory	74	48	- 35.1	2,312	1,609	- 30.4
Jacksonville	47	34	- 27.6	4,252	3,641	- 14.3
Kinston	48	49	2.0	4,184	3,693	- 11.7
New Bern	53	35	- 33.9	3,074	1,295	- 57.8
Rocky Mt.	26	34	30.7	1,508	642	- 57.4
Wilmington	88	105	19.3	7,558	9,485	25.4
Winston Salem	71	41	- 42.2	6,593	801	- 87.8
SOUTH CAROLINA						
Florence	75	73	- 2.6	3,316	1,761	- 46.8
Myrtle Beach	65	149	129.2	4,660	9,136	96.0
Greenville/ Spartanburg	188	270	43.6	15,321	23,336	52.3
TENNESSEE						
Clarksville	12	18	50.0	528	126	- 76.1
Jackson	85	33	- 61.1	1,269	594	- 53.1

Source: Office of Essential Air Service, U.S. D.O.T. November, 1986 Report.

of subsidies paid to carriers for serving small communities. To foster this objective the Section 419(a) program was implemented. Table 10 illustrates the results of this program.

It is noteworthy that total subsidy requirements for these states has fallen from a high of $13,172,038 under the former Section 406 program in 1981 to only $1,474,197 Section 419(a) monies in 1986. Of particular interest is the fact that Florida, Kentucky, Mississippi and South Carolina no longer receive any subsidies.

It is also of interest to attempt to project the number of EAS points in this sample that can exist in the free market when the 419(a) program ends in 1988.

	Subsidy Free Points	(1986)	Subsidized Points
Alabama	4		1
Florida	5		0
Georgia	7		2
Kentucky	3		0
Mississippi	10		0
North Carolina	8		2
South Carolina	3		0
Tennessee	1		1
Totals	41		6

It is encouraging to observe that only six EAS points remain on subsidy. In many cases the trend has been less subsidy than prior years (e.g., Gadsden, AL; Athens, GA; Winston-Salem, NC). It appears therefore that relatively few markets in this region of the U.S. are candidates for loss of service in 1988.

H. Interline Fares/Ticketing/Baggage Transfer Issues

One common concern of air service to small communities is the relatively high air fares in these short haul markets. Additional concerns are the availability (or lack thereof) of joint/through fares, tickets and baggage checking.

In general the state aviation officials were quite satisfied with interline services. Most agreed with the concern of relatively high short haul fares in intrastate markets. (a notable exception was Florida). Some gave examples of passenger driving to larger hubs to take advantage of lower super-saver fares. However, most expressed the idea that higher costs in short haul, low density markets warrant higher fares. None of the state officials expressed concern over levels of complaints by air passengers.

I. Post-1988 Funding and Participation in NCCSAS Legislation Proposals

As noted earlier the EAS program guarantees for service and funding are scheduled to end in 1988. In addition the National Committee of Cities and States for Airline Service (NCCSAS) is attempting to develop legislation that will ensure the continuation of the EAS program beyond 1988. Therefore this study asked for the opinions of the surveyed state officials as to whether they felt the EAS program would continue beyond 1988 and if their state supported the NCCSAS efforts.

The findings were almost unanimous. None of the respondents felt that the EAS program and Section 419(a) monies would be extended beyond 1988. Also none supported or were even aware of the NCCSAS efforts.

J. State Efforts to Promote/Develop Air Service at EAS Points

Levels of activity varied widely between states. Some officials were actively involved while others

TABLE 10

Comparison of 406 vs. 419 subsidies 1981 - 1986

State	1981 406 Subsidy	1982 419 Subsidies	1983	1984	1986
ALABLAMA	$ 1,920,405	N/A	$ 295,929	$ 232,986	$ 111,409
FLORIDA	$ 860,144	N/A	N/A	N/A	N/A
GEORGIA	$ 1,419,036	$ 710,920	$1,127,357	$1,127,357	$ 654,273
KENTUCKY	$ 241,214	$ 250,460	$ 250,460	$ 291,243	N/A
MISSISSIPPI	$ 3,619,456	$ 456,422	$ 626,009	$ 867,636	$ 114,681*
NORTH CAROLINA	$ 3,319,590	$ 270,766	$ 270,766	$ 402,229	$ 294,119
SOUTH CAROLINA	$ 1,086,403	N/A	N/A	N/A	N/A
TENNESSEE	$ 705,790	$ 116,174	$ 125,239	$ 125,239	$ 299,715
TOTALS	$13,172,038	$1,804,742	$2,695,760	$3,046,690	$1,474,197

Sources: Summary of Tables 1-8
*Mississippi - level down to zero as per late 1986 (MS., D.O.T.)

were content to let the "test of the market" prevail. Examples of state efforts are listed below.

- No states provided direct subsidies to airlines or communities
- Aid communities in solicitation of new carriers to a particular city
- Make marketing studies for small communities
- Local community funding for commuter airline advertising
- Statewide study/survey of commuter air services
- Reduced fuel taxes for commuter airlines

VIII. CONCLUSIONS AND SUMMARY

This section summarizes the findings concerning EAS in eight Southeastern States over the nine research questions/topics.

1) The vast majority of EAS communities have experienced increased flight frequencies (departures) at the expense of fewer weekly seats as commuters replace larger carriers. Most consumers would prefer an increased number of flights in a market.
2) Passenger enplanement figures are mixed. However, negatives outweigh positives. Many subsidy free markets exist and can continue beyond 1988. Some markets demonstrate little hope for survival and could lose all service in the near future.
3) Service interruptions have been very infrequent at EAS points. Only in extremely rare cases were replacements not found or incumbents "held-in". Georgia and South Carolina experienced more carrier turnover problems than other states.
4) No state has experienced the "bumping-right" procedure. In fact no cases have existed to date in the U.S. (D.O.T. office of Essential Air Service).
5) Georgia was the only state in this sample that attempted to develop Section 419(b) cities. (Rome, Waycross). Neither cases were successful and do not have service today.
6) Subsidy levels for these eight states have declined dramatically from a high of $13.2 million in 1981 (Section 406) to $1.5 million in 1986 (Section 419) (See Table 10). This is an 89% reduction in federal funding.
7) Interline ticketing, baggage service and fares/rates are acceptable. The most common complaint is relatively high fares for short-haul markets (usually intra-state). State officials generally agree that this is the result of higher costs and lower demand markets.
8) None of the respondents felt that the EAS program and Section 419 subsidies will continue beyond 1988. Also none participated or were even aware of the new legislative efforts of the NCCSAS group.
9) Efforts by state officials range from great to almost none. Some respondents gave the impression of no interest or very limited (even erroneous) knowledge of their state EAS programs. No state provided any funding to carriers. Most state aid consisted of surveys, aid to small cities and work with commuter carriers.

REFERENCES

Addus, Abdussalam A., "Essential Air Service Determination for Small Communities," *Transportation Quarterly,* Vol. 38, No. 4, October, 1984, pp. 559-574.

Ahmed, Samir A., "Air Transportation to Small Communities: Passenger Characteristics and Perceptions of Service Attributes," *Transportation Quarterly,* Vol. 38, No. 1, January, 1984, pp. 15-21.

"Airline Subsidy Payments," by the National Association of State Aviation Officials, Washington, D.C., March 19, 1981, 7 pages.

"Deregulation: Increased Competition is Making Airlines More Efficient and Responsive to Consumers," U.S. General Accounting Office Report GAO/RCED-86-26 dated November 6, 1985. "DOT's View of Airline Deregulation Challenged by Small Cities, Labor," *Traffic World,* June 20, 1983, pp. 16-17.

"Ellison, Anthony P., "The Structural Change of the Airline Industry Following Deregulation," *Transportation Journal,* Vol. 21, No. 3. Spring 1982, pp. 58-69.

"Essential Air Service Program: Post 1988," Regional Airline Association Times, November-December, 1986, pp. 17-18.

Graham, David R. and Daniel P. Kaplan, "Competition and the Airlines: An Evaluation of Deregulation," Staff Report, Office of Economic Analysis, Civil Aeronautics Board, Washington, D.C., December, 1982 (especially Chapter Five "Small Community Service").

Junod, Thomas and E. F. Cudworth, "Majors Speak Out on Commuters' Future," *Commuter Air,* Vol. 5, No. 12, December 1983, pp. 24-26.

Kilman, Scott, "Airlines' Links to Regionals Spark Battles," *The Wall Street Journal,* May 16, 1984, p. 31.

Langland, Rolf, "Airline Deregulation: An Upper Midwest Perspective," report prepared by the Upper Midwest Council, Minneapolis, MN, June, 1982.

Meyer, James S., "Section 419 of the Airline Deregulation Act: What Has Been the Effect on Air Service to Small Communities?" *Journal of Air Law and Commerce,* Vol. 47, No. 1, Fall, 1981, pp. 151-185.

"More Flexible Eligibility Criteria Could Enhance the Small Communities Essential Air Service Subsidy Program," U.S. General Accounting Office Report GAO/RCED-83-97 dated May 18, 1983.

Nammack, John A., "EAS": Facing Reality," *Commuter Air,* Vol. 5, No. 11, November, 1983, pp. 20-22.

Oster, Clinton V. and C. Kurt Zorn, "Deregulation and Commuter Airline Safety," *Journal of Air Law and Commerce,* Vol. 49, No. 2, Winter, 1984, pp. 315-335.

"Regional/Commuter Airline Industry Annual Report, 1982," Prepared by the Regional Airline Association, Washington, D.C., June, 1983 (especially section entitled, "Summary of U.S. Essential Air Service," pp. 133-149).

"Regional/Commuter Airline Industry Annual Report, 1983," Prepared by the Regional Airline Association, Washington, D.C., June, 1984 (especially section entitled, "Summary of U.S. Essential Air Service," pp. 94-111).

"Regional/Commuter Airline Industry Annual Report, 1985," Prepared by the Regional Airline Association, Washington, D.C., June, 1986 (especially section entitled, "Summary of U.S. Essential Air Service," pp. 103-120).

"Regional/Commuter Airline Industry Annual Report, 1986," Prepared by the Regional Airline Association, Washington, D.C., June, 1987 (especially section entitled, "Summary of U.S. Essential Air Service," pp. 86-103).

Replacement Carrier Procedures, Federal D.O.T., Washington D.C., Essential Air Services Division, January 1, 1985.

Rose, Warren, "Three Years After Airline Deregulation in the United States: A Report Card on Truckline Carriers." Transportation Journal, Vol. 21, No. 2, Winter, 1981, pp. 51-58.

"Small Airlines, in Competitive Squeeze, Are Linking with Big Carriers' Routes," *The Wall Street Journal,* July 31, 1984, p. 10.

Stephenson, Fred J. and Fred Beier, "The Effects of Airline Deregulation on Air Service to Small Communities," *Transportation Journal,* Vol. 20, No. 4, Summer 1981, pp. 54-62.

"The Changing Airline Industry: A Status Report Through 1980," U.S. General Accounting Office Report GAO/CED-81-103 dated June 1, 1981.

"The Changing Airline Industry: A Status Report Through 1980," U.S. General Accounting Office Report GAO/CED-92-94 dated June 24, 1982.

Warren, William D., "Changing Air Transportation Services for Smaller Metropolitan Regions: 1980-82," *Transportation Quarterly,* Vol. 38, No. 2, April 1984, pp. 245-266.

Williamson, Kenneth C., Lawrence F. Cunningham and Mark G. Singer, "Scheduled Passenger Air Service to Small Communities: A Role for State and Local Governments," *Transportation Journal,* Vol. 21, No. 4, Summer, 1982, pp. 25-34.

Yoe, Daniel G., "Airline Deregulation Essential Air Transportation for Small Communities," *Journal of Air Law and Commerce,* Vol. 48, No. 2, Winter, 1983, pp. 435-450.

ENDNOTES

* Department of Purchasing, Transportation Operations, College of Business, Arizona State University.

** School of Business Administration, California State University, Stanislaus, CA.

Value Capture in Transit

*by W. Bruce Allen**

ABSTRACT

Value capture, in this case, is the taking of some or all of the enhanced value of property which is attributable to the government's investment in a transit system. Various studies have suggested that value capture be used to help finance transit systems. Recent studies have alleged that the value to be captured is small and tightly concentrated around the investment.

An interdependent set of modal choice, station choice, and savings models are developed using the economic law of market areas. These models are enhancements of earlier models developed by the author. Such models spatially separate the auto users from the transit users, spatially separates the users of various stations, and spatially identifies locii of equal savings, i.e., auto cost minus transit cost (where costs are a generalized function of out-of-pocket costs and time).

The station choice model is tested by using auto access data for all seven suburban stations of the Philadelphia-Lindenwold High Speed Line for a morning rush hour (13,000 observations). The savings model is tested by postulating that residential sales price is a function of the characteristics of the property, neighborhood characteristics, distance to the Philadelphia CBD, and savings. Value capture is tested by examining to what degree the savings are capitalized into the selling price of the property. Over 1,300 real estate transactions are utilized to test the model. The results demonstrate that approximately one third of the capital cost of building a completely new Lindenwold Line could be captured from the increased property value.

I. INTRODUCTION

Value capture, as used in this context, is the takingn of some or all of the enhanced value of land which is caused by the government's investment in a transit system.

Various studies have suggested that value capture can be used to help finance transit systems. The average transit system in the US can only cover 30-40% of its operating costs, with even the most successful systems, e.g., Philadelphia's Lindenwold Line, only covering 80% of operating costs.

Since systems require subsidies to cover a portion of operating costs, all of the capital costs must be subsidized. Currently, 75-85% of the capital costs come from the Federal government (Congressional Budget Office, 1985, p. 45), with the balance from state and local governments. The Federal government would like to see greater aid coming from the local governments where the majority of the beneficiaries are likely to live. However, recent studies have suggested that the value to be captured is very small (Anas, 1979, 1980, 1983, Anas and Lee, 1980) and is very tightly concentrated around the investment (Dewees, 1976, Damm, Lerman, et al., 1980, Ferguson, Goldberg, and Mark, 1984).

An interdependent set of models and modal choice, station choice, and travel savings is developed using the economic law of market areas (Fetter, 1924 and Hyson and Hyson, 1950). The application of the market area analysis to transport choice was first introduced by Mohring (1961) in the continuous access case and by Moses (1965) in the discrete access case. It was extended to station choice and savings locii by Boyce, Allen, et. al. (1972, 1973), Boyce and Allen, (1974), Boyce, Allen, and Tang (1976), and Allen et. al. (1986).

The models herein spatially separate two sets of commuters to the Central Business District (CBD)—auto users and transit users (the modal choice model); spatially separates the users of one station of the transit line from the users of another station (the station choice model); and spatially denotes the locii of all points where the users save an equal amount of money from using transit over auto (the savings model). All of these models (with the assumptions used herein) generally yield hyperbolic locii which bend around the stations on the transit system.

The purpose of the paper is to utilize the savings model to test whether savings are capitalized into the property values in the transit market area. If so, value would seem to exist to be captured. Given the existence of such value, the question of how much value is relevant. Finally, the legality and practicality of a value capture scheme needs to be examined.

The paper first introduces the three models and shows their interdependence. The setting of the empirical work is then described. The empirical tests and the data set are then explained, followed by the empirical results. Lastly, conclusions are drawn concerning the use of value capture as a funding mechanism.

II. THE THEORY

It is first necessary to delineate an area of impact for the transit investment. The area is defined as the area where individuals would choose to utilize the transit system over their previous alternative—the automobile.

The model assumes that individuals are located throughout a circular city. Work takes place only within the CBD, while residences surround the CBD. With no distinguishing characteristics associated with the landscape and individuals of identical tastes and incomes, the city would develop as a perfect circle with a declining rent gradient from the CBD. These are the typical Von Thunen assumptions and results. While the above assumptions would never be met, they provide a starting point for the value capture analysis.

The model assumes that an individual's workplace and homeplace are specified and that an individual commutes one roundtrip per day. Since the consumption of work, housing, and quantity of work trips are fixed, the individual will maximize utility by minimizing his/her roundtrip transportation cost. Transportation cost is defined here to represent a "full" cost, i.e., dollar costs plus value of time costs. Access/egress costs in the CBD are ignored. A straight line transit system is constructed running from a suburban station G to the CBD station H. Driving can occur on a straight line basis as per Von Thunen (actual road networks can be explicitly considered in more complex versions of the model). The following notation applies:

F_i = the fare on the transit system from station i to the CBD (assumed to increase with distance)

W = expected wait time for the transit system (assumed to be the same at all stations)

T_i = line haul time on the transit system from station i to the CBD (assumed to increase with distance)

V = value of the commuter's time (assumed to be the same for all individuals)

c = auto driving cost per mile (assumed to be the same for all drivers regardless of the distance driven, type of car, direction driven, time of day, etc.)

m = auto miles per hour (assumed to be the same for all drivers regardless of distance driven, type of car, direction driven, time of day, etc.)

P_A = parking costs associated with auto in the CBD (assumed to be the same for all)

B = other pure monetary costs incurred by auto not related to distance, e.g., bridge tolls

D_{ix} = straight line distance from location X to station i = G,H

D_{GH} = straight line distance from G to H

TC_j = full cost by j = A, Ti; where A = auto and Ti = transit from station i

Full transit cost TC_{Ti} and auto cost TC_A are as follows:

(1) $TC_{TG} = 2F_G + 2V(T_G + W) + 2cD_{GX} + 2(V/m)D_{GX}$

(2) $TC_A = 2cD_{HX} + 2(V/m)D_{HX} + 2B + P_A$

Equating equations (1) and (2) yields the locus of all points x where an individual will be indifferent between using automobile and using transit.

(3) $$D_{HX} - D_{GX} = \frac{F_G - B - (1/2)P_A + V(T_G + W)}{c + (V/m)} = K$$

With knowledge of all of the items on the right hand side of (3), the right hand side would be a constant, K.

Equation (3) gives the equation of a hyperbola which bends around G if K is positive and around H if K is negative. The larger K is in absolute value, the closer the vertex of the hyperbola is to the points G or H and the "tighter" the bend of the hyperbola around G or H. The hyperbola collapses to a horizontal line and its vertex is at G or H when D_{HG} = [K]. The hyperbola is the perpendicular bisector of the line when K = 0. No hyperbolas exist when D_{HG} > [K], i.e., either all trips are by auto or all trips are by transit.

The model resulting from (3) is the modal split model. Its form appears as Figure 1 assuming O > K > D_{HG}. Equation (3) represents the locus of all points X where a person located at X is different between driving or using the transit system. All persons residing to the right of the hyperbola will use the transit line (because it is cheaper than auto, i.e., $TC_{TG} > TC_A$), and all individuals residing to the left of the hyperbola will use auto because $TC_A > TC_{TG}$.

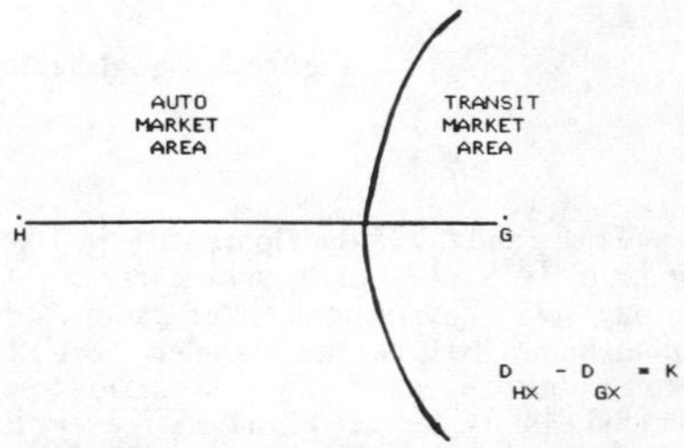

Figure 1. Simple Auto/Transit Market Areas (Modal Split Model)

More stations are easily added to the analysis. This yields station choice models analogous to (3) with new stations beyond G diverting traffic from old stations and diverting traffic from auto if fares have a distance taper. In addition, the impact of different V's over space can be seen by taking (dK/dV). Since this can be positive or negative (since auto time is required both for direct driving and driving to the transit system), the transit market area can either expand or contract as V increases.

Consider a level of savings, S, where savings are defined as transit cost minus auto cost. The area where S ≤ O is obviously the transit market area and S = O where $D_{HX} - D_{GX} = K$. Since S is defined as $TC_A - TC_{TG}$ for the one station case, it can be rewritten as:

(4) $$D_{HX} - D_{GX} = K + \frac{S}{2(c + [V/m])} = K'$$

Thus, when S = O, (4) is the modal split boundary. Since K′ cannot exceed D_{HG}, $S_{MAX} = 2(c + [V/m])(D_{HG} - K)$ and by varying S between zero and $2(c + [V/m])(D_{HG} - K)$, the locus of all points X which incur equal savings can be traced by (4). Thus, the beneficiaries of a transit investment can be spatially identified by savings level by the method above. For multiple stations, the savings locii appear as in Figure 2.

III. THE SETTING OF THE EMPIRICAL STUDY

The Lindenwold High Speed Line opened for business between Lindenwold, Camden County,

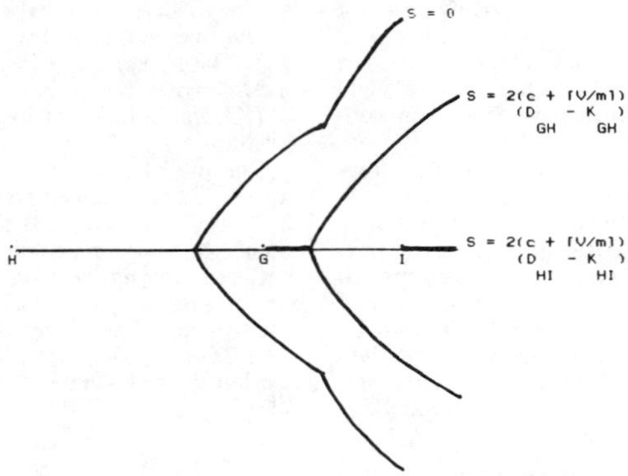

Figure 2. Equal Savings Locii When Two Stations Exist

New Jersey and the Philadelphia CBD in 1969. The Line runs 14.5 miles and provides service 24 hours per day, seven days a week. Four stations serve the Philadelphia CBD, two the Camden, NJ CBD, and seven suburban, park and ride stations exist in Camden County. Service frequency is every three to four minutes during the rush hours and no worse than ten minute headways exist inbound from 5:45 AM and outbound until 7:20 PM. Approximately 12,000 parking places currently exist at the seven suburban stations. New Jersey Transit buses feed the stations, bike racks exist at the stations, and the stations are pedestrian accessible.

A trip from Lindenwold to Philadelphia currently costs $1.60 one way and takes approximately 24 minutes for an average speed of 36 miles per hour including station stops. This contrasts with a 45 minute drive, the auto out-of-pocket operating cost, bridge tolls (to cross the Delaware River), and CBD parking costs.

The Line is perceived to be a reliable, fast, safe, and relatively inexpensive way to travel to the Philadelphia CBD during rush hour. Approximately 20,000 riders use the system each workday (making two trips each). Most of the trips are peak oriented, i.e., work trips, and most of the trips are to the Philadelphia CBD. The number of intra-New Jersey trips and the number of contra-peak trips are small. The peak hour users of the system tend to be well dressed white collar workers.

The 1980 Census of Population shows 14,191 workers from Camden county that list their workplace as the Philadelphia CBD. In Burlington County (the contiguous county to the north), 5,684 Philadelphia CBD workers are found, while in Gloucester County (the contiguous county to the south), the comparable total is 3,312. Thus, in the three county area, 23,187 residents work in the Philadelphia CBD. The Census also reports that 14,037 workers state that rail transit is their primary means of transportation in their journey to work (11,941 from Camden County, 1,601 from Burlington County, and 495 from Gloucester County). It seems certain that the Line is moving a significant number of the South Jersey Philadelphia CBD workers into Philadelphia each day. Other users include shoppers, students, sightseers, etc.

IV. THE DATA

The basic data for the empirical analysis comes from three sources. The first source is a license plate survey of all rush hour (6:30 AM-9:30 AM) auto feeder users of the Line on Census day 1980 (about 78% of the Line's rush hour users either park and ride, i.e., leave their car parked at the station all day, or kiss and ride, i.e., are dropped off at the station by automobile). By matching the license plate at each station with the address to which that car is registered (according to the New Jersey Division of Motor Vehicles' files), the likely origin of the trip is found. This data set is used to develop the modal split, station split, and savings (value capture) models.

The second data set consists of all arm's length residential real estate transactions conducted by the members of the Camden County Board of Realtors (the line bisects Camden County). While non-member realtors and out-of-county realtors do make sales in Camden County and transactions take place without the use of realtors, most transactions in the county will be covered by the data from the Board. For the year 1980, 1,341 useable transactions exist.

Table 1 gives a summary of the license plate data collected by station. The dominance of auto feedership is shown by these statistics. Total ridership was calculated by subtracting the 6:30 AM gate counts at each station (from the automatic counter on each turnstile) from the 9:30 AM gate count. Walk-ons are a residual estimate assuming that each park and ride vehicle carries one individual (a likely underestimate and hence walk-ons will be slightly overestimated).

The real estate data was gathered from the Comparable Sales books of the Camden County Board of Realtors. These books are basically the same as the MLS "picture" books which are available for inspection by customers at any MLS member real estate office, except that they contain the final transaction price and the number of days that the house was on the market in addition to the information normally available to the prospective buyer, e.g., ask price, lot and structural information, etc., from the MLS listing. The variables listed in Table 2 were collected.

TABLE 1
April 1, 1980 Speed Line Rush Hour Traffic

	6:30-9:30 Entrances	Kiss & Ride	Bus	Rail	Park & Ride	Maximum Walk-Ons +
Lindenwold	4,225	709	482@	200	2,426	408
Ashland	2,109	300	91#	—	1,411	307
Woodcrest	393	89	28	—	325	−49
Haddonfield	2,194	417	313	—	1,103	361
Westmont	1,391	195	10	—	964	222
Collingswood	1,111	170	0	—	548	393
Ferry Ave.	2,004	279	77	—	1,482	166
	13,427$	2,159	1,001*	200	8,259	1,808

+ includes bike riders and multiple riders from park and ride
@includes apartment busses in addition to New Jersey Transit
#includes Echelon Mall shuttle in addition to New Jersey Transit
$excludes pass entrances, i.e., only revenue entrances counted
*small bus ridership transferred at station to another bus and thus was need feeding the Line. There was no way to eliminate such riders, but the number is small.
ˆerror in the data set

TABLE 2
Real Estate Variables

Location Of Property	Final Sales Price	Lot Size (Sq. Ft.)
(street address, town)	Days Property On Market	Property Taxes
Number Of Bedrooms	Number Of Bathrooms	Number Of Stories
Style Of House	Construction Of House	Age Of House
(single, duplex, etc.)	(frame, brick, etc.)	Existence Of Garage
Type of Heat	Existence Of Basement	Type Of Floors
Existence Of Fireplace	Existence Of Porch/Deck	Type Of Plumbing
Census Tract Location	Existence Of Family Room	Type Of Walls
Existence of Insulation	Existence Of Dining Room	
Quarter And Year Sold	Existence Of Utility Room	

Each transaction was mapped into its appropriate Census Tract. For each Census Tract, the information in Table 3 was collected.

TABLE 3
Census Tract Variables

Average Age Of Resident	# Blacks Residing	# Whites Residing
Total # Of Residents	# Persons/Family	% High School Grads
# Enrolled In College	# Enrolled In School	# Of Workers
# Using Rail As Primary Mode	# Using Public Trans Mode	# Who Work In Phila CBD
For Work Trip	For Work Trip	Per Capita Income
# Who Work In Camden	# Managerial Workers	# Vacant House Units
Median Household Income	Mean Household Income	# Housing Units
# Homes Owner Occupied	% Families Below Poverty	
Median Value Of Owned Home		

This information was merged with the property transactions file.

V. THE EMPIRICAL STUDY

The auto cost equation was estimated in the following manner. The c was chosen to be the auto cost/mile as used by the local MPO (Metropolitan Planning Organization)—the DVRPC (Delaware Valley Regional Planning Commission). This is an out-of-pocket cost of 7.87 cents per mile.

D_{PX} was chosen to be the distance from the centroid of the Census Tract to the corner of Broad and Market (the center of the Philadelphia CBD) via either the Ben Franklin or the Walt Whitman Bridges (whichever yields the minimum distance). The m was chosen to be the miles per hour as used by the local MPO. This ranged from 20 to 30 miles per hour with a mean and mode of 23.5.

The value of time for auto V_A was calculated on the basis of the journey to work analysis by McFadden, Talvitie, et. al. (1977) which gave values of time as a percentage of the wage rate for in-vehicle auto time (178% of the wage rate), in-vehicle transit time (74% of the rate rate), transit expected wait time (165% of wage rate), and transit access/egress walk time (338% of wage rate).

Since the wage rate was not available, the mean family income for each Census Tract was divided by 2,080 hours to yield an hourly income rate. It was assumed that a commuter from the Census Tract would use such a figure as the basis on which to value his/her time. This figure was then multiplied by 1.78 to give the estimate of V_A for each Census Tract.

The B was chosen to be the cost of obtaining a commuter sticker for crossing either the Ben Franklin or the Walt Whitman Bridges. This costs $6 (in 1980) for a monthly sticker plus a payment of $.25 for each crossing. Assuming a 22 day work month, the average commuter would have 44 bridge crossings. This would then cost $.3864 per crossing.

The P was determined in the following fashion. No formal records were found of Center City Philadelphia parking costs. An established Center City parking operator estimated that the rates varied between $90 and $150 per month depending on the quality (enclosed, lot, etc.) and location in 1980. The simple average ($120) was taken and divided by 22 work days to yield a daily CBD parking cost of $5.4545.

The cost of driving from each Census Tract centroid X to the Philadelphia CBD P was estimated as:

$$(5)\ \text{AUTO COST} = \$6.2273 + \$.1574\, D_{PX} + .0851063\, V_A D_{PX}$$

The costs of using each one of the stations (ST) is formally stated as:

$$(6)\ \text{TRANSIT COST} = 2cD_{STX} + 2(V_{T1}/m)D_{STX} + 2F_{ST} + P_{ST} + 2V_{T2}T_{ST} + 2V_{T3}W_{ST} + 2V_{T4}T_{WST}$$

where: c and m are defined as above

- V_{T1} is auto in-vehicle value of time for a transit user, i.e., $1.78V_A$
- V_{T2} is transit in-vehicle value of time, i.e., $.74V_A$
- V_{T3} is transit expected wait value of time, i.e., $1.65V_A$
- V_{T4} is transit access/egress walk value of time, i.e., $3.38V_A$
- D_{STX} distance from the centroid of the Census Tract to the station (as calculated by the MPO over the regional highway network)
- F_{ST} fare from station ST to the CBD
- P_{ST} expected parking cost at station ST
- T_{ST} ride time from station ST to the CBD
- W_{ST} expected wait time for a train at station ST
- T_{WST} average walk time from station ST parking lots to station ST

The F_{ST} is the applicable fare for 1980 (see Table 4). The P_{ST} is calculated by multiplying the pay parking spaces used times $.25 (the pay lot rate) divided by the total used spaces (both pay and free). The results of these calculations are shown in Table 4.

T_{ST} is taken from the Line's timetable and the W_{ST} is calculated by dividing the rush-hour schedule headway in half. T_{WST} is calculated by taking the distance from each parking lot centroid at a given station, weighting them with respect to cars using the lot, and calculating the average distance walked by a park and ride patron at the station. This was then converted to minutes by assuming a walk speed of 2.5 miles per hour. These results are shown in Table 4.

Converting the times from minutes to hours, the McFadden et. al. study was utilized to give values of time for transit users. The value of time was then calculated in the same fashion as with the auto users where it was assumed that transit users valued their in-vehicle access time at 178% of their hourly in-

TABLE 4

Basic Data For Station Costs

	One Way Fare	Average Station Parking Cost	Line Haul Time	Wait Time	Walk Time To Station
Ferry Ave.	.75	.1250	12	2	1.95
Collingswood	.95	.1960	14	2	1.27
Westmont	.95	.1698	16	2	1.40
Haddonfield	.95	.1639	18	2	1.42
Woodcrest	1.15	.2500	21	2	1.41
Ashland	1.15	.1472	22	2	1.53
Lindenwold	1.15	.1464	24	2	2.52

come rate. The total transit costs are then calculated for each station in the form of (6).

The preferred station for each Census Tract is the mode station actually chosen by the users from each Tract. Given the preferred station, the Tract's transit costs are calculated using one of the seven stations used cost formulas of the form (6). Savings for each Tract are then calculated by subtracting the transit costs from the centroid of said Tract to the Tract's preferred Line station from the auto costs from the centroid of the Tract to the Philadelphia CBD by the minimum distance bridge. This savings variable is then used in the value capture analysis.

However, some adjustments must be made. The data collected represents the sales price of improved land—not change in sales price (discounted change in site rent) of unimproved land. Thus the dependent variable in the analysis is the transaction price of single family residential homes. Since improvements have been made to the land, in order to control for such improvements, they are entered into the analysis as independent variables. A list of such variables appears above. Such inclusion yields hedonic prices for the characteristics of the property.

Since the properties are influenced by the Philadelphia/Camden rent gradient, distances from both the Ben Franklin and Walt Whitman Bridges are included in the analysis to register the presence of the urban rent gradient. The analysis herein is designed to measure how that gradient has been perturbed as a result of the Line.

Characteristics of the neighborhood will also influence housing prices. Characteristics of each Census Tract were taken from the 1980 Census and are listed above. These characteristics were taken from the 1980 Census and are listed above. These characteristics are added as independent variables to control for the neighborhood effects.

Finally, the savings variable, as described above is calculated. It becomes the independent variable of interest in the value capture analysis.

Characteristics of the neighborhood will also influence housing prices. Characteristics of each Census Tract were taken from the 1980 Census and listed above. These characteristics are added as independent variables to control for the neighborhood effects.

Finally, the savings variable, as described above is calculated. It becomes the independent variable of interest in the value capture analysis.

Equation (7) reports the results of the regression for the variables with a significant effect:

The equation is highly significant as measured by the F.

The regression run searched the independent variables for the optimal entrants in terms of contribution to R^2. Obviously, many of the characteristics of the property listed above are not significant and none of the neighborhood characeristics listed above are significant.

The mean sales price is $62,428 (while the mean price in previous work was $21,260; see Boyce, Allen, and Tang, 1976). While the intercept is not statistically significant, it is presented as a reference point. The two bridges are used as reference poitns for the urban rent gradient since heavy traffic on each bridge is oriented to the Philadelphia CBD. The sum of the two coefficients states that a property equidistant from the two bridges (approximately the exact route of the Line) experiences a ceteris parabus drop in housing price of $1,117 for each mile located further from Philadelphia. Property above the Line falls slower in value as one gets further from Philadelphia, while property below the Line falls faster as one gets further from Philadelphia. These results seem plausible given the perceived quality of the areas above the Line.

That larger lot sizes command higher prices is not surprising. Despite the intuition that higher property taxes should be negatiively capitalized into sales prices, the value of the house and the property tax on the house are highly positively correlated. This is because property taxes are based on property values and also relate to the degree of municipal services which the property enjoys. Variables and combinations of variables which were proxies for municipal services did not show any statistically significant results. The hedonic prices for garages and fireplaces are positive and strong. The bathroom variables are dummies related to the two bathroom and larger house. As can be seen, houses with only one bathroom or a bath and a half are strongly discounted in the market relative to houses with more bathrooms.

Finally, the savings variable is significant with a positive coefficient. It states that a dollar's worth of savings is worth $443 of sales price for a house. A dollar's worth of transportation "full" cost savings is worth $250 per year (assuming 250 work days per year). This indicates a discount raate of 56.4% assuming that the savings rate would accrue to the property forever, i.e., the pure capitalization hypothesis. Obviously, this seems to be a very high discount rate.

(7) SALES PRICE = 33.053 (1.53) − 5,392 (−3.06) DISTANCE TO THE BEN FRANKLIN BRIDGE

+ 4,275 (1.98) DISTANCE TO THE WALT WHITMAN BRIDGE + 1.568 (4.79) LOT SIZE

+ 13.18 (6.32) PROPERTY TAX PAID + 7,492 (3.18) IF HOUSE HAS A GARAGE

+ 8,277 (3.51) IF HOUSE HAS A FIREPLACE—15,510 (−3.96) IF HOUSE HAS ONLY ONE BATHROOM—13,518 (−4.17) IF HOUSE HAS ONLY ONE AND ONE HALF BATHROOMS

+ 443 (2.30) SAVINGS $R^2 = .4177$ $F = 5.27$

(where figures in parentheses are t statistics)

However, there are multiple reason why the pure capitalization hypothesis would not apply. The first is that which discount rate to use, in general, is difficult to determine. Because of risk and uncertainty of the future, it should likely be a high one. Gasoline prices have doubled in one year in the past and such changes would alter the future savings rate, i.e., the level S would not prevail at location X forever. In addition, there is obviously much unexplained variance in the sales price which is not explained by the data set. In addition, the pure capitalization effect must be modified downward by the adjustment factor described in Allen et. al. (1986).

However, elastic demands for land and population growth will violate assumptions on which the model is built. The assumptions of an elastic demand and population growth will, at worst, hold the city's shape in the non-transit impacted area and cause the city to grow even more in the impacted area. This will tend to hold rents up in the non-impacted area and cause them to grow more than indicated in a pure Von Thunen/Mohring city (See Allen et. al., 1986). If the demand is elastic enough, rents could grow area-wide (albeit moreso in the impacted area). Thus supply and demand effects are also working in the area. Likewise, the Von Thunen center city dominance does not hold in the case herein. While the Philadelphia CBD is the single largest concentration of employment in the Philadelphia region, it does not hold the majority of the jobs. Sub-regional centers exist and these centers may cause the rent gradient to rise and fall rather than monotonically fall as one gets further from the CBD ceteris parabus.

The average value of savings and the rest of the value capture analysis was calculated in the following fashion. Since 4% of the journeys to work in the nation are by transit, those Census Tracts which had 4% or more of their journeys to work by the Line were counted as transit Census Tracts. (In the three county area, from the Census Tracts which had both transit users and property transactions, the transit usage averaged 4.34%. Using 4 or 4.34 yields virtually the same results). If all Tracts which had both Line users and real estate transactions were included in the analysis, the value capture number would be 66% higher than that given below. The analysis had already cut some Tracts before the transit Census Tracts were defined, i.e., if no users wre found to exist from a Tract, real estate data was not used from the Tract; likewise, some Tracts with users were not entered into the analysis because no real estate transactions existed for that Tract.

VI. CONCLUSION

The research described herein developed simple modal choice, station choice, and savings models, all from the economic law of market area analysis. The simple theoretical models described herein can all be made more complicated by the addition of actual road networks and congestion thereon. In fact, the actual empirical calculations of distances and times from Census Tracts to stations were made by the DVRPC on their highway network of the region. The interdependence of all three models can also be demonstrated, e.g., the linkage of the modal split K's, the station choice K's, and the savings K's.

The models were then tested using data collected showing where the autor users of each station originated their trips and the transactions' prices of single family houses in the Line's capture area.

The value capture regression showed that savings were capitalized into the property values such that each dollar's worth of daily savings added $443 to the value of the property. The average savings was $10.34 implying that $4,581 is available to be captured per single family home. With 61,021 single family homes in the transit Census Tracts, the level of value to be captured is of the magnitude of $279.5 million. This figure is 117.9% of the construction costs of the Line (such costs inflated to represent 1980 dollars). Such construction costs were very low relative to other systems in that tunnels in Philadelphia and Camden and a Delaware River bridge crossing did not have to be built and the right of way was already in place. Had such costs been incurred, the Line would have cost approximately $820 million in 1980 dollars. This would place the above value capture percentage in the one third range. This is still a high number relative to the results of Anas. Of course, in this study, the fall of rents elsewhere are not included in the analysis.

It seems likely that the impacts of "what might have been" in non-impacted areas are probably quite small for any individual property and very difficult to substantiate. Thus such rent decreases or lack of rent increases as high as they might have been do not provide a strong case for disbenefits. On the other hand, the property in the transit corridor is likely to experience measurable benefits whose measurement can be made by acceptable (and applied) statistical techniques, i.e., econometric assessment.

Since multiple family dwelling units, commercial property, vacant land, and single family residential units in non-transit Census Tracts (but still with transit use) have been excluded from the analysis, higher value to be captured would seem to exist. In addition, a level of savings which valued auto access time at 74% of the income rate for transit users (just as they value their in-transit vehicle time) yielded value capture levels over 30% higher than those used herein and a statistically significant savings coefficient of $622 for every dollar's worth of savings. Such a value of time assumption leads to appreciable higher savings levels for most Census Tracts.

However, the model suffers from a lack of an internally generated value of time and only explains 42% of the variance in sales price (typical of such models). Thus the results herein should not be judged on a strict quantitative basis. Rather qualitative statements are appropriate—such as a series of models exist which can be used to yield value capture magnitudes if the data is available and that, subject to the data which was available for this case, significant value was available to be captured.

A policy of value capture for the Lindenwold Line at this time would cause great equity problems. If a property-holder has not transacted his/her property from before the announcement/opening of the Line, then the holder of the appreciated property is the beneficiary and could be taxed. However, if the property has transacted, it is the seller, who held the property from before the announcement/opening and sold after the announcement/opening, who has captured the benefit. The current occupant, if required to pay a value capture tax, would be paying for the same benefit twice. Another problem is that the gain is an unrealized capital gain until the property is sold. For a property-holder with no other assets, a value capture tax on the property may require the sale of one's home.

This does not mitigate the conclusion that there is value to be captured. It only states that the current occupants are only a subset of the beneficiaries. Thus the time to implement a value capture tax is when the beneficiaries are occupying the property. This would be between the announcement and the opening of the system. This still provides some difficulty because announcement still entails uncertainty about the level and surety of savings (as well as a discount effect because the savings will only be realized in the future). Opening provides certainty, but some property may already have been transacted. While some researchers have investigated these impacts, more work needs to be done.

Likewise, more work should be done on value capture in general. It appears to have the ability to raise significant levels of money.

BIBLIOGRAPHY

(1) Allen, W.B. and D. Boyce, "Impact of a High Speed Rapid Transit Facility on Residential Property Values," Proceedings of the International Conference on High Speed Ground Transportation, High Speed Ground Transportation Journal, Vol. 8, No. 2, 1974, pp. 53-60.

(2) Allen, W.B., K. Chang, D. Marchetti, and J. Pokalsky, "Value Capture in Transit: The Case of the Lindenwold High Speed Line," Report PA-11-0031, Urban Mass Transportation Administration, Washington, DC, April, 1986.

(3) Anas, A., "The Impact of Transit Investment on Housing Value: A Simulation Experiment," Environment and Planning A, Vol. 11, 1979, pp. 239-255.

(4) Anas, A., "Evaluating the Effects of Transportation-Land Use Policies on Housing Values and Household Welfare," Environment and Planning A, Vol. 12, 1980, pp. 747-764.

(5) Anas, A. and G. Lee, "The Potential for a Value Capture Policy to Finance Rapid Trnasit Projects on Chicago's Southwest Side: An Empirical Simulation Analysis," Research in Urban Economics, Vol. 2, 1980.

(6) Anas, A., "The Effects of Transportation on the Tax Base and Development of Cities," United States Department of Transportation, DOT/OST/P-30/85/005, Washington, DC, April, 1983.

(7) Boyce, D., W.B. Allen, et. al., "Impact of Rapid Transit on Suburban Residential Property Values and Land Development," Phase One Report to the Office of the Secretary, United States Department of Transportation, Washington, DC, November, 1972.

(8) Boyce, D., W.B. Allen, et. al., "Impact of Access Distance and Parking Availability on Suburban Rapid Transit Station Choice," US Government Clearing House, Springfield, VA, PB-220694, 1973.

(9) Boyce, D. and W.B. Allen, "Impact of Rapid Transit on Suburban Residential Development: Analysis of the Phildelphia-Lindenwold High Speed Line, in T. Norman (ed.), New Jersey Trends: Population and Economy, Natural Resources, Housing, Transportation, Energy, Institute for Environmental Studies, Rutgers University, New Brunswick, NJ, 1974, pp. 478-497.

(10) Boyce, D., W.B. Allen, and F. Tang, "Impact of Rapid Transit on Residential Property Sales Price," in M. Chatterji (ed.), Space, Location, and Regional Development, Pion Limited, London, 1976, pp. 145-153.

(11) Congressional Budget Office, "The Federal Budget for Public Works Infrastructure," US Government Printing Office, Washington, DC, July, 1985.

(12) Damm, D., S. Lerman, et al., "Response of Urban Real Estate Values in Anticipation of the Washington Metro," Journal of Transport Economics and Policy, September, 1980, pp. 315-336.

(13) Dewees, D., "The Effect of a Subway on Residential Property Values in Toronto," Journal of Urban Economics, Vol. 3, No. 4, 1976, pp. 357-369.

(14) Ferguson, B., M. Goldberg, and J. Mark, "Thee Pre-Service Impacts of the Vancouver Advanced Light Rail Transit System on Single Family Property Value," unpublished paper, University of British Columbia, Vancouver, December, 1984.

(15) Fetter, F., "The Economic Law of Market Areas," Quarterly Journal of Economics, Vol. 38, 1924, pp. 520-529.

(16) Hyson, C. and W. Hyson, "The Economic Law of Market Areas," Quarterly Journal of Economics, Vol. 64, 1950, pp. 319-325.

(17) McFadden, D., A. Talvitie, et. al., "Urban Travel Demand Forecasting Project-Final Report," Institute of Transportation Studies, University of California-Berkeley, Berkeley, 1977.

(18) Mohring, H., "Land Values and the Measurement of Highway Benefits," Journal of Political Economy, Vol. 79, 1961, pp. 236-249.

(19) Moses, L., class lectures in C-53, Location Theory, Northwestern University, Department of Economics, Evanston, IL, 1965.

(20) Von Thunen, J., Der Isolirte Staat In Beziehung Auf Landwirthschaft Und National Okonomie, Wiegandt, Berlin, 1836.

ENDNOTE

*The Wharton School, University of Pennsylvania, Philadelphia, PA.

Scale Economies Among United States Bus Transit Systems

*by Michael Keough**

ABSTRACT

To provide insight into the question of economies of scale among United States bus transit systems, a literature review and a multiple regression analysis of data for more than 70 United States bus transit systems were conducted.

The literature review proved inconclusive. Various authors have reported increasing, constant and decreasing returns to scale. Seven reasons are cited that may, in part, explain why conflicting findings have been reported.

The objective of the quantitative analysis was to determine the extent to which selected unit operating expense variables are a function of various measures of scale and other nonfinancial system definition variables (e.g., spare ratio). The analysis was based for 73 United States large and medium-size bus transit systems.

The major findings of the analysis include: (1) Diseconomies of scale are present among United States bus transit systems; (2) Variation in unit operating expense variables which is associated with variations in scale and nonfinancial system definition variables (primarily population density) ranges from 6.1 to 26.6 percent; and (3) Most other nonfinancial system definition variables (e.g., spare ratio) play little or no role in explaining variation in the unit operating expense variables.

These findings challenge the widely-held belief that economies of scale are present among United States bus transit systems. This suggests the need for a reappraisal of the institutional arrangements used to deliver bus transit in the nation's metropolitan areas.

While this research demonstrates that diseconomies of scale are present among United States bus transit systems, it leaves unanswered many questions. Priorities for future research include: (1) replicating this study using unit measures of total, rather than operating, expense as the dependent variables; (2) conducting a time-series analysis of selected systems; and (3) performing case study analyses of individual transit systems.

I. INTRODUCTION

The concept of economies of scale is a relatively straightforward one. Suppose that an organization, such as a bus transit system, increases the amount of all its inputs by the same proportion. What happens to the output? If the output increases by a larger proportion than each of the inputs, there are increasing returns to scale—economies of scale. If output increases by a smaller proportion than each of the inputs, there are decreasing returns to scale—diseconomies of scale. If output and input change by the same proportion, there are constant returns to scale.

The presence of either economies of scale or diseconomies of scale among United States bus transit systems would have significant implications. The presence of economies of scale would support the continuation of the delivery system which is presently most common: transit service for an entire metropolitan area provided directly by one public entity. The presence of diseconomies of scale would suggest that different institutional arrangements should be considered. This might include, for example, large-scale contracting with private companies to provide publicly-funded bus transit service.

II. LITERATURE REVIEW

As Berechman and Guiliano (1985) report in an excellent review of evidence regarding economies of scale in bus transit, several empirical studies have been conducted during the past two decades. Some utilized traditional cost functions while many of the more recent studies utilized generalized cost functions. These studies have produced conflicting findings: economies of scale, diseconomies of scale and constant returns to scale have all been reported. Moreover, some individual studies have reported increasing returns to scale under certain circumstances and constant or decreasing returns to scale under others.

A. Studies Reporting Increasing Returns to Scale

Several cross-section studies have reported increasing returns to scale (i.e., economies of scale). Veatch (1983), in a study of 37 bus systems, found increasing returns to scale among small systems. In this study, vehicle miles served as the output measure (independent variable) while total cost was used as the dependent variable.

Williams and Dalal (1981), in a study of 22 systems, found increasing returns to scale among the larger ones. In this study, vehicle miles served as the output measure while operating cost was the dependent variable. Viton (1981) conducted a study of 54 systems which examined the relationship between vehicle miles and operating cost. He found short-run increasing returns for the entire sample but long-run economies of scale only among the smaller systems. Williams and Hall (1981), in a study of intercity carriers which used passenger miles as the output measure and operating cost as the dependent variable, found increasing returns to scale.

Two time-series studies have reported economies of scale. Berechman (1981), in a study of the relationship between passenger revenue and total cost among Israeli bus systems, found increasing returns to scale. More recently, Berechman and Giuliano (1984) studied the relationship between revenue passengers and total cost in one Israeli bus company over a 28-quarter period and found increasing returns.

B. Studies Reporting Constant Returns to Scale

Three cross-section studies have reported constant returns to scale. Koshal (1972) studied ten bus companies and found constant returns to scale between vehicle miles and total cost. Nelson (1982) examined data for 45 systems for 1960 and 40 systems for 1968. He found constant returns to scale between vehicle miles and the natural log of total cost. Finally, Veatch (1973), using a smaller sample size than that for which the aforementioned increasing return to scale finding was reported (29 versus 37), found constant returns to scale between vehicle miles and total cost per vehicle mile.

C. Studies Reporting Decreasing Returns to Scale

Three studies have reported decreasing returns to scale. Veatch (1973), in his study of 37 systems, found decreasing returns to scale among larger systems. Wabe and Coles (1975), in a cross-section study of British municipal bus systems, found slightly decreasing returns to scale between vehicle miles and operating cost per bus. Finally, Berechman and Giuliano (1984), in the aforementioned time-series study of one Israeli bus system, found decreasing returns to scale between vehicle miles and total cost.

D. Possible Reasons for Conflicting Findings

There are several possible reasons for these conflicting findings. One, the data bases varied considerably. For example, Wabe and Coles (1975) studied British municipal systems, Berechman (1981) and later Berechman and Giuliano (1984) studied Israeli systems while Williams and Hall (1981) and others examined United States bus systems. Two, the number of cases examined varied widely, from ten (Koshal) to 76 (Wabe and Coles). Three, some studies were cross-section while others were time-series.

Four, although most authors used supply variables (e.g., vehicle miles) as the sole output measure, several (Williams and Hall, Berechman, and Berechman and Guiliano) used consumption variables such as passengers or passenger miles. Five, several different measures served as dependent variables. Some were amounts (e.g., total cost and operating cost) while others were rates (total cost per vehicle mile and operating cost per bus).

Six, the cross-section studies did not adjust system-specific cost data to reflect intercity income or cost of living variations. Such variations could be expected to have a profound effect on the relative cost of providing bus service since the industry is very labor-intensive.

Finally, most of these studies did not control for intersystem variations in operating characteristics which may affect the cost of providing bus transit service. These factors include the spare ratio (ratio of total buses in the revenue fleet to buses operated in maximum scheduled service), peak-base ratio (ratio of buses operated in maximum scheduled service to buses operated during the base period) and average operating speed.

For the aforementioned seven reasons, it is not surprising that conflicting findings have been reported.

E. Utility of Past Research

As described above, past research in this area has produced conflicting results. While several reasons for these conflicting findings can be identified, these studies varied so much that it is not possible to control for interstudy methodological variations and discern a trend. Consequently, past research in this area provides little guidance to transit policymakers and managers.

III. METHODOLOGY

The methodology for this analysis has four distinguishing features: (1) number of cases examined; (2) exclusion of selected multimodal systems; (3) controls for interarea variation in income; and (4) controls for intersystem variations in system definition variables (operating characteristics). As described below, these features should enhance the validity of the analysis.

A. Systems Examined

To provide insight into the question of scale economies among United States bus transit systems, multiple regression analysis was performed on data for 73 United States (and Puerto Rico) single mode bus and predominantly bus multimodal transit systems. Except as noted below, the analysis covered all United States public transit systems with fleets of 250 or more buses and a random sample of approximately 50 percent of those in the 40–249 bus range.

Several systems were excluded because key variables for these systems were missing or obviously wrong. For example, systems reporting average operating speeds such as 1.2 or 120 miles per hour were excluded.

A few others were excluded because a significant portion of their regular route service was purchased rather than directly operated. It is the norm for a United States transit system to directly operate all or virtually all of the regular route bus transit service that it provides. Inasmuch as such contracting may affect a system's cost efficiency, those few transit systems for which purchased transportation represented more than five percent of total bus transit operating expense were excluded.

Selected multimodal public transit systems were also excluded from the analysis. In multimodal operations (e.g., Southeastern Pennsylvania Transportation Authority) many administrative expenses are distributed among modes using system-specific expense allocation formulae. Because there is no uniform, industry-wide basis for allocating such expenses, it would not be possible to ascertain the extent to which intersystem unit expense variations among the bus operations of multimodal systems are due to variations in these allocation formulae. Consequently, multimodal systems where the bus mode accounted for less than 85 percent of systemwide operating costs were excluded from this analysis.

B. Data Sources

The primary source of the data for this analysis was the *National Urban Mass Transportation Statistics—FY 1983 Section 15 Annual Report* (United States Department of Transportation, 1984). Data in this report are for individual transit system fiscal years which ended between January 1 and December 31, 1983.

Another source of data was *Transportation Planning Data for Urbanized Areas Based on the 1980 Census* (United States Department of Transportation, 1985). This document provided demographic, economic and travel behavior data for the nation's urbanized areas.

C. Model Construction

Attempts were made to specify simple and multiple regression equations which would explain the following unit operating expense variables: (1) Adjusted Operating Expense Per Vehicle Hour (ADEXPVH); (2) Adjusted Vehicle Operations Expense Per Vehicle Hour (ADVOPVH); (3) Adjusted Other Operating Expense Per Vehicle Hour (ADOTHPVH); (4) Adjusted Vehicle Maintenance Expense Per Vehicle Hour (ADVMPVH); and (5) Adjusted General Administration Expense Per Vehicle Hour (ADGAPVH). The key relationships among these variables are as follows: (1) ADEXPVH = ADVOPVH + ADOTHPVH; and (2) ADOTHPVH > ADVMPVH + ADGAPVH (because ADOTHPVH also includes non-vehicle maintenance).

Note that all unit operating expense variables were adjusted to control for intercity variations in income levels (median family income). if income-adjusted variables were not used, intercity income variations would tend to exaggerate diseconomies of scale (should they exist). This is because larger bus systems tend to be found in larger urbanized areas, which, in turn, tend to have higher incomes and living costs. For the same reason, if adjusted expense variables were not used, intercity income variations would tend to mask economies of scale (should they exist). Finally, should constant returns to scale exist, the use of unadjusted expense variables would tend to give the appearance of diseconomies of scale where, in fact, they did not exist.

Nine trial independent variables—three scale variables and six nonfinancial system definition variables—were tested in various combinations. The scale variables were: (1) Buses Operated During the Base Period (BASE); (2) Buses Operated in Maximum Scheduled Service (MAXPEAK); and (3) Total Buses in the Revenue Fleet (REVBUS). The nonfinancial system definition variables (operating characteristics) were: (1) Population Density (DENSITY); (2) Average Number of Passengers On-Board (HOWFULL); (3) Ratio of Buses Operated in Maximum Scheduled Service to Buses Operated During the Base Period (PEAKBASE); (4) Ratio of Total Buses in the Revenue Fleet to Buses Operated in Maximum Scheduled Service (SPARE); (5) Average Operating Speed (SPEED); and (6) Proportion of Urbanized Area Workers Making Work Trips by Public Transportation (SPLIT). Descriptive statistics for the dependent and trial independent variables are shown in Exhibit 1.

IV. EXPECTED RELATIONSHIPS

A. Scale Variables

No a priori assumption was made regarding the nature of the returns to scale that would be found among United States bus transit systems. In other words, no hypotheses were made as to the expected relationships between each scale variable—Buses Operated During the Base Period (BASE), Buses Operated in Maximum Scheduled Service (MAXPEAK) and Total Buses in the Revenue Fleet (REVBUS)—and each dependent variable—Adjusted Operating Expense Per Vehicle Hour (ADEXPVH), Adjusted Vehicle Operations Expense Per Vehicle Hour (ADVOPVH), Adjusted Other Operating Expense Per Vehicle Hour (ADOTHPVH), Adjusted Vehicle Maintenance Expense Per Vehicle Hour (ADVMPVH), and Adjusted General Administration Expense Per Vehicle Hour (ADGAPVH). As described in the Literature Review, various studies have reported increasing, constant and decreasing returns to scale.

B. Population Density (DENSITY)

It was hypothesized that Population Density (DENSITY) would be positively correlated with the unit operating expense variables. Systems operating in more densely populated areas may tend to operate more slowly and be used more intensively than systems in less densely populated areas. To the extent that this is true, such systems would be expected to incur higher unit maintenance costs (due to wear and tear on the vehicles) and, to a lesser extent, higher unit vehicle operations costs (due to lower fuel economy). Moreover, to the extent that some administrative expenses are a function of ridership and that there is a relationship between service area density and ridership, systems operating in densely populated areas would be expected to have somewhat higher unit administrative costs than those operating in less densely populated areas.

C. Average Number of Passengers On-Board (HOWFULL)

It was hypothesized that Average Number of Passengers On-Board (HOWFULL) would be positively correlated with the unit operating expense variables. To the extent that there is a relationship between vehicle weight and fuel economy, systems which have fuller average loads would be expected to experience lower fuel economy and, therefore, incur slightly higher unit vehicle operations expenses. Similarly, to the extent that there is a relationship between intensity of use and vehicle wear and tear, systems which have fuller average loads would be expected to experience higher unit maintenance expenses. Finally, to the extent that some administra-

EXHIBIT 1
Descriptive Statistics

Variable	Mean	Standard Deviation	Minimum	Maximum	Label
EXPVH	$36.207	7.146	16.32	54.19	Operating Expense Per Vehicle Hour
ADEXPVH	$37.250	7.196	20.51	66.34	Adjusted Operating Expense Per Vehicle Hour
ADVOPVH	$22.692	3.812	12.80	34.70	Adjusted Vehicle Operations Exp. Per Veh. Hr.
ADOTHPVH	$14.525	4.146	7.19	31.71	Adjusted Other Operating Exp. Per Vehicle Hour
ADVMPVH	$ 7.883	2.468	4.07	16.96	Adjusted Vehicle Maintenance Exp. Per Veh. Hr.
ADGAPVH	$ 5.766	2.078	2.02	14.33	Adjusted General Admin. Exp. Per Vehicle Hour
BASE	101.836	88.527	14	387	Buses Operated During the Base Period
MAXPEAK	197.575	193.879	29	821	Buses Operated in Maximum Scheduled Service
REVBUS	273.836	263.079	42	1249	Total Buses in the Revenue Fleet
DENSITY	3061.425	1216.994	1145	7208	Population Density
HOWFULL	11.608	5.417	2.91	28.96	Average Number of Passengers On-Board
PEAKBASE	.875	.619	.03	2.53	Peak-Base Ratio (equals [MAXPEAK/BASE] - 1.0)
SPARE	.465	.332	.15	2.55	Spare Ratio (equals [REVBUS/MAXPEAK] - 1.0)
SPEED	13.131	2.225	7.12	20.46	Average Operating Speed
SPLIT	.081	.077	.010	.307	Proportion of Work Trips Made By Transit

tive expenses are usage-related, systems which have fuller average loads would be expected to experience higher unit administrative expenses.

D. Ratio of Buses Operated in Maximum Scheduled Service to Buses Operated During the Base Period (PEAKBASE)

It was hypothesized that Ratio of Buses Operated in Maximum Scheduled Service to Buses Operated During the Base Period (PEAKBASE) would be positively correlated with the unit operating expense variables. Variation in demand for bus transit service by time of day triggers the need for more vehicles to be in service during the early morning and late afternoon than during the midday or other periods. This could be expected to adversely affect unit costs in both the vehicle operation and maintenance areas. At most transit systems, drivers whose work assignments (runs) are split in two (or more) pieces with several intervening off-duty hours (e.g., several hours in the early morning and several hours in the late afternoon) are paid a premium to compensate them for the inconvenience of the long span between when their work day first begins and finally ends.

In the maintenance area, peaking creates inefficiencies in vehicle servicing. It also creates inefficiencies in maintenance manpower deployment as the ratio of vehicles in the garage to maintenance personnel fluctuates throughout the day.

E. Ratio of Total Buses in the Revenue Fleet to Buses Operated in Maximum Scheduled Service (SPARE)

A positive correlation between Ratio of Total Buses in the Revenue Fleet to Buses Operated in Maximum Scheduled Service (SPARE) and selected unit operating expense variables was also hypothesized. SPARE could have a significant effect on unit operating expense in the maintenance area. Too large a proportion of spares could result in lower vehicle productivity (fewer miles per vehicle per period), higher unit materials expenses (assuming that all vehicles are kept road-worthy) and higher unit labor expenses (assuming that some work done on the vehicles is done at time rather than mileage intervals). SPARE, however, would not be expected to have any significant effect on unit expenses in either the vehicle operations or administration areas.

F. Proportion of Urbanized Area Workers Making Work Trips by Transit (SPLIT)

It was hypothesized that Proportion of Urbanized Area Workers Making Work Trips by Transit (SPLIT) and selected unit operating expense variables would be positively correlated. In the context of this study, SPLIT is a surrogate for union strength. It is assumed that where SPLIT is high that union strength is great and, in turn, where union strength is great, higher unit expenses can be expected. The vehicle operations and maintenance areas of most transit systems are unionized. Consequently, union strength—through the surrogate measure SPLIT—is expected to be positively correlated with the unit expenses of these functions.

G. Average Operating Speed (SPEED)

It was hypothesized that Average Operating Speed (SPEED) would be negatively correlated with selected unit operating expense variables. The relationship between vehicle speed and fuel economy resembles an inverted-V. Fuel economy is lowest at extremely low and high speeds. Average operating speeds for buses operated in a transit environment are clustered in the teens, well below the cruising speeds at which a bus realizes optimum fuel economy. Since the range of expected values for SPEED would be along the left half of the inverted-V, the expected correlation between SPEED and unit vehicle operations expense would be negative. A similar relationship would be expected to exist between SPEED and unit maintenance expense.

V. FINDINGS

The major findings of the analysis are as follows:

1. *Diseconomies of scale are present among United States bus transit systems.* For most unit expense variables studied, Buses Operated During the Base Period (BASE) is the scale variable most highly correlated with the dependent variable. For the unit vehicle maintenance expense variable (ADVMPVH), the most highly correlated scale variable is Total Buses in the Revenue Fleet (REVBUS). (See Exhibit 2.) The proportion of variation in a given unit operating expense variable that is associated with variation in a given scale variable varies from case to case but is generally in the six to twelve percent range.

2. *Population Density (DENSITY) plays a key role in explaining variation in selected operating expense variables.* As expected, the correlations between Population Density and unit operating expense variables are positive. In many cases, the correlation between Population Density and a given dependent variable is stronger than that between any of the three scale variables and that dependent variable. (See Exhibit 2, rows one through four.)

3. *In general, the other nonfinancial variables—* Average Number of Passengers On-Board (HOWFULL), Ratio of Buses Operated in Maximum Scheduled Service to Buses Operated During the Base Period (PEAKBASE), Ratio of Total Buses in the Revenue Fleet to Buses Operated in Maximum Scheduled Service (SPARE), Average Operating Speed (SPEED) and Proportion of Urbanized Area Workers Making Work Trips by Transit (SPLIT)—*play little or no role in explaining variation in the dependent variables.* (See Exhibit 2, last five rows.)

4. *Strong relationships exist between selected pairs of independent variables.* Very strong relationships (r>.9) exist among Buses Operated During the Base Period (BASE), Buses Operated in Maximum Scheduled Service (MAXPEAK) and Total Buses in the Revenue Fleet (REVBUS). (See Exhibit 3.) These are to be expected since REVBUS is always greater than or equal to MAXPEAK which, in turn, is always greater than or equal to BASE. More significantly, the analysis shows a strong relationship between DENSITY and SPLIT (r = .750). (See Exhibit 3.) This finding supports the generally accepted notion that transit usage and demand is largely a function of the density of the service area in which it operates.

5. *Variation in unit expense variables which is associated with variation in scale and nonfinancial system definition variables ranges from 6.1 to 26.6 percent:*

- More than one-fifth (21.4 percent) of the variation in Adjusted Operating Expense Per Vehicle

EXHIBIT 2
Correlation Matrix
Independent—Dependent Variable Pairs
(pairs where /r/ > .3 are underlined)

Independent Variable	Dependent Variable				
	ADEXPVH	ADVOPVH	ADOTHPVH	ADVMPVH	ADGAPVH
BASE	<u>.363</u>	.271	<u>.381</u>	<u>.315</u>	<u>.323</u>
MAXPEAK	<u>.312</u>	.220	<u>.339</u>	<u>.303</u>	.251
REVBUS	<u>.341</u>	.249	<u>.364</u>	<u>.330</u>	.258
DENSITY	<u>.402</u>	.265	<u>.457</u>	<u>.494</u>	.279
HOWFULL	.111	.034	.155	.296	-.016
PEAKBASE	.045	-.005	.079	.149	-.049
SPARE	.101	.172	.014	-.032	-.022
SPEED	-.024	.020	-.060	-.151	.053
SPLIT	.152	.066	.204	<u>.374</u>	-.051

EXHIBIT 3
Correlation Matrix: Independent Variables
(pairs where /r/ > .3 are underlined)

	BASE	MAXPEAK	REVBUS	DENSITY	HOWFULL	PEAKBASE	SPARE	SPEED	SPLIT
BASE	1.000	.921	.925	.246	.304	.124	-.207	.058	.173
MAXPEAK	.921	1.000	.987	.135	.283	.421	-.245	.035	.163
REVBUS	.925	.987	1.000	.127	.244	.384	-.151	.050	.126
DENSITY	.246	.135	.127	1.000	.439	-.109	.014	-.312	.750
HOWFULL	.304	.283	.244	.439	1.000	.114	-.331	-.273	.549
PEAKBASE	.124	.421	.384	-.109	.114	1.000	-.285	-.092	.108
SPARE	-.207	-.245	-.151	.104	-.331	-.285	1.000	.090	-.063
SPEED	.058	.035	.050	-.312	-.273	-.092	.090	1.000	-.407
SPLIT	.173	.163	.126	.750	.549	.108	-.063	-.407	1.000

Hour (ADEXPVH) is associated with variation in Population Density (DENSITY) and Buses Operated During the Base Period (BASE).

- More than one-quarter (26.6 percent) of the variation in Adjusted Other Operating Expense Per Vehicle Hour (ADOTHPVH) is associated with variation in Population Density (DENSITY) and Buses Operated During the Base Period (BASE).
- More than one-quarter (26.6 percent) of the variation in Adjusted Vehicle Maintenance Expense Per Vehicle Hour (ADVMPVH) is associated with variation in Population Density (DENSITY) and the Ratio of Buses Operated in Maximum Scheduled Service to Buses Operated During the Base Period (PEAKBASE).
- Less than one-tenth (9.2 percent) of the variation in Adjusted General Administration Expense Per Vehicle Hour (ADGAPVH) is associated with variation in Buses Operated During the Base Period (BASE).
- Less than one-tenth (6.1 percent) of the variation in Adjusted Vehicle Operations Expense Per Vehicle Hour (ADVOPVH) is associated with variation in Buses Operated During the Base Period (BASE).

Specification and analysis of regression equations which explain the five dependent variables, as functions of scale and other system definition variables, is shown in Exhibit 4.

VI. IMPLICATIONS FOR TRANSIT POLICYMAKERS AND MANAGERS

As noted at the outset, the presence of diseconomies of scale among United States bus transit systems would warrant a reconsideration of the institutional arrangements presently used for the delivery this service. This analysis has demonstrated that such diseconomies of scale exist.

There is a need, as a result, for policymakers to consider alternate institutional arrangements which would promote cost-efficiency (without sacrificing quality). Greater use of private providers is one option. As reported by Teal, Giulianno, et. al. (1984) there are numerous examples of increased cost-efficiency resulting from private companies providing publicly-supported bus transit service. There are, however, major institutional and market obstacles to widespread use of purchase-of-service contracting.

Further, public transit agencies might consider more extensive use of contractors to provide administrative and maintenance services on their behalf. Many transit systems presently contract with private companies to provide selected administrative or maintenance services. However, these contracts generally represent a very small fraction of expenses in these functional areas. Consideration should be given to using private contractors for non-traditional assignments such as telephone information service, vehicle servicing and preventive maintenance inspections.

VII. NEED FOR ADDITIONAL RESEARCH

While this research demonstrates that diseconomies of scale are present among the nation's medium and large bus transit systems, it leaves unanswered many questions. Priorities for future research, therefore, include:

- replicating this study using unit measures of total, rather than operating, expense as the dependent variables;
- replicating this study using measures of labor and material expense and productivity as the dependent variables;
- conducting time-series analyses of selected systems;
- performing case study analyses of individual systems; and
- conceptualizing alternative institutional arrangements for the delivery of transit service in the nation's major metropolitan areas.

At the very least, there is a need for an on-going investigation of the causes of and possible remedies for the inefficiencies which plague the nation's larger bus transit systems. The status quo—the monopolistic, regional transportation authority—does not appear to be a particularly cost-efficient method of delivering bus transit service.

EXHIBIT 4

Specification of Explanatory Variables
(t scores are shown in parentheses)

```
ADEXPVH = $28.896 + $.002 DENSITY + $.023 BASE
          (13.93)       (3.09)          (2.60)
Adjusted r2 = .214   Statistically significant at .0001 level


ADVOPVH = $21.502 + $.012 BASE
           (32.5)      (2.38)
Adjusted r2 = .061   Statistically significant at .02 level

ADOTHPVH = $9.128 + $.001 DENSITY + $.013 BASE
           (7.90)       (3.71)          (2.75)
Adjusted r2 = .266   Statistically significant at .0001 level

ADVMPVH = $3.959 + $.001 DENSITY + $.819 PEAKBASE
          (4.98)       (5.09)          (2.02)
Adjusted r2 = .266   Statistically significant at .0001 level

ADGAPVH = $4.993 + $.008 BASE
          (14.08)      (2.88)
Adjusted r2 = .092   Statistically significant at .005 level
```

REFERENCES

Berechman, J. (1983). *"Costs, economies of scale and factor substitution in bus transport. An analysis"*, Journal of Transport Economics and Policy XVII(1):7-24.

Berechman, J. and Giuliano, G. (1984) *"Analysis of the cost structure of an urban bus transit property"*, Transportation Research.

Berechman, J. and Giuliano, G. (1985) *"Economies of scale in bus transit: A review of concepts and evidence"*, Transportation 12:313-332.

Koshal, R.K. (1972). *"Economies of scale. II. Bus transport: Some United States experience"*, Journal of Transport Economics and Policy 6(2):151-153.

Nelson, G.R. (1972). *An Econometric Model of Urban Bus Transit Operations*. Paper P-863. Arlington, VA:Institute for Defense Analysis.

Teal, R.F., Giuliano, G., Brenner, M.G., Rooney, S.B. and Jacobs, J. (1984). *Private Sector Options for Commuter Transportation*. Prepared for United States Department of Transportation, Urban Mass Transportation Administration. Washington, DC. UMTA-CA-11-0022-2.

United States Department of Transportation, Transportation Systems Center (1984). *National Urban Mass Transportation Statistics—FY 1983 Section 15 Annual Report*. Washington, DC: United States Department of Transportation DOT-TSC-UMTA-87-37

United States Department of Transportation, Federal Highway Administration (1985). *Transportation Planning Data for Urbanized Areas—Based on the 1980 Census*. Washington, DC: United States Department of Transportation DOT-I-85-20

Veatch, J.F. (1973). *Cost and Demand for Urban Bus Transit*. Unpublished doctoral dissertation. University of Illinois, available from Xerox University Microfilms as No. 74-5723.

Viton, P.A. (1981). *"A translog function for urban bus transit"*, Journal of Industrial Economics 29(3):287-304.

Wabe, J.S. and Coles, O.B. (1975) *"The short and long-run cost of bus transport in urban areas"*, Journal of Transport Economics and Policy 9(2):127-140.

Williams, M.L. and Dalal, A. (1981). *"Estimation of elasticities of factor substitution in urban bus transportation: A cost function approach"*, Journal of Regional Science 21(2):263-275.

Williams, M.L. and Hall, C. (1981). *"Returns to scale in the United States intercity bus industry"*, Regional Science and Urban Economics 11:573-584.

ENDNOTE

* New York Institute of Technology.

Labor-Management Cooperation in Mass Transit: Conceptual and Applied Considerations

by Ken Jennings, Jay A. Smith,** and Earle C. Traynham****

ABSTRACT

The research suggests several broad conclusions for labor-management cooperative activities in mass transit: (1) there is a direct relationship between mutually successful collective bargaining and successful cooperative activities; (2) "canned" approaches to labor-management cooperation do not work; (3) each successful cooperative venture is tailor made for that transit system; and (4) more and better communication about labor-management cooperation can improve the climate for productive, positive results.

I. INTRODUCTION

Although the resolution of important workplace issues in mass transit continues to be most commonly addressed within the traditional system of collective bargaining, there is an increasing use of cooperative arrangements such as quality circles, productivity sharing plans, labor-management committees, etc. All of these programs are structural interventions which attempt to generate greater worker interest, involvement, and effort toward achieving organizational goals. This renewed interest in cooperative ventures offers an opportunity for a long-term reshaping of many transit bargaining relationships. The success of cooperative efforts will likely determine whether these endeavors become a permanent part of transit's industrial relationship systems or merely a temporary interruption in traditional adversary collective bargaining.

There has been a recent, somewhat urgent, consideration of labor-management cooperation (LMC) in the United States. Members of the Reagan administration, noted academics and some union and management executives have observed the increase in cooperative initiatives and have urged the continuation of this situation.

LMC has become a prominent aspect of the mass transit industry as witnessed by the First National Transit Labor-Management Conference in Miami (1982). Today, the mass-transit industry faces severe challenges in the forms of reduced ridership, fiscal austerity, and reduced government subsidies. Union and management officials will need to marshal cooperative attitudes and efforts to approach these challenges successfully.

This paper has two major purposes. It first briefly explores the varied dimensions and related confusion over the concept, "labor-management cooperation." It then formulates and discusses various pre-conditions and other variables which affect this concept's *measured success*. This discussion should be directly applicable to the mass transit industry since it draws related insights from interviews with 67 mass transit union and management officials. A model of successful labor-management cooperation will then be derived by meshing practitioner insights with an extensive literature review.

II. A MODEL OF LABOR-MANAGEMENT COOPERATION

Labor-Management Cooperation: The Dependent Variable

Defining "labor-management cooperation" (LMC) is relatively easy; measuring this concept becomes more difficult. A classic definition of LMC was provided by Harbison and Coleman in 1959.

> "Union management cooperation" describes a relationship in which the parties through joint action attempt to reduce unit costs of production, increase efficiency, and improve the competitive position of the firm. This kind of relationship is rare."

Yet, this definition suggests that LMC can represent a goal, an approach, attitude or activity. If LMC is to be examined at mass transit properties, appropriate measures need to be formulated to determine relative involvement and success rates. The following model has been developed to focus on cooperative activities that are more clear cut and susceptible to results measurement than such vague outcomes as a goal or an attitude. To be included, LMC activities also must have the following dimensions:

—*Voluntary initiation* by union management officials. Collective bargaining issues carry sanctions (strikes and lockouts) and definite time frames (duration of the labor agreement). On the other hand, union and management officials are free to enter and leave the cooperative arrangement at any time.

—*Risk* takes place when management and union representatives have to give up some of their traditional prerequisites in approaching a LMC activity. Management officials have to be willing to share power as well as profits by relinquishing some of their former unilateral decision-making authority (Schlesinger & Oshry, 1984). Union officials have to take a more active role in problem solving—there has to be a sense of joint ownership in the activity instead of the traditional pattern "management acts an the union reacts through the grievance procedure" (Batt & Weinberg, 1978, Raskin, 1982). Union members might believe that union leaders following the more active approach have "sold out," becoming "partners with management." Union officers faced with this situation might lose subsequent reelection bids.

—*Joint definition and resolution of a problem topic represents the focal point* for union and management participants. Related examples found at our investigated mass transit properties are union-management design of a training program to reduce employee stress at a particular facility; union-management committees to determine if a bus driver's driving accident should be labeled "chargeable" or "non-chargeable" for subsequent employee discipline; and, productivity improvement programs.

These and other related topics would likely receive more limited treatment in formal contract negotiations and/or administration (grievance procedures found in the labor agreement) for several reasons. Contract negotiation issues are often discussed in an argument-counter argument fashion. Each party stresses their respective positions on the issue instead of specific issue aspects which can be agreed upon by both.

Also, collective bargaining topics are often discussed in potential, instead of eventual, terms. "Trade-off" value of a bargaining issue can become a consideration, particularly if the topic pertains to a relatively small proportion of the bargaining unit employees. Little, if any, serious attention will be given to a collective bargaining issue if one or both parties feel it will be dropped in the final labor agreement settlement.

There are many potential LMC activities. Some researchers and practitioners might find that a classification system orders their thinking and helps in predicting differential success rates. Our research and literature review suggests the following classification system.

1. *Employee Input Programs* which can pertain to all employees (e.g., suggestion programs), selected union and management officials across the organization (e.g., labor-management committees or boards), or to employees and managers from a common work area (e.g., quality circles).
2. *Safety and Health Programs* such as safety committees or employee assistance programs designed to counsel employees about several problems.
3. *Performance Incentive Programs* which can range from recognition of above average performance (e.g., banquets, trophies and so forth) to productivity gains as justification for salary increases.
4. *Training Programs* where both union and management officials play an active part in determining and delivering appropriate topics.

Once defined, LMC activities need related success measures. One of the most common approaches is to declare a program "successful" if it meets established targets of change—an attendance incentive program might be considered successful when 20 percent of the employees meet. Another success measurement approach is less concerned with the LMC activity's results and more concerned with the process of reaching the goals. "A scorecard is kept to record meetings cancelled, attendance at meetings, number of projects completed, number of employees involved and so on" (Clark, Warren & Greisinger, 1983, p.5).

Cost-benefit calculations might also be considered in some or all of these success measurements. However, these calculations, even for performance incentive programs, are not always agreed upon by all interested parties. Assume, for example, that a LMC activity results in the reduction of 50 employees in a particular classification. These employees are transferred to other classifications for which new employees would have been hired. However, the resulting "savings" might be disputed on several dimensions such as the amount of time needed to implement or start the program; the time periods it would have taken to recruit and develop replacements and train transferred employees in their new positions; the wage rate of entry level employees; and, related fringe benefit costs such as pension contribution versus similar cost categories for the transferred employees.

In many cases, hard line productivity measures are not used to operationalize possible benefits of LMC activities (Jennings, Smith & Traynham, 1986). This situation might be explained by several reasons. First "cause and effect" is difficult, if not impossible, to establish. For example, is reduction in employee turnover due to a LMC activity or to poor economic conditions that make it difficult to leave their jobs at a particular facility? Secondly, it might also be difficult for management to isolate diverse performance and personnel statistics. Thirdly, a hard line result might have both positive and negative implications. To illustrate, an employee assistance program might be associated with a reduction of disciplinary grievances. This statistic might reflect a better understanding between labor and management officials or it might mean that supervisors, fearing "second guessing" and an increased watchdog function of the personnel department, ignore the situation entirely.

Finally, a cost-benefit analysis no matter how extensive can overlook necessary but vaguely defined, long range benefits such as "high level of participation," "improved morale/teamwork," and "increased enthusiasm." There is a real possibility that labor-management cooperation is a progressive situation where "quick hitters" or somewhat minor problems that are easily resolved pave the way for subsequent, more difficult efforts. Thus, precise cost-benefit figures might run contrary to longer run goals of the LMC activity.

Preconditions Affecting Labor-Management Cooperation

Collective bargaining behaviors and results suggest the extent that union and management officials will engage in labor-management cooperation. A facility might have successful bargaining (both sides are pleased with resolution) and no LMC efforts. But it is doubtful that a property will have unsuccessful collective bargaining and successful LMC efforts.

The negotiated labor agreement seldom exerts a large positive influence on collective bargaining activities. Contract language typically restricts discretion of both union and management officials. For example, contractual provisions indicating the bidding procedure for job transfers tells union and management officials what they cannot do since these individuals are not free to substitute their own procedures. Thus, a comprehensive detailed labor agreement probably restricts the potential for subsequent LMC activities.

III. ORGANIZATIONAL AND PARTICIPANT CHARACTERISTICS

Declining financial/competitive aspects of the organization can prompt one or both parties to seek cooperative alternatives. Mass transit services, for example, have been recently confronted with declining ridership, competitive services (paratransit, for example), increased costs of operations, and reduced level of governmental financial assistance. This situation might prompt union-management interest in some cooperative categories, such as employee input programs and performance incentive programs. Employees in various mass transit classifications might prefer these programs when they perceive them to result in higher wage increases than those formerly obtained under more traditional arguments such as wage comparability.

There are at least three participant characteristics which can affect the success of a particular LMC activity. The first category pertains to negotiation experience as measured by number of formal labor agreements negotiated, number of years as union or management official, years of experience in dealing with officials from other party, and perceived security in remaining in union and management organization at the same or higher level capacity. This latter dimension is particularly significant for union officials who are elected on a periodic basis (Ramquist, 1982) by members typically desiring leadership capability and credibility. Experienced negotiators also offer the advantage of knowing whether a possible cooperative issue is best pursued through collective bargaining or through a more informal, flexible approach such as a union-management task force. Previous research at mass transit properties has found that mass transit negotiators have rather extensive experience in the mass-transit industry and in collective bargaining.

Each party involved in labor-management cooperation should also recognize the legitimate rights and interest of the other party (Scearce, 1977). Management officials have to recognize that the union is more than the legally sanctioned exclusive bargaining agent for the employees (Holley & Jennings, 1984; Gray, 1984). Union officials in a successful cooperative activity will likely believe that management has the right to determine layout of equipment; financial policies and practices, source of materials; and determination of products and services to be rendered (Perline, 1971 and Derber, 1961). The two major labor organizations in mass transit, ATU and TWU, have encouraged local unions to consider and possibly implement cooperative activities that might be uniquely tailored for the particular location.

Finally, both union and management officials need to be committed to the particular LMC activity. This characteristic helps insure the consistent potential of cooperative efforts even if there are interruptions in these activities. For example, there were elections of new union officers at a couple of the interviewed properties. Union and management officials who are committed to LMC activities at this property at least might believe in the resumption after internal difficulties are resolved.

"Commitment" might be difficult to measure in terms other than attitudinal reflections. However, there are some concrete measures, such as quality and promptness of responses to employee suggestions, amount of money spent on trips to send employee input groups such as quality circles to other locations for related insights; remaining with employees (if appropriate) in program orientation and feedback sessions; and returning telephone calls concerning the program's content and direction. Our model and interviews with mass transit practitioners suggests that commitment is likely when each party believes that the advantages of the particular LMC activity outweigh the disadvantages. Some obvious disadvantages of LMC activities include the possibility that managers' decision making and leadership skills are challenged/second guessed by their subordinates possibly in front of higher level executives; requiring subsequent alternatives (Siegel and Weinberg, 1982), and union member charges that participating union officials "sold out" (trying to obtain a management position through LMC efforts, for example). Top managers might contend that LMC efforts extend bargaining over the labor agreement 365 days a year while some other participants might fear retribution from their supervisors for their critical remarks made during LMC discussions.

There are also several possible advantages of LMC activities which might accrue to some or all LMC participants. For example, a LMC activity might have little effort placed on "face saving," a significant aspect of collective bargaining and related labor relations considerations. Union and management participants are not wearing their respective positions on their sleeves since cooperation focuses on "win-win" situations and there is no "victory" labeling or posturing at stake. Management participants might also be pleased if their subordinates involved in LMC learn to appreciate better their difficult decision making roles. Many employees claim managers take the easy way out by responding "no" to a particular decision. This situation might change when employees become involved in problems design and solving—they realize that once viewed, "easy" problems become more complicated and difficult to resolve. At a minimum, a LMC activity might reduce organizational hostility toward managerial decisions and resultant changes (Juechter, 1982). Also LMC activities might facilitate experimentation and innovation (U.S. Department of Labor, 1982) which might allow level managers to obtain additional perspectives and responsibilities from higher level management officials.

LMC participation might also benefit the union officer concerned with representing "significant minority membership factions." An example might be maintenance employees, who are politically active but do not have a large number of votes to warrant spending large amounts of time in collective bargaining and related labor agreement provisions/grievances. Quite possibly, labor-management cooperative activities such as safety committees or quality circles can benefit specialized employee groupings while not detracting from collective-bargaining efforts. This means that mass transit union can escape the change that they are "driver dominated" by enabling maintenance employees to resolve their problems throughout the year.

MODEL OF A LABOR-MANAGEMENT COOPERATIVE (LMC) ACTIVITY AT A FACILITY LEVEL

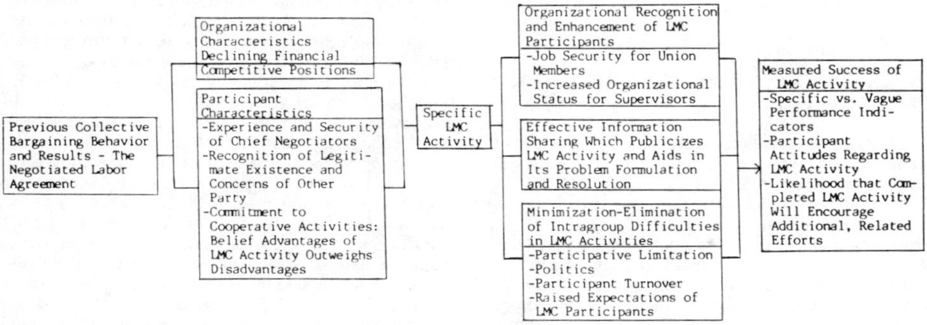

IV. OTHER VARIABLES AFFECTING LMC AND RELATED RESULTS

A. Organizational Recognition and Enhancement of LMC Participants

Job security represents the most fundamental collective bargaining issue that must be addressed and sustained if labor-management cooperative activities are to succeed. We uncovered no specific guarantees or related programs to maintain job security. This might be due to the fact that all but possibly one of the sites anticipated consistent if not expanded service and related employment levels. Union officials at this possible exception doubted whether broadly based cooperative efforts would occur there. Union officials represent a subset of hourly employees; in addition to their job security concerns they are also interested in their political survival. Related guarantees at our interviewed mass transit properties include express prohibitions against having any previously attained labor agreement provision as a subject of a particular LMC activity. Some organizations also indicate that a particular LMC activity has to be unanimously approved before it is initiated; and a single negative vote by a participant can terminate the LMC activity in subsequent stages. Of course, the right to file grievances over any aspect of the LMC activity furnishes union officials with yet another political safeguard.

Lower level management participants are perhaps the most publicized individuals in a LMC activity. Unfortunately, most of this publicity is negative. One representative research finding noted, "the biggest problem, as other industries have discovered in trying the participation approach, is convincing first line supervisors that they must change their management styles and listen to the suggestions of others instead of barking orders" (Siegel & Weinberg, 1982, Zahra, 1984).

This problem has many sources, but few solutions. Negative reactions from management participants can occur when the LMC activity is suddenly sprung on them, when they are expected to change their former supervisory role with little attention spent on formulating their new rights and responsibilities (Fenn & Yankelovich, 1972; and Ozley & Ball, 1982).

B. Information Sharing

This variable is initially necessary to publicize the LMC activity. Our survey found that all employee-input activities had at least one printed newsletter sent to all employees of the facilities to inform them of the program's purpose and process. The initial written publicity for quality circles typically explained the organization, operation, typical projects, and related problem solving techniques used in quality circles. Employee Assistance Programs (EAPs) very likely rely on initial publicity more than other labor-management cooperative efforts. The emphasis on sensitive, personal problems, such as substance and alcohol abuse, requires precise statements pertaining to the rights and responsibilities of employees/potential clients.

Publicity can also occur during the operation of the labor management cooperative activity. The most common information exchange mechanism among members are minutes of meetings which serve two purposes: a follow-up for the next meeting and an indicator of the various items acted upon by management.

This situation might be partly due to the subject matter of labor-management cooperative activities. For example, many Accident Review Boards (ARBs) do not include the vote totals on a chargeable accident. This incomplete publicity lets union and management members vote their convictions without running the risk of being second-guessed/harassed by affected drivers or other employees who believe union ARB members should vote the "party line." Similarly, the sensitive nature of Employee Assistance Plans (EAPs) preclude union and management officials from receiving detailed status reports. However, at least one EAP enables these individuals to receive at least some of the information about the participating employee on a need-to-know basis. Sometimes union stewards can obtain information concerning an employee to counter possible misinformation supplied by the employee. For example, one employee claimed that a supervisor had it "in for him," although a review of his records revealed nine disciplinary write-ups, only two of which were from the supervisor in question.

Information sharing's second purpose is to aid in the formulation and resolution of the problem. Employee suggestion systems often have two unique information-sharing sources. The first item is a rather detailed manual that specifies the rights and responsibilities of various individuals in the suggestion process, as well as proper considerations to be used in making asuggestion. A second source is the evaluation written report that details the reasons for acceptance or rejection of the suggestion.

C. Minimization-Elimination of Intragroup Difficulties

Space limitations preclude a detailed discussion of variables within this category; although some examples will be given for illustrative purposes. Employee-input programs such as quality circles can have participative limitations such as *deference patterns* or situations that give those in higher organizations "more air time" and their opinions more weight. Only two interviewed locations reported such patterns. Perhaps the relative lack of deference patterns is due to the large number of participants who are union officers who are ". . . experienced in being vocal and standing their ground."

Politics, a second intragroup difficulty, can refer to the resolution of internal differences within the union where union officials attempt to strengthen, or at least maintain their positions in the organization. In order for labor-management cooperative activities to be successful, the union must be strong, and members must trust leaders (Guest, 1979). A union leader is typically concerned with reelection. Related attempts to balance labor-management cooperative considerations with the traditional role of the union leader is not always easy. Union officials at two locations who were identified with LMC activities were defeated in subsequent union elections. In one of these cases, the newly elected union president has not formally endorsed the employee-input program that was encouraged by the defeated union president. According to one interviewee, the current union president functions as a "gatekeeper." This individual wants to receive all members' problems, then get back to each member on a one-to-one basis. This leadership style is not particularly conducive to labor-management cooperative activities. In addition to leadership style, the extent that union officers are politically affected by LMC activities could depend on the type of activity, and the impact of "peer pressure." Labor-management cooperative activities that "pass judgment" on a potentially wide number of union members on a fairly frequent basis might be politically risky for union officers. Employee-Assistance Programs (EAPs) represent one such activity. Perhaps one reason no union officials reported any political problems over EAPs is that union officials are not directly involved in the judgment of other employees. Also, an employee who is a substance abuser can affect employees' safety, and safety is a mutually desirable goal capable of propelling union and management officials into cooperative action. However, one management official did reveal an unanticipated EAP political disadvantage to union officials. An EAP program indicates that management officials are very concerned with giving an employee a second chance, thereby saving this individual's job. This concern might preclude union leaders from being identified with the traditional goal of employee protection from "arbitrary and capricious" managerial actions.

A second LMC activity that calls political considerations into play is the Accident Review Board (ARB). ARBs have prompted political adjustments, such as not indicating specific votes (even totals) associated with union and management participants. Other political considerations of ARBs were reported at two interviewed properties. At one location, politics and resultant tensions among union members resulted in a structural change of the ARB. Originally, there were four members, two union representatives and two management representatives. A coin would be flipped to get a fifth, possibly tie-breaking, voting member for the ARB's decision. If a union member-bus driver lost a three-two vote with three union representatives on the ARB, the membership, according to one interviewed union respondent, would raise "holy havoc with the union leadership." The ARB's structure has since been changed to five permanent members: two union representatives, two management representatives, and one "professional" (a former state highway patrol employee, for example), who only acts to break a tie vote.

An ARB at another location witnessed some political considerations on at least one occasion. Even though the name of the driver is left off the report and specific ARB members are not identified with vote total, partisan actions can be taken. In one case, a union steward's son was involved in an accident; the resulting 5-5 deadlock vote from the 10-member board was changed to a nonchargeable accident.

Some experiences with LMC activities indicate that "peer pressure" can work against labor-management cooperation team members. It is incorrect to assume that all employees will embrace programs. For example, one estimate suggests that only 25 percent to 50 percent of the rank and file support employee-input programs (Steel, 1983). At a minimum, negative rank-and-file attitudes would confuse union officials; these attitudes could even make union leaders wary of cooperative efforts.

Peer pressure negatively affected quality circles at only two interview locations. Management and union respondents indicated that this peer pressure was actually short-lived skepticism about the quality circle's potential. A union official at another property perhaps explained the relative absence of negative attitudes from other employees: "All employees at least believe in having a say in their working conditions instead of having orders shoved down their throat."

Loss of key persons and turnover was not suggested by the literature review; however, several interviewed properties reported the existence and at least potential impact of turnover in two general directions. First is the loss of key management or union officials. At five locations, the agencies had witnessed a change in the highest management position within a year of the interview. At a sixth location, there had been a turnover of three union presidents within a short period of time. While the precise impact of these employment shifts on labor-management cooperative efforts cannot be determined at this time, the probable impact, at least in terms of delay and in terms of reestablishing previous commitments, appears major. Perhaps turnover of top union and management officials explains the relatively short-lived nature of many employee-input programs.

Turnover of quality-circle members was regarded as a problem at three locations. Management and union respondents furnished the following four reasons for this turnover:

> Many circle members have been promoted to managerial positions. This reason was reported at the largest property interviewed. Perhaps this problem is more applicable to larger properties.
> Many circle members are afraid to give related, formal presentations before management officials.
> Some circle members get "burned out" or tired after the project is over and are, therefore, reluctant to continue in other quality circles.
> Once a year people have the opportunity to bid to other jobs, and they may go to locations not covered by the quality circle.

The fourth reason appears to at least harm the potential of employee-input programs. One location's "status report" noted that "the new line-up presents the potential for a serious disruption and possible discontinuation of QC activities if significant reassignment of QC members occurs."

One possible way to minimize turnover is to carefully structure the quality circle. However, as noted at one location, this strategy can backfire. Initially QCs were started in work areas with only one shift where employees were centrally located. These selection criteria failed to generate sufficient volunteers. The eligible list was later broadened to remedy this problem. However, some employees who were declared originally "ineligible" (in the wrong location) refused to join.

Risk of raising expectations can occur in LMC activities. At one location, union officials contended that management raised employee expectations in some farming out (subcontracting) projects, such as a car overhaul program, and then did not provide employees with adequate parts. For example, bargaining-unit employees performed the work and had to rework old door frames, which, of course, takes more time than that involved by the subcontractor who simply replaces door frames with new frames.

At another location, the third-party facilitator of the labor-management board indicated, "We have received a lot of 'no's' from management, a situation which is likely to occur when potential ideas are not (and will not be) screened by the facilitator." Turndowns have a negative effect, particularly if the participants believe management is turning down an idea for cost reasons. Management officials can also raise employee expectations by delaying their response to employee suggestions. A "no" received after a particularly lengthy delay with no communication as to viable status or outcome of the suggestion will probably increase membership bitterness.

Driver recognition programs represent one labor-management cooperative effort that can raise employee expectations. Management first attempted to meet employee expectations through its long-range achievement plan: rewarding drivers who have met performance criteria for 10 years. This plan ensures follow through because employees know it will be in operation for at least a decade.

Yet driver-recognition programs typically have two administrative difficulties that are, in turn, related to employee expectations. One problem has to do with maintaining incentives for good performance over a long period of time. Most of the programs reward safe driving, perfect attendance, and so on for a period of one or two years. Employees who extend their good performance beyond the initial time period may expect greater rewards; however, if not rewarded, driver morale suffers. Another problem concerns the use of a rigid calendar for judging good performance. For example, suppose a driver is rewarded for a perfect driving record such as no chargeable accidents for 12 months beginning July 1. If the driver has a chargeable accident in early July, there is no incentive to make the extra effort for safe driving for the remainder of the year. One agency has tried to adjust to this by going to a rolling calendar; an employee is rewarded for 12 consecutive months of safe driving. Of course, this creates some difficulties with annual awards banquets in that a substantial amount of time might elapse between an employee's accomplishment and the subsequent recognition.

A management official at one location suggested that labor-management cooperative efforts, particularly employee-input programs, might raise management's expectations more than those of the participants. Some of the management officials seeking quick solutions and not receiving them have asked, "Where's the money we have spent for training?"

V. SUMMARY

Hopefully our model of LMC has illustrated several applications/ implications for the mass transit industry. First, LMC is not an easily defined concept; although it should be operationalized in rather concrete terms so that related success measures can be gauged. Second, previous collective bargaining activity and results will not always translate into subsequent LMC efforts although negative collective bargaining indicators will make LMC unlikely. Third, industrial and participant characteristics found in most of our interviewed mass transit locations have been conducive to many LMC efforts. Finally, the extent these activities are successful often depends on a) organizational recognition and enhancement of LMC participants; b) effective information sharing; and c) minimization-elimination of related intragroup difficulties.

REFERENCES

Batt, W. and Weinberg, E. Labor-management cooperation today. *Harvard Business Review*, 1978 (January-February), 96-104.

Clark, S., Warren, K. and Greisinger, G. *Assessment of Quality-of-WorkLife Program for The Transit Industry Research Report No. 5,* Washington, D.C.: Transportation Research Board, 1983.

Derber, M., Chalmers, W., and Edleman, M. Union participation in plant decision making, *Industrial and Labor Relations Review*, 1961 (October), 83-101.

Fenn, D. and Yankelovich, D. Responding to the employee voice, *Harvard Business Review*, 1972 (May, June), 83-91.

Gray, L. Union-management cooperation: A passing fad or permanent change? *Labor Studies Journal*, 1984 (Winter), 209-220.

Guest, R. Quality of work life-learning from Tarrytown. *Harvard Business Review*, 1979 (July-August), 76-87.

Harbison, F. and Coleman, J. *Goals and Strategy in Collective Bargaining*. New York: Harpers, 1951.

Holley, W. and Jennings, K. *The Labor Relations Process*. Chicago: Dryden Press, 1984.

Jennings, K., Smith, J., and Traynham, E. *Labor Management Cooperation in a Public Service Industry*. New York: Praeger Publishers, 1986.

Juechter, W. The pros and cons of participative management. *Management Review*, 1982 (September), 44-48.

Ozley, L. and Ball, J. Quality of worklife: Initiating successful efforts in labor-management organizations. *Personnel Administrator*, 1982 (May), 27-33.

Perline, M. Organized labor and managerial prerogatives: An empirical study. *California Management Review*, 1971 (Winter), 46-50.

Ramquist, J. Labor-management cooperation—the Scanlon plan at work. *Sloan Management Review*, 1982 (April), 49-55.

Raskin, A. Can management and labor become partners? *Across the Board*, 1982 (July-August), 12-16.

Scearce, J. Labor-management cooperation: Myth or reality? In Loftus, J. and Walfish, B., eds. *Breakthrough in Union-Management Corporation*. New York: Work in American Institute, 1977.

Schlesinger, L. and Oshry, B. Quality of work life and the manager: Muddle in the middle. *Organizational Dynamics*, 1984 (Summer), 4-19.

Siegel, I. and Weinberg, E. *Labor Management Cooperation: The American Experience*. Kalamazoo: W. E. Upjohn Institute, 1982.

"Steel Listen to Workers and Likes What it Hears," *Business Week*, December 19, 1983, p. 92.

U.S. Department of Labor. *Report on the Secretary of Labor's Symposium on Cooperative Labor-Management Programs*. Washington, D.C.: 1982.

Zarha, S. What supervisors think about quality circles. *Supervisory Management*, 1984 (August), 27-33.

ENDNOTE

* Industrial Relations, College of Busniess, University of North Florida, Jacksonville, FL.

* * Labor Economics, College of Business, University of North Florida, Jacksonville, FL.

* * * College of Business, North Texas State University, Denton, TX.

Privatizing Bus Transit: Cost Savings From Competitive Contracting

*by Edward K. Morlok**

ABSTRACT

This paper identifies and defines the various organizational arrangements for providing public bus transit, including the competitive contracting form which is the form most actively pursued now in the U.S. It then describes the results of a recent research project which surveyed all U.S. transit systems that are competitively contracting bus service, with an emphasis on the cost savings from privatization. Based on this survey a model for estimating cost savings was developed, and the estimated national savings from increased competitive contracting, as is being mandated by the U.S. Department of Transportation, are presented.

I. INTRODUCTION

The purpose of this paper is to summarize the results of a recent survey of U.S. transit systems which use the technique of competitive contracting of bus service operation to private organizations. Since the primary motivation for increased use of competitive contracting is cost reduction, one focus was on cost savings. A model for estimating the cost savings from contracting was developed using these data, and applied to all U.S. Transit systems with 100 or more buses to estimate the savings they would experience from contracting. These are then summed to obtain national savings.

Because privatization has so many different meanings, it is important to define competitive contracting within the context of these various forms. Broadly speaking, there are four alternative institutional arrangements for the provision of transit service (with some variations):

1. *Public Monopoly*: A publicly owned transit authority or government depart ment plans and operates the service, with any deficit (all U.S. and virtually all systems in developed nations operate at a loss) being made up by govern ment subsidies. Government grants cover virtually all capital purchases (vehicles, garage construction, subway construction, etc.).

2. *Competitive Contracting*: The public transit authority plans the service routes, schedules, fares, etc. but contracts on a competitive basis with others for the operation of some or all of the service. Most bidders are private firms, but the operating division of the transit authority itself or other public agencies may bid as well.

3. *Private Monopoly*: Transit service is planned and operated by one or more private, for-profit companies, which are regulated. Each firm typically has an exclusive area monopoly, although sometimes areas overlap.

4. *Free Market*: Private firms plan and operate their own services, without any restrictions on entry, exit, routes or service, fares, etc. Three important variants of these (numbered to facilitate association with the parent form) are:

 1A. *Public Monopoly with Private Financing*: Operation by a transit authority as in No. 1, but with the deficit or funding for capital purchases covered partly by government funds and partly by private contributions. Usually such private sector financing is through ownership of particular facilities, with the motivation being increased profit either from tax credits (e.g., lease-back of vehicles) or greater accessibility of other businesses (e.g., stores, offices).

 1B. *Public Monopoly with Sole Source Contracting*: Service planned as in No. 1, but with contracting on a sole source or negotiated basis to private companies for some or all operations, for system management or for maintenance.

 3A. *Private Monopoly with Public Funding*: Service provision as in No. 3, but with government funding of deficits and often low-price leasing of government owned vehicles or other capital stock.

Historically, public transit in the U.S. was primarily of form No. 3 from the early 1900's through the 1950's and 1960's, when most systems were purchased by public monopolies (yielding forms Nos. 1 and 1B). In some areas private ownership continued under No. 3A. At the present time there is considerable interest in increased privatization following patterns No. 2 and 1A. The major emphasis in this paper is of course on competitive contracting (No. 2). The free market arrangement has not been used for urban mass transit in the U.S. for decades, although the deregulation of interstate bus service in 1983 had the unintended effect of allowing free entry into a few urban markets (e.g., New Jersey suburbs to New York City). Various efforts at the state and local levels have been under way to bring such service under regulatory control.

In the context of competitive contracting of transit service, it is important to distinguish clearly between two different roles. These are the roles of service sponsor and of service operator. A *service sponsor* decides what services to be provided and its characteristics, such as routes, schedules, and fares, and arranges for provision of the service. The sponsor's role is essentially one of policy making, planning, facilitation, and financing. The *service oper-*

ator, on the other hand, actually produces the service—operates the vehicles, maintains them, hires the drivers, and so forth. There is no reason why the roles of sponsor and operator must be fulfilled by the same organization, and in many public services the two roles are separate—for example, in the provision of roads, where governments contract with construction firms for road building and often for maintenance.

The interest in privatization in the form of competitive service contracting has been fostered by the desire to reduce costs substantially, while at the same time retaining governmental control of service. The concern for cost reduction stems from the rapid escalation of transit cost per vehicle mile (or other service output unit) and simultaneously rapid increases in the deficits of service in the past two decades. Figure 1 presents relevant data in the U.S. for the period from 1950 to 1980. All costs and revenues have been adjusted for inflation using the consumer price index. The average cost for all modes of transit have generally increased much more rapidly than inflation. While there have been some changes in transit service which might have lead to a slight increase in this average, such as improved vehicle quality (e.g, air conditioning), and an increase in the peakedness of traffic and hence increases in average operating costs, a cost increase of 78% from 1960 to 1980 surely can not be attributed solely to these changes. Furthermore, in the same period, the revenue to cost situation changed from one in which revenues covered all operating costs to one in which revenues covered less than forty percent of operating costs and virtually no capital expenditures. This situation, and pressing other demands for government funds at all levels, has led to considerable interest in competitive contracting.

II. ISSUES IN COST SAVING

The cost saving from competitive contracting of service operation is not an unambiguous concept, and therefore consideration must be given to exactly what type of cost saving is relevant. Three issues are important: (1) the type of cost basis (e.g., fully allocated costs vs. marginal costs) for the saving, (2) what cost items, in the sense of cost to whom, are to be included, and (3) the time period for savings to be realized.

Turning to the first question, the most relevant cost saving would seem to be the *change in total cost* of providing transit service resulting from com-

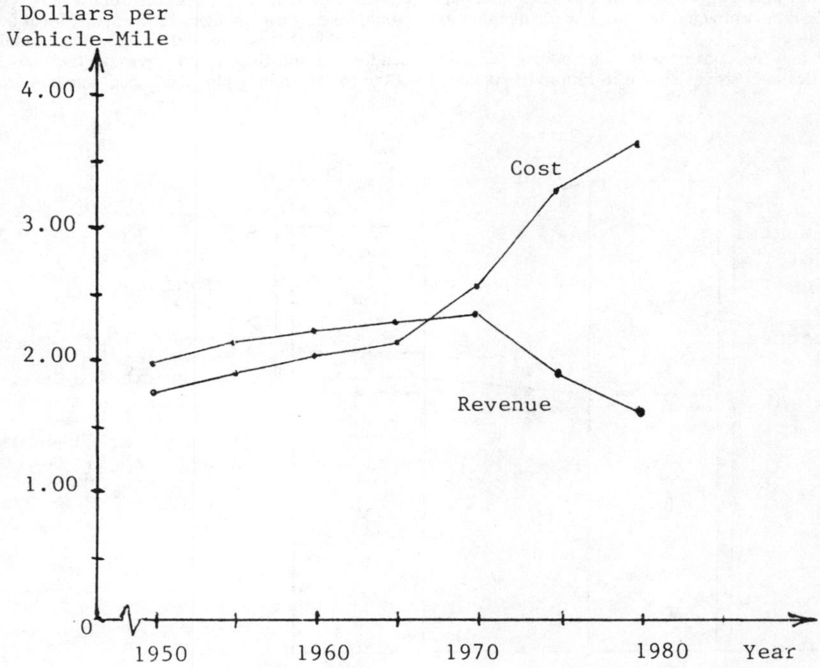

Source: Statistical Abstract of the U.S. (various years)

Figure 1. Trends of average mass transit revenue and expense per vehicle-mile (including bus, streetcar, and rail rapid), in 1983 dollars, 1950-1980.

petitively contracting some or all of an agency's service. The cost saving would be composed of two elements. The first is the reduction in cost to the authority of operating the service itself, and secondly the additional cost represented by the contract amount paid to the private operator plus any additional contract management or other costs resulting from the competitive contracting arrangement. In the former category, there would be the savings in drivers, maintenance of vehicles, fuel, etc., and possibly other costs such as administrative costs associated directly with the operation of the service. In the second category would be not only the contract amount (i.e., the winning bid) but also the costs associated with preparing the request for quotes, evaluating those, negotiating the contract(s), and monitoring the performance of the contractor(s). The categories of cost where a change in amount would result from contracting, and the amount of that change, will depend on which tasks are provided by the contractor and which are continued by the operator. While vehicle operation and provision of drivers are invariably the responsibility of the contractor, some variation exists in other tasks, such as provision of the vehicles, making passenger traffic counts, and purchase of insurance.

The concept of change in total cost is conceptually illustrated in Figure 2. As indicated there, many costs of the authority would continue, including the cost of non-contracted service and general and administrative costs. Also, the authority's efficiency in providing non-contracted service may change due to contracting.

There may be considerable "lumpiness" or discontinuities in costs, as shown in Figure 3. As contracting increases, it may not be possible to adjust all resources used. A good example might be a bus garage, which might not be eliminated until almost all routes using it are contracted out, illustrated by the curve in Figure 3a.

Another example consists of drivers. Contracting out some service or routes may eliminate entirely the need for drivers whose assignments yield relatively few bus-miles (e.g., the extra drivers needed for peak period runs), resulting in savings more than proportionate to bus-miles contracted out (illustrated by Figure 3b). In other situations, contracting of, say, a route, may eliminate some but not all the work of a driver, and hence that labor cost would not be saved. Even after reallocation of runs to drivers, the decrease in efficiency (hours actually driving to hours paid) may remain (see Figure 3c). Finally, and by no means least in importance, some managements may simply be aggressive in reducing in-house costs as service is contracted, and others may be quite lax in eliminating unnecessary jobs, combining jobs, or renting out or selling unneeded portions of facilities.

Turning to the second question, cost responsibility for paying transit costs is shared among the different governmental units involved in transit, and this impacts cost and cost saving. Transit authorities are typically responsible for all of the operating and management costs associated with service, but generally do not pay any of the capital costs of vehicles or facilities. Therefore, one measure of cost saving would be the cost savings as seen by the local transit agency. Another would add to these the capital expenditures, and this would represent the viewpoint of the transit agency plus the local, state and federal

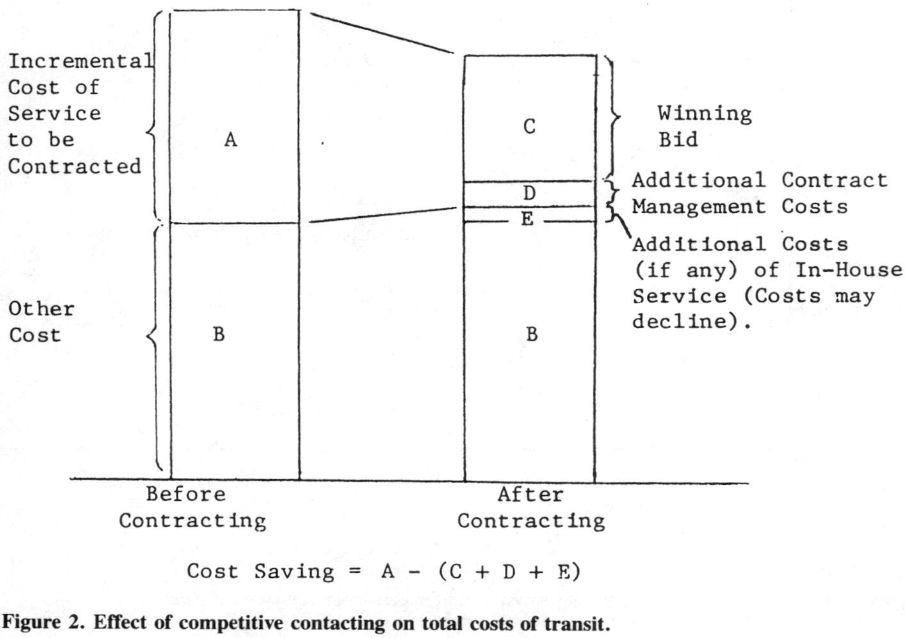

Figure 2. Effect of competitive contacting on total costs of transit.

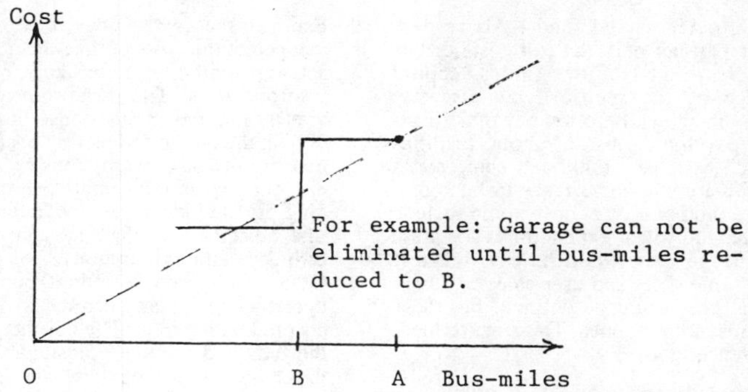

a. Discontinuous cost curve ("lumpy" costs).

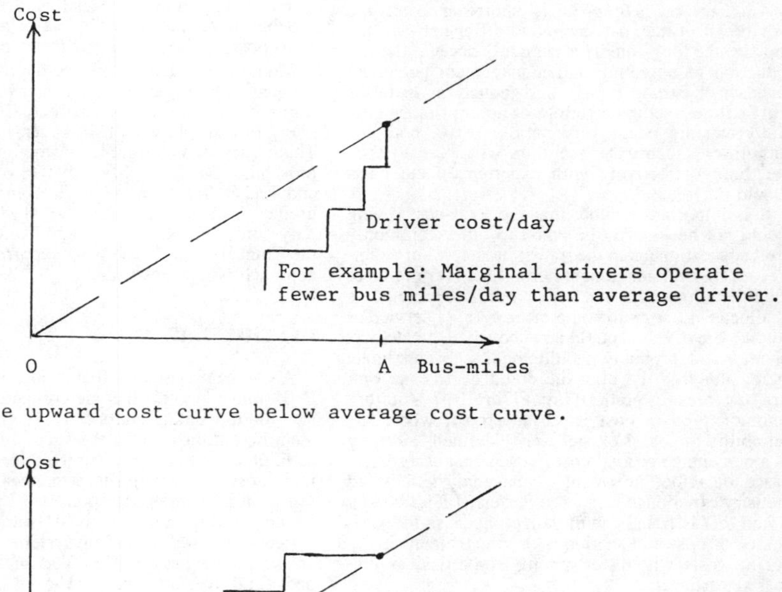

b. Concave upward cost curve below average cost curve.

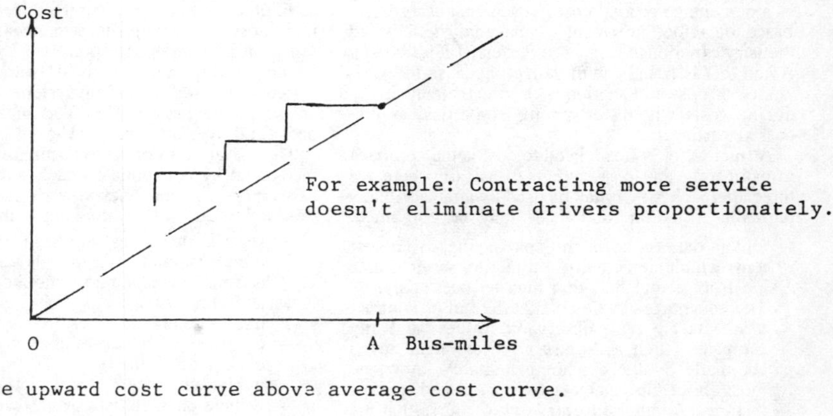

c. Concave upward cost curve above average cost curve.

Figure 3. Three types of cost relationship applicable to cost reduction from contracting. (Initial bus-miles before contracting is A).

agencies which provide capital funds. Above this, private operators pay not only the full costs of bus operation, vehicles, and facilities (unless capital stock is provided by the sponsor), but also pay general taxes on income and profits, pay other business taxes, and pay for the use of public facilities and services such as streets (although some states and municipalities forgive some taxes to bus companies). Thus yet another type of cost would include an estimate of the equivalent expenditures for these items that transit agencies would have to make if subjected to the same taxes and user charges. Other measures of total cost could be defined, but these seem to be the most relevant ones. These are termed, for ease of reference:

1. Transit Agency Cost (operating costs only)
2. Government Costs (operating plus capital costs); and
3. Total Market Costs (operating plus capital costs plus other direct costs)

Another issue is long-run vs. short-run costs. One would expect the cost savings to differ between the shortvs. the long-run. This is mainly because the full reduction in operating and management personnel (at least of certain types), and reductions in buses and facilities (selling, renting, or not replacing) generally can not occur very rapidly. Also, both the authorities and private operators will learn to contract more effectively with experience, and costs should decline.

It is important to note that the total cost saving would not necessarily be equal to other commonly used cost estimates in the transit industry. For example, for many management purposes it is common to use fully allocated costs, where the basic principle is to allocate *all* costs to some measure(s) of service or output. Clearly, this could erroneously estimate cost savings and probably would tend to overestimate them, although the opposite could be true as well (for the reasons underlying Figure 3b). Another common type of cost is variable cost, where the variability of cost is exogeneously defined—usually as a percentage of total cost in each cost category—based on either judgment of the analyst or broad industry experience (e.g., see Ferreri, 1979, 240-1). Again this is fraught with danger, because the variability of cost and vigilence of management in reducing costs will differ among properties, as discussed earlier.

A final set of issues related to cost is the approach to estimating any cost savings. Basically there are three approaches that may be used by analysts in any particular situation, and these may be combined:

1. Obtain data on costs and cost savings from systems which are actually contracting service; data are from actual bids and internal cost analyses. The advantages of this is that the bid or contract cost data are real, observable values, and any estimates (such as contract management costs) are made by those who presumably know the most about the particular operation and its cost features. Another advantage is of course that the data are from managers, and such data would presumably have the greatest credibility among other managers. The main disadvantages are that the basis of each bid or estimate may be unknown, and it may not be possible to obtain estimates of non-bid costs from some systems.
2. Estimate the cost savings, based on models developed of the costs of the transit authority and of actual potential or, more likely, hypothetical private operators. The disadvantage of this is obviously that the analysts have limited knowledge of both the public and private operators, and must make numerous assumptions regarding cost levels (e.g., wage rates and work rules of the winning private bidder), cost variability, and profits. The advantage is that the analyst controls the estimates and assumptions.
3. Obtain bids from potential public and private operators and make estimates of the costs not covered by the bids. The advantage of this is that the bids and costs are presumably estimated by those who know them best, and if the situation is constructed such that they are likely to have to abide by the preliminary bids or estimates, then each is likely to be done carefully. But, as in approach no. 1, it is not clear what the basis of each estimate is. Another disadvantage is of course that in many cases it is difficult or impossible to get potential bidders or actual operators to provide the bids or estimates needed.

Methods 1 and 2 seem to be the most appropriate for general policy study situations, while 3 is largely limited to situations where actual contracting is likely to take place. All, however, have been used. This paper is an example of approach no. 1. Approach no. 2 is exemplified by the work by Guiliano and Teal in Teal et al. (1987). Approach no. 3 was used by McKnight and Paaswell (1984) in a preliminary study of potential savings from contracting in the Chicago area. In the work reported in this paper, approach No. 1 was used.

III. THE SURVEY

A companion study (Teal et al., 1987) identified U.S. transit systems that are competitively contracting. In the entire nation, 41 such systems were identified at the time of the survey. (Since then, 11 additional systems that contract have been identified; most are in California which was already represented in the survey). Responses to a detailed survey on contracting were received from 31 of the 41 systems surveyed, including systems in metropolitan areas from the largest (New York area) to very small ones, and over most of the U.S.

The survey was conducted primarily by mail, supplemented by telephone contact, although in a few cases personal interviews were conducted. Data were collected on the following items:

— service operating and traffic characteristics
— number and amount of bids, and winning bid
— contract management, monitoring and other additional costs
— due to contracting
— estimated or bid costs of contracted operation of service by the
— public authority
— vehicle characteristics and ownership
— cost savings from contracting

We now turn to the cost saving results.

While information on bids and operating characteristics was received from all respondents, the additional detailed cost data were needed to calculate savings was received from only 17 systems. These data enabled calculation of the percentage savings from contracting, equal to:

$$PSO = ((WB + IM)/AO)100 \quad (1)$$

where: PSO = percent savings in costs to local sponsor
WB = winning bid
IM = incremental other costs resulting from contracting
AO = incremental cost of authority operation of contracted service

The distribution of savings, as a percent of the cost of in-house operation the service (or of contracting to the local transit operating agency if different from the sponsor) is shown in Figure 4. The average (or mean) saving is 28.7% and the standard deviation is 17.6%. This range is from essentially zero (0.1%) to just over 50%. Especially noteworthy is the substantial variation in savings, and the relatively uniform distribution over the entire range.

The very small savings are for small systems in small communities. These savings are all net of the additional contract management costs, which in almost all cases were less than ten percent of the total cost of the contracted service. But they represent the savings that apply to local agencies and governments, and are not adjusted to reflect, for example, federal grants for vehicle purchase as described earlier. Thus these underestimate the Total Market Cost savings, as defined earlier.

IV. DISCUSSION OF SAVINGS

Two issues related to these cost savings bear discussion: (1) the reasonableness of the savings, and (2) the effect of contracting on service quality.

Since the number of transit systems that are competitively contracting fixed route bus service is small, and cost saving estimates are available for only some of these, it is especially important to ascertain the degree to which these rather impressive levels of savings are corroborated on the basis of other evidence. If other evidence supports such savings, then the expectation of similar savings as contracting expands to other cities is reinforced. If the other evidence is inconsistent with these savings, then there is greater uncertainty about realizing such savings.

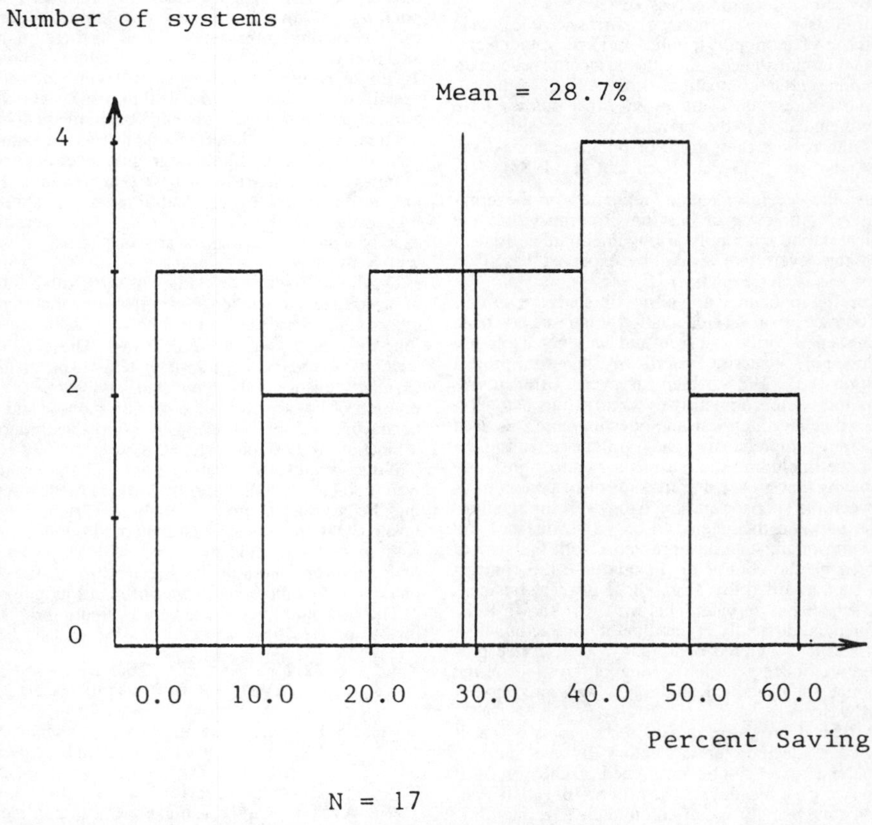

Figure 4. Distribution of actual local cost savings from contracting.

The other evidence on cost savings that might results from competitive contracting is summarized as follows:

1. A review of the experience with contracting other public services which, like transit, have a product that is easily specified revealed similar (about 30%) savings.
2. An examination of the costs of private firms that might be service contractors. Since bus transit is very labor intensive, this focused on labor costs. A detailed study of the labor costs (wages plus benefits) of the various "over-the-road" carriers of passengers in the Philadelphia area revealed that small operators in competitive lines of business had driver costs as low as one-half those of the regional transit agency (Morlok and Krouk, 1983), and similar relationships were found in other less detailed studies in other cities. All this suggests that substantial savings would results from competitive contracting.
3. As described earlier, *after* adjustments for inflation, from 1960 to 1980 the average operating costs per transit vehicle-mile increased by 78%. Allowing for a 30% increase in costs above inflation to account for service improvements and any structural changes, this alone would suggest that costs could be cut by 27% ((1.78 − 1.30)/1.78) by eliminating the excess cost.
4. If costs of private monopoly firms were less than those of monopoly public authorities under current circumstances, then the costs of private operation under competitive contracting would almost surely be even lower. Most studies have concluded that the private costs are either the same or less than those of public agencies (see Anderson, 1983, and Pucher et al., 1983).

Thus, the levels of saving reported for the competitive contracting of bus transit compared to the typical public monopoly arrangement are consistent with the levels that would be expected based on other less direct evidence.

Turning to the question of possible adverse effects of contracting on service quality, this issue has been addressed in prior research and was not explicitly included in the survey. Specifically, in a prior project (Burgin et al., 1985) which surveyed 11 transit systems that were competitively contracting bus service, a detailed interview and questionnaire was used to obtain comprehensive data on the contracting experience, including the reasons for contracting, any problems that arose, the division of responsibilities between the sponsor and operator, working relationships between them, etc. In only one instance did the sponsor indicate any problems with the service offered by the contractor. In this case the contract had been awarded to a firm whose primary business was school bus service. This firm was known to be in financial difficulty at the time of contractor selection and later entered bankruptcy. The service provided was inadequate in the view of the sponsor, the contract was cancelled, and a new award to another firm was made.

A typical provision in service contracts is that if service is deemed unsatisfactory by the sponsor, then the contract can be terminated (usually on short notice, e.g., a few days). The criteria for satisfactory service are usually specified quite clearly, including factors such as condition of the buses, on-time performance, and appearance and courtesy of the drivers. As discussed earlier, the nature of transit service lends itself to quantitative measures for such criteria, facilitating the inclusion of performance specifications in contracts. In addition, contractors are typically chosen considering ability to provide good service as well as price (recall that the lowest bidder was rarely chosen), helping to ensure good service quality.

While the more recent cost-oriented survey did not include specific questions on quality problems, if there had been any serious ones it is likely the respondents would have mentioned this in the typically rather lengthy phone conversations. Thus there is no indication from the surveys that service quality is likely to suffer as a result of contracting. However, if transit agencies were required to make awards to the lowest bidder, then service quality could become a serious problem.

V. ESTIMATING SAVINGS

To estimate national savings accruing to transit authorities from competitive contracting, a model was developed based on the data from systems that are competitively contracting. The detailed data on bids and costs indicated clearly that a model incorporating random phenomena is essential for realism, for the factors influencing actual savings in any situation can not be treated with absolute certainty. To do so would be analogous to trying to predict exactly who would become ill tomorrow, with what ailment, and how much it would cost in medical bills for treatment, etc. Nevertheless, just as we know a great deal about the likely aggregate level of illness in our society tomorrow or next year, we should be able to determine a great deal about aggregate savings even though we can not say with certainty exactly what the savings in any one transit system will be.

On the basis of prior studies of the cost structure of bus transit and studies of savings from contracting other public service, factors likely to influence the long run cost savings were identified. These are: (1) area characteristics, (2) transit system characteristics, (3) characteristics of the contracted service, and (4) the division of tasks between the service sponsor and the contractor. The percent savings was hypothesized to be a function of these characteristics.

Nine variables describing these characteristics were obtained for all systems. Various models relating the savings to groups of these variables were posited, the parameters estimated through regression, and the models evaluated with respect to *a priori* and goodness-of-fit criteria. From over 100 models, three emerged as superior in our judgement.

The particular model used for estimating the savings to be presented later is:

$$PSO = -22.101 + 0.593(10(BBP) - 5)^2 + 0.91(AVS) + 4.011(\ln(POP)) \quad (2)$$

where: PSO = percent saving in transit agency cost
BBP = ratio of buses operated in base period to buses operated in peak, in contracted service
AVS = average vehicle speed in contracted service, mph
POP = area population, thousands

This model has an R2 of 63.36% and a standard error equal to 11.8% of the mean. The F-value is 7.494 and the tail area is 0.004. The largest partial tail area, for the variable BBP, is 0.08. In addition, this model has the feature that the range of the variable PSO is between zero and about 45% for the range of possible independent variable values (i.e., the values for U.S. transit systems), and the signs of the coefficients are as would be expected on *a priori* grounds.

It is noteworthy that the data never supported the inclusion of one service characteristic which we thought would be a significant factor: the percent of contract service vehicles provided by the service sponsor. Given the federal capital grant program, it had been assumed this would be an important cost factor, increasing the savings if the sponsor provided the vehicles. Naturally the cost disadvantage of private firms providing vehicles may be decreased by opportunities to used the vehicles in other services, and such opportunities may enhance the ability to share drivers and other costs as well. And the complexities of tax and depreciation rules may offset much of the superficial public advantage, as recently has been revealed for public works facilities (see Au and Hendrickson, 1985). More research is clearly needed on this subject.

The model is then used to estimate savings on individual systems, based on values of the three variables identified earlier for the contracted service and the amount of contracted service. The result is an estimate of the expected value (or average) savings and of the standard deviation of such savings, either as a percentage or as a dollar amount.

Before applying any model, it is important to consider sources of variance in the application context that do not appear in the data on which it is based. This variance, or variability, is in additional to the internal variance built into the model, and accounting for those additional factors is important in providing realistic estimates, even though this usually must be done by judgement (see, e.g., Mosteller and Tukey, 1977, 129-131).

Three characteristics of the sample data stand out as especially relevant to such supplementary variance: (1) the sample was small and not necessarily random, (2) competitively contracting systems are not randomly distributed among all transit systems (in particular, the first systems to contract might have features that lead to the largest savings), and (3) features that were not included in the analysis could influence the savings (e.g., the local political climate toward government cost reduction).

To take these into consideration clearly requires the use of judgement, and different researchers will have different conclusions regarding the proper magnitude of the adjustment. Our conclusion was to adjust the model by reducing the estimate of mean value by 10%, i.e., a mean of 25.5% would be reduced by 1/10 to 23.0%, and by increasing the standard deviation by 15% (or the variance by 32%). These clearly err on the side of conservatively estimating the savings.

Now the saving can be estimated for each system in which contracting is contemplated, and these can be summed to yield national savings. These savings will naturally depend on exactly which specific services are competitively contracted, and a myriad of combinations of specific services and systems are possible.

The scenario to be considered is contracting 5% of all service in systems of 100 or more buses. Such systems represent the vast majority of bus service in the U.S.A. by any measure, representing (in 1983) 89.5% of total bus operating expenditures, 75.8% of federal operating assistance, and 84.4% of total public operating assistance reported in 1983 (according to the Section 15 data report). Contracting a small percentage of service in small systems would not make sense, as too few vehicles would be involved to cover even a single route. The contracted service in each system possessed operating characteristics identical to the system averages.

The approach is to estimate the dollar value of savings (to the authority) in each system i (SSi), and the standard deviation of those savings (DSi), and then sum to yield the corresponding national values. That is, the expected value of national savings (NS) and its standard deviation (DNS) are:

$$NS = \sum_i SS_i \tag{2}$$

$$DNS = (\sum_i DS_i^2)^{1/2} \tag{3}$$

As one might expect from the data on savings from systems already contracting, considering especially the range of bids obtained and the ability of any one system to choose a high or low bid, the spread of likely values of savings for any one system is very large. The expected value of the savings in each of these larger systems was typically 20% to 35%, with a standard deviation of 20% to 40% of the mean. But when the savings are summed over all systems, the standard deviation decreases (in relative terms) and the distribution becomes quite "tight".

The expected value of national savings of Transit Agency Costs (i.e., operating expenses) is 22.7%, and the standard deviation is only 13.6% of this value. The appropriate distribution to be used for these savings is the Normal distribution, and with the distribution specified we can determine the probability of various levels of savings. The results for savings in Transit Agency Costs are:

— a 98% probability that national operating costs savings will be at least 16.4%,

— a 90% probability that savings will be at least 18.8%, and

— a 50% probability that savings will be at least 22.7%.

(Other scenarios were evaluated, such as contracting services which had characteristics favorable to larger savings, and calculation of savings including capital costs and other market costs, but space precludes coverage here).

This type of information is usually presented in what is known as a cumulative probability density function curve, and the curve for this scenario is shown in Figure 5. On the horizontal axis are the various possible levels of savings (as a percentage of the full cost of the contracted service). The vertical axis on the left goes from 100% at the bottom to 0% at the top. Its interpretation is as follows. For any given percentage saving (on the horizontal axis), the corresponding value on the left axis is the probability that the savings will be at least the given amount. Here can be seen the 90% probability that the savings will be at least 18.8%, for example.

Thus the use of a formal model which provides statistically meaningful results for each system has

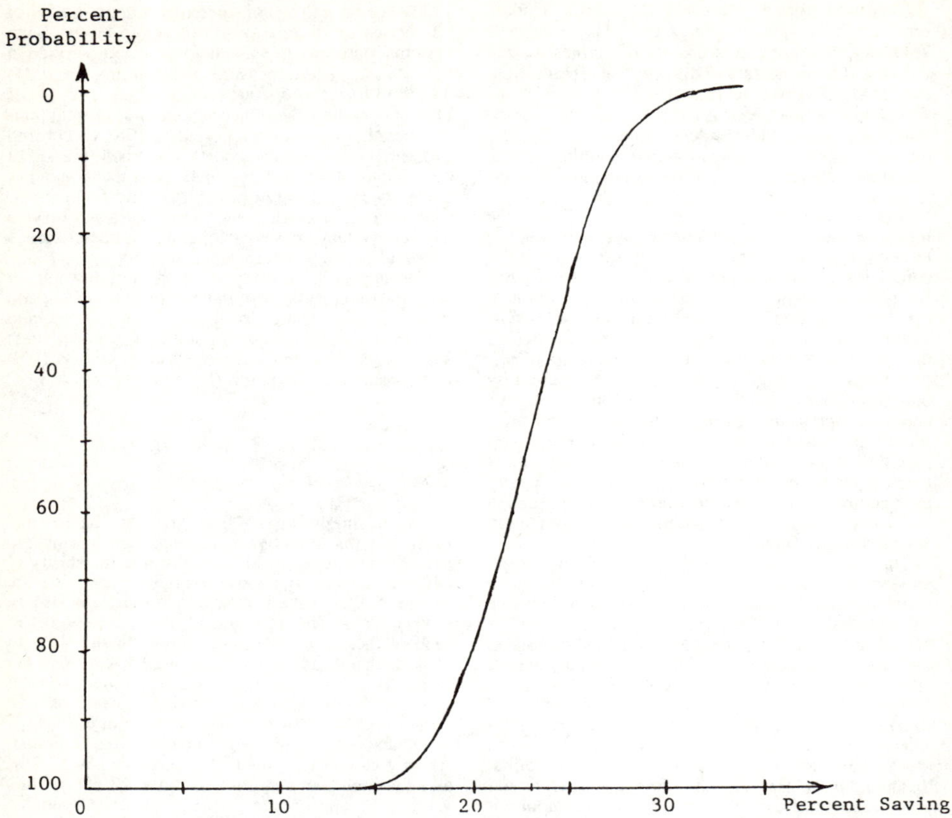

Figure 5. Probability of percentage national saving from contracting at least as great as indicated. (For example, the probability of a saving of at lesat 20% is approximately 80%).

enabled us to make strong statements about overall savings even though savings in any one system are highly uncertain. This is exactly analogous to the example of illness among individuals discussed earlier. While we can not predict exactly who will be ill next year, we do have a good idea about total illness in the population. Similarly, while the savings in any one city are of highly uncertain magnitude, the level of national savings can be bracketed within a small range.

VI. CONCLUSIONS

Thus the potential national saving from competitive contracting was estimated using a model developed from data on systems that are actually contracting. This model takes into account characteristics of the metropolitan area and of the contracted service. Consideration was given to the fact that the systems currently competitively contracting are just "the tip of the iceberg" relative to the potential number of competitively contracting systems. To reflect the uncertainty inherent in such a situation, we used our judgement, based on the experience in contracting other public services, to further adjust savings to be more conservative.

Of course, future savings can not be predicted with certainty, as a variety of unpredictable factors will influence savings, such as the amounts of the bids and which bid is chosen for each contract. The model explicitly takes this into account, estimating the probability of different levels of savings. Contracting of services without regard to special characteristics favorable to savings yield savings of at least 19% in authority costs with a 90% probability. Choosing services carefully would simply increase the savings. Hence substantial savings are virtually certain with careful implementation of competitive contracting. Competitive contracting clearly is a powerful management tool for cost reduction in U.S. bus transit systems.

REFERENCES

Anderson, Shirley, C. (1983). The Effect of Government Ownership and Subsidy on Performance: Evidence from the Bus Transit Industry, Transportation Research A, 17, 3, 191-200.

Au, Tung and Chris Hendrickson (1985). Public Versus Private Ownership of Constructed Facilities, Manuscript, Civil Engineering Department, Carnegie-Mellon University, Pittsburgh, PA.

Burgin, Lynn G., Donald R. Ellerman and Edward K. Morlok (1985). Initial Survey of Competitive Contracting of Bus Transit, University of Pennsylvania, Philadelphia, PA.

Cox, Wendell (1984). The Growing Involvement of Private Providers in Local Urban Public Transportation. Paper presented to the 63rd Annual Meeting of the Transportation Research Board, Washington, DC.

Ferreri, Michael F. (1979). Comparative Costs of Transit Modes, in Lester A. Hoel, Ed., *Public Transportation*, Prentice-Hall, Englewood Cliffs, NJ, 226-252.

McKnight, Claire E. and Robert E. Paaswell (1984). Cost Analysis: Selected Routes, Urban Transportation Center, University of Illinois, Chicago, IL.

Morlok, Edward K. (1984). Economics of Private Operator Service, Transportation Research Record 980, 55-59.

Morlok, Edward K. and Steven K. Krouk (1983). Variations in Driver Wage Rates and Opportunities for Transport Cost Reduction, *Proc.* Trans. Res. Forum, XXIV, 1, 108-115.

Morlok, Edward K. and Philip A. Viton (1985). Comparative Costs of Publicly-Provided and Privately-Provided Urban Mass Transit, in Charles A. Lave, ed., Urban Transit, Ballinger Publishing Co., Cambridge, MA.

Mosteller, Frederick and John W. Tukey (1977). *Data Analysis and Regression*, Addison-Wesley, Reading, MA.

Pucher, John, Anders Markstedt, and Ira Hirshman (1983). Impact of Subsidies on the Costs of Urban Public Transport, Journal of Transport Economics and Policy, 17, 2, 155-176.

Teal, Roger F., et al. (1987). Estimating the Cost Impacts of Transit Service Contracting Alternatives. Institute of Transportation Studies, University of California, Irvine, CA.

ENDNOTE

* Department of Systems, University of Pennsylvania, Philadelphia, PA.

This research was supported by grants from the Urban Mass Transportation Administration and the Public Policy Initiatives Fund. This support is gratefully acknowledged but implies no endorsement of the findings. The assistance of Donald R. Ellerman and Frederick A. Moseley, Research Fellows at the University of Pennsylvania, in preparing the regressions, and of Professor Ashish K. Sen, of the University of Illinois—Chicago, in reviewing the data analyses, is gratefully acknowledged.

The Dynamics of West Coast Container Port Competition

*by Robert L. Hanelt and Daniel S. Smith**

ABSTRACT

Competition among U.S. container ports has broadened and intensified in the last decade. New port service offerings have expanded the scope of competition. At the same time, changes within the container trades have made the port choice decision more critical to ocean carrier business strategies, and raised the stakes in port competition. The dynamics of this competition, and the shifting priorities of ocean carriers, are illustrated by changes in coastal and port area shares of U.S. foreign trade. The paper uses a ten-year time series of containerizable foreign trade to demonstrate the impact of intermodal services, ocean carriers' increasing concern with land transportation, and other maritime industry developments. Carrier and shipper port selection criteria (from recent interviews) are discussed.

I. BACKGROUND

Maritime trade has been a vital part of West Coast economic activity for over a century. Superb natural harbors, abundant resources, and a position on the Pacific guarantee important roles for West Coast ports. As Pacific liner trade has grown, it has also been containerized. The emergence of intermodal transportation has allowed the West Coast ports to serve the growing Pacific trade for the entire nation.

West Coast ports compete vigorously for containerized traffic. The economics of containerization favor large ships, mechanized terminals, and few port calls. The decision of a major ocean carrier to call at one port rather than another, or to direct discretionary cargo through one port rather than another, affects port cargo volumes, employment, and revenues. This paper addresses the competition among three West Coast port regions:

- *San Francisco Bay,* including San Francisco and Oakland.
- *Pacific Northwest,* including Seattle, Tacoma, and Portland.
- *Southern California,* including Los Angeles and Long Beach.

Other West Coast ports, some of which handle small volumes of containerized cargo, were grouped with Alaska and Hawaii as "Other West Coast." We also compared the West Coast with other U.S. port regions, including the East Coast, the Gulf Coast, and the Great Lakes.

We began by documenting the international containerizable cargo flows through West Coast and U.S. port regions for the last ten years (1976-1985). The results of port competition are apparent: the data show substantial cargo shifts between Southern California, the Bay Area, and the Pacific Northwest. To obtain an understanding of the factors affecting port choice, the ranking of those factors, and the degree of port control over the most important factors, we drew on previous proprietary studies, contacts with ports, and semi-structured interviews with both carriers and shippers. Our results were reviewed with key respondents to ensure that our interpretations were realistic. This approach allowed us to address issues for which statistical data do not exist, but which lie at the core of container port competition.

II. CONTAINERIZABLE CARGO VOLUMES AND SHARES

A. Source

The data base used for this analysis is derived from U.S. Bureau of the Census import and export data (SA305/705). It gives tonnage and value of cargo by commodity (Schedules A & E), by service (liner and non-liner), by commodity containerizability and density, by foreign country of origin or destination, and by U.S. port. Our ten-year time series for this analysis included only *containerizable, liner* cargo. We estimated the equivalent container volumes, by commodity, in two stages. First, we used proprietary containerizability factors to estimate the portion of each commodity likely to be containerized in each trade. Second, we converted containerizable tonnages to twenty-foot equivalent units (TEUs) using proprietary density factors.

The proprietary factors for both containerizability and density are empirically derived, and periodically updated with the cooperation of West Coast ports and container operators. The containerizability factor is derived from total and containerized tonnage figures within Schedule A and E commodity categories, and gives the percentage of cargo of each commodity group historically carried in containers. These factors vary from only a few percent (for heavy equipment) to 100 percent (for consumer goods in major trades). The density factors give the tons per TEU for each commodity group, allowing for stowage practices that typically result in less than complete space utilization. It was critical that individual density factors were used for each commodity, since densities range from 3 to 18 tons per TEU. Different commodity mixes can thus yield significantly different container volumes for the same tonnage, making the use of overall averages unreliable and misleading.

B. TEUs and Shares

Table 1 gives the ten-year time series for West Coast containerizable liner trade in TEUs, and the corresponding regional shares. The time series shows very strong growth in both Southern California and the Pacific Northwest since 1980, a reflection of both increased trade and increased intermodal traffic. San Francisco Bay TEUs grew by 67 percent between 1976 and 1985, but Bay Area share dropped from 24.5 to 17.0 percent. As the Pacific Northwest has the smallest of the three local markets, that port region is apparently drawing cargo to and from inland points.

The West Coast accounted for roughly 68 percent of the total U.S. growth in all liner trades (Table 2). The West Coast share grew from 28.2 percent in 1976 to 42.8 percent in 1985, and probably exceeded the East Coast share in 1986-1987. The growth of West Coast cargo has apparently shifted much of the containerized traffic from the Gulf, and absorbed much of the growth that would otherwise have occurred at East Coast ports. The dramatic increase in West Coast share is consistent with an influx of discretionary intermodal cargo.

C. Imports and Exports

Table 3 gives TEUs and TEU shares for Far East Asia and Total U.S. imports and exports for shown separately for 1976-1980 and 1980-1985. Two points are immediately apparent.

First, growth in U.S. trade with Far East Asia accounts for much of the total growth, for much of the increased West Coast share, and for much of the changes in shares within the West Coast. Trade with Far East Asia grew much faster than other trades, and exports to Far East Asia grew 25.6 percent between 1980 and 1985 when U.S. exports declined 15.6 percent overall. Thus, the ability of the West Coast ports to attract intermodal Far East Asia imports destined to inland points accounts for much of the West Coast growth.

Second, the logistics of discretionary imports and exports account for much of the change in West Coast shares. The San Francisco Bay ports have traditionally handled a large share of West Coast exports, and their export volume grew from 1976-1980. Since 1980, however, the Pacific Northwest ports have shown exceptional growth in Far East Asian and total exports, and the Bay Area share

TABLE 1
WEST COAST CONTAINERIZABLE LINER
TEUS AND TEU SHARES

	1976	1977	1978	1979	1980	1981	1982	1983	1984	1985
TEUs (000)										
San Francisco Bay	263	307	346	367	394	384	411	384	455	438
Southern California	481	564	636	680	760	801	824	962	1,146	1,435
Pacific Northwest	265	303	338	354	406	433	407	470	573	617
Other West Coast	64	74	74	72	81	74	66	82	93	87
TOTAL	1,072	1,247	1,394	1,474	1,641	1,691	1,709	1,898	2,267	2,578
TEU Shares (Percent)										
San Francisco Bay	24.5	24.6	24.8	24.9	24.0	22.7	24.1	20.2	20.1	17.0
Southern California	44.9	45.2	45.6	46.2	46.3	47.3	48.2	50.7	50.6	55.7
Pacific Northwest	24.7	24.3	24.3	24.1	24.7	25.6	23.8	24.8	25.3	23.9
Other West Coast	5.9	5.9	5.3	4.9	4.9	4.4	3.9	4.3	4.1	3.4

TABLE 2
U.S. CONTAINERIZABLE LINER TEUS AND
TEU SHARES

	1976	1977	1978	1979	1980	1981	1982	1983	1984	1985
TEUs (000)										
West Coast	1,072	1,247	1,394	1,474	1,641	1,691	1,709	1,898	2,267	2,578
East Coast	1,956	1,809	2,130	2,200	2,181	2,178	2,059	2,192	2,508	2,664
Gulf Coast	724	681	785	851	863	859	775	761	812	754
Great Lakes	51	60	64	57	46	28	24	20	24	24
TOTAL	3,803	3,797	4,373	4,581	4,731	4,757	4,566	4,871	5,610	6,019
TEU Shares (percent)										
West Coast	28.2	32.9	31.9	32.2	34.7	35.6	37.4	39.0	40.4	42.8
East Coast	51.4	47.6	48.7	48.0	46.1	45.8	45.1	45.0	44.7	44.3
Gulf Coast	19.0	17.9	18.0	18.6	18.3	18.1	17.0	15.6	14.5	12.5
Great Lakes	1.3	1.6	1.5	1.2	1.0	0.6	0.5	0.4	0.4	0.4

Source: Manalytics' Waterborne Trade Data Base.

has declined as a result. Increases in local production for export do not account for the Pacific Northwest growth; it is apparently due instead to increased flows of discretionary intermodal exports.

The major source of growth for Southern California ports was containerizable imports. From 1976 to 1985, Southern California container ports increased their overall import market share from 51.7 to 65.3

TABLE 3

CONTAINERIZABLE LINER IMPORT AND EXPORT TEUS AND TEU SHARES

	1976		1985		1976-80 Growth		1980-85 Growth	
Far East Asia Imports	TEU (000)	Share (%)	TEU (000)	Share (%)	Total (%)	Annual (%)	Total (%)	Annual (%)
S.F. Bay	75	16.8	133	10.0	36.0	8.0	30.4	5.5
So. Calif.	242	54.3	896	67.3	38.8	8.6	166.7	21.7
Pac. N.W.	120	26.9	294	22.1	36.7	8.1	79.3	12.4
Other W. Coast	9	2.0	8	0.6	-44.4	-13.7	60.0	9.9
West Coast	446	61.2	1,331	73.7	36.1	8.0	119.3	17.0
East Coast	260	35.7	462	25.6	0.4	0.1	77.0	12.1
Gulf Coast	22	3.0	12	0.7	-50.0	-15.9	9.1	1.8
Great Lakes	1	0.1	0	0	0	-20.5	0	0
TOTAL	729	--	1,805	--	20.6	4.8	105.3	15.5
Total U.S. Imports								
S.F. Bay	119	19.9	219	12.9	33.6	7.5	37.7	6.6
So. Calif.	309	51.7	1,108	65.3	44.7	9.7	147.9	19.9
Pac. N.W.	150	25.1	347	20.5	38.7	8.6	66.5	10.7
Other W. Coast	20	3.3	22	1.3	-35.0	-11.0	75.0	11.8
West Coast	597	30.2	1,696	43.8	38.4	8.5	105.1	15.4
East Coast	1,136	57.3	1,834	47.4	-2.2	-0.6	65.1	10.5
Gulf Coast	224	11.3	332	8.6	0.9	0.2	46.9	8.0
Great Lakes	24	1.2	7	0.2	-60.8	-20.9	-22.3	-4.9
TOTAL	1,981	--	3,869	--	9.7	2.3	78.0	12.2
Far East Asia Exports								
S.F. Bay	89	31.1	152	23.5	64.0	13.2	4.1	0.8
So. Calif.	113	39.5	244	37.7	86.7	16.9	15.6	2.9
Pac. N.W.	66	23.1	216	33.4	104.5	19.6	60.0	9.9
Other W. Coast	18	6.3	35	5.4	55.6	11.7	25.0	4.6
West Coast	286	76.3	647	74.9	81.8	16.1	24.4	4.5
East Coast	66	17.6	195	22.6	103.0	19.4	45.5	7.8
Gulf Coast	23	6.1	22	2.5	47.8	10.3	-35.3	-8.3
Great Lakes	0	0	0	0	0	0	0	0
TOTAL	375	--	864	--	83.5	16.4	25.6	4.7
Total U.S. Exports								
S.F. Bay	144	30.1	220	24.9	62.9	13.0	-6.1	-1.2
So. Calif.	173	36.1	328	37.2	80.9	16.1	4.8	0.9
Pac. N.W.	115	24.2	270	30.6	71.3	14.3	36.9	6.5
Other W. Coast	46	9.5	65	7.4	56.5	11.9	-9.2	-1.9
West Coast	478	26.2	884	41.1	71.0	14.4	8.2	1.6
East Coast	818	44.9	826	38.4	30.4	6.9	-22.6	-5.0
Gulf Coast	499	27.4	423	19.6	28.3	6.4	-33.9	-7.9
Great Lakes	26	1.4	20	0.9	38.3	8.5	-44.4	-11.1
TOTAL	1,821	--	2,153	--	40.6	8.9	-15.9	-3.4

Source: Manalytics' Waterborne Trade Data Base

percent, and their share from Far East Asia from 54.3 percent to 67.3 percent. San Francisco Bay and Pacific Northwest shares were correspondingly reduced.

The statistics show cargo share shifts within the West Coast, and to the West Coast from other U.S. coasts. The available cargo statistics do not, however, capture the numerous influences at work or reveal anything about the degree of port control.

III. FACTORS IN PORT CHOICE

To explore the factors in port choice and cargo routing, we began with a "laundry list" of potential influences based on our own experience and on the maritime industry literature. That list was incorporated in semi-structured interviews with a number of major carriers and large shippers. The sample was not random: a number of potential interviewees were identified to obtain the views of participants with different traffic volumes and business strategies. From our interviews, we derived the rankings shown in Table 4. These are grouped according to the degree of port control. These findings suggest strongly that major factors in port choice and cargo routing are beyond direct control of the ports. In ranking these factors, we contacted ports, carriers, exporters, and importers, and found substantial agreement. Different parties are influenced by different factors, and Table 5 regroups the factors accordingly.

TABLE 4
RANKING OF PORT CHOICE FACTORS AND DEGREE OF PORT CONTROL

Major Factors

Outside Port Control

1) Local market size
2) Domestic transport costs

Under Port Influence

1) Port labor productivity
2) Railroad connections and transit times
3) Port access
4) Total transit time
5) Ocean carrier intermodal networks

Under Port Control

1) Terminal gate capacity
2) Channel and berth depth
3) Indirect port charges
4) Port charges on cargo
5) Port charges on carrier
6) Cargo handling equipment

Secondary Factors

Outside Port Control

1) Directional container balance
2) Domestic backhaul opportunities
3) Corporate relationships between carrier, shipper agent, or terminal operator

Under Port Control

1) Rail/water interface
2) Terminal container storage
3) Distribution center availability
4) Industrial park or foreign trade zone availability
5) Availability and price of bunker fuel
6) Reputation of port management and level of port service

Minor Factors

Under Port Control

1) Availability of an automated customs clearance service
2) Port-provided cargo tracking system
3) Port-provided cargo statistics system

TABLE 5
MAJOR FACTORS INFLUENCING USERS IN PORT REGION SELECTION

Exporters

1) Domestic transport cost
2) Total transit time
3) Indirect port costs
4) Local export market size
5) Port charges on cargo
6) Port access

Importers

1) Local import market size
2) Total transit time
3) Port access
4) Access to proprietary distribution networks
5) Availability of container storage
6) Port charges on cargo

Large, Intermodal Ocean Carriers

1) Local market size
2) Rail/water interface
3) Port access
4) Port charges (on carrier and cargo)
5) Railroad connections and transit time
6) Carrier intermodal network
7) Channel and berth depth

Smaller Container Carriers

1) Local market size
2) Port charges on cargo
3) Port charges on carrier
4) Terminal gate capacity
5) Labor productivity
6) Railroad connections and transit time
7) Availability and price of bunker fuel

A. Exporters

Table 5 ranks six major factors influencing the exporters' port choice. Only indirect port costs and charges to the cargo are under port control. Exporters choose ports first to minimize total transport cost, and second to minimize total transit time. Where the shipper is responsible for domestic transport, such as a California cotton exporter, domestic transport cost is the first criterion in port choice. A shipper who is not responsible for domestic transport, such as an exporter using an intermodal bill of lading, will emphasize transit time.

B. Importers

Six major factors influence the importers' port choice (Table 5). Only container storage space and port charges are under port control. The local import market is the major reason for importers selecting the Southern California ports. This import market leads transpacific ocean carriers to call at Southern California first. The carriers' routings satisfy the importers' second criterion: shortest transit time.

C. Large Intermodal Ocean Carriers

The large intermodal container operators typically require their own terminal, rely upon proprietary intermodal networks, and need efficient intermodal transfer. Table 5 lists seven major factors that influence large intermodal operators in port choice and intermodal routing. Their first port choice criterion is the size of the local import market. Although the majority of their cargo is discretionary, intermodal capabilities do not influence carriers in their *initial* port choice: once port rotation has been established by local market size, the discretionary cargo follows.

D. Smaller Ocean Carriers

Smaller container operators are less concerned about intermodal capabilities. Table 5 lists seven major factors that influence their port choice. Except for local market size, railroad connections, and transit time, the factors are under port influence or control. Since a large proportion of their cargo originates or terminates in the local hinterland, the size of the local market is of first importance. All aspects of cost are important to these smaller carriers. They are more likely to shift discretionary cargo between ports than are the large carriers with established intermodal networks.

How then do these factors and preferences operate to yield the observed cargo shares and cargo share shifts?

IV. LOCAL AND INTERMODAL MARKETS

A. Local Imports

In the Pacific trades, import containers outnumber export containers by a wide margin (Table 3). Ocean carrier logistics, including port choice, are thus dictated primarily by import flows. The size of the local import market, roughly defined as the local "hinterland" accessible by truck, is the greatest single influence on port choice.

The size of the Southern California hinterland market, which can include portions of Arizona and Nevada and the lower San Joaquin Valley, accounts more than any other factor for direct Los Angeles/Long Beach port calls in all trades. The size of this market is reflected in both population and economic activity. Table 6 shows the estimated population changes for Southern California, Northern California, and the Pacific Northwest over the past ten years. Southern California has the largest share and

TABLE 6
LOCAL MARKETS FOR WEST COAST PORTS

POPULATION (000)

	1970	1980	1983	Average Annual Growth 1970-1980	Average Annual Growth 1980-1983
Los Angeles Basin	9,981	11,498	12,191	1.5%	2.0%
LA/LB Hinterland	2,276	4,536	4,964	6.6%	3.1%
San Francisco Bay Area	4,754	5,368	5,624	1.3%	1.6%
SF/Oakland Hinterland	1,994	2,518	2,719	2.6%	1.6%
Seattle-Tacoma	1,837	2,093	2,187	1.4%	1.5%
Sea/Tac Hinterland	1,521	1,890	1,935	2.4%	0.8%

RETAIL AND WHOLESALE ACTIVITY ($000,000)

	1977	1982	Average Annual Growth 1977-1982
Los Angeles Basin	112,524	189,899	13.8%
LA/LB Hinterland	29,410	51,946	15.3%
San Francisco Bay Area	56,288	86,285	10.7%
SF/Oakland Hinterland	19,484	30,443	11.2%
Seattle-Tacoma	21,192	33,803	11.9%
Sea-Tac Hinterland	15,887	24,921	11.4%

Los Angeles Basin SMSAs: Anaheim, Los Angeles, Oxnard, Riverside.
LA/LB Hinterland SMSAs: Bakersfield, San Diego, Santa Barbara, Visalia, Las Vegas, Phoenix

San Francisco Bay SMSAs: San Francisco, Oakland, San Jose, Santa Cruz, Santa Rosa, Vallejo
SF/Oak Hinterland SMSAs: Fresno, Modesto, Sacramento, Salinas, Stockton

Seattle-Tacoma SMSAs: Seattle, Tacoma
Sea/Tac Hinterland SMSAs: Portland, Spokane

Sources: U.S. Statistical Abstract, 1985.
Metropolitan Statistical Abstract, 1985.

the highest growth rate of the three areas. Table 6 also compares economic activity for the three regions. Again, Southern California has the largest share, and a high growth rate.

B. Intermodal Imports

Major container carriers move much of their traffic to and from inland points. In effect, carriers are using intermodal services to serve a second, more distant hinterland through a single port of call. Such intermodal cargo is *discretionary*, in that ocean carriers can usually route the cargo through different West Coast ports with comparable efficiency. Chicago cargo, for example, can be readily shifted among West Coast ports as their relative advantages change. This discretionary cargo has become the prime focus of competition among container ports.

As of May, 1987, about 28 weekly double-stack trains originated in Southern California, 17 weekly trains in the Pacific Northwest, and 2 weekly trains in the San Francisco Bay Area. The volume of such traffic suggests that ocean carriers have routed most of their discretionary cargo movements through Southern California and the Pacific Northwest.

C. Local Exports

A large local export market leads carriers to make that region the *last* port of call. The availability of local exports depends on the extent of local production. On the West Coast, a large portion of the containerized exports consists of agricultural products that are containerized in the area of production and exported through the nearest port region. Most commodities, such as Pacific Northwest forest products and California rice, are thus economically limited to only one port region.

D. Intermodal Exports

With the expansion of intermodal services, exports from the East and Midwest (but not the Gulf) can move with comparable efficiency through any of the West Coast ports. As the strong growth in exports through the Pacific Northwest indicates, discretionary intermodal exports tend to move through the last U.S. port of call. This tendency can be altered, however, by the use of domestic freight to fill empty westbound containers, since the large local markets for imports also attract westbound domestic freight. A carrier may choose to combine westbound domestic freight with intermodal exports in a single train movement to Southern California, and to connect that train movement to a Southern California port call.

V. TRANSIT TIME AND VESSEL LOGISTICS

Vessel logistics dominate ocean carrier costs. Modern containerships are very large, up to 4,200 TEUs, and 70-80 percent of their costs are fixed. These costs, which can reach $30,000-$40,000 per day, lead carriers to minimize the number of port calls and the time in each port. Vessel logistics have encouraged the growth of intermodal movements: by using rail to serve the Gulf, for example, a transpacific carrier can turn the vessel on the West Coast instead of sailing through the Panama Canal. Ocean carriers prefer to unload import cargo at the first port of call to minimize transit time and maximize backhaul capacity. Carriers thus call first where the most import cargo (local *and* intermodal) can be unloaded, usually Southern California. The opposite is true for exports, which the ocean carrier seeks to load at the last port of call, usually San Francisco Bay or the Pacific Northwest.

Once first and last ports of call are established, they tend to attract the cargo of shippers seeking faster transit. The advantage of using the first port of call outweighs any small difference in rail transit time for intermodal imports. The last port of call gives the fastest possible transit time for exports, since they are not delayed while the vessel calls at other U.S. ports. Los Angeles and Long Beach have become important intermodal ports because the import market guarantees that they will usually be the first ports of call on the West Coast. Thus, intermodal cargoes transferred to rail at the first port of call in Southern California have shorter *total* transit times than cargoes discharged at later calls in the Bay Area or the Pacific Northwest, even though port and rail transit times are similar. The reverse is true for intermodal exports, and has contributed to the growth of Pacific Northwest cargo volumes.

In Table 7, we compare total transit times to Chicago through all three port regions. We assume that the foreign port is the last foreign port of call, and the U.S. port region is the first U.S. port of call. We also assume a 12-hour port time (from ship to rail), although less time may be required at some terminals. From the Far East, the Pacific Northwest has an advantage over San Francisco Bay, and San Francisco Bay has a similar advantage over Southern California. From Yokohama, for example, total transit time to Chicago is 265-273 hours through the Pacific Northwest, 278-290 hours through San Francisco Bay, and 292-300 hours through Southern California. As the inland destination is shifted south, the advantage gradually shifts to Southern California.

A. Ocean Steaming Time

Table 7 includes the steaming time for various foreign/West Coast port combinations. The Pacific Northwest is closer to the Far East, Southern California is closer to other points, and the Bay Area is in the middle of the range. At an average speed of 22 knots, a vessel from Yokohama could take 13 hours longer to reach San Francisco Bay than to reach the Pacific Northwest, and 14 hours longer yet to reach Southern California.

B. Port Time

If a vessel is prevented from docking because of berth congestion, voyage schedules are disrupted, cargo is delayed, and costs mount rapidly. Guaranteed availability of berth space to match vessel schedules is a minimum condition for calling at any port. Once docked, the vessel may be delayed due to terminal congestion. The San Francisco and the Pacific Northwest both have adequate berth and terminal capacity, and additional capacity under develop-

TABLE 7
TOTAL INTERMODAL TRANSIT TIMES TO
CHICAGO FROM PACIFIC RIM PORTS
(Hours from Last Foreign Port)

Via First U.S. Port		Yokohama	Pusan	Sydney	Hong Kong
Bay Area (SF/Oakland)	Ocean	206	224	293	275
	Port	12	12	12	12
	Rail*	60-72	60-72	60-72	60-72
	TOTAL	278-290	296-308	365-377	347-359
Southern Calif. (LA/LB)	Ocean	220	239	296	290
	Port	12	12	12	12
	Rail	60-68	60-68	60-68	60-68
	TOTAL	292-300	311-319	368-376	362-370
Pacific Northwest (Seattle/Tacoma)	Ocean	193	209	310	262
	Port	12	12	12	12
	Rail	60-68	60-68	60-68	60-68
	TOTAL	265-273	281-289	382-390	334-342

* Wider range for Bay Area rail reflects 6-12 hours longer time for San Francisco.

Assumptions: Ocean transit at 22 knots; port times of 12 hours from vessel arrival to train departure.

Source: Distances Between Ports, 1965.

ment. Congestion is becoming serious in Southern California. Carriers may encounter congestion within the marine terminals, on port access roads, at truck gates, on freeways, and in Customs clearance. In this respect, the Ports of Los Angeles and Long Beach are the victims of their own success in attracting discretionary traffic. While the Southern California ports have plans for expansion and revision of marine terminals, they have less latitude, and much less influence, to improve port access. Further growth of Southern California container traffic will increase the congestion, and likely lead some ocean carriers to shift discretionary cargo to other ports.

C. Rail Transit Time

The effective rail transit time includes more than just the train movement. Rail transit begins when the last containers are accepted for loading onto the train. Rail transit ends with the transfer of containers onto chassis, when the containers are released by the railroad to the consignee. An examination of rail distances suggests that the Southern California ports have an advantage to the Gulf, and the Pacific Northwest ports have a small advantage to Chicago. The San Francisco Bay Area ports can be competitive to all midwest points, but they do not have a compelling advantage or disadvantage anywhere. At 50-70 MPH, a 100-mile route difference makes only a 1.5-2.0 hour transit time difference. Larger differences are more a function of contract terms and operating practices than of route distances. Chicago remains the major mid-continent rail destination and interchange, and transit times to Chicago are comparable for all three West Coast port regions. Our discussions with ocean carriers found a minimum scheduled transit time of *60-62 hours* to Chicago from Oakland, Seattle, or Los Angeles. Transit times for lower priority intermodal service are 64-72 hours, depending on route.

VI. PORT CAPACITY AND TERMINAL PRODUCTIVITY

Port capacity and terminal productivity influence the choice of ports or terminals within a region, and the choice of port for discretionary intermodal cargo that is not tied to a particular region. The number and productivity of container cranes, the terminal space available, and labor productivity are directly related to cargo throughput capacity vessel turn time, and underlying unit cost.

Vessel turn time depends largely on the rate at which container cranes can move containers on and off the ship. The prime considerations are the container moves per crane-hour, and the number of cranes available (usually two or three). The provision of container cranes can be a significant issue in securing an agreement from an ocean carrier to use a particular terminal. All the West Coast ports have responded by installing new cranes as needed. Crane productivity, however, is a function of labor productivity, as discussed below.

Terminal space requirements are affected by the type of terminal operation. Some of the larger carriers calling at West Coast ports, APL and Sea-Land for example, favor chassis operations, and thus have an exceptionally high demand for terminal space. All of the West Coast ports have room for expansion, although expansion is more difficult and expensive where the port areas are highly developed. This is especially true in Southern California, where expansion plans entail extensive filling of San Pedro Bay. In the San Francisco Bay Area and the Pacific Northwest, port expansion is being accomplished by converting older facilities to container terminals.

Labor and terminal productivity are closely linked, since labor accounts for most of the terminal operating costs. Hard data on labor productivity are uncommon and inconsistent, so carrier routing decisions are based on the *perceived* terminal productivity. To assess this factor, we interviewed representatives of five ocean carriers regarding the *perceived* rankings of West Coast ports in terminal productivity. The rankings we obtained are given below.

to the influx of intermodal discretionary cargo into Southern California and the Pacific Northwest. Competition for this cargo among port regions remains very keen. Factors in this competition include facility characteristics and transfer charges.

A. Intermodal Transfer Facilities

Current West Coast transfer facilities range from "on-dock" transfers in or adjacent to the marine terminal, to separate rail facilities 20 miles or more away. The merits of one arrangement over another is less an issue of distance than an issue of jurisdiction and transfer cost. The container typically passes from the ocean carrier's jurisdiction to a drayman at the marine terminal gate, and from the drayman to the railroad at the rail yard gate. Two inspections are performed in this procedure, with two sets of documents and attendant delays. Since the drayage cost is based on time, the ocean carrier sees delay as a direct cost. One alternative to the double-transfer system is to eliminate drayage by bringing the rail transfer "on-dock." In such a system, the container passes directly from the ocean carrier to the railroad, with only one inspection and only one set of paperwork.

Other things being equal, the ability to provide on-dock intermodal transfer and eliminate drayage is a competitive advantage for discretionary cargo. Other things are seldom equal, however, and savings in drayage may be offset by gate fees or terminal productivity differences. On-dock transfers now take place in Tacoma, Portland, San Francisco, and Long Beach. The Ports of Seattle and Oakland are in

Ocean Carrier Productivity Rankings of West Coast Ports

	A	B	C	D	E
Highest	Oakland	Tacoma	Tacoma	Tacoma	Tacoma
	Tacoma	LA/LB	Oakland	Seattle	Oakland
	Seattle	Oakland	Seattle	Portland	Seattle
	Portland	Seattle	Portland	Oakland	Portland
Lowest	LA/LB	Portland	LA/LB	LA/LB	LA/LB

Tacoma's high ranking was attributed to management and labor cooperation. Carriers generally agree that the favorable labor climate in Tacoma results from the union's aggressive approach to getting and keeping traffic, and thus maintaining employment for its members. Oakland's productivity was ranked differently by different carriers. The attitude and ability of the Bay Area labor force were generally praised, while port and stevedore management policies and practices received some criticism. Carriers consistently rank the Southern California ports low on terminal productivity, and attribute that ranking to a poor labor/management climate, as well as to increasing congestion. The one carrier who ranked Los Angeles/Long Beach high in productivity cited superior management at a particular terminal.

VII. INTERMODAL CAPABILITIES

Ten-year time series suggests that a large portion of the shift in West Coast cargo shares is attributable

the process of adding on-dock rail transfer at some terminals. Ports and railroads that do not offer on-dock transfer have pursued other strategies to remain competitive.

Increased traffic, especially the advent of double-stack container trains, has strained the capacity and efficiency of older intermodal yards. New facilities have been built in Seattle, Tacoma, San Francisco, and Los Angeles. Existing facilities are being upgraded at Seattle, Oakland, and Los Angeles. Upgrading and new construction have set a higher standard of performance: railroads and ports with older facilities are perceived to be disadvantaged.

B. Intermodal Transfer Charges

Transfer charges may include drayage; chassis positioning or per diem; and gate charges at the rail terminal. Overall, the Pacific Northwest ports have the lowest transfer costs and the Southern California ports the highest, with San Francisco Bay ports in between. However, the charges are determined by a mix of port and railroad policies, and can change quickly.

The Pacific Northwest has a mixed system. For its contract container traffic, BN absorbs the drayage costs (if any) for loaded movements at Seattle, Tacoma, and Portland. Otherwise, the drayage cost in Seattle is $30-$75 per container, depending on marine terminal delay and the need for chassis repositioning. UP customers pay similar drayage costs. In Tacoma, most intermodal moves use the on-dock transfers. At the North Intermodal Terminal, operators charge $33 per container for this transfer (as of May, 1986). The South Intermodal Terminal is operated under contract for Sea-Land traffic. The port-operated Portland rail facility is on-dock, and served by BN and UP.

In Oakland, drayage charges have been roughly $35 per container for the SP and UP yards, and Santa Fe has historically equalized drayage charges to its Richmond Yard. The San Francisco ICTF is operationally on-dock, requiring no drayage, and is served by SP.

Drayage is the rule in Southern California. "K" Line uses on-dock UP service at Long Beach. All other carriers dray containers to UP's City of Commerce Yard, Santa Fe's Hobart Yard, or SP's new Intermodal Container Transfer Facility (ICTF). The UP and Santa Fe yards are over 20 miles from the marine terminals, and drayage averages $75-$80 per container. Highway congestion and other delays can raise the drayage cost to $120 per container. The Santa Fe yard has a chassis pool, enabling ocean carriers to reduce empty chassis repositioning. The SP ICTF is about 4 miles from the ports, reducing drayage costs to about $45-50 per container. There is a $30 per container gate charge at the ICTF, however, raising the total cost to about $75-80 per container.

VIII. PORT CHARGES

Within each of the three port regions, differences in port charges could influence a carrier to call at one port over another. Traditionally, the ship was charged for the use of the dock (dockage), and the cargo for the use of the wharf (wharfage). Other port charges were separate. Today, the various charges are being replaced by contractual, all-inclusive rates.

In choosing among regions, however, port charges are not a significant factor. Wharfage is expressed as a dollar charge per revenue ton, or per TEU. Wharfage tends to be 10-20 percent of total port costs, and the variations between ports are small. Dockage is based on vessel characteristics and time in port. For medium-sized ocean carriers, dockage on the order of $5 per container might be typical. In comparison to other ocean carrier costs, wharfage and dockage are minor considerations on a per-container basis.

IX. CONCLUSIONS

Our findings suggest that long-term cargo flows are mainly influenced by factors outside the control of the ports. Exporters consider domestic transportation costs and total transit time the two most important criteria in choosing among ports. Importers consider the local market size and transit time the two most criteria. The size of the local market unquestionably the single biggest reason why carriers elect to call at a particular port region. The local market is also a major factor influencing port rotation, and port rotation determines the routing of discretionary intermodal cargo. Following the size of the local market, carrier port routing criteria vary by the size and strategy of the various carriers. Large intermodal operators look for efficient, large-scale, intermodal connections; smaller container operators look for port cost and service efficiencies.

The Southern California cargo volumes have grown, and continue to grow, because the large local market attracts the first U.S. port calls. Those first calls lead carriers and importers to route discretionary cargo through Southern California. The Pacific Northwest cargo volumes have grown faster than the local market size would imply because of favorable intermodal capabilities and the attraction of being the last port of call for exports. While the San Francisco Bay cargo volumes have grown, their share has declined due to factors beyond their control that draw discretionary cargo north or south.

ENDNOTE

*Manalytics, Inc.

Assessment of Container-on-Barge Service on the Mississippi River System

by James Crew and Kevin Horn***

ABSTRACT

The only sector of the water transportation industry that has not been affected by containerization and intermodalism is the Mississippi River system inland barge and towing industry. With one of the world's most developed inland waterways system and towing industry, the U.S. lags behind other nations in developing container-on-barge (COB) services. In spite of several attempts by towing firms to institute COB service on the Mississippi River System, only coastal feeder barge movements and container movements on the Columbia-Snake River system have met with any long-term success.

An evaluation of the economic and operating characteristics of COB service was performed to assess the potential for the Mississippi River system. An inland/ocean transportation cost analysis was conducted to determine the relative competitive position of the Port of New Orleans for COB service on the Mississippi River. The cost analysis indicated limited potential for cargo diversion away from coastal ports to a COB service between major river cities and New Orleans. Marine container flows handled by rail between major river cities and the Port of New Orleans were analyzed as the primary markets for potential diversion to COB service.

COB linehaul costs were developed for general towing. Different COB operating patterns and service schedules were examined to determine minimum cost service relative to volume and competitive conditions. COB rates were developed to be competitive with macro-bridge rail rates.

The results of this evaluation indicate that Mississippi River COB service is not feasible except for every large volume movements of low value merchandise or empty marine containers. Inclusion of shippers inventory holding costs erases any significant margin of savings attributable to COB. Moreover, even when savings exist, COB service is perceived to be too infrequent and unreliable by steamship lines that control domestic routings of containers under intermodal rates. While COB appears intuitively appealing, given the relatively low linehaul costs for barge, the viability of any innovative service on the Mississippi River is highly speculative under the most favorable assumptions.

I. INTRODUCTION

Since 1980 several attempts have been made to institute container-on-barge (COB) service on the Mississippi River. The growth in containerized cargoes at the Port of New Orleans and trend toward intermodalism appear to be preconditions associated with successful COB ventures on the Columbia-Snake River system and Rhine River. The depressed Mississippi River towing industry has essentially reduced rates to variable costs. Declining fuel prices have improved the cost competitiveness of shallow draft barge compared to rail. It would appear that COB could be price competitive with rail intermodal service.

The most recent COB service was operated by Leaseway Transportation Corporation for five months in 1984. In 1986 the Maritime Administration (MARAD) sponsored a conference as part of a feasibility study of domestic COB services. The COB concept is still perceived to be viable.[1]

The purpose of this paper is to examine the underlying presuppositions for a successful Mississippi River COB service. The essential determinants of COB will be: (1) time value of inventory in transit; (2) dependability of barge service; and (3) minimum traffic volumes to breakeven.

II. TIME VALUE OF INVENTORY

COB linehaul transit times on the Mississippi River between major midwest load centers and New Orleans would be approximately 14 days for Chicago, 10 days for St. Louis and 5 days for Memphis.[2] Rail transit time between these cities and New Orleans is normally two days for Chicago, one day for St. Louis and overnight (next day) for Memphis.[3] Slower COB service will increase shippers linehaul in-transit inventory between New Orleans and midwest load centers as follows: Chicago–12 days; St. Louis–8 days and Memphis–4 days.

The speed disadvantage of COB does not include allowances for reduced service frequency (bi-weekly or weekly). It is assumed that shippers are able to meet COB sailings without incurring any delay compared to daily rail service. At a minimum COB rates would have to compensate shippers for the linehaul speed disadvantage. Three daily inventory carrying costs are used to represent low, medium and high value merchandise: (1) \$5; (2) \$10; and (3) \$20. The COB linehaul inventory cost disadvantage for midwest service is contained in Table 1. The in-transit inventory costs associated with COB represent the minimum rate savings that must be offered to shippers.

III. DEPENDABILITY OF SERVICE

Mississippi River barge transportation is subject to unavoidable delays caused by river conditions.

TABLE 1

Shippers Increased Linehaul Inventory Costs Using COB Service Between Midwest and New Orleans: Dollars Per Container Trip

Interior Port:	Inventory Carrying Costs Per Day		
	Low ($5)	Medium ($10)	High ($20)
Chicago	60	120	240
St. Louis	40	80	160
Memphis	20	40	80

Fluctuating water levels and fog can disrupt normal towing operations on the open river south of St. Louis. The slack water portion of the waterway between St. Louis and Chicago is vulnerable to delay due to lock congestion and ice. Unfortunately, none of these circumstances can be avoided by barge operators except by lengthening schedules to incorporate anticipated delays.[4]

The average transit times used to develop Table 1 incorporate normal delays, however, disruptions to linehaul operations are excluded. COB transit times would have to be increased in the fall for fog conditions on the lower river and lengthened during the winter navigation season on the upper river. In some instances arrangements for substitute rail service would be necessary due to river closure, for example ice on the Illinois River or lock closures.

COB linehaul service will be provided in mixed or general tows of bulk commodities. Schedules may have to be increased to reflect seasonal delays encountered in linking the container barge with linehaul river tows. Some delays will also be experienced at major fleeting areas, such as St. Louis and Cairo, Illinois.

Efforts to insure dependable service will increase shippers in-transit inventory costs or increase COB operator's costs insofar as securing substitute service. Both of these problems characterized the Leaseway COB service. Leaseway's brief service was reliable, however, good performance did not develop a steady balanced flow of repeat customers. The lack of repetitive shippers was a primary reason why Leaseway management abandoned COB. It appeared that shippers were experimenting with COB, but not committing to the service, in spite of reliable operations.[5]

IV. MINIMUM VOLUMES

The breakeven number of containers for a COB service is calculated as follows: (1) COB maximum linehaul rates are derived from rail intermodal rates by subtracting in-transit inventory carrying costs developed in Table 1; (2) Container loading and unloading costs at origin and destination ports are subtracted from COB rates; (3) COB net linehaul rates per container are compared to linehaul towing costs and anticipated overhead expenses.

Negotiated rail intermodal contract rates for Plan III ramp- to-ramp movements of 40-foot marine containers between New Orleans and the midwest were developed for Chicago, St. Louis, and Memphis. The macro bridge rates between these cities and New Orleans are $400, $350, and $250, respectively.6 Shippers increased in-transit inventory costs for slower COB service (Table 1) were subtracted from rail rates. COB linehaul (dock-to-dock) rates for 40-foot containers are shown in Table 2 for three levels of inventory carrying costs. COB linehaul rates include container loading and unloading at ports.

TABLE 2

COB Linehaul Dock-To-Dock Rates Between Midwest and New Orleans: Dollars per 40-Foot Container

Interior Port:	Inventory Carrying Costs Per Day		
	Low ($5)	Medium ($10)	High ($20)
Chicago	340	280	160
St. Louis	310	270	190
Memphis	230	210	170

Container loading and unloading costs are highly variable because of fluctuating volumes and labor costs. Lift-on or lift- off charges per box at public terminals were quoted at $100 in Memphis and $125 in New Orleans. These charges are based on I.L.A. labor rates and gang sizes. Port costs of this magnitude would be prohibitive for a COB service. Therefore, lift-on and lift-off charges were estimated for non-union labor using crawling cranes at interior ports and efficiently sized I.L.A. gangs using a toplift crane at New Orleans. Loading and unloading costs were estimated to be a minimum of $30 per container movement at ports. Depending on gang productivity and volume, costs per box could increase to $55 per movement, however.

Table 3 contains COB container revenue after subtracting loading and unloading costs from linehaul dock-to-dock rates (Table 2). The data indicate that the speed disadvantage of COB results in container revenue that is not proportional to distance. For high value merchandise the container revenue is virtually the same for all three midwest load centers.

Linehaul costs for COB were estimated based on quotations from operators for towing jumbo river barges (195 × 35) between the midwest cities of Chicago, St. Louis, and Memphis and the Port of New Orleans. Linehaul towing costs incorporate fleeting, switching, and tow-make-up and break-up expenses of $900 per barge to or from Chicago, and $600 per barge for St. Louis and Memphis. On the lower Mississippi, south of St. Louis, towing charges vary by direction. Linehaul costs reflect average one- way expenses for a round trip barge

TABLE 3

COB Linehaul Revenues Per Container After Port Loading and Unloading Costs

Interior Port:	Inventory Carrying Costs Per Day		
	Low ($5)	Medium ($10)	High ($20)
Chicago	280	220	100
St. Louis	250	210	130
Memphis	170	150	110

movement. Linehaul costs per barge are estimated to be: $7,800 for Chicago; $5,100 for St. Louis; and $3,000 for Memphis. The costs do not include insurance or any non-linehaul expenses such as chassis rental or COB administration and overhead.

Barge linehaul cost breakeven volumes, excluding any administration and overhead, are shown in Table 4. Capacity of a jumbo barge (195 × 35) is twelve 40-foot boxes per tier. Leaseway projected that four tiers of containers could be safely lashed on a jumbo box barge. Maximum barge carrying capacity would be 48 containers. The breakeven volumes in Table 4 suggest that COB is potentially viable.

TABLE 4

COB Barge Linehaul Cost Breakeven Volumes of 40-Foot Containers Excluding Administration and Overhead

	Inventory Carrying Costs Per Day		
Interior Port:	Low ($5)	Medium ($10)	High ($20)
Chicago	28	35	78*
St. Louis	21	25	39
Memphis	18	20	27

*exceeds barge capacity

COB administration and overhead costs were estimated assuming a manager, coastal port captain and interior port directors and clerical personnel. COB minimum administration and overhead costs were estimated to be $6,000 per week for service between one interior port and New Orleans. Service to each additional interior port would require a port director and clerk, increasing administration costs by $1,700 per week. COB breakeven volumes of containers for linehaul and administration costs are shown in Table 5 for direct weekly service between one, two, or three midwest load centers and New Orleans. Administration costs are divided by two, assuming weekly COB service in each direction to and from the ports and added to average one-way barge towing costs. Direct weekly COB service between each midwest load center and New Orleans would require at least 39, 32 and 36 containers of low value merchandise per barge trip between Chicago, St. Louis, Memphis and New Orleans, respectively, to breakeven. Round trip weekly breakeven volumes would be 78 boxes for Chicago, 64 boxes for St. Louis and 72 boxes for Memphis.

Table 5 indicates COB breakeven volumes when threshold administration costs are allocated to more than one interior port. Breakeven volumes are not particularly sensitive to direct service between multiple inland ports and New Orleans. Instituting single barge service between Chicago or St. Louis and New Orleans with intermediate port calls at St. Louis and Memphis to handle additional containers is not feasible. The high breakeven volumes for any individual port leave very little excess capacity to serve intermediate ports, considering port costs associated with tow make-up, break-up, switching and fleeting ($600 at St. Louis and Memphis) and additional delay for in-transit containers.

Table 5 also indicates that COB service would have to be oriented to low value merchandise to be feasible. High value cargoes cannot be profitably handled other than to fill excess capacity beyond breakeven. The breakeven volumes for direct service to three ports are 77 percent of capacity for Chicago; 65 percent of capacity for St. Louis; and 69 percent of capacity for Memphis.

TABLE 5

COB Total Cost Breakeven Volumes of 40-Foot Containers

Direct Service to One Interior Port

	Inventory Carrying Costs Per Day		
Interior Port:	Low ($5)	Medium ($10)	High ($20)
Chicago	39	49*	108*
St. Louis	32	39	63*
Memphis	36	40	55*

*exceeds barge capacity

Direct Service to Two Interior Ports

	Inventory Carrying Costs Per Day		
Interior Port:	Low ($5)	Medium ($10)	High ($20)
Chicago	38	48	105*
St. Louis	32	37	60*
Memphis	34	38	52*

*exceeds barge capacity

Direct Service to Three Interior Ports

	Inventory Carrying Costs Per Day		
Interior Port:	Low ($5)	Medium ($10)	High ($20)
Chicago	37	47	103*
St. Louis	31	37	59*
Memphis	33	37	50*

*exceeds barge capacity

V. COB Outlook

The potential for a viable Mississippi River COB service is very limited. The data in Table 5 represent an optimistic scenario in which shippers are given no real rate savings over relatively dependable daily rail service. In Table 6 COB breakeven volumes are projected for a direct weekly COB service between midwest load centers and New Orleans assuming real savings to shippers. COB linehaul rates were developed by discounting rail rates 25 percent or $50 per container and subtracting shippers increased in-transit inventory costs. After deducting container loading and unloading expenses, the data indicate that COB is not feasible with 25 percent real rate savings. The $50 per box incentive would necessitate that COB be operated at nearly 100 percent of capacity (48 containers) to breakeven.

Nominal savings of $50 per container leave no margin for cargo insurance, chassis rental, etc., that could range upwards to $10 per day. Moreover, the rail rates used do not include any special volume incentive discounts or rebates for large steamship lines and freight forwarders. The breakeven volumes for COB are also quite large considering the weak

TABLE 6
COB Total Cost Breakeven Volumes of 40-Foot Containers: Incentive Rate Savings for Shippers

		COB Linehaul Rate After Deductions:			
Interior Port:	Rail Rate:	Less 25% COB Reduction:	Less In-Transit Inventory Cost:[1]	Less Port Costs:[2]	Breakeven Volume:[3]
Chicago	400	300	240	180	60*
St. Louis	350	260	220	160	51*
Memphis	250	190	170	110	55*

		COB Linehaul Rate After Deductions:			
Interior Port:	Rail Rate:	Less $50 per box Reduction:	Less In-Transit Inventory Cost:[1]	Less Port Costs:[2]	Breakeven Volume:[3]
Chicago	400	350	290	230	47
St. Louis	350	300	260	200	41
Memphis	250	200	180	120	50

*exceeds barge capacity

[1]Based on low value carrying costs of $5 per day.

[2]Based on container lift-on and lift-off costs of $30.

[3]Based on towing costs and administrative expenses for weekly service in each direction.

competitive position of Gulf Coast ports for midwest traffic.[7] Available evidence indicates that COB would have to capture a significant proportion of the marine container traffic handled by rail between the midwest and New Orleans. The estimated annual number of marine containers handled by rail between the midwest and the Port of New Orleans is as follows: Chicago–5,400; St. Louis–14,200; and Memphis–13,700.[8] COB would have to operate at nearly full capacity to be potentially viable. This would require approximately 5,000 40-foot containers a year for each load center (48 containers per barge trip times two to reflect round-trip weekly service times 52 weeks). COB would require nearly 100 percent of the Chicago market, however, only one-third of the market would be required at St. Louis and Memphis.[9]

It is doubtful that shippers could be induced to use COB service given very nominal rate savings. Moreover, steamship line representatives are very skeptical of the quality of a COB service irrespective of real rate savings. Liner operators calling at New Orleans indicate that they do not handle sufficient volume of low value cargo that would be susceptible to slower COB service.

VI. CONCLUSIONS

The potential for a viable Mississippi River COB service is virtually nonexistent. Although COB is intuitively appealing from the standpoint of low barge linehaul costs, the time factor and relatively low rail rates for dedicated intermodal trains preclude any successful COB venture. The unique determinants of successful COB services on the Columbia-Snake and Rhine Rivers are not transferable to the Mississippi River.

REFERENCES

Brown Associates, Inc., *Shipper and Towing Company Responses to Lock Congestion* (Fort Belvoir, Virginia: Institute for Water Resources, 1984).

Crew, J. G. and Horn, K. H., "The Evolving Functions of U.S. Gulf Coast Ports," *Proceedings of the Transportation Research Forum*, Vol. XXVI, No. 1, (Washington, D.C.: Transportation Research Forum, 1985) pp. 152-160.

Illinois Central Gulf Railroad Company, "Contract for Movement of Exempt Freight" (Chicago, Illinois: Intermodal Department).

Illinois Central Gulf Railroad Company, *System Timetable No. 1*, January 12, 1986.

Ports and Waterways Institute, *Container Transport By Inland Waterways* (Washington, D.C.: Maritime Administration, February 1987).

ENDNOTES

* Transportation Research and Analysis Center

** Department of Management, Marketing and Logistics, University of North Florida, Jacksonville, FL.

1. Ports and Waterways Institute (1987).
2. Upriver navigation speeds will be slower than downriver speeds south of St. Louis. These average transit times reflect a composite of upward and downward movements.
3. Illinois Central Gulf Railroad (1986).
4. Brown Associates, Inc. (1984).
5. Jerome D. Krassenstein, Group Vice President Intermodal, Leaseway Transportation Corporation, *Proceedings, Container-On-Barge Conference*, April 29, 1987.
6. Illinois Central Gulf Railroad.
7. James G. Crew and Kevin H. Horn (1985).
8. Ports and Waterways Institute (1987).
9. COB breakeven market shares, expressed in 40-foot containers, are underestimated because of the existence of a large volume of 20-foot containers in the Port of New Orleans midwest hinterland. For example, approximately 60 to 70 percent of the Chicago rail marine container traffic consists of 20-foot containers. The percentages of 20-foot containers in the port's St. Louis and Memphis hinterlands are approximately 60 and 50 percent, respectively.

The Intermodal Equipment Distribution Model

*by Kenneth C. K. Chih and Carl D. Van Dyke**

ABSTRACT

The Intermodal Equipment Distribution Model (IEDM) is a set of powerful computer programs designed to assist managers of an intermodal equipment fleet in the distribution of the fleet's trailers and/or containers. The model's goal is to efficiently utilize the fleet's equipment and improve shipper service and reliability. To achieve this goal the model uses a real-time multicommodity transshipment optimization algorithm and a dynamic real-time approach. The model can be easily modified to handle many other types of equipment such as rail cars or airline containers.

The IEDM can be broken into five components: 1) input, 2) forecasting of supply and demand, 3) network generation, 4) optimization, and 5) post-processing. The primary user interactions are with the input, forecasting, and post-processing sections. The input and forecasting sections feed the network generator which creates a network in space and time over which loaded and empty equipment can be moved. The heart of the system is an optimization algorithm which uses a multi-commodity temporal-spatial network approach to effectively produce real-time optimal results for equipment distribution. The post-processor converts the optimization results into specific trailer distribution orders.

I. INTRODUCTION

The demand for intermodal service is unbalanced. As a result, it is often necessary to reposition empty equipment over substantial distances and time periods. The task of managing these empty movements is usually called trailer distribution. The trailer distribution manager must make decisions relating to the railroad's current and future equipment supply, what to do with equipment as it becomes empty, how much empty equipment to hold in inventory at each location, and how much equipment will be required by shippers by location. These decisions will affect thousands of trailers at any point in time, both on and off a particular railroad.

Most trailer distribution managers can only consider the demand and supply of equipment on a local basis. The demand for equipment is generally expressed in terms of an aggregate demand at a ramp for a multiple day period, broken out by size of trailer, but without regard to where the equipment will be sent in the loaded move. As a result most distribution managers focus on the inventory level at each location, keeping each ramp supplied with enough trailers to prevent a "stock-out" situation. As a result of this approach a railroad may tend to hold too much equipment in inventory, resulting in a requirement for a larger fleet. In addition, non-optimal choices may be made about how to replenish the inventories, resulting in excessive empty miles, empty days, and equipment leasing costs.

The Intermodal Equipment Distribution Model (IEDM) is a powerful railroad network optimization and costing model for supporting the day to day operating decisions that intermodal distribution managers have to make. The purpose of the IEDM is to simplify this complicated problem by helping the trailer distribution manager more efficiently manage the railroad's fleet of equipment. The IEDM assists the distribution staff by providing simple recommendations of when and where to send each type of equipment to meet current and future demand. It makes these recommendations while fully taking into consideration the railroad's total system needs for equipment on both a global and local basis over the coming one to two week period.

Generally, the IEDM is intended to be run once per day. This run should be made late at night or in the early morning so that the latest information will be available for the next day's distribution decisions. The model could be run more or less frequently, depending on the railroad's needs and how frequently the input data can be collected. The results of each run lists when, and what kind of trailers need to be sent from which origins to which destinations.

The IEDM was developed by ALK Associates Inc. as part of the efforts of the Association of American Railroad's Intermodal Productivity Task Force under the supervision of the Subcommittee on Equipment Distribution. It is designed to be installed at the same location as a railroad's central equipment distribution staff. The model is programed in FORTRAN, and is designed to run on an IBM mainframe computer.

II. MODEL INPUTS AND OUTPUTS

The primary points of user interaction with the software are the IEDM's inputs and outputs. While the inputs are mostly generated automatically, they can be reviewed by the distribution staff prior to running the model. To use the Intermodal Equipment Distribution Model, the user supplies a number of input files:

1. *Network Description and Operating Rules*—The network defines all of the relevant locations in the system such as ramps, junctions, classification points, and storage points. Links define how equipment can be moved between these points and what the associated distances and transit times are. Operating rules define how equipment moves between the various locations, what types of equipment can be substituted for other types of equipment, and where equipment can be interchanged with other carriers.
2. *Status Information*—Status information includes the current empty equipment inventories, the

known demands for empties, and what loaded and empty equipment is in the pipeline. This information is usually taken directly from an on-line car movement system. If available, information on off-line equipment and expected/known demand can be incorporated into this data.

3. *Empty Equipment Supply Information*—The empty supply data consists of two parts: a status component and a forecast component. Based on the status information, current and future known supply is computed. Based on a constantly updated computer study of past supply levels, and the known equipment supply, the total equipment supply is forecasted. The user can modify the supply forecast on a location by location basis.
4. *Empty Equipment Demand Information*—The forecast of future demand is based on past loadings plus any user supplied demand forecast and/or equipment order data. Ideally the demand forecast contains the destination of the loaded move, in addition to the demand location. In the absence of destination data for known demand, past destination patterns are used.

Based on the input information outlined above, the IEDM automatically generates a time-space network and optimally solves the distribution problem for the forecasted situation. Based on the results, a number of reports are generated. There are four primary reports:

1. *Recommended Empty Distribution Orders*—This includes orders of empty equipment already dispatched to a destination, equipment not currently at the "correct" location, and equipment that is expected to become available at on-line locations.
2. *Expected Equipment Inventories*—This report shows the expected levels of empty equipment on-hand at each location for each day in the planning period.
3. *Demand/Supply Yard Line-Ups*—The line-up report shows each actual or forecasted demand, the characteristics of the demand, and the source and characteristics of the supply satisfying the demand. Demand information includes:
   a) Location
   b) Date/Time
   c) Destination (Projected or Actual)
   d) Equipment Type Required
   e) Quantity of Equipment Required

   Supply information includes:
   a) Source
   b) Current Location (On-line or Off-line)
   c) Current Status (in a train, inventory, on the street)
   d) Earliest Expected Availability
4. *Foreign Equipment Sent Off-Line*—This report details those pieces of equipment that the IEDM recommends you should send off-line.

III. SUPPLY/DEMAND FORECAST

Using information on the current status, past movement history, and past supply history, a supply forecast is generated for each location for each day, broken out by equipment type and ownership. The basic supply forecast is made up of three components:

1) On-line empty equipment
2) On-line loaded equipment
3) Off-line equipment

The current on-line empty equipment not destined for off-line points is added directly to the supply forecast. Historical transit times, un-load times, and return rates are used to project when and if the on-line loaded equipment will become available for reloading. A projection of equipment expected to be received from off-line points is made based on past levels of equipment received. This is done using a forward weighted rolling average for the previous 28 to 35 days. This projection includes the mix of loaded and empty equipment, as well as the destinations of the loaded equipment. It is added to the supply forecast in the same manner as the on-line loaded and empty equipment.

Given this projected equipment supply, the "Supply File Generator" creates the actual supply file to be used by the IEDM. This supply file can be reviewed and edited by the user prior to use by the model.

The level of demand is also forecasted for each location for each day, broken out by equipment type and destination. The demand forecast is based on historical loading patterns taken from the railroad's trailer movement system. A projection of equipment requirements is created using a forward weighted rolling average for the previous 28 to 35 days. The forecast takes into account the destination of the equipment and the size of the equipment required. As with the supply forecast, the demand forecast can be edited/reviewed by one's distribution staff prior to use by the model.

If agent supplied demand information and/or known equipment orders are available on a consistent basis, these can be substituted for the computer generated demand or incorporated into the forecast system at user's discretion. If user demand information is not destination specific, destinations can be assigned based on historical destination patterns.

There is some uncertainty in the level of demand, supply, transit times, and all other factors affecting the matching of empty equipment to loads. This uncertainty can be alleviated by imposing a minimum inventory goal for each location, called a buffer stock. The level of this "buffer stock" can be determined by the amount of variation in the level of supply and demand, and the level of acceptable unsatisfied demand.

IV. NETWORK GENERATION

As described above, the network defines all of the relevant times and locations on the system, and the rules for how to move between these locations including intermediate blocking points, running times, costs, and distances.

There are three components of the operating rules that are very important:

1) equipment substitution
2) cost modelling
3) revenue modelling

Each is described below.

1. *Equipment Substitution*—Railroads use numerous types of trailers and containers. In some

cases this equipment can be freely substituted, in other cases substitution is allowed only in one direction (upgrades are allowed, downgrades are not). For example, a 45 ft trailer can be used to satisfy a demand for a 40 ft trailer; however, a 40 ft trailer cannot be used to satisfy a demand for a 45 ft trailer. The substitution rules are defined by the railroad. These rules are used to create groups of trailers that can be treated in as interchangeable, and to define the relationship between these equipment groups. These groups can be further subdivided by their cost structure, based on ownership and equipment type.

2. *Cost Modelling*—In order to minimize cost or maximize profit, the cost of using and moving equipment must be modelled. In the IEDM the cost of moving a piece of equipment is a function of distance, ownership, load/empty status, and type of equipment. The level of detail of the cost model is a function of how much information is available. In general, for foreign-owned and leased equipment the ownership cost will be the equipment's per diem rate. The rate for railroad owned equipment should reflect both the cost of the equipment to the railroad and the priority of using this equipment relative to foreign and leased equipment. A fixed cost can also be applied to each move, reflecting the initial and final handling cost for the move. Thus, the cost formula will usually be:

   Costs = Fixed Cost +
   Miles x (Cost/Mile) +
   Hours x (Cost/Hour)

The model recognizes that sending foreign and leased equipment off-line will save the railroad per diem charges, and that empty foreign and leased equipment can only be sent off-line at specific locations as a function of ownership. The IEDM also recognizes that system equipment sent off-line can earn the railroad per diem revenue. The model will take all of these factors into account when deciding when and where to send equipment.

3. *Revenue Modelling*—To properly model the trade-offs of holding a trailer in inventory for a future load versus sending it to some other location, and the proper handling of foreign-owned equipment, the revenue associated with each demand must be modelled. The model uses an average revenue per mile approach. This can be broken out by equipment type, ownership, distance, origin, and destination. The basic formula is:

   Revenue = Fixed Revenue +
   Miles x (Revenue/Mile)

   Other revenue functions can be easily implemented, depending on one's needs. For example, to model an operating environment in which loads are never turned down, a very high revenue can be used for all demands so that revenue will always exceed the costs of repositioning empty equipment.

V. STRUCTURE OF THE IEDM

The IEDM is divided into six modules including two input modules, three main processing routines, and one post-processor. Each module prepares data for the next by passing on its results via a set of disk files. These disk files make the inner workings of each program unimportant to the rest of the IEDM as long as the correct set of output files is produced. Thus, any program in the IEDM can be changed without having to worry about the impact of these changes on the rest of the IEDM, as long as the input and output files remain the same. The layout of these modules is depicted in Figure 1.

Modules 1 and 2 are the demand and supply file creation routines. They take current and historical car movement data as input and generate the supply and demand forecast files required by the network generator.

Module 3, the "Network Generator," takes the supply and demand information, physical network definition, and operating rules as input and generates a network in space and time. This network is used to define the costs and revenue associated with different types of equipment when it is kept in inventory, assigned to a load, repositioned to a different point on the system, or sent off-line. This network takes into account numerous factors including those mentioned above plus factors such as equipment substitution and travel time.

Module 4 is a network optimization program that determines when and where to:

—hold empty trailers in inventory,
—assign a trailer to a particular demand,
—send equipment off-line, and
—reposition empty equipment for loading at a different point on the system.

This optimization program uses a resource-directive multi-commodity network transshipment algorithm detailed in Section VII.

Module 5 takes the output of the network optimization program, which consists of individual link volumes between various points in space and time, and transforms them into origin-destination movements. This is also called the "reverse A-B pairs process."

Module 6 is the post-processor program that generates the distribution orders. The module takes the reverse A-B pairs output and transforms it into a form readily used by a railroad's trailer distribution staff. The volumes assigned by the optimization and reverse A-B pairs algorithm determine what the trailer distribution orders should be. From these volumes orders can be generated to:

a) hold trailers in inventory
b) send trailers empty from one location to another
c) send trailers off-line

In addition, the solution indicates what trailer ownerships and equipment types should be used to satisfy the various demands based on the destination of the demands.

VI. THE TEMPORAL-SPATIAL NETWORK

The key to the optimization techniques used by the IEDM is the temporal-spatial network. Figure 2 depicts a simplified temporal-spatial network with 2 locations and 2 commodities. It gives a simplified graphical representation of the opportunities the equipment distribution manager faces. The links in

Figure 1. Overall Structure of Intermodal Model

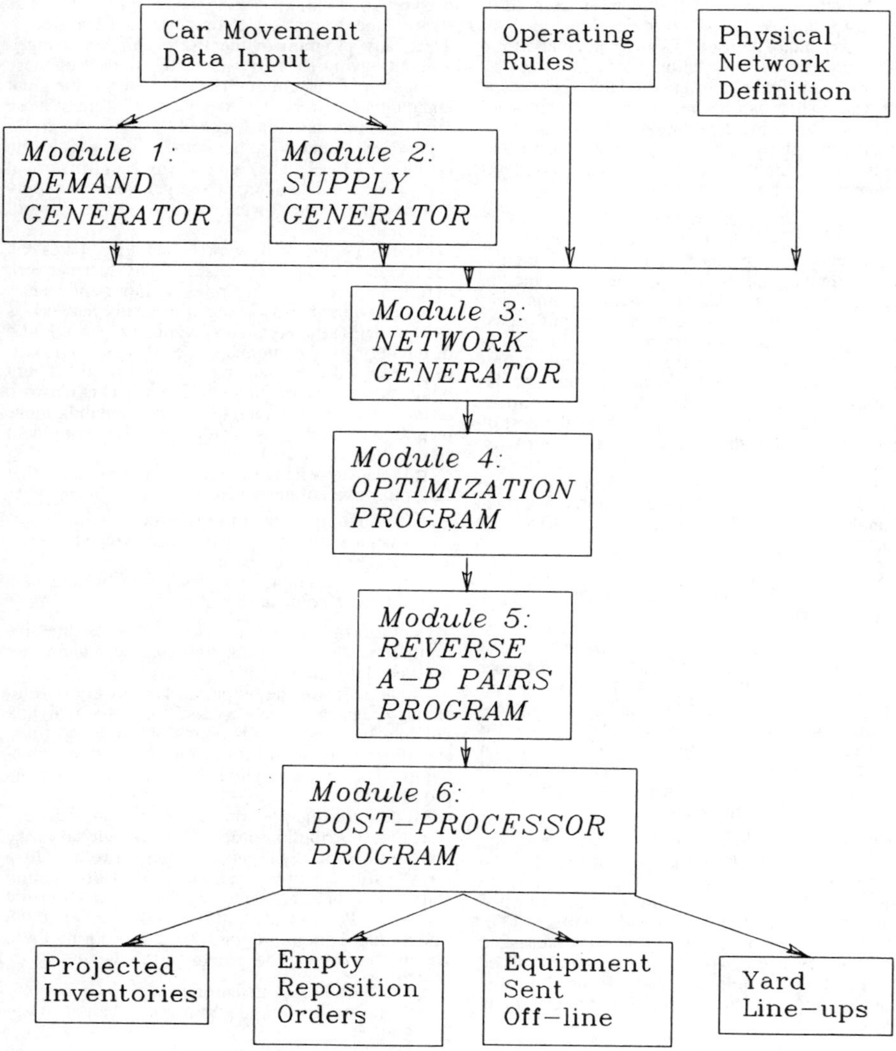

this network represent the three basic decisions that the equipment manager can make:

1. Inventory links—the manager can keep a trailer of a certain type at the current location. As shown in figure 2 an inventory link goes between the same location at sequential time steps. The cost of assigning a trailer to this link would be the the cost of keeping that trailer in inventory for one time step.
2. Empty movement links—the manager can send an empty trailer from one location to another (point a to point b). As shown in figure 2 the elapsed time depends on the time to move the trailer. The cost of assigning a trailer to this link is the cost of moving the trailer from a to b.
3. Loaded movement links—the manager can assign the empty trailer to a revenue movement. The time and location attributes of the loaded links are the same as the empty movement links.

The loaded links are capacitated, with the capacity equal to the level of demand. The link costs are the loaded movement's revenue minus the car movement costs.

It should be noted that in the temporal-spatial network, links can not be constructed between two points in the same time step because you cannot move a trailer with infinite speed (Chih [2]).

VII. THE OPTIMIZATION ALGORITHM EMPLOYED BY THE IEDM

The fourth module of the IEDM uses a resource-directive decomposition approach to solve the multi-commodity temporal-spatial network flow problem created by the third module of the IEDM. An extensive survey of the solution algorithms for solving multicommodity network flow problems can be found in Assad [1], and Kennington and Helgason [6]. These problems have been extensively studied because of their numerous applications and their underlying network structure.

In the context of the IEDM each commodity represents a particular trailer size and owner combination. These "commodities" must be "flowed" over a network in space and time in an optimal fashion. The resulting optimal flows represent how empty equipment should be repositioned to satisfy the demands for reloading.

A resource-directive decomposition approach consists of distributing the link capacity among the individual commodities in such a way that solving K subproblems yields an optimal flow for the original problem (Kennington and Shalaby [7]). At each iteration, a subgradient allocation (Polyak [8], Held et al. [4]) is made with a nearest point projection (Chih

Figure 2. An Example of a Temporal Spatial Network with Two Commodities

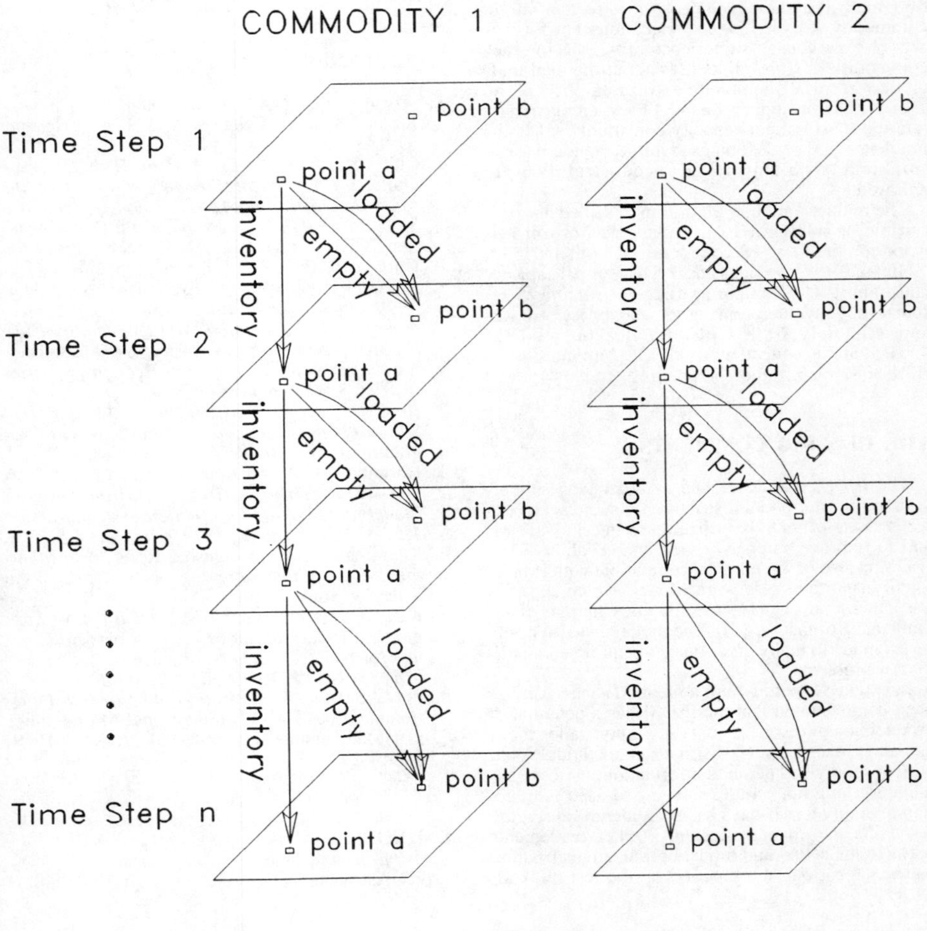

et al. [3]) and K single commodity flow problems are solved. The sum of the capacities allocated to a link over all commodities is equal to the link capacity in the original problem. Therefore, the combined flow from the solutions of the subproblems provides a feasible flow for the original problem. Optimality is tested and the procedure either terminates or a new link capacity allocation is developed.

Letting K denote the set of commodities from 1 to k, and A denote the set of links in the network, the multicommodity minimum cost network flow problem can be stated as follows (Chih [2]):

$$\text{Minimize } \Sigma_{k\varepsilon K}\Sigma_{(i,j)\varepsilon A}\, c^k_{ij}x^k_{ij}$$

subject to

$$(1)\ \Sigma_j x^k_{ij} - \Sigma_l x^k_{li} = b^k_i, \text{ for all } k, \text{ all } (i,j)\varepsilon A,$$

$$(2)\ \Sigma_{k\varepsilon K} x^k_{ij} \leq D_{ij}, \text{ for all } (i,j)\varepsilon A,$$

$$(3)\ x^k_{ij} \geq 0, \text{ for all } k\varepsilon K, \text{ all } (i,j)\varepsilon A.$$

where x^k_{ij} is the number of units of commodity k over link(i,j) and c^k_{ij} is the cost of moving a unit of commodity k over link(i,j). Equations (1) form K sets of flow conservation constraints, one for each commodity k. Constraints (2) restrict the total units of flow from all commodities over link(i,j) be no greater than its upper bound. They are generally referred to as mutual capacity constraints, since the problem could be decomposed into K standard transshipment problems if these constraints were removed.

The multicommodity algorithm descibed above is capable of solving a 19 day planning horizon network composed of 19 commodities, with 16,556 x 19 links, 989 x 19 nodes and 7,810 multual capacity constraints. The computing time for running 5 resource-directive iterations of the above test case was approximately 61.8 CPU minutes on an IBM 4361-5, and is equivalent to 7.7 CPU minutes on an IBM 3081 mainframe computer.

VIII. CLOSING COMMENTS

The IEDM was developed as a decision support system for intermodal distribution managers. Its objective is to efficiently utilize one's intermodal fleet and to improve shipper service and reliability. The IEDM is also designed to be compatible with Princeton Transportation Network Model, which supports the generation of computer graphics outputs (Hornung and Kornhauser [5]). The model can also easily be extended to many other types of equipment distribution problems.

In extensive testing with a major US carrier it has been demonstrated that the IEDM has a potential of producing significant savings in empty trailer miles and days over current distribution methods. With perfect supply and demand information, up to a 25% reduction in empty trailer miles, and a 50% reduction in empty trailer-days has been identified. Figure 3 depicts the change in empty trailer movements between the actual and model optimized distribution patterns for a one week period on the test railroad. The map shows total savings of 465,622 empty trailer miles and 13,140 empty trailer days. In the figure one can see that total empty miles has decreased, and the model has selected the optimal mix of gateways for the disposal of empty foreign equipment.

Numerous attempts have been made to develop computerized distribution models. Among them, the IEDM is unique for the following reasons:

1. It takes a pratical and realistic approach by considering both spatial and temporal imbalance, taking into full consideration the when and where of the supply and demand for empty equipment.
2. It is cost and price sensitive to different trailer types and ownerships in distributing the equipment.
3. It takes into consideration the rules relating to the substitution of different types of equipment.
4. While it uses forecasting to project future supply and demand of equipment, the frequency of its use makes errors in these forecasts self-correcting.
5. Its development was at the request and under the supervision of several railroads.

REFERENCES

1. Assad, Arjang A., *Multicommodity Network Flows—A Survey,* Networks, Vol. 8, 1978, pp. 37-91.
2. Chih, Kenneth C. K., *A Real Time Dynamic Optimal Freight Car Simulation Model of the Multiple Railroad, Multicommodity Temporal Spatial Network Flow Problem*, Ph.D. Dissertation, Civil Engineering Department, Princeton University, Princeton NJ, June 1986.
3. Chih, C. K., Powell W., Kornhauser A., *A Simple Procedure for Finding Nearest Point Projections for the Capacitated Multicommodity Flow Problem*, Princeton University, Civil Engineering Department, Engineering Economics Series EES-86-8, March 1986.
4. Held, M., Wolfe, P., Crowder, P., *Validation of Subgradient Optimization*, Mathematical Programming, Vol. 6, 1974, pp. 62-88.
5. Hornung, M. A., Kornhauser, Alain L., *PTNM/GIS—The Princeton Transportation Network Model and Graphic Information System, Version 4.5*, ALK Associates, 1987.
6. Kennington, J. L., Helgason, R. V., *Algorithms for Network Programming*, New York: John Wiley & Sons, 1980.
7. Kennington, J., Shalaby M., *An Effective Subgradient Procedure for Minimum Cost Multicommodity Flow Problems*, Management Science, Vol. 23, No. 9, 1977, pp. 994-1004.
8. Polyak, B. T., *Minimization of Unsmooth Functionals*, U.S.S.R. Computational Mathematics and Mathematical Physics, Vol. 9, No.3 1969, pp. 14-29.

ENDNOTE

*ALK Associates, Inc.

Figure 3

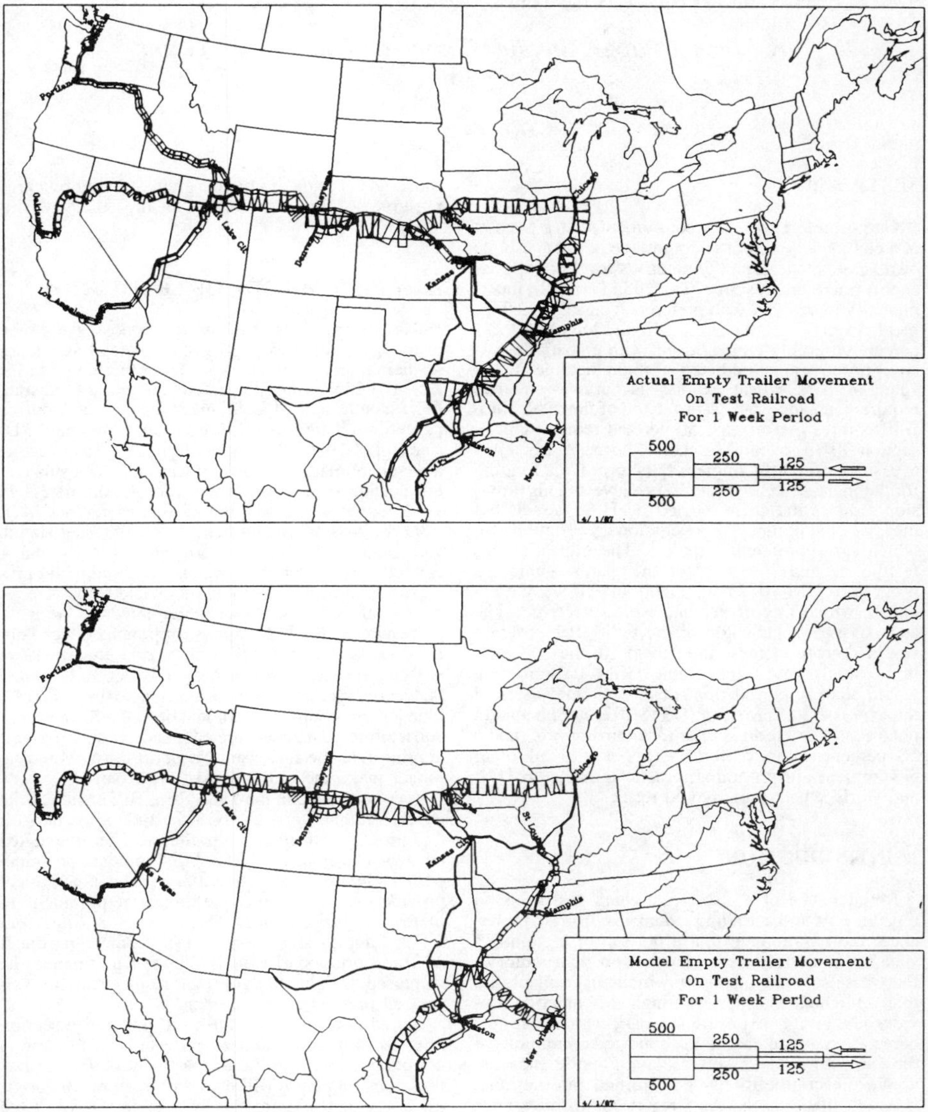
Actual Empty Trailer Movement
On Test Railroad
For 1 Week Period
500
250
125
125
250
500
Model Empty Trailer Movement
On Test Railroad
For 1 Week Period
500
250
125
125
250
500

Cargo Preference: Its Impact on the U.S. Wheat Sector

*by Mechel Paggi, Stephen Fuller and Warren R. Grant**

ABSTRACT

Cargo preference, the requirement that a portion of a country's waterborne commerce be carried in its own vessels, can affect a country's level of exports, export prices and revenue. The 1985 Farm Bill incrementally increases cargo preference on PL 480 shipments from 50 to 75 percent. Simulation of 50 to 75 percent cargo preference levels with a spatial price-equilibruim model indicate a slight decline in the value of U.S. wheat exports as cargo preference requirements increase. A provision of the 1985 Farm Bill requires the original 50 percent requirement of section 901(b) of the Merchant Marine Act of 1936 to resume effect if Congress fails to provide funding for the increased cargo preference levels. This provision reduces the share carried by U.S. vessels but increases the number of concessionary programs for which cargo preference applies. The impcat of this feature depends on whether the export guarantee program (GSM-102) is included and, if so, the extent to which import demand would decrease. The analysis shows inclusion of the GSM-102 program would sharply increase the subsidy to the US. maritime industry. An inoperable GSM-102 program would sharply curtail this subsidy. Should users of the export credit program (GSM-102) not be able to obtain another credit source for their imports, and a 50 percent cargo preference applied to all other government aid programs, drastic impacts on U.S. and world wheat trade would result.

I. INTRODUCTION

Maintenance of a strong merchant marine is an objective of most trading countries. Cargo preference, the requirement that a portion of a country's waterborne commerce be carried on that country's flag vessels, is one means by which governments aid their merchant marine. Although this practice provides less budget exposure for a government, it may come as an added cost to the exported commodity if the rates for the vessels receiving preference are above the competitively determined international rates of other vessels. As a result, cargo preference may shift the competitiveness of a country's traded commodities. Current U.S. legislation requires that a percentage of all PL 480 commodities be shipped on American flag vessels. Since rates for U.S. flag vessels are higher than non-U.S. flag vessels, changes in cargo preference requirements are of concern to the U.S. maritime industry, agricultural exporters, commodity industry groups, and the taxpayers. This study examines recent changes in U.S. cargo preference legislation and estimates its impact on the United States maritime industry, United States wheat industry, and international wheat trade. Since wheat is the dominant grain affected by cargo preference, that commodity is the focus of this study.

II. A CHRONOLOGY OF LEGISLATION

The governments of most nations possessing merchant fleets offer some form of financial assistance to their maritime industries. The Merchant Marine Act of 1928 was the first statute that granted substantial financial aid to U.S. maritime interests during peacetime. This was followed by the Merchant Marine Act of 1936. The intent of this Act was to create a merchant marine capable of transporting a substantial portion of the U.S.'s waterborne commerce. In 1954 Congress passed the Cargo Preference Act. This Act was an amendment to the 1936 legislation and dictated that at least 50 percent of the cargo generated by government-sponsored foreign aid programs be allocated to privately-owned ships of U.S. registry if available at reasonable rates. As a result, 50 percent of the U.S.'s grain shipments under Public Law 480 had to be transported in American vessels. In early 1985 a U.S. District Judge ruled that cargo preference also applied to the USDA's blended credit program. In reaction, the Secretary of Agriculture suspended the blended credit program, arguing that the increased transport costs would no longer make the program cost effective.

A compromise in the 1985 Farm Bill excluded the export credit guarantee (GSM) and other non-PL 480 programs from cargo preference, but mandated an increasing share of PL 480 shipments be transported aboard U.S. flag vessels. The amendment provides for the cargo preference requirement to increase to 60 percent for fiscal 1986, to 70 percent in 1987 and to 75 percent in 1988. The Department of Transportation (DOT) is required to finance the increased freight charges that result from the increased preference requirement.

In addition, the 1985 farm bill includes a section referred to as the snapback provision. If funding is not obtained through the Department of Transportation, the 1985 Farm Bill provision requires the original 50 percent requirement of section 901(b) of the Merchant Marine Act of 1936 to resume effect, i.e., to snapback. Because of the 1985 U.S. District Court ruling on the blended credit program, the USDA's export credit guarantee program (GSM) could be challenged in the Courts for inclusion under the 50 percent cargo preference requirement.

III. LITERATURE REVIEW

In the mid-1970's Vanags (1977) concluded that costs and benefits of cargo preference schemes are analogous to the cost and benefits of protection in other industries. More recently, Castillo-Manuel (1980), using a spatial equilibrium model, analyzed the impact of various cargo preference schemes on the world and U.S. grain trade. Results show cargo preference is a protection mechanism for the U.S. maritime industry and a trade barrier to the U.S. grain industry. Paarlberg (1984) addressed the consequences of requiring various proportions of U.S. concessional and commercial wheat sales be shipped on U.S. flag vessels. In general, Paarlberg's analysis shows relatively modest impacts from increased cargo preference requirements. U.S. exports decline between 1-2 percent, while U.S. prices are slightly lower and foreign prices are slightly higher.

IV. CONCEPTUAL AND MATHEMATICAL MODEL

The relationship between the United States, other world exporters, and the total world is shown by the simplified model in Figure 1. In the absence of cargo preference, the equilibrium CIF price is OA. The United States would export quantity OG while other exporters would export OH. In total, exporters would export quantity OI (OG + OH) to world importers. Imposing cargo preference on the United States will shift the U.S.'s excess supply. This is facilitated in Figure 1 by shifting the excess supply function to the left (Xsus to X'sus) which subsequently affects a decline in world supply (Xsw to X'sw). This yields higher delivered prices (OF versus OC) and a reduction in quantity traded (OL versus OI). The United States exports less, OJ versus OG, at a higher delivered price, OD versus OA. Because the increased price only partially offsets the increased U.S. flag vessel cost the U.S. FOB price must be lowered to remain competitive. Other world exporters increase exports from OH to OK and receive the benefit of the higher price with no change in their transportation rates. World importers would import less at the higher delivered price. The extent price and quantity shift is determined by the excess supply and demand elasticities in impacted countries.

A spatial price-equilibrium model was used in this study to evaluate the impact of changing cargo preference levels on the U.S. and world wheat trade. The competitive, spatial price equilibruim model describes the pricing and allocation of a commodity between exporting and importing countries separated by transportation costs. The assumptions of this model are (1) the markets are competitive, (2) each country represents a distinct market, (3) a homogenous commodity is traded, (4) supply and demand can be expressed in a functional form, (5) transportation costs are constant, and (6) the countries may be specified a priori (Holland, 1985). These models are usually static with the solutions providing values given alternative policies or shocks. They do not provide information on the transition from one solution to another The competitive spatial price equilibrium model can be stated mathematically as a surplus supply/demand maximization model or as a set of equilibrium conditions (Holland, 1985). The model used in this study is an adaptation of an equilibirum wheat model developed by Holland and Sharples (1984). The model solutions were obtained on the IBM-XT personal computer using the GTP (Generalized Transportation Problem) program developed by Holland (1985).

A price linkage function in the spatial price equilbirum model depicts the price relationship between exporters and importers. The price linkage in this study assumes no tariff and all trading and transportation costs are expressed in dollars per metric ton. No conversion to internal country prices is made. Exchange rates are set at 1. Thus, the difference in export and import price for a given two country trade is the transportation costs between the two countries. The GTP program links importing and exporting countries by choosing the least cost supplier for each importing country. However, trade flow constraints can override the least cost choice; but once the constraints are met, the least cost criteria is followed to distribute remaining grain supplies (Holland, 1984). The excess supply and demand schedules are included in the GTP program in the following functional form:

$$\text{Quantity} = \text{Alpha} + \text{Beta} * (\text{Price})^{\text{Gamma}}$$

This expression allows scalar, linear, or nonlinear schedules.

The Vector Sandwich Method (VSM) algorithm is used in the GTP program to find a competitive, spatial price equilibrium (Holland, 1984). The VSM algorithm is one of a class of algorithms referred to as a fixed point, equilibria, homotopy, or path-following algorithms. All in this class are path-follow-

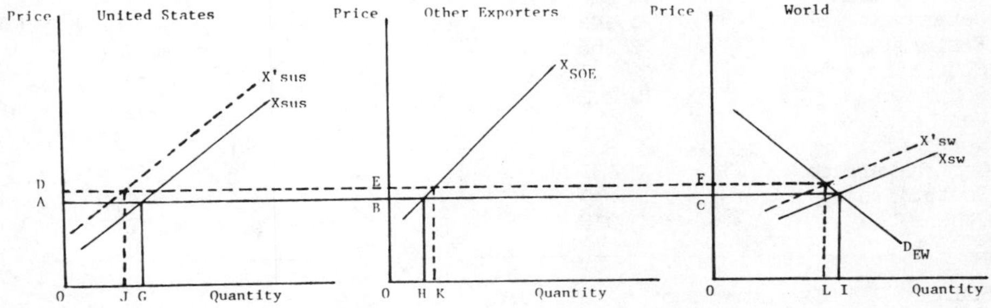

Figure 1. Cargo Preference imposed on the United States.

ing in the sense that given an initial point they follow a path to an equilibrium point (Holland, 1985). The VSM algorithm originally developed by Kuhn and Mackinnon (1975) to solve general equilibrium problems was applied to the generalized transportation problem by MacKinnon (1975).[1] Holland and Sharples (1984) applied the VSM algorithm to world wheat trade. We modified their world wheat trade model to reflect current conditions and cargo preference requirements..

V. PROCEDURE

The first step in the estimation process was to construct a partial equilibruim wheat trade model incorporating excess supply and demand equations for a base period. To this we added a matrix of wheat shipping costs and a set of contraints representing bilateral trade agreements existing in the base period. The final step was to vary the assumptions on cargo preference (transportation cost) in order to obtain alternative equilibrium (Holland and Sharples, 1984).

The model was specified to include 9 exporting countries/regions and 17 importing country/regions. Excess supply and demand elasticities for the world exporting and importing country/regions are given in Table 1.[2] Excess supply functions based on these elasticities and defined to pass through the base period price-quantity were derived for each country/region. Each of the 17 importing regions in the model has an excess demand function defined to pass through the base period price-quantity.

United States excess supply was separated into three groups: U.S. commercial, U.S. PL 480, and U.S. GSM. Excess supply elasticities for commercial and GSM sectors were assumed to be identical. The GSM program is primarily an export finance program. Thus, U.S. wheat exporters would be in-

TABLE 1

International Wheat Exporters and Importers and Their Assumed Excess Supply/Demand Elasticities, 1985/86-1986/87 Average

Exporters	Net Exports/Imports (a)	Excess Supply/Demand Elasticities (b)
	---------- 1000 Metric Tons ----------	
Net Exporters:		
Canada	17,690	.50
U.S. Commercial	16,271	1.92
Argentina	5,361	.10
EEC-10	15,250	1.00
O.W. Europe	1,302	1.00
Australia	5,179	.25
U.S. PL 480	4,410	1.00
U.S. GSM	6,569	1.92
Net Importers:		
Central America	1,475	-0.2
Brazil	1,750	-0.2
O.S. America	3,630	-0.4
EEC-10	2,575	0.0
O.W. Europe	772	0.0
East Europe	3,425	0.0
Soviet Union	15,850	0.0
Rep. of China	6,800	-0.8
Japan	5,445	0.0
East Asia	7,253	-0.4
Southeast Asia	600	-0.6
South Asia	3,216	-0.8
West Asia	8,705	0.0
North Africa	12,959	0.0
Central Africa	1,800	-0.8
South Africa	470	0.0
Others	10,026	0.0

[a] Source: U.S. Department of Agriculture, a,b, and c.

[b] Source: Holland and Sharples, 1984.

TABLE 2
Estimated Base Period Transportation Costs[a]

Importer	Exporter								
	Canada	United States	Argen-tino	EEC-10	West Europe	Aus-tralia	Other	U.S. PL 480[b]	U.S. GSM
					$ /MT				
Central America	15.62	13.14	20.31	16.18	20.23	22.25	16.00	19.97	13.14
Brazil	16.99	13.20	12.14	16.18	20.23	22.25	20.00	30.36	13.20
Other S. America	18.61	12.86	16.59	17.56	20.23	20.88	19.00	28.41	12.86
EEC-10	10.68	11.73	20.47	–	9.71	26.87	10.00	26.98	11.73
West Europe	12.46	16.10	20.23	9.71	–	32.04	10.00	37.03	16.10
East Europe	12.95	15.15	20.72	9.95	10.11	22.66	8.00	34.84	15.15
Soviet Union	13.19	16.02	20.31	12.14	12.14	15.94	8.00	36.85	16.02
Peoples Rep. China	22.82	23.25	28.48	24.28	28.32	18.77	25.00	53.48	23.25
Japan	16.43	12.25	21.12	21.69	24.28	14.73	20.00	28.18	12.25
East Asia	21.52	15.29	27.51	24.28	28.32	18.77	25.00	34.47	15.29
Southeast Asia	31.40	26.70	22.25	28.40	32.37	17.15	28.00	61.41	26.70
South Asia	29.94	27.12	21.69	31.56	32.37	22.66	32.00	63.96	27.12
West Asia	19.02	11.19	25.73	14.65	16.18	28.56	18.00	22.82	11.19
North Africa	21.44	21.35	21.85	11.81	12.14	26.14	16.00	50.63	21.35
Central Africa	25.25	31.12	23.06	25.57	24.28	26.62	26.00	76.03	31.12
South Africa	28.65	15.00	25.97	28.32	28.32	20.88	30.00	57.61	15.00
Others	20.00	22.00	23.00	19.00	19.00	25.00	24.00	50.60	22.00

[a] Compiled from unpublished data furnished by Sharples.

[b] Average (blended) transportation cost with 65 percent cargo preference requirement.

different as to whether to make commercial or GSM sales. However, the PL 480 surplus supply elasticity is different. Each fiscal year, the U.S. government allocates limited funds for PL 480 wheat exports. If the price of wheat varies, the fund remains unchanged, indicating an opposite proportional variation in quantity.[3] Thus, the PL 480 excess supply elasticity was assumed at 1.0 (or unity). PL 480 freight costs were assumed to be paid from U.S. funds allocated for shipments. Interactions of the model's excess supply and demand in the world impacted slightly on the PL 480 sector, i.e. total cost for PL 480 (grain plus transportation costs) varies slightly across simulations.

Transportation costs between exporters and importers were furnished by Sharples (1986) (Table 2). However, transportation costs associated with shipping on American flag vessels are higher than on foreign flag vessels. Reported 1982-1985 costs of shipping on foreign and U.S. flag vessels between U.S. ports and the 17 importing regions were analyzed to estimate the ratios of U.S. to foreign flag vessel cost by importing region. Blended transportation costs between U.S. and foreign ports were then estimated for each assumed cargo preference level (Table 3). The weighted average ratio of U.S. to foreign shipping coasts was used to estimate costs for missing data. We assumed foreign flag vessels would not change rates as the level of U.S. cargo preference shifts.

VI. CARGO PREFERENCE ASSUMPTIONS

Recent legislation stipulates that increasing increments of PL 480 sales (>50 percent) be shipped on American vessels. This level rises to a maximum of 75 percent in 1988. The Department of Transportation (DOT) is required to finance the increased freight charges incurred in any fiscal year resulting from the increased preference requirement. If funding is not budgeted and the DOT cannot fund the increased costs, a snapback provision will take effect. That provision requires that the pre-1986 cargo preference level, 50 percent, become applicable. Exports under the CCC Export Guarantee Program (GSM) would not fall under this requirement. However, in the event the courts or Congress includes the GSM program under the cargo preference requirement, importers would have to pay the additional cost of shipping half of their purchase on U.S. vessels—a requirement that may offset the benefits of the GSM program, rendering it inoperable. Because there are several possible outcomes resulting from recent legislation, this study focuses on six different cargo preference scenarios:

1) 50 percent cargo preference requirement on PL 480 sales.
2) 65 percent cargo preference requirement on PL 480 sales.
3) 75 percent cargo preference requirement on PL 480 sales.
4) 50 percent cargo preference requirement on PL 480 and GSM sales.
5) 50 percent cargo preference requirement on PL 480 sales, GSM program inoperable, and no assumed change in importers demand.
6) Same as (5) but importer's demand is assumed to decline by average GSM sale levels.

These six cargo preference arrangements were programmed assuming U.S. and world wheat trade levels consistent with 1985/86-1986/87 excess supply elasticities.

TABLE 3

Estimated Transportation Rate from U.S. Ports to World Markets.

Country/Region	Foreign Vessels (1)	American Vessels (2)	Average Transportation Costs 50% cargo preference	65% cargo preference	75% cargo preference	1982/85 Ratio American to Foreign Rate
		$/Metric Ton				
Central America	13.14	23.65	18 40	19.90	21.02	1.80
Brazil	13.20	39.60	26 40	30.36	33.00	3.00 a
Other South America	12.86	36.78	24 82	28.41	30.80	2.86
EEC Importers	11.73	35.19	23 46	26.98	29.32	3.00 a
Other W. Europe Importers	16.10	48.30	32 20	37.03	40.25	3.00 a
Eastern Europe	15.15	45.45	30 30	34.84	37.83	3.00 a
Soviet Union	16.02	48 06	32 04	36.85	40.05	3.00 a
China	23.25	69.75	46 50	53.48	58.12	3.00 a
Japan	12.25	36.75	24 50	28.18	30.62	3.00 a
East Asia	15.29	44.80	30 04	34.47	37.42	2.93
South East Asia	26.70	80.10	53 40	61.41	66.75	3.00 a
South Asia	27.12	83.80	55 46	63.96	69.63	3.09
West Asia	11.19	29.09	20.14	22.82	24.62	2.60
North Africa	21.35	66.40	43.80	50.63	55.14	3.11
Central Africa	31.12	100.21	65 66	76.03	82.94	3.22
South Africa	15.00	80.55	47 78	57.61	64.16	5.37
Others	22.00	66 00	44 00	50.60	55.00	3.00 a

a No American vessel rates reported. Ratio based on weighted average of rates reported for both foreign and American vessels.

VII. RESULTS

The results involve two main sections—one which includes outcomes where the level of cargo preference is varied on PL 480 exports only and the other where the GSM program is part of the cargo preference requirement. Each main section includes subsections. Because recent legislation requires increasing increments of PL 480 sales to be shipped on American vessels, the assumptions which involve a 50, 65, and 75 percent cargo preference requirement are contrasted. This comparison is included in one subsection. Concern has been expressed by some that the GSM program may eventually be included under the cargo preference program. An evaluation of this event with three assumptions on GSM importer responses is presented.

A. Cargo Preference on PL 480 Sales

PL 480 wheat exports averaged 4.4 million metric tons during 1985/86-1986/87. Prior to implementation of the 1985 farm bill 50 percent of these quantities were shipped on American flag vessels. This proportion increases incrementally through time to 75 percent by 1988. The 50 percent level of cargo preference is used as a basis for comparison with the other scenarios.

50 Percent Cargo Preference:

Including the 50 percent cargo preference requirement on PL 480 exports increases revenues to the U.S. maritime industry $94 million above the cost of shipping the wheat on foreign flag vessels. The model estimates world wheat trade at 85,569,000 metric tons and a CIF price of $137.17 per metric ton (Table 4). An estimated 4,669,000 metric tons would be exported under PL 480 while other U.S. and world exports total 80,900,000 metric tons. With these assumptions U.S. wheat export value totals $3,121 million. It would cost $517 million to transport U.S. wheat to importing countries (Table 4). The value of U.S. PL 480 wheat exports and shipping costs equal $721 million.

50 vs. 65 Percent Cargo Preference:

Expanding the cargo preference requirement to 65 percent on U.S. PL 480 exports increases revenues to the U.S. maritime industry $121 million over the charges associated with shipping the wheat on foreign flag vessels, a 29 percent increase over the 50 percent cargo preference requirement. This increase in cargo preference lowers total world exports to 85,558,000 metric tons and raises average world CIF price to $137.56 per metric ton. Non-U.S. wheat exports increase 21,000 metric tons. The decline in PL 480 exports, 101,000 metric tons, more than offsets the increase in U.S. commercial (47,000 metric ton) and GSM (22,000 metric ton) exports. The shift in U.S. export prices was not enough to offset the overall drop in U.S. exports, accordingly, the value of U.S. wheat exports declined $11 million. Government expenditures on PL 480 rise slightly with the increase in cargo preference.

50 vs. 75 Percent Cargo Preference:

Increasing the cargo preference requirement from 50 to 75 percent on PL 480 exports lowers total U.S. exports 38,000 metric tons but increases transportation cost of these exports from $190 to $234 million. The U.S. maritime industry receives $139 million above the charges associated with shipping the wheat at competitive foreign flag rates. Commercial and GSM exports increase, but not enough to offset reductions in PL 480 exports. The added cost of American vessels lower PL 480 exports. Value of U.S. wheat exports drop $12 million. Transportation costs for all U.S. wheat exports increase from $517 million to $562 million. Government program costs increase $20 million. Exports from other countries increase 25,000 metric tons, absorbing most of the

TABLE 4
Impact of Alternative Cargo Preference Policies.

Item	Cargo Preference on PL 480 at 50%	Cargo Preference on PL 480 at 65%	Cargo Preference on PL 480 at 75%	Cargo Preference on PL 480 and GSM, GSM Inoperable: Demand Unchanged	Cargo Preference on PL 480 and GSM, GSM Inoperable: Demand Lower	GSM Operable
	----------1,000 Metric Ton----------					
EXPORTS:						
United States						
Commercial	14,446	14,493	14,496	21,105	17,203	14,572
PL 480	4,669	4,568	4,558	4,669	4,415	4,692
GSM	6,659	6,681	6,682	-	-	6,414
Total	25,774	25,742	25,736	25,774	21,618	25,678
O. Exporters	59,795	59,816	59,820	59,795	58,820	59,855
World	85,569	85,558	85,556	85,569	80,438	85,533
	----------Dollars/Metric Ton----------					
PRICE:						
United States						
Commercial	122.69	122.90	122.91	122.69	110.30	123.95
PL 480	113.74	111.27	111.04	113.74	106.98	114.30
GSM	122.69	122.90	122.91	-	-	117.43
World FOB	119.64	119.71	119.71	119.64	110.24	119.75
World CIF	137.17	137.56	137.78	137.17	127.76	138.54
	----------Million Dollars----------					
VALUE OF WHEAT:						
United States						
Commercial	1,773	1,781	1,781	2,590	1,897	1,796
PL 480	531	508	506	531	472	536
GSM	817	821	822	-	-	753
Total	3,121	3,110	3,109	3,121	2,369	3,085
O. Exporters	7,116	7,132	7,133	7,116	6,498	7,157
World FOB	10,237	10,242	10,242	10,237	8,867	10,242
World CIF	11,738	11,179	11,738	11,738	10,272	11,850
	----------Million Dollars----------					
TRANSPORTATION COSTS:						
United States						
Commercial	215	215	216	327	253	217
PL 480	190	216	234	190	185	191
GSM	112	112	112	-	-	216
Total	517	543	562	517	438	623
World	1,501	1,527	1,546	1,501	1,409	1,608
SUBSIDY TO U.S. MARITIME INDUSTRY:						
PL 480	94	121	139	94	92	94
GSM	0	0	0	-	-	107
Total	94	121	139	94	92	201

loss in U.S. exports. FOB world price increases slightly yielding a modest rise in quantities supplied by other exporting countries.

B. Cargo Preference on PL 480 and GSM Programs

There is concern that GSM exports may eventually be included under the cargo preference requirement. Because of uncertainties regarding the effects of expanding the cargo preference requirement to include the GSM program, three different scenarios are examined.

50 Percent Cargo Preference on GSM:

If 50 percent of GSM sales had to be shipped on American flag vessels, the U.S. maritime industry would receive payments which exceed the charges foreign flag vessels by about $201 million—$94 million from PL 480 and $107 million from GSM exports. The additional shipping cost of PL 480 exports would be paid by the importers using the program. U.S. wheat exports total 25,678,000 metric tons and are valued at $3,085 million (Table 4). World exports were estimated at 85,533,000 metric tons with an export value of $10,242 million. Government program cost plus value of PL 480 sales and transportation cost equals $741 million. Should the advantage of the GSM program to importers be less than the added cost of the assumed 50 percent cargo preference requirement, then importers probably would not use the GSM program, rendering it inoperable. If this happens, the question of whether these importers could find credit for their purchases is relevant. Two scenarios concerning an inoperable GSM program and resulting import demand are contrasted below with an operable GSM program.

Inoperable GSM, Demand Unchanged:

If the 50 percent cargo preference level is required on all governmentaided exports and the GSM program is rendered inoperable due to its added cost, the U.S. maritime industry would lose returns compared to an operable GSM program regardless of what happens to the demand of the former users of the GSM program. However, the U.S. wheat industry gains provided the demand for GSM exports can be shifted to effective commercial exports. Total U.S. wheat exports increase 96,000 metric tons with an increased value of $36 million. Government program costs decline $6 million, $727 to $721 million. Total value of world wheat exports decline $5 million. Importers benefit with reduced cost ($107 million) and total CIF wheat value declines from $11,850 million to $11,738 million.

Inoperable GSM, Demand Lowered:

An inoperable GSM program under the cargo preference requirement could trigger a lower world demand for wheat if importing countries that use the GSM program cannot find another credit source. U.S. wheat exports drop 16 percent as compared to the scenario with a 50 percent cargo preference on PL 480 and GSM sales. However, exports from non-U.S. exporters decline only 2 percent. Most of the decline in world exports would be borne by the United States wheat industry through a decline in demand and lower wheat prices. Thus, the sharp decline in U.S. wheat exports coupled with a decline in prices lowers the value of U.S. wheat exports 23 percent, $3,085,000 to $2,369,000. Subsidy to the U.S. maritime industry would decline from $201 million to $92 million. Thus, if the GSM program becomes inoperable and countries historically participating in this program are unable to secure alternative funding sources, the U.S. wheat and maritime industries are both dramatically impacted. Including the GSM program under the cargo preference requirement is to the advantage of the U.S. maritime industry provided the GSM program remains viable. If the GSM program is not viable because of the added cost of shipping on American vessels, then the U.S. maritime industry subsidy falls to a much lower level. The impact of an inoperable GSM program on the U.S. wheat industry depends upon what happens to demand for wheat imports in the GSM importing nations. Should these countries not be able to purchase wheat, the U.S. wheat industry and the world wheat exporters face a sharp decline in exports, price, and export value. The U.S. would absorb a larger portion of these declines than non-U.S. wheat exporters.

VIII. CONCLUSIONS

Cargo preference, the requirement that a portion of a country's waterborne commerce be carried in its own vessels, can affect a country's level of exports, export prices, and revenue from exports if that country's vessel costs are high. Shipping costs on U.S. vessels are 2.5 to 3 times of many other countries, accordingly U.S. wheat exports and wheat producers are unfavorably impacted, as are wheat importing nations. However, other exporting countries benefit when the United States imposes cargo preference requirements.

Under the 1985 farm bill, the Secretary of Transportation is required to finance increased ocean freight charges above the 50 percent cargo preference level. If funding is not obtained, the 1985 farm bill requires the original 50 percent requirement of section 901(b) of the Merchant Marine Act of 1936 to resume effect. This provision reduces the proportion of grain moving under cargo preference, but opens the door for possibly other programs coming under cargo preference requirements. Revenues to the U.S. maritime industry are maximized when a portion of all government-aided exports are required to move on U.S. vessels provided these program remain viable. If the GSM program is included under cargo preference policy, then U.S. wheat exports, export prices and total value of wheat exports would decline compared to that situation where only PL 480 sales are included in the cargo preference requirement.

If the GSM program is included under cargo preference requirements, but becomes inoperable, and import demand is unchanged, then U.S. exports and export value increases, but revenue to the U.S. maritime industry is reduced. However, should users of the GSM program not be able to obtain credit for their imports (i.e., effectively lowering their excess demand) the value of world wheat trade declines sharply. The U.S. wheat industry would bear a major portion of this negative impact.

ENDNOTES

* Department of Agricultural Economics, Texas A & M University, College Station, TX.

1. More details on the VSM algorithm can be obtained from reports by Holland and Sharples; Holland; Kuhn and MacKinnon; Zangwill and Garcia.
2. Elasticity of excess supply expresses the percentage change in quantity supplied for export in response to a one percent change in export price, other factors held constant. In the short run it may be estimated as follows:

$$Ex = \frac{(Qc)(Ec) + (Qd)(Ed)}{-Qx}$$

Where Ex = elasticity of excess supply
Ec = carryover or ending stacks
Qd = quantity of domestic use
Ed = elasticity of demand for domestic use
Qx = quantity exported

3. The Government, through a political process, can change the level of funds available for PL 480 wheat exports. This would impact on the model solutions. However, we assumed the allocated funds would not change appreciably for this shortrun comparative analysis.

REFERENCES

Castillo-Manuel, P. 1980. *A Spatial Equilibrium Analysis of the Impact of Cargo Preference in the World and U.S. Grain Trade,* unpublished Ph.D. dissertation, University of Idaho.

Holland, Forest D. 1985. *GTP: A Microcomputer Program for the Spatial Equilibrium Problem,* Economic Research Service, U.S. Department of Agriculture, Staff Report AGES 850514.

Holland, Forrest D., and Jerry A. Sharples. 1984. *World Wheat Trade: Implications for U.S. Exports,* Department of Agricultural Economics, Purdue University, Staff paper No. 84-20.

Kuhn, H.W., and J.G. MacKinnon. 1975. "Sandwich Method for Finding Fixed Points," *Journal of Optimization Theory and Applications,* 11:189-204.

Mackinnon, J.G. 1975. "An Algorithm for the Generalized Transportation Problem," *Regional Science and Urban Economics,* 5:445-464.

Sharples, Jerry A. 1986. Unpublished spatial price equilibrium model.

Paarlberg, Philip L. 1984. *Cargo Preference and U.S. Wheat Exports,* Economic Research Service, U.S. Department of Agriculture, Staff Report No. AGES 840216.

U.S. Department of Agriculture, a. *Export Markets for U.S. Grain and Feed Commodities,* Foreign Agricultural Service, (selected issues).

U.S. Department of Agriculture, b. *Wheat Outlook and Situation Report* Economic Research Service, (selected issues).

Vanags, A.H. 1977. "Flag Discrimination: An Economic Analysis," *Advances in Maritime Economics,* edited by R.O. Goss, Cambridge University Press, Cambridge, England.

Zangwill, W.I. and C.B. Garcia. 1981. *Pathways to Solutions, Fixed Points, and Equilibrium,* Englewood Cliffs, New Jersey: Prentice-Hall.

A Statewide Network Model for Rationalizing the Grain and Livestock Transportation System in Rural Nebraska

*by Azzeddine Azzam and Dean Linsenmeyer**

ABSTRACT

The purpose of this paper is to report on a statewide transportation model. The model, a multicommodity, multimodal network is designed to identify: a) current multicommodity flows by highways, rural roads and rail lines, b) anticipated impacts on commodity flows over specific rural roads and highways of rail line abandonment, bridge/road upgrading or closure, shifts in the agricultural production base, structural changes in rural commodity markets, and changes in transportation costs. The model also calculates public and private-user costs to aid planners in assessing the costs and benefits of alternative network configurations.

The network model is programmed in Pascal and can be used on an IBM-PC. It is user friendly and allows county engineers to experiment graphically with alternative investment/disinvestment scenarios.

I. INTRODUCTION

To assess the impact of decisions by the private rail and motor carriers, the rural roads administration and the private rural shipper/receiver, the Nebraska Department of Roads (NDR) needed a statewide, multicommodity, multimodal network model. The commodities of interest to NDR were corn, sorghum, soybeans, wheat, fat cattle, feeder cattle, fertilizer and bulk petroleum. The information needed was private user costs by road type, road quality, and truck type; public costs incurred from accomodating the total flow of the above commodities; and a mechanism to calculate the costs under current and alternative road, bridge and rail configurations. The purpose of this paper to report on the procedures used in developing this model and will be orgainized in five sections: (1) conceptual formulation of the model, (2) empirical specification, (3) description of the network algorithm, (4) a policy example and sample output, and (5) summary and conclusions.

II. CONCEPTUAL FORMULATION

To formulate the problem, denote the state area by A, and the area of its counties by a_i, for i = 1,2,...N. Subdivide the a_i's further into j components say square miles, such that $A = \epsilon_i \epsilon_j a_{ij}$, and let P_{ij} be the total production in the square mile a_{ij}.

For a given county i, consider each a_j as a source. Associated with each source is a net nonnegative P_j to be shipped to potential sinks (e.g. elevators). If the objective of the shipper is to maximize revenue received at a_j associated with the flow of P_j to potential sinks S_k, for k = 1,2,...,K, then there exists a least cost path L^*_j linking the source a_j and a sink S* S_k. The minimum total transportation cost per unit C* for goods P_j can be written as:

$$C^* = c \sum_{r=1}^{S} t_r \qquad (1)$$

where t_r denotes the distance in miles of each branch and S the number of branches on the least cost path L^*_j, and c is a constant unit cost per volume/mile assuming a given vehicle type and a given quality factor for all S branches. If the branches are of unequal quality, as is usually the case, then associated with each t_r and vehicle type v is a unique cost, c_{vr} and C* becomes

$$C^* = \sum_{r=1}^{S} c_{vr} t_r \quad \text{for } v = 1,2,\ldots,V \qquad (2)$$

C* is the private shipper cost of moving one unit of farm produce from a source to a sink, assuming a specific vehicle type, v. For M shippers, the total private costs (TPV) are simply

$$TPV = \sum_{v=1}^{V} \sum_{j=1}^{M} \sum_{r=1}^{S} C_{vr} t_r P_j \qquad (3)$$

This is illustrated in Figure 1. The diagram represents a hypothetical road network in a hypothetical county. Consider the large square as a county in the state and the numbered nodes in the center of each square as sources of the flows P_j in the square with area a_j. Assume all roads connecting the squares are unpaved rural roads. If S* is the sink associated with a least cost path by moving west from each source, then south on AS*, then L^*_1, can be written as:

$$L^*_1 = \{2,1,3,5\}$$

with j=4 links. The associated cost is:

$$C^*_{v1} = C_{v1}^{2} + C_{v2}^{1} + C_{v3}^{3} + C_{v4}^{5}$$

Where the superscripts denote the location of a particular segment on L^*_1 and the subscripts denote the number of the sequential segments on the diagram followed by vehicle type v.

Denote the capacity of truck v by Q_v, then the proportion P_j hauled by truck type v is written as

$$H_{vj} = \frac{R_{vj}(CAP_v)}{\Sigma_{v=1}^{V} R_v(CAP_v)} \qquad (4)$$

Figure 1. Hypothetical county with six sources and one sink.

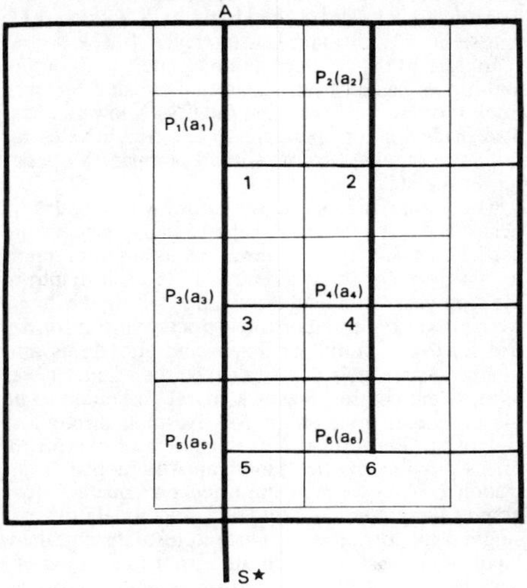

where R_{vj} is the probability of truck type v hauling P_j to market. The actual volume hauled by truck type v is

$$T_{vj} = P_j * H_{vj} \tag{5}$$

and the number of trips by truck type v to haul T_{vj} to market is

$$N_{vj} = \frac{T_{vj}}{CAP_v} \tag{6}$$

with appropriate conversion of T_{vj} to 18 kip equivalents[1] (KIP_{vj}), the total number of kips applied to each road segment r on the least cost path L^*_j is

$$F_{jr} = \sum_{v=1}^{V} (N_{vjr})\ (KIP_{vjr}) \tag{7}$$

$$\text{for } r = 1,2,\ldots,S$$

Since a particular road segment r may also be part of other least cost paths L^*_j for $j=2,\ldots,M$, the aggregate number of kips from the M sources is

$$F_r = \sum_{j=1}^{M} \sum_{v=1}^{V} (N_{vjr}) \tag{8}$$

$$\text{for } r = 1,2,\ldots,S.$$

If the public cost per road segment PC_r are functionally related to F_r, then the total public cost for the region is simply to the total of PC_r over the existing r's segments which constitute all possible least cost path from the M sources.

The link between traffic dynamics on rural roads and rail lines is formulated as follows. Consider Figure 2 where a grid system is imposed on a hypothetical county. The lighter squares again represent the sources (P_j) of the flow and the darker lines

Figure 2. Hypothetical county with rail line, country and state roads

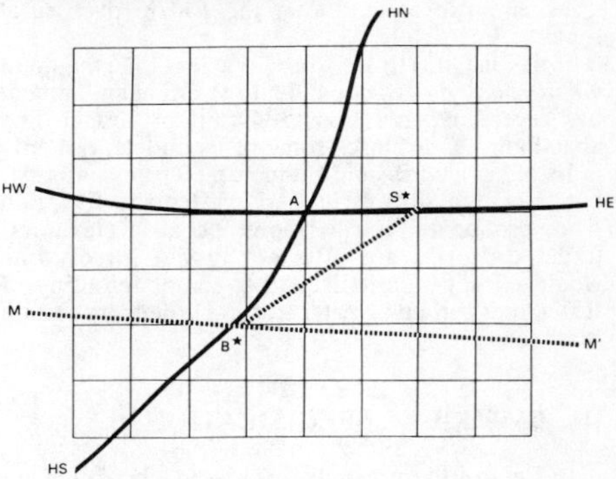

represent rural roads. Superimposed on the grid are two state highways, HN-HS and HW-HE, a main rail line, MM′ and a branch line, S*B*. Assume S* is a country elevator and B* is a subterminal elevator. S* may ship grain to B* and/or to an external market. The flow from the sources surrounding the two elevators may go to either elevator. The size of each elevator's trade territory hinges on its relative bid price and road accessibility. S* has the option of shipping to B* either through the rail link, S*B*, or along the road link, S*AB*, depending on the relative transport costs.

Suppose the load out capacity constraint at S* and the quality of branch line S*B* are such that the rail company owning S*B* and MM′ decides to close the branch line. The impact on roads hinges on the interaction of many variables.

First, using the traditional model of rail/truck competition (Figure 3), if the intersection of the rate schedules of the two modes happen to coincide with or exceeds the distance from S* to B*, then for distances less than S*B* trucking rates per cwt. are consistently lower than for rail rates. Therefore, the only effect of rail abandonment on roads would be the flow diverted to the state highway moving west

Figure 3. Model for rail truck competition

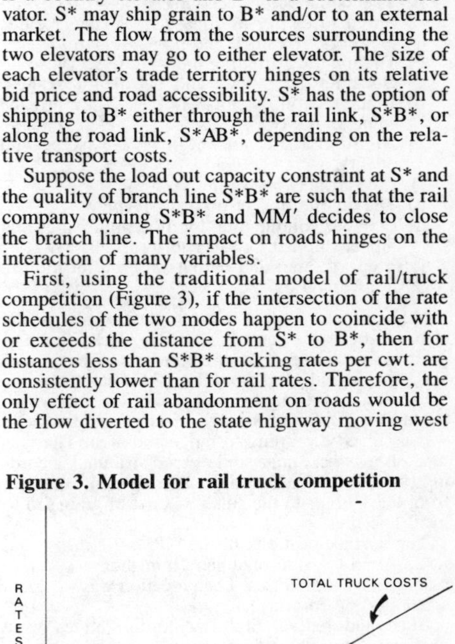

on HW-HE and south on AB*. Hence, rail abandonment will have no effect on trade territory boundaries and subsequently on the traffic over rural roads.

If the distance to B* from S* is less than the point where the two rates coincide, then rail abandonment has several repercussions. First, if by branch line abandonment, S* loses some of its outlets, not only is its bid price affected by increased transport cost, but also by a shift in its derived demand for grain from producers. Furthermore, because elevators' trade territories are affected by bid prices, one would expect a shift in market share, rerouting of flows, and perhaps greater traffic burdens on certain rural roads.

III. EMPIRICAL SPECIFICATION

In light of the previous discussion, the following items need to be identified:

a) The commodity producing units (CPU)s and their respective shipment volumes (P_j)s which include corn, sorghum, wheat, soybeans, fat and feeder cattle.
b) The commodity receiving units (CRU)s referred to earlier as sinks (S_k)s. They include elevators, livestock and hog sale barns.
c) The input distributing units (IDU)s. These include fertilizer, feed and petroleum dealers.
d) Input receiving units (IRU)s. They are the same as the (CPU)s.
e) All rural roads and highways by type and quality.
f) The rail network in the state.
g) The distribution of trucks by type.
h) Cost per volume mile for all commodities and inputs shipped by rail and by truck.
i) Prices of input and outputs at each location in items b and c.

With respect to item a, each square mile in the state has been designated as a CRU. The volume of grain assigned to each square mile was based on area specific soil and yield conditions.[2] Fat cattle volume was based on the number and size of feedlots which location is public knowledge. Feeder cattle volume was assigned according to rangeland area. The volume of hogs per square mile was distributed according to feed grain availability. All shipments of grain from the sources to the sinks was net of grain fed to local livestock.

The spatial locations of the CRUs (item b), and IDUs (item c) were obtained from their respective association's directories. Each location was assigned a unique xy coordinate.

State and federal highway configurations were obtained from the official state map and were arranged in separate arcs connecting various nodes (towns). Each node was assigned an xy coordinate. Rural roads, on the other hand, were entered as an empty grid system. This way, the county engineer can alter the configuration and quality of the roads under his jurisdiction.

The rail network (item f) was also arranged as arcs linking various nodes (connection points such as towns). Each node was assigned an xy coordinate. Each arc was assigned its actual distance in miles. As for item g, the probability of using a particular type to haul grain and livestock was estimated using multivariate logit analysis of farm survey data. The principal covariates used in the logit model are distance to market and grain or livestock shipment size. The trucks considered are single axle, tandem axle and semis (Azzam and Linsenmeyer, 1987).

In item h, the cost per volume mile for all output and inputs hauled by truck were estimated for each truck type by road type and quality. Allowance was also made for turn penalties.[3] The freight rates for rail were obtained from railroad companies operating in the state.

The actual bid prices for corn, wheat and soybeans at each of the 62 subterminal elevators are used.[4] Each subterminal elevator was asked to report its bid price for the last Thursday of each month in the crop year. Those subterminals retailing fertilizers were also asked to report their prices per ton for dry bag fertilizer, liquid fertilizers and anhydrous ammonia during their peak sales season. Omaha was used as the central livestock market assumed to be the dominant price setter for livestock throughout Nebraska. Preliminary investigation of petroleum prices provided by the American Automobile Association (AAA) Service indicated no apparent geographical pattern. Consequently, the model utilizes a single price for the state. Once items a through i are inputted, a least-cost-path algorithm (discussed below) is used in a three step optimization process. The first step transmits prices from the subterminal (wholesale) centers to the country (retail) markets by discounting central market prices by the estimated transportation costs along the least-cost routing. This determines the prices quoted by the local grain elevator, fertilizer or bulk petroleum retailer, or livestock collection station. Step two utilizes that price to estimate the trade area and affiliated volume of business for each local collector (retailer). In doing so, the least-cost-path algorithm estimates the routing used to transport rural production from the farm (source) to the collection centers (sinks). This step also calculates producer's hauling costs by truck type using the logit functions and determines road impacts caused by the freight flow.[5] Step three then attributes the estimated volume transacted by each rural collector (retailer) to the least-cost routing connecting it with its respective terminal (wholesale) center identified in the first step. By eliminating or adding links to the road and rail network, costs and benefits of subsequent optimal routings are calculated and compared to alternative network configurations.

Specifically, the model defines subterminals as those facilities located on rail lines with 50-car unit train loading capacity or greater. Those elevators not located on the rail or with less than 50-car unit train loading capacity comprise the country collectors. The procedure for transmitting grain prices from subterminal centers to country collectors is described below.

Let elevators "A" and "B" in figure 4 consist of 50-car capacity subterminals whose prices were discussed earlier. Let elevator "C" be a country elevator with a rail load-out capacity less than 50-cars. The bid price at "C" is determined by comparing the market options that would yield the largest sales revenue per bushel net of transportation costs. In this illustration four separate calculations would be compared: (a) The price at "A" minus the rail costs over the "AC" rail routing; (b) The price at "A" minus the trucking costs over the "AOC" highway routing;

Figure 4. Illustration of subterminal-county linkage

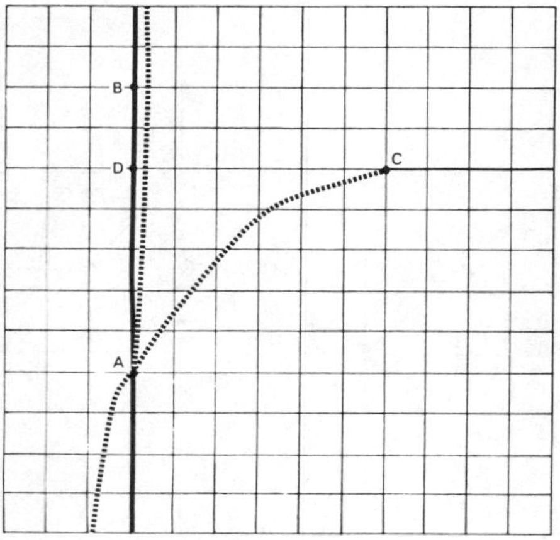

and (c) the price at "B" minus the trucking costs over the "BOC" highway routing. The rail rate is differentiated by the rail shipping capacity at the shipping point. The truck rate is differentiated by the quality of the road network for both on-rail and off-rail country elevators. If option (b) yields a higher net sales revenue at "C" than either option (a) or (c), then the bid price at "C" is set as the net sales revenue per bushel from trucking grain to subterminal "A". After the rural trade territory is established, the tonnage of grain acquired at "C" is then added to the estimated freight flow over the "AOC" road segment with its associated impact on road quality.

In transmitting fertilizer prices, the procedure is the mirror image of that for determining grain prices at country elevators. The same locations, "A" and "B", in Figure 4 are assumed to be the dominant local pricing sources at which fertilizer ingredients are received from distant manufacturers/processors. Fertilizer prices at these subterminal locations are compounded by the rail and/or truck charges for transporting fertilizers from the source to the country outlets. Once again a least-cost algorithm is used, starting at the subterminal locations and sequentially adding the transportation costs for each segment of the optimal path road linkages in "AOC" and "BOC" as well as the rail linkage "AC". The source, "A" or "B", which can deliver the fertilizer ingredients to "C" at the least total delivered cost per ton is the sole supplier of "C" and in effect determines the retail price at "C".

In the case of cattle (fat and feeder cattle) and hogs, prices at each auction barn and/or collection point are determined from the average livestock prices received by farmers at the central market in Omaha. The total delivery cost is computed by multiplying the transportation charge per hundredweight per mile by the distance between Omaha and each respective auction barn/livestock collection point. This delivery cost per hundredweight is subtracted from the Omaha livestock price to establish local prices.

Once prices at the local collector/distributor points are calculated, they become an essential input in determining the outcome of the economic interplay between merchandisers and rural producers.

As the country grain merchandiser attempts to obtain the largest sales revenue net of transportation costs, the rural producer aspires to do the same. The merchandiser discounts the grain bids from numerous subterminals, processors, and so forth, by the transportation cost incurred in accessing each market and then comparing their equivalent "alongside" prices. In like manner, the producer discounts commodity bids from country collectors by his cost of delivery to each in arriving at their respective "farm-gate" prices. The same parallelism applies to producers' economic behavior in the purchase of inputs except that alternative retail prices are compounded by transportation costs while attempting to minimize total delivered costs.

In attempting to minimize the transportation cost, the producer will attempt to choose that road system which (a) has the least distance, (b) has the greatest proportion of better quality roads, and (c) has the least constraint on load capacity, safety hazard. Other factors such as past social-cultural ties to a particular market or the presence of related business firms located at a particular center will also be important in determining the specific market.

IV. ALGORITHM OF COUNTRY MARKET-PRODUCER LINKAGE: DESCRIPTION

A least cost algorithm was built to approximate the interplay described above. A rectangular grid of roads intersecting at one-mile intervals, representing the rural road system was assumed and is illustrated in Figure 5. This assumption can be relaxed by deletion of roads or by specifying diagonal roads. Each potential intersection of two mile-roads was identified as a vertice or node. For example from any one node, "A", a maximum of four roads lead outward in different directions connecting that node

Figure 5. Rectangular grid system: An approximation of diagonal roads

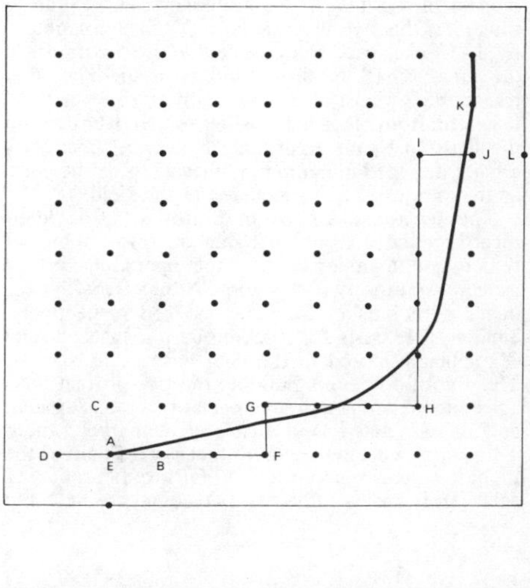

with four other potential nodes, "B", "C", "D" and "E". Data stored describing each of these interconnecting links must be site specific. Two matrices of identifications were used to locate a specific segment being considered: one matrix identifying the east-west roads and one the north-south. Both matrices use the intersection of the 40th parallel and the 6th Principal Meridian as the origin from which all mile segments are identified.

Diagonal roads were approximated by rectangular movements toward that intersection closest to diagonal road. For example, a diagonal section of major highway "AK" would be approximated by the rectangular roads "AFGHIJK". The rectangular approximation of any diagonal section overestimates the actual distance. Therefore each of the separate links, "AF", "FG", etc. In the rectangular approximation, the links are adjusted proportionally so that their sum equals the distance of the diagonal road itself. (Turn penalties, normally assigned to the rectangular corners, are eliminated when it represents a diagonal road.) To reflect the shift of traffic off the diagonal, a turn penalty is imposed for traffic going straight when the approximated diagonal turns. Traffic movements on "IJL" would experience a turn penalty at "J"; while no turn penalty would be imposed on "IJK" traffic. The disadvantage of this method is that the road quality, transportation costs, etc. of the diagonal is superimposed on its rectangular road substitute and the actual conditions of the original rectangular road are lost.

The least-cost path algorithm is run for each truck type and each commodity type since the cost per mile of a specific road type is unique to the truck type and the commodity being transported. Each mile road segment carries the tonnage produced/required by the quarter section (160 acres) bordering on each side of the road. Production from areas "A" and "B" are attributed to the road segment "AB" in Figure 6. In addition any segment may also carry the cumulative tonnage of other roads which feed onto it enroute to market. Road segment "AC" provides transport service for the tonnage originating from "C" and "D" as well as "E" and "F" which is acquired from road segment "CD".

When areas larger than one square mile are treated as an aggregate, imaginary roads with an artificially high cost of transport are introduced to drain the interior of the area to the boundary road. Figure 6 illustrates this by the area "AEHD". The imaginary road "CG" is used to transport the production from the area "GH" to the boundary road. The high transportation cost of these imaginary roads prohibits traffic from other roads being routed over them in the least-cost-path algorithm. However, their costs are not included in computing private or public costs as they are essentially incurred in the field.

The least-cost-algorithm builds what is technically called a depth-first tree traversal. Figure 7 will assist in understanding the operations of the model. Assume two elevators, A and B, with producer corn bids of $2.57 and $2.58 respectively. Tandem-axle costs for trucking corn over each road segment are stored in the data base as illustrated. The algorithm begins building the tree of road linkages one branch at a time connecting a road segment and its associated land area with a market center. This starts with the segment that would deliver the highest net farm gate price (NFGP) to its producing acres. In frame "i", the first two segments are 1 and

Figure 6. Production areas contributing to rural frieght flows

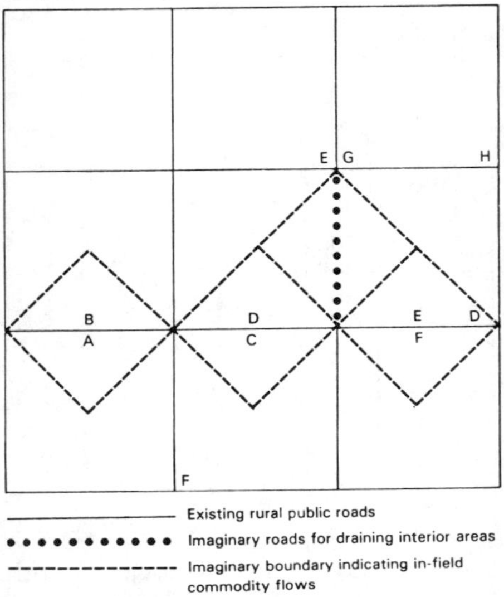

2 since they originate from elevator "B" which has the highest bid price combined with their lowest transportation cost of 1.0 cents per bushel per mile. This yields the highest possible NFGP of $2.57 per bushel for the four quarter sections bordering segments 1 and 2. All other segments from either "A" or "B" would produce lower NFGPs.

In frame "ii", the model now considers all possible linkages radiating from the nodes currently included. After comparing the costs for all of these linkages, the model selects sequentially those segments which result in NFGPs of descending magnitude: Linkage 3, $2.569; Linkage 4, $2.568; and Linkage 5, $2.56. This technique simulates the producers' evaluation of elevator bids and transportation costs which in combination will give him his highest net sales revenue. This is also equivalent to identifying the least-cost routing from each producing area to its respective market center.

The same procedure is repeated in from "iii". When a comparison is made of all remaining linkages spreading out from the included nodes, elevator "A" now commands the highest net farm gate price of $2.559 on segment 6. It appears initially that both segment 6 and segment 7 in frome "iv" produce the same NFGP. However, a turn penalty of $.0005, reflecting the cost of deceleration and acceleration results in a NFGP for segment 7 of $2.5585. Turn penalties are indicated by diagonal lines traversing the relevant corners. The node between segments 6 and 7 begins to identify the trade area boundary between "A" and "B". Segment 8 is added next in the series.

Segments 9, 10, and 11 are added in frame "v". Both segments 9 and 10 have turn penalties and both result in exactly the same NFGP of $2.5575. In the program they would then be added simultaneously. This also applies to links 12, 13, and 14 which share a common NFGP of $2.5555 in frame "vi". The

Assumes:
Price @B = $2.58/bu Price @A = $2.57/bu Turn Penalty = $.0005/bu

Link #	Farm Gate Price	Farm Gate Price	Link #	Farm Gate Price	Farm Gate Price
1	$ 2.57	$	13	$ 2.5555	$
2	2.57		14	2.5555	
3	2.569		15	2.5550	
4	2.568		16	2.5545	
5	2.56		17	2.5545	
6		2.559	18	2.5465	
7	2.5585		19	2.5465	
8		2.558	20		2.5455
9	2.5575		21		2.5455
10	2.5575		22	2.5445	
11		2.557	23		2.5440
12	2.5555		24	2.5425	

Note: Small sized numbers represent cents/10 bushel/mile trucking cost for each respective segment. Large sized numbers represent order in which segment is added to least-cost routing.

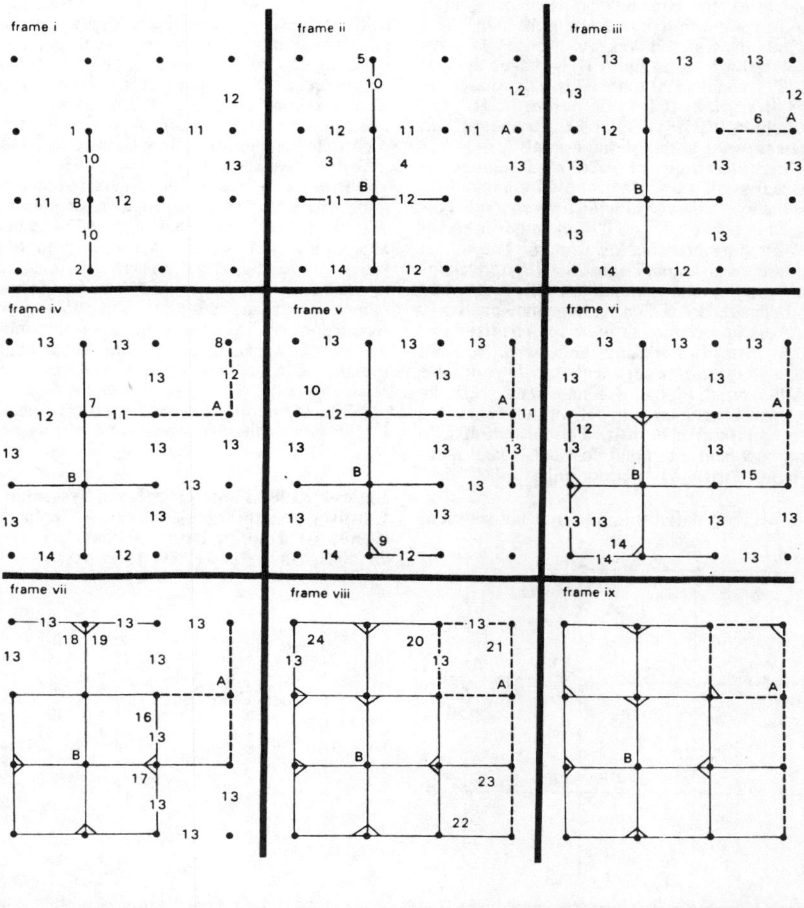

Figure 7. Hypothetical ordering of least-cost algorithm for routing corn in a tandem axle truck

program proceeds through frame "vii" and is completed in frame "viii". The completed tree spanning the entire region is presented in frame "ix". The least-cost path and the resulting trade area in frame "ix" is unique to the specific truck type and commodity type considered. Therefore the algorithm is repeated for each. It is evident that any change in bid price, or road condition would also alter the flow and the final trade area configuration. The NFGPs at a given point in the boundary of the trade territory are approximately equal with respect to either elevator. However the NFGPs differ when comparing different points on the boundary.

Now that the least-cost routing is established, the model begins to accumulate tonnage as illustrated in Figure 8. The algorithm traverses the tree, to the end of a branch first. It then carries that tonnage to the first junction, deposits it, goes to the end of the alternate branch, returns to the same junction, and proceeds until the tonnage from all branches are gathered at the main junction before moving to the next major limb. For example, assume a uniform corn surplus above livestock requirements of 130 bushels per acre over the entire area in figure 8. This is equivalent to approximately 560 tons per quarter section. The algorithm moves to link "DE" and gathers 560 tons which it deposits at node "D". An additional 560 tons from link "DF" makes the toal at node "D" equal 1120 tons. This is increased by 560 tons from link "CD", 560 from "CH" and 1120 from links "CGF" to total 3920 at node"C" to be carried forward to the elevator at "B".

The freight tonnage for each road segment is summed across all truck sizes and all commodities to reveal the total tonnage moving a given road. This is indicated in figure 8 for each of the roads immediately surrounding both market centers. The model converts this total tonnage into l8 KIP equivalents which are used in later sections to estimate the road wear and changes in annual maintenance costs.

The reverse procedure is used for fertilizer and petroleum. Instead of starting with the highest bid price, the algorithm begins with the lowest sales price at the retail Figure 8 centers and adds the trucking costs in an attempt to deliver the input at the minimum net farm gate cost. This also results in input routings being minimal cost to all land areas and trade territories drawn accordingly.

Figure 8. Accumulated tonnage by road segment

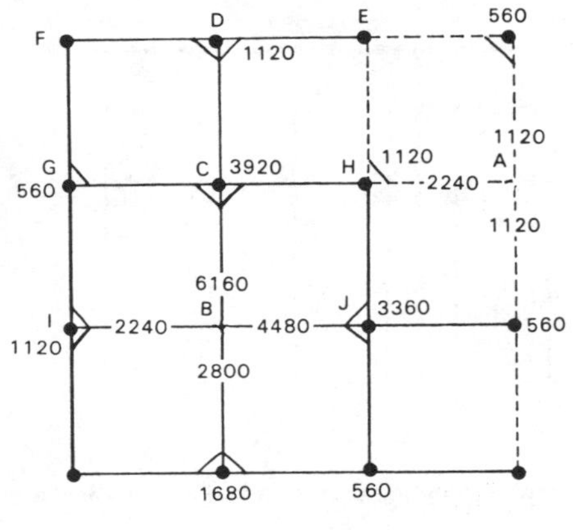

V. POLICY EXAMPLE AND SAMPLE OUTPUT

Saline County, Nebraska was selected as a pilot test of the impact of reclassification of low-volume, public dirt roads to permit private access only. In an attempt to reduce public maintenance costs without incurring significant increases in private transportation costs, this policy may be considered by county road officials.

The base run of RTM with the public dirt roads in their current condition is partially illustrated in Figure 9. This flow of commodities over the rural road network resulted in an estimated private trucking costs of approximately $225,287 annually to transport all fertilizer, bulk petroleum, feedgrains, wheat, soybeans, feeder cattle, fat cattle, and slaughter hogs. In order to provide these transportation services while maintaining the current road qualities, the public sector paid approximately $6,230 in annual maintenance costs. Figure 9 estimates the 1000 18-Kips of freight over each road segment in Saline County in 1984 given the base run conditions.

In the alternative run, all public dirt roads in Saline County were reclassified as restricted access so that they were essentially private access only. All other conditions of the base run, such as truckers not crossing underweight bridges and not considering freight needs in the perimeter outside the county, were held constant. Figure 10 provides the user with the estimated 1000 18-Kips of commodity flow over each road segment in Saline County in 1984 under the alternative scenario.

The impact of reclassification of public dirt roads as private access roads was to decrease public expenditures for road maintenance by $254 annually or approximately 4 percent. While one might expect the public savings to be greater, one must consider that the same tonnage must still be transported over other public roads, the closing to public access has forced some redirection of flows possibly increasing the ton-miles hauled, and the dirt roads chosen for reclassification were already very light-density roads.

While the public sector experienced a net savings by the policy, the private sector saw transportation

Figure 9. 1984 Flows Over Road System in Saline Country Assuming a) a Zero Mile Parameter Zone, b) Trucks Don't Cross Underweight Bridges. *Base Run.* (1000 18 Kips)

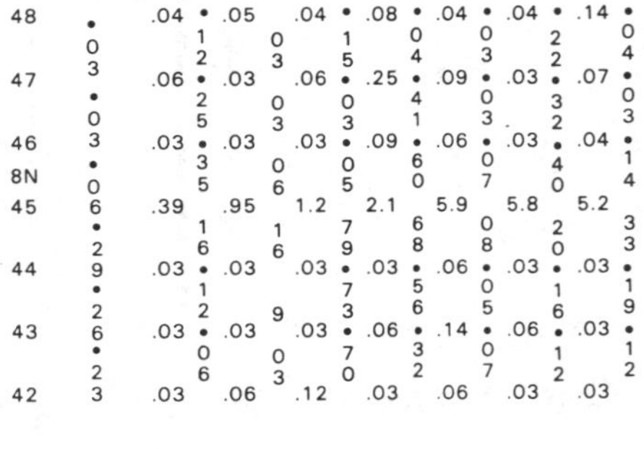

Figure 10. 1984 Flows Over Roads System in Saline County. *Simulation Run* Assuming: a) A Zero Mile Parameter Zone, b) Trucks Don't Cross Underweight Bridges, and c) All Dirt Roads are Private Access Only.

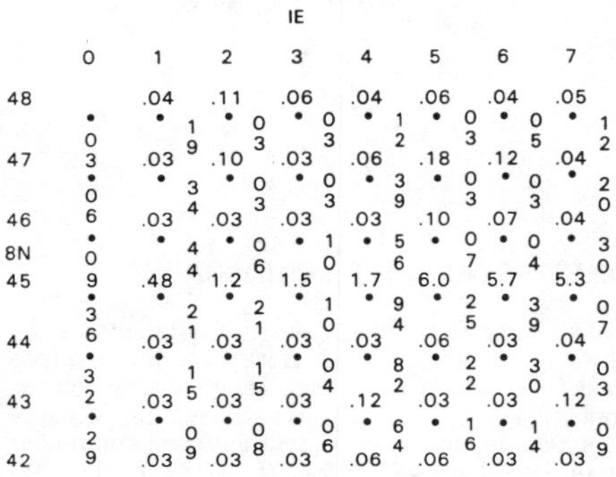

costs rise by $3,706. This change would need to be taken into consideration in evaluating the net gains to the local economy as a whole of reducing the public revenue expenditure.

VI. SUMMARY AND CONCLUSIONS

This paper highlighted a statewide network model developed for the purpose of assisting county and state planners in approximating the private and public costs of rural traffic. The model is updatable and interactive. By displaying the network on the computer screen, the model allows county engineers to a) graphically witness optimal routings and flows, and b) maintain a graphical road inventory which configuration and quality can be altered at the engineer's discretion.

The model is an improvement over existing rural transportation models (Baumel et al., 1973; Trychniewicz and Tosterud, 1975) in three important ways. First, it encompasses the entire state and bordering counties. This allows the user to observe and assess the impact of his/her area-specific network configuration on networks beyond the immediate area. Second, it distributes grain and livestock production according to area-specific agronomic conditions. Since planned expenditures are based on projected flows, public dollars will likely be misallocated if production is distributed equally across the county or state.[6] Third, the model considers several agricultural commodities; all existing roads, rail lines, bridges and their respective classifications. This way, rationalizing the grain and livestock transportation system in rural Nebraska is conducted in a multicommodity multimodal framework.

REFERENCES

Azzam, A.M. and D. Linsenmeyer. "The Impact of Volume of Marketing and Distance to Market on the Choice of Truck Size." *North Central Journal of Agricultural Economics*. 8(1):135-144.

Baumel, P.C., T. P. Drinka, D.R. Lifferth and J. Miller, *An Economic Analysis of Alternative Grain Transportation Systems: A Case Study.* Nat. Tech. Info. Rep. No. 224819. Springfield, Virginia, Nov. 1973.

Trychniewicz, E.W. and R.J. Tosterud. "A Model for Rationalizing the Canadian Grain Transportation and Handling System on a Regional Bisis." *American Journal of Agricultural Economics*. 55(1973):805-13.

Zaniewski, J.P., B.C. Butler and G. Cunningham. "Vehicle Operating Costs, Fuel Consumption, and Pavement Type and Condition Factors." Texas Research Research Development Foundation, FWHA-PL-82-001. 1982.

ENDNOTES

* Department of Agricultural Economics, University of Nebraska, Lincoln, NE.

1. A kip is equivalent to 1,000 pounds.
2. The soil types or Soil Association Names (SAN) per square mile we obtained from the Agricultural Stabilization and Conservation Service (ASCS) maps. Associated with each SAN are 4 qualitative categories: 1) farmland suitability; 2) irrigation suitability; 3) slope; and 4) water erosion potential. Each qualitative category is in turn broken down into 5 categories: high, good, fair, low and very low. The method for using the above information to assign production per square mile is too lengthy to discuss. Interested readers should contact the author(s).
3. Turn penalties are costs of vehicle acceleration and deceleration to turn right or left at a road intersection.
4. Initial data for prices and quantities reflected 1984 conditions. The model is updated for use in subsequent years.
5. In calculating the road impacts, we considered the distribution and total kip equivalents by truck-type and by commodity. The weight was distributed over the front axle, tractor drive, and tractor back wheels.
6. This is especially true for a State like Nebraska where agronomic conditions and practices in the west are radically different from the east.

The Effect of Elevator Competition Upon Market Areas

*by Frank J. Dooley and Wesley W. Wilson**

ABSTRACT

The purpose of this paper is to develop a theoretical model which determines the market boundary of a particular elevator based upon economic conditions. Specifically, a conceptual framework is developed which considers the production in a region, the prices offered by elevators, and assembly and shipping costs. Farmers are assumed to ship grain to the elevator offering the highest net price. Therefore, the boundary for a particular elevator's market area depends upon arbitrage conditions such that farmers are indifferent between shipping grain to various elevators. As a result, the model can be applied to study competition between elevators and the effect of such competition upon elevator bid prices. In particular, the model can be used to analyze policy implications of varying transportation rate structures upon interfirm elevator competition.

I. INTRODUCTION

Market area efficiency, the study of the optimal organization of an industry operating in spatial markets, has been an important and enduring research problem for agricultural economists. In recent years, several market area efficiency studies have been conducted to study the effect of recent changes in transportation technology and rate structures upon the restructuring of the country grain elevator system.[1] This research has primarily dealt with the optimum number, size, and location of marketing facilities. "Although such information may be of general value in formulating both public and private goals, the results are apt to be sterile in the absence of some central planning authority" (French 1977, p. 159). In other words, a major weakness of the market area efficiency models is the inability of the method to address the effects of change at the firm level.

The purpose of this paper is to develop a theoretical model which determines the market boundary of a particular elevator based upon economic conditions. Using arbitrage conditions, the model is applied to determine the market boundary and hence market areas of particular elevators in a region. The specific objectives are: 1) to briefly review the conceptual framework of assembly area models; 2) to develop a theoretical model to analyze assembly areas based upon economic considerations; and 3) to illustrate and identify potential applications of the method as a policy tool to evaluate competition between elevators.

II. CONCEPTUAL FRAMEWORK

Assembly areas have been estimated in most grain elevator market area efficiency studies using a model developed by French (Dooley). French assumed that processing plants operate at the origin of a circular supply region with a uniform, continuous production density (1960, p. 769.) The model states that total quantity supplied is equal to the product of area and production density, or:

(1) $Q = D\pi r^2$

where: Q = quantity of product handled;
D = density of supply per unit of area;
r = the radius of the circular supply area;
π = 3.1416.

Solving for r yields the average distance from the farm to the elevator,

(2) $r = Q^{1/2}(D\pi)^{1/2}$

While appropriate for grain industry market restructuring studies, the French model is not suitable for the firm level analysis of elevator competition. The French model assumes that the entire volume of grain for a supply region goes to one centrally located elevator. However, farmers are generally not bound to a single centrally located elevator within the supply region, but rather have options in their marketing of grain. For example, in seeking to maximize profits, farmers may bypass nearby small local elevators and haul grain directly to multiple-car rail elevators (MCRs) because of higher bid prices at the latter.

In this paper, elevator competition is added by including more than one elevator in the supply region. Thus, profitmaximizing farmers may choose among competing elevators when merchandising their grain. In such a framework, elevator pricing decisions endogenize the market boundary and hence, throughput to the economic model. Since economic variables such as bid prices and assembly costs are reflected in the model, explicit consideration of changing economic and legal conditions in the industry can be analyzed.

In recent years, there have been significant changes in several of the factors affecting elevator bid prices and farmer assembly costs. For example, MCRs have been able to expand their market areas because multiple-car and contract rail rates have lowered shipping costs for MCRs. Market areas also vary as assembly costs change. Farmer assembly costs of moving grain to elevators have fallen because new farm trucks have greater payload capacities and higher fuel efficiencies. Lower assembly costs may allow farmers to haul their grain to more distant elevators.

In the next section, a model is discussed which considers the effect that economic factors such as bid price and assembly costs have upon market areas. This model can be applied to analyze competition between elevators and the effect of such competition upon market areas, elevator margins, and elevator bid prices to farmers.

III. THEORETICAL MODEL

Assume, as with the French model, that the production of agricultural commodities is distributed uniformly over a geographic region. Elevators may not discriminate among farmers with bid prices. Thus, all farmers receive the same price offered by a particular elevator. Farmers are also assumed to ship their grain to the elevator offering the highest bid price net of a constant transport assembly cost per mile. Further assume that the elevator does not have a storage capacity constraint in the sense that they can serve all farmers in the relevant region.

Given these assumptions, the market boundary of an elevator is dictated by the elevator bid price, the distance the farmer ships to the elevator and the farmer's transport cost such that the farmer is indifferent between shipping to elevator i or any other elevator k.[2] The assembly area for an elevator depends upon farmer arbitrage conditions which are characterized by:

(3) $P_i - td_i = P_k - td_k$ for all k

where P = the bid price at elevator i and elevator k;

d = the distance to elevator i and elevator k; and

t = linear transportation assembly cost.

Equation (3) holds only for farmer shipments emanating from the market boundary. For shipments off of the market boundary, the farmer's decision does not reflect indifference; rather it reflects a preferred alternative.

To complete the model, the elevator's decision framework is added. Without loss of generality, let the location of the set of elevators be denoted by a Cartesian coordinate system. Each elevator's location (l_i) can be represented by a coordinate pair, $l_i = (x_i, y_i)$. The ith firm's market area (M_i) is characterized by:

(4) $M_i = \{(x_i, y_i) \mid P_i - td_i \geq P_k - td_k\}$ for all k

Thus, the market area for the ith firm is the set of Cartesian coordinates such that each farmer within the area ships to the ith elevator. The market area boundary is characterized by the locus of points such that the farmer is indifferent between shipping to elevator i and any other elevator k. In other words, the boundary of the ith firm, $B(M_i)$, is a direct result of the farmer's arbitrage condition, or:

(5) $B(M_i) = \{(x_i, y_i) \mid P_i - td_i = P_k - td_k\}$ for all k

Given production is characterized by a distribution f(·) over the market area, the ith elevator's expected throughput, Q_i, is given by:

(6) $$Q_i = D\int_{\mu \epsilon M_i} d\mu$$

where D is production per unit area and M_i is the ith elevator's total market area. Given (4), (5), and (6), the price paid to farmers by each elevator represents the choice variable for an elevator seeking to maximize profits. In such a framework, the elevator trades off the benefits from a larger market area associated with a higher bid price against the cost of a lower margin.[3]

IV. MODEL APPLICATION

In this section, the spatial competition notion of market areas is illustrated with a hypothetical example. The model is applied to five cases to study the pricing behavior of one of the elevators in a fixed region as shipping costs, assembly costs, and the competing elevator's prices vary. Before discussing the results, the features of elevator competition in a fixed region are discussed.

Suppose that there are two elevators competing in a 100 square mile production region. Production points in the region are denoted by discrete Cartesian coordinate pairs (x, y) taking values in the set ($x = \pm 5$, $y = \pm 5$). Consequently, the production region is square with 121 discrete production points. Let elevator 1 be located at (-3.5, 0) and elevator 2 be located at (3.5, 0).

By equation (5), the market boundary is characterized by:

(7) $P1 - td_1 = P2 - td_2$

where: P_i is the elevator bid price offered to all farmers by elevator i = 1, 2;

t is the transportation cost per mile faced by all farmers; and

d_i is the distance from a given production point to elevator i.

In (7), there are several cases to consider. First, when both elevators offer the same bid price, the market boundary reduces to a perpendicular line equidistant between the two elevators. Second, if one elevator offers a price greater than the farmer's assembly cost between the two elevators, $\{|P1 - P2| > 7t\}$, then the entire region is serviced by the elevator with the highest price. In this case, the elevator is paying a price high enough relative to the other elevator to induce farmers to drive by the other elevator. Third, when the difference in elevator bid prices is less than the farmer's transportation cost but nonzero, the market boundary is a hyperbola bowing away from the highest paying elevator and intersecting the x-axis at x = (P1 - P2)/2t. Finally, the farmer's assembly cost also plays a critical role in this model. As t goes to zero, the market boundary at the limit is characterized by P1 = P2. In this case, farmers are indifferent between which elevator services them and a small change in price results in one elevator serving the entire region. On the other hand as t increases, elevator competition is "less intense" in the sense that elevator price changes have less of an impact on the farmers' selling decision. In short, the market boundary is the result of elevator pricing policies and assembly costs.

TABLE 1
Starting Values for Elevator 1's Pricing Problem

Variable	Abbreviation	Starting Value
Market Price	Pm	$3.00/bushel
Elevator 1's shipping cost	T1	$0.38/bushel
Elevator 2's shipping cost	T2	$0.38/bushel
Farmer's assembly cost	t	$0.01/mile
Elevator 2's margin	π2	$0.10/bushel
Storage and handling costs	C2	$0.15/bushel

At the outset, behavioral assumptions pertaining to the two elevators are made. There are several alternative pricing strategies available to elevators. In this case, it is assumed that one of the elevators follows a fixed price rule. In other words, this elevator (elevator 2) observes the terminal market price (Pm), and then determines its bid price to farmers (P2) by subtracting from Pm the shipping cost to the terminal market (T2), the related storage and handling costs (C2), and a fixed margin per bushel ($\pi 2$). Elevator 1 is assumed to be more aggressive in its marketing practices. Given elevator 2's bid price (P2), elevator 1 chooses its bid price (P1) so as to maximize its profits.

The algorithm used to solve elevator 1's pricing problem follows three steps. First, elevator 2's offering price, P2, is determined. Given P2, the minimum price elevator 1 can offer and still provide service (P1MIN) and the maximum price elevator 1 needs to offer such that it serves the entire region (P1MAX) are calculated. Second, the difference between the two prices (P1MAX - P1MIN) is divided by 100 to yield the step size of each search over all possible prices used in the algorithm. Third, profit for each step taken is compared to determine the maximum profit yielding price for elevator 1. As a starting point, a summary of the beginning values for the algorithm are presented in Table 1.

The effect of changing elevator 1's shipping cost to the terminal market is the first case investigated (Table 2). The beginning rate of 38 cents per bushel approximates the single car rate for moving wheat from Rosalia, Washington to Portland, Oregon, a distance of approximately 350 miles (Borris et al.). The rate for a 26-car shipment is about 30 cents per bushel. It is assumed that the 26 cent per bushel rate is reflective of a contract rate.[4]

The results suggest that as elevator 1's shipping costs are lowered, some of the savings are passed on to farmers (Table 2). Elevator 1's shipping costs fell 8 cents per bushel as it shifted from a single car to a multiple-car rail shipper (MCR). However, the savings are not completely passed on to farmers as elevator 1 retains some of the lower shipping costs in the form of a slightly higher margin. Some of the higher margin may be regarded as the return on investment for any equipment required to convert a country elevator to a MCR.[5] The difference in bid

TABLE 2
Elevator 1's Profit Maximizing Results With Varying Shipping Costs

Elevator 1's Shipping Rate ($/bu)	Elevator 1's Bid Price ($/bu)	Elevator 1's Profit[a]	Farmers Served by Elevator 1 N=121
.38	2.3714	6.5076	66
.35	2.3914	8.5004	79
.33	2.4114	10.4448	102
.30	2.4274	13.7312	112
.26	2.4274	18.2112	112

[a]Profit is an index which needs to be multiplied by the number of bushels moved. It is the margin per bushel multiplied by the number of farmers shipping to elevator 1.

TABLE 3
Elevator 1's Profit Maximizing Results with Varying Shipping Costs for Elevator 2

Elevator 2's Shipping Rate ($/bu)	Elevator 1's Bid Price ($/bu)	Elevator 1's Profit[a]	Farmers Served by Elevator 1 N=121
.38	2.3714	6.5076	66
.35	2.4014	4.5276	66
.33	2.4088	3.3660	55
.30	2.4192	1.9304	38
.26	2.4438	0.6026	23

[a]Profit is an index which needs to be multiplied by the number of bushels moved. It is the margin per bushel multiplied by the number of farmers shipping to elevator 1.

prices to farmers is 5.6 cents, while elevator 1 increased its margin by 2.4 cents per bushel. The number of farmers hauling grain to elevator 1 increased from 66 to 112 as the shipping rate fell to 30 cents per bushel.

The second case analyzed is analogous to the first, except that elevator 2's shipping costs are allowed to vary instead of elevator 1's. The results suggest that an elevator responds to its competitor becoming a MCR by paying a higher price to farmers (Table 3). Recall that the transport shipping savings realized by elevator 2 are completely passed on to farmers since elevator 2 practices a fixed pricing rule. As a result, elevator 2's bid price increased by 8 cents per bushel. There is a negative impact on elevator 1's market area, with the number of farmers served falling from 66 to 38. Elevator 1 reacts by increasing its bid price. However, its price does not increase as much as elevator 2's. For example, as elevator 2 moves from the single rate of 38 cents per bushel to the multiple-car rate of 30 cents per bushel, elevator 1 increases its offer price by only 4.8 cents per bushel.

The third case investigated considers changes in the terminal market price received by both elevators. The results suggest that changes in the market price are passed on completely by elevator 1 to the farmers (Table 4). In this case, elevator 1's pricing behavior is equivalent to elevator 2's fixed price rule. Profit margins, market area, and the number of farmers served are unaffected by varying the terminal market price. The only difference is the change in revenue received by farmers.

Allowing elevator 2 to change its fixed per bushel profit margin so as to vary elevator 2's offering prices to farmers is the fourth case investigated. This

TABLE 4
Elevator 1's Profit Maximizing Results With Varying Market Price

Terminal Market Price ($/bu)	Elevator 1's Bid Price ($/bu)	Elevator 1's Profit[a]	Farmers Served by Elevator 1 N=121
3.25	2.6214	6.5076	66
3.15	2.5214	6.5076	66
3.00	2.3714	6.5076	66
2.85	2.2214	6.5076	66
2.70	2.0714	6.5076	66

[a]Profit is an index which needs to be multiplied by the number of bushels moved. It is the margin per bushel multiplied by the number of farmers shipping to elevator 1.

TABLE 5
Elevator 1's Profit Maximizing Results With Varying Profit Margins for Elevator 2

Elevator 2's Profit Margin ($/bu)	Elevator 1's Bid Price ($/bu)	Elevator 1's Profit[a]	Farmers Served by Elevator 1 N=121
.13	2.3624	8.5004	79
.10	2.3714	6.5076	79
.07	2.4014	4.5276	66
.04	2.4188	2.8160	55
.01	2.4292	1.5504	38

[a]Profit is an index which needs to be multiplied by the number of bushels moved. It is the margin per bushel multiplied by the number of farmers shipping to elevator 1.

situation is similar to the case in which elevator 2's shipping costs decreased. The results suggest that elevator 2 becomes more competitive as it lowers its profit margin (Table 5). Once again, elevator 1 responds to an increase in competition by increasing its offer price to farmers and allowing its market area to shrink.

The final competitive situation analyzed considers the farmer's assembly cost. The results suggest that when farmer transport costs are low, the pricing decisions of the aggressively marketing elevator 1 are more sensitive than when transport costs are high (Table 6). In other words, elevator 1 must offer a higher incentive to attract grain when assembly costs are high. On the other hand, only a slight increase in elevator 1's price is necessary when assembly costs are low. Indeed, elevator 1 effectively prices elevator 2 out of the market when transport costs in this example fell to 0.4 cents a mile.

V. SUMMARY AND CONCLUSIONS

The pricing policy of an aggressive marketing elevator has important ramifications upon market areas. The results of the hypothetical example sug-

TABLE 6
Elevator 1's Profit Maximizing Results With Varying Farmer Assembly Costs

Farmer Assembly Cost ($/mi)	Elevator 1's Bid Price ($/bu)	Elevator 1's Profit[a]	Farmers Served by Elevator 1 N=121
.0130	2.3709	5.3994	66
.0100	2.3707	6.5538	66
.0075	2.4010	6.6264	96
.0065	2.4010	7.0441	102
.0060	2.4044	7.3427	112
.0040	2.3980	8.1720	121[b]
.0010	2.3770	11.2530	121[b]

[a]Profit is an index which needs to be multiplied by the number of bushels moved. It is the margin per bushel multiplied by the number of farmers shipping to elevator 1.
[b]Reflect elevator 1 serving the entire region.

gest than an elevator's bid price will increase when it becomes a multiple-car shipper because profits increase as it passes on some of its lower outbound shipping costs to farmers. Bid prices will also rise when a competitor becomes a MCR shipper as the single car shipper uses higher prices to maintain some of its market area. Similarly, an elevator will also raise its bid price in response to lower margins by a competitor. An elevator's market area and profitability does not vary with changes in the terminal market price because each elevator changes its bid price by the same amount. Finally, assembly costs are an important determinant of an elevators market area. Increases in the farmer's assembly costs limit the effectiveness of using bid prices to expand the market areas of elevators.

While the results seem plausible, it is important to remember that this research analyzes a hypothetical case. A weakness of the analysis is the failure to consider other factors such as elevator storage capacity constraints or the element of risk in pricing. Nevertheless, the model provides a useful framework for analyzing elevator pricing and market area decisions for a fixed region. Policy-makers may find the model useful when analyzing how the benefits and costs of lower transportation rates are allocated among shippers, railroads, and farmers. As the next step, the model will be applied to study the effect of rail contract rates upon elevator bid prices and market areas.

REFERENCES

Borris, Bonnie, et al. *A Preliminary Assessment of the Impact on Washington Highways of Railline Abandonment and Other Transportation Changes in the Washington Grain Industry.* Washington State Transportation Center, Olympia, 1985.

Casavant, Kenneth L. and Frank J. Dooley. *Transportation Needs of Washington Agriculture Phase II: Present and Future Needs.* Washington Dept. of Agr., Olympia, 1983.

Dooley, Frank J. "The Theory and Economics of Multiplant Firms Applied to Washington Grain Elevators." Ph.D. dissertation, Dept. of Agr. Econ., Washington State Univ., Pullman, 1986.

French, Ben C. "Some Considerations in Estimating Assembly Cost Functions for Agricultural Processing Operations." *J. of Farm Econ.,* 42(4):767-778, 1960.

______________. "The Analysis of Productive Efficiency in Agricultural Marketing: Models, Methods, and Progress."In *A Survey of Agricultural Economics Literature, Volume 1: Traditional Fields of Agricultural Economics, 1940's to 1970's,* pp. 93-206, edited by L.R. Martin. Minneapolis: Univ. of Minnesota Press for Amer. Agr. Econ. Assoc., 1977.

Fuller, Steven W., et al. "Alternative Wheat Collection and Transportation Systems for the Southern U.S. Plains." *Western J. of Agr. Econ.,* 6(1):91-101, 1981.

Greenhut, Melvin, George Norman, and Chao-Shun Hung. *The Economics of Imperfect Competition: A Spatial Approach.* Cambridge: Cambridge Univ. Press, 1987.

Harris, James M. "The Impact of Unit Train Rates Upon Spatial Price Differences for Corn." Ph.D. dissertation, Dept. of Agr. Econ., Univ. of Illinois at Urbana-Champaign, 1980.

Hilger, Donald A., Bruce A. McCarl, and J. William Uhrig. "Facilities Location: The Case of Grain Subterminals." *Amer. J. of Agr. Econ.,* 59(4):674-82, 1977.

Ladd, George W. and Dennis R. Lifferth. "An Analysis of Alternative Grain Distribution Systems." *Amer. J. Agr. Econ.,* 57(3):420-30, 1975.

Martin, F. Larry, D. Grant Devine, and Surendra N. Kulshreshtha. "Centralized Prarie Grain Collection: Savings Related toMarket Efficiency." *Canadian J. Agr. Econ.,* 26(2):18-34,1976.

Wilson, Wesley W. and Frank J. Dooley. *Discrete Country Elevator Investment Decisions.* Unpublished manuscript, Dept. of Agr. Econ., Washington State Univ., Pullman, 1987.

ENDNOTES

* Research Associate at the Upper Great Plains Transportation Institute, North Dakota State University and Assistant Professor in the Department of Agricultural Economics, Washington State University, respectively. Authorship seniority is not designated. The authors give special thanks to Drs. Ken Casavant, Ron Mittelhammer, and Mike Kallaher, as well as the TRF reviewers, for their insights and assistance on this paper.

1. For example, see Fuller et. al. (1981); Harris,; Hilger, McCarl, and Uhrig; Ladd and Liffereth; and Martin, Devine, and Kulshreshtha.
2. This type of approach is analogous to the spatial equilibrium models for firms selling products (See Greenhut, Norman, and Hung for an in-depth review).
3. For further detail about the model see Wilson and Dooley.
4. Rail industry officials suggest that close to 50 percent of the nationwide grain traffic moves under contract rates.
5. In 1984, the average investment cost of upgrading a single car elevator to a multiple-car elevator in Washington was approximately $200,000 (Casavant and Dooley).

Optimal Cross-Country Shipment of Perishable Commodities: An Economic Approach

*by Jeffrey L. Jordan, S. E. Prussia, R. L. Shewfelt and W. A. Flanders**

ABSTRACT

The purpose of this paper is to predict the shelf-life of lettuce due to temperature management during transportation from California to Georgia. The economic feasibility of investments to increase shelf-life are examined. A shelf-life model is specified and related to the price at which the commodity is sold.

I. INTRODUCTION

The transportation system directly affects the nature and extent of postharvest losses of perishable commodities—in quantity, quality and nutrition. Since transporting represents one-half or more of the packer-to-retailer time period in the distribution process, the level of transportation services is an important determinant of the final quality of perishable commodities.

Postharvest losses often occur due to poor management of temperature during shipment of perishables. Temperature control is required to keep perishables at a set temperature to maximize shelf-life and maintain quality. Extensive research efforts have determined optimum physical environments for the transport and storage of fresh fruits and vegetables. These studies have resulted in lists of recommended transit temperature, desired relative humidity, and the best combination of oxygen and carbon dioxide.

Most of the research on the cause and extent of loss of lettuce during shipping has focused on temperature. Barger et al. (1958) studied the relation of temperature to market quality of lettuce in rail shipments from California to New York. When lettuce was shipped at average temperatures ranging from 1°C to 6°C, differences could not be detected upon arrival at the market in general appearance, marginal browning, decay, or russet spotting. Lettuce shipped at temperatures in the lower part of this range had a longer shelf-life than that shipped above 5°C.

Harris and Harvey (1978) observed the transit temperatures and arrival conditions of California lettuce shipped to the Far East in van containers. The temperatures throughout the loads averaged 2°C and varied up to 1.5°C from the average. Morris et al. (1974) demonstrated that as much as 18 percent of the postharvest deterioration of lettuce could occur in transit if the temperature was allowed to rise to 10°C for a transit period of 10 days. Deterioration was only 8 percent when the temperature was maintained at 0°C for a transit period of 8 days.

A study by MacLeod et al. (1976) on the effect of temperature on the quality and storage life found that lettuce became unsalable in only 12 days at 15°C, but not until 35 days at 0°C. Finally, Bush (1982) found that lettuce arrival quality declined exponentially whenever temperatures exceeded 41°F.

The literature on the environments for the transportation of fresh fruits and vegetables, and especially lettuce, leads to the conclusion that refrigeration at optimum temperature for maintenance of quality is an important loss-reducing factor in the postharvest system. What is lacking in most of these studies is an examination of the economic costs and benefits of achieving optimum temperature control.

In a study (Jordan, et. al, 1985) on the effect of stacking patterns on temperature control in lettuce shipments from California to Georgia it was found that stacking pattern did not affect shelf-life, and thus economic returns. However, that study had two limitations:

1. Moderate ambient road temperatures in the spring. Thus, the effect of high temperature during transportation was not measured, and;
2. Little knowledge about the preloading handling practices in the production/shipping regions.

Consequently, this study extends that research to the summer months, and includes observations of preloading conditions in California and unloading conditions in Georgia.

II. SCOPE AND APPROACH

The purpose of this study is to predict the shelf-life of California Iceburg lettuce due to temperature management during transportation from California to Georgia. The optimal economic temperature conditions were evaluated using the following approach:

1. Transit times and temperatures of 15 straight loads of lettuce from California to the sales point in Atlanta (GA) were monitored. Three of these loads were paired shipments with known harvesting and preloading data. The departing temperatures of the three measured loads were 35.3°F, 37.7°F, and 38.8°F.

   For the other 12 loads, the mean of the three known departure temperatures (37.2°F) was used in the shelf-life model. The known and mean temperatures conform to the observation that most lettuce in California is adequately pre-cooled to between 350F and 40°F.
2. The arrival conditions and prices at the Atlanta retail warehouse for each load were also monitored. The pulp temperatures of twoheads in each of 36 selected boxes throughout the load were recorded.
3. The arrival temperatures from the 36 boxes in each load, alongwith other temperature and time data, were used in a lettuce degradation model to estimate shelf-life.

4. The shelf-life estimates were used to predict gross and net revenues based on the location of the boxes in the truck.
5. A regression equation was specified to determine the most significant variables that affect expected shelf-life. Also estimated will be the arrival temperature above which economic returns decline.

Lettuce was selected as the commodity for this project because of it's popularity as a salad vegetable and it's commercial importance; its commercial value in the U.S. is exceeded only by potatoes and tomatoes. Lettuce is susceptible to damage and reduced shelf-life when temperature is not properly controlled during shipment. The arrival pulp temperature then is related to changes in the length of shelf-life and consequently, to the value of the commodity.

The Atlanta-area was the study location since it is the distribution and integration point for the Southeast. Truckloads of lettuce arriving from California at a major grocery warehouse were inspected in order to determine optimal economic shipping conditions.

The shelf-life model is from Jordan et. al (1985 and 1986). The model employs a time-temperature profile of lettuce from harvest to retail, and can be expressed as:

$$SL = \frac{-\ln C_3}{-A\,e^{MT_3}} - \frac{t_1}{M(T_1-T_o)}\,\frac{e^{MT_1} - e^{MT_o}}{e^{MT_3}} - (t_2 - t_1)\,\frac{e^{MT_2}}{e^{MT_3}} \quad (1)$$

where:
- C_3 = quality when lettuce is unsalable;
- T_o = departure temperature from California °F;
- T_1 = arrival temperature in Atlanta °F;
- T_2 = storage temperature at warehouse °F;
- T_3 = storage temperature at grocery store °F;
- t_1 = days in transit from California;
- t_2 = day lettuce leaves storage;
- A = 0.0150769 (l/day);
- M = 0.0327645 (1/°F).

The values for A&M were determined by the method of least squares based on test data at constant and different storage temperatures (Pratt, et al.). The model predicts the number of days lettuce will remain at a quality for which it can be sold at full price, at half-price, and when it should be discarded.

Interviews with grocery produce managers indicated that lettuce is sold for its full price until it deteriorates to approximately 75 percent of grocery arrival quality. Perceived quality at the grocery store is related to leaf discoloration and wilting. Those heads not sold are then reduced to half-price until they deteriorate to 50 percent of grocery arrival quality, at which point the remaining heads are discarded. Under such a system, approximately 80 percent of a load of lettuce is sold at full price, 15 percent at half-price and five percent is discarded. To achieve these results, the lettuce must have a shelf-life of two to three days at high quality (before going below 75 percent of arrival quality) and two to three more days of shelf-life before falling below the unsalable level. The gross revenue for each lettuce load can then be calculated as:

$$GR_i = 0.8\,H_i\,P + 0.15\,H_i\,\frac{P}{2} \quad (2)$$

where: GR_i = gross returns, load i;
H_i = total number of lettuce heads in load i;
P = retail price (cents) per head;

and net revenue would be:

$$NR_i = GR_i - \text{shipping cost}. \quad (3)$$

If the lettuce does not last two days before deteriorating to 75 percent of arrival quality (SL_{75}) then the retail outlet will not be able to sell 80 percent of the lettuce at full price. Thus, equation 2 must be revised to reflect a shorter selling period. Also, if shelf life before deteriorating to 50 percent of arrival quality (SL_{50}) is less than four days, then 15 percent cannot be sold at half price. For those cases where SL_{75} is below 2 days, and SL_{50} is below 4 days, the percent sold at full and half price will be reduced proportionate to the percentage of time below the cut-off points:

$$GR_i = \frac{SL_{75}\text{ (days)}}{2\text{ days}} \cdot 0.8\,H_i + \frac{SL_{50}\text{ (days)}}{4\text{ days}}\,0.15\,H_i\,\frac{P}{2} \quad (4)$$

III. RESULTS

Table 1 shows the predicted shelf-lives when load means were used in the lettuce degradation formula. In the model, the departure temperature in California (T_o) included the three known temperatures and the mean of those three for the other 12 loads.[1] The arrival temperature in Atlanta (T_1), the storage temperature in the warehouse (T_2), and the transit time (t_1), were as observed. The storage time in the warehouse (t_2-t_1) was constant at two days, and the grocery storage display temperature was set constant at 60°F, as observed in local stores. This table presents the predicted shelf-life for lettuce until it reaches 75 percent of grocery arrival quality (the point at which the retail price is reduced) and 50 percent of grocery arrival quality (unsalable level). From the time the lettuce arrives in the grocery store, the sample lettuce was predicted to remain above 75 percent arrival quality for approximately two days. This conforms generally to the observation that lettuce is displayed about two days on the grocery shelf for sale at full price.

In six shipments, the predicted shelf-life was below the two days required for sale at full price, but in all cases, less than a half day below. However, that will reduce the amount of lettuce to be sold at full price and will be reflected in revenue estimates.

It was observed that to sell the remaining lettuce at half-price, the shelf-life would have to be over four days after arrival in the grocery store before the lettuce is unsalable. In all cases, the model predicted lettuce would last the required time.

In 4 of the 6 cases where shelf-life is predicted to be less than two days to 75 percent quality, the transit time was five days. Only in loads 6 and 15 did a four day transit time produce a shelf-life less than two days, and that by less than 6 minutes.

IV. SHELF-LIFE REGRESSION RESULTS

To further explore the variables that affect shelf-life, the following empirical model was specified:

TABLE 1
Predicted grocery shelf-life (days) until quality declines to 75 and 50 percent.

Load no.	Shipping time from California to Atlanta	Shelf-life: 75 percent of arrival quality	Shelf-life: 50 percent of arrival quality
	- - - - - - - - - -	- - - - - days - - - - -	- - - - - - - - - -
1	5	1.7	5.5
2	5	1.7	5.5
3	4	2.4	6.2
4	5	1.8	5.5
5	4	1.9	5.6
6	4	2.2	6.0
7	4	2.3	6.1
8	4	2.3	6.1
9	4	2.2	6.0
10	4	2.2	6.0
11	4	2.2	6.0
12	4	2.2	5.9
13	5	1.6	5.3
14	4	2.1	5.9
15	4	1.9	5.7

$$SL_i = \beta_0 + \beta_1 T_{o_i} + \beta_2 \text{Time } 1_i + \beta_3 HR_i + \beta_4 IT_i + \beta_5 BOX_i + \beta_6 T_{1_i} + \beta_7 T_{2_i} + \varepsilon_i \quad (5)$$

where:

SL_i = Mean shelf-life (days), $load_i$;
T_{o_i} = California departure temperature °F, load i;
Time 1_i = Transit Time (days), load i;
HR_i = Highest temperature °F, shown on Ryan Recorder, load i;
IT_i = Ambient truck temperature °F, when opened in Atlanta, load i;
BOX_i = Number of boxes in load i;
T_{1_i} = Mean arrival temperature °F, load i;
T_{2_i} = Warehouse temperature °F, load i.

The model was estimated for shelf-life to 75 percent arrival quality and unsalable level separately. The expected signs for all variables are negative since an increase in any will decrease shelf-life. The parameters of equation (5) were estimated using OLS procedure, and the results are shown in Table 2.

The parameter estimates of departure temperature, transit time, arrival temperature and warehouse temperature, were all highly significant and of the correct a priori sign. The estimates for high Ryan temperature and number of boxes were not significant at any meaningful level.

As was expected from the previous results, the most significant factor in determining shelf-life was transit time, followed by warehouse temperature, arrival temperature and departure temperature.

V. SHELF-LIFE BY BOX LOCATION

Exploring the relationship between the boxes' position on the truck and shelf-life and returns, the mean arrival temperature for each of the 36 locations were calculated and used in the shelf-life model and returns equation. Table 3 shows the results.

Of the 36 locations, ten had shelf-lives below the required two days to sell at full price. Figure 1 shows the location of those boxes. Seven of the ten were located on the bottom of the truck, indicating airflow was not adequately reaching those boxes. Fur-

TABLE 2
Regression results of effects of transit variables on shelf life.

Variable	Parameter-estimate
Departure temperature	-0.028* (-7.469)[a]
Transit time	-0.496* (-98.876)
High Ryan temperature	-0.0004 (-1.515)
Number of boxes	.00004 (1.223)
Arrival temperature	-0.035* (-44.314)
Warehouse temperature	-0.038* (-39.797)

a. Values in parenthesis are the calculated t-statistics.
*Significant at the .01 level.
R^2 = .99.

TABLE 3
Shelf-life and revenues over all quarters, sides, and positions.[a]

Obs.	Qtr.	Side	Position	Shelf75	Shelf50	Arrival temp °F	Gross Rev.	Cost	Net rev.
								-$-	
1	1	1	1	2.1	5.8	40.0	18.85	9	9.85
2	1	1	2	2.0	5.8	41.3	18.49	9	9.49
3	1	1	3	2.0	5.8	41.1	18.53	9	9.53
4	1	2	1	2.1	5.9	39.5	18.88	9	9.88
5	1	2	2	1.9	5.8	42.0	18.24	9	9.24
6	1	2	3	2.0	5.8	40.8	18.65	9	9.65
7	1	3	1	2.1	5.8	40.4	18.64	9	9.64
8	1	3	2	2.0	5.8	41.5	18.50	9	9.50
9	1	3	3	1.9	5.8	42.1	18.21	9	9.21
10	2	1	1	2.1	5.9	39.8	18.79	9	9.79
11	2	1	2	1.9	5.8	42.0	18.36	9	9.36
12	2	1	3	1.9	5.7	42.3	18.08	9	9.08
13	2	2	1	2.1	5.9	38.7	18.95	9	9.95
14	2	2	2	2.1	5.8	40.6	18.82	9	9.82
15	2	2	3	2.0	5.8	41.5	18.51	9	9.51
16	2	3	1	2.1	5.8	39.8	18.56	9	9.56
17	2	3	2	1.9	5.7	42.7	17.98	9	8.98
18	2	3	3	1.8	5.5	45.1	16.09	9	7.09
19	3	1	1	2.1	5.9	38.8	18.99	9	9.99
20	3	1	2	2.1	5.8	40.6	18.72	9	9.72
21	3	1	3	2.0	5.8	41.9	18.47	9	9.47
22	3	2	1	2.2	5.9	37.7	19.09	9	10.09
23	3	2	2	2.1	5.9	39.5	18.90	9	9.90
24	3	2	3	2.0	5.8	41.5	18.58	9	9.58
25	3	3	1	2.1	5.8	40.0	18.58	9	9.58
26	3	3	2	2.0	5.8	41.3	18.64	9	9.64
27	3	3	3	1.9	5.8	42.5	18.37	9	9.37
28	4	1	1	2.1	5.9	38.1	19.12	9	10.12
29	4	1	2	2.0	5.8	41.5	18.55	9	9.55
30	4	1	3	1.9	5.8	42.6	18.48	9	9.48
31	4	2	1	2.1	5.9	38.1	19.14	9	10.14
32	4	2	2	2.1	5.8	40.4	18.90	9	9.90
33	4	2	3	1.9	5.8	42.2	18.61	9	9.61
34	4	3	1	2.1	5.9	38.8	19.08	9	10.08
35	4	3	2	2.0	5.8	40.9	18.69	9	9.69
36	4	3	3	1.9	5.7	42.8	18.23	9	9.23

a. Quarter 1 was at the back door of the truck, 2 was the back middle, 3 front middle and 4 front wall, as shown in figure 1. Side refers to the row numbers 1, 2, 3 and position refers to boxes stacked top, middle and bottom, 1, 2, 3 respectively.

Fig. 1. Positions where arrival temperature produced shelf-lives less than two days.

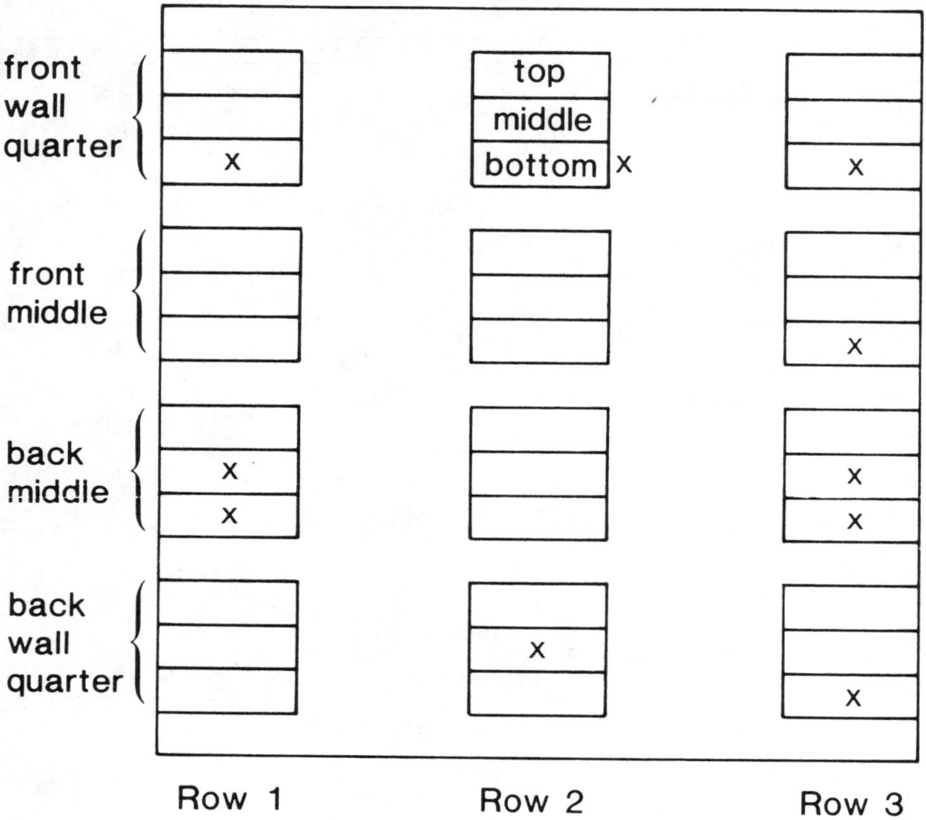

ther, five of the ten are located on the right-side wall. Also, four of the ten boxes (and two of the three that were in the middle layer) were above the rear wheelwell. None of the boxes on the top layer had estimated shelf-lives less than two days.

All ten boxes with shelf-lives below two days had arrival temperatures of 42°F or higher. Those were the only boxes with temperatures above that level. This conforms to the observation noted earlier by Bush.

The average net revenue[2] for the 26 locations with shelf-lives above two days was $9.75 per box. The average net revenue for the ten boxes with shelf-lives below two days was $9.06. Consequently, lettuce boxes that are not kept adequately cooled, where arrival temperature was 42°F or higher, resulted in an average net revenue loss of $.69 per box. In a truck with 900 boxes, each of the 36 positions would represent 25 boxes of lettuce. Thus, over an entire load, the loss in average net revenue due to poor temperature management would be $172.50.

Looking at the 540 boxes individually, all of the boxes in 5 loads (loads 3, 7, 8, 10, and 14) had shelf-lives to 75 percent of arrival quality (SL_{75}) over two days, and transit times of four days. In the four loads with transit times of five days (loads 1, 2, 4, and 13) all boxes had shelf-lives (SL_{75}) of under two days. In four loads (loads 6, 9, 11, 12, and 15) there were a total of 15 boxes with shelf-lives (SL_{75}) under two days. In load 5, 18 boxes had shelf-lives (SL_{75}) under two days. In all these cases, the majority were boxes on the floor.

VI. IMPLICATIONS

Most of the boxes of lettuce that exhibited low expected shelf-lives were on the floor of the trucks. This indicates that refrigerated air was not reaching those boxes to reduce the arrival temperatures below the estimated cut-off level of 42°F.

The implications of these findings are:

1. Better ways to ensure bottom boxes are kept below 42°F are needed. Bottom air delivery may be considered along with the normal cooling method. More immediately, the outfitting

of trucks with pressurized bulk-heads would move the air through the bottom of the truck.

2. Much of the apparent temperature problems appear to be related to exposure to external heat at the rear wheel well and on theright-side wall of the truck. The latter appears to occur because roads are crowned to the right to aid water run-off. Thus, over the transit time, boxes shift to rest against the right-side wall and are exposed to external heat (Kasmire, 1982). Also, as trucks travel East from California to Georgia, the right side of the truck faces more directly toward the sun.
3. Over ten round trips throughout a summer between California and Atlanta, the loss in lettuce revenues due to temperature management would be $1,725.00. Thus, any attempt to better insulate the side walls and wheel wells or to stop the shift ofboxes to the right would be economically feasible if the cost were below that loss. For example, pressurized bulk-heads can be custom made and fitted in each truck for approximately $300. The loss in lettuce shelf-life in just two trips more than pays for this investment. Also, duct-T floors can be outfitted for a modest investment.

One area of concern to transportation managers dealing with perishable commodities is the control of temperature during shipment. Previous studies on optimal shipping temperatures rarely focused on economic costs and returns and thus set the optimal temperatures for lettuce shipment near 35°F. This study indicated that optimal economic returns are derived when the arrival temperature of a head of lettuce is below 42°F. The study also found that the problem areas for temperature management are primarily on the floor of the truck. Therefore, investments in transportation equipment to better control temperature on the truck floor would be economically feasible.

REFERENCES

Barger, W. R., J. K. Stewart, J. M. Harvey, M. J. Ceponis, Leonard L. Morris, and R. F. Kasmire. 1958. Transit Temperatures of California Lettuce. U.S. Department of Agriculture, AMS, Marketing Research Report No. 285, Washington, D.C.

Bush, Paul. 1982. California Lettuce Industry Confronts Transport Ills. *The Packer*. July 24.

Harris, Max and John M. Harvey. 1978. Transit Temperatures and Arrival Condition of California Lettuce and Table Grapes Shipped to the Far East. U.S. Department of Agriculture, AMS, Marketing Research Report, No. 1083. Washington, DC.

Jordan, J. L., D. Mann, S. E. Prussia, and C. N. Thai. 1985. Managing the Transportation of Perishable Commodities: Temperature Control and Stacking Patterns. *Trans. Res. Forum*. 26(1):585- 589.

Jordan, J. L., R. L. Shewfelt, C. N. Thai, and S. E. Prussia.1986. Transporting Perishable Commodities: The Economic Impacts of Separating Ethylene-Generators from Ethylene-Sensitive Produce. *Trans. Res. Forum*. 27(1):229-234.

Kasmire, R. F. 1982. Factors Affecting Transit Temperatures in Truck Shipments of Fresh Produce. Perishables Handling Transportation Supplement No. 1. Davis, CA: University of California.

MacLeod, R. F., A. A. Kader, and L. L. Morris. 1976. Damage to Fresh Tomatoes Can Be Reduced. *California Agriculture*,30(12): 10-12.

Morris, L. L., J. A. Klaustermeyer, and A. A. Kader. 1974. Postharvest Requirements of Lettuce to Control Physiological Disorders. 26th International Conference on Handling Perishable Agricultural Commodities, Michigan State University.

Pratt, K. Harlan, L. L. Morris and C. L. Tucker. 1953. Temperatureand Lettuce Deterioration. University of California, Department of Vegetable Crops, Davis, California.

ENDNOTES

* Associate Professors, Department of Agricultural Economics, Agricultural Engineering, and Food Science, and former Research Coordinator, Agricultural Economics, respectively, University of Georgia, Georgia Experiment Station.

This research was conducted under a cooperative agreement with the USDA-AMS, number 12-25-A-3129, coordinated by Mr. Robert Mongelli.
The authors wish to acknowledge Mr. Joseph Garner, Research Engineer, Postharvest Systems Laboratory, Mr. Tim Campbell, Research Coordinator, Department of Agricultural Engineering, Ms. Sue Ellen McCullough, Mr. Larry Hitchcock, and Mr. Bob Flewellen, Research Technicians, Department of Food Science, and Mr. Jerry Davis, Research Technician, Department of Agricultural Engineering, University of Georgia Experiment Station, for their assistance. Also, Dr. Chi N. Thai, Assistant Professor, Agricultural Engineering for his assistance in developing the lettuce shelf-life model. The authors also wish to thank Dr. Hal Ricker, USDA-AMS, for his helpful comments.

1. Loads number 3, 6, and 10 were the paired shipments where data was collected on the truck in both California and Atlanta. For those three loads, departure temperatures were measured. For the other 12 loads, the mean of the three known departure temperatures were recorded. The known and mean temperatures conform to the observation that most lettuce in California is adequately pre-cooled to between 35°F and 40°F.
2. For the 15 loads sampled, the shipping and purchase price of the lettuce was constant as shown in Table 3. The $5.00 per box shipping cost was a contractual amount with lettuce shippers from-California in the summer of 1986. The price at which the retail chain bought lettuce was based on a set formula. If the market price of the lettuce was between $5 and $7.50, the retailer paid that price for the full load. If the market price was above $7.50 per box, the retailer paid the market price for half the boxes, and $7.50 for the other half. If the market price was below $5.00, the retailer paid the market price for half the boxes and $5.00 for theother half. Throughout the study period, the market price of California lettuce averaged $3.00 per box, so the purchase price was set at $4.00. Thus, the purchase price and shipping cost amounted to $9.00.

A Role of Transportation Costs in Agricultural Product Trade

*by Won W. Koo**

ABSTRACT

The traditional Heckscher Ohlin Theorem of International Trade has been successful in describing the relationship between trade and factor endowments. This theorem was re-examined for the world wheat trade with the inclusion of transportation costs between exporting and importing countries.

Two models are developed on the basis of Heckscher-Ohlin Theorem; one is a trade model which determines trade direction of a country and the other is a trade flow model which determines flows of a commodity from i^{th} country to all other countries. The trade model indicates that over 75 percent of trade direction (imports or exports) of wheat is determined by resource endowments. On the other hand, this study reveals that trade flows of wheat are not largely influenced by resource endowments. It is also found that transportation costs for agricultural products play a very important role in determining the quantities of wheat moved from exporting countries to importing countries.

I. INTRODUCTION

Agricultural production has been highly specialized based on the principle of comparative advantage because it requires certain weather conditions, soil types, and specialized farming technology. As a result, distribution of agricultural products among countries is an important element in stabilizing the world food market. For example, in 1980, five major grain exporting countries supplied approximately 85 percent of the grain traded in the world market and over 20 percent of the grain consumed in the world. Transportation costs are one of the most important factors affecting the distribution of agricultural products. Transportation costs of grains occupy a substantial portion of their prices in the world market. Substantial price differences between exporting and importing countries are, therefore, largely attributed to transportation costs, which could serve, like tariffs, as a trade impediment.

A large body of literature concerning trade impediments in agricultural products includes studies by Bale and Lutz, Carter and Schmitz, and Shei and Thompson. Bale and Lutz theoretically discussed impacts of trade intervention on price instability. Carter and Schmitz analyzed the distribution of welfare under optimal tariff rates. Shei and Thompson focused on spatial price determination in importing and exporting countries under alternative trade restrictions and policies. In contrast with literature in both theoretical and empirical examination of artificial trade barriers such as tariffs and quotas, the study of transport costs was rather neglected until recently. Among other things, the exclusion of transport cost is one of the shortcomings of the Orthodox Theory of International Trade (Gary). Bryan argued that the neglect of transport costs in the theory is largely attributed to the postwar preoccupation with the Heckscher-Ohlin theorem of trade that requires the assumption of zero transport costs to be logically correct.

The traditional Heckscher-Ohlin theory of international trade describes a country's trade patterns in terms of its factor endowments. In the simple traditional model with assumptions of zero cost of transport and no other impediments to trade,[1] trade is a linear function of the endowments. A country's trade relation with the rest of the world depends on its endowment of production factors. This paper empirically examines the relationship between endowments and trade when the assumption of no transportation costs and of no other impediments is altered in order to ascertain whether transportation costs are important in determining trade patterns of agricultural products subject to an economy's factor endowments. Specifically, this paper is an empirical work focusing on wheat trade.

The results of the study have policy implications for United States wheat exports. If it is shown that a functional relationship between trade and factor endowments is preserved in the presence of transportation costs, policy options for an increase in wheat exports would include a means to find the most cost-effective way to transport United States wheat overseas or to actively pursue transportation rate reduction. If the relationship cannot be preserved, on the other hand, policy recommendations should focus on factors affecting the production structure of the United States wheat industry.

II. THE MODEL

International commodity trade has been explained by the national endowment of major factors of production, such as capital, labor, and land, and/or their intensity in the production process on the basis of the Heckscher-Ohlin theorem of international trade (Balassa 1979, 1983; Bowen; Williams; Bergstrand; Neary; and Clifton and Marxsin). The theorem is employed to examine agricultural trade relationships among countries.

Based on the Heckscher-Ohlin theorem, the relationship between outputs and factor endowments can be expressed as follows[2]:

$$AX_i = V_i \qquad (1)$$

where

A is the n X n matrix of factor input coefficients in which each element is equal to the amount of factor k used to produce one unit of commodity j,

X_i is the vector of outputs of n commodities produced in country i, and

V_i is the vector of endowments of n resources in country i

The matrix A in Equation 1 can be inverted[3] to obtain the following relationship:

$$X_i = A^{-1} V_i. \quad (2)$$

This equation expresses output in country i as a linear function of the endowments.

The total world outputs (X_W) can be expressed as a function of the total world endowments (V_W) under an assumption that factor intensity (A) remains constant[4] as follows:

$$X_w = A^{-1} V_W. \quad (3)$$

The vector of net exports in country i (T_i) is the difference between production (X_i) and consumption (C_i) in the country as follows:

$$T_i = X_i - C_i, \quad (4)$$

and it is assumed that[5]

$$C_i = sXw \quad (5)$$

where s is the country's consumption share of the world output, and C_i is its consumption vector.

Combining equations (2), (3), (4), and (5) yields

$$T_i = A^{-1}(V_i - sVw). \quad (6)$$

This equation was derived on the assumption that the number of commodities included in the model is equal to the number of factors used in the production of the commodities (A is assumed to be a square matrix). The assumption was violated in developing the empirical model because the number of commodities used in the model is not equal to the number of factors. This means that the matrix in equation (1) is not square, and, thus, cannot be inverted. Equation (7) is, however, obtained while equation (6) cannot be derived because factor intensity is implicitly assumed to be constant across countries in this paper as follows:

$$AT_i = V_i - sVw. \quad (7)$$

This equation indicates that the factors embodied in the net exports are equal to the excess factor supply, which indicates that trade is an exchange of factors, not an exchange of commodities.

Equation (7) shows a relation between trade volumes (T_i) and factor intensities (A) and/or factor endowments (V_i − Vw). Thus, the proper application of the theory begins with independent measures of these three concepts.[6] Most studies of trade have focused on the relationships between trade and either factor intensity or factor endowments and have inferred the other. A factor content study by Leontief used measures of trade (T) and factor intensities (A) and inferred the factor abundance vector (V − sVw). Chenery used measures of trade (T) and endowments (V) to infer the matrix of intensities (A). This study follows the latter tradition in the sense that measures of trade (T) and factor endowments (V) are used and, thus, factor intensity (A) is implicitly inferred.

The empirical model developed for wheat[7] on the basis of equation (7) is as follows:

$$T_i = \beta_0 + \beta_1 L_i + \beta_2 LD_i + \beta_3 K_i + e_i \quad (8)$$

where

T_i is imports or exports of wheat in country i,
L_i is labor endowment in country i,
LD_i is land endowment in country i,
K_i is capital endowment in country i, and
e_i is the random error term.

The main hypothesis of this empirical examination is that agricultural product trade (exports or imports) for a country is a linear function of its factor endowments, such as labor, land, and capital. Because equation (8) is based on production characteristics of agricultural products, it is expected that the parameter β_2 will have a positive sign and other parameters will have either a negative or a positive sign depending upon contributions of these factors in agricultural production.

In addition to determining a country's trade volume, another important concern in agricultural trade centers around the factors that determine trade flows of agricultural products among countries. This paper is especially concerned with impacts of transportation costs in determining trade flows of agricultural products. Because the dependent variable in equation (8) represents the quantity of wheat imported or exported by country i, transportation costs for grain shipments cannot be incorporated in the model. A trade flow model is developed by identifying flows of wheat from each exporting country to each importing country (T_{im}). Also, explanatory variables (L, LD, and K) are measured in a relative term—a ratio of resource endowments in exporting country i to that in importing country m. The empirical model used is, therefore, specified as follows:

$$T_{im} = \alpha_0 + \alpha_1 L_{im} + \alpha_2 LD_{im} + \alpha_3 K_{im} + \alpha_4 t_{im} + e_{im} \quad (9)$$

where T_{im} is the quantity of wheat traded between exporting country i and importing country m, L_{im} is a ratio of labor endowment in country i to that in country m, LD_{im} is a ratio of land endowment in country i that in country m, K_{im} is a ratio of capital endowments in country i to that in country m, t_{im} is the unit transportation costs of wheat between exporting and importing countries, and e_{im} is the random error term. It is expected that the parameter α_4 in equation (9) will have a negative sign based on economic theory of spatial equilibrium and α_2 will have a positive sign because wheat is a land-intensive product. The sign of other parameters can be either positive or negative depending upon the degree of contributions of these factors in producing and marketing wheat in exporting and importing countries.

Sample countries used in this study are 5 major wheat exporting and 20 major importing countries (table 1). The 5 major exporting countries handle about 85 percent of the wheat traded in the world market, while the 20 importing countries import about 51 percent of the wheat traded in the world

TABLE 1
Quantities of Wheat Traded in the World Market in 1975 and 1980

Country	1975 Quantity (MT)	1975 Percent	1980 Quantity (MT)	1980 Percent
Exporting				
United States	31,753,243	43.13	36,861,680	37.20
Canada	11,701,566	15.89	17,376,361	17.53
Argentina	1,920,003	2.61	4,537,621	4.58
Australia	8,200,507	11.14	14,955,305	15.09
France	7,838,195	10.65	11,967,433	12.08
Other	12,206,622	16.58	13,399,087	13.52
TOTAL	73,620,136	100.00	99,097,487	100.00
Importing				
Japan	5,654,378	7.78	5,682,425	5.78
United Kingdom	3,634,691	5.00	2,264,111	2.30
Algeria	1,576,853	2.17	3,000,580	3.05
Korea	1,550,988	2.13	1,878,894	1.91
Brazil	2,108,850	2.90	4,758,501	4.84
Egypt	3,404,813	4.68	5,423,105	5.51
Germany	1,544,374	2.12	1,326,120	1.35
Morocco	1,383,248	1.90	1,693,170	1.72
India	7,182,919	9.88	374,274	.38
Nigeria	407,626	.56	1,176,395	1.20
Bangledesh	2,047,000	2.82	2,199,000	2.24
Indonesia	730,000	1.00	1,503,000	1.53
Iran	1,473,000	2.03	1,574,000	1.60
Iraq	568,000	.78	1,698,000	1.73
Pakistan	1,183,000	1.63	604,000	.61
Italy	900,000	1.24	1,936,000	1.97
Netherlands	551,000	.76	644,000	.65
Phillipines	541,000	.74	798,000	.81
Peru	802,000	1.1	823,000	.84
Other	35,337,665	48.78	58,467,517	59.98
Total	73,620,136	100.00	99,097,487	100.00

market (table 1). Our major concern was whether or not these 20 importing countries are sufficient samples to get a robust estimate of the model. The main criterian for choosing the 20 countries as importers are as follows: (1) the countries are continuous wheat importers and (2) some data are available for only these countries.

III. DATA

Data used for this study are both time-series data from 1975 to 1980 and cross-section data for 25 countries (5 exporting and 20 importing countries). Wheat exports in the 5 exporting countries, wheat imports in the 20 importing countries,[8] and quantities of wheat moved from exporting countries to importing countries for the time period from 1975 to 1980 were obtained from the various issues of *World Wheat Statistics* (International Wheat Council). Wheat imports are differentiated from wheat exports by identifying imports as a negative number and exports as a positive number. Capital endowment for a country is defined as the accumulated domestic investment flows in the country assuming a 15-year average asset life. The capital endowments are calculated for each country from 1975 to 1980. The annual domestic capital investments for 25 countries from 1961 to 1980 were obtained from *World Tables* (World Bank). Labor endowment in a country is defined as the economically active population in the country (Leamer) as presented in *Production Yearbook*[9] (FAO).

Land endowment in a country is defined as the total area of arable land in the country. The aggregate area of arable land for each country was obtained from *Production Yearbook* (FAO). Land is subdivided into two categories,[10] Land 1 is defined as that in humid and tropical climate, and Land 2 is defined as all other arable land in dry, humid mesothermal, and humid microthermal climate. The total area of classified land for each country was obtained from Leamer.

Ocean transportation costs from each exporting country to importing countries were obtained from various issues of *World Wheat Statistics* (International Wheat Council) and *Chartering Annuals* from 1976 to 1980 (Maritime Research, Inc.). An average of all contract rates for a particular shipment was estimated for this study.

IV. EMPIRICAL RESULTS

Equations (8) and (9) are estimated using multiple regression. Since cross-section data for each year from 1975 to 1980 are not large enough to get a robust estimate, both time-series and cross-section data were combined to estimate the models. It was expected that the disturbance terms would likely be time-series-related disturbances, cross-section disturbances and a combination of both. This study, however, found that the model does not have any complexity with the disturbance term—no heteroskedasticity and/or no autocorrelation.

The data set used for this study is subject to chronic measurement errors as is any type of "data." Thus, the question is whether measurement errors are so severe that these data by themselves are useless. The treatment of chronic errors in variables begins with the computation of the direct and the reverse regressions in an econometric tradition. This approach to the treatment of chronic measurement errors has been extended to the multiple regression setting by Klepper and Leamer (1984). If there are k explanatory variables, then there are k + 1 ways of estimating the regression, depending upon the direction in which the residuals are measured. If these k + 1 regressions have the same sign patterns, then any weighted average of these k + 1 regressions is a maximum likelihood estimate. But if these k + 1 regressions have sign changes, then the set of maximum likelihood estimates is unbounded.

Direct and reverse regressions for equations (8) and (9) show that all reverse regressions have the same sign patterns, and the set of maximum likelihood estimates remains bounded. The reverse regressions indicate that there are no reasons to suspect the existence of measurement errors in the regression estimates.

A. Impacts of Factor Endowments on Wheat Trade

Table 2 presents estimated parameters of the corresponding variables in equation (8). Two models are estimated, Model 1 does not include the dummy variables representing the years from 1975 to 1979 under the assumption that there are no major changes in export and import volumes over time due to changes in resource endowments, while Model 2 includes the dummy variables. The F test is conducted to test whether the dummy variables representing the years from 1975 to 1979 should be included in the model. The F test concludes that the dummy variables are not statistically significant at the 1 percent critical level by accepting the null hypothesis H_o: $\beta_5 = \beta_6 = \beta_7 k = \beta_8 = \beta_9 = 0$. This implies that Model 1 is statistically better than Model 2 in terms of a model specification. It should be noted, however, that the estimated coefficients for Land 1, Land 2, labor, and capital are almost the same between the two models. Adjusted R^2_s $(\bar{R}^2)$ are 0.78 and 0.77 for Models 1 and 2, respectively. This indicates that over 77 percent of variations in export and import volume can be explained by resource endowments as specified in equation (8).

TABLE 2

Estimation of Wheat Trade Models

Variable	Trade Model 1	Trade Model 2
Constant (β_0)	-2,883.507 (7.55)**	-2,729.993 (3.46)**
Land 1 (β_1)	-0.002 (1.05)	-0.002 (1.03)
Land 2 (β_2)	0.022 (16.42)**	0.022 (16.11)**
Labor (β_3)	-0.003 (2.09)*	-0.003 (2.06)*
Capital (β_4)	0.004 (6.87)**	0.004 (6.71)**
Dummy 1 (β_5)		-174.362 (0.16)
Dummy 2 (β_6)		-17.909 (0.02)
Dummy 3 (β_7)		-130.83 (0.12)
Dummy 4 (β_8)		-160.18 (0.15)
Dummy 5 (β_9)		-436.07 (0.42)
$\bar{R}^2$	0.78	0.77
S	3,644.31	3,705.88

Note: t-statistics are in parentheses.

* and ** indicate statistical significance of the corresponding variables at 0.05 and 0.01 critical level, respectively.

The estimated parameters for Land 2 and capital are statistically significant at the 1 percent level and that of labor is significant at the 5 percent level. This indicates that these variables are highly correlated to the quantities of wheat traded and trade direction. On the other hand, Land 1 is not statistically different from zero, indicating that its effect on wheat exports and imports is not substantial mainly because Land 1 is more appropriate for other crop production such as rice than wheat production.

The estimated coefficients for Land 1 and labor have a negative sign, indicating that those countries which have abundant Land 1 and labor have a comparative disadvantage over other countries in producing and exporting wheat. On the other hand, the estimated coefficients for Land 2 and capital have a positive sign, indicating that those countries which have abundant Land 2 and capital have a comparative advantage over other countries in producing and exporting wheat. This implies that wheat is a capital- and Land 2-intensive product assuming that factor intensity in the production of wheat is constant across countries.

Beta coefficients are estimated to evaluate the relative importance of the independent variables in Model 1 (table 3).[11] The beta coefficients are 0.0468, 0.586, 0.0762, and 0.3623 for Land 1,

TABLE 3
Estimated Beta Coefficients and Elasticities for the Variables in Trade Model 1

Variable	Beta Coefficient[a]	Elasticity[b]
Land 1	-0.0468	-0.0944
Land 2	0.586	2.0150
Labor	-0.0762	-0.0988
Capital	0.3623	1.049

a The beta coefficient for variable X_i is calculated as follows:

$$\overset{*}{\beta}_i = \hat{\beta}_i \; (Sx_i/Sy_i)$$

b Elasticity of variable X_i is calcualted at mean values of X_i and Y.

Land 2, labor, and capital, respectively. These beta coefficients indicate that Land 2 and capital endowments in a country are the most important factors influencing agricultural production specialization and trade, and Land 1 is the least significant variable.

Elasticity for each variable in the model is estimated at mean values of dependent and independent variables (table 3). The elasticities associated with Land 1 and labor are inelastic whereas that for Land 2 is very elastic (2.015). In addition, elasticity for capital is 1.049. This implies that Land 2 and capital endowments are the most significant economic variables in agricultural production specialization and trade. On the other hand, Land 1 and labor appear to be less significant economic variables.

B. Interdependency Between Trade Patterns and Transportation Costs

Covariance pooling procedures of cross-section and time-series data were used to estimate equation (9). Two models are estimated (table 4); Model 1 includes two different types of dummy variables (the variables representing importing countries and those representing years from 1975 to 1980), whereas Model 2 excludes dummy variables representing years under the assumption that there are no major changes in trade flows of wheat due to changes in resource endowments and transportation costs over time. The F-test accepts the null hypothesis H_o: $\alpha_{22} = \alpha_{23} = \alpha_{24} = \alpha_{25} = \alpha_{26} = 0$ indicating that the dummy variables representing years are not statistically significant at the 5 percent critical level. This implies that Model 2 is statistically better than Model 1 in terms of model specification. The estimated coefficients in Model 1 are similar to those in Model 2 except transportation costs.

The adjusted R^2s ($\bar{R}^2$) are 0.26 in both models. This indicates that about 26 percent of the variations in wheat trade flows can be explained by ratios of resource endowments in exporting countries to those in importing countries, and the transportation costs. The R^2s for the flow models are much smaller than those for the trade model. This implies that the resource endowments influence more import or export volume in a country (direction of trade) than quantities of wheat traded between importing and exporting countries (trade flows).

Most variables in flow Models 1 and 2 (table 4) are significant at the 1 percent critical level except Land 2. Land 2 is not statistically different from zero at the 20 percent level, indicating that Land 2 is not an important variable affecting trade flows among countries. Note that Land 2 was the most important variable explaining a country's trade potential of wheat. The insignificance of Land 2 in the flow model is mainly because trade flows are influenced more by transportation costs rather than factor endowments.

Beta coefficients for the independent variables in the model are presented in table 5. The beta coefficients are 0.103, 0.079, 0.193, and 0.138 for transportation costs, Land 2, capital and labor, respectively, indicating that capital is statistically more important than other variables and that Land 2 is statistically least important variable. The beta coefficients for transportation costs and labor are not the highest, but they are statistically significant in determining trade flows among countries.

The quantity of wheat traded between importing and exporting countries is negatively related to transportation costs between the two countries. This is consistent with economic theory of spatial equilibrium. For example, an increase in transportation costs results in a rise in wheat prices at importing countries and a decrease in wheat prices at exporting countries. The changes in wheat prices reduce the total quantity of wheat traded between two countries. On the other hand, the quantity of wheat traded among countries is positively related to the capital ratio which reflects differences in farming technology between two countries. This indicates that exporting countries are more capital and technology abundant relative to importing countries. The labor ratio has a positive sign in the flow models in table 4. This indicates that the labor ratio between exporting and importing countries contributes to an increase in the quantity of wheat traded between the countries. Primary reasons are: (1) agricultural sector is a labor intensive relative to other sectors of the economy such as electronic and computer, (2) some importing countries have a large labor population employed in wheat production yet still import some wheat (e.g., India and Brazil), and (3) labor population in the most importing countries in the sample is smaller than that in exporting countries (e.g., Korea, Japan). For example, the labor ratio of the United States to India is 0.39 and is almost constant over the time period. India's imports from the United States gradually declined; 4300 thousand ton in 1975, 4178 in 1976, 2053 in 1977, 405 in 1978, 271 in 1979, and 222 in 1980. On the other hand, the labor ratio of the United States to Korea is about 9 over the six-year period which is much larger than that of the United States to India. Korea's imports from the United States in 1980 totaled 1,791 thousand ton, which is about 8 times larger than India's imports from the United States. This results in a positive relationship between trade volume and the labor ratio.

TABLE 4

The Estimated Trade Flow Model

Variable	Model 1		Model 2	
Constant (α_0)	326.852	(1.411)	-63.684	(0.455)
Ocean rate (α_1)	-17.265	(2.875)	-7.599	(2.984)
Land 2 (α_2)	0.547	(0.606)	0.809	(0.904)
Capital (α_3)	1.301	(3.275)	1.299	(3.273)
Labor (α_4)	10.444	(2.113)	12.138	(2.490)
D1 (α_5)	380.902	(2.181)	366.771	(2.102)
D2 (α_6)	916.873	(5.462)	888.559	(5.311)
D3 (α_7)	245.197	(1.506)	231.054	(1.420)
D4 (α_8)	937.081	(5.219)	986.114	(5.540)
D5 (α_9)	334.735	(1.905)	243.800	(1.433)
D6 (α_{10})	971.633	(5.380)	878.917	(5.021)
D7 (α_{11})	549.203	(3.169)	487.089	(2.858)
D8 (α_{12})	412.336	(2.405)	336.519	(2.008)
D9 (α_{13})	1619.702	(9.815)	1605.068	(9.737)
D10 (α_{14})	602.480	(3.887)	561.213	(3.650)
D11 (α_{15})	475.146	(2.763)	414.871	(2.447)
D12 (α_{16})	228.964	(1.423)	258.497	(1.613)
D13 (α_{17})	513.919	(3.195)	521.460	(3.243)
D14 (α_{18})	248.745	(1.443)	229.493	(1.334)
D15 (α_{19})	528.222	(3.264)	558.703	(3.467)
D16 (α_{20})	-167.959	(1.407)		
D17 (α_{21})	-276.476	(1.886)		
D18 (α_{22})	-280.741	(2.034)		
D19 (α_{23})	-252.535	(1.832)		
D20 (α_{24})	-237.320	(1.946)		
$\bar{R}^2$	0.262		0.263	
S	577.835		465.560	

Note: t-statistics are in parentheses.

TABLE 5
Estimated Beta Coefficients and Elasticities for the Variables in Trade Flow Model 2

Variable	Beta[a]	Elasticity[b]
Transportation cost	0.103	-0.356
Land 2	0.079	0.001
Capital	0.193	0.118
Labor	0.138	0.111

[a] The beta coefficient for variable X_i is calculated as follows:

$$\beta_i^* = \hat{\beta}_i \; (Sx_i/Sy_i)$$

b Elasticity of variable X_i is calcualted at mean values of X_i and Y.

Elasticity of each variable is estimated at mean values of dependent and independent variables (table 5). The elasticities are all inelastic ranging from 0.356 for transportation costs to 0.001 for Land 2 in the absolute value. The elasticities for capital and labor are 0.118 and 0.111, respectively. The elasticities indicate that transportation costs are the most significant variable in determining trade flows among countries. On the other hand, Land 2 appears to be the least significant economic variable.

V. CONCLUDING REMARKS

The traditional Heckscher-Ohlin theorem of international trade has been successful in describing the relationship between trade and factor endowments. This theorem was re-examined for the world wheat trade with the inclusion of transportation costs between exporting and importing countries.

This study reveals that the wheat trade of a country is positively related to the country's land and capital endowments and is negatively related to the country's labor endowments. This indicates that wheat is a land- and capital-intensive product. When transportation costs are included in the flow model, the same relationship holds except for labor. The labor ratio is positively related to the quantities of wheat moved from exporting countries to importing countries. The flow model also reveals that wheat trade is negatively related to transportation costs and the relationship is statistically significant at the 0.01 percent level.

In summary, factor endowments influence wheat export or import volume in a country more than trade flows among trading partners. This implies that a country can increase its trade potential by focusing its policy on reallocating of factor endowments in the country. It is also found that transportation costs for agricultural products play a very important role in determining the quantities of wheat moved from each exporting country to each importing country. It should be recognized that transportation costs serves as a trade-resisting factor in the international wheat trade. Improvement in physical efficiency of grain distribution system influences the quantities of wheat traded among trading partners more than the trade potential of a country.

REFERENCES

Balassa, Bela. 1979. The Changing Pattern of Comparative Advantage in Manufactured Goods. *Review of Economics and Statistics* 61:259-66.

Balassa, Bela. 1983. Comparative Advantage in Manufactured Goods: A Reappraisal. *Review of Economics and Statistics* 65:315-18.

Bale, Malcolm D., and Ernsst Lutz. 1979. The Effects of Trade Invervention on International Price Instability. *American Journal of Agricultural Economics* 61:512-16.

Bergstrand, Jeffrey H. 1986. The Gravity Equation in International Trade: Some Microeconomic Foundations and Empirical Evidence. *Review of Economics and Statistics* 68:474-81.

Bowen, Harry P. 1983. Changes in the International Distribution of Resources and Their Impacts on U.S. Comparative Advantage. *Review of Economics and Statistics* 65:402-14.

Bryan, I. A. 1974. The Effect of Ocean Transport Costs on the Demand for Some Canadian Exports. *Review of World Economics* 110:642-62.

Carter, Colin, and Andrew Schmitz. 1979. Import Tariff and Price Formation in World Wheat Market. *American Journal of Agricultural Economics* 61:517-22.

Chang, W. 1979. Some Theorems of Trade and General Equilibrium with Many Costs and Factors. *Econometrica* 7:709-26.

Cheneny, H.B. 1960. Patterns of Industrial Growth. *American Economic Review* 50:624-54.

Finger, J. M. and A. J. Yeats. 1976. Effective Protection by Transport Cost and Tariffs: A Comparison of Magnitude. *Quarterly Journal of Economics* 90:169-76.

FAO (Food and Agriculture Organization). 1975-1981. *Production Yearbook.* Rome.

Gray, H. Peter. 1976. *A Generalized Theory of International Trade*. New York: MacMillan Press Limited.

Hufbauer, G. C. 1970. The Impact of National Characteristics and Technology on the Commodity Composition of Trade in Manufactured Goods. *The Technology Factor in Intermediate Trade,* ed. R. Vernon, pp. 145-231. New York: Columbia University Press.

International Wheat Council. 1982. *World Wheat Statistics*. London.

Johnson, Paul R., Thomas Grennes, and Marie Thusby. 1977. Evaluation, Foreign Trade Controls and Domestic Wheat Prices. *American Journal of Agricultural Economics* 59:619-27.

Klepper, S., and E. E. Leamer. 1984. Sets of Maximum Likelihood Estimates for Regressions with All Variables Measured with Error. *Econometrica*.

Leamer, E. E. 1978. *Specification Searches: Ad Hoc Inference with Nonexperimental Data*. New York: Wiley.

______. 1984. *Sources of International Comparative Advantage: Theory and Evidence*. Cambridge, Mass.: MIT Press.

Leontief, W. W. 1953. Domestic Production and Foreign Trade: The American Capital Position Reexamined. *Proceedings of the American Philosophical Society.*

Maritime Research Inc. 1976-1981. *Chartering Annual*. Parlin, New Jersey.

Neary, J. Peter. 1985. Two-by-two International Trade Theory with Many Goods and Factors. *Econometrica* 53:1233-47.

Shei, S., and Robert L. Thompson. 1977. The Impacts of Trade Restriction on Price Stability in the World Wheat Market. *American Journal of Agricultural Economics* 59:628-38.

Vanek, J. 1968. The Factor Proportion Theory: The N-Factor Case. *Kyklos* 21, 4.

Williams, James R. 1970. The Resource Content in International Trade. *Canadian Journal of Economics* 3:111-22.

World Bank. 1980. *World Tables,* 2nd edition, Washington, D.C.

ENDNOTES

* Agricultural Economics, North Dakota State University, Fargo, ND.

1. Other assumptions of the model include equal number of goods and factors, international factor intensity, constant return to scale, no factor intensity reversals, and no demand bias.
2. In a world where there are only two goods (X_1 and X_2) and two factors (K and L), the equilibrium condition in the factor market can be described as follows:

$$A_{L1} X_1 + A_{L2} X_2 = L$$

$$A_{K1} X_1 + A_{K2} X_2 = K$$

$$\begin{bmatrix} A_{L1} & A_{L2} \\ A_{K1} & A_{K2} \end{bmatrix} \begin{bmatrix} X_1 \\ X_2 \end{bmatrix} = \begin{bmatrix} L \\ K \end{bmatrix}$$

where A represents input requirements in producing one unit of output.

3. The inverse exists as long as the relative input intensities among commodities are equal.
4. $X_w = \sum_i A^{-1} V_i = A^{-1} \sum_i V_i = A^{-1} V_w$
5. Since the commodity prices are given internationally, each country consumes commodities in the same proportion.
6. Hufbauer measured the three concepts based on the Leontief-type calculation. A recent attempt was made by Balassa (1986).
7. Since importing and exporting countries are different for each crop, the empirical model was developed for each crop (e.g., wheat) rather than for the aggregate level.
8. Some wheat importing countries such as the People's Republic of China and the Soviet Union are not included because the data used are not available for these countries.
9. Labor used in the production process varies greatly in quantity and its divisibility based on the product produced and the nature of labor's contribution. Typically, labor is divided into at least three categories: skilled, semi-skilled, and unskilled, based upon the educational background of the labor force in a particular country. This study aggregates labor endowments of each country across all skill levels because all skill levels are employed in the agricultural sector. Although this study focuses on wheat trade, labor endowment is defined as economically productive population employed in the economy because labor can be moved from one sector to another.
10. Wheat, by its naure, is a weed and can be grown in diverse conditions. However, land under humid and tropical climate is more suitable for rice production than wheat, while the other land is more suitable for wheat production.
11. The beta coefficients adjust the estimated parameter in the model by the ratio of the standard deviation of the independent variable to the standard deviation of dependent variable. In other words, the beta coefficient is the same as the estimated parameter for the normalized variable, indicating that the rescaling associated with the normalized regression makes it possible to compare beta coefficient directly.

Quantifying the Costs & Uncertainties of Transporting Grains From Farms to Assembly Point

*by Neil Meyer and Gordon Sparks**

I. INTRODUCTION

The impact on producers of adjustments in the Canadian grain collection systems depends heavily on how the system evolves. Economics dictate that substantial rationalization/centralization in terms of fewer but larger elevators is needed and that trucks should play a substantially more significant role in grain assembly from farm to elevator prior to transfer to the rail mode and movement to export position.

Previous research has demonstrated that the unit costs of transporting grain by truck from the farm to the elevator can be reduced even when haul distances are increased as a result of centralization of elevators [1]. This was accomplished by introducing larger, more productive trucks combined with increased annual utilization.

Deterministic costing models similar to that used in the previous work provide estimates of the expected (mean) values for costs incurred in assembling grain by truck under specific operating scenarios. While these values provide significant insight into the potential impact on producers as a whole, they provide little insight with respect to the range of costs likely to be experienced by producers.

This paper is an extension of previous work [1] and presents the results of the application of a probabilistic modelling methodology to the assembly of grain from the farm to the elevator on the Canadian Prairies. It is done under a range of scenarios varying from 2 axle farm trucks on one-way hauls of 10 kilometers to 7 axle commercial grain trucks on 80 kilometer delivery hauls. This approach provides estimates of expected (mean) values for the costs incurred by producers as well as quantifying the range in costs likely to be experienced. That is, the methodology estimates both the expected values and the variance thereby quantifying the uncertainty related to these estimates.

Cases analyzed include a farmer transporting his own grain in a 2 or 3 axle farm truck, a custom hauler using a 3 axle farm truck plus pup transporting the production from seven farmers (i.e., a farmer hauling his own grain plus that of six neighbors) and a commercial trucker using a seven axle A-Train (5 axle semi plus pup) hauling for as many farmers as time will permit. Costs are determined for haul distances of 10, 40, and 80 kilometers.

Larger loads and longer hauls decrease per unit costs because they increase efficiency or utilization. This is the incentive for producers to custom haul grain for other farmers. This model quantifies the expected values for cost of grain transport for different sized trucks hauling different distances.

If expected (mean) values for each input parameter are substituted into the costing model, an (single value) estimate of the costs for assembling grain by truck is obtained. While these single value estimates provide useful insight, they do not represent the costs incurred by any specific producer. Or, put another way, there are as many values for the costs of transporting grain as there are producers involved because they each have a somewhat different cost base (i.e., different values for at least some of the determining input parameters).

To illustrate, one farmer may have a ten year old (essentially depreciated) farm truck while another has a new one representing a capital investyment of \$20-30,000. Obviously, these two individuals operate from a substantially different cost base. In a similar manner, many of the other values for determining input parameters are producer specific. In this context, it is of considerable interest to not only have estimates for the expected (mean) cost for various scenarios (i.e. degrees of rationalization) but also to have estimates of the variability and/or range of uncertainty associated with unit costs as a function of variabiliity of input parameters.

II. DETERMINISTIC COSTING MODEL

The costing model developed as part of the previous work and used as a starting point here, is illustrated conceptually in Figure 1. Fixed costs are for capital (depreciation), interest on investment, licenses and insurance, and tarps. Variable costs include fuel, tire wear, repairs and maintenance, and labor. Capacity/utilization variables include annual kilometers, capacity per trip, tonnes transported per year, distance to delivery point, on-farm loading time, travel time to delivery point, and unloading time at the delivery point. As noted, unit costs (\$/tonne Km or #/tonne) are derived from a combination of parameters quantifying the fixed and variable costs of owning and operating a truck.

III. SENSITIVITY ANALYSIS

Sensitivity analysis carried out using the deterministic costing model provides some insight into the range of expected end costs resulting from variations/uncertainties of values for input parameters. A sensitivity analysis was performed by developing minimum, maximum, and most likely estimates for the values for each of the input parameters for the deterministic costing model for each of a number of scenarios. These ranges can be thought of either as representing the uncertainty in the value for the parameter (i.e. fuel consumption) or representative of the range of values found in the population (i.e. distribution of the capital cost for farm trucks). Either type of variability manifests itself as a distribu-

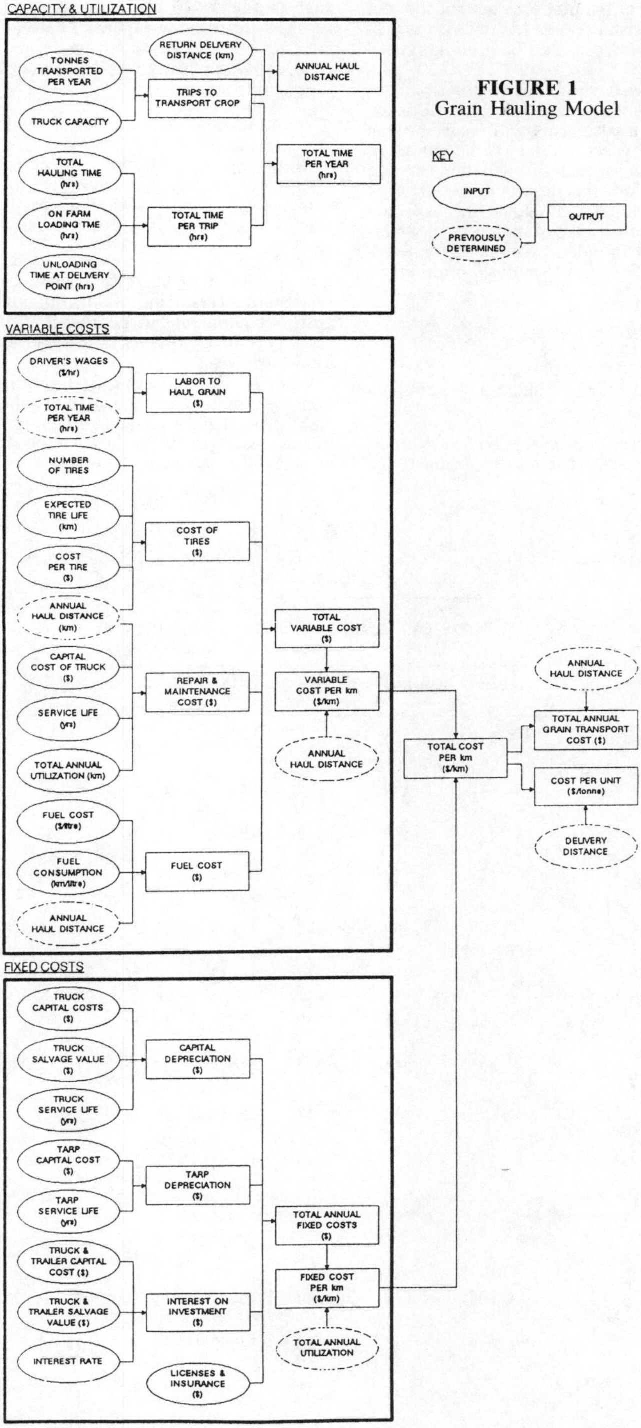

FIGURE 1
Grain Hauling Model

tion over end cost. For the purposes herein, it is not necessary to distinguish between the two types: they are both of interest and can both be accommodated in the analysis.

A sensitivity analysis can be carried out by substituting the most likely (mean) value for all of the parameters into the model. This will result in what we have termed the expected value. If the minimum and maximum values for each of the parameters are in turn substituted into the model while all other parameters of the model are held at the expected (mean) values, the sensitivity of model results relative to the variability in individual input parameters can be determined. Sensitivity analysis of this type was carried out for:

(1) a 2 axle farm truck,
(2) a 3 axle farm truck,
(3) a Custom 3 axle truck plus pup,
(4) a Commercial 7 axle A-Train (5 axle semi plus 2 axle pup)

Typical input and results for a 2 axle farm truck are summarized in Table 1. The results of sensitivity analysis for the 40 km haul distance for each truck type are illustrated in Figures 2 through 5. It is noted that the costs of transporting grain by truck are particularly sensitive to variability/uncertainty in the following parameters:

(1) utilization—the number of kilometers a truck is driven annually,
(2) delivery time—minutes per trip
(3) truck capacity—tonnes
(4) purchase price—capital cost of the vehicle
(5) service life—life of the truck in years

All types of operations were found to be particularly sensitive to utilization, purchase price, and service life. This points out the importance of capital costs of equipment and the impact of spreading these fixed costs over a larger number of kilometers and tonnes per year.

Costs for 2 axle trucks used only in farm operations were particularly sensitive to annual kilometers and the purchase price and service life of the truck. In this case, variability in costs per tonne were as great as 100 per cent. In the case of 3 axle farm

TABLE 1

Deterministic Sensitivity for Operating Costs of a 2-Axle Farm Grain Truck Hauling 40 Kilometers (25 Miles)

INPUTS	TYPICAL VALUE			PRODUCTIVITY (cost / tonne)			PERCENT VARIATION
	MIN	LIKELY	MAX	MIN	EXPECTED	MAX	(%)
Utilization (km / yr)	5,040	6,780	20,000	18.41	15.34	9.47	58
Delivery time (min / trip)	85	130	200	14.00	15.34	17.42	22
Truck capacity (tonnes)	4.5	5.6	6.7	19.09	15.34	12.82	41
License & Insurance ($)	147	147	147	15.34	15.34	15.34	0
Purchase Price ($)	10,000	17,400	43,500	11.68	15.34	28.26	108
Service life (years)	6	10	30	19.37	15.34	11.31	53
Salvage value ($)	1,500	2,610	6,525	15.46	15.34	14.97	3
Tarp cost ($)	800	1,000	1,200	15.30	15.34	15.38	1
Tarp service life (years)	6	10	17	15.48	15.34	15.25	1
Fuel consumption (km / liter)	1.7	2.1	2.5	15.88	15.34	14.97	6
Fuel cost ($ / liter)	0.27	0.34	0.55	14.86	15.34	16.77	12
Tire cost ($)	202	252	350	15.29	15.34	15.45	1
Tire life (km)	64,000	80,000	130,000	15.41	15.34	15.24	1
Interest rate (%)	9	11	15	14.92	15.34	16.18	8
Drivers wages ($ / hr)	8	10	12	14.57	15.34	16.11	10

FIGURE 2
Deterministic Sensitivity 2 Axle Farm Grain Truck
40 km Haul Distance

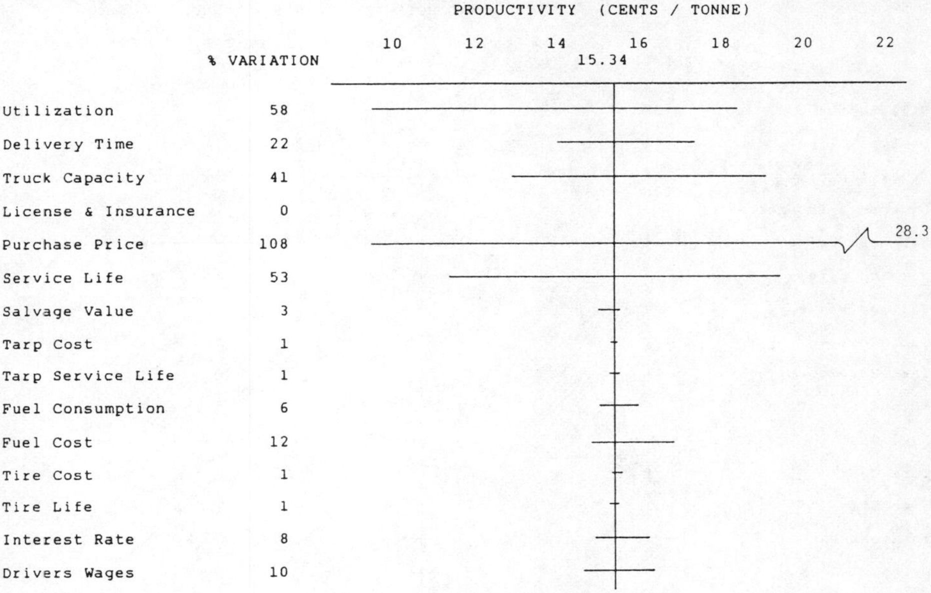

FIGURE 3
Deterministic Sensitivity 3 Axle Farm Grain Truck
40 km Haul Distance

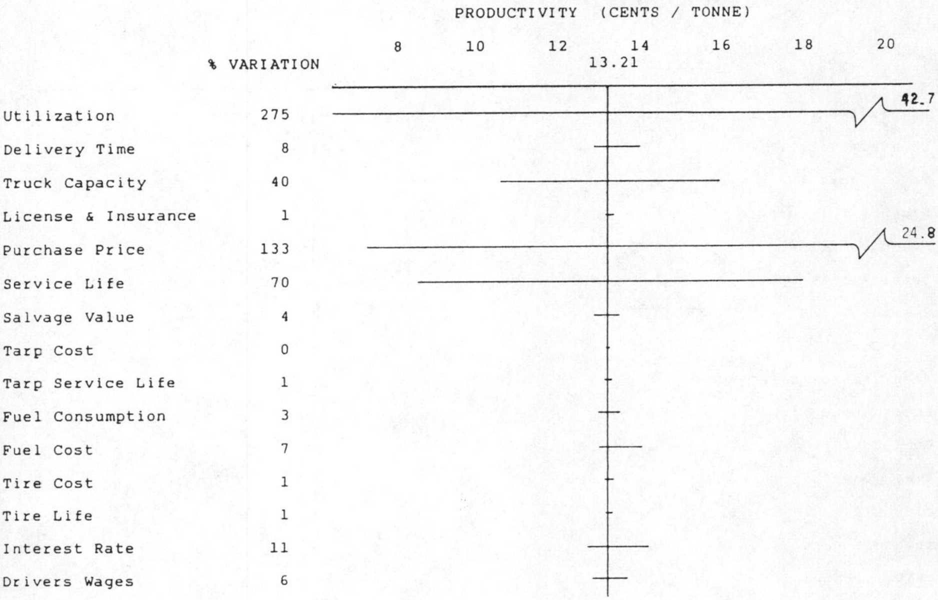

FIGURE 4
Deterministic Sensitivity Custom 3 Axle Grain Truck with Pup
40 km Haul Distance

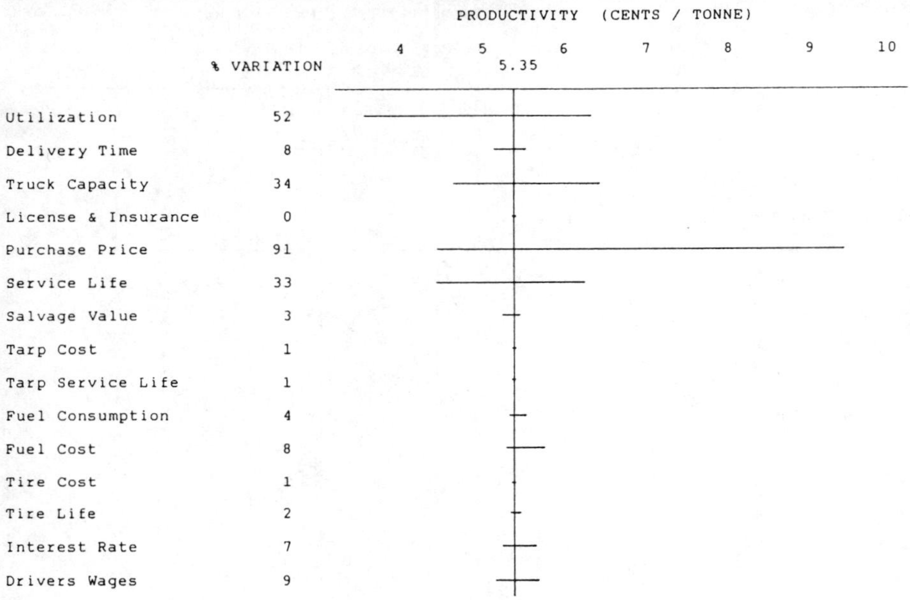

FIGURE 5
Deterministic Sensitivity Commercial 7 Axle Grain Truck (A Train)
40 km Haul Distance

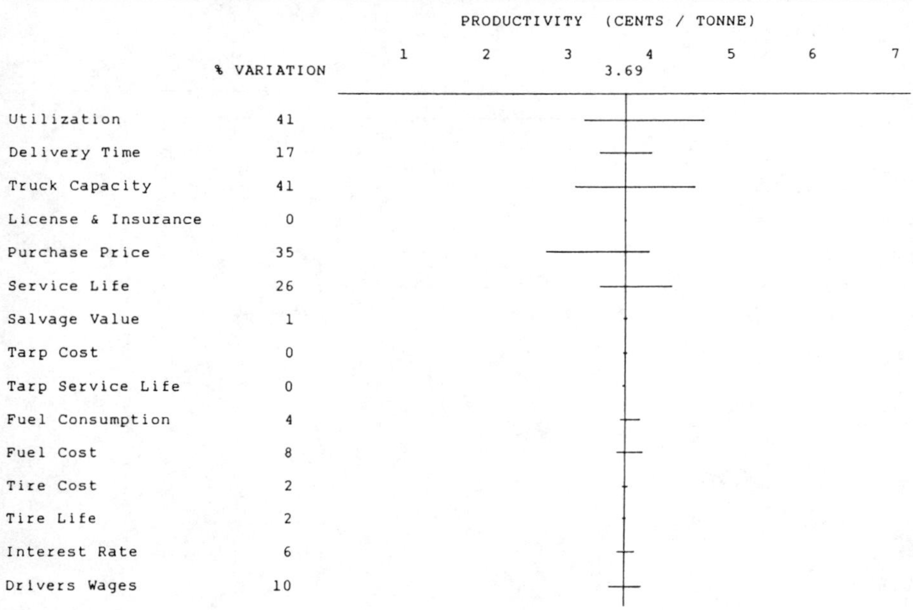

trucks, utilization and purchase price were most sensitive. In this case, the model showed 100-300% variation in the per tonne costs. This is because of the larger cost and greater variation in price which producers can pay when purchasing various trucks. This also explains why service life is a source of considerable variation; larger cost means annual depreciation and interest charges are greater. Variation in vehicle capacity showed a 40% variation in per tonne costs. This explains why producers tend to overload trucks used for delivering grain. They directly gain the benefits of lower per unit costs.

Custom operations were defined as units hauling for individual producers and six neighbors. This larger volume permitted allocating the fixed costs over more tonnes. The results of this analysis are shown in Fig.4. Utilization and purchase price were associated with the greatest variation in per tonne cost. Other important parameters included truck capacity and service life. As illustrated in Figure 5, commercial trucking operations were about equally influenced by utilization, truck capacity, purchase price, and service life.

IV. PROBABILISTIC MODEL

The sensitivity analysis undertaken and summarized above illustrates how the end costs of transporting grain vary relative to individual variation/uncertainties in single parameters. It is of interest to also quantify the overall uncertainty resulting from the effects of the combined variations/uncertainties in all of the input parameters. While this could conceivably be done, it is extremely computationally intensive. (That is, the lights would dim when the computer was turned on).

The computational intensity of the problem can be substantially reduced without significantly compromising the quality of the results by exploiting the structure of the problem. That is, the majority of the uncertainty of the end results (i.e. $/tonne of grain transported) will be associated with the variation/uncertainties associated with the variables to which the end results are most sensitive. (This sensitivity can be associated with either the fact that the model is very sensitive to a particular parameter and/or that there is considerable variability/uncertainly with respect to the value of the parameter). Variability/uncertainty associated with insensitve parameters on the other hand will add little to the variation in end results. As such, it is reasonable to restrict the probabilistic modelling effort only to the parameters to which the results are sensitive. This has in fact been done herein by fixing all insensitive parameters at their expected (mean) values and performing the stochastic analysis over the ranges for only the sensitive parameters. The details of the stochastic modelling methodology are presented in a previous paper [2]. The results obtained are illustrated in Figures 6 through 13.

FIGURE 6
Probabilistic Model Two Axle Grain Truck

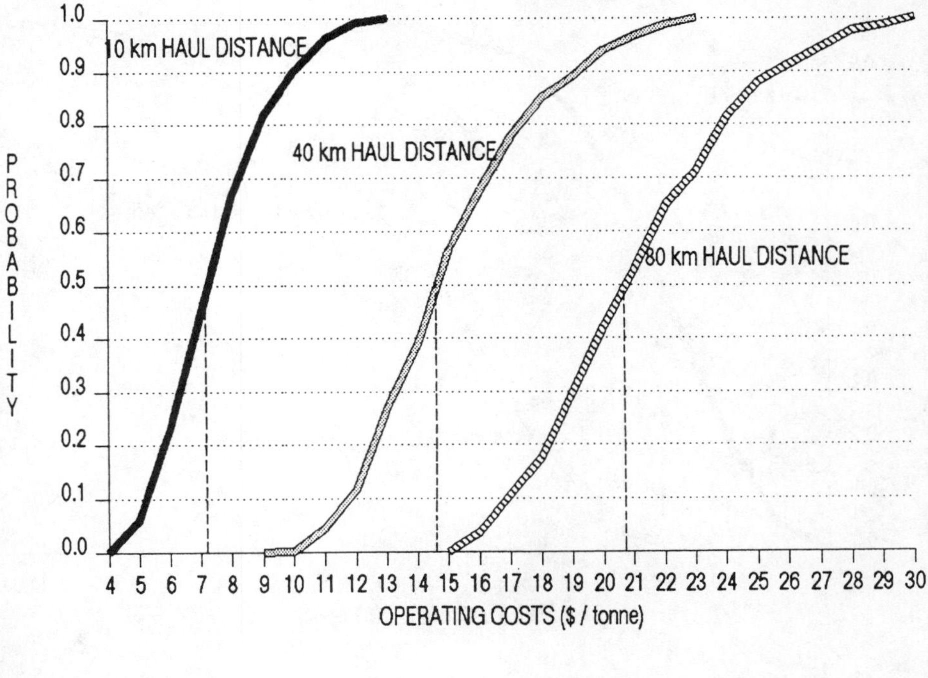

FIGURE 7
Probabilistic Model Three Axle Grain Truck

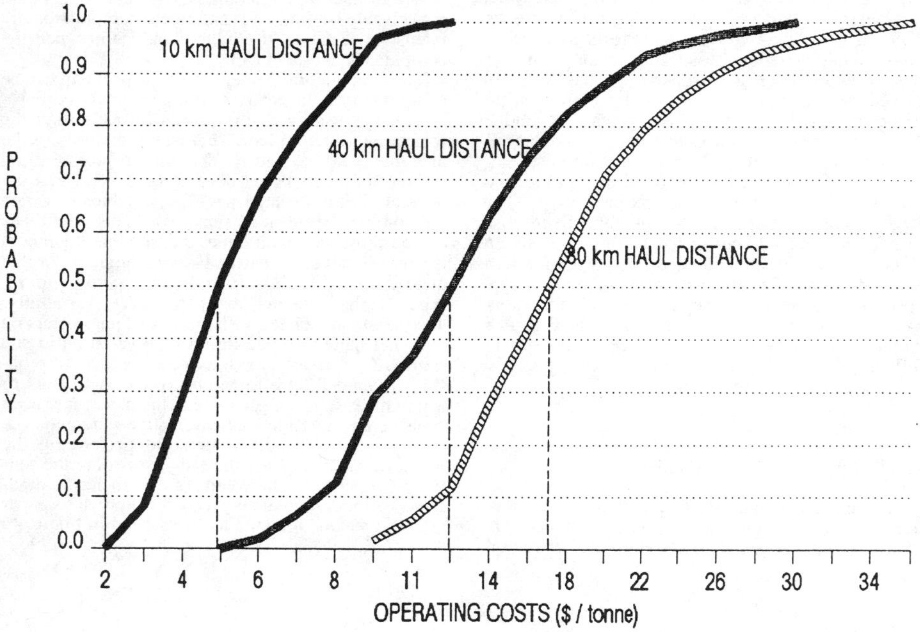

FIGURE 8
Probabilistic Model Custom Three Axle Truck With PUP

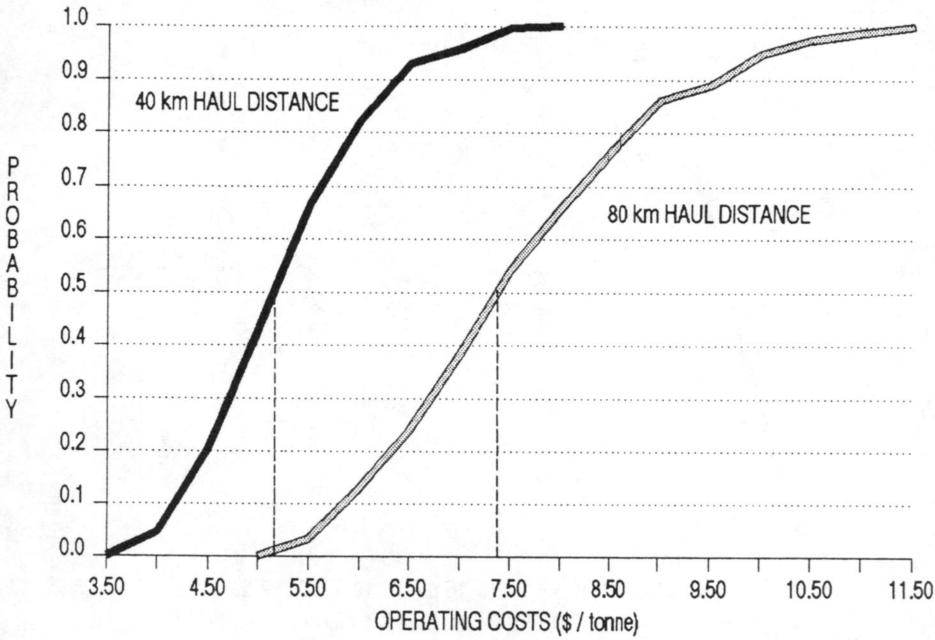

V. CONCLUSION

Figures 6 through 9 illustrate (as would be expected) that costs for a particular truck type increases as haul distance increases. It is interesting to note, however, that expected costs do not increase in proportion to haul distance (i.e. expected costs for a 2 axle truck hauling 10 km is about $7.25/tonne, for 40 km (4 times further) is about $14.50/tonne and 80 km is $20.75/tonne—that is, it is possible to haul 4 and 8 times further and costs increase by 2 and 3 times respectively— similar observations apply to the 3 axle plus pup and 7 axle units. This is attributable to increased utilization (i.e. distance) per year for longer hauls—thereby spreading fixed costs over greater distance.

Also from Figure 6 through 9 it is noted that some of the distributions "overlap" each other. That is the distribution for the 10 km haul overlaps the distribution for the 40 km haul for the 2 axle truck. This "overlapping" of distributions suggests that some producers would experience the same costs even if they hauled different distances. This is attributable to different cost bases. It is also noted that the distributions "tighten-up" as the truck size increases. That is the variation/uncertainties tend to reduce in relative terms with the larger trucks. This again is attributable to greater utilization as well as less relative variability/uncertainty with respect to the values for determining parametrers.

All of the plots illustrated in Figure 6 through 9 are illustrated together in Figure 10. This plot illustrates that a large truck with high utilization can haul 80 km (curve 10 in Figure 10) for the same cost a smaller truck with lower utilization can haul 10 km (curve 4). Other similar conclusions are possible for various truck sizes and haul distances.

Figure 10 also illustrates how the distributions (i.e. variations/ uncertainties) in costs associated with the larger trucks are "tighter" than those for the smaller vehicles. That is curves 7, 8, 9 and 10 demonstrate less variations/uncertainties in costs than do curves 1, 2, 3, etc.

Figures 11 through 13 illustrate the probability density functions associated with the cumulative probability functions previously illustrated. These Figures illustrate the "overlapping" of distributions associated with various vehicle types for particular haul distances. In this context the "overlapping" can be interpreted to mean that producers with different vehicle types would be experiencing the same costs for that particular haul distance. That is, for a haul distance of 10 km (Fig. 11) one would expect to see farmers using both 2 and 3 axle vehicles because they could have the same associated costs in many cases.

On the other hand, Figure 12 and 13 illustrate that one would expect custom 3 axle units with pups and commercial 7 axle units to dominate if the haul distances increased to 40 or 80 kms. This is because the associated costs are substantially less than those associated with 2 and 3 axle units (i.e. there is little overlap).

FIGURE 9

Probabilistic Model Commercial 7-Axle A-Train

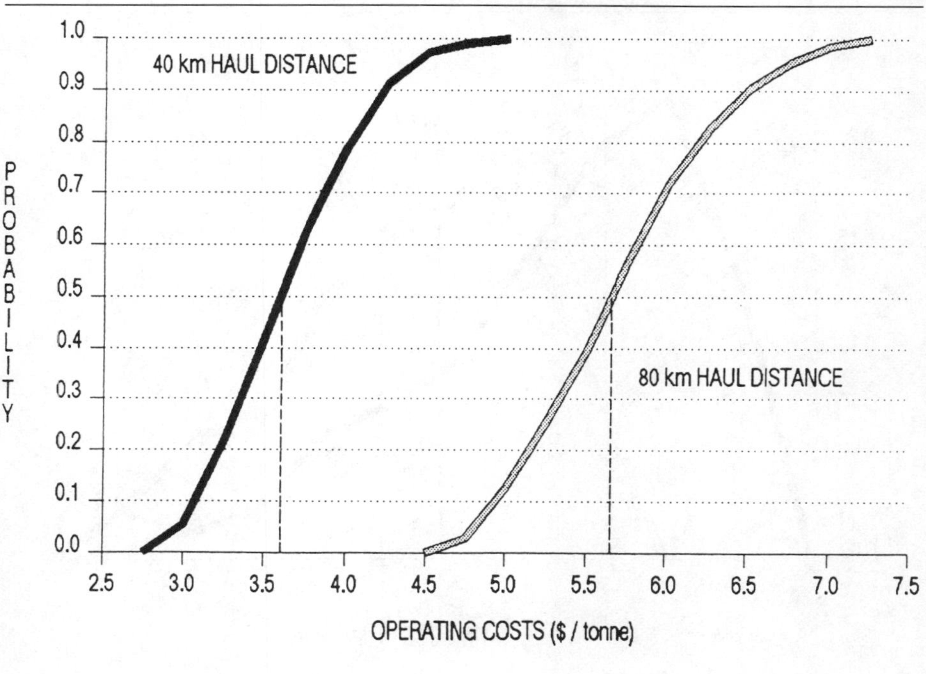

FIGURE 10
Probabilistic Model Various Truck Types

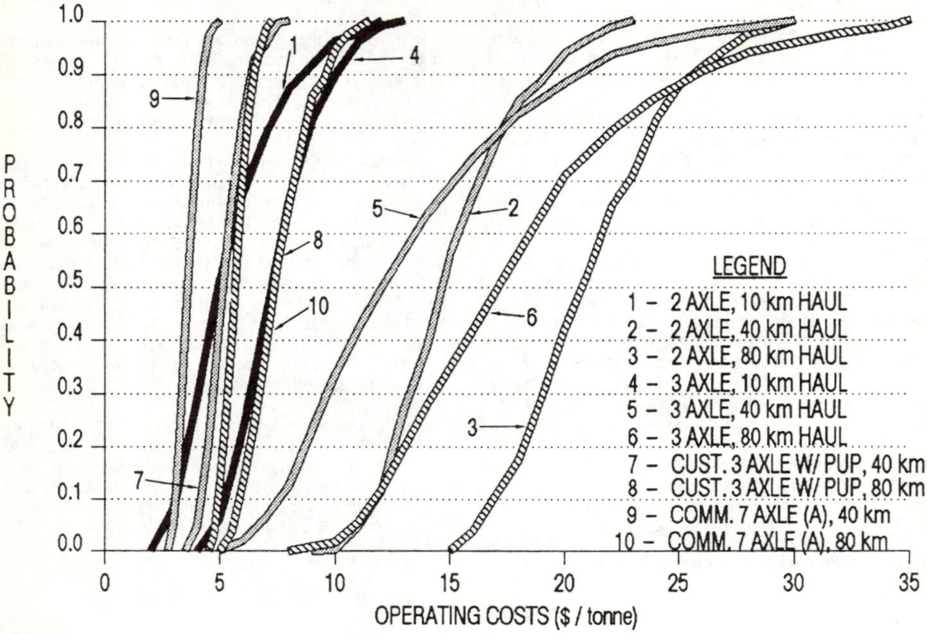

FIGURE 11
Probabilistic Model 10 km Haul Distance

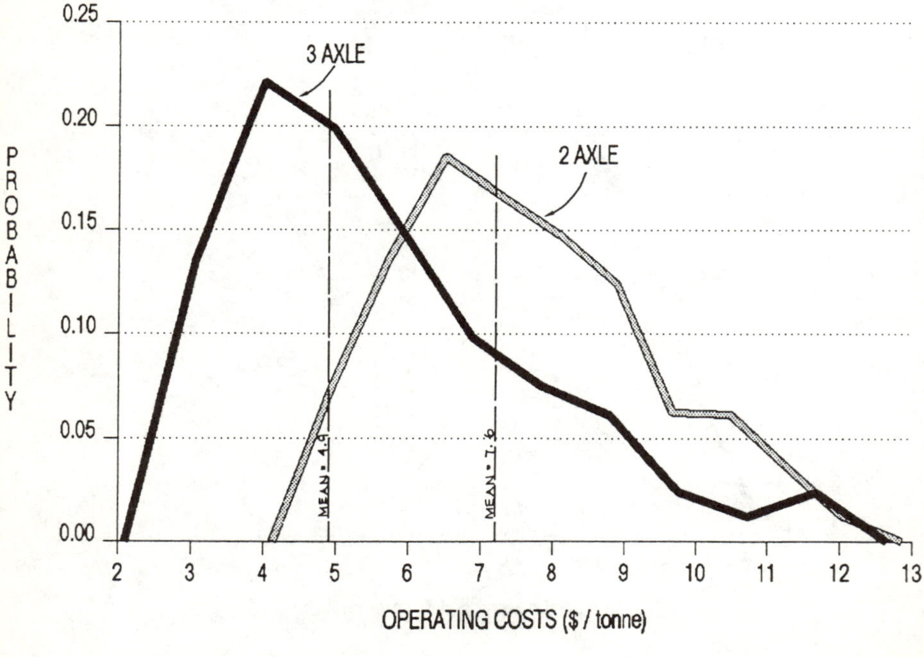

FIGURE 12
Probabilistic Model 40 km Haul Distance

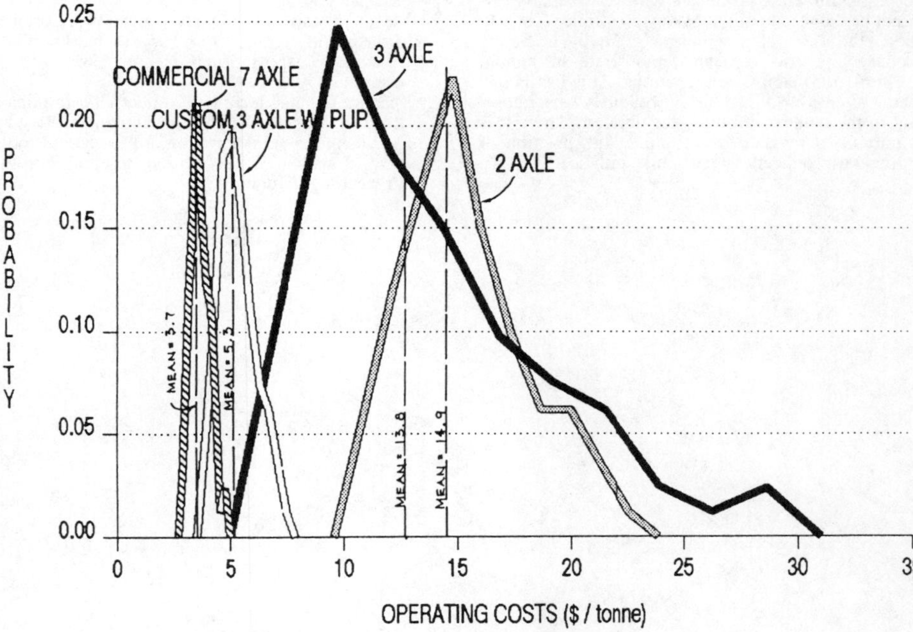

FIGURE 13
Probabilistic Model 80 km Haul Distance

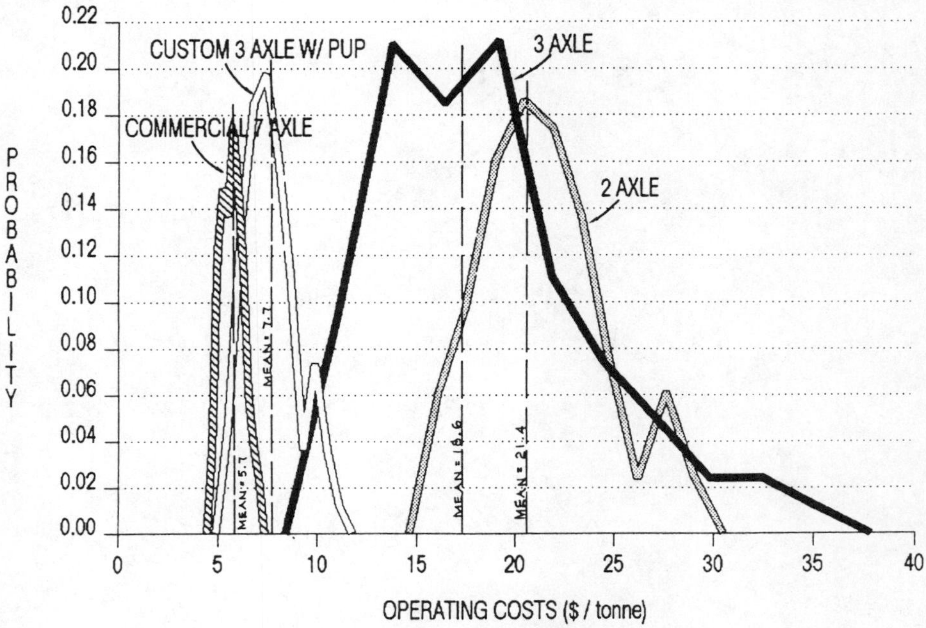

In summary, the above analysis has demonstrated how different producers with similar haul distance can experience substantially different costs for transporting grain while producers with substantially different haul distance might have very similar costs. In short, the costs of assembling grain by truck are producer specific and can demonstrate substantial (i.e. order of magnitude) variations. This fact is one that is often overlooked but is obviously very important in the context of policy formulation where it is important to be able to anticipate the reaction of individual producers to particular policies.

REFERENCES

* Department of Civil Engineering, University of Saskatchewan.

[1] Meyer, N, Sparks, G. The Economic Cost of Transporting Grains from Farm to Market, Canadian Transportation Research Forum Proceedings, June 1987.

[2] Sparks, G, Neudorf, R. Productivity Implications of Vehicle Weight and Dimension Regulations in the Transportation of Petroleum Products, Canadian Transportation Research Forum Proceedings, June 1987.

Impact of Rail Mergers on Kansas Export Wheat Rail Rates and Costs

*by Kenneth Lemke and Michael W. Babcock**

ABSTRACT

In the 1980's the U. S. railroad system is experiencing a rapid rate of consolidation. Proponents of rail mergers argue that they benefit shippers through access to new markets, faster delivery time, and reduced rates. Opponents argue that mergers increase market power and rail rates. This paper provides some empirical perspective on the impact of rail mergers on rail costs and grain rates. Specifically the study measures the effect on western Kansas export wheat rates of actual and proposed mergers of western railroads.

It is assumed that railroads pursue a profit maximizing rate strategy. However the study measures the maximum constraint on railroad ability to raise rates. This is accomplished by assuming all rival carriers employ variable cost pricing. Profit maximizing rates are initially measured for each railroad in each county assuming no mergers. Rates and rail costs are re-estimated as a result of the Union Pacific-Missouri Pacific merger and a hypothetical Santa Fe-Southern Pacific merger.

A Santa Fe-Southern Pacific merger would result in a significant increase in the ability of the merged system to raise rates in southwest Kansas. The Union-Pacific-Missouri Pacific merger produces no increase in market power. Both mergers result in lower rail costs due to more direct routing to the Gulf.

The study indicates that competitive rail rates will occur if shippers have access to at least two railroads.

I. INTRODUCTION

In 1950 railroads accounted for 56% of U. S. freight ton miles. By 1985 rail market share had declined to 37%. Many factors played a role in the decline of railroads but regulation was the most significant. Rate regulation prevented railroads from responding quickly to changing market conditions. Any significant rate change proposal was greeted by protests from affected shippers, carriers, and regional interests. The resulting slow, cumbersome rate review process had a substantial negative effect on rail market share.

The Staggers Act of 1980 greatly reduced regulation of the rail industry. The Act recognized the high degree of intermodal competition in most transportation markets. Also in 1980 there were 40 Class I railroads, indicating potential intramodal competition. Thus competition would regulate rail rates.

The Staggers Act also contained provisions designed to expedite railroad mergers. For example merger decisions by the ICC must be made within 30 months of the date of application. Although the impact of deregulation is difficult to measure, railroad merger activity has accelerated in recent years. The following major combinations have occurred in the 1980's.

1. In April, 1980 the Burlington Northern merged with the St. Louis, San Francisco to form the nation's largest railroad.
2. In October 1980 the Chessie System combined with the Seaboard System to form the CSX railroad (22,000 miles).
3. The Southern railroad and the Norfolk & Western merged in March, 1982 to form the Norfolk Southern railroad (18,000 miles.)
4. In June, 1982 the Union Pacific, Missouri Pacific, Western Pacific merger was approved.

The proposed Santa Fe-Southern Pacific merger has been rejected once by the ICC but is currently being reviewed. The Union Pacific has proposed a merger with the MKT railroad.

Proponents of mergers cite several potential advantages. Some of these would benefit shippers while others would assist merging carriers. In end-to-end mergers, shippers have direct access to more markets on the merged system. Reduced need for interlining also results in faster delivery time and less administrative costs for shippers. In parallel mergers the elimination of duplicate facilities reduces the costs of the combined system. Mergers will reduce rail costs and strong intermodal competition will turn rail cost reductions into reduced rates.

Market power was the major reason for regulating rail rates. The recent rapid consolidation of the rail industry concerns many observers including many who initially supported the Staggers Act. Critics of rail mergers argue that the potential cost savings are overstated and that the primary effect will be an increase in rail rates. Mergers will not produce resource savings. They merely transfer income from shippers to railroads.

There have been many studies of the impact of rail deregulation on agricultural transportation markets (1, 3, 4, 5, 6, 7, 8, 11, 14, 19, 21, 22, 23, 25, 27). However there have been comparatively few studies of the effects of rail mergers. The few that exist have examined railroad efficiency gains, market share changes, and the effect of merger configuration on efficiency gains (i.e. end-to-end vs. parallel mergers). Virtually no studies exist which directly address the impact of railroad mergers on rail grain rates and the distribution of any efficiency gains.

Gallamore (10) investigated nine railroad mergers that occurred between 1957 and 1964. He found that only four of the nine mergers resulted in significant cost savings when compared to a control group of railroads with similar characteristics. The cost reduction did not appear to be related to the geographical nature of the merger.

Sloss et al. (24) found that merged railroads performed only slightly better than selected control firms. The authors compared the performance of merged railroads vs. the control group, using fifteen different performance measures. The merged railroads failed to do better on any of the performance measures at the .05 level of significance. At the .10 level of significance, the merged firms did slightly better on five of the performance criteria.

Levin and Weinberg (18) investigated the effect of the geographic configuration of railroad mergers on rail cost savings. Using market share as a measure of efficiency gains, they found that firms engaging in parallel mergers did not increase their market share. However, an important defect of the study is that the analysis did not permit the merging firms to abandon parallel routes. The authors found that firms engaging in end-to-end mergers were able to raise their market share.

Harris and Winston (12) concluded that cost savings associated with end-to-end mergers are larger than those of parallel mergers. However, the conclusion is weakened by the assumption of a fixed network (i.e., no abandonment). The main benefit of end-to-end mergers is said to be an improvement in service quality.

Levin's study (17) of the impact of rail deregulation on rail rates and profitability sheds some light on the effect of rail concentration on rail grain rates. With a modest degree of intra rail competition, rail grain rates remain relatively stable. However, without intra rail competition, they increase by 40-80% compared to the regulated rate base case.

MacDonald (19) has recently completed a study that examines the impact of mergers on rail corn, soybean, and wheat rates. The study measures the effect of a Norfolk Southern-Conrail merger on corn and soybean rates in Indiana and Ohio. Impacts of the proposed Santa Fe-Southern Pacific merger on Texas, Oklahoma, and Kansas wheat rates are also examined. In general the size of any merger induced rate increase depends on the pre merger level of rail concentration, the increase in concentration caused by the merger, and the distance of the railroads from water competition.

The overall objective of the paper is to provide some empirical perspective on the impact of rail mergers on rail costs and grain rates. Specific objectives include the following:

1. Estimate the impact on western Kansas export wheat rates of the merger of the Union Pacific and Missouri Pacific.
2. Measure the potential impact on western Kansas export wheat rates of the proposed merger of Santa Fe and Southern Pacific.
3. Measure the potential impact on western Kansas export wheat rates of an abandonment of the Kyle railroad.
4. Measure the rail cost savings in western Kansas of the Union Pacific-Missouri Pacific merger.
5. Measure the rail cost savings in western Kansas of the proposed Santa Fe-Southern Pacific merger.

II. THE STUDY AREA AND THE MODEL

The grain origin area is the three crop reporting districts of western Kansas which comprise about one-third of the state's land area. See Figure 1. In 1983, the area produced 178.3 million bushels of wheat (40% of the state's production) of which 62% was exported through the Gulf of Mexico. The area also produced 54.4 million bushels of corn (64% of state output) and 43.2 million bushels of sorghum. Kansas is the third largest cattle feeding state and 70% of the state's cattle on feed are in the study area.

The northern third of the study region is served by the Burlington Northern, the Kyle, and the Union Pacific railroads. In the central portion, rail service is provided by the Santa Fe and the Missouri Pacific. The southern part is served by the Santa Fe and the Southern Pacific. In general the Union Pacific System dominates the northern part of the area and the Santa Fe dominates the south.

The area has about 250 country elevators but only one unit trainloading facility (Colby, Kansas).

The grain origin area is only part of the export wheat logistics system. In Kansas most of the wheat is stored in country elevators at harvest. During the marketing year, wheat is shipped from country elevators to large inland terminals. Ultimately the grain is shipped from inland terminals to export elevators at the Gulf (or domestic milling locations). The model contains seven inland terminal locations. These are Atchison, Hutchinson, Kansas City, Salina, and Wichita, Kansas; Catoosa, Oklahoma, and Amarillo, Texas. Atchison, Kansas City, and Catoosa are served by rail, truck, and water carriers. Only rail and truck transport is available at all other inland terminal locations. The model incorporates all major export areas on the Gulf of Mexico including New Orleans, Houston, Galveston, and Corpus Christi.

In the past, linear programming has been used successfully for solving the "transportation problem." However in recent years network models have more often been employed for solving large scale grain logistics problems. This is because the lower cost of the network model facilitates the handling of more complex logistics systems and permits more spatial disaggregation. For details of the network model see (9).

This study uses a network flow model employing out-of-kilter algorithm to obtain least cost grain flows. The objective function minimizes total shipping and handling costs for the movement of wheat from production origins to Gulf terminals. If other non-Gulf destinations had been included, the least cost solution would indicate that some production areas do not ship wheat to Gulf ports. This is undesirable for three reasons. First, the objective of the study is to determine the ability of railroads to increase rates on shipments to the Gulf. Therefore, it is necessary to find the least cost route from each production site to the Gulf. Second, rail service to the Gulf is available from every elevator site in the study, therefore, railroads have to establish rates for these movements. Third, over 60 percent of the study area's 1983 wheat crop moved to Gulf ports for export, and significant amounts moved from each production area. Therefore, a model that shows wheat flows to the Gulf from *each* production origin is more realistic.

Exclusion of non-Gulf port areas will not prevent achieving the objectives of the study. This is because those ports are served by other wheat supply areas which have a locational advantage. Thus the Pacific Northwest ports are served by Washington, Mon-

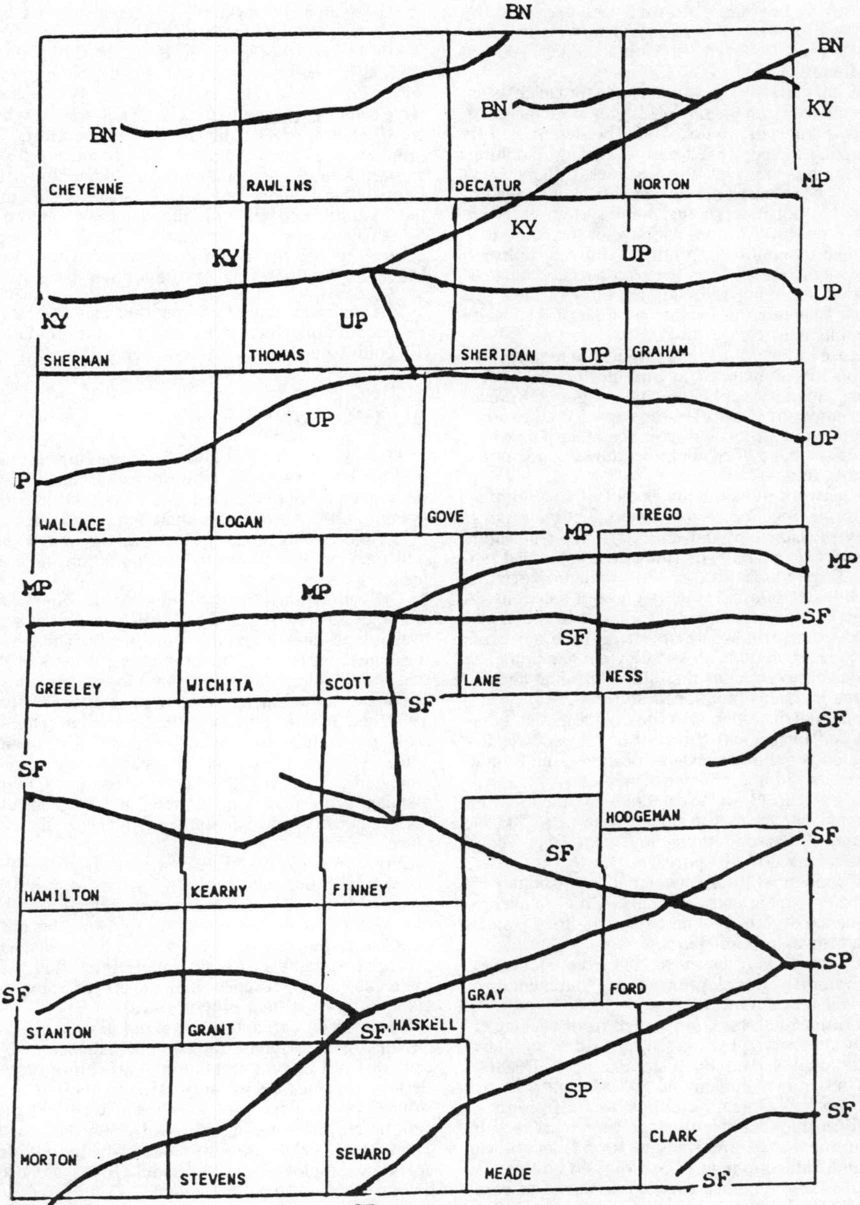

Figure 1. Study Area

tana, and the Dakotas. California ports are served by Utah, California, and Arizona. The western Great Lakes ports are served by Montana, the Dakotas, and Minnesota (15).

Elevator capacity constraints and storage costs are also excluded from the model. If they were included, least cost transport routes would be determined by minimizing the combined cost of storing, handling, and shipping. In effect, this assumes a "benevolent dictator" who forces all participants in the logistics system to cooperate so that total system costs are minimized. However the objective of the study is to determine the ability of profit maximizing railroads to raise their rates. Thus least cost routes must be identified assuming railroads act only in their self interest. This requires exclusion of elevator capacity constraints and storage costs.

The use of the out-of-kilter algorithm requires that total quantity demanded at final destinations equal the total quantity supplied at the origins. However the assumption of no elevator capacity constraints means that the presence or absence of surplus wheat or excess demand does not affect the selection of the least cost routes.

The network model includes all types of shipping and storing locations such as production origins, country elevators, inland terminals, river elevators and export terminals. The study area is divided into 20-by-20 mile areas to form 70 production origins. The model includes 57 country elevator locations. In general, country elevator sites closest to the center of production origins are selected. In some cases origin areas shared the closest elevator location. The model also incorporates the inland terminal and export area locations mentioned above.

The model includes all types of potential intramodal and intermodal competition. After wheat is assembled at country elevator locations, the following types of shipment patterns may occur:

1. Multi car (3 or 5 car) rail shipments from country elevators to inland terminals. 50, 54, and 60 car unit train movement from inland terminals to Gulf ports.
2. Commercial truck movement from country elevators to inland terminals. Unit train shipments of 50, 54, and 60 cars from inland terminals to Gulf ports.
3. Commercial truck movement from country elevators to river elevators. Barge shipment from river elevators to the Gulf.
4. Commercial truck movement from country elevators served by a given railroad to an inland terminal served by a competing railroad(s). Unit train shipment of 50, 54, and 60 cars from these inland terminals to Gulf ports.
5. Farm truck shipment to the train loading subterminal at Colby, Kansas. 50, 54, and 60 car unit train shipment from Colby to Gulf ports.

Railroads in the study area are assumed to compete with each other and do not interline traffic. Implicit or explicit collusion is possible but not likely. Two provisions of the Staggers Act have induced rivalrous railroad pricing behavior. These are (1) prohibition of discussion of single line rates in rate bureaus and (2) secret rail/shipper contracts. In the four years after the Staggers Act, nominal rail rates in Kansas declined by 34% which is inconsistent with effective cartel pricing (5).

Under certain circumstances, transport rates from other wheat supply areas could restrain rail rates in the study area. For example rail rates from western Kansas to the Gulf can be no higher than rates from South Dakota to the Gulf. However this type of alternative source competition is not a binding constraint on rail rates in the study area. This is because large alternative supplies of hard red winter wheat are located at greater distances from the Gulf. The principal constraint on rail ability to raise rates in western Kansas is competition from other study area railroads and competing modes. Stated differently, the binding constraint on the Santa Fe rate from Scott City, Kansas is the Missouri Pacific rate from Scott City not the Burlington Northern rate from Mitchell, South Dakota. As Levin has stated:

> "Thus inter source competition can at best be an imperfect substitute for intramodal point to point competition." (16)

III. THE DATA

The objective of the study is to measure the impact of rail mergers on rail costs and ability to raise rates. No attempt is made to measure total logistics system costs. Accordingly there are no elevator capacity constraints or storage costs in the model. The only data inputs required are transportation and grain handling costs.

The estimation year for the model is 1983. This was the latest year for which all data inputs were available when work on the study began. There have been no significant structural changes since 1983 that would alter the results of the study. Kansas intrastate motor carrier rates were regulated before 1983 and remain so today. Water carrier rates on bulk commodities have been exempt from regulation since 1940. Railroads and interstate motor carriers were deregulated in 1980. There have been no major changes in the location, number, and types of elevators and no rail abandonment in the study area since 1983.

Farm trucking costs are obtained from a 1983 economic engineering study of 300 and 450 bushel farm trucks (7). The estimated costs include variable costs (with distance) fixed costs, and transfer costs which are a function of loading/unloading costs. The principal variable costs are labor time, fuel, and maintenance. Fixed costs include depreciation, insurance, interest, and property taxes.

Commercial agricultural trucking has the characteristics of a perfectly competitive industry and is exempt from federal regulation. The ability of carriers to cut rates below fully allocated cost is constrained by a high proportion of variable costs. Competition prevents pricing above fully allocated costs. Therefore it makes little difference whether rates or costs are employed in the model. Rates are used because they are readily available (13).

The barge industry has the characteristics of a competitive industry with many small carriers, low barriers to entry, homogeneous service, and regulation exemption of bulk commodity movements. Thus long run barge rates will approximate barge cost. The model employs barge rates. The last published rates expired in February, 1976 and there is no publicly available source of more recent rates. The 1976 rates are inflated to the 1983 price level (2).

A rail cost program developed at North Dakota State University is used to measure the variable costs of rail movements (26). The program is based on Phase II, Worktable E of the Uniform Railroad Cost System (URCS). It can provide cost estimates for single car, multi car, single origin trainload, multiple origin trainload, and unit train movements. All cost estimates are in 1983 dollars.

The estimated 1974/75 grain handling costs at country elevators, inland terminals, and Gulf port terminals are inflated to the 1983 price level (28). These *cost* estimates are close to Kansas elevator *rates* for handling grain in 1983 (20). Handling costs are those associated with receiving and loading out grain.

IV. EMPIRICAL RESULTS

The rail transportation market in western Kansas is an oligopoly. Accordingly there are many possible pricing strategies depending on how rivals react to each other's initiatives. In order to measure the maximum constraint on rail market power in the region we assume that *motor carriers and water carriers set rates equal to variable cost*. When trucks haul grain to an inland terminal for export movement by a rival railroad (rival to another railroad attempting to increase rates) we assume the *rival railroad sets rates equal to variable cost*.

To measure the ability of originating railroads to raise rates, the following three step procedure is employed.

1. The model determines least cost routes and rail carriers for each production origin. Each originating railroad sets rates equal to variable cost.
2. A railroad (for example the Santa Fe) is selected to determine its ability to raise rates. Variable cost rates of the selected railroad are increased to the level of its best competitor. This is nearly always truck shipment to an inland terminal and rail shipment by a rival railroad from the inland terminal to the Gulf.
3. Simulation (1) is compared to simulation (2). The cost differences on various routes served by the selected railroad represents its cost advantage over its best competitor. *This is also the railroads ability to raise rates without losing any traffic*.

The above procedure is executed for each of the railroads serving the region. The results are in Table 1. The table contains the ratio of rail rates to rail variable cost for the least cost railroad in each county of the study area. An examination of Table 1 reveals wide variation in the ability of railroads to raise rates. Santa Fe dominance of the southwest area of the region is quite evident. The 140% of

TABLE 1

Railroad Rate/Variable Cost in Western Kansas[1]

County	Rail Rate/ Variable Cost × 100	Originating Railroads[2]
Cheyenne	100	Burlington Northern
Rawlins	118	Union Pacific/Burlington Northern
Decatur	108	Burlington Northern/Kyle
Norton	107	Burlington Northern/Kyle
Sherman	113	Kyle
Thomas	123	Union Pacific
Sheridan	109	Union Pacific
Graham	111	Union Pacific
Wallace	119	Union Pacific
Logan	119	Union Pacific
Gove	105	Union Pacific
Trego	107	Union Pacific
Greeley	109	Missouri Pacific/Santa Fe
Wichita	103	Missouri Pacific
Scott	104	Missouri Pacific/Santa Fe
Lane	104	Missouri Pacific/Santa Fe
Ness	104	Missouri Pacific/Santa Fe
Hamilton	115	Santa Fe
Kearny	114	Santa Fe
Finney	116	Santa Fe
Hodgeman	121	Santa Fe
Stanton	140	Santa Fe
Grant	130	Santa Fe
Haskell	124	Santa Fe
Gray	133	Santa Fe
Ford	129	Santa Fe
Morton	139	Santa Fe
Stevens	115	Santa Fe
Seward	106	Southern Pacific
Meade	112	Santa Fe
Clark	128	Santa Fe

[1]Assumes all railroad competitors charge variable cost rates.
[2]These originating railroads are the least cost carriers in each county.

variable cost rate in Stanton county and 139% in Morton county are the highest in the region. Vigorous competition between the Santa Fe and the Missouri Pacific occurs in the central part of the area. The 104% of variable cost rates in Scott, Lane, and Ness counties are among the lowest in the region. Perhaps the greatest amount of variation in ability to raise rates occurs in the northern counties. The Burlington Northern has no ability to raise rates above variable cost in Cheyenne county. However the Union Pacific can achieve a 123% rate in Thomas county.

The impact of a merger on railroad rates is simulated in a three step procedure.

1. The ability of *each* study area railroad to raise its rates is measured as described above.
2. The Santa Fe and Southern Pacific (or the Union Pacific-Missouri Pacific) are combined and the costs on the merged system's routes are raised to the level of the merged railroad's best competitor.
3. Simulation (1) is subtracted from simulation (2). The result is the ability of the merged system to raise rates without losing any traffic.

The results of simulation for the Santa Fe-Southern Pacific merger are in Table 2. The analysis indicates that the merged system has the ability to significantly raise rates in the southwest part of the study area. The rate increases due to the merger range from 13.4% in Stanton county to 34% in Stevens county. The rate changes are substantial because the Southern Pacific is the Santa Fe's best competitor in this region. Conversely the Santa Fe is the Southern Pacific's best competitor.

The Union Pacific-Missouri Pacific merger produced no increase in market power. The combined system has no ability to raise rates that the individual railroads cannot attain separately. This is because the Union Pacific and Missouri Pacific are not each others most effective competition. The Santa Fe is the Missouri Pacific's best competitor and constrains Missouri Pacific rate increases with or without the merger. Similarly the Kyle-Santa Fe combination restricts Union Pacific. This is evident from an examination of Table 3 which contains Union Pacific potential rate increases if the Kyle is abandoned. These range from a 12.5% increase in Sherman county to a 24.4% increase in Thomas county.

TABLE 2

Market Power Increase Due to Santa Fe-Southern Pacific Merger

County	Pre Merger Rail Rate/ Variable Cost x 100	Post Merger Rail Rate/ Variable Cost x 100	Percent Increase in Rates Due to Merger
Stanton	140	159	13.4%
Grant	130	154	18.4%
Haskell	124	148	19.5%
Morton	139	166	19.3%
Stevens	115	154	34.0%
Seward	106	139	31.3%
Meade	112	134	19.2%

TABLE 3

Increase in Union Pacific Rates if Kyle Railroad is Abandoned

County	Union Pacific Rate/Variable Cost x 100 With Kyle	Union Pacific Rate/Variable Cost x 100 Without Kyle	Percent Increase Due to Kyle Abandonment
Norton	107	121	12.8%
Sherman	113	127	12.5%
Thomas	123	153	24.4%
Sheridan	109	128	17.3%
Graham	111	126	13.2%

TABLE 4

Percentage Reductions in Rail Variable Cost Due to Union Pacific-Missouri Pacific and Santa Fe-Southern Pacific Mergers

County	Variable Cost Reduction
Union Pacific-Missouri Pacific	
Rawlins	10.6%
Decatur	1.4%
Sherman	5.9%
Thomas	17.3%
Sheridan	8.8%
Graham	4.3%
Wallace	8.1%
Logan	11.7%
Gove	5.6%
Santa Fe-Southern Pacific	
Ford	1.5%
Seward	5.0%
Meade	6.1%
Clark	5.8%

The above simulations also yield estimates of cost savings due to mergers. The model reveals cost savings if the merged carrier has shorter routes to the Gulf. Prior to the 1982 merger with Missouri Pacific, the Union Pacific had no direct route to the Gulf. In the pre merger simulations we assume study area railroads compete with each other and do not interline traffic. As a result, Union Pacific originated grain moves east to the Atchison-Kansas City area and then to the Gulf on the MKT railroad. However after the merger the Union Pacific can interline with the Missouri Pacific at Salina—a much shorter route to the Gulf. *Thus the primary benefit to the merged system is cost savings on Union Pacific routes*. These are displayed in Table 4. The cost reductions occur all across the northern part of the study area, ranging from a 1.4% decline in Decatur county to a 17.3% decrease in Thomas county.

As stated above, in the pre merger simulation we assume that no study area railroad interlines with any other study area railroad. The Southern Pacific has very circuitous routes from Kansas to the Gulf. Excluding potential connections with other study area railroads, the Southern Pacific's first opportunity to interline traffic for Gulf movement is at Herington, Kansas. This is much farther north and east of Santa Fe routes to the Gulf. After the merger, the Southern Pacific can interline with the Santa Fe at Pratt or Hutchinson, Kansas. These are shorter routes and are the source of cost saving for the merged system. As Table 4 indicates, the cost savings for the Santa Fe-Southern Pacific merger are not as large or pervasive as those of the Union Pacific-Missouri Pacific. The cost declines range from 1.5% in Ford county to 6.1% in Meade county.

V. CONCLUSION

Several types of potential and actual transport competition are present in the study area. Thus railroad pricing latitude is contingent on pricing strategy of rival railroads and modes. This study has measured the *maximum constraint* on the ability of merged railroads to raise rates by assuming that all rivals charge variable cost rates. However during high demand periods, competing modal capacity may be reached. In this environment, railroad ability to raise rates will increase. However this may be mitigated by elevator postponement of shipment until additional motor carrier capacity occurs.

Analysis reveals that the Santa Fe-Southern Pacific can significantly raise rates in southwest Kansas. This is because each railroad is the others best competitor in the region. The merged system can also realize some cost reductions on traffic originated in southwest Kansas. This is due to more direct routing of Southern Pacific traffic. Since rates can increase and costs decline the impact of the merger is higher rail profits on study area grain traffic. Shippers in part of the area would pay higher rates.

The Union Pacific-Missouri Pacific has no ability to raise rates in the northern half of the study area. Competition from the Santa Fe prevents the Missouri Pacific from increasing rates. The Kyle railroad restricts the Union Pacific. The constraint on Union Pacific rates is the major benefit of the Kyle to the area's grain shippers. The merged system realizes substantial cost reduction in northwest Kansas due to more direct routing of Union Pacific traffic. Thus the merger would not lead to higher rates. Indeed lower rates could be the result. Since the Union Pacific realizes lower costs it could choose to increase market share by lowering rates. The success of this strategy would depend on the reaction of the Kyle. In fact both railroads have reduced rates since 1983.

We compared the rates predicted by the model to actual 1983 Santa Fe rates on traffic originated in southwest Kansas. Santa Fe multi car rates ranged from 5% above to 8% below the rates predicted by the model. Since 1983 the rate structure in the area has changed dramatically. By 1986 average Santa Fe Gulf export rates had declined 15.4%. This could be due to a variety of factors including declining wheat prices, depressed wheat exports, increases in rail capacity, and increasing motor carrier competition. In this environment a Santa Fe-Southern Pacific merger would not result in increased rates for Kansas grain shippers. However the merged system would realize somewhat greater profits due to cost reductions. A recovery in the grain markets would yield results similar to those indicated by the model.

The current level of intra rail competition in Kansas is sufficient to prevent major rail rate increases when an agricultural recovery occurs. This is predicated on the ability of shippers in western Kansas to truck grain to inland terminals where several railroads compete. The study also indicates that competitive rail rates will occur if shippers have access to at least two railroads. In northwest Kansas the Kyle railroad constrains Union Pacific rates. In west central Kansas, the Missouri Pacific and Santa Fe contain each others rates. In southwest Kansas the Southern Pacific provides some restraint on Santa Fe rates. The model indicates that a regional monopoly will raise rates substantially.

REFERENCES

* Economics Department, Kansas State University, Manhattan, KS.

1. B. D. Adam and D. G. Anderson "Implications of the Staggers Rail Act of 1980 for Level and Variability of Country Elevator Price Bids," *Proceedings of the Transportation Research Forum-1985,* 26 (November, 1985) 357-363.
2. Arrow Transportation Company, *Guide to Published Barge Rates on Bulk Grain,* Supplement No. 3, Schedule No. 8, Sheffield, Alabama, 1975.
3. M. W. Babcock, "Potential Impact of Rail Deregulation in the Kansas Wheat Market," *Journal of Economics,* 7 (St. Louis, 1981) 93-98.
4. ——————, *Efficiency and Adjustment: The Impact of Railroad Deregulation,* Cato Institute (Washington, D.C.: 1984).
5. ——————, et al. "Impact of the Staggers Rail Act on Agriculture: A Kansas Case Study," *Proceedings of the Transportation Research Forum-1985,* 26, No. 1 (November, 1985) 364-372.
6. K. L. Casavant, "An Examination of the Relationship Between the Staggers Rail Act and Agriculture in the United States," *Proceedings of the Transportation Research Forum-1985,* 26 (November, 1985) 373-376.
7. M. H. Chow, *Economic Impacts of Structural Changes in the Wheat Logistics System: Exporting Winter Wheat From Twelve Counties in Northwest Kansas*. Unpublished Ph.D. dissertation, Kansas State University (1984) 90.
8. S. Fuller, L. Makus and M. Taylor, "Effect of Railroad Deregulation on Export Grain Rates, *North Central Journal of Agricultural Economics,* 5 (January, 1983) 51-63.
9. S. W. Fuller and C. V. Shanmugham, "Network Flow Models: Use in Rural Freight Transportation Analysis and A Comparison with Linear Programming," *Southern Journal of Agricultural Economics,* 10, (1978) 183-88.
10. R. E. Gallamore, "Railroad Mergers: Cost, Competition, and the Future Organization of the American Railroad Industry," Ph.D. dissertation, Harvard University, 1969.
11. H. W. German and M. W. Babcock, "The Impact of Rail Deregulation on the Movement of Fresh Fruits and Vegetables," *Logistics and Transportation Review,* 18/4 (December, 1982) 373-384.
12. R. G. Harris and C. Winston, "Potential Benefits of Rail Mergers: An Econometric Analysis of Network Effects on Service Quality," *The Review of Economics and Statistics,* LXV, No. 1 (February, 1983) 32-40.
13. Kansas Corporation Commission, *Kansas Motor Carrier Association, Motor Freight Tariff No. 50-I.* Issued March 15, 1983.
14. K. A. Klindworth, et al. *Impacts of Rail Deregulation on Marketing of Kansas Wheat,* U. S. Department of Agriculture, Office of Transportation (Washington, D.C.: 1985).
15. M. N. Leath, L. D. Hill, and S. W. Fuller, *Wheat Movements in the United States,* North Central Regional Research Publication No. 274, University of Illinois at Urbana-Champaign (January, 1981).
16. R. C. Levin, *Verified Statement of Richard C. Levin, ICC Finance Docket No. 30,000 (Sub-No. 34) Union Pacific-Control-Missouri Pacific* (February 11, 1981) 18.
17. R. C. Levin, "Railroad Rates, Profitability, and Welfare Under Deregulation," *Bell Journal of Economics,* 12 (Spring, 1981) 1-26.
18. —————— and D. H. Weinberg,

"Alternatives for Restructuring the Railroads: End to End or Parallel Mergers?" *Economic Inquiry,* 17 (July, 1979) 371-388.

19. J. M. MacDonald, "Competition and Rail Rates for the Shipment of Corn, Soybeans, and Wheat," *Rand Journal,* 18 (Spring, 1987) 151-163.
20. National Grain and Feed Association, *Grain Storage and Merchandising in 1986* (Washington, D.C., 1986) 103-104.
21. Office of Transportation, U. S. Department of Agriculture, *An Assessment of Impacts on Agriculture of the Staggers Rail Act and the Motor Carrier Act of 1980* (August, 1982).
22. Office of Transportation, U. S. Department of Agriculture, *Effects of the Staggers Act on Grain Marketing* (July, 1984).
23. Sarwar and D. G. Anderson, "Impact of the Staggers Act on Variability and Uncertainty of Farm Product Prices," *Proceedings of the Transportation Research Forum-1986,* 27 (September, 1986) 144-150.
24. F. Sloss, T. J. Humphrey, and F. N. Kruttner, "An Analysis and Evaluation of Past Experience in Rationalizing Railroad Networks," 16 *Studies in Railroad Operations and Economics* (MIT: 1975).
25. L. O. Sorenson, "Some Impacts of Rail Regulatory Changes on Grain Industries," *American Journal of Agricultural Economics* 66 (December, 1984) 645-650.
26. D. Tolliver and W. Wilson, "Railroad Costing Software for the Analysis of Agricultural and Bulk Commodity Shipments: Documentation and User's Guide," Upper Great Plains Transportation Institute, North Dakota State University, September 14, 1985.
27. W. W. Wilson, "Railroad Pricing and Deregulation," *Proceedings of the Transportation Research Forum-1985,* 26 (November, 1985) 105-109.
28. U. S. Department of Agriculture, Economic Research Service, "Costs of Storing and Handling Grain in Commercial Elevators, 1971-72 and Projections for 1974/75," (Washington, D.C., 1973).

Effect of Deregulation on Export-Grain Rail Rates in the Plains and Corn Belt

by Stephen Fuller, David Bessler,* James MacDonald,** and Michael Wohlgenant****

ABSTRACT

The purpose of this study is to measure the impact of deregulation on export-grain rates with use of a procedure which attempts to correct for shortcomings of earlier research studies. In particular, the specified procedure controls for the impact of events which may be coincident with deregulation and, in addition, circumvents the problem associated with inaccessible rate information included in the confidential contracts.

Results show export-rail rates in the central and south Plains to decline dramatically while rates linking eastern Corn Belt areas with East Coast ports declined modestly. Deregulation seems to have had little impact on corridors linking Iowa and Illinois with Gulf ports. Even though price spreads on these corridors declined during the deregulated era the analysis suggests this was due to declining barge rates rather than deregulation.

I. INTRODUCTION

The Staggers Rail Act of 1980 represents one of the most dramatic changes in federal policy toward railroads since the Interstate Commerce Act of 1887, the Act responsible for creation of the Interstate Commerce Commission (ICC). This historic legislation was based on the premise that the railroad industry was a monopoly which required ICC regulation. In contrast the Staggers Act is based on the notion that competition is generally adequate to constrain any potential abuses by railroads.

Some agricultural interests have questioned the workability of the competitive transportation environment, even with the competitive provisions included in the Staggers Act. As a consequence, there is great interest in knowing the effect of deregulation, in particular its impact on railroad prices. Because the theoretical expectations regarding the effect of deregulation are indeterminant, the impact on rates must be addressed as an empirical issue. Several insightful studies have attempted to identify the impact of deregulation by measuring and contrasting rate trends in the preand post-Staggers era. Unfortunately, there are several recognized problems associated with direct measurement of published tariff rates. First, because much of the rail-transported grain (63 percent) moves under confidential contracts, the published tariffs provide incomplete information, i.e., accurate rate information is not accessible. Second, coincident events make it difficult to segregate the effect of deregulation from other happenings. As an example, grain exports have declined nearly 50 percent from the peak years in 1980-81 and presumably this has led to declines in the demand for transportation and a downward pressure on rates. Clearly, a method which controls for these effects is necessary to accurately measure the impact of deregulation on rates.

The purpose of this study is to measure the impact of deregulation on export-grain rates with use of a procedure which attempts to correct for shortcomings of earlier research studies. In particular, the specified procedure controls for the impact of events which may be coincident with deregulation and, in addition, circumvents the problem associated with inaccessible rate information included in the confidential contracts. The procedure is applied to major export-grain transportation corridors linking the central and south Plains and Corn Belt regions with their respective port areas.

II. REVIEW OF LITERATURE

Limited efforts were made to extrapolate the likely consequences of deregulation prior to enactment of the Staggers Act. Most noteworthy among efforts which attempted to evaluate the likely effects of deregulation were those by Friedlaender and Spady, and Levin. Friedlaender and Spady found regulated rail rates were below marginal costs for manufactured products and above marginal costs for bulk products such as grain. They conclude that deregulation would yield a decline in bulk rates and an increase in manufactured product rates if a competitive transportation environment evolved. Levin developed a model which included transportation demands and truck and rail costs for purposes of simulating the effect of deregulation under various assumptions regarding the extent of interrailroad competition. The analysis shows the social desirability of deregulation is closely related to the degree of interrailroad competition and indicates grain transportation markets are probably most accurately represented by an intermediate degree of railroad competition. Therefore, Levin believes grain rates are unlikely to increase much in a deregulated environment.

Despite problems associated with the empirical analysis of deregulation useful evidence has been collected. Several Kansas studies measured and contrasted rate trends during the preand post-Staggers era (Babcock, et al., Klindworth et al., and Sorenson). They report published rail rates between 14 Kansas elevator sites and Gulf ports increased 39 cents in the four years preceding the Staggers Act and declined 37 cents during the four years following deregulation. Adam and Anderson investigated corn and soybean price spreads for a sample of

Nebraska elevators from September 1978 through August 1984. They conclude price spreads declined in the post-Staggers period by large and statistically significant amounts. The Kansas and Nebraska studies imply similar impacts attributable to deregulation.

III. PAPER ORGANIZATION

The paper includes five additional sections. First, attention is given to describing the selected study regions and their historic grain transportation patterns. This is followed by a section that outlines the conceptual model which directs the analysis and a statistical model designed to isolate the effect of deregulation on rates. An additional section provides discussion of variables included in the statistical model and data sources. The results section focuses on that portion of the statistical model which summarizes the effect of deregulation and elaborates on secondary issues associated with deregulation. Finally, a summary and implications are offered.

IV. STUDY AREA GRAIN MARKETING PATTERNS

The study area comprises the states of Kansas, Iowa, and Indiana as well as the Texas Panhandle and portions of Illinois. Kansas and the Texas Panhandle regions are surplus producers of hard red winter wheat, a wheat class which has historically comprised about half of the U.S.'s annual wheat production and about 63 percent of its wheat exports. A study by Leath, Hill and Fuller showed the Texas study regions export from 60 to 81 percent of their wheat shipments, while the Kansas region exported slightly over half of its shipments. Gulf ports receive over 99 percent of the Kansas and Texas regions' export shipments with the Houston-Galveston-Beaumont complex being the principal export location. All study regions rely heavily on rail transportation—about ninety percent of the western Kansas wheat outflow has been rail-carried, whereas northeast Kansas and the Texas Panhandle have shipped about 75 percent by rail. The states of Iowa, Illinois, and Indiana have historically produced nearly 50 percent of the U.S.'s annual corn production with Illinois and Iowa each producing 1.0 to 1.7 billion bushels, while Indiana's annual production ranges between .4 and .7 billion bushels. About 30 percent of the U.S.'s annual corn production typically moves to export markets. All Corn Belt states shipped export grain to at least three of the four export coastal areas (Great Lakes, Atlantic, Gulf, Pacific); however, for Illinois and Iowa the majority of export shipments (88-90 percent) were to Gulf ports, principally Mississippi River ports. In contrast, Indiana shipped about 75 percent of its corn exports to Atlantic ports. Iowa and Illinois shipped up to 90 percent of their Gulf shipments by barge, while all of Indiana's corn movement to Atlantic ports were rail transported (Hill, Leath and Fuller).

V. PROCEDURE TO ESTIMATE EFFECT OF DEREGULATION ON RATES

To identify the effect of deregulation on export-grain railroad price structures, the analyses focuses on the price spread between port and associated hinterland region. In general, grain prices in an export-grain transportation corridor reflect the port price and the transportation and marketing service charges necessary to market the grain to the respective port area. That is, a hinterland's farm level price can be approximated by subtracting from the exporter's port price the transportation and marketing charges necessary to market the hinterland's grain to the export location. In which case, the analyzed price spread includes transportation charges. And, by controlling for changes in other marketing service charges, and by focusing on price spreads rather than published rates, the procedure attempts to isolate the impact of deregulation on transportation charges.

The analyses assumes the grain handling industry is sufficiently competitive so that any change in transportation charges that result from railroad deregulation will be reflected in the geographic price spread. This assumption is supported by several Kansas studies which found price spreads to closely follow changing rail rates (Klindworth, et al., Sorenson). Further, Klindworth, et al. shows the transportation charge to be the largest component of the farm-to-port price spread; for 14 Kansas elevators the rail rate averaged 84 to 91 percent of this geographic price spread.

The model shown in Figure 1 was developed to aid in conceptualization of the procedure. Price relationships on major grain transportation corridors are assumed to be determined under the derived demand framework represented by this model. The grain price spread between a U.S. port and an associated hinterland producing region is determined by the interaction of the export grain demand curve at the port (Dg), the farm-level supply function (Si), the demand for export marketing services derived by subtracting these two schedules (Dm), and the supply of marketing services on the transportation corridor which links the hinterland and the port (Sm). The price spread (m) is the price difference, Pg-Pi, where Pg is the grain price at a port, e.g., a Gulf port, and Pi is the price at an interior location, e.g., a Kansas producing region. To illustrate the economic model, consider the case involving a decline in export demand. This is represented by a leftward

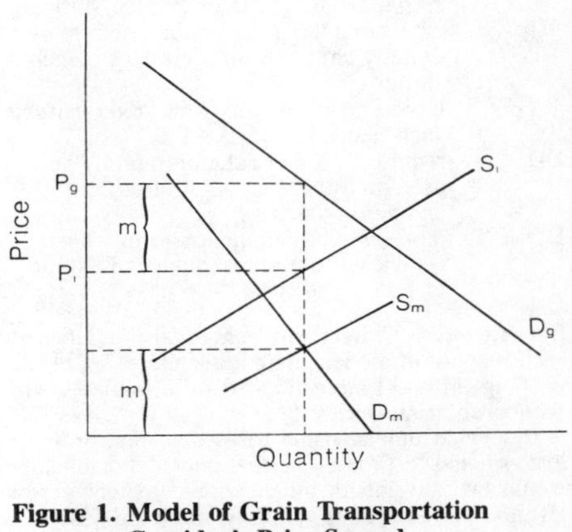

Figure 1. Model of Grain Transportation Corridor's Price Spread

adjustment in Dg and a subsequent and similar adjustment in the derived demand for marketing services (Dm). If farm-level (Si) and marketing service (Sm) supply schedules were unchanged, the price spread (m) would narrow.—Ceteris paribus,—an increase (decrease) in export demand (Dg), or farm-level supply (Si) or a decline (increase) in the supply of marketing services (Dm) will tend to widen (narrow) the price spread, m. And, if deregulation accompanies these shifts, it would be impossible to segregate the impact of the various events through the observation of railroad prices.

In general, the adopted procedure involves the estimation of regression equations which include hinterland's farm-to-port price spread (m) as the dependent variable and independent variables which (1) control for shifts in the above-noted demand (Dg) and supply schedules (Si and Sm); (2) region and time dummies; (3) a time trend; and (4) a dummy and an interaction term to isolate changes in rates that may have resulted from the 1980 deregulation.

The effect of deregulation on rates is accomplished by analyzing the monthly price spread between port and selected hinterland regions from 1976 through 1985. For purposes of determining this effect, the following general model is estimated:

$$P_{it} = a + \sum_{i=1}^{k} b_i D_{it} + \sum_{i=1}^{h} C_i S_{it} + \sum_{i=1}^{l} d_i MS_{it} + \sum_{j=1}^{m} e_j RD_j + \sum_{j=1}^{n} f_j SD_j + gT + hDD + i\ (DD \times T).$$

where,

Pit: represents monthly average differences between the port price and the associated hinterland region's farm level price;

Dit: represents the i *th* export demand shifter of a class of k such shifters;

Sit: represents the i *th* farm level supply shifter of a class of h such shifters;

MSit: represents the i *th* marketing service supply shifter from a class of l such shifters;

RDj: represents the j *th* regional dummy shifter from a class of m such shifters;

SDj: represents the j *th* monthly (seasonal) dummy shifter from a class of n such shifters;

T: represents a monthly time trend variable such that $t = 1,2,\ldots, 120$

DD: represents a deregulation dummy variable such that DD is one for $T > 60$ and zero otherwise;

DD X T: represents an interaction term whose coefficient measures changes in slope of time trend for $T > 60$.

Because several hinterland regions are included in each regression model, the parameters a, bi, ci, di, ei, fi, g, h, and i are estimated from pooled cross-section and time-series data.

It seemed unreasonable to assume that ordinary least-squares estimates of the intercept and slope would be constant for all hinterland regions across all time periods. Thus, dummy variables which allow for the intercept term to vary over time (SDj) and to vary across regions (RDj) were introduced. The region dummies attempt to control for cross-region price spread determinants not formally incorporated in the model.

The effect of deregulation on export-grain rail rate structures is accomplished by controlling for the impact of various variables on the geographic price spread and then isolating changing rates with a dummy variable (DD) designed to estimate any adjustment in the regression equation's intercept that may result from deregulation and an interaction term. The deregulation dummy (DD) is interacted with the time trend variable (T) to create an interaction term (DD x T) whose estimated coefficient measures any change in slope of the time trend variable that may occur as a result of deregulation.

VI. REGRESSION MODEL VARIABLES AND DATA SOURCES

Port hinterland areas were identified so meaningful geographic price spreads might be analyzed. For example, to analyze a Kansas-East Coast price spread would be meaningless since Kansas is not linked to the East Coast by an active grain transportation corridor. Perusal of previous studies, and 1981-85 railroad weighbill and barge grain flow data showed (1) Kansas and Texas to be closely linked with Gulf ports; (2) Indiana to be linked with East Coast ports; (3) Illinois to be linked with Gulf and East Coast ports; and (4) Iowa to be linked with Gulf ports , but to an increasing extent with Pacific Northwest ports since 1980.(Hill, Leath and Fuller; Leath, Hill and Fuller). Unfortunately, it was impossible to analyze the Iowa-Pacific Northwest price spread because export price information was lacking prior to 1980. However, monthly price spread data were available for all other transportation corridors. Monthly price spread information was collected for 30 subregions. These included: nine Indiana regions, three Illinois regions, six Iowa regions, nine Kansas regions, and three Texas Panhandle regions (U.S. Department of Agriculture, 1976-1985a, 1976-1985d).

Several types of variables were identified as potential shifters of export demand, farm-level grain supply and marketing services supply. National export levels and international grain prices were considered as potential measures of export demand. However, due to the lag between the grain sale and its subsequent shipment, the quantity exported per month was predetermined with respect to the current month's price. Thus, quantity rather than price was assumed exogeneous. Furthermore, inclusion of international grain price (ex., Rotterdam price) as a demand shifter may have created a spurious regression problem, i.e., the error term may not have been independent of the explanatory variable. For these reasons, monthly national exports was selected as the appropriate proxy for export demand (U.S. Department of Agriculture, 1976-1985a). Port area demand should be closely related to the national export level.

Several variables were tenatively selected as proxies for the farm-level supply. An effort was made to collect annual grain supply for each hinterland study region, but for several states, the production and price data were not available on corresponding geographic units. It was assumed that the quantity of produced grain was predetermined, so as an alternative, state-level and national grain supply data were the selected variables (U.S. Department of Agriculture, 1976-1985b, 1976-1985c).

During the study period there were substantial changes in factor prices. These changes may have impacted spreads between hinterland farm-level prices and port area export prices. For purposes of this analysis, the real cost of holding grain was represented by the difference between the nominal interest rate and the percent change in the wholesale price index (International Monetary Fund). A simultaneity problem would arise if percent change in grain value were substituted for the wholesale price index.

Because changes in real wage rates may alter the analyzed price spread an index of wages was included (International Monetary Fund). In addition, an index of fuel prices was considered as a potential shifter, since these prices adjusted dramatically during the study period (Association of American Railroads, 1979-1986). Any real increase in wages or fuel prices will have shifted the farm-level and the marketing services supply function (Figure 1). In the case of the farm-level supply function an increase in input prices will tend to narrow the price spread. Increasing input prices will tend to shift the marketing services supply function to the left and increase the price spread. In which case, increasing input prices may create off-setting effects on the analyzed price spread and make it impossible to predict signs. However, for purposes of this study, this outcome does not represent a problem since interest is only in controlling for such effects. To measure the potential effect of changing rail costs on rates and the subsequent impact on price spreads, two rail cost indices were selected. One index measures materials prices and wages, whereas the other includes fuel, material prices and wages (Association of American Railroads, 1979-1986).

A rail utilization index was constructed from the AAR's Rail Grain Traffic by Quarter and Rail Grain Fleet data (Association of American Railroads, 1987). The numerator of the index includes quantity of grain railtransported per unit of time and the denominator reflects railroads' car supply or its grain carrying capacity.

To reflect a widening price spread which may be attributable to shortages in hinterland storage capacity, a ratio variable which measures unused storage capacity relative to total storage capacity was constructed. Due to limited data, it was necessary to estimate the ratio on a statewide basis. The ratio was calculated with use of grain stocks and storage capacity data and exhibits the expected intra-year pattern (U.S. Department of Agriculture, 1976-1985d).

Export barge rates on the inland waterway system were included in the analysis (Illinois Department of Agriculture). The barge mode is the principal carrier of corn from Iowa and Illinois to Gulf ports and is a major competitor of the railroad industry. Table 1 identifies and briefly describes variables included in the regression models. VII. Results

Five regression models were estimated to measure the effect of deregulation: a model was estimated for each state (Indiana, Illinois, Iowa, Kansas and Texas). The Durbin-Watson statistics showed the residuals in each model to be positively correlated or indeterminate; therefore, all models were reestimated and corrected for autocorrelation. The adjusted R-squared for the estimated models range from .69 to .78. With few exceptions, the signs and magnitudes of the statistically significant parameters generally appear plausible, given the outlined theoretical framework.

Tables 2 and 3 includes parameter estimates for models (Texas, Kansas and Indiana) whose deregulation variables are statistically significant. The effect of rail deregulation on rates is measured with the deregulation dummy (DD) and the interaction term (DSLOPE) which was created by interacting the time trend variable (TIME) with DD. The statistical significance of the DD and DSLOPE variables imply an impact attributable to deregulation. The coefficients on these variables may be used to estimate preand post-Staggers rate trends. For example, the Texas model's rate trend during the pre-Staggers months (1-60) is calculated with the INTERCEPT coefficient (-\$0.2017) and the TIME coefficient (\$0.0035) which is multiplied by month. Based on these coefficients the pre-Staggers (1976-1980) real rate increase is estimated to be \$.21 per bushel (\$.0083 \$.2017 = \$.21). The post-Staggers trend (months 61-120) is estimated by aggregating the DINTER coefficient (\$0.5092) with the INTERCEPT coefficient and the DSLOPE coefficient (-\$0.0091) with the TIME coefficient (-\$.0091 + \$.0035 = -\$.0056) and multiplying the resulting coefficient by month. And, by subtracting the estimated rate trend value for month 60 (pre-Staggers) from the estimated value for month 120 (post-Staggers), the post-Staggers real rate change in Texas is estimated to be -\$.37 per bushel (\$.0083 \$.3645 = \$.3728).

The analysis shows deregulation to have had a dramatic effect on rail transportation corridors linking the Plains states (Texas and Kansas) with Gulf ports and a relatively modest impact on the corridor linking Indiana with East Coast ports. The Texas and Kansas corridors had estimated real rate declines of \$.37 and \$.31 per bushel during the 1981-1985 period, while deregulation is estimated to have decreased rates on the Indiana corridor by \$.08 per bushel. The estimated decline in Kansas closely parallels Sorenson's observation regarding the impact of deregulation at 14 Kansas elevator locations.

Additional analysis was carried out with the five regression models to identify state subregions where the effect of deregulation may have differed from the estimated state-wide effect. This showed northwest Kansas to have a \$.37 per bushel post-Staggers rate decline rather than the statewide average of \$.31, while rail corridors linking northeast and southeast Illinois with the East Coast were impacted by deregulation and had estimated rate declines of \$.01 and \$.02 per bushel, respectively.

The Iowa and Illinois regression models show deregulation to have had little sustained effect on rates (at usual levels of statistical significance) linking these regions with Gulf ports. And, in contrast to the findings in the Plains (Kansas, Texas), barge rates (BRAT) have an extremely important effect on corridor price spreads (Table 2). In particular, in the Illinois and Iowa models a dollar decline in rates serves to reduce the price spread by \$.51 and \$.73 per bushel, respectively, while the impact in Indiana is a more modest \$.38 per bushel (Table 2). It seems that much of the decline in Corn Belt price spreads during the post-Staggers era can be attributed to declining barge rates, not to a reduction in rail rates.

TABLE 1

Definition of Variables Included In Regression Equations

PSFD:	Monthly hard red winter wheat or corn price spread between port location and hinterland study region ($/bu.) (1975=100)
EXPD:	Monthly U.S. hard red winter wheat or corn exports (1,000 bu.)
FSUP:	Region's annual hard red winter or corn production (1,000,000 bu.)
WAGE:	Monthly real wage rate index (1975=100)
T-Bill:	Monthly 3-month T-Bill rate
EXPECT:	Monthly percent change in wholesale price index
RUI:	Monthly rail equipment utilization index
STOR:	Monthly ratio of unused storage capacity to total storage capacity in study region
TIME:	Monthly time trend variable
DD:	Variable to measure changes in intercept due to rail deregulation (pre-Staggers = 0, post-Staggers = 1)
DSLOPE:	Variable to measure change in TIME variable's slope due to rail deregulation (DD x TIME)
BRAT:	Monthly real barge rates linking mid-Illinois River with Gulf port locations ($/bu.) (1975=100)
JAN-NOV:	Month dummies
TXTRI:	Texas Triangle region
TXNOC:	Texas Canadian River region
TXSOL:	Texas South of Plainview-Muleshoe Line (base region).
KSNW-KSSE:	Kansas regions. Last two letters identify geographic location of region in state. KSSE was the base region.
INNW-INEC:	Indiana regions. Last two letters identify geographic location of region in state. INSE was the base region.

The exceptions are in those corridors linking Indiana and east Illinois with East Coast ports where, after controlling for barge rates, modest rail rate declines were identified. These findings are expected since it is generally acknowledged that water transportation has a substantial cost advantage for movement of low-valued bulky commodities and empirical evidence verifies that barge carriage is dominant for those grain-surplus regions with access to the Mississippi River system. The dominance of barge is further supported by Fuller, et al. whose work show barge rates to affect rail rates in the Corn Belt, but not vice versa.

Some have argued that the post-Staggers decline in rates was primarily the result of declining export demand in the 1980's. To test this notion, attention was focused on the export demand variable (EXPD) which was significant at usual levels in the Texas, Kansas and Indiana models. The EXPD variable controls or accounts for the effect of changing export levels on the observed price spread. To identify the sensitivity of the price spread to changing export levels, an export elasticity was calculated for each model. When calculated at the means of the data, the estimated export elasticities for Texas and Kansas were .06 and .07, respectively, while Indiana's estimated elasticity was .13. In which case, a 1 percent change in hard red winter wheat exports alters the price spread on the Kansas' corridor by .06 percent. This insensitive relationship implies that the decline in export demand had little effect on price spreads. To test the notion that railroads became increasingly sensitive to changing demand in the post-Staggers era, an interaction term (EXPD x DD) was constructed and included in reestimated models. The results imply an equally insensitive relationship between the price spread and export levels during the preand post-Staggers period. This outcome offers no support to the notion that increasing export levels or demand would lead to dramatic and substantial rail rate increases or that the reduced rates were attributable to the declining export demand in the 1980's.

TABLE 2

Estimated Linear Regression Coefficients for the Non-Binary and Deregulation Variables Included in the Kansas, Texas and Indiana Models.

	Kansas		Texas		Indiana	
Variable	coefficient	t-ratio	coefficient	t-ratio	coefficient	t-ratio
INTERCEPT	-0.1379	-0.871	-0.2017	-0.981	-0.0811	-0.691
EXPD	0.6999-8	5.549	0.6309-8	3.916	0.2036-8	2.825
FSUP	0.0002	5.280	0.0001	3.029	-0.7292-9	-3.195
WAGE	0.0055	4.463	0.0059	3.696	0.0046	4.578
T-BILL	0.0025	1.935	-0.0017	-0.992	-0.0069	-4.883
EXPECT	0.0052	2.971	0.0014	0.662	-0.4460	-1.979
RUI	-0.0196	-1.081	-0.0587	-2.415	0.0092	0.394
STOR	-0.1150	-2.951	0.0038	0.108	-0.1025	-2.533
TIME	0.0005	1.817	0.0035	9.443	-0.0012	-4.155
DD	0.1334	4.179	0.5092	12.593	0.1332	4.005
DSLOPE	-0.0040	-8.536	-0.0091	-15.244	-0.0018	-3.431
BRAT	--	--	--	--	0.3832	7.973
Adj. R-squared	.7359		.6928		.7648	
N	1080		360		1080	

Additional analysis was attempted with the rail cost index to determine whether this variable had a similar effect on the price spread during the preand post-Staggers era. This was accomplished by replacing the labor cost index with a rail cost index in the Kansas, Texas and Indiana models and adding an interaction term designed to measure the sensitivity of the price spread to changing rail costs in the post-Staggers period. The labor cost index was removed because of its high correlation with the rail cost index. The rail cost index and its associated interaction term were found statistically significant at usual levels in all models and when converted to an elasticity indicated the existance of a positive and elastic relationship (1.2 to 2.7) between rail costs and price spread in the pre-Staggers period. But, in contrast, a negative and inelastic relationship (-.3 to -.9) in the post-Staggers era. This outcome shows a decline in the sensitivity of the wheat price spread to changing rail costs in the post-Staggers era. The extreme sensitivity of rates to costs during pre-Staggers era seem reasonable in view of the cost-based rate increases which were authorized by the ICC during this regulated period. In the post-Staggers period, rates were shown to decline even though rail costs increased through 1983. Increased railroad productivity and interrailroad competition have probably been largely responsible for the altered relationship between the cost index and rate levels. Railroad productivity was undoubtedly enhanced as a result of the dramatic increase in use of multicar and unit train shipments that were shown to occur in wheat export corridors following deregulation (MacDonald).

The results provide strong evidence that the Staggers Rail Act of 1980 led to a restructuring of export-grain rates. There appears to have been a substantial decline in export-rail rates linking central and south Plains wheat producing regions with Gulf ports, and a more modest impact on rates linking the eastern Corn Belt with East Coast ports. In contrast, deregulation seems to have had little or no effect on corridors linking Iowa and Illinois with Gulf ports. Furthermore, the declining rates do not seem to be attributable to the diminished foreign demand for U.S. grain or shifts in the marketing services supply functions. VIII. Summary and Implications

This study measures the impact of deregulation on export-grain rate structures through use of a procedure which attempts to control for events that were coincident with deregulation and circumvent the problem associated with the unavailability of contract rate information. The procedure is applied to export-grain transportation corridors linking central and south Plains and Corn Belt regions with their respective port areas.

Results show export-rail rates in the central and south Plains to decline dramatically while rates linking eastern Corn Belt areas with East Coast ports declined modestly. Deregulation seems to have had little impact on corridors linking Iowa and Illinois with Gulf ports. Even though price spreads on these corridors declined during the deregulated era the analysis suggests this was due to declining barge rates rather than deregulation. These results seem plausbile and are corroborated by findings at Kansas (Sorenson) and Nebraska (Anderson, et al.) and research at the University of Illinois (Hauser). Further, the rate declines cannot, in general, be attributed to coincident events, in particular, the diminished demand for U.S. grains. This outcome is corroborated by recent analysis which shows that published wheat rates from Kansas elevators began their decline prior to the drop in exports (MacDonald).

The Staggers Rail Act offers contracting and a reduced role for rate bureaus as means of creating rivalrous railroad pricing behavior. In view of these regulatory modifications and notions regardng the

TABLE 3
Estimated Linear Regression Coefficients for the Binary Variables Included in the Kansas, Texas and Indiana Models.

	Kansas		Texas		Indiana	
Variable	coefficient	t-ratio	coefficient	t-ratio	coefficient	t-ratio
JAN	0.0005	0.081	-0.0125	-1.659	-0.0228	-2.531
FEB	-0.0142	-2.014	-0.0238	-2.716	0.0088	1.004
MAR	-0.0016	-0.197	-0.0243	-2.618	-0.0167	-2.009
APR	-0.0197	-2.065	-0.0229	-2.274	-0.0232	-2.730
MAY	-0.0182	-1.788	-0.0263	-2.596	-0.0200	-2.240
JUN	-0.0349	-3.251	-0.0475	-4.651	-0.0323	-3.038
JUL	-0.0153	-1.335	-0.0599	-5.702	-0.0399	-3.197
AUG	0.0087	0.940	-0.0293	-2.929	-0.0523	-3.837
SPT	-0.0167	-2.148	-0.0288	-2.929	-0.0247	-1.696
OCT	-0.0255	-3.847	-0.0189	-2.044	0.0804	5.370
NOV	-0.0089	-1.536	-0.0152	-2.025	0.0782	8.190
KSNW	0.1182	14.658	--	--	--	--
KSWC	0.0949	11.766	--	--	--	--
KSSW	0.0678	8.412	--	--	--	--
KSNC	0.0252	3.120	--	--	--	--
KSC	0.0093	1.156	--	--	--	--
KSSC	-0.0026	-0.318	--	--	--	--
KSNE	-0.0800	-9.290	--	--	--	--
KSEC	-0.0254	-3.158	--	--	--	--
TXTRI	--	--	-0.0256	-3.783	--	--
TXNOC	--	--	0.0005	0.069	--	--
INNW	--		--	--	0.0092	1.177
INWC	--	--	--	--	0.0093	1.195
INSW	--	--	--	--	-0.0583	-7.435
INNC	--	--	--	--	0.0292	3.726
INC	--	--	--	--	0.0116	1.485
INSC	--	--	--	--	-0.0015	-0.189
INNE	--	--	--	--	0.0464	5.921
INEC	--	--	--	--	0.0461	5.887

ICC's pre-Staggers behavior, the explanation for the observed results come into perspective. Apparently rate bureaus were allowed to function as a cartel and the ICC permitted railroads to offer rates whose upper bound was determined by competitive forces, i.e., the value-of-service pricing theory had no role in the pre-Staggers export-grain rate structure. In which case, the lack of intermodal competition in the Plains led to high rates, whereas the strong intermodal competition on those corridors linking the Corn Belt with Gulf ports created relatively low rates. With the passage of Staggers, interrailroad competition was apparently facilitated. And, in the Plains region, which had little intermodal competition and the highest pre-Staggers rates, there was a substantial opportunity for rates to decline as a result of rivalrous pricing behavior. Accordingly, rates declined precipitously in the Plains as compared with the Corn Belt where rates declined modestly or not at all. Strong water competition in the Corn Belt had kept pre-Staggers rail rates at low levels, thus the limited opportunity for rate reductions.

Several implications of the results seem important. Evidence supports the notion that rail regulation served to allow cartel pricing (Hilton). Rates have fallen in the Plains region where a cartel would have been effective, but, in the Corn Belt, where intermodal competition would have made for a less effective cartel, rates have declined little or not at all. In addition, this study's findings seem to support Levin's earlier work. Levin showed the social desirability of deregulation to be closely related to the existence of interrailroad competition. It seems that removing the immunity of rate bureaus and contracting have generated interrailroad competition in those regions of the U.S. where railroads have historically enjoyed monopolistic power and relatively high rates.

REFERENCES

Adam, Brian G., and Dale G. Anderson. "Implications of the Staggers Rail Act of 1980 for Level and Variability of Country Elevator Price Bids." *Transportation Research Forum* 26(1985):357-363

Association of American Railroads. *The Grain Book, 1986,* Washington, D.C.: Information and Public Affairs Department, 1987.

Association of American Railroads. *Railroads Cost Recovery Index*. Washington, D.C.: Information and Public Affairs Department, various issues 1979-1986.

Babcock, Michael W., L. Orlo Sorenson, Ming H. Chow, and Keith Klindworth. "Impact of the Staggers Rail Act on Agriculture: A Kansas Study." *Transportation Research Forum* 26(1985):364-372.

Friedlaender, Ann F., and Richard H. Spady. *Freight Tranport Regulation: Equity, Efficiency, and Competition in the Rail and Trucking Industries*. Cambridge, MA: The MIT Press, 1981.

Fuller, Stephen, James MacDonald, David Bessler and Michael Wohlgenant, *Effect of Railroad Reregulation on Export-Grain Transportation Rate Structures,* Texas Agricultural Experiment Station, in press.

Hauser, Robert J. "Regional Measures of Competition in the U.S. Inland Transport Industry for Export Grain, 1978-1983," Unpublished final report to U.S. Department of Transportation, University of Illinois, 1986.

Hill, Lowell D., Mack N. Leath, and Stephen W. Fuller. *Corn Movements in the United States,* Urbana: Illinois Agricultural Experiment Station, January 1981.

Hilton, George W. "The Basic Behavior of Regulatory Commissions, *"American Economic Review Papers and Proceedings,* 62(1972): 47-54.

Illinois Department of Agriculture, Division of Marketing, "Transportation Situation," Springfield, Ill., various issues 1976-1985.

International Monetary Fund, *International Financial Statistics,* Washington, D.C.: Monthly Summary Statistics, various issues 1976-1985.

Klindworth, Keith A., L. Orio Sorenson, Michael W. Babcock, and Ming H. Chow. *Impacts of Rail Deregulation on the Marketing of Kansas Wheat*. Washington, D.C.: U.S. Department of Agriculture, Office of Transportation, 1985.

Leath, Mack D., Lowell D. Hill, and Stephen W. Fuller. *Wheat Movements in the United States,* Urbana: Illinois Agricultural Experiment Station, Bulletin 767, January, 1981.

Levin, Richard C. "Railroad Rates, Profitability, and Welfare Under Deregulation." *Bell J. Econ*. 12(1981):1-26.

Sorenson, L. Orio. "Some Impacts of Rail Regulatory Changes on Grain Industries." *Amer. J. of Agri. Econ*. 66(1984):645-650.

MacDonald, James, "Railroad Deregulation, Innovation, and Competition: Effects of the Staggers Act on Grain Transportation" unpublished manuscript, June 1987.

U.S. Department of Agriculture, Agricultural Marketing Service, "Grain and Feed Market News," Washington, D.C., various issues 1976-1985a.

U.S. Department of Agriculture, Economic Research Service, "Wheat Situation and Outlook," Washington, D.C., various issues 1976-1985b.

U.S. Department of Agriculture, Statistical Reporting Service, "Agricultural Statistics," Washington, D.C., various issues 1976-1985c.

U.S. Department of Agriculture, Statistical Reporting Service, Crop Reporting Board, "Grain Stocks," Washington, D.C., various issues 1976-1985d.

ENDNOTES

* Department of Agricultural Economics, Texas A&M University, College Station, TX.

** USDA-ERS-NED-FCME

*** Department of Economics, North Carolina State University, Raleigh, NC.

Use of Rail Rate Contracting on Varying Transportation Competitive Environments and its Impact on the Competitive Structure of Grain Merchandising

by Henry M. Bahn, Kenneth L. Casavant, Gene C. Griffin and Daniel L. Zink[1]

ABSTRACT

This study compares the impact of rail rate contracting in three adjacent geographical areas. The areas share common production and market characteristics but have different transportation competitive environments. The Northern Great Plains is served by two competitive railroads and ships to Pacific Northwest (PNW) and Great Lakes destinations. Montana is dependent upon a single railroad which carries 75 percent of the state's grain, and truck backhaul. Montana serves PNW and Great Lakes market destinations. The Pacific Northwest is served by two railroads, truck and barge via the Columbia Snake River system. Grain flows to North Coast millers and export destinations.

Grain handlers and merchandisers were surveyed to determine the impact of rail rate contracting on the competitive position of their firms. Analysis suggests that the impact of rail rate contracting is dependent on the competitive environment for transportation, firm size and business structure.

I. INTRODUCTION

The Staggers Rail Act of 1980 gave rail carriers the authority to enter into contracts with shippers. Provisions of the act provided opportunities to railroads to increase earnings by allowing more flexible rate establishment procedures and cost reduction mechanisms. These changes have affected the U.S. grain merchandizing industry.

Rail rates for grain have not increased appreciably since late 1981 due to increased competitiveness of railroads, reduced costs associated with rail line abandonment, more efficient pricing (including contracting) and decreased demand for rail transportation due to poor grain market conditions. However, monthly fluctuations in the rail rate index indicate that rail rates change quickly due to the increased rate-making flexibility provided by the Staggers Act, (Figure 1). The extent of rail rate contracting and its impact varies within and between geographic regions. Depending on the competitive environment for transportation, rail rate contracting may have positive or negative impacts on the competitive structure of grain merchandizing. The impact on individual firms varies according to firm size and business structure.

Kindworth, Sorenson, Babcock and Chow (1985) found a reduction in rail rates following deregulation. They noted that rate reductions were disproportionate between areas and shipping points within the state of Kansas. Furthermore, they found that price spreads between county elevators and terminal markets decreased. Other studies have explored price variability after the passage of the Staggers Act. Adams and Anderson (1985) found that basis variability increased for Nebraska corn following passage of Staggers, but further research by Sarwar and Anderson (1986) found no significant change in cash price variability in Nebraska between pre and post Staggers periods.

There has been speculation regarding the impacts of rail rate contracting on the competitive structure of the grain merchandizing industry, but little is known about the impacts in different transportation competitive environments. This study compares the impacts of rail rate contracting in three adjacent wheat and barley producing areas.

The goal of the study is to contrast the impacts of rail rate contracting under different transportation competitive environments and grain marketing structures. Specifically, this paper addresses three objectives:

a) Develop a conceptual framework of supply and demand for rail rate contracts;

b) Define a study area with similar market characteristics and different transportation competitive structures, and;

c) Evaluate the use and impact of rail rate contracts.

The remainder of the paper will discuss a theoretical framework of rail rate contracts, and present a structural review of the study area, survey methods, results and implications.

II. THEORETICAL FRAMEWORK

The degree to which rail rate contracts become utilized in the grain transportation marketplace depends on the legality of the contracting procedure and the perception of contracting parties regarding mutual or relative benefits. The legal precedent was established in the Staggers Rail Act of 1980, after having railroad contracting rules established by the Interstate Commerce Commission (ICC) in Ex Parte No. 358, Sub. 1. The incidence of rail rate contracts depends on a number of variables which affect the magnitude of net benefits received by the contracting parties.

Figure 1. Rail Freight Rate Index for Grain
January 1979–December 1986
(December 1984 = 100)

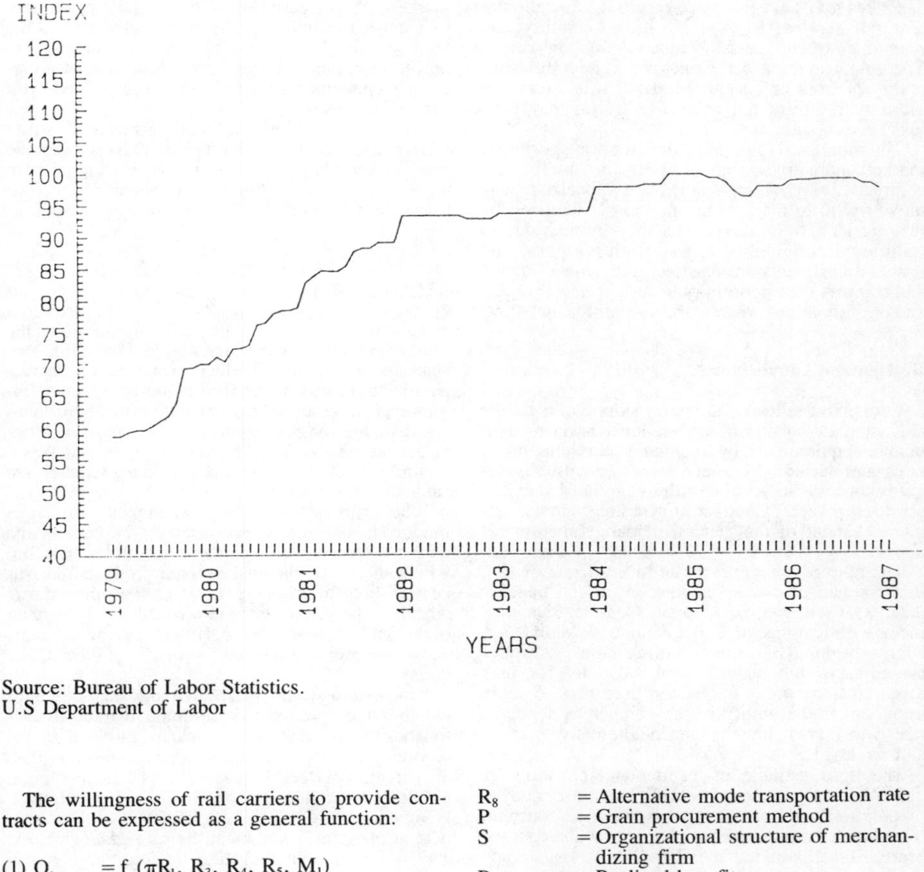

Source: Bureau of Labor Statistics.
U.S Department of Labor

The willingness of rail carriers to provide contracts can be expressed as a general function:

$$(1)\ Q_s = f\left(\frac{\pi R_1}{\pi R_2}, R_3, R_4, R_5, M_1\right)$$

Where Q_s = Number of contracts supplied

$\frac{\pi R_1}{\pi R_2}$ = Profitability ratio between contract and existing traffic rates
R_3 = Truck transportation rate
R_4 = Truck-barge transportation rate
R_5 = Competing rail carrier transportation rate
M_1 = Transportation market alternatives

The desire for contracts by grain merchandisers can be similarly expressed:

$$(2)\ Q_d = f\left(\frac{\pi R_6}{R_7}, R_8, P, S, B, M_2\right)$$

Where Qd = Number of contracts demanded

$\frac{R_6}{R_7}$ = Ratio of contract rate to existing rail tariff rate
R_8 = Alternative mode transportation rate
P = Grain procurement method
S = Organizational structure of merchandizing firm
B = Realized benefits
M_2 = Grain market alternatives

A. Supply Considerations

The overall goal of rail carriers entering into contracts centers primarily on desired increases in volume or cost efficiencies while maintaining a specified level of service. The variables delineated in general functional form (1) above serve to characterize the competitive transportation environment faced by a rail carrier in any given market.

The profitability ratio, $\pi R_1/\pi R_2$, is the relationship between the profitability of the contract rate, R_1, (the contract rate minus additional carrier costs) to the profitability of the existing base tariff, R_2, (tariff rate minus existing carrier costs). The existing base tariff is the relevant tariff for each shipper; for grain merchandisers it could represent single car, or 3, 5, 10, 26 or 52 multiple car rates. This variable evaluates the increase in net returns, if any, per unit of traffic handled. It reflects the carrier's desire for increased volume as well as the attempt to achieve internal cost effectiveness.

The next three variables measure competitive intensity within a region or to a specific merchandiser. The rates for truck, R_3, truck-barge, R_4, and alternative rail service, R_5, specify the availability and magnitude of competition to the contracting carrier. The truck and truck-barge rates also reflect the availability of front or backhaul traffic in the area. Regions with substantial back haul traffic may have lower internodal rates.

The final variable, M1, defines transportation market alternatives, enhanced by the use of rate contracts. By providing contracts, a rail carriers may move traffic to markets not previously serviced by that carrier. These alternatives vary by area due to traditional commodity flows. Both temporal and spatial adjustments are feasible and provide buyers and shippers the opportunity to seek atypical origins and or destinations when market conditions warrant.

B. Demand Conditions

Contracting allows individual shippers to fit the characteristics of the transportation market to their unique requirements by seeking a combination of rates and services. However, contracts usually require concessions to obtain the negotiated rate and service package. Contract concessions are an implicit element of the general demand function (2) above.

The ratio of the contract rate to tariff rate, R_6/R_7, is a comparison of concessions given and benefits received by a specific shipper. Contract rate may include obligations such as guaranteed volume, orderly shipment, limited loading times, shipper owned cars, etc., which, when accounted for, may result in a net ratio of greater than one. Equally important is the ability of the shipper to negotiate rates which are below the rates of alternative carriers and modes, R_8.

The third variable in the demand equation, P, describes the grain procurement method employed by the merchandisers. If grain is purchased outright (merchandiser takes title) the benefits of the contract accrue directly to the merchandiser, and a strong incentive to negotiate exists. When the elevator provides only storage or throughput (title does not pass to the merchandiser), less direct benefits accrue to the negotiating party and there is correspondingly less incentive to the shipper to seek negotiated rates.

The organizational structure of the firm, S, can be a deterministic factor in contract usage. In regions where multi elevator firms exist managers may routinely transship between elevators and volume or equipment requirements pose little or no logistical constraints. Large, multi-plant firms may be disposed to negotiate single contracts for multiple origins and destinations. This reduces negotiation costs and eliminates the direct involvement of individual elevators and their managers.

Realized benefits of the contract are defined by B. The realized benefit is that portion of the net contracted rate which is retained by the contracting firm. Contract obligations such as volume or time commitments may require more aggressive buying techniques by shippers. In order to ensure that adequate product is available to ship, merchandisers will bid higher cash prices to farmers, effectively passing on a portion of the contract benefits. The retained or realized benefits to the contracting firm are compared with the costs and other benefits (if any) of contracting.

The final variable, M_2 defines alternative grain markets which influence the desire of merchandisers to enter into contracts. As with rail carriers, the temporal flexibility associated with contracts may enable merchandisers to capture a portion of a previously unattainable market either on a temporary or permanent basis.

The supply and demand functions reveal the structural characteristics of individual markets and the resultant market power of each participating party in a contract. Final resolution is the incidence of contracts by firms within a region or a market area.

III. STRUCTURAL REVIEW OF THE STUDY AREA

The study area is a contiguous region along the northern grain growing tier of the U.S. The area shares common grain production and market characteristics but different transportation competitive environments. Three subregions are defined: Northern Great Plains (North Dakota and Minnesota), Montana, and the Pacific Northwest (Oregon and Washington). Production and market characteristics are summarized in Table 1.

Wheat and barley are the predominant grain crops produced. The region produced 28 percent of the nation's wheat crop during 1982 and 1983, but within that total, the study region produced 83 percent of the durum, 79 percent of the other spring wheat, 70 percent of the white wheat and 14 percent of the winter wheat (USDA, 1984). During the same period the region produced 59 percent of the nation's barley.

Grain storage and merchandizing methods vary but the study area shares common market characteristics. Grain is stored primarily on-farm in the Northern Great Plains and Montana, with less than 30 percent of Dec. 1, 1986 stocks held by mills, elevators, warehouses, terminals and processors. However, nearly 75 percent of PNW stocks were held at off-farm locations at the same date (USDA, 1987). Throughout the study region, large multi-elevator grain merchandizing firms are dominant. In the PNW cooperatives are most common, while commercial noncooperative firms predominate in Montana and the Northern Great Plains. Cooperatives play an important, through more limited, role in Montana and the Northern Great Plains. Throughout the study area single elevator, independent firms are declining.

Grain procurement methods vary, but firms can be generally categorized as brokers or buyers, depending on whether they take title to the grain they handle. Active buyers include firms which purchase for their own use (millers, feed processors, cattle feeders) and firms which purchase for resale. The latter may include cooperatives and independent private firms as well as major commercial merchandisers.

Much of the wheat produced on the region is exported. Virtually all of the white wheat is exported through PNW terminals, although some white wheat is retained for domestic feed if export price levels are low. Most of the winter and spring wheat produced in Montana is exported through PNW facili-

TABLE 1

Grain Production and Market Characteristics—Upper Great Plains, Montana, Pacific Northwest (PNW)

	UPPER GREAT PLAINS	MONTANA	PACIFIC NORTHWEST
Grain Production	Dark Northern Spring Wheat Hard Red Winter Wheat Durum Wheat Feed Barley Malting Barley	Dark Northern Spring Wheat Hard Red Winter Wheat Durum Wheat Soft White Wheat (Limited) Feed Barley Malting Barley	Soft White Wheat Soft Red Wheat (Limited) Feed Barley Malting Barley
Grain Storage	On-Farm Commercial Warehouse	On-Farm Commercial Warehouse (Limited)	On Farm (Limited) Commercial Warehouse
Grain Merchandising	Major Firms – Multi-elevator Independent Firms Cooperatives – Multi-elevator (Limited)	Major Firms – Multi-elevator Independent Firms (Limited) Cooperatives – Multi-elevator (Limited)	Major Firms – Multi-elevator (Limited) Cooperatives – Multi-elevator Independent Firms (Limited)
Domestic Markets	Interior Millers Plains Feeders (Barley) Southwest US (Barley) Central US Malsters	Local Millers (Limited) PNW Millers Interior Millers Local Feeders (Barley) PNW Feeders (Barley) Southwest U.S. (Barley) PNW Maltsters	PNW Feeders (Barley) Southwest U.S. (Barley) PNW Maltsters
Export Terminals	PNW Great Lakes Gulf (Limited)	PNW Great Lakes (Limited) Gulf (Limited)	PNW
Transportation	2 Class – 1 Railroads 1 Shortline Railroad Truck Backhaul Truck-Barge (Limited)	1 Class – 1 Railroad 1 Shortline Railroad Truck Backhaul Truck-Barge (Limited)	3 Class – 1 Railroads Truck Backhaul Truck-Barge

ties although some export shipments are loaded at Great Lakes and Texas Gulf ports. Northern Plains wheat is exported through the Great Lakes, the PNW, and the Texas Gulf. Barley is consumed locally, shipped to domestic feeders in the Central Plains and Southwest U.S. and exported from PNW and California ports. Barley malting takes place in the PNW and at Central US locations.

The mix of available transportation service is the most important distinction between the subregions of the study area. Rail shipment is the dominant mode of grain transportation, accounting for an estimated 65 percent of PNW shipments, 74 percent in the Northern Great Plains and 75 percent in Montana. However, the PNW is served by three major railroads (Burlington Northern, Union Pacific and Southern Pacific), the Northern Great Plains are served by the Burlington Northern and the Soo Line, and Montana is served by the Burlington Northern.

Truck transportation is primarily backhaul throughout the study area. In the PNW it is primarily short haul while the other two subregions utilize more extensive long distance backhaul. Availability of backhaul provides competition to rail carriers but does not present a threat to their market share.

Truck-barge combinations compete directly with rail carriers in the PNW and also draw grain from Montana and the Northern Great Plains. Data verifying truck-barge transloadings are not available but elevators at Lewistown, Idaho, Pasco, Washington, and other Columbia-Snake River ports report receiving grain from as far east as Minnesota.

IV. SURVEY METHODS

Grain merchandisers were surveyed as part of an extensive ICC sponsored study (ICC 1987). For the regional study the authors surveyed firms to determine extent of contracting activity, volume of grain shipped and the impact of rail rate contracting on relative competitive positions and prices paid to farmers. A typical survey document is presented in the appendix.

A total of 756 elevators received mail surveys. Of those returned, 376 (49.7 percent) were deemed usable. In addition to the mail survey, telephone interviews were utilized to elicit qualitative responses. An effort was made to obtain responses from county elevators as well as parent firms. Because of the difference between firm organizational structure and individual procurement procedures and survey methods, attempts to determine specific types of contracts (own firm versus another firm's origin, destination) were not completely successful. A high percentage of firms responding had experience with rail rate contracting, but the survey results indicated differences between the subregions of the study area.

V. RESULTS

In general, survey results were predictable: large volume, multiplant firms with alternative transportation modes had greater contracting experience and experienced positive net impacts. Contracts have changed the competitive environment and benefits have been partially passed on to farmers in the form of higher bid prices. Nearly three quarters of the responders had some contract experience. (Table 2).

In the Upper Great Plains and Montana, the majority of respondents with contract experience used other firm's contracts (either destination contracts or those written by the parent firm) rather than their own. While PNW respondents had greater experience with their own contracts, this difference may not be significant, due to the large number of cooperatives in the PNW. The cooperatives may not have considered contracts negotiated by the parent office to be "other firm's" contracts.

Use of rail rate contracts increased with annual volume of grain shipped (Table 3). While about 30 percent of the smallest elevators (those which ship less than 750,000 bushels annually) used contracts, over three quarters of the largest volume shippers did. Closer examination of Table 3 indicates that all sizes of shippers were proportionately more dependent on other firm's contracts, but contract use increased with annual volume shipped. Thus the smallest shippers' use of their own contracts was low (21 percent of survey respondents) and the largest shippers' use of other firms contracts was high (80 percent).

Utilization of their own contracts was lowest for smaller firms (those shipping less than 1.5 million

TABLE 2

ELEVATOR USE OF RAIL RATE CONTRACTS—NORTHERN GREAT PLAINS, MONTANA, PACIFIC NORTHWEST (PNW)

SUB REGION	OWN CONTRACT	OTHER FIRM'S CONTRACT	NO EXPERIENCE	TOTAL*
No. Great Plains	116	204	62	382
Montana	5	22	10	36
PNW	16	12	29	43
TOTAL	137	238	101	461

*total not additive due to multiple answers by some respondents

TABLE 3
ELEVATOR USE OF RAIL RATE CONTRACTS,
BY ANNUAL VOLUME SHIPPED, NORTHERN GREAT PLAINS, MONTANA,
PACIFIC NORTHWEST (PNW)

	ANNUAL VOLUME SHIPPED (000 BUSHELS)							
SUB REGION	< 750		751 - 1,500		1,501 - 3,000		> 3,000	
	OWN FIRMS CONTRACTS							
Northern Great Plains	33	(31%)	33	(43%)	26	(57%)	24	(80%)
Montana	3	(16%)	1	(25%)	0		1	(33%)
PNW	4	(17%)	2	(29%)	6	(75%)	4	(100%)
	OTHER FIRMS CONTRACTS							
Northern Great Plains	57	(50%)	54	(61%)	39	(70%)	29	(91%)
Montana	13	(62%)	2	(40%)	3	(100%)	2	(67%)
PNW	1	(6%)	2	(22%)	4	(57%)	5	(83%)

NOTE: % indicates percentage of survey respondents in each subcategory who have used rail rate contracts.

Northwest. The low usage of own firm's contracts across all sizes of shippers in Montana suggests an inferior bargaining position for individual elevators, due to a lack of transportation alternatives, and heavier dependence on contracts negotiated by firms located elsewhere (destination contracts) or by multi plant firms such as major grain merchandisers which meet volume or equipment utilization obligations. This is consistent with the demand conditions presented in Section II. Economics of size and scale transfer of grain by cooperatives in the Pacific bushels annually) in Montana and the Pacific Northwest. This could be due to the lack of rail competition in Montana and to more extensive intra firm affect the demand for contracts by affecting market alternatives.

The survey tested three impacts of rail rate contract rates. They were impacts on the elevator's competitive position (positive, negative or no charge), prices paid to farmers (higher, lower or no change) and annual grain volume shipped (higher, lower or no change).

Impacts varied with the use of contracts and the type of contract utilized. Firms negotiating their own contracts (Table 4) generally reported that their competitive position was enhanced, and their stronger competitive position resulted in higher prices paid to farmers and increased annual volume shipped. However, no definite competitive impact could be established in the PNW, and 59 percent of the respondents indicated that their annual volume of shipments declined. Soft white wheat export shipments declined during the past several years, however, so the PNW experience is not necessarily inconsistent with theoretical expectations.

Respondents utilizing other firm's contracts reported a negative competitive impact and a lower proportion of them bid higher prices to farmers. Merchandisers shipping on destination contracts of other firms did not receive direct reductions on rates and thus would passed benefits on to farmers. The realized benefits to these respondents may have been negligeable despite positive realized benefits to the parent firm.

PNW shippers which utilized other firm's contracts increased shipment volume. Parent firm contracts encourage transshipment between elevators if distances between the facilities are not too great. Shipment between cooperative elevators or other firms with a multiplant organizational structure can result in efficiencies which benefit the organization and its patrons.

Volume shipped by elevators utilizing other firm's contracts decreased or stayed the same in the Northern Great Plains. Destination contracts for these firms resulted in a marginal reduction of transportation costs which is insufficient to maintain market share relative to firms negotiating their own contract rates. The realized benefits would, in these cases, be negligeable.

Elevators not utilizing contracts noted negative competitive impacts. This is not a surprising result since the realized benefits associated with contracting would obviously be denied to those firms unable or unwilling to negotiate favorable rates. Particularly in Montana, survey respondents not using contracts noted both an erosion of their competitive position and lower prices bid to farmers. For grain which is priced basis terminal market less transportation, lower producer bids imply relatively higher transpor-

TABLE 4
IMPACTS OF RAIL RATE CONTRACTS ON CONTRACT USERS ON COMPETITION, PRICES PAID AND VOLUME SHIPPED, NORTHERN GREAT PLAINS, MONTANA, PACIFIC NORTHWEST (PNW)

SUB REGION	COMPETITIVE IMPACT			PRICES PAID TO FARMERS			VOLUME SHIPPED		
	Positive	Negative	No ▲	Higher	Lower	No. ▲	Higher	Lower	No. ▲
	OWN FIRM CONTRACTS								
Northern Great Plains	63 (56%)	21 (18%)	29 (26%)	75 (66%)	8 (7%)	31 (24%)	52 (45%)	20 (17%)	44 (32%)
Montana	4 (80%)	0	1 (20%)	4 (80%)	0	1 (20%)	N/A	N/A	N/A
PNW	5 (31%)	5 (31%)	6 (38%)	N/A	N/A	N/A	5 (29%)	10 (59%)	2 (12%)
	OTHER FIRM CONTRACTS								
Northern Great Plains	63 (35%)	75 (42%)	42 (23%)	96 (54%)	20 (11%)	61 (34%)	52 (28%)	70 (38%)	62 (34%)
Montana	6 (30%)	12 (60%)	2 (10%)	6 (32%)	6 (32%)	7 (36%)	N/A	N/A	N/A
PNW	5 (45%)	5 (45%)	1 (10%)	N/A	N/A	N/A	7 (47%)	3 (20%)	5 (33%)

NOTE: % indicates percentage of survey respondents in each sub category who have used rail rate contracts.
N/A = not available

tation costs. Lower bid prices ultimately result in loss of market share, reduced returns and a decline in profitability. Impacts on non contract shippers are summarized in Table 5.

VI. IMPLICATIONS

The overall finding of the mail survey and telephone followup discussions is that larger shippers with transportation alternatives had greater experience utilizing rail rate contracts (both their own and those of other firms). Those utilizing their own contracts experienced more positive impacts (enhanced overall competitiveness, increased volume shipped and higher prices paid to farmers) than users of other firm's contracts. Although the incidence of contracting was similar across the study area, Northern Great Plains survey respondents more often experienced positive benefits. Montana experienced negative, eroded competitive positions and lower prices paid to farmers.

Elements of firm size, organizational structure and the competitive environment for transportation help define the impact of rate contracting.

A. Firm Size

Larger firms used more contracts and received more direct benefits from them. Farmers selling to these firms received higher prices for their grain.

Larger firms possess economies of size and scale which provide a greater incentive to negotiate contracts and which simultaneously enhance their bargaining position. Examples of the former include storage capacity, rail sidings adequate to service multiple car or unit trains and a multi-plant corporate structure (private or cooperative) which allows pooling of equipment and intra-plant transfer of grain. An example of the latter is sufficient annual volume or market share to meet contract obligations.

TABLE 5
IMPACTS OF RAIL RATE CONTRACTS ON FIRMS NOT USING CONTRACTS ON COMPETITION, PRICES PAID AND VOLUME SHIPPED, NORTHERN GREAT PLAINS, MONTANA, PACIFIC NORTHWEST (PNW)

SUBREGION	COMPETITIVE IMPACT			PRICES PAID TO FARMERS			VOLUME SHIPPED		
	POSITIVE	NEGATIVE	NO ▲	HIGHER	LOWER	NO ▲	HIGHER	LOWER	NO ▲
Northern Great Plains	59 (23%)	148 (58%)	48 (19%)	76 (31%)	66 (2%)	101 (42%)	32 (12%)	149 (58%)	78 (3%)
Montana	2 (25%)	4 (50%)	2 (25%)	0	6 (75%)	2 (25%)	N/A	N/A	N/A
PNW	5 (99%)	19 (33%)	33 (58%)	N/A	N/A	N/A	8 (14%)	40 (72%)	8 (14%)

NOTE: % indicates percentage of survey respondents in each sub category who have not used rail rate contracts.
N/A = Not Available

Negative impacts experienced by smaller shippers may be attributable to their competitor's size and scale economies resulting from transport scale economies (Casavant and Brinkley 1983) rather than to the existence or use of rate contracts. However, because of their general visibility (and the confidentiality of specific details) contracts are a ready target. Dissatisfaction with rail rate contracts may be symptomatic of broader structural changes in the grain marketing industry.

B. Firm Structure and Grain Procurement Methods.

Firm structure and organization affect the magnitude and distribution of contract benefits. Multiplant firms centralize or coordinate their merchandizing activities and may be in a superior position to respond to changing market conditions. Centralized firms tend to be more aggressive in their merchandizing and usually purchase for resale. They are thus more likely candidates for contracts.

The combination of large size, centralized merchandizing and purchase for resale provides both the opportunity and the incentive to service alternative markets. The temporal flexibility of negotiated contracts permits short term (arbitrage) adjustments such as shipment of Northern Great Plains and Montana wheat to Texas Gulf export facilities when market conditions justify shipment to atypical destinations. Over the longer term, contracts enhance the development of new market or enable merchandisers to penetrate previously unteachable markets. Merchandisers in Montana and the Northern Great Plains cite rail rate contracts as instrumental in developing the Southwest U. S. barley market. One PNW merchandiser noted an erosion of market share for the same market was caused by increased competition from Montana and the Dakotas. This is consistent with the demand agreements of Section II and underscores the fact that rate contracting has interregional as well as intra-regional competitive impacts.

C. Competitive Environment

The grain merchandizing industry displays a high degree of competition. Regional use of and satisfaction with rail rate contracts is affected by the transportation alternatives facing each firm. Merchandisers in Montana, regardless of size and organizational structure, have only one rail carrier to negotiate with. Opportunities to transship to competing lines are extremely limited. In the Northern Great Plains and to a lesser extent the PNW, competing rail carriers offer east-west service, and merchandisers have the incentive to negotiate contracts between carriers. For their part, the railroads are motivated to negotiate contract rates at or below the marginal cost of transshipment between competing lines if they are to maintain volume or market share. Competing transportation rates have a more explicit impact on the rail carriers supply function. Successful contractors in Montana, on the other hand, are generally shippers with substantial truck backhaul opportunities.

Despite dissatisfaction with contracts and their negative impact, Montana merchandisers did utilize them. This suggests a high level of competition between merchandisers who reluctantly adopt contracting in order to maintain their share of the market and is consistent with buyer behavior in a bilateral oligopoly (Knutson, 1968). The expected result would be stable or declining elevator margins[5] and higher cash bids to farmers. However, many Montana survey respondents noted lower prices were bid to farmers, suggesting that little or no contract benefits were passed on to farmers.[6] This apparent contradiction has not been resolved but is clearly worthy of further study.

The dominance of destination contracts suggests regional differences in the competitive environment. Firms with multiple elevators throughout the study area have the ability to negotiate contracts which cover many or all of their facilities. The merchandiser buys and ships grain from a variety of country locations on an FOB elevator basis in fulfillment of contract obligations. County elevators receive FOB track bids and are not aware of the rail rate. The manager compares the FOB bid with other selling alternatives. Lack of information and a reduced participatory role in the marketing process are a related concern for county elevators. However, shipping under destination contracts transfers the risk associated with contract commitments to the merchandiser. Elevators located in areas with exclusive access to the commodity specified in contracts may gain price advantages and favorable transportation rates.

VII. SUMMARY AND CONCLUSION

Grain merchandisers in North Dakota and Minnesota, Montana, and Oregon and Washington were surveyed to determine the impact of rail rate contracting. Large elevators in regional with alternative transportation modes and those directly involved in contracting generally favored the use of contracts, citing increased competitiveness and increased volume as benefits. Furthermore, they noted a positive price impact for farmers. Smaller shippers generally perceived contracts as discriminatory, noting an erosion of competitive position and reduced volume of shipments. Smaller shippers felt prices paid to farmers declined because of rail rate contracts.

Northern Great Plains and PNW elevators were more positively impacted than those in Montana. Both the Northern Great Plains and the PNW have service from two or more railroads, while rail service in Montana is largely limited to a single carrier.

Telephone discussions with the largest firms verified a preference for contracts, although respondents were not unanimous in their support. Some firms felt compelled to participate in contracting in order to maintain market share or volume of shipments. Large contractors utilized destination contracts and passed benefits on to country elevators and farmers. Firms negotiating their own contracts were generally more favorable than those who utilized other firm's contracts.

Declining price levels during the 1980's and technological changes such as development of unit train loading facilities make it difficult to quantify the benefits of rail rate contracting. However, firms with sufficient size, volume or market power to directly negotiate contracts from more than one carrier perceive rail rate contracts to be beneficial to the industry and to farmers. Small firms which face limited rail competition perceive a reduction of competition and correspondingly lower prices to grain producers.

REFERENCES

Adams, B.D. and D. G. Anderson. "Implications of the Staggers Rail Act of 1980 on the Level and Variability of Country Elevator Bid Prices," *Proceedings, 26th Annual Meeting*, Transportation Research Forum.

Casavant, K. L. and J. K. Binkley. "Transportation Changes and Agricultural Marketing Research," in Farris, Paul, ed. *Future Frontiers in Agricultural Marketing Research*, Iowa State University Press, 1983.

Interstate Commerce Commission. *Contract Rate Competitive Report*. Report to Congress, Ex Parte 387 (Sub. No. 959) February 18, 1987.

Klundworth, K. A., L. O. Sorenson, M. W. Babcock, and M. W. Chow. *Impacts of Rail Deregulation on Marketing of Kansas Wheat*, USDA, Office of Transportation, September 1985.

Knutson, R. D. Buyer "Strategy in Bilateral Oligopoly," American Journal of Agricultural Economics, 50, (1968): 1507-1511.

Sarwar, G. and D. G. Anderson. "Impact of the Staggers Act on Variability and Uncertainty of Farm Prices," *Proceedings; 27th Annual Meeting Transportation Research Forum.*

USDA, Statistical Reporting Service *Field Crops*, Statistical Bulletin 708, June 1984.

__________ Agricultural Statistics Board. *Grain Stocks*, GrLg 11-1, January 1987.

ENDNOTES

1. Henry Bahn is Assistant Professor of Economics at Montana State University. Kenneth Casavant is Professor of Agricultural Economics at Washington State University. Gene Griffin and Daniel Zink are respectively, Director and Transportation Economist, Upper Great Plains Institute.
2. Montana has limited eastbound Soo Line service in the extreme northeast corner of the state and limited southbound service on Union Pacific from the Southwest portion of the state. However approximately 77 percent of Montana wheat is shipped westbound to domestic and export destinations.
3. These firms are generally merchandisers which take title to the grain which they purchase for resale.
4. Larger capacity elevators may have lower variable costs. However, per unit loading or handling costs (elevator margins) are not necessarily lower depending on the total cost structure facing the firm. For a more thorough discussion, see Nelson (1983), p124.
5. The impact of contract rates on elevator margins has not been verified. Thorough analysis is virtually impossible without access to contracted rates. However, stronger local basis has been observed during the post Staggers period but it should be noted that grain basis tends to strengthen during periods of low prices as merchandisers must bid grain of or away from government loan programs.
6. Large firms surveyed by telephone indicated that contracting generally had a positive impact on cash prices bid to farmers. Declining cash prices levels due to poor market conditions over the past several years may have masked or offset the benefits passed on to farmers by contracting merchandisers.

APPENDIX

1. Have you shipped grain under rail contract rates?
   A. Your firm's contract Yes _____ No _____
   B. Another firm's contract Yes _____ Believe I have _____ No _____

2. What is the impact of rail contract rates on your elevator's competitive position?
   Positive _____ Negative _____ No change _____

3. What is the impact of rail contract rates on your price to farmers?
   Higher _____ Lower _____ No change _____

4. What is your average <u>annual</u> number of carloads shipped over the past two years? _____

5. What percent of your annual rail shipments are:
   A. Single car _____ %
   B. 26 car multiple rates (single origin) _____ %
   C. 26 car multiple rates (multiple origin) _____ %
   D. 52 car unit trains _____ %

Economic Pressure, Long Distance Trucking, and Safety

by Richard Beilock and Russell B. Capelle, Jr.***

ABSTRACT

The safety-related practices of long distance drivers are examined. The results reveal that a large proportion of drivers have delivery schedules that require violations of speed limits and/or Hours of Service Regulations. Also indicated are widespread violations of speed limits and Hours of Service regulations and use of stimulants while driving. There was a general concensus among drivers that safety conditions have deteriorated since 1980.

I. INTRODUCTION

In recent years concerns have increased regarding motor carrier safety. Reasons for this include uncertainty about evidence of increasing highway speeds by all classes of drivers, raised public awareness of the hazards of driving under the influence of alcohol or drugs or while excessively fatigued, and the safety of larger and multiple trailer configurations. These concerns have been underscored by the US DOT announcement that in 1984 truck accidents in the U.S. increased by 16.5 percent over the previous year. In 1984, there were 36,854 accidents involving trucks, resulting in 2,721 deaths, 29,149 injuries and over $404 million dollars in property damage (Strah).

In this paper the safety-related practices of long distance drivers are examined. Three major areas are addressed: 1. speeding and violation of Hours of Service regulations, 2. drug and alcohol usage, and 3. equipment age. Perceived changes in highway safety since 1980 are also examined.

II. DISCUSSION

It has been proposed by various individuals, both in and outside of the industry, that there is a negative relationship between the level of economic pressure and attention to safety. In particular, some contend that the increased economic pressures attendant the partial deregulation of the early 1980's have resulted in some carriers compromising on safety as a cost-cutting and revenue enhancing strategy.[1]

At the heart of the safety problem is the fact that firms may be able to shift or externalize some costs, at least for a time, by reducing their safety standards. This may result in nonoptimal levels of safety from the point of view of the society as a whole. For example, a motor carrier may reduce maintenance on brakes. This action saves the carrier money, unless discovered by an insurance agent, but shifts a cost to the public in the form of increased risk of property damage and bodily injury. In a world without information and transactions costs, many of those sharing the same roadways would be willing to pay the carrier to reduce this risk.[2]

If the goal of profit maximization is assumed for all carriers, then at one level it could be argued that regardless of the amount of economic stress, all should ruthlessly seek ways of shifting safety-related costs to the public. However, this reasoning does not account for the expected costs of accidents. Settlements can be extremely costly, particularly if negligence can be shown. Moreover, if there is negligence, the managers or owners of a firm may be directly subject to litigation, regardless of the firm type (i.e., corporation, partnership, etc.). If it is assumed that the value of the assets and time of a firm's managers and owners is positively correlated with the net worth of the firm; then the higher the firm's net worth, the more that the firm and its owners and managers risk from cutting corners regarding safety, *ceteris paribus*. This effect may be particularly evident with firms teetering on the edge of bankruptcy. For them, there are few assets for an injured party to acquire. It should, however, be stressed that the existence of a linkage between the financial state of carriers and their safety practices has never been demonstrated empirically and will not be in this article. But, as this discussion has hopefully demonstrated, the existence of such a relationship is plausible.

Economic pressure to lower safety standards may be divided into two categories: 1. carrier-induced pressures to complete jobs earlier or cheaper to enhance net revenues, and 2. pressures to complete jobs earlier to meet the service requirements of the shipper/receivers. This distinction can be subtle, but important. The carrier may elect to lower safety standards to increase net revenues over a given time period (for example, by reducing maintenance expenditures or by speeding). As argued above, a carrier's propensity to do this may be negatively related to its net worth. Another source of pressure for reduced safety standards may be the shipper/receiver. This would occur when shipper/receivers require transit times not possible without violating speed or Hours of Service Regulations (HSR). If the carrier refuses the load, those revenues are forgone, but if it is accepted, the carrier normally bears full responsibility for the results of the illegal transit.[3] If such shipper/receiver demands are widespread, it is difficult to imagine how any carrier could survive economically without acquiescing.

It should be recognized that economic pressure is not the only reason for speeding or violating HSR. Another reason is driver pleasure or convenience. A trip is over sooner if one drives faster or for longer hours, and there is more time for nonwork activities. Moreover many people simply enjoy driving faster than posted speed limits.[4] Therefore, even if sufficient time has been alloted to ensure safe operations during a trip, speed and HSR violations may occur.

A major focus of the current study is determining the extent to which driver schedules are sufficiently tight to force violation of speed limits or HSR. Such evidence is indicative of economic pressure, carrier- or shipper/receiver-induced, rather than pleasure or convenience as a motivation for such infractions. Tight schedules may be dictated by the carrier to increase equipment utilization or by the shipper/receiver to reduce cargo transit times or by both. It should be noted that while tight driver schedules are indicative of economic pressure, loose driver schedules do not necessarily indicate a lack of such pressure. More work may be available to the driver after the current job is completed. The schedule for the current job may be loose, but the driver has an economic incentive to speed or operate for longer hours to maximize earnings per unit of time. This is likely to be the case for company drivers of firms which are lax in monitoring speeds and hours worked, and for owner-operators. The methodology employed to determine schedule tightness is described along with the results.

Another safety-related indication of economic pressure is the age of the equipment. Firms in poor financial condition, *ceteris paribus*, would be more likely than financially healthy firms to defray equipment replacement.[5] Well-maintained older vehicles may be as safe as newer vehicles. However, given similar efforts regarding maintenance, the older vehicle would seem to be more likely to develop safety-related problems, and older equipment usually are not retrofitted with recently developed or improved safety-related components. In this study equipment ages are examined as well as the estimated total mileage on the tractor. Of particular interest is determining the differences between fleet carriers and owner-operators and between those specializing in hauling regulated commodities and those specializing in exempt commodities. If differences among these groups exist, it may be indicative of linkages between safety, on the one hand, and firm type or economic regulation, on the other hand.

Driving under the influence of drugs and alcohol is a growing safety concern. Alcohol is generally acknowledged to impair driving ability. Therefore, its use is not an indicator of economic stress. The use of stimulants, however, is consistent with economic stress. A driver needing to work long hours may resort to a drug, such as amphetamines, to stay alert, albeit dangerously so.

III. DATA

Little data exist regarding safety-related practices of truckers. This is in large part due to difficulties in sampling a representative group of truckers and to difficulties in securing information regarding illegal practices. Perhaps the most singular feature of the current study is the methodology employed to mitigate these problems.

To sample a representative group of truckers, interviews were carried out with truckers passing through Florida Agricultural Inspection Stations as they exited from the Florida Peninsula. Three sites were chosen: US I-10 (westbound), US I-75 (heading northwest), and US I-95 (heading northeast). These sites account for approximately 85 percent of all the motor carrier traffic out of the Peninsula. There is an inspection station on all other roads out of the Peninsula. All of these stations are open 24 hours per day, 7 days per week. All trucks are inspected. Therefore, there is no reason to suspect that the trucks passing through the three stations selected are a typical of those traveling along the other routings.[6]

The problem of response bias to avoid self-incrimination was handled in a number of ways. To lessen intimidation, the enumerators all dressed casually, their attire included orange safety vests with conspicuous University of Florida seals, and they introduced themselves as students. From the onset, the respondents were assured of anonymity. To underscore that there was nothing to hide, whenever possible, the enumerators positioned the questionnaires so that the respondents could read the questions as they were asked and the answers as they were recorded. Finally, virtually all potentially incriminating questions were phrased in terms of the respondent's perception of his/her fellow drivers, rather than his/her own actions. The success of these measures is suggested by the low refusal rates (see below). Moreover, at no time did a driver state that he/she was declining to be interviewed due to suspicions regarding our motives. Rather, insufficient time was almost always given as the reason for a refusal.

The data for this study are from two major survey efforts. In order of importance to the study they are:

Survey name	Time period	Number completed interviews	Refusal rate
Nonproduce	9/86	1,319	13%
Produce	11/84-6/85	1,836	3%

The higher refusal rate for the Nonproduce Survey is thought to be due to the fact that almost all of the respondents were not hauling agricultural products. As such they would normally spend only a brief time at the agricultural inspection station, and usually would not have to leave their cabs. Therefore, the 3 to 5 minute survey represented a significant increase in the time at the station. On the other hand for the produce survey, the drivers expected to be stopped for several minutes. Frequently the surveys could be carried out during the agricultural inspection or while the driver waited for inspection.

The selection criteria for the surveys are described in the Appendix. While the samples cannot technically be described as random, they are believed to be representative of all drivers exiting Florida during the time of the surveys with respect to the characteristics of interest for the study. The sole exception to this is that those refusing to be surveyed were likely to have tighter schedules on average.

IV. RESULTS

A. Schedule Tightness

To measure schedule tightness, the average speed (SCHSPD) was calculated necessary to meet the next scheduled drop or pickup (after the interview) without violating HSR. If this speed exceeds the speed limit or, indeed is fairly close to it, then either the trucker must be late or be in violation of the law.

To make this calculation, the following data were collected: interview time and date, location from which the driver came to the interview site[7], location of the next stop, time and date after which the driver considered him/herself to be late at the next stop,[8] and presence or absence of a co-driver. Three basic assumptions were made: that the driver(s) had been off-duty for several days prior to this trip, that no rests were taken on the journey to the interview site, and that those reporting no due time for their next destination had very loose schedules.

The first assumption is extremely conservative. In virtually all cases the driver(s) would have logged several hours of driving and on-duty time during that and immediately preceding days. However, there was no practical way of determining the number and timing of these driving and on-duty hours. The second assumption is probably accurate in almost all cases. The surveys were conducted from 6:00PM to 1:00AM, and the average distance from the origin to the interview sites was 182 miles. It seems unlikely that drivers would have stopped, other than briefly, prior to reaching the interview sites. It should be noted that this would not have been the case if the interviews had been conducted from dawn to early afternoon. In that case, there likely would have been a significant proportion of drivers who had pulled over to sleep prior to arriving at the interview site. The final assumption is also probably accurate in almost all cases. Some drivers indicated that anytime they get to their destination was alright or that no due time had been specified. Clearly, these perceived drivers that they were under little or no time pressure.

Over one quarter of the respondents in the 1984/85 Produce survey and 18 percent of those in the 1986 Nonproduce survey had SCHSPD in excess of 55 MPH (the study was completed prior to relaxation of the 55 MPH national speed limit). However to not exceed 55 MPH, a vehicle would have to average a somewhat lower speed. The gap between and attainable average speed and the 55 MPH maximum would vary according to the characteristics of the vehicle (eg., distance between refueling and ability to maintain speed on an upgrade), the types of terrain, and the proportion of the route that is along congested roads and those with lower speed limits. The percentages of respondents with SCHSPD in excess of various speeds are presented below and in Figure 1:

	Percent of respondents with SCHSPD in excess of:			
	40 MPH	45 MPH	50 MPH	55 MPH
Produce Survey	55	44	35	28
Nonproduce Survey:				
All days	37	27	23	20
Omitting Friday	45	35	25	20

Figure 1. Schedule Tightness for Produce and Nonproduce Long Distance Motor Carriers, 1986

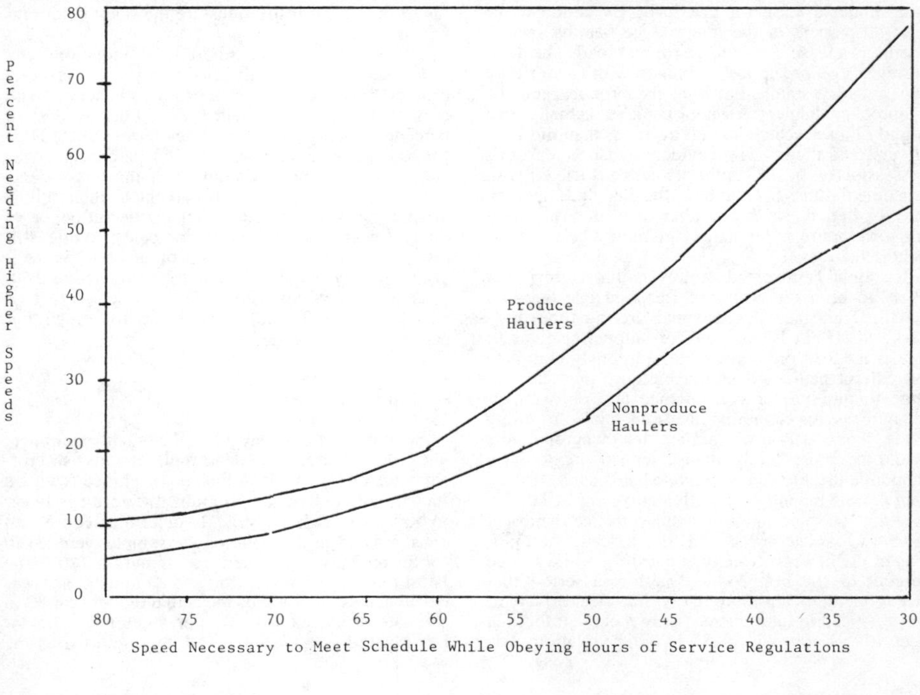

Regardless of the cutoff used for SCHSPD, the percentage of drivers with violation-inducing schedules is disturbingly high. This conclusion is strongly echoed by Transportation Research and Marketing. Using the 45 MPH cutoff, 44 percent of the produce haulers and 27 percent of the nonproduce haulers had to speed and/or violate HSR to avoid being late. The Nonproduce survey was conducted on a Monday, Wednesday, and Friday. It was hypothesized that for nonproduce loads schedules would be looser on Fridays as many drivers would have to wait till Monday to unload. This effect can be seen in the above data and in Figure 1. A larger proportion of the nonFriday subset has schedules permitting average speeds below 45 MPH (differences statistically significant at the .01 level).

Even omitting the Friday responses and regardless of the cutoff speed chosen, a much higher proportion of produce than nonproduce respondents would have had to violate HSR or speed limits to maintain their schedules. The most likely reason for this is the nature of the commodities involved. Produce has a high value and is perishable. Both of these factors contribute to high carrying costs and, hence, to a premium on transit time. Moreover, produce is often sold at terminal markets that have definite cycles of activity. A load that is one or two hours late may not be salable at all or only at a sharp discount.

Another possibility is that there may be a linkage between the regulatory status of the carrier and/or the carrier type and schedule tightness. The produce haulers interviewed were exempt from economic regulation, at least for the duration of that haul, whereas virtually all of the nonproduce haulers were subject to regulation for their current haul. Moreover, 48 percent of the produce haulers were owner operators versus 37 percent for the nonproduce haulers. As the produce was exempt, virtually all of the produce hauling owner-operators were not then operating under a lease for that load. By contrast, for over 80 percent of the nonproduce hauling owner-operators a lease was in force for that load. The data, however, do *not* support a linkage with carrier type or regulatory status. For both the produce and the nonproduce haulers, owner-operators actually had slightly looser schedules, on average, than did fleet drivers, and there was no evidence that the carrier's possession of an ICC authority was associated with schedule tightness. Therefore, the data does not support the hypothesis that smaller firms or firms operating exempt from regulation are more likely to have tight schedules.

It should be stressed, however, that there is not likely to be a one-to-one correspondence between SCHSPD and actual speed and driving time practices. SCHSPD is an index of shipper/receiver or carrier-induced pressures to expeditiously complete the current haul. It does not measure pressures related to pleasure or convenience and only imperfectly measures economic pressures related to future loads.[9] Previously it was argued that owneroperators would be more likely than fleet drivers to have economic incentives to complete hauls early. If true, then despite having essentially equivalent SCHSPD, they may be more likely in actual practice than fleet drivers to violate speed or HSR. Indeed, the opinions of the drivers strongly suggest this. Sixty-three percent of the nonproduce haulers asserted that owner-operators are more likely than fleet drivers to violate speed limits, whereas only 5 percent felt that fleet drivers are more likely to be in violation. The results were similar with respect to HSR. Of particular note is the fact that owner-operators and fleet drivers are in agreement that the former is more likely to violate speed and HSR (Table 1). Table 1.

One third of the trucks of the nonproduce sample had no loads. Differences with regard to SCHSPD between empty and full trucks were slight. There was some indication that drivers of empty trucks are more likely to have very loose schedules. However for higher SCHSPD, there were similar percentages of drivers of empty and full trucks.

B. Frequency of Violations of Speed and Hours of Service Regulations and Drug and Alcohol Usage

Drivers in the nonproduce sample were asked to estimate the percentage of their fellow drivers that **sometimes** drive under the influence of alcohol or under the influence of uppers, such as amphetamines, and **regularly** violate HSR. They were also asked for the average speed currently maintained by truckers on nonmetropolitan sections of US Interstates. This approach was taken to avoid obvious response bias problems that would occur if the respondents had been directly asked if they violated these laws.

Less than 3 percent of the respondents felt that average truck speeds on nonmetropolitan sections of US Interstates are 55 MPH or less. The mean response was 65 MPH, with 92 percent of the respondents estimating between 60 MPH and 70 MPH (Table 1). This result is consistent with that for the recent study by Transportation Research and Marketing. Curiously, the mean desired speed limit for these roadways was also 65 MPH and the distribution of the responses was similar (Table 1). This suggests, perhaps, that driver or firm preferences rather than legal restrictions are the main determinants of speed.

The results with regard to HSR violations and drug usage were equally disturbing. Fifty-six percent of those surveyed expressed satisfaction with current HSR. Yet, over half believed that at least 70 percent of their fellow drivers regularly violate HSR (the average response was that 63 percent are violators), see Figure 2. Over half of the respondents also believe that at least 30 percent of their fellow drivers periodically operate under the influence of drugs (mean response of 37 percent). While still high, the perceived incidence of alcohol figure 2 usage is only about half that for drugs (Figure 2). This pattern is consistent. Drivers who must or choose to drive for long periods may use drugs, but not alcohol, to stay alert.

C. Equipment Age

Economic stress may be reflected in equipment age, because firms may defer replacements as a cost-saving strategy. Unless this is combined with an increased maintenance program, the result is likely to be reduced safety levels. To determine equipment ages, drivers in the nonproduce sample were asked for the model years of their tractor and trailer(s). The latest model year was 1987, and equipment age was calculated as 1987 minus the actual model year.[10] An estimate of the total number of miles that a tractor has been driven was derived by multiplying the

TABLE 1
Carrier Type of Respondent[1]

	O/O	For-hire Fleet	PC	All
	percent of responses			
More likely to violate speed limits				
O/O	54.9	65.3	72.0	63.1
Fleet	7.5	4.0	4.0	5.3
Same or unsure	37.6	30.7	24.0	31.6
More likely to violate Hours of Service				
O/O	58.3	67.0	69.6	64.4
Fleet	6.2	5.4	4.0	5.3
Same or unsure	35.5	27.6	26.4	31.2
More likely to drive under the influence of drugs or alcohol				
O/O	24.3	40.0	48.3	36.3
Fleet	12.8	6.2	6.1	8.6
Same or unsure	62.9	53.8	54.4	55.1

Miles Per Hour[2]

	52.5 or less	52.5-57.5	57.5-62.5	62.5-67.5	67.5-72.5	Over 72.5	Average
	percent of responses						
Motor carrier average or rural Interstate	.1	6.1	11.5	62.5	22.8	3.2	65.0
Desired rural Interstate speed limit	.1	3.5	23.5	52.2	17.3	3.4	65.4

Equipment Age (years)[1,2,3]

	0-1	1-3	3-5	5-7	over 7	Average Age
	percent of responses					
Tractor:						
O/O	9.9	24.4	9.9	13.2	42.6	6.5
For-hire Fleet	19.5	38.0	16.7	8.9	16.9	4.2
PC	21.0	39.0	11.3	9.1	19.6	4.4
All	16.3	33.3	12.8	10.6	27.0	5.1
Trailer:						
O/O	12.0	22.7	11.2	11.0	43.1	6.3
For-hire Fleet	16.5	32.6	12.5	8.0	30.4	4.4
PC	12.2	26.2	12.8	10.7	38.1	5.4
All	16.6	30.9	13.6	11.1	27.8	5.3

Notes: 1. O/O = Owner/Operators
For-hire Fleet = For-hire fleet drivers
PC = Private carrier drivers
2. Greater than or equal to lower end of ranges
3. Age = 0 for 1987 model year equipment.

Figure 2: Drug, Alcohol, and Hours of Service Regulations : Perceived Percentages of Drivers in Violation

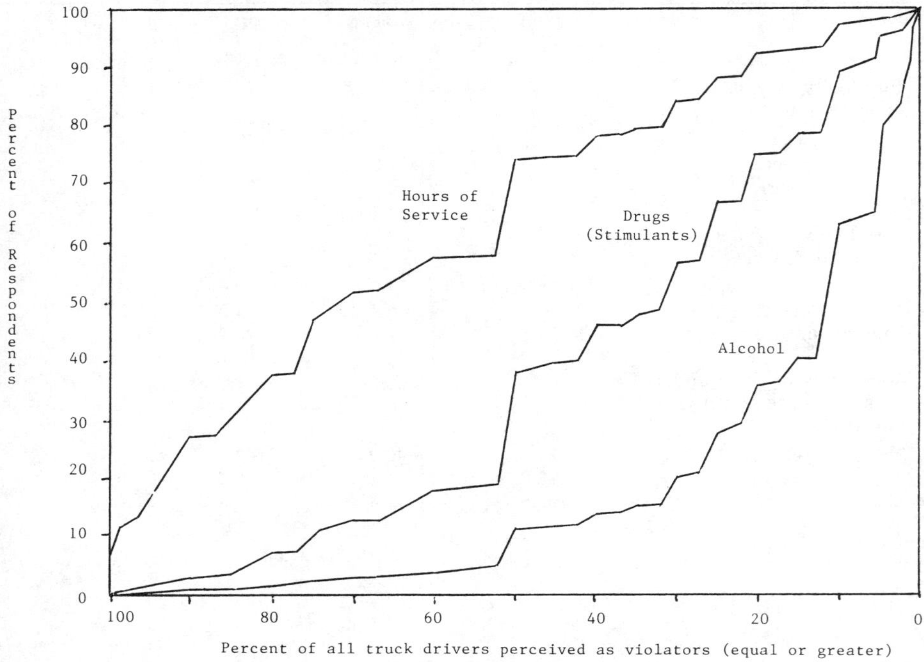

driver's estimated annual mileage times the tractor's age.

The average tractor age was 5.1 years. There were marked differences in tractor ages between owner-operators and fleets, 6.5 and 4.3 years, respectively (significant at the .001 level). Only 44 percent of owner-operator tractors were aged 5 years or less compared to 73 percent for the fleets (Table 1). The average estimated total number of miles per tractor is 565,884. Reflecting the differences in average ages,[11] the average total miles for owner-operator tractors is almost 50 percent more than for fleets (709,650 versus 481,568 miles).

The average trailer age was 5.4 years. As might be expected, there were considerable differences across trailer types:

Trailer type	Average age (years)	Number in sample
Dry van	5.5	630
Refrigerated van	4.8	437
Flatbed	5.9	149
Tanker	7.7	51
Twin vans	2.9	45
Auto carrier	8.6	26
Bulk	4.5	24
Livestock	5.3	22
Flatbed for container	5.4	16

The much lower age for twin vans follows as these have only been legal in most Eastern states since 1983. It also seems reasonable that refrigerated vans, with their complex and breakable temperature control equipment, would have a lower average life-span than dry vans and flatbeds.

As with the tractors, there are significant differences between the ages of trailers of owner-operators and of fleets, 6.3 and 4.8 years, respectively, significant at the .001 level.[12] Forty-three percent of the trailers of owner-operators are aged 5 years or less, compared to 53 percent of those for the fleets. Controlling for trailer types, these differences persist. The mean ages in years for the three most common trailer types are as follows:

Trailer type	Owner-operator	Fleet
Dry van	6.3	5.1
Refrigerated van	5.7	4.1
Flatbed	7.1	5.0

It might be noted that the results of the *RCCC Age of Assets Survey,* Transportation Research and Marketing, and the current study are consistent. All three studies indicate similar overall average equipment ages as well as older equipment for smaller firms.

D. Perceived Changes Since 1980

Perceived changes in safety conditions may offer clues as to possible linkages between economic regulatory reform and safety. For example, if drivers feel that safety conditions have deteriorated since 1980 (and the reforms embodied in the Motor Carrier Act of 1980), then there is circumstantial evidence of a linkage between the two. It should be stressed strongly, however, that such evidence is only circumstantial. Since 1980 there have been a host of developments unrelated to motor carrier regulation which likely also impact upon safety conditions faced by motor carriers. Among these are reduced fuel costs, the economic recovery, and the generally recognized trend for higher speeds by drivers of private automobiles. Any of these may overshadow the effects of a reform-safety linkage.

Drivers in the nonproduce sample were asked about changes since 1980 in the percent of unsafe truckers, the percent of poorly maintained trucks, and the difficulty in driving as safely as they would like to. If changes with regard to overall difficulty were noted, the driver was asked to give the main reason for this.

Fifty-five percent of the respondents expressed the view that the proportion of unsafe truck drivers has increased since 1980, three times the number who felt that the incidence of unsafe drivers has decreased. This result was consistent across the sample. That is, no subset (based, for example, on carrier type or possession of an authority) could be identified that had markedly different opinions in this regard.

The balance of opinion was much different with respect to equipment maintenance. Somewhat more drivers saw decreases in the incidence of poorly maintained vehicles than saw the reverse (44 to 32 percent, respectively). An interesting but puzzling result was that 46 percent of the union drivers saw deteriorations in maintenance, versus 22 percent who saw improvements (union drivers, however, comprised only 7 percent of the total nonproduce sample).

The general consensus that truck maintenance levels are not deteriorating is consistent with the responses to two later questions in the survey. Drivers were asked about changes in overall difficulty in driving safely since 1980. Only 3 percent of those who perceived increased difficulty gave poorer maintenance or older equipment as a major reason. In contrast, 16 percent of those who perceived improved conditions gave better maintenance and equipment as a reason. One of the last questions in the survey queried drivers about the main reason for the recent increase in accidents involving trucks. Only 4 percent gave equipment failure as a reason, while two thirds indicated human error (truck or auto driver).

The surprising degree of driver satisfaction with equipment condition may be quite realistic given improvements over time in truck designs and maintenance techniques. Moreover, there is empirical evidence indicating that human error far overshadows equipment failure as a contributing cause for accidents. For example, in a study of 346 accidents involving motor carriers, mechanical problems were noted as a possible contributing cause in 18 percent versus human error in 84 percent (Bureau of Motor Carrier Safety). The preceding, however, should not be seen as reason for complacency regarding truck maintenance. Some recent findings suggest disturbingly high percentages of mechanical deficiencies in trucks. For example, in 1985 29 percent of the 366,400 trucks inspected under the Federal Motor Carrier Safety Assistance Program did not pass, and in 1986 the failure rate was expected to approach 35 percent (Hanley). Finally, work by the American Trucking Associations indicates higher incidences of unsatisfactory equipment safety ratings among motor carriers with smaller profit margins (Glaskowsky).

With regard to overall conditions influencing a driver's ability to drive safely, 57 percent indicated that these have worsened since 1980, while only 10 percent have seen improvements. The large majority of the reasons given for worsening safety conditions fell into four areas: 1. increased traffic (45%), 2. deregulation and competition (25%), 3. poor driver skills (20%), and 4. traffic laws and enforcement (19%).[13] Increased (truck) traffic and poorer (average) truck skills may be partially explainable by increased entry attendant regulatory reform. Therefore, there is reason for concern regarding a regulatory reform-safety link.

V. SUMMARY AND CONCLUSIONS

The main empirical results of the study are:

1. Schedules are sufficiently tight for many drivers to induce them to violate speed or HSR. For example, if HSR were obeyed, 35 percent ofthe produce haulers and 23 percent of the nonproduce haulers would haveto exceed an average speed of 50 MPH in order to maintain their schedules.
2. There is no indication that schedules are tighter for owner-operators than for fleets. In fact, there is weak evidence that the reverse is true. However, the overwhelming opinion of all drivers, including owner-operators, is that owner-operators are more prone to violate speed limits and HSR than are fleet drivers.
3. Less than 3 percent perceived that currently-maintained average truck speeds on nonmetropolitan portions of US Interstates are 55 MPH or less. Over 90 percent estimated average speeds between 60 and 70 MPH, with almost half estimating 65 MPH.
4. Perceived frequencies of violations of HSR, and usage of drugs and alcohol are alarmingly high. The average respondent expressed the view that 63 percent of truckers regularly violate HSR and 37 and 17 percent, respectively, sometimes drive under the influence of drugs and alcohol.
5. Owner-operator equipment is markedly older than that for fleet carriers. On average, owner-operator tractors are 50 percent older and trailers are 30 percent older than those for fleets.
6. Nearly six times as many drivers perceive that since 1980 it has become more difficult to drive safely as believe that it is now easier to drive safely. Also, nearly three times as many see increases than

see reductions in the percentage of unsafe drivers. Opinions are divided, however, regarding changes in the incidence of poorly maintained vehicles.

It appears clear from the results that there is a linkage between economic pressure and safety. For example, carriers hauling time—sensitive produce are more likely to have violation-inducing schedules, *ceteris paribus*, than those hauling non-agricultural commodities. There is also reason to suspect that linkages exist between regulatory reform and safety. That is, a side effect of increasing the intensity of economic competition through regulatory reform may be increased economic stress which, in turn, results in a lowering of safety standards among highly stressed firms. It should be emphasized, however, that safety problems were not absent prior to regulatory reform. For example, in 1970 Fellmeth reported propensities to speed, violate HSR, and to use drugs that are comparable to those reported in this study. Rather, it is the recent rise in motor carrier accident rates and the proportions of trucks failing roadside safety inspections, and the near-consensus reported in this study that safety conditions have worsened since 1980 that offers *circumstantial* evidence of a safety-reform linkage.

If such a linkage exists, then presumably, if it were possible to undo regulatory reforms and thereby reduce economic stress on carriers, improved safety standards would result. However by this logic, the same result could be achieved if carriers were given assets subject to the stipulation that the would not have a mishap.[14] For produce haulers, perhaps we should promote irradiation as a safety measure. With less perishable produce, shipper/receivers would have less reason to pressure carriers to reduce transit times. Better yet, to promote highway safety, perhaps we should ban fresh produce altogether in favor of canned fruits and vegetables. Even if any of these were possible, they would be absurdly indirect, inefficient, and improper responses to safety problems.

The system has simply lost legitimacy and broken down. Many of the rules are not viewed as being equitable. For example, drivers (and carriers) frequently expressed the view that they feel trapped into violating rules to meet otherwise impossible schedules. Those dictating the schedules most often are essentially exempt from prosecution or, at the very least, not at physical risk. Moreover, current enforcement efforts are not credible deterrents. This is why, for example, violating speed and HSR has been able to become a way of life for a large segment of long distance truckers. A corollary to this is that a high proportion of truckers use stimulants.

There will always be economic incentives on carriers and drivers to cut corners on safety. Making carriers wealthier, either by protective economic regulations or by some sort of grant, may lower the propensity to reduce safety levels, but, almost surely, at an unreasonable economic cost. Moreover, this strategy does not address pleasure or convenience-associated safety violations. Instead, the society must determine the levels of safety it requires *and* is prepared to enforce. What financial, insurance, and maintenance standards must carriers meet to be allowed to use public roads? What are the minimum physical and safety/violations records for professional drivers, and should they undergo drug testing? Should shipper/receivers share liability with carriers for violationinducing schedules? What speed limits are proper? These are difficult and complex questions to answer. But without the answers and the dedication to enforce the decisions, the toll of death, injury, and property damage seems destined to rise.

REFERENCES

Beilock, R. *Motor Carrier Safety* report prepared for the Regular Common Carrier Conference, 1987a (forthcoming).

Beilock, R. *Produce Transportation: the Case of Florida* ERS Bulletin, 1987b (forthcoming).

Bureau of Motor Carrier Safety *Analysis and Summary of Accident Investigations, 1977-1979* Bureau of Motor Carrier Safety, U.S. Department of Transportation, 1981.

Fellmeth, R. *The Interstate Commerce Omission: The Public Interestand the ICC* Grossman Publishers, New York, 1970.

Glaskowsky, N. *Effects of Deregulation on Motor Carriers* Eno Foundation for Transportation, Westport, CT, 1986.

Hanley, R. "60% of Trucks Inspected in New York Area Fail" *The New York Times*, October 8, 1986, pp. B1 and B4.

NOVA "Why Do Planes Crash?" NOVA, Public Broadcasting Stations, March 1987.

Regular Common Carrier Conference *Regular Common Carrier Age of Assets Survey* 1986.

Strah, T. "Substantial '84 Accident Rise Confirmed: BMCS Reports Truck Accidents Increased 16.5% *Transport Topics*, September 1, 1986, pp. 1 and 32.

Transportation Research and Marketing *A Study of Current Truck Driver Practices* report prepared for the Motor Common Carrier Association, 1986 (forthcoming).

ENDNOTES

* University of Florida, Gainesville, FL.

** American Trucking Associations, Inc.

1. It is interesting to note that similar concerns are being raised regarding the airlines (NOVA).
2. A driver would be willing to pay unsafe carriers up to the point where marginal increases in these payments overtook resulting marginal decreases in insurance payments.
3. The carrier may be able to secure a rate premium for assuming the risks of operating illegally. However, that one party may, essentially with impunity, pay another party to commit an unlawful and dangerous act seems to be at variance with our society's legal philosophy.
4. This rationale may also be put in terms of maximization of economic benefit by recognizing that there is a value to leisure time and to pleasure, such as that which may be derived from fast driving.
5. This follows both because firms in poor finanacial conditions are less able to raise capital for equipment. Moreover, these firms have shorter future expected lifespans which lowers the economic rationale for equipment replacement.

6. The exception to this is that trucks passing out of the Peninsula on secondary roads may be more likely to have shorter total trip distances.
7. For the produce survey, the origin point of the journey was not obtained. Rather, it was assumed that the driver had just traveled 130 miles. As Miami is almost three times this distance, 130 miles is thought to be a very conservative estimate.
8. The importance of using the driver's perceived due time at the next stop should be stressed, for it is this time, upon which the driver will base operational decisions.
9. This would be measured if the carrier communicated to the driver a due time that took into account pressures to reposition the vehicle for ensuing loads, rather than the due time required by the shipper/receiver currently being served.
10. It could be argued that using "1987" somewhat overstates equipment ages. However, this effect is likely to be modest, at most 1 year, and the focus of the analysis is on differences across groups in equipment ages, rather than absolute age levels.
11. The average for total miles driven annually were virtually identical across carrier types.
12. Unfortunately, owner-operators were not asked if they owned the trailer they were then pulling. A portion of these trailers likely belonged to fleets (80 percent of the owner-operators were under lease). Unless the owner-operators systematically receive their lessee's oldest trailers, this would have resulted in some underestimation of the age of trailers actually owned by owner-operators.
13. The numbers in parentheses are the percentages of those respondents viewing worsening conditions who gave a reason that fit under each category. Some respondents gave two reasons. Therefore, the numbers do not total to 100.
14. The use of a carrot to deter nondesired actions is sometimes used in this and other countries. For example, farmers are sometimes paid to not plant crops and, via lower taxes, landowners are sometimes paid to not plant houses.

APPENDIX

The selection criteria and rationales for these for the Nonproduce survey were as follows:

Criteria		Reason
1.	Only combination rigs	To avoid interviewing drivers of essentially local vehicles
2.	Every third empty truck and all full trucks	A high proportion of the trucks exiting the Peninsula are empty. This rule ensures that a high proportion of those interviewed had loads. This was done on the premise that trucks with loads would be more likely than empty trucks to be operating under very tight schedules. No other differences with respect to the items of interest in the study were anticipated. Nevertheless, for all of the analyses differences between full and empty trucks were tested for and are reported if statistically significant.

The selection criteria and rationales for these for the Produce Survey was as follows:

Criteria		Reason
1.	Only combination rigs	To avoid interviewing drivers of essentially local vehicles
2.	Only those hauling exempt produce or ornamentals	Information regarding the transport of these commodities was the main focus of these efforts.

Note: For more detail regarding the survey methodology, see Beilock (1987a) and Beilock (1987b)

What if Everything We Know About Highway Safety is Wrong?

*by John Semmens and Dianne Kresich**

ABSTRACT

Public policy over the last two decades has proceeded on the basis that compelling improvements in vehicles and driver behavior is the best means of enhancing highway safety. While narrow focus analyses of the specific effects of specific regulations in specifically defined circumstances produces positive inferences on the efficacy of compulsory safety measures, a broader focus casts some doubt on the over all results.

The presence of a long term decline in the traffic accident rate confounds a simple before-and-after comparison of accident statistics. The question remains as to whether the number of accidents or fatalities per passenger mile of travel would have declined anyway. Further, there is the prospect that the expenditure of resources on a specific safety measure may merely transfer the risk rather than simply reducing risk.

This paper will examine some of the critical evidence from a broader perspective in order to better determine what the actual effects of compulsory safety regulations are and whether more voluntary and market-based approaches would produce a safer environment for those who value safety highly.

I. INTRODUCTION

Safe is preferable to dangerous. This is one of those choices for which we easily find universal concurrence. No one would seriously argue on behalf of increasing peril on our roadways. Merely being in favor of safety, though, does not answer the very pertinent question of how safer roads are to be obtained. From an engineering standpoint we could try to improve the road itself, the vehicle using the road, or the driver operating the vehicle. All three approaches have been employed in the attempt to lessen the danger. How effective have the safety measures promoted by public policy been?

While many appear to be satisfied that the assorted safety mandates promulgated by various levels of government are, indeed, making roads safer, the fact is we simply do not **know** whether public policy is helping or hurting the effort to enhance safety. How can this be? Doesn't logic tell us that installing traffic signals should reduce collisions, that wearing seat belts saves lives, and that driver education improves operator skill? As long as we deal with an abstract conception of an environment in which nothing else but the safety-enhancing improvement changes, logic would seem to dictate an obvious conclusion: intervention increases safety. However, the reality is that there is a complex interaction between safety-enhancement measures and human behavior that makes simplistic extrapolation from intent to results prone to err. As some of the data we shall review show, actions intended to improve safety may not have the desired effects.

That intended safety improvements might fail is not to be interpreted as an indictment of the good motives of those who have designed or promoted such measures. Likewise, neither should the questioning of the effects of various safety mandates be judged a neanderthal opposition to progress. The issue is whether chosen means to a desired end (safer roads) are suitable and cost-effective. Even though we may agree that it is difficult, if not impossible, to place a value on a human life, the issue of cost is still relevant. Resources available for meeting human needs are limited. Those expended on ineffective safety programs are not available for more urgently needed purposes. So, the notion that any money spent on safety is warranted because "no price is too high to pay for saving a life" ignores the very real opportunity cost entailed in any employment of scarce resources.

Over a decade ago, economist Sam Peltzman authored an article questioning the efficacy of mandatory safety improvements to vehicles (Peltzman 1975). His suggestion that regulations meant to improve safety might be useless or, worse, counterproductive shocked the safety establishment. His methodology was attacked on the grounds that is failed to *prove* that the criticized safety rules didn't work. Since his methodology was an examination of accident statistics in the post-safety regulation period, there never was any prospect for proving the inefficacy of the regulations. At the same time, though, the proponents of mandatory safety rules have little to offer other than similarly derived statistics in support of their case. They, too, have failed to *prove* that the safety measures are effective.

It may be that it will never be possible to conclusively prove whether specific safety measures do or don't work. Perhaps the best we can hope for is to weigh conflicting evidence and decide what we want to believe about the advisability of mandating specific safety regulations. It is the purpose of this paper to observe some of the evidence contradicting the conventional wisdom on safety. After all, none of the promotional efforts on behalf of any mandatory safety rules shows any awareness of conflicting evidence. Many of those making decisions on safety legislation or on how to expend highway resources have not been exposed to evidence that the rules or measures they contemplate may be ineffective. It is hoped that by creating some exposure and provoking a more robust debate on safety issues, better and more productive outcomes will result.

II. LONG TERM AGGREGATE TRENDS

The long term decline in traffic fatalities per vehicle mile of travel has been a source of pride for many safety advocates. The plunge from 18 fatalities per 100 million vehicle miles in 1925 to a current figure around 3 fatalities per 100 million miles (see table 1) would appear to reflect a successful safety program. That is, at first glance and without much thought it would seem to be evidence of successful regulation. However, the inexorable progress of the decline raises some doubt as to whether specific safety measures could really be the significant factors. The absence of any substantial deviations from the long term trend could be taken as indicative of the *lack* of consequence of specific safety measures.

If we convert the long term data to a logarithmic scale (see table 2) we cannot help but note that the trend fluctuates around a straight line. This line approximates a constant 3% per year decline in fatalities per vehicle mile of travel. If legislated safety measures had had a significant impact on traffic fatalities, wouldn't we expect to see dramatic changes in the rate of decline? While the absence of dramatic changes in the rate of decline in traffic fatalities doesn't prove that legislated safety has no impact, neither does the long term decline prove that legislated safety is the cause of the salutary trend.

Similar long term declining trends in fatalities per vehicle mile of travel have been observed in other parts of the world. In general, there appears to be an inverse relationship between degree of motorization and fatality rates. As societies become more motorized, the fatality rates decline. This phenomenon has been dubbed "Smeed's Law" (Adams 1985). Interestingly, the phenomenon cuts across time and cultural differences such that modern, safety-equipped cars in relatively unmotorized third World nations achieve fatality rates comparable to those in the United States when Model T Fords were the dominant vehicle (Adams 1987). Perhaps the long term decline in fatalities per mile would have occurred regardless of whether government mandated specific safety measures. If this is the case, then billions of dollars have been expended on devices that have had no significant effect on safety.

A further examination of long term traffic fatality statistics produces some interesting information. One of Peltzman's contentions was that drivers might choose to consume intended safety benefits in other forms (Peltzman 1975). For example, improved crashworthiness might induce many drivers to opt for more speed rather than more safety at previous speeds. Another possibility is that the perception of improved safety would contribute to a greater propensity to travel. That factors such as these might be at work can be observed from the long term data on traffic fatalities per capita (see table 3). Unlike the fatalities per vehicle mile data, there is no long term salutary trend. The 1985 death rate of 19 per 100,000 persons is right back where it was in 1925. The up and down fluctuations give no hint of assurance that the future will produce improved safety.

If traffic death rates have shown no improvement over several generations we are inevitably led to the question of whether all the effort and expense on behalf of specific safety measures was worth it. That people are traveling further and faster is a value in

TABLE 1

U.S. Motor Vehicle Fatalities

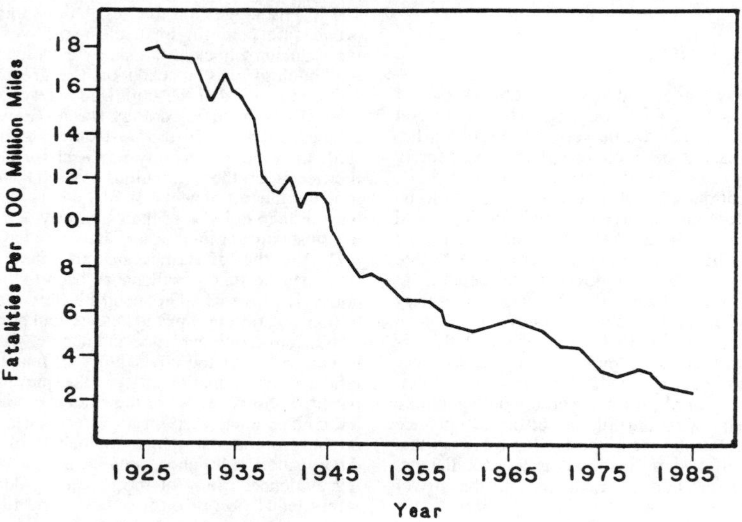

Sources:
Statistical Abstract of the United States
Historical Statistics of the United States

TABLE 2
U.S. Motor Vehicle Fatalities
(on Logarithmic Scale)

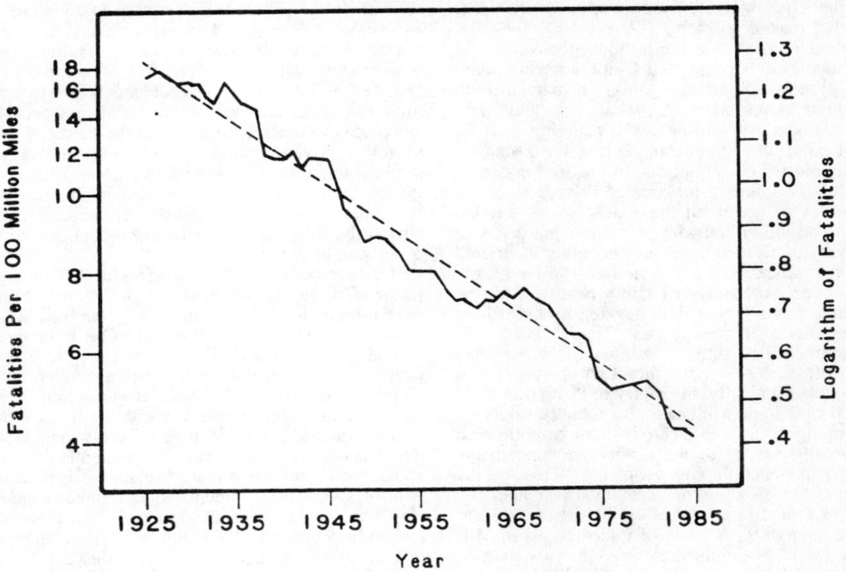

Source:
Statistical Abstract of the United States
Historical Statistics of the United States

itself. However, the appeal that "if even one life is saved, the effort is worthwhile," doesn't have the same impact when actual human behavior effectively converts it to "if even one more mile is traveled . . . " Further, when we consider that there are threats to life and limb other than traffic hazards, it is entirely possible that funds expent on safety measures with negligible impact may lower overall safety. Consider that death by homicide has more that doubled over the last generation. Given this decidedly unsalutary trend, can we be confident that diverting police resources to the writing of citations for motorists not wearing seat belts, for example, will actually improve the net safety of the environment?

III. AMBIGUOUS EVIDENCE

While there are certain propositions about which there can be little argument—for example the damage done by collisions at high speeds is greater than at low speeds— this knowledge does not lead directly to feasible safety strategies in the real world. Obviously, the number and severity of collisions could be reduced to nil by prohibiting movement. However, we prefer the values to be achieved, albeit at some higher levels of risk, from movement. Safety is not, then, an all-consuming objective. We purposefully choose less safety in order to pursue other ends. The debate that exists cannot be couched in simplistic safe vs. unsafe terms, despite emotional rhetoric (i.e., "unsafe at any speed") that may be employed. There are trade-offs of safety vs. other objectives. Improvements that can be made are more properly considered as gains at the margin.

Most practical approaches to the safety issue aim at limiting or reducing negative consequences rather than imagining that risk is going to be eliminated. Even with this more realistic perspective, the directional effect of any measure intended to enhance safety is not as certain as many advocates of these measures might like to believe. Below we survey some of the research on the impacts of various programs intended as safety improvements. It is by no means clear that the intent is realized in practice.

The issue of drunken driving is a major concern in traffic safety. Enforcement personnel estimate that up to 50% of the fatal accidents on our highways involve an intoxicated driver. Coming up with methods of reducing this hazard could contribute to significant improvements in road safety. A currently popular proposal along these lines entails the restriction of packaged liquor sales. Specifically, many communities are banning the simultaneous vending of gasoline and packaged alcoholic beverages. The intent is that it will be more difficult for persons to both drink and drive. Recent evidence indicates that contrary to intent, the actual effect of banning the sale of alcoholic drink at gas stations may be to increase drunken driving. In a study of 400 California communities, economists Patrick McCarthy and John Umbeck of Purdue University found that cities with the bans experienced increases in alcohol-related accidents ("Driven..." 1987).

While the ban does make it more inconvenient to obtain alcoholic beverages, it is not deterring determined drinkers. Some are going the extra mile to find liquor stores where they can buy their beverages. Others are settling for consuming their liquor at bars rather than transporting the packaged goods home for consumption. In the first instance, the policy is inducing intoxicated drivers to put in more miles of travel on the roads. In the second instance, an individual who previously got drunk in the privacy of his own home now becomes inebriated in a public place from which he has to drive home. In either case, the intended safety benefit from the regulation may be more than negated by the behavioral response of road users.

Securing traffic signals and painted crosswalks are favorite aims of neighborhood groups concerned about safety. Typically, the injury or death of a child moves people to demand action to enhance safety by installing lights, or painted crosswalks, or both. And we have all seen the public service announcements on television encouraging people to "cross at the green and not in-between." Obviously, signals and crosswalks must improve safety. Statistics, though, indicate otherwise. A study of intersections in Milwaukee failed to demonstrate any improvement in safety from the installation of traffic lights. In the three-year period prior to signalization, the intersections studied experienced 520 accidents. In the three years after signalization, there were 522 recorded accidents. Injuries sustained in the two periods were 134 and 154, respectively. Right angle collisions were down by 34%, but rear-end, head-on, and vehicle-bicycle collisions were up by 37% to 41% (Short et al. 1982).

The statistics on painted crosswalks are even more disconcerting. There are six times as many accidents in marked crosswalks than in unmarked areas. Of course, crossing pedestrian volume is higher for marked crosswalks, but even after adjusting for this fact we find that the rate of accidents is still twice as high for painted crosswalks (Herms). Some pedestrians place too much confidence in the power of paint to induce drivers to stop. Perhaps we have oversold the idea that crosswalks are safe places. Thus, even though the law holds the driver responsible for stopping, the installation of marked crosswalks is not having the fully intended effect of reducing vehicle-pedestrian collisions.

If we are convinced that driver attitude is a major factor affecting the accident rate, the natural course would seem to call for improved training. Better driver knowledge and skill ought to be an unassailable aid to safer roads. Even here, though, the evidence is mixed. Insurance companies give premium reductions to drivers who have had driver education courses (Cushman 1986). Since these firms are in business to make money, they have a strong incentive to accurately assess risk. Consequently, it would appear well established that mandating driver education for high school students should enhance safety. How, then, do we explain the results of a Connecticut study (Robertson 1981) that showed the elimination of mandatory driver education classes was followed by lower accident rates for students of those schools? While it is likely that on an individual basis driver education improves safety, on an aggregate basis this may not be the case. Forcing students to take driver education courses may encourage earlier driving by those who would otherwise delay this

TABLE 3
U.S. Motor Vehicle Fatality Rate

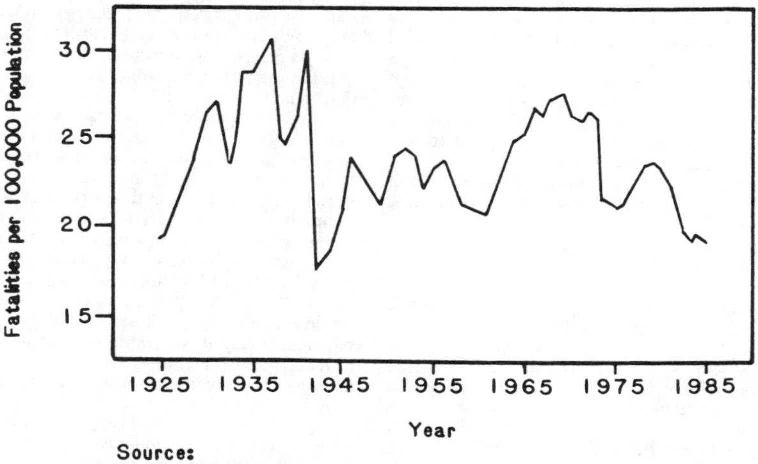

Source:
Statistical Abstract of the United States
Historical Statistics of the United States

undertaking. As a result, the streets could be made more dangerous by hastening teenagers into operating motor vehicles.

Another example of how we can mislead ourselves with unquestioning devotion to safety dogma is in the case of the 55 mph speed limit. Public transportation agencies have sported signs with the familiar "55 saves lives" slogan. Many attribute the decline in fatalities per vehicle mile of travel since 1974 to the reduction in maximum speed limits from 70 or 75 mph to 55 mph. The mechanism by which the 55 mph speed limit is supposed to work may be simple in theory, but in practice it is very mysterious. It is not that we challenge the idea that vehicles crashing at higher speeds inflict greater damage. Rather, we challenge the notion that an unobserved limit can be expected to have a significant impact on road safety. Before the 55 mph limit was relaxed for rural Interstate highways, surveys revealed that over 70% of the vehicles traveled at speeds in excess of 55 mph (Mancuso 1986). Observation of highway speeds in Arizona showed an average rural Interstate speed of 68 mph in 1973, the year before limits were lowered to 55 mph. In 1986, these same routes showed an average speed of 66 mph, even though the posed limit was 55 mph (Miller 1987). How can the mere posting of signs be credited with saving lives if the speed limit is ignored? It would seem that the vehement defense of the 55 mph speed limit as a boost to safety is more an act of faith that a demonstrated reality.

Perhaps the biggest current controversy in auto safety is the issue of mandatory seat belt use laws. A combination of factors has brought the focus of occupant restraint to compulsory belt wearing. Seat belts themselves became standard equipment for cars back in the 1960s. Thus, purchasers of new autos were compelled to pay for the addition of seat belts to vehicles. However, only a minority of drivers actually use the belts in their cars. The "flaw" in this safety equipment is that it requires action on the part of the user to be effective. The failure of users to regularly take action to save their lives spawned the campaign for passive restraints: those that could be effective even if the driver neglects to take the precaution of fastening his own seat belt. The mandated mechanism for achieving passive restraint is the so-called "air bag": a device that deploys automatically when the vehicle crashes.

The problem with the "air bag" is that it protects only against a limited variety of crashes for only an initial impact and costs a considerable amount of money. Quite logically, the automobile manufactures would prefer a solution to the lack of seat belt use that does not add extra cost to their products. In order to give a lower cost option an opportunity, Secretary of Transportation Elizabeth Dole decreed that if states representing two-thirds of the population enacted mandatory belt use laws by April 1, 1989 vehicle manufacturers, would not be required to install "air bags" in all new cars. It is by no means clear that the alternative of enacting compulsory belt use laws will avert the cost of mandatory "air bags." Neither is it clear that the enactment of seat belt use laws will make highway travel safer.

Of course, the evidence is very persuasive that in any given crash vehicle occupants are more likely to survive if they are wearing seat belts. However, this does not prove that compulsory belt use laws will make the traffic environment safer. There are several ways in which the intended safety improvement can break down. First, mere enactment of a mandatory use law is not the equivalent of attaining higher rates of use. The experience with non-observance of the 55 mph speed limit is reminder enough of this possibility. Second, resources devoted to enforcement of mandatory belt use laws are not simultaneously available for other purposes. For example, the time and money used to apprehend, cite, prosecute, and punish those not wearing seat belts is not available for the apprehension, arrest, prosecution, and punishment of potentially more dangerous threats to public safety (for example, drunken driving). Third, the imposition of compulsory seat belt use may have some undesirable side-effects that detract from, or even outweigh, the saving of lives for vehicle occupants.

A study of the post-mandatory belt use law in Great Britain raises some disturbing implications. As expected, driver and front-seat passenger fatalities were down by 18% and 25%, respectively, in the post-law period. At the same time, though, fatality rates for back-seat passengers, pedestrians, and cyclists were up by 27%, 8%, and 13%, respectively (Harvey and Durbin 1986). If one considers non-occupants killed only by those vehicles covered by the belt use law, we find pedestrian death rates up by 14% and cyclist deaths up by 40% (Hamer 1985c).

There is some controversy over the above statistics. While statistics cannot prove the cause and effect relationship we might like to have in order to be certain, they can suggest possible cause and effect sequences. John Adams of University College in London has gained some notoriety by suggesting a link between mandatory belt use laws and increased risk to pedestrians and cyclists (Adams 1985). The instantaneous reaction may be to reject the idea that a driver would purposefully risk a crash merely because a seat belt gives him better protection. Yet, calm reflection tells us that the willingness to take greater risks is influenced by perceptions of enhanced safety: non-swimmers more willingly venture into water with life-jackets, high-wire acts are more daring when there is a safety net, firemen equipped with oxygen and asbestos suits venture into burning buildings, etc. In traffic safety terms, statistics have consistently shown higher crash rates for larger cars (Evans (a)). For its part, Great Britain's Department of Transportation predicted a potentially alarming increase in pedestrian deaths if mandatory seat belt use because law (Hamer 1985c). Shamefully, though, this accurate prediction was officially suppressed until two years after the mandatory belt use law went into effect (Hamer 1985 a). We do not do the cause of better safety a service by refusing to contemplate inconvenient data or arguments. If forcing people to use seat belts is such a meritorious undertaking, it ought to be able to prevail over all contrary evidence and arguments.

IV. REAL SAFETY ENHANCEMENT

Most transportation practitioners are engineers. It stands to reason that an engineering approach to safety would predominate in highway agencies and in the advice transportation professionals give to those who legislate safety rules. Unfortunately, the engineering approach does not allow for offsetting human behavioral responses to intended safety improvements. All other things being held constant, the removal of a hazard or the provision of safety device ought to result in less danger. The problem is that in the real world, few things are held constant for very long. Human beings are one of the most dynamic elements in any prospective safety equation. We cannot expect to have any lasting or significant impacts on safety if we ignore the impacts of changes in human behavior in response to our intended safety improvements.

A dispassionate examination of the record in intended safety improvements cannot fail to convince us that we do not really know whether the wide array of measures taken have produced net gains. Even if we can be sure that a specific device, like a seat belt, makes vehicle occupants safer, this is not the same as making the roads safer. If all, or even part, of the result of mandatory safety rules is a mere shifting of the burden of risk, can we really justify the use of compulsion to achieve such mixed and/or questionable ends?

The current focus of most safety programs is too narrow. Better brakes, bumpers, or belts are the physical products of intended safety improvements. Even if the physical product is improved, though, there is no assurance that the environment is safer. It is not the means of a person's demise that is so crucial as the fact of his demise. Does it really matter that fewer people die as the result of brake failure, if in the end the traffic fatality toll is unchanged in total? That the traffic death rate per capita is the same today as it was six decades ago should inspire more restraint in our willingness to *impose* mandatory safety regulations on the traveling public. This should especially be the case when we presume to save people from the consequences of their own negligence.

The notion that there is no downside to public mandates aimed at increasing safety is a cruel self-deception. Everything has its cost. Over the last decade some $1.3 billion spent on intended safety improvements by the federal government produced "little demonstrated effect in reducing the traffic crash toll" according to the General Accounting Office ("GAO . . ." 1980). Perhaps this money could have been employed to better effect in some other manner. For example, while statistics do not show an improvement from government efforts in traffic safety, they do show a deteriorating trend in crimes of violence. A person's chance of being murdered has doubled since the 1960s. Resources that many states are now committing to enforcement of seat belt use laws are resources that could be devoted to better efforts to prevent crimes of a more serious nature.

So, the opportunity cost of mandating intended safety improvements has been an overlooked aspect in the overwhelming majority of safety promotional campaigns. This is all the more critical when it becomes apparent that the "obvious" safety improvements do not necessarily produce the intended results. It is one thing to tread on individual freedom armed with the rationale that the inflicted compulsion is for the individual's "own good." As the evidence shows, though, we cannot be sure that the safety rules government imposes *are* for the individual's own good. Less enthusiasm for coercion and more respect for individual autonomy would appear to be the scientifically warranted direction for those truly interested in enhancing safety to take.

REFERENCES

Adams, John. *Risk and Freedom*. Cardiff, Great Britain: Transportation Publishing Projects, 1985.

Adams, John. "Risk Homeostasis and the Purpose of Safety Regulation." Presented to CEC Workshop on Risky Decision Making in Transport Operations, TNO Institute for Perception, Soesterberg, The Netherlands, November 9, 1986.

Adams, John. "Smeed's Law: Some Further Thoughts." *Traffic Engineering & Control,* February 1987.

Bedard, Patrick. "If We Really Want to Save Lives, Let's Make Dying Against the Law." *Motor Trend,* May 1984, p. 183.

Block, Walter. "Theories of Highway Safety." Unpublished manuscript, circa 1980.

Chaddar, Himmat Paul Schoufeld. "Are Pedestrians Safe at Right-Turn-on-Red Intersections?" *Journal of Transportation Engineering,* January 1985, pp. 1-15.

Cirillo, Julie and Forrest Council, "Highway Safety: Twenty Years Later." *Transportation Research Record 1068,* pp. 90-95.

Copulos, Milton, "The High Cost of the 55 MPH Speed Limit." *Backgrounder,* Heritage Foundation, September 9, 1986.

"Curfews Reduce Crashes of 16-Year-Olds." *Status Report,* Insurance Institute for Highway Safety, October 5, 1982, pp. 1-2.

Cushman, William. "Driver Education," *Traffic Safety,* May/June 1986, pp. 6-7, 25-27.

"Driven to Driving," *Reason,* July 1987, pp. 17-19.

Evans, Leonard. "Accident Involvment Rate and Car Size." *Accident Analysis and Prevention,* Vol. 16, pp. 387-401 (a).

Evans, Leonard. "Human Behavior Feedback and Traffic Safety." *Human Factors,* October 1985, pp. 555-576 (b).

Evans, Leonard. *Estimating Fatality Reductions from Increased Safety Belt Use*. General Motor Research Laboratories, May 21, 1986 (c).

Evans, Leonard, "Risk Homeostasis Theory and Traffic Accident Data." *Risk Analysis,* Vol. 6, No. 1, 1986, pp. 81-94 (d).

Evans, Leonard, and Paul Wasielewski. "Risky Driving Related to Driver and Vehicle Characteristics." *Accident Analysis and Prevention,* Vol. 15, No. 2, 1983, pp. 121-136.

Flint, Jerry. "Back to the Drawing Board." *Regulation,* September/October 1986, pp. 50-51.

"GAO Questions Effectiveness of Safety Grants," *Status Report* Insurance Institute for Highway Safety, November 21, 1980, pp. 1-2.

Graham, John. and Steven Garber, "Evaluating the Effects of Automobile Safety Regulation," *Journal of Policy Analysis and Management,* Vol. 3, No. 2, 1984, pp. 206-224.

Hamer, Mick, "Report Questions Whether Seat Belts Save Lives." *New Scientist,* February 7, 1985, p. 7(a).

Hamer, Mick. "Belted Drivers Are Killing More Pedestrians." *New Scientist,* April 25, 1985, p. 8(b).

Hamer, Mick. "Britain to Decide on Whether to Keep Seat Belts." *New Scientist,* December 19, 1985, p. 9(c).

Harvey, A. C. and J. Durbin. "The Effects of Seat Belt Legislation on British Road Casualties." Royal Statistical Society, March 19, 1986.

Herms, B. F. "Pedestrian Crosswalk Study," *Highway Research Record* 406, pp. 1-13.

"Inflator Explosion." *Status Report,* Insurance Institute for Highway Safety, August 21, 1979, pp. 11-12.

"Influence of Truck Size and Weight on Highway Crashes." *Status Report,* Insurance Institute for Highway Safety, March, 1983.

Mancuso, James. "Finding the Truth About Highway Safety." *Heartland Perspective,* August 8, 1986.

McKenna, Frank. "Do Safety Measures Really Work?" *Ergonomics,* Vol. 28, No. 2, pp. 489-498.

McKenna, Frank. "The Human Factor in Driving Accidents." *Ergonomics*. Vol. 25, No. 10, pp. 867-877.

Miller, Charles. Director, Arizona Department of Transportation, "Comments to the House Public Works Committee on the 55 MPH Issues," 1987.

"NHTSA Estimates Cost of Auto Safety Regulations." *Status Report,* Insurance Institute for Highway Safety, April 27, 1982, pp. 2-3.

"No Signs in the Window." *Today's Traffic,* November/December 1986, p. 5.

"Not All See 21 as a Cure-All." *Traffic Safety,* March/April 1986, pp. 8-9.

"Patterns of Teenager Crash Involvement are Studied." *Status Report,* Insurance Institute for Highway Safety, November 21, 1980, p. 3.

Peltzman, Sam. "The Effects of Automobile Safety Regulations." *Journal of Political Economy,* Vol. 83, No. 4, 1975, pp. 677-725.

Pisarski, Alan, "Deep-Six 55." *Reason,* November 1985, pp. 3-35.

Raub, Richard. "Removal of Roof-Mounted Emergency Lighting from Police Patrol Vehicles: An Evaluation." *Transportation Research Record 1047,* pp. 83-88.

Robertson, Leon. "Pattern of Teenaged Driver Involvement in Fatal Motor Vehicle Crashes." *Journal of Health Politics, Policy and Law,* Summer 1981, pp. 303-314.

Semmens, John and Dianne Kresich. "Coercion is No Cure for Collision," *Heartland Perspective,* January 26, 1987.

Short, Michael S., and Gerald A. Woelfl and Chia-Juch Chang. "Effects of Traffic Signal Installation on Accidents." *Accident Analysis and Prevention,* Vol. 14, No. 2, 1982, pp. 135-145.

Smith, Kevin. "Drop in Highway Deaths Mystifies Experts." *Car and Driver,* May 1984.

Summers, Brian. "Is It Safer to Put Uncle Sam at Wheel of Your Car?" *Orange County Register,* October 21, 1984.

Sussman, E. D. H. Bishop, B. Madnick and R. Walter. "Driver Inattention and Highway Safety." *Transportation Research Record 1047,* pp. 40-48.

Trucking Deregulation in 1986. U. S. Department of Transportation, 1986.

"Two Studies Cite Failure of Drunk-Driving Programs." *Status Report,* Insurance Institute for Highway Safety, April 9, 1979, pp. 7, 9.

"Wet Pavement Found to Multiply Fatal Crash Risks." *Status Report,* Insurance Institute for Highway Safety, May 20, 1980, p.4.

Wilde, Gerald. "Evidence Refuting the Theory of Risk Homeostasis?" *Ergonomics,* Vol. 27, No. 3 pp. 297-304.

Williams, Allan F., Paul L. Zador, Sandra S. Harris, and Ronald S. Karpf. *The Effect of Raising the Legal Minimum Drinking Age on Fatal Crash Involvement*. Insurance Institute for Highway Safety, June 1981.

ENDNOTE

* Arizona Department of Transportation

Motor Carrier Cost Analysis

*by Russell S. Elliott**

ABSTRACT

Motor carrier rates historically have been compared on a per hundred weight (cwt) basis which does not accurately reflect distances as a factor in the cost. A Defense Logistics Agency (DLA) Defense Depot with shorter average less-than-truckload (LTL) distances for example, could have a low cwt cost when compared to other depots. However, when distances are considered, the same defense depot could be spending more on a cost per mile basis than other defense depots. The purpose of this study was to provide insight into motor carrier costs on a per cwt per mile (cwt/mile) basis and to compare these costs across defense depots. One year of historical data were used to develop the mean cwt/mile rates for each defense depot. These cwt/mile rates were subdivided into weight bracket and mileage groups. Graphic representation by weight bracket and mileage groups was used to compare the defense depots and to observe any obvious trends. An analysis of variance (ANOVA) tests which blocked weight brackets and mileage brackets along with multiple classification analysis (MCA) was used to determine differences in mean cwt/mile across the defense depots.

I. INTRODUCTION

A requirement existed to examine motor carrier cost trends on a cost per mile basis from Defense Logistics Agency (DLA) Defense Depots (hereinafter referred to as 'depot(s)'). A cost per mile study will enhance the interpretation of depot cost trends because the traditional cost per hundred pound (cwt) charges do not accurately reflect distance as a factor in the cost. A depot with shorter average less-than-truckload (LTL) distances for example, could have a low cwt cost when compared to other depots. However, when distances are considered, the same depot could be spending more on a per mile basis than other depots.

The primary study objective was to determine the motor carrier costs on a per mile basis for each DLA depot and to compare those costs across the depots. The study was limited to interstate motor carrier shipments with DLA depots as origins and all continental United States (CONUS) points as destinations. Dedicated truck service was excluded from the analysis due to the special nature of this service. The time period of the data was fiscal year 1985 (FY85).

II. METHODOLOGY

A. Input Data

The input data consisted of records from the Freight Information System (FINS) file for the fiscal year 1985. For a record to be selected for the analysis, the transportation mode field had to indicate motor carrier shipments. The data used from each record selected consisted of the origin depot Standard Point Location Code (SPLC); the origin depot Government Bill of Lading Location Code (GBLOC); the total weight; the billed weight; the total charges; the destination state; the destination city; and the SPLC assigned to the destination activity. The mileages used for the study were generated by an algorithm which approximated mileages based on the longitude and latitude of the origin and destination SPLCs.

B. Data Preparation

A computer program was written and run to select the appropriate records from the FINS file. A second program was run which used an algorithm to approximate mileages and appended these mileages to the selected records from the FINS file. Finally, a third program was run for each depot to select the data and make computations as follows: eliminate dedicated truck service; eliminate obviously erroneous data; convert weights to the billed weight; select records that were interstate shipments only; aggregate the transportation charges according to weight bracket and mileage group; and compute the average transportation charge per cwt per mile for each weight bracket and mileage group. Table 1 shows the weight brackets used and Table 2 shows the mileage groups used for this analysis.

TABLE 1

WEIGHT BRACKETS

0- 499 lbs	15000-19999 lbs
500- 999 lbs	20000-24999 lbs
1000- 1999 lbs	25000-29999 lbs
2000- 4999 lbs	30000-34999 lbs
5000- 9999 lbs	35000-39999 lbs
10000-14999 lbs	40000 lbs and over

TABLE 2

MILEAGE GROUPS

1– 50-100 miles	10–1000-1200 miles
2–100-150 miles	11–1200-1400 miles
3–150-200 miles	12–1400-1600 miles
4–200-300 miles	13–1600-1800 miles
5–300-400 miles	14–1800-2200 miles
6–400-550 miles	15–2200-2600 miles
7–550-700 miles	16–2600-3000 miles
8–700-850 miles	17–3000 miles and over
9–850-1000 miles	

III. ANALYSIS

A. Data Trends

The data tend to behave as expected on an individual depot basis. The motor carrier rates per cwt per mile tend to decrease with respect to increases in distance and/or increases in weight. Carrier costs follow the tapering principle such that as distance and/or weight increases, total cost increases at a slower rate. Thus, on a per unit basis, the average rate per cwt per mile decreases.

B. Cost Comparisons Across Defense Depots

There tends to be a greater disparity in rates per cwt per mile between the shorter distances (550 miles or less), particularly for the 50-100 mile group. Above the 550-700 mile group, the rates tend to converge along a narrow band for LTL shipments under 10,000 pounds. Figure 1 shows a typical rate comparison for LTL shipments.

There are two major occurrences at the 10,000 pound break point. First, shipments greater than 10,000 pounds are normally considered as volume or truckload (TL) shipments. Each depot defines its own LTL and TL weight break point. Three depots define TL as 10,000 pounds, one depot defines TL as 15,000 pounds, one depot defines TL as 20,000 pounds and one depot defines TL as 25,000 pounds. The significance of this is that TL rates tend to be lower than LTL rates.

Second, the DLA Guaranteed Traffic Program (GTP) generally starts at 10,000 pounds for TL shipments. The purpose of the TL GTP is to reduce the cost and transit time of TL shipments. Figure 2 shows a typical rate comparison for TL shipments.

The behavior of the curves in the figures were expected. LTL rates were expected to be higher than TL rates because LTL rates contain terminal and line haul costs whereas TL rates contain mainly line haul costs. TL rates tend to be more linear than LTL rates. The low mileage groups have an effect upon both TL and LTL rates. This effect can be caused by the minimum charges assessed by motor carriers.

C. Statistical Analysis of Data

The GTP has been implemented for all TL shipments throughout DLA. DLA depots have also implemented GTP for LTL shipments with the exception of DDF. Since one of the primary purposes of the GTP is to reduce rates, a reasonable expectation is that DDF rates for LTL traffic will be higher than those of the other defense depots. Another expectation is that a geographical difference in rates should be shown. The general nature of motor carrier rates published by rate bureaus tend to reflect the different general operational characteristics of the geographical areas of CONUS. For example, the

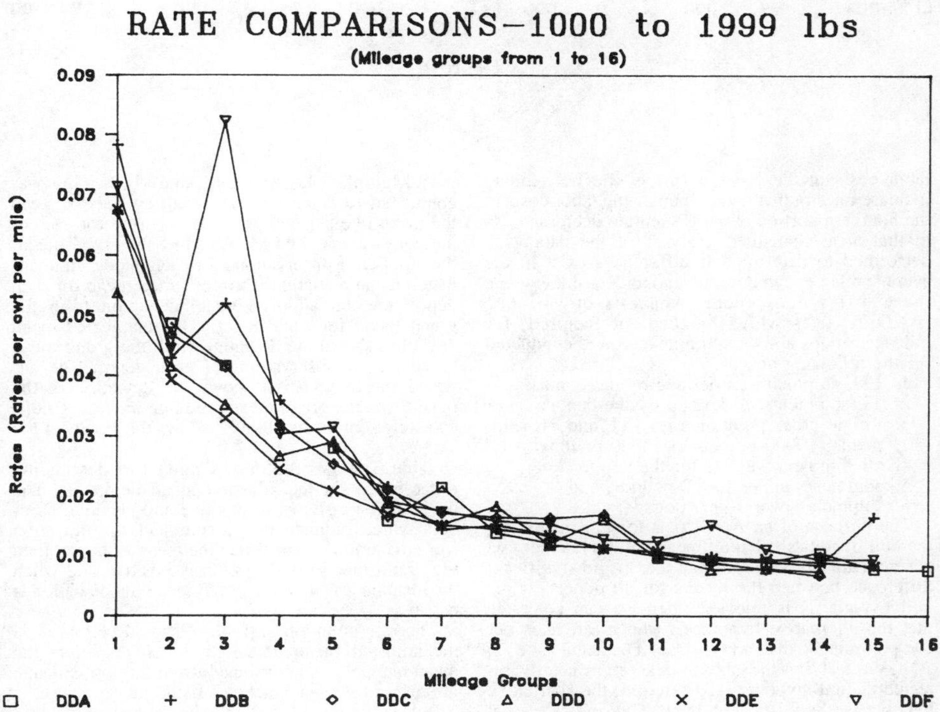

Figure 1

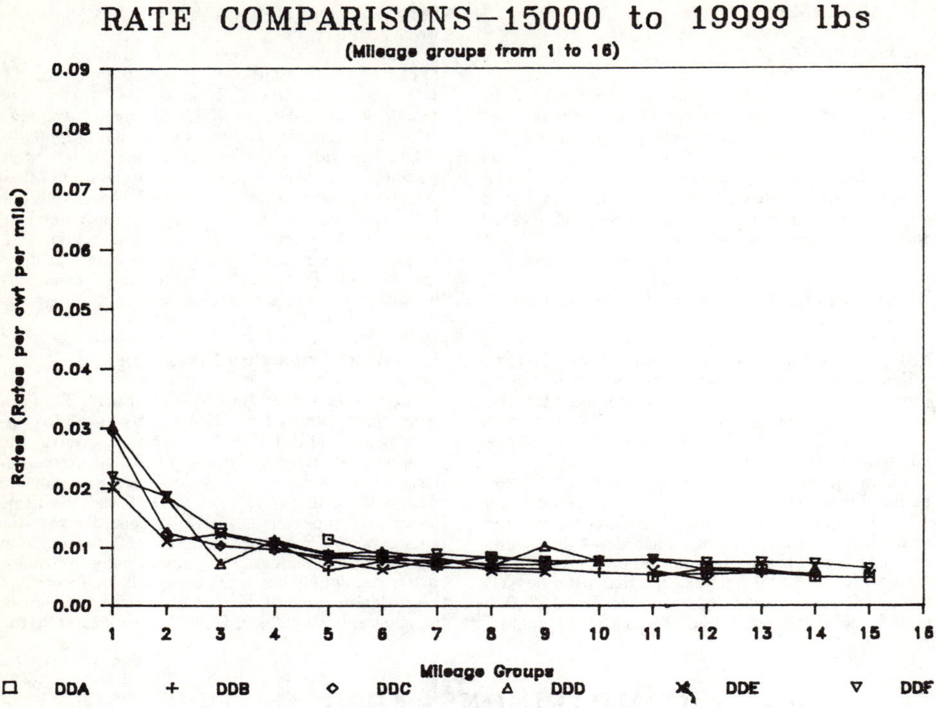

Figure 2

labor costs in the Eastern part of the US tend to increase rates in that area, whereas the labor costs in the Southern portion of the US tend to decrease rates in that area. Statistical analyses of the data were performed to determine if differences exist in the average rate per cwt per mile and to determine where these differences occur. Analysis of variance (ANOVA) tests which blocked, or factored, for mileage groups and weight brackets were conducted for the following areas:

1. LTL shipments as defined by the depots;
2. TL shipments as defined by the depots;
3. weight break point area for LTL and TL shipments (10000 pounds to 25000 pounds);
4. all shipment weights for the depots;
5. shipments under 10000 pounds; and
6. shipments over 10000 pounds.

The purpose of an ANOVA is to statistically test the null hypothesis that all means of the rate per cwt per mile are equivalent or that there is no significant difference between the means for all depots. If the null hypothesis is rejected, then we can conclude that the alternative hypothesis (the mean rates per cwt per mile are different) is true. The results of each ANOVA test for the above areas significantly rejected the null hypothesis. Therefore, the alternative hypothesis should be accepted. This indicates that the mean rates per cwt per mile are not equivalent for all depots.

A Multiple Classification Analysis (MCA) was conducted to determine where differences between the mean rate per cwt per mile occurred for each of the above areas. The MCAs show the grand mean, the number of observations for each depot, an unadjusted mean deviation from the grand mean for each depot, and an adjusted mean deviation from the grand mean for each depot. The unadjusted mean deviation shows the deviation from the grand mean based on all observations of each depot. The adjusted mean deviation shows the deviation of the mean from the grand mean based on mileage groups and weight brackets. Table 3 shows the results of the MCAs.

Table 3 shows the depots ranked in descending order based on the adjusted mean deviation. The deviations are shown in positive and negative numbers which indicate the direction of the deviation from the grand mean. When the mean is greater than the grand mean the deviation is positive and when the mean is less than the grand mean the deviation is negative.

The interpretation of Table 3 should be based on the adjusted mean deviation which considers the mean rate per cwt per mile adjusted for both mileage groups and weight brackets. If the interpretation is based on the unadjusted deviation the results would be misleading. For example, the results of the MCA for TL shipments, as defined by the depots show that

Table 3
Multiple Classification Analysis Results

LTL Shipments as defined by the depots
GRAND MEAN = 0.1275

DEPOT	OBSERVATIONS	UNADJUSTED DEVIATION	ADJUSTED DEVIATION
DDB	39633	0.0097	0.0560
DDE	75976	0.0558	−0.0007
DDA	29276	−0.0694	−0.0077
DDF	36893	0.0012	−0.0087
DDC	88855	−0.0240	−0.0120
DDD	26201	−0.0194	−0.0212

TL Shipments as defined by the depots
GRAND MEAN = 0.0075

DEPOT	OBSERVATIONS	UNADJUSTED DEVIATION	ADJUSTED DEVIATION
DDF	4041	0.0009	0.0007
DDA	772	−0.0031	0.0007
DDD	494	−0.0016	0.0005
DDC	5546	−0.0019	−0.0001
DDE	2737	0.0051	−0.0004
DDB	2427	−0.0016	−0.0009

Weight break point area for LTL and TL shipments (10000 pounds to 25000 pounds)
GRAND MEAN = 0.0099

DEPOT	OBSERVATIONS	UNADJUSTED DEVIATION	ADJUSTED DEVIATION
DDD	869	0.0001	0.0013
DDF	2008	0.0006	0.0006
DDC	3318	−0.0013	0.0000
DDA	676	−0.0025	−0.0004
DDB	1412	−0.0024	−0.0006
DDE	1783	0.0045	−0.0008

All shipments
GRAND MEAN = 0.1228

DEPOT	OBSERVATIONS	UNADJUSTED DEVIATION	ADJUSTED DEVIATION
DDB	42060	0.0082	0.0536
DDE	78713	0.0560	−0.0014
DDA	30048	−0.0646	−0.0068
DDF	40934	−0.0046	−0.0097
DDC	94401	−0.0236	−0.0107
DDD	26695	−0.0151	−0.0198

Shipments under 10,000 pounds
GRAND MEAN = 0.1301

DEPOT	OBSERVATIONS	UNADJUSTED DEVIATION	ADJUSTED DEVIATION
DDB	39633	0.0086	0.0561
DDE	75976	0.0547	−0.0010
DDA	28600	−0.0693	−0.0080
DDF	36896	0.0001	−0.0089
DDC	87471	−0.0236	−0.0120
DDD	25524	−0.0179	−0.0212

Shipments over 10,000 pounds
GRAND MEAN = 0.0079

DEPOT	OBSERVATIONS	UNADJUSTED DEVIATION	ADJUSTED DEVIATION
DDD	1171	0.0009	0.0090
DDF	4041	0.0005	0.0070
DDA	1448	−0.0020	−0.0000
DDC	6930	−0.0013	−0.0002
DDE	2737	0.0048	−0.0005
DDB	2427	−0.0020	−0.0007

DDE has the largest positive unadjusted deviation of 0.0051. DDE's adjusted deviation is -0.0004, which is one of the lowest adjusted deviations. The majority of DDE shipments are in the lower mileage groups with the highest rates per cwt per mile. When the deviation is adjusted for weight brackets and mileage groups, the adjusted deviation drops to -0.0004 for DDE. Therefore, the results will be explained in terms of the adjusted deviations.

DDF tends to have a high rate for depot defined TL shipments and shipments over 10000 pounds and lower than average for shipments under 10000 pounds and depot defined LTL shipments. This indicates that there is an inconsistent relationship between TL and LTL rates for all depots. It would be expected that if TL rates were high, then LTL rates would be high for the same depot which is not always the case.

DDD also tends to have a high rate for depot defined TL and shipments over 10000 pounds. This result was unexpected because it indicates that the geographical differences in motor carrier rate bureaus do not exist for TL GTP traffic. If there was a geographical difference in TL GTP rates, then it would be expected that DDE would have higher rates than DDD.

DDA was expected to have higher mean rates than the results show. One explanation is that the analysis did not include intrastate shipments. The elimination of intrastate rates can decrease the number of low mileage shipments, which in turn, tends to produce lower rates on a per mile basis. DDA's lower mean rate may be explained by the fact that the rates from DDA may reflect backhaul rates from the west coast area to the east cost are. Backhaul rates are normally lower than regular rates because most carriers will charge a lower rate to cover the cost of returning his equipment to his base territory (i.e., fuel costs and driver costs only).

The results indicate that DDB tends to have the highest rates for depot defined LTL shipments, shipments under 10000 pounds and all shipments. This result was expected due to the geographical location of DDB. DDB has the lowest rates for depot defined TL and shipments over 10000 pounds. This was initially unexpected. The reason for the low rates may be explained by backhaul rates.

DDE tends to have a high mean rate per cwt per mile for depot defined LTL, all shipments, and shipments under 10,000 pounds. The majority of DDE shipments tend to be in the lower weight brackets and the lower mileage groups, which will cause a higher mean rate. DDE also tends to ship in the motor rate bureau areas, which tend to have the highest rates of the rate bureaus.

DDC tends to have a low mean rate per cwt per mile (lower than the mean rate). This may indicate that the GTP, for both TL and LTL traffic, is meeting its objective to decrease motor carrier transportation rates.

DDD mean rates tend to vary between LTL and TL traffic. DDD tends to have the lowest mean rate for LTL traffic, one of the highest mean rates for TL traffic, and the lowest mean rate for all shipments. The reason for this is the low number of TL shipments (26201 LTL shipments vs. 494 TL shipments).

General observations of the results indicate that a majority of motor carrier shipments from the depots are LTL. The weight brackets of 0 to 499 pounds tend to contain the majority of the LTL traffic. The LTL shipments have a great impact on the mean rate per cwt per mile for all depot shipments. The grand mean for all shipments is 0.1228, 0.1301 for LTL (shipments under 10000 lbs), and 0.0079 for TL (shipments over 10000 lbs). This indicates that the LTL rate per cwt per mile strongly influences the overall rate per cwt per mile for all the depots. This is expected because the overwhelming majority of the shipments are LTL for each depot. The differences in the mean rate per cwt per mile for each depot are relatively small (i.e., from under six cents to under one cent), yet these differences are statistically significant. The results did not show strong geographical rate differences. The reason is that all of the defense depots have interregional shipments which would tend to decrease the impact of the differences in the general operational characteristics of the geographical areas.

D. LTL GTP versus LTL NON-GTP

The purpose of GTP for LTL traffic is to reduce transit times, transportation costs, and administrative handling on an individual depot basis. Several studies analyzing GTP indicate that GTP does reduce the transit times and transportation rate on a per depot basis. When a rate is reduced, then the rate per cwt per mile is also reduced. DDF has not implemented the LTL GTP, thus it was expected that DDF's LTL mean rates per cwt per mile would be higher than the other depots. The first analysis does not support this assumption because all shipments were analyzed.

The first analysis showed that the majority of the shipments are LTL and under 500 pounds therefore a second analysis was conducted to compare LTL GTP versus LTL non-GTP mean rates per cwt per mile. Shipments under 200 pounds were excluded from the additional analyses. The weight of 200 pounds was selected because the GTP agreements establish a minimum charge based on a minimum shipment weight of 200 pounds.

The additional analyses were conducted in the following categories: depot defined LTL shipments; LTL shipments under 10000 pounds; and all shipments. The ANOVA test results rejected the null hypothesis (the mean rates per cwt per mile are equivalent for each depot) for each category; therefore, the alternative hypothesis was accepted (the mean rates per cwt per mile for each depot are statistically different). The test statistic, which indicates whether to accept or reject the null hypothesis, was reduced for each ANOVA test category. This indicates that by eliminating shipments under 200 pounds some variability in the mean rates per cwt per mile was eliminated.

The MCA results are shown in Table 4. These results are interesting for three reasons. First, the ranking of the depots are the same in each of the categories. Second, DDB and DDF are the two depots with the highest mean rates per cwt per mile. The differences between DDB and DDF adjusted deviations of mean rates per cwt per mile are at least twice the mean rate per cwt per mile of the third ranked depot. Third, the grand means decrease by at least .0991.

The results show that DDB mean rates per cwt per mile are the highest. This is expected due to the

Table 4
MCA Results for LTL Shipments Over 200 Pounds

LTL Shipments as defined by the depots
GRAND MEAN = 0.0254

DEPOT	OBSERVATIONS	UNADJUSTED DEVIATION	ADJUSTED DEVIATION
DDB	19850	-0.0053	0.0023
DDF	23574	0.0055	0.0020
DDC	44602	-0.0018	0.0007
DDD	13890	-0.0002	0.0005
DDA	17918	-0.0094	0.0004
DDE	36776	0.0060	-0.0038

LTL Shipments under 10000 pounds
GRAND MEAN = 0.0257

DEPOT	OBSERVATIONS	UNADJUSTED DEVIATION	ADJUSTED DEVIATION
DDB	19850	-0.0056	0.0024
DDF	23574	0.0052	0.0020
DDC	43218	-0.0016	0.0008
DDD	13213	0.0060	0.0005
DDA	17242	-0.0093	0.0004
DDE	36776	0.0058	-0.0038

All shipments
GRAND MEAN = 0.0237

DEPOT	OBSERVATIONS	UNADJUSTED DEVIATION	ADJUSTED DEVIATION
DDB	22277	-0.0052	0.0022
DDF	27615	0.0038	0.0016
DDC	50148	-0.0021	0.0008
DDD	14384	0.0011	0.0007
DDA	18690	-0.0082	0.0004
DDE	39513	0.0064	-0.0038

geographical location of DDB. DDB and DDF adjusted mean rates per cwt per mile are out of line when compared with the other depots. The LTL GTP for DDC, DDD, DDA and DDE appears to be working properly. DDE has an extremely low mean rate per cwt per mile which indicates that their LTL GTP is excellent. DDF mean rates are the second highest for each of the above categories. This was expected since DDF has not implemented GTP for LTL traffic.

IV. CONCLUSION

The results of this analysis show that the mean rates per cwt per mile for all the depots are not equivalent. The ANOVA test used for each of the nine areas under consideration resulted in significant rejection of the null hypothesis. The largest absolute deviations from the grand means occurred for LTL traffic. This indicates that greater savings can probably be achieved with GTP for LTL traffic. Since the largest percentage of shipments are LTL, more cost savings attention to LTL traffic should be given.

The minimum charges that are established by the LTL GTP agreements are based on the minimum weight of 200 pounds. The effect of the minimum charges become apparent when shipments under 200 pounds are eliminated from the analyses. The ranking of the defense depots changes and the grand mean is reduced. DDF had a below average mean rate per cwt per mile when all shipments were analyzed. However, when shipments over 200 pounds were analyzed, DDF's mean rates per cwt per mile jumped to the second highest. This shows that the minimum charge for LTL GTP can have an adverse effect on mean rates per cwt per mile.

The results also show that closer attention should be paid to the minimum charges during the negotiation of GTPs. High minimum charges affect the low mileage group rates and this effect is shown in the figures for all weight brackets. Lower minimum charges would reduce the rate per cwt per mile for the lower mileage groups, which in turn would reduce the mean rate per cwt per mile for LTL shipments. This could result in a more cost effective GTP. The results also show that each depot is unique in its shipment patterns. DDA and DDB mean rates appear to be influenced by backhaul rates, DDE mean rates appear to be influenced by its short distance shipments and rate bureau territory. Each depot's mean rate is influenced by the type of commodity shipped. This uniqueness of each depot increases the difficulty of comparing the mean rate per cwt per mile across the depots.

ENDNOTE

*Defense Logistics Agency, Operations Research and Economic Analysis Management Support Office

A Model of Firm Behavior and Costs in Exempt Motor Carriage

*by Wesley W. Wilson**

ABSTRACT

In this paper an empirical model of exempt motor carrier costs is developed formally from an explicit model of firm behavior. There are a variety of cost models employed in the literature to obtain cost estimates, to estimate the parameters of the cost function, to evaluate competitiveness and performance, and to ascertain optimum industry structure. In general, these models have little behavioral components short of cost-minimizing behavior. Here the output decision and critical input decisions are explicitly modelled.

Theoretically, it is maintained carriers choose the desired number of miles they travel and the number of trucks to operate subject to a capacity constraint on each truck employed. The capacity constraint enters in the problem because of road weight limitations, truck configuration constraints (e.g., no double bottoms), driver hours constraints, etc. In the decision process there is a trade-off between higher overhead costs (e.g., insurance and license fees) associated with additional trucks driven and reduced repair costs. When higher price levels and/or increases in the truck capacity constraint (e.g., increased speed limits) obtain the firm will choose a higher desired output level and operate a greater number of trucks. When insurance fees, license fees, and/or interest expenses increase the firm will respond by reducing the number of trucks operated, reduce the number of desired miles operated, and drive its trucks harder.

Empirically, considerable structure is placed on the cost structure. Due to the manner in which the data were collected, costs are not observed in the data. However, the data do provide information allowing a proxy for costs to be defined. While a proxy for costs can be defined in the data, it is defined by known formulae. Using that information results in more precise cost estimates, reduces the number of parameters to estimate, and greatly simplifies estimation. The empirical model of costs is obtained by parameterizing overhead components of cost as a function of exogenous firm characteristics, the number of trucks employed, and the utilization of equipment. The latter two of these variables are the result of endogenous decisions of the agent. Hence, estimation of the cost model by OLS may result in biased coefficients. Unlike predecessors to this study, a Haussman test is employed to test for the presence of bias. The results suggest variable and fixed overhead cost parameters are significantly different from zero and no bias is suggested by using endogenous variables as explanatory variables in the estimation.

I. INTRODUCTION

Transportation of agricultural commodities to market is provided by air, railroad, motor carriers, and barge. This study focuses on the truck portion of the grain transportation infrastructure. Currently, this sector is confronted with a variety of issues. These include dramatic increases in overhead costs (e.g., the insurance crisis), growing competitive pressures (e.g., a depressed farm economy, deregulation of transport), and changing constraints on operations (e.g., weight restrictions, speed limits, etc.).[1] It has been suggested that increases in overhead costs, growing competitive pressures, and/or increases in equipment sizes have been the source of increasing safety hazards in the industry.[2] Safety issues are not directly addressed in this paper. Rather a framework for examining firm responses and/or benefits to increasing overhead costs, growing competitive pressures, and changing constraints on operations is provided and applied to "exempt" motor carriage.

Generally, firm responses and/or benefits associated with these changes are identified from a combination of behavioral considerations. By a specification of firm behavior, the benefit accruing to the motor carrier from relaxing constraints on operation (speed limit) can be estimated. Firm tendencies toward driving older equipment is addressed by developing and estimating a cost function indexed by firm specific characteristics (age of equipment). In this regard, there have been a number of analyses focussing on the costs of the agricultural motor carrier[3] and/or the supply of the exempt trucker and regulated trucking firms.[4] Such analyses have been used to evaluate performance of the industry, in network flow analyses, and to evaluate the effectiveness of the mode in constraining railroad prices.

In general, recent empirical models used to characterize the structure of costs in trucking have been based on flexible functional forms. Such forms are taken to approximate an arbitrary technology. In this paper, the fact that previous research has found little if any support for non-constant returns to scale in TL motor carriage along with a priori knowledge concerning the structure of costs in exempt trucking is used to specify an empirical model. In short, the resulting model is greatly simplified from a "flexible form" and captures heterogeneity in costs emanating from differences in equipment characteristics such as age of equipment in the cost parameters. The "indexed" cost function is then used to characterize benefits to the firm from driving older equipment.

In Section II the behavioral and resulting empirical model of costs for the firm is posited. Data characteristics are presented in Section III, followed by empirical results and application in Section IV. Implications and relevance of the findings conclude the paper.

II. CONCEPTUAL FRAMEWORK

In this section a brief discussion of the environment within which the agricultural carriers (used in this analysis) operate is provided, a conceptual model of the firm is presented, and the empirical model formulated. The resulting model provides a method to estimate the cost per mile of operating constraints on trucks and to partially explain firms driving older equipment.

The agricultural carriers analyzed in this analysis provide transportation from North Dakota origins to two Minnesota destinations (Minneapolis and Duluth). These carriers typically serve the same elevator(s) regularly for their outgoing or fronthaul movements of agricultural products. On incoming or backhaul movements these firms generally haul commodities subject to ICC jurisdiction when such loads can be obtained. In this regard Duluth is considered the low potential backhaul market while Minneapolis (in a relative sense) is considered the high potential backhaul market.[5] However, despite the differences in potential returns with Minneapolis, Duluth was the most common truck market with 64 and 77 percent of total truck movements in 1976 and 1979, respectively [Griffin (1980)]. In explaining this paradox, Wilson (1986) developed a market model finding that, in equilibrium, prices adjusted such that expected marginal profits equate across markets. That result forms a maintained hypothesis of this model. Namely, in equilibrium, the firm maximizing expected profit is indifferent to the market served.

At this point a brief overview of cost and revenue characteristics is provided, followed by a model of firm behavior. Costs are divided into three categories including assignable costs, fixed overhead, and variable overhead. Assignable costs are those costs which vary with output and can be directly assigned to output. These costs include labor, fuel, usage based depreciation, tire, and management expenditures. Note that this specification does not allow for substitution effects in producing output. For example, in providing round trips, the firm cannot trade off labor for fuel. The rationale for the specification follows from definition of cost components (data) and from the fact that in most cases there is no reason to expect that such trade-offs can be made in this industry (a priori consideration). Fixed overhead costs are those costs that, given the equipment decision, are invariant to output levels. These costs include license and insurance fees as well as interest expenditures and a portion of repair costs. Finally, variable overhead costs are those costs which vary with output measures but cannot be directly assigned to a given trip. Here such costs are represented by a portion of repair expenditures.

In terms of the revenue function, recall that the carriers analyzed serve multiple markets. As noted above, a maintained hypothesis of the analysis is that in equilibrium prices adjust such that the carrier is indifferent between markets served. Hence, any trip to a given destination is randomly assigned to the trucker by the shipper. This type of behavior is consistent with Cosgriff's (1978) finding that loads in these markets are commonly set up by the firm calling the elevator with the added caveat that the trucker will accept the load regardless of the destination given expected price is equal to or exceeds expected marginal costs.[6]

In providing round trips the trucker sequentially serves two markets—the fronthaul market and the backhaul market. On the fronthaul leg of the round trip the trucker received revenue with certainty while on the backhaul the firm is randomly loaded yielding backhaul revenue as uncertain.

Let P_h, M, and m_h represent the expected price per mile associated with each destination (h), total miles, and the proportion of total miles to the h destination. Total revenues (R) may then be written as $R = \Sigma_h P_h m_h M$. Under the assumption that the probability of being loaded on a backhaul is independent of miles traveled to each destination, expected revenues may be written as $ER = M \, EP_h \, Em_h$.

With the above specification of cots and revenue, the behavioral component, firm decisions, is now discussed. A formal mathematical representation of the economic model is provided in Appendix A.

Firms are taken to choose the number of miles to travel and the number of trucks to operate over the time period. In making these decisions, there is a theoretical and/or operational constraint on the number of miles a truck can travel over the time period (e.g., speed limits, driver hour limits, waiting to unload, etc.). Also, since assignable costs can be directly assigned to trips and therefore miles, they are incorporated simply by marking down the expected revenue. In terms of the choice variables remaining costs (fixed and variable overhead), repair costs are taken as a function of both the number of trucks and the utilization (miles/trucks) of those trucks, while remaining overhead costs are taken as a function of the number of trucks.

As shown in the Appendix, the profit maximizing number of miles traveled equates the expected price per-mile (net of assignable costs) with marginal repair costs per truck plus the opportunity cost per mile associated with the truck capacity constraint. The profit maximizing number of trucks equates the marginal increase in fixed overhead with the marginal repair cost savings per truck (driving trucks less intensively results in cost savings) plus the added profit associated with increasing the truck capacity of the firm (the marginal profit per truck associated with increasing truck capacity multiplied by the number of miles per truck).

The usual comparative statics emanate from this model. When p^e increases, the firm responds by buying more equipment and traveling more miles. When factor prices increase, it responds by driving fewer miles and buying fewer trucks. Finally, when the capacity of truck increases, the number of trucks and miles traveled each increase.[7]

The above framework is used for two purposes. First, given estimates of the repair cost function and remaining fixed overhead cost function (license, insurance), the value associated with increasing truck capacity can be estimated. Second, as all cost categories may depend on the age of equipment, incorporation of that information enables a comparison of costs and benefits of driving older equipment. The remainder of this section describes the methodology used to obtain estimates of the non-assignable cost functions—repair costs and other overhead costs (license, insurance, interest).

In parameterizing repair costs, $C^R(\frac{M}{U}, U)$ and overhead costs, $C^o(U)$ a linear specification is employed. The particular specification employed is given as

$$C^R(\frac{M}{U}, U) = K^R + a\frac{M}{U} + bU + \epsilon_R$$
(1)
$$C^o(U) = K^o + cU + \epsilon_o$$

In (1), a, b, and c are the marginal cost parameters to be estimated, $K^i(i = R,O)$ are the constant terms, M is miles traveled, U is trucks operated, and ϵ_i are the error terms. Such marginal cost parameters are expected to vary systematically across firms. To account for such variability a and b are taken to be a linear function of average equipment age (A) while c is taken as a function of not only average equipment age but also the price of equipment (P^E).[8] By parameterizing these parameters the cost function is indexed by firm specific equipment characteristics—an indexed cost function. Finally, as two years of data (1976 and 1979) were employed intercept and interaction dummy variables (D) were added. The final model to be estimated is as follows.

$$C^R = (r_o + r_1D) + (r_2 + r_3D)\frac{M}{U} + (r_4 + r_5D)A\frac{M}{U} + (r_6 + r_7D)U + (r_8 + r_9D)AU + \epsilon_r$$
(2)
$$C^o = (v_o + v_1D) + (v_2 + v_3D)U + (v_4 + v_5D)UP^E + (v_6 + v_7D)UA + \epsilon_o$$

where: C^R = repair costs;
C^o = overhead costs[9];
D = 1 if 1979, D = O if 1976;
A = average equipment age;
U = the number of equipment units; and
M = miles traveled.

Under the assumptions of the model, both M and U are the result of decisions made at the beginning of the year. If this is the case then estimation of (2) proceeds by either ordinary least squares (OLS) or seemingly unrelated regressions (SUR). However, if firms adjust over the year, then M and U are endogenous, and the use of OLS or SUR can result in biased coefficient estimates. The detection of such bias was tested using a Hausman (1978) specification test.

III. DATA

There are three data sets used in this paper. All were collected from surveys or from published sources by the staff of the Upper Great Plains Transportation Institute (UGPTI) located at North Dakota State University. Two data sets were the result of surveys of "exempt" motor carriers providing transportation of North Dakota's agricultural commodities to two primary destinations—Minneapolis/St. Paul, Minnesota and Duluth, Minnesota/Superior, Wisconsin. These surveys contain firm-specific annual cost and operating statistics for the years 1976 and 1979. The 1976 data are reported in Cosgriff (1978) and the 1979 data are reported in Wilson, Griffin, and Casavant (1982). An important advantage of these data is that they contain data allowing market specific measures of firm output to be constructed.

The Cosgriff (1978) survey covered 600 "exempt" motor carriers from which 84 surveys were returned. Wilson, Griffin, and Casavant (1982) surveyed 694 firms of which 144 surveys were returned. As discussed in Cosgriff (1978), there were an estimated 600-1000 firms in the industry when his sample of 600 was drawn. Wilson, Griffin, and Casavant (1982) sampled all firms (694) known to exist in the industry at that time.

Of all the data available, a total of sixty observations used were in this study, representing 32 observations from 1979 and 28 observations from 1976. The remaining observations were omitted for a variety of reasons including missing values, serving markets other than Minneapolis and Duluth, and/or the data represented a regulated carrier which sometimes hauled agricultural products.

The data provide little or no information regarding truck rates. Fronthaul rates were constructed following the methodology described in Wilson (1986). In short, partially available 1976 truck rates were fitted to a function of distance for each market. Missing data were completed using the predicted values. There were no data available corresponding to 1979. These data were constructed as follows. First, rail rate schedules (rate as a function of distance) were estimated for 1976 and 1979. The truck and rail rates for 1976 were used to specify an equation fitting truck to rail rates. The results provided parameter estimates enabling a truck rate for 1979 to be constructed.

As discussed in the previous section, costs are divided into the three categories of assignable, variable overhead, and fixed overhead. Assignable costs include labor, fuel, depreciation, tire, and management expenditures. Based on Cosgriff (1978) truck interview results, a rate plus \$10 method was used to calculate per mile labor costs. For example, a rate of 55 cents per hundred weight yields labor cost of \$65. Fuel costs per mile were calculated directly from the survey data. The 1976 data provided fuel expenditures and miles traveled. The 1979 data provide both the price of fuel and miles per gallon loaded and unloaded. Vehicle and tire depreciation per mile calculations were based on usage depreciation methods.[10] For equipment, depreciable value of tractors was set at 75 percent of initial cost while for trailers this value was set at 70 percent.[11] While the data are quite limited regarding the approximate lives of tractors and trailers, the average values of 700,000 and 950,000 were used here.

Tire depreciation per mile follows analogously with the exception that no salvage value was included. The depreciable miles used were 142,000, 85,500, and 118,500 for radial, recapped and standard tires, respectively. Finally, management fees were also not observed in the data. Cosgriff (1978) suggests that using the brokerage fee of 8 percent of the gross revenue per trip is a reasonable approximation of these costs. This approach was used here.

Variable overhead and fixed overhead costs include repair, license, insurance, and interest expenditures. All of these costs are directly observed in the data except for interest expenditures. These latter costs are approximated in the application of the model using assumed 10 and 18 percent interest rate for 1976 and 1979. These values parallel those used by Cosgriff (1978) and Casavant et al. (1982). Investment costs are then constructed by the product of the interest rate and the average investment in the tractors and trucks over the period.

A summary of all revenue and cost variables employed is provided in Table 1.

IV. EMPIRICAL RESULTS AND APPLICATION

In this section a brief overview of the empirical results is provided then applied to the conceptual model to obtain estimates of the marginal profit associated with increasing truck mileage capacity and to examine firm decision rules pertaining to the vintage of truck driven.

TABLE 1
Summary of Observed and Constructed Data

	Duluth		Minneapolis	
Variable	1976	1979	1976	1979
Fronthaul Rate[a]:	.425	.527	.377	.50
Assignable Costs[a]:				
Labor	.099	.119	.088	.114
Fuel	.086	.225	.091	.231
Depreciation	.044	.077	.044	.077
Tire	.028	.031	.028	.031
Management	.034	.042	.030	.040
Overhead:			1976	1979
Repairs			9051.82	8563.25
License			1061.071	2023.438
Insurance			3574.643	5917.469
Interest			3496.475	13923.000
Output:				
Miles Traveled			156,099.071	153,962.906
Number of Trucks			1.536	1.781
Average Backhaul Proportion			.194	.19
Proportion of Trips to Duluth			.731	.758
Estimated Backhaul Probability - Duluth[b]			.042	.105
Proportion of Trips to Minneapolis			.269	.242
Estimated Backhaul Probability - Minneapolis[b]			.767	.446
Other:				
Average Price of Equipment			35,071	66,775
Average Age of Equipment			4.3	4.7
Utilization (miles/trucks)			103,275	87,584

a Per round trip mile.

b Estimates taken from Wilson (1986).

At the outset, the results of the Hausman test suggest no significant bias is introduced by treating either miles traveled or trucks driven as exogenous.[12] Hence all results reported are based on the SUR estimator.

All coefficient estimators, standard errors and related statistics are reported in Appendix B (Tables B-1 and B-2) and summarized below. The estimated repair equations for 1976 and 1979 are

$$C^R(1976) = -666.11^{**} + .135^* \frac{M}{U} - .008^* A \frac{M}{U} + 17.46\ U + 753.09^*\ AU$$

(3)

$$C^R(1979) = 811.17 + .047 \frac{M}{U} - .006\ A \frac{M}{U} - 586.16\ U + 839.25^*\ AU$$

where a (*) indicates significance at the five percent level and a (**) indicates significance at the ten percent level. In each year utilization had an increasing impact on repair costs but the impact was lower on older equipment suggesting marginal repair costs per mile per truck full with the age of equipment. With regard to the fixed overhead component of repair costs, in each year these costs were positive over the range of data and increasing with the age of equipment. In short, older equipment at the margin has lower marginal repair costs but the fixed component of repair costs increases with equipment age.

The empirical results corresponding to the remaining fixed overhead costs (insurance, license, interest) are given for each year by

$$C^0(1976) = 173.57 + 2172.36\ U + .108576^*\ UP^e - 103.11\ AU$$

(4)

$$C^0(1979) = 1783.11^* + 4059.26^*\ U + .129^*\ UP^e + 102.57\ AU$$

As expected, in each year over the range of the data these fixed overhead costs increase with the number of trucks driven. However, the indexing parameters suggest the rate at which these costs increase depends significantly on the value of the equipment. That is, as the value of equipment increases, the rate at which these costs increase, increases. In the indexing function mixed results occur with regard to age of equipment but in no case is significant.

Treating the estimated cost schedules as first order linear approximations to the true cost function and using the mean values of observables in Table 1 and equation B-6, two different values of the marginal profit associated with increasing truck capacities may be calculated for each year—estimates based on the mile decision of the firm and on the number of trucks decision of the firm. These values are provided in Table 2.

Except for the value calculated in 1979 based on the equipment decision the values are less than 10 cents a mile. With regard to the 1979 equipment based decision, it is noted the empirical results suggested considerable increases in not only the relevant parameters of the fixed overhead cost schedule, but the average indexing variable price of equipment almost doubled compared with 1976. Hence, the evidence suggests fixed overhead costs increased dramatically between 1976 and 1979, a finding analogous to the current "insurance crisis." Therefore, based on the equipment decision values in Table 2, the value of truck mileage capacities is regarded as increasing with increasing overhead expenditures.

A final issue addressed in this paper concerns firm responses to growing competitive pressures and/or increasing overhead costs. On the benefit side, there are savings associated with driving older equipment from lower variable overhead costs. In addition, added benefits from driving older equipment include facing lower equipment costs which impact no only assignable costs (e.g., lower usage-based depreciation) but also lower fixed overhead (e.g., lower insurance schedules). Consequently, for this purpose the correlation between age of equipment and price of equipment need be included. To this end, equipment prices were fit to equipment ages via OLS, yielding the following:

$$P^E = \underset{(4964.87)}{48795.87^*} + \underset{(7293.85)}{47848.12^*D} - \underset{(938.83)}{3156.42^*A} - \underset{(1363.88)}{3207.07^*AD}$$

(13)

$$R^2 = .6815$$

where a (*) denotes significance at the five percent level and standard errors are in parentheses.

The salient features of the decision rule are summarized by

Benefits:	**1976**	**1979**
Lower Variable Overhead	\$814.04	\$537.27
Lower Fixed Overhead (from lower P^E)	526.52	1459.63

TABLE 2

Estimated Marginal Profits Associated with Relaxing Truck Mileage Constraints

	Added Profits Based on:[a]	
Year	M	U
1976	9.01 cents/mile	3.34 cents/mile
1979	6.37	32.16

[a]Since backhaul prices and associated costs are not observed, these values are based on the no backhaul firm.

Costs:		
Higher Fixed Overhead	649.99	941.82
Net Benefit	\$690.57	\$1055.08

The results suggest moderate savings from driving older equipment of about \$700 and \$1055 per year.

V. SUMMARY AND CONCLUDING REMARKS

This paper provided a model of firm behavior which was used to address two issues facing exempt from regulation motor carriages—the impact of changing constraints on truck operations and the impact of increasing overhead costs. Increases in capacity constraints translates into increases in profit for exempt motor carriage while increases in overhead costs result in carriers driving older equipment.

In addressing these issues a formal model of firm behavior was posed, and an indexed cost function yielding firm specific marginal cost parameters was developed and estimated under differing assumptions. While much of the literature has suggested constant returns to scale, the approach used here suggests scale parameters vary across firms systematically. Using these results in conjunction with an optimality assumption enable point estimates of the marginal profit associated with increasing truck capacity constraints. Finally, the indexed cost formulation enabled a cost-benefit analysis of firm strategies to drive older equipment. The results suggest moderate savings associated with such a strategy.

While the data employed are somewhat dated, changes through time reported are analogous to current changes. That is, insurance schedules are increasing, and firms are driving older equipment. Furthermore, the profits added to truckers form changing constraints on truck operations is still an issue. These results are suggestive of the hypothesis that with growing competitive pressures and in particular rising overhead costs have a significant effect on the value of relaxing constraints on truck capacity.

REFERENCES

Buxton, Freeman K., "Preliminary Findings of the 1985 Motor Carrier Insurance Crisis," Presented at the 27th Annual Meeting of the Transportation Research Forum.

Casavant, K. C., W. W. Wilson, and G. C. Griffin, "Changing Costs and Characteristics of Operating Motor Carriers: A Case Study of Grain Trucking Firms in North Dakota," *Transportation Research Forum Proceedings,* Vol. 23, No. 1, (1982).

Cosgriff, J. G., "The Costs and Operations of Exempt Motor Carriers in North Dakota," Upper Great Plains Transportation Institute Report No. 33, North Dakota State University (1978).

Griffin, G. C., "North Dakota Grain and Oilseed Transportation Statistics," Upper Great Plains Transportation Institute Report No. 36, North Dakota State University (1980).

Hausman, J. A., "Specification Tests in Econometrics," *Econometrica,* vol. 46, pp. 1251-1271 (1978).

Labich, Kenneth, "The Scandal of Killer Trucks," *Fortune,* March 30, 1987.

McGinley, Laurie, "Hands On: Federal Regulation Rises Anew in Matters that Worry the Public," *Wall Street Journal,* April 21, 1987.

Strah, Thomas M., "Judge Strikes Down N.Y. City Restrictions on Twins," *Transport Topics,* March 9, 1987.

Wilson, William W., "Railroad Pricing and Deregulation," *Transportation Research Forum Proceedings,* Vol. 26, No. 1 (1985).

Wilson, William W., Wilson, W. W., and Koo, W. W., "Modal Competition and Pricing in Grain Transportation," Staff Paper Series, AE 85009, Department of Agricultural Economics, NOrth Dakota Agricultural Experiment Station, North Dakota State University (1985).

Wilson, Wesley W., "Transport Markets and Firm Behavior: The Backhaul Problem," unpublished dissertation, Washington State University (1986).

Wyckoff, D. D. and Mainster, D. H., *The Owner Operator: Independent Trucker* (Lexington: Lexington Books, 1977).

ENDNOTES

* The author is an Assistant Professor of Agricultural Economics at Washington State University. Comments from Greg Duncan, Ken Casavant, Fred Inaba, William Hallagan and reviewers are gratefully acknowledged. Also, the author wishes to express his gratitude to Gene Griffin and the staff of the Upper Great Plains Transportation Institute for providing the data employed in this analysis.

1. See Buxton (1986), Labich (1987), Strah (1987), and McGinley (1987).
2. Discussed in Buxton and Labich.
3. For example, see Cosgriff (1978), Casavant, Wilson and Griffin (1982).
4. See Wilson (1985) and Wilson, Wilson and Koo (1985).
5. Using Wilson's (1986) results the average estimated probabilities from each destination are 4 and 77 percent for Duluth and Minneapolis, respectively. And, 10.5 and 45 percent for Duluth and Minneapolis, respectively, in 1979.
6. It is noted here, reandom assignment of loads is all that is needed here as opposed to the equilibrium argument, a somewhat stronger assumption.
7. This result requires that $Pe > CR_{21}$ or the expected price per mile is larger than the change in marginal overhead repair costs (C_2 per truck from increased utilization levels. *A priori* C_{21} could be considered equal to zero yielding the result. That is, marginal overhead costs per truck does not vary with utilization levels.
8. These relationships follow a discussion with the North Dakota State Highway Licensing Division.
9. Interest expenditures were not available in the data and are approximated as discussed later.
10. Virtually all truckers in the 1976 survey indicated they used some notion of depreciation in their pricing divisions.
11. These figures were taken from Wycoff and Mainster (1975).
12. The chi-square statistics of 1.23 and 7.64 asso-

ciated with the repair equation and the fixed overhead equationwith 10 and 8 degrees of freedom were less than the critical values of 18.3 and 15.5 needed for detection of such simultaneity bias.

13. Thus, a portion of these costs are considered fixed given the equipment decision.

14. It should be noted that when multiplied by expected payload capacity the interpretation then becomes the opportunity cost associated with increasing ton-mile capacities one unit (e.g., an increase in weight restrictions).

APPENDIX A

In this appendix, the mathematical model of firm behavior discussed in Section II is provided.

Let P_h, M, and m_h represent the expected price per mile associated with each destination (h), total miles, and the proportion of total miles to the h destination. Total revenues (R) may then be written as

$$\text{(B-1)} \quad R = \Sigma P_h M m_h$$

Under the assumption that the probability of being loaded on a backhaul is independent of miles traveled to each destination, expected revenues may be written as

$$\text{(B-2)} \quad ER = M\, EP_h\, Em_h$$

The firm's optimization problem is now provided after some notational definitions. First, the expected price per mile to a given destination (EP_h) is redefined net of expected assignable costs (fuel, depreciation, management, labor, and tires) per mile yielding a net expected price per mile associated with destination h. This expected price net of assignable costs is given by Ep_h. This allows expected revenues net of assignable costs (r) to be written as

$$\text{(B-3)} \quad Er = M\, \Sigma_h\, Ep_h\, Em_h$$

The firm is taken to maximize expected profits by choosing the desired number of miles to travel each year and the number of trucks to operate at the beginning of the year. Repair costs are assumed to be a function of not only the number of trucks but also of the utilization of those trucks.[13] Let $C^{\$}(M/U,U)$ represent these overhead costs, where M is the number of miles traveled and U is the number of truck units. Fixed overhead costs except for the fixed repair costs (i.e., license, insurance, and interest expenditures) are assumed to depend only on the number of trucks $C^o(U)$. In this framework, expected profits is given by

$$\text{(B-4)} \quad E\Pi = P^e M - C^R\left(\frac{M}{U}, U\right) - C^o(U)$$

where $P^e = \Sigma p_h m_h$ which has the interpretation as the expected price net of assignable costs per round trip mile.

It should be noted here that the number of trucks and desired number of miles may not be independent decisions. That is, given speed limits, hourly constraints on drivers, etc. there is, in principle, a given number of miles a truck can operate in a year. Let K represent this theoretical truck capacity in miles. The Lagrangian associated with (4) is then given by

$$\text{(B-5)} \quad L = MP^e - C^R\left(\frac{M}{U}, U\right) - C^o(U) + \lambda\left(K - \frac{M}{U}\right)$$

It is noted also that the constraint can be redefined in terms of ton-miles by multiplying the constraint by the payload capacity of a truck. The multiplier x has the interpretation as the expected marginal profit per mile associated with increasing the maximum number of miles a truck can operate in a year (K).[14]

The first order conditions associated with (5) are given by

$$\text{(B-6)} \quad \begin{aligned} &\text{(i)} \quad \frac{\partial L}{\partial M} = P^e - C_1^R \frac{1}{U} - \lambda \frac{1}{U} \le 0 \\ &\text{(ii)} \quad \frac{\partial L}{\partial u} = C_1^R \frac{M}{U^2} - C_2^R - C_1^o + \lambda \frac{M}{U^2} \le 0 \\ &\text{(iii)} \quad \frac{\partial L}{\partial \lambda} = K - \frac{M}{U} \ge 0 \end{aligned}$$

where C^j_i represents the derivative of the jth (j = R,O) function with respect to its ith argument.

Interpretation of (B-6) is as follows. First, firms choose a desired mileage level such that expected price equates with marginal repair costs per mile per truck ($C_1^R \frac{1}{U}$) plus the opportunity cost associated with truck mile capacity constraint per truck. Second, the firm chooses a truck up to the point where the marginal repair savings per truck plus the opportunity cost of truck capacity per truck equates with fixed overhead costs.

APPENDIX B
Empirical Results

TABLE B-1
SUR Estimation Results.

Coefficient	Variable	Estimate[a]	Standard Error
r_0	1	-6666.11**	3512.28
r_1	D	7477.278	5225.116
r_2	M/U	.13488*	.041556
r_3	(M/U)D	-.08743	.0572
r_4	A(M/U)	-.00801*	.00371
r_5	AD(M/U)	.001795	.00859
r_6	U	17.4636	1535.385
r_7	DU	-603.627	2898
r_8	AU	753.0969*	288.263
r_9	DAU	86.1523	492.265
v_o	1	173.565	548.6745
v_1	D	-1956.67*	789.9666
v_2	U	2172.359	2019.473
v_3	DU	1886.898	2368.73
v_4	UP^e	.108576*	.03355
v_5	DUP^e	.02025	.035402
v_6	UA	-103.109	149.6223
v_7	DUA	205.6761	177.8839

[a] A (*) and a (**) indicate significance at the five and ten percent levels, respectively.

Weighted R-square for system = .9297 Fixed Overhead OLS R-Square = .9541
Repairs OLS R-Square = .7058

Table B-2
1979 Marginal Cost Parameter Estimates.

Coefficient	Variable	Estimate[a]	F-Statistic
$r_0 + r_1$	Intercept	811.168	.0451
$r_2 + r_3$	M/U	.04745	1.4911
$r_4 + r_5$	A(M/U)	-.006215	.6597
$r_6 + r_7$	U	-586.1634	.0584
$r_8 + r_9$	AU	839.2492*	4.5403
$v_0 + v_1$	Intercept	1783.105*	9.7149
$v_2 + v_3$	U	4059.257*	10.6111
$v_4 + v_5$	UP^e	.128826*	127.9516
$v_6 + v_7$	UA	102.5671	1.1217

[a] A (*) indicates significance at the five percent level.

Identification of Plant Rationalization Strategies Using Optimization Techniques

*by George List**

ABSTRACT

This paper presents a methodology that can be used to analyze the plant rationalization problems typically encountered by railroads. It also presents the results of its application to a case study environment. An optimal network design model is combined with demand sensitivity analysis to develop optimal networks for increasingly more restrictive traffic scenarios. These results, when augmented by more qualitative measures of network value are then used to select a final, rationalized network.

I. INTRODUCTION

Plant rationalization is one of the more challenging problems facing the railroad industry today. Mergers and changing traffic bases make it important to ask which lines can be cut, which branchlines can be let go, and which mainlines can be shared to eliminate redundant trackage.

The problem is complex. Analysts must at once deal with changes to the operating plan, the system's resource requirements and its traffic base. The number of possible solutions is immense for networks the size of typical railroads. To aid the decision-making process, a methodology is needed to enumerate the possible solutions and identify those that are best.

This paper describes an approach that can be used to analyze such problems. An optimal network design model is combined with demand sensitivity analysis to develop optimal networks for increasingly more restrictive traffic scenarios. These results, when combined with more qualitative measures of network value are used to select a final, rationalized network. Section II describes the methodology itself, Section III presents results of its application, and Section IV contains some concluding remarks.

A. Network Design

Categorically, plant rationalization falls under the heading of network analysis (see Friesz, 1985) and more specifically, network design. Notable reviews can be found in Magnanti and Wong (1984) and Boyce (1984); Golden and Magnanti (1977) provide a comprehensive, but somewhat dated, bibliography of relevant literature.

The basic problem is (1) to choose an optimal group of facilities to add to a given network or (2) to determine their optimal capacities. In some cases, both are relevant. Facility selection typically involves discrete choices (the inclusion or exclusion of specific links and nodes) and integer variables are used. Capacity determination typically involves gradations in investment; hence, continuous variables are often used (e.g., for the width of a highway). In some cases, although the capacity additions are discrete (e.g., the addition of classification tracks in a hump yard), they can be treated in a continuous fashion because the capacity (and cost) increments are relatively small.

Mathematically, the problem can be described as follows. Assume a network consists of links I and nodes N. Also, assume a set of links S is being considered for addition to the network; B is the budget limit for adding links; and b_i is the fixed cost of building link i. In addition, assume f_i is the flow crossing link i and $c_i(f_i)$ is the average cost of transportation on link i given f_i. Further, assume y_i is a variable equal to one if link i is added to the network, and zero otherwise. Finally, let $f = \{f_i\}$ satisfy an equilibrium flow pattern based either on Wardrop's first or second principle (see Wardrop, 1952). Then, from LeBlanc (1975), the link-addition formulation of the problem can be described as follows:

(P1): minimize $$\sum_i c_i(f_i)\, f_i, \quad (1)$$

subject to:

$$\sum_{i \in S} b_i\, y_i \le B, \quad (2)$$

$$f_i - M\, y_i \le 0 \quad \text{for all } i, \quad (3)$$

$$y_i = (0,1) \quad \text{for all } i, \quad (4)$$

$$f \text{ is an equilibrium flow pattern,} \quad (5)$$

where M is a very large constant.

If the flow pattern $f = f_i$ follows Wardrop's first principal, then no single unit of flow passing across the network can improve its travel cost (in time, dollars or other suitable measure) by changing its path. This is quite appropriate for highway studies where individual drivers (the users) control the vehicle routing, but for railroad-oriented problems, where the carrier has that control, Wardrop's second principal should be followed: the flow pattern should maximize or minimize some objective function (e.g., total cost, profit, or net consumer surplus).

Based on Abdulaal and LeBlanc (1979), we can also describe an alternate formulation of the problem that involves the continuous addition of capacity:

(P2) minimize $$\sum_i c_i(f,y)\, f_i + w \sum_i b_i(y)$$

subject to:

$$y \ge 0$$

$$f \text{ is an equilibrium flow pattern}$$

where y_i is the capacity enhancement of link i; $c_i(f,y)$ is the unit cost of transportation on link i as a function of both the vector of flows f and the vector of network enhancements y; $b_i(y)$ is the cost of improving link i as a function of the vector of improvements y; and w is a dual variable for the budget constraint d_i $b_i(y)$ s B, where B is the fixed budget for capacity enhancement (see Friesz, 1985).

B. Plant Rationalization

Plant rationalization is a variant on these two formulations. Facilities are to be pruned from rather than added to the network. For example, in the formulation (P1), S would represent the set of facilities to be reviewed; b_i, the fixed cost of retaining link i; and B, the budget available for operating and maintaining the retained links. S is typically some subset of the existing links, but it may also contain new links that make it possible to streamline the network (e.g., create connections between links that run parallel to each other or that cross without intersecting).

C. Railroad-Oriented Research

Railroad-oriented research on plant rationalization has been sparse, unlike the highway mode, where there is a well developed body of literature. Only a dozen or so articles have appeared over the past 20 years. Clearly, analysis methodologies do exist because the industry has been trimming its network ever since 1916 when the total miles of line reached 254,037, almost double what it is today, but the methodologies employed have largely not been reported in the formal transportation literature.

Those methodologies that have been reported are examples of either branchline analysis, network simulation or network optimization. Branchline analysis is a carefully tailored form of cost/benefit analysis that assesses the economic viability of a given network link. Assessments are performed singly, ignoring for the most part the system as a whole, and carry no guarantee of systemwide optimality. Revenues (or benefits) are attributed to the link as are costs, typically on a mileage basis, and if certain viability criteria are met (e.g.; link revenues exceed link costs), a link is retained. The technique has been used for many years in abandonment proceedings before the Interstate Commerce Commission and more recently to analyze various branchlines prior to the creation of ConRail (see United States Railway Association, 1976) and to study the railroad network in Iowa (see Baumel et. al., 1976), among others. Harris (1980) has translated the process into a formal mathematical construct, making it possible to analyze many links simultaneously by computer. Grimm (1986) has updated Harris' work so that the results are more reflective of current conditions.

Network simulations focus on the routing of traffic and the subsequent analysis of link volumes, revenues, and costs. Plans based on this approach include the final system plan for ConRail (see United States Railway Association, 1975) and many of the merger plans submitted to the Interstate Commerce Commission. Hornung and Kornhauser (1979), Wood and Williamson (1980), Bronzini (1980), and Victor and Stewart (1982) provide descriptions of the methodologies typically employed. Many use a shortest path algorithm (minimum impedances) to route the traffic across a pre-specified network. The impedance metric is typically mileage, time, cost or a combination of these three. Impedances are typically pre-specified, so the volumes for any given run have no effect on link performance. Equilibrium assignments, balancing traffic against capacity, can only be achieved through an iterative process. Assigned volumes are taken as a measure of link value and lightly-trafficked segments are targeted for elimination. The process iterates until an acceptably pruned network is identified. The main problem with this approach is that its results depend heavily on the talents of the analyst. As Morlok (1970) points out, if the alternatives he or she develops are sub-optimal, then only a sub-optimal result can be obtained; better solutions may exist if a means were available to identify them.

Network optimization employs some form of mathematical programming to both generate and select an optimal network. Friesz and Harker (1983) and List and Clausing (1983) present possible modeling frameworks. Friesz and Harker describe a multicriteria spatial price equilibrium model that is capable of accounting for the various interrelationships and impacts of network investment. Separate objectives are used to represent the interests of consumers, producers, transportation firms, and the government. Weights are used to combine these into a single, albeit non-linear, objective function and two techniques, one an iterative optimization-equilibrium algorithm and the other a pattern search, are used to find non-dominated solutions. The approach is data intensive, and as the authors acknowledge, capable of generating models which are very difficult to solve. Separately, on an international scale, Steenbrink (1974) reports that similar models have been developed by the French (see Barbier, 1966) and the Dutch (see Haubrich, 1972). List and Clausing present a somewhat simpler approach that involves mixed-integer linear programming. Their model minimizes total transportation cost, taken from the carrier's perspective, and uses integer variables to represent link additions and continuous variables to represent further capacity additions. It thus represents a melding of problems (P1) and (P2) above.

II. METHODOLOGY DESCRIPTION

In work that followed logically from that presented in List and Clausing (1983), the following network rationalization methodology was developed. Assume the network consists of links I and nodes N and that some nodes are yards K where incoming traffic is classified to various destinations. Let f_{rj} be the flow (cars per year) moving on route r (from origin o to destination d) in service type j (e.g., unit train, general merchandise, intermodal); y_i be a (0,1) variable indicating whether line segment i is retained or not; y_{im} be the stepwise level of capacity (m = 1,2,3,..) to be provided on segment i; and y_{ik} be a (0,1) variable indicating whether yard k is open or closed.

Assuming the demand is constant, the objective is then to minimize cost subject to constraints for (a) routing the traffic, (b) determining the line segments to retain, and (c) providing sufficient line and yard capacity to handle the traffic. Mathematically, the problem can be stated as follows:

$$\text{(P3)} \quad \text{minimize} \quad \sum_r \sum_j c_{rj} f_{rj} + \sum_i \left[c_i y_i + \sum_m c_{im} y_{im} \right] + \sum_k c_k y_k \qquad (9)$$

subject to:

$$\sum_{r \in R_{odj}} f_{rj} = F_j^{od} \quad \text{for all } o,d,j \qquad (10)$$

$$\sum_j \sum_{r \in R_i} f_{rj} \le M_i^* y_i \quad \text{for all } i \qquad (11)$$

$$\sum_j \frac{1}{q_j} \sum_{r \in R_i} f_{rj} \le K y_i + \sum_m K_m y_{im} \quad \text{for all } i \qquad (12)$$

$$\sum_k \sum_{r \in R_k} f_{rj} \le M_k^* y_k \quad \text{for all } k \qquad (13)$$

$$y_i,\ y_{im},\ y_k \in (0,1) \qquad (14)$$

$$f_{rj} \ge 0 \qquad (15)$$

The objective function (9) reflects costs for moving the traffic, retaining the line segments, and retaining the yards. Coefficient c_{rj} is the cost of transporting a single car of traffic type j on route r (i.e., from o to d via route r), including not only the variable "above the rail" costs of transportation and equipment ownership and maintenance, but also the variable portion of the "below the rail" costs, such as the variable cost associated with line segments in the path (e.g., dollars per gross ton mile), the variable costs of classification at the intervening yards (e.g. cost per car switched), and any variable, yard-related "below the rail" costs. Coefficient c_i is the fixed cost associated with retaining line segment i; c_{im} reflects the cost of the mth increment of capacity for segment i (e.g. various track classes); and c_{ik} is the fixed cost associated with retaining yard k.

Equations (10) ensure that for each OD pair, the traffic is assigned to one or more of the routes r available for each service type j. R_{odj} is the set of routes r available for traffic type j moving from o to d and F_j^{od} is the total volume of traffic type j moving from o to d.

Equations (11) ensure that segments are retained if the traffic passes over them and equations (12) match volume to capacity. R_i is the set of routes using line segment i and $M_i^*$is a number chosen large enough so that no reasonable level of traffic will be precluded. In equations (12), coefficient q_j is the gross ton miles per car for service type j; K_i is the initial capacity in gross ton miles per year made available by retaining line segment i; and K_{im} is the incremental capacity provided by retaining the mth incremental expansion of capacity on line segment i.

In some cases, only equations (12) are required. If the costs of adding capacity (c_i, followed by c_{im}, c_{im}, . . .) are monotonically increasing, equations (12) are sufficient. However, if the cost of any one of the c_{im} is smaller than c_i, as is the case in the example described later, then equations (11) and (12) are both necessary.

Equations (13) ensure that yards are retained if traffic passes through them. If yard k is used, y_{ik} has a value of one, otherwise it is zero. R_i is the set of routes passing through yard k and M_k^* is the maximum volume that can flow through yard k. If the data suggest that yard capacity costs rise non-linearly with volume, then equations much like (12) should also be included. However, if the relationship is linear, they can be omitted, with the variable cost of classification being included in c_{rj}.

A. Demand Sensitivity Analyses

To investigate the effects of changes in demand, two approaches can be taken. The first is to add "demand" constraints to (P3); the second is to conduct a series of runs, one for each "traffic scenario". Since direct inclusion of demand is very difficult (see Friesz, 1985), the latter approach is more tractable.

The approach to the demand sensitivity analysis is shown in Figure 1. First, future traffic is estimated using historic flows and an economic forecasting technique, such as an input-output model (see List et. al., 1982). Second, various traffic scenarios are developed, based on traffic retention criteria. Two logical possibilities are: volume and commodity. If, for a particular scenario, a given flow is of sufficient size to be retained (e.g., 10 cars per week), and if it is for a commodity of interest, the flow is included in the retained traffic. The third step is to use (P3) to develop networks for each traffic scenario. To reduce the computational burden, the traffic can be screened before implementing (P3) to identify facilities which must be included; for example, dead-end branchlines leading to traffic origins or destinations; they can be marked for automatic inclusion. Equations and variables relevant to their selection in (P3) can be omitted. The fourth step is to examine the networks for sensitivity to various changes in the future economic climate. Line segments and yards can be flagged, for example, if they are dependent on only one commodity or shipper. These facilities might represent significant investment risks if the shipper were to go out of business or if the market for the product were to disappear. Finally, based on all the above, an acceptably pruned network can be identified.

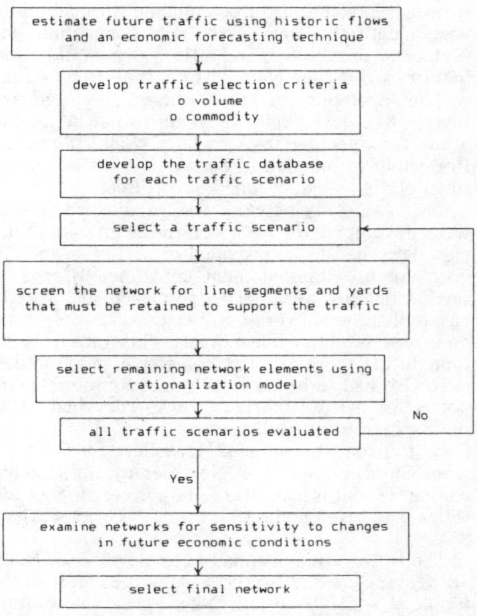

Figure 1. Plant Rationalization Methodology

III. EXPERIENCE IN THE USE OF THE METHODOLOGY

An application of the methodology was performed for the state of New Jersey (see List et. al., 1982). The state's goal was to establish a "core rail network" that could support the state's future economic growth. Separately, there was a desire to determine whether any minor investments (e.g., in connections) could make the network more efficient. It was assumed that any line segment could be excluded if it was not needed. Moreover, since the focus was on freight service, segments with heavy passenger service (e.g., the Northeast Corridor) had to be justified based on freight use alone.

A. Problem Specification

Establishing the scope of the network to study was one of the first tasks. It was decided that all rail facilities within the state should be included (see Figure 2). The Federal Railroad Administration data showed there were 235 line segments, but for purposes of the study, this was trimmed to 118, by judicious simplifications. To arrive at the network in Figure 3, segments on dead-end branch lines were put in a special file for inclusion if and only if the traffic on them was included. (The dead-end segments shown in the figure are there only for purposes of clarity; e.g., segments 122, 164, and 506). Segments that abutted one another on line segments between nodes were consolidated. In some cases, however, if a shipper or a cluster of shippers was located at a point between two nodes, the line segment was subdivided (regardless of whether it had originally been comprised of one or more segments or not) so that only "half" of the segment would have to be retained (e.g., one example is segments 110 and 111 between Monmouth Junction and Browns). In addition, 14 potential network additions were identified. These were new connecting line segments, such as 810 and 801, which would allow industries along the Northeast Corridor to be served without operating on the Corridor; and segments such as 812 (technically speaking two switches and a short connecting track) which would allow the Bridgeton/Norma area to be served via Clayton instead of via Winslow Junction or Elmer.

Once the study network was established, routes were developed for each origin-destination (OD) pair. They had to be provided explicitly since (P3) does not use transshipment equations. Moreover, several options were needed for each OD pair to ensure that the network optimization was not somehow inadvertently constrained. This meant being thoughtful, thorough and somewhat imaginative. Care also had to be taken to select appropriate end points for the out of state flows. They had to be places of routing indifference (e.g., Buffalo for flows to/from Chicago and Harrisburg for flows to/from Saint Louis). For New Jersey, this meant adding 18 out-of-state links (all those in the 900 series) and seven nodes (55, 56, 57, 67, 73, 82, and 83).

The list of yards also had to be developed. Here, 18 were included. Also, because of the nature of the problem, while 16 were in New Jersey, two were in Pennsylvania; Allentown and Abrams were viewed as possible substitutes for some of the in-state yards.

Rail Form A costs were used for the "above the rail" portions of c_{rj} (e.g., for transportation and equipment-related costs), with costs per car by commodity and car type being based on car-miles, car-hours, net ton-miles, empty return ratio, tare ton-miles, and the number of classifications involved. Commodity mix was then used to develop an average cost per car. In the interest of saving space, more details on developing these costs can be found in List et. al. (1982).

The "below the rail" costs (e.g., for operating and maintaining the line segments and for building the proposed connections) were developed separately. Track-related costs were based on normalized maintenance practices appropriate for the conditions found in New Jersey. The cost relationships, presented in Figure 4, show annual track costs as a function of track class and tonnage. These cost relationships can be superimposed to develop an overall relationship, as illustrated in Figure 5. Annual track-related costs per route-mile, with the step changes in cost occurring at points where, based on engineering judgement, the carrier would transition from one track class to another (which in this case is different from the envelope formed by the long- run marginal cost curve because of the improvements in performance, primarily speed, that can be achieved, which are not shown). The piecewise linear nature of this relationship made it possible to separate the stepwise cost increments (i.e., c_i and the c_{im}) from the linearly increasing component, with the latter being incorporated into c_{rj}.

Yard-related track costs included a fixed and a variable component. About two miles of track are required to make a yard functionally useful, and that the yard's size increases at about 16 miles of track for every additional 1,000 cars per day that pass through the yard. These facts yielded the yard track cost function portrayed in Figure 6.

B. Demand Sensitivity Analysis

The demand sensitivity analysis followed the procedure outlined in Figure 1. Traffic scenarios were developed first on a volume basis using breakpoints for traffic into and out of nodal clusters. The cluster volumes ranged from a low of 0 to a high of 513 cars per day. The cut-offs actually employed were 0 cars per day (to retain everything), 5, 20, and 50.

The resulting networks were overlaid to identify a pattern of network contraction. The networks showed great stability, with line segments of lesser density being dropped logically as the cut off volume was raised. Also, the original traffic flows were reviewed to see if lesser volume flows could be added back into the traffic base, given the selected network. This added significant percentage points to the traffic retained.

As Figure 7 shows, at the 50 car per day cut-off, 29 percent of the network was needed; at 20 cars per day, 38%; at 5 cars per day, 65%; and at 0 cars per day; 80%. Moreover, with the lesser traffic flows added back in, the 50 car per day network supported 80 percent of the traffic; the 20 car per day network, 90 percent; and the 5 car per day network, 99 percent. These are staggering results. At the time, they suggested that significant portions of the New Jersey network could be shed without significantly affecting the amount of service provided. In a more general sense, they may imply that we can shed significant portions of the national network, with significant savings in cost, without having a similar effect on total traffic. We may have confirmed the findings of both Harris (1980) and Grimm (1986) through a completely independent means.

Commodity-based networks were also developed. The purpose here was to determine what network elements should be retained for the State's "vital" industries (e.g., coal, petrochemicals, lumber). Clearly, from a carrier's perspective, this could be translated into a consideration of contribution to revenues or profits. From a governmental perspective, we used an input-output model to identify those production sectors responsible for the top 60 and 80 percent of the State's "GNP" and develop networks for each of these industrial groups. Our experience, however, was that this did not result in much line shedding. Since the origins and destinations for the retained traffic flows were well dispersed, almost all portions of the network had some traffic. Hence, little insight was gained from this analysis and the results were not used heavily.

C. Selecting the Retained Network

Once the sensitivity analyses were complete, the final task was to determine the retained network. Typically, some hybrid network is selected, because of externalities that cannot be reflected in the modeling framework. This is perhaps as it should be. Optimization studies are not necessarily intended to find "the right answer" but rather to indicate what the "good" choices would be. New Jersey opted to make minor adjustments, basically to the 20 cars per day solution, to develop the network shown in Figure 8.

Several factors played major roles in the adjustment process, routing flexibility being one of them. It was felt that if the network was trimmed too much, not enough routing options would be available. A maintenance problem with one segment might affect the use of all the others. Another important factor was the unknowns about shippers and receivers. The State was hesitant about including line segments with only one shipper, or even one dominant commodity. It was felt these segments could become strong abandonment candidates if something went awry. Finally, the economic interests of the carriers and other parties played a role. The State cautiously added some segments to the network because it felt they would be protected by one or more of these interested parties, even if the State could not help out. The State thereby assumed it would have direct responsibility for only those portions that no other party opted to protect.

IV. CONCLUDING REMARKS

The ideas presented here can be viewed as a first step in the development of a more sophisticated plant rationalization technique. One possible extension is the direct incorporation of demand, perhaps in a simple sense so that traffic levels at least increase and decrease in response to the capacity provided. This would then make it possible to treat profit maximization in a meaningful way. As the methodology presently stands, this can only be addressed through sensitivity analysis.

Another possible extension is a heuristic that can find nearly optimal solutions in a time-effective manner. We found, as many others have, that even the best commercially available mixed-integer linear programming code is slow and cumbersome in solving these types of problems.

We also think the cost portion of the model can be improved. More detailed empirical research might show that the track cost relationship is non-linear; or that it can be approximated better with a different functional form. Separately, it may be possible to use two separate objective functions within the model; one for major carriers and one for shortline operators. In line with this, perhaps a multi-carrier version of the model can be developed where retention decisions are made separately for each carrier's network.

However, even in its present state of development, the methodology can have a significant value in evaluating ways to prune networks down to an efficient and manageable size. Certainly, other future investigations will provide opportunities to learn more about the plant rationalization process and the way in which it should be addressed.

ACKNOWLEDGEMENT

The work on which this paper is based was supported by the Freight Division, New Jersey Department of Transportation (NJDOT). The author would like to thank NJDOT for providing the opportunity to work on this very interesting problem and particularly Kenneth Kyte for his invaluable support as NJDOT's project manager. Thanks should also go to colleagues Ernest Clausing and Joel Koppelman for their help with the details of the methodology.

REFERENCES

1. Abdulaal, M. and L.J. LeBlanc (1979), "Continuous Equilibrium Network Design Models," *Transportation Research*, Vol. 13B, pp. 19-32.
2. Barbier, M. (1966), "Le Future Reseau de Transports en Region de Paris,"*Chaiers de l'Institut d'Amenagement et d'Urbanisme de la Region Parisienne*, Vol. 4-5, No. 4, pp. 5-20.
3. Boyce, D.E. (1984), "Urban Transportation Network Equilibrium and Design Models: Recent Achievements and Future Prospects," *Environmental Planning*, A16, pp. 1445-1474.
4. Bronzini, M.S. (1980), "Evolution of a Multimodal Freight Transportation Network Model," *Transportation Research Forum Proceedings*, Vol. 21, No.1, pp. 486-498.
5. Friesz, T.L. and P.T. Harker (1983), "Multicriteria Spatial Price Equilibrium Network Design: Theory and Computational Results,"*Transportation Research*, Vol. 17B, No. 5, pp. 411-426.
6. Friesz, T.L. (1985), "Transportation Network Equilibrium, Design, and Aggregation: Key Developments and Research Opportunities," *Transportation Research*, Vol. 19A, No. 5/6, pp. 413-427.
7. Golden, B.L. and T.L. Magnanti (1977), "Deterministic Network Optimization: A Bibliography," *Networks*, Vol. 7, pp. 149-183.

8. Grimm, C.M. (1986), "Excess Branch Line Capacity in the U.S. Railroad Industry: A Simulation Model Approach," *Logistics and Transportation Review*, Vol. 22, No. 3, pp. 223-239.
9. Harris, R.G. (1980), "Economic Analysis of Light Density Rail Lines,"*Logistics and Transportation Review*, Vol. 16, No. 1, pp. 3-31.
10. Haubrich, G.Th.M. (1972), De optimzlisering van het spoorwegnet in Nederland. Korte beshrijving van methode en resultaten," *Tijdschrift voor Vervoerswetenschap*, Vol. 8.
11. Hornung, M.A. and A.L. Kornhauser (1979), *An Analytical Model for Railroad Network Restructuring*, Working Paper 79-TR-11, Transportation Program, Princeton University, Princeton, NJ.
12. LeBlanc, L.J. (1975), "An Algorithm for the Discrete Network Design Problem," *Transportation Science*, Vol. 9, No. 3, pp. 183-199.
13. List, G.F., E.L. Clausing, H. Ramp, and J.M. Koppelman (1982), *New Jersey Rail Needs Study*, technical report, Transportation and Distribution Associates, Philadelphia, PA.
14. List, G.F. and E.L. Clausing (1983), "Coal Transport Planning by Network Analysis," *Transportation Research Forum Proceedings*, Vol. 24, No. 1, pp.638-648.
15. Magnanti, T.L. and R.T. Wong (1984), "Network Design and Transportation Planning: Models and Algorithms," *Transportation Science*, Vol. 18, No. 1,pp. 1-55.
16. Morlok, E.K. (1970), "A Goal-Directed Transportation Planning Model,"*Transportation Research*, Vol. 4., pp. 199-213.
17. Steenbrink, P.A. (1974), "Transport Network Optimization in the Dutch Integral Transportation Study," *Transportation Research*, Vol. 8, pp. 11-27.
18. United States Railway Association (1975), *Final System Plan for Restructuring Railroads in the Northeast and Midwest Region: Pursuant to the Regional Rail Reorganization Act of 1973*, Washington, DC.
19. United States Railway Association (1976), *Viability of Light Density Lines*, Washington, DC.
20. Victor, P.M. and H.G. Stewart (1982), "Strategic Planning at the ICG," *Transportation Research Forum Proceedings*, Vol. 23, No. 1, pp. 162-167.
21. Wardrop, J.G. (1952), "Some Theoretical Aspects of Road Traffic Research," *Proceedings of the Institute of Civil Engineers*, Part 2, Vol. 1, No. 2, pp. 325-378.
22. Wood, C.F. and W.E. Williamson (1980) "Rail Operations Analysis in Operations and Cost Model," *Transportation Research Forum Proceedings*, Vol. 21, No. 1, pp. 453-457.

ENDNOTE

* Department of Civil Engineering, Rennesselaer Polytechnic Institute, Troy, NY.

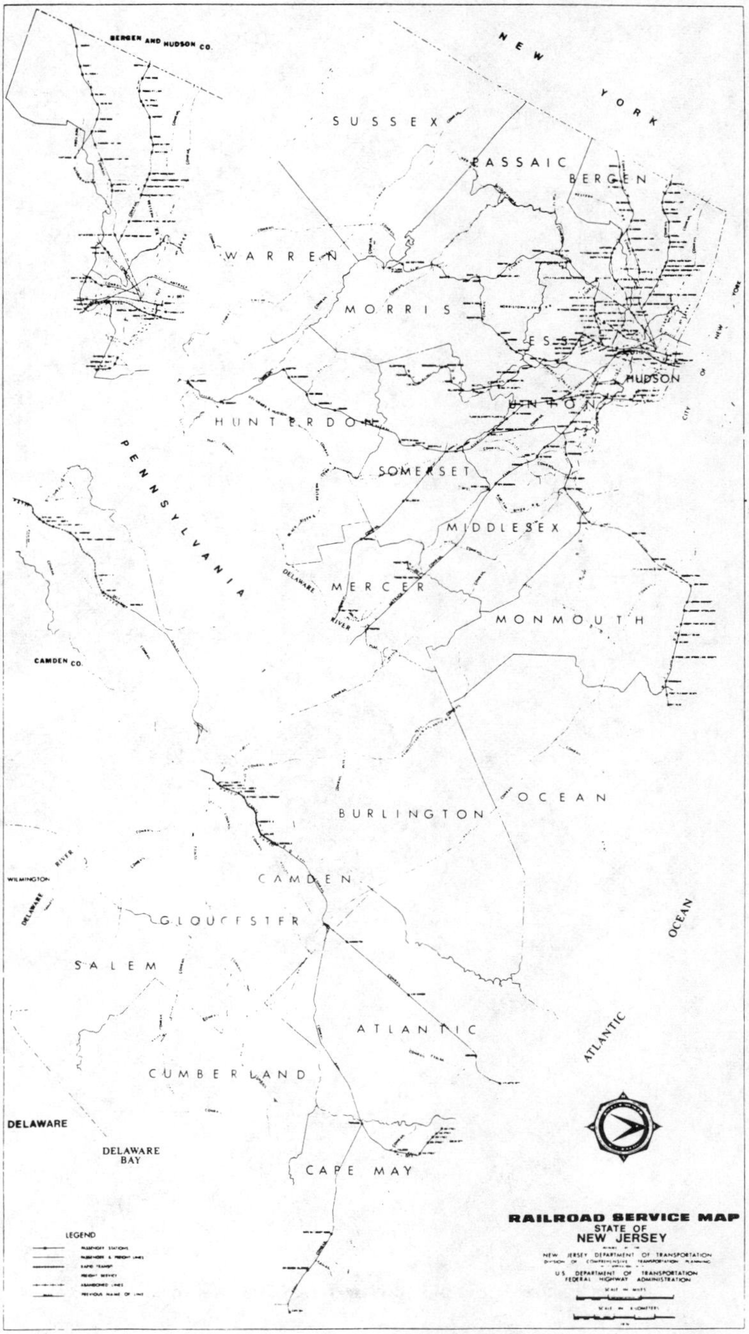

Figure 2. New Jersey Rail Network

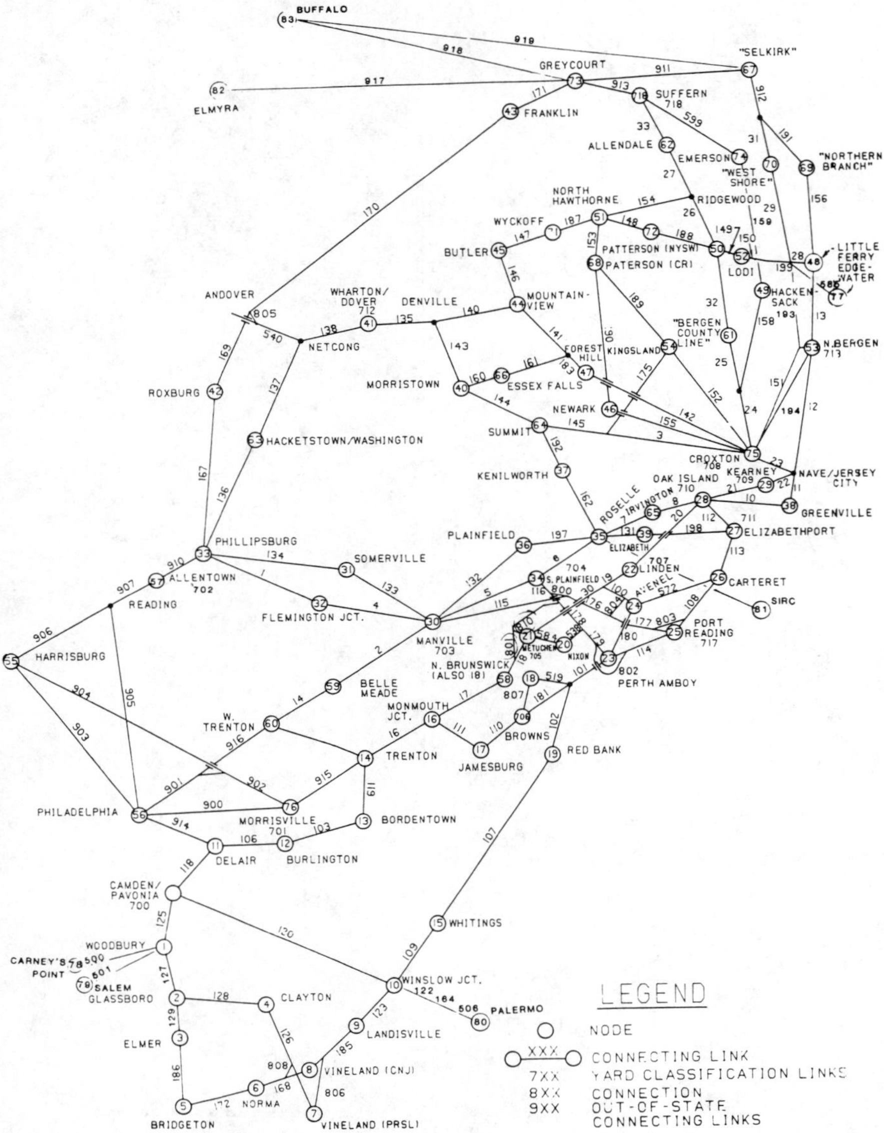

Figure 3. Simplified New Jersey Rail Network

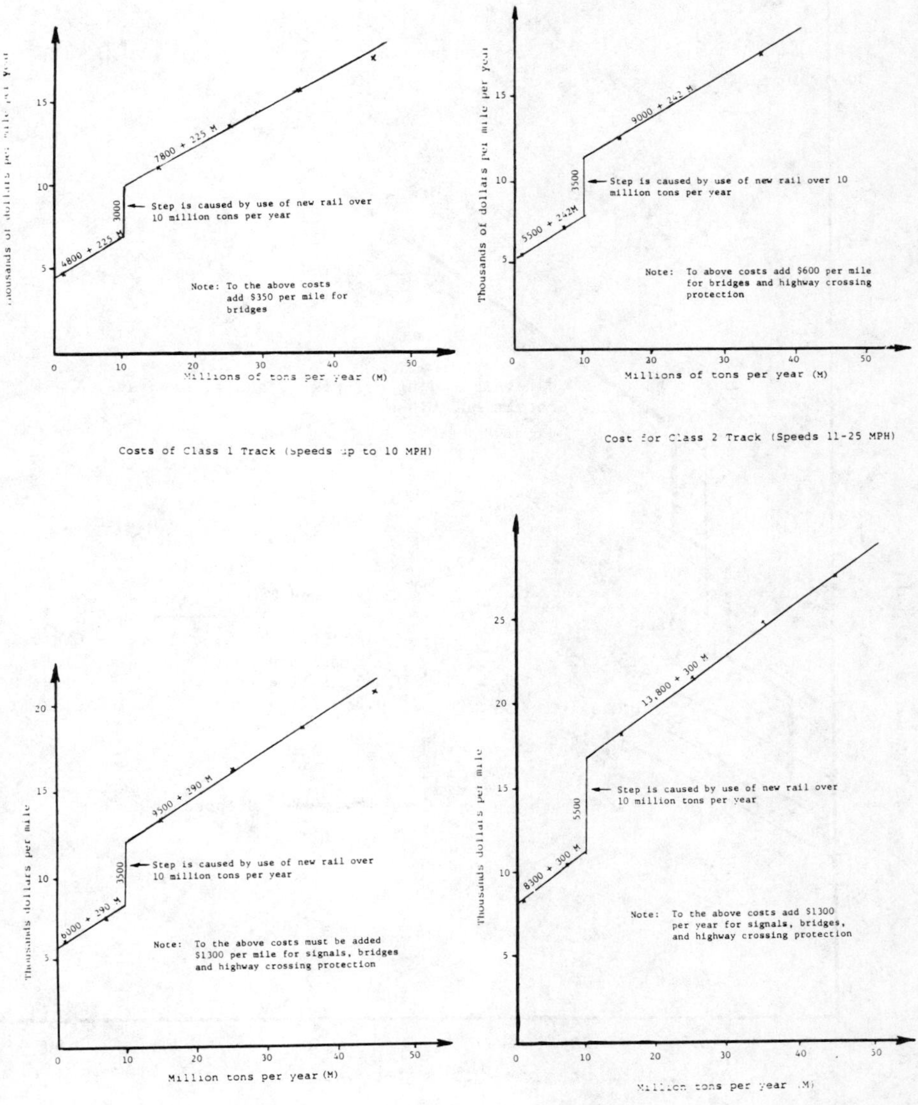

Costs of Class 1 Track (Speeds up to 10 MPH)

Cost for Class 2 Track (Speeds 11-25 MPH)

Costs for Class 3 Track (Speeds 26-40 MPH)

Costs for Class 4 Track (Speed Greater Than 40 MPH)

Figure 4. Line Segment Cost Relationships

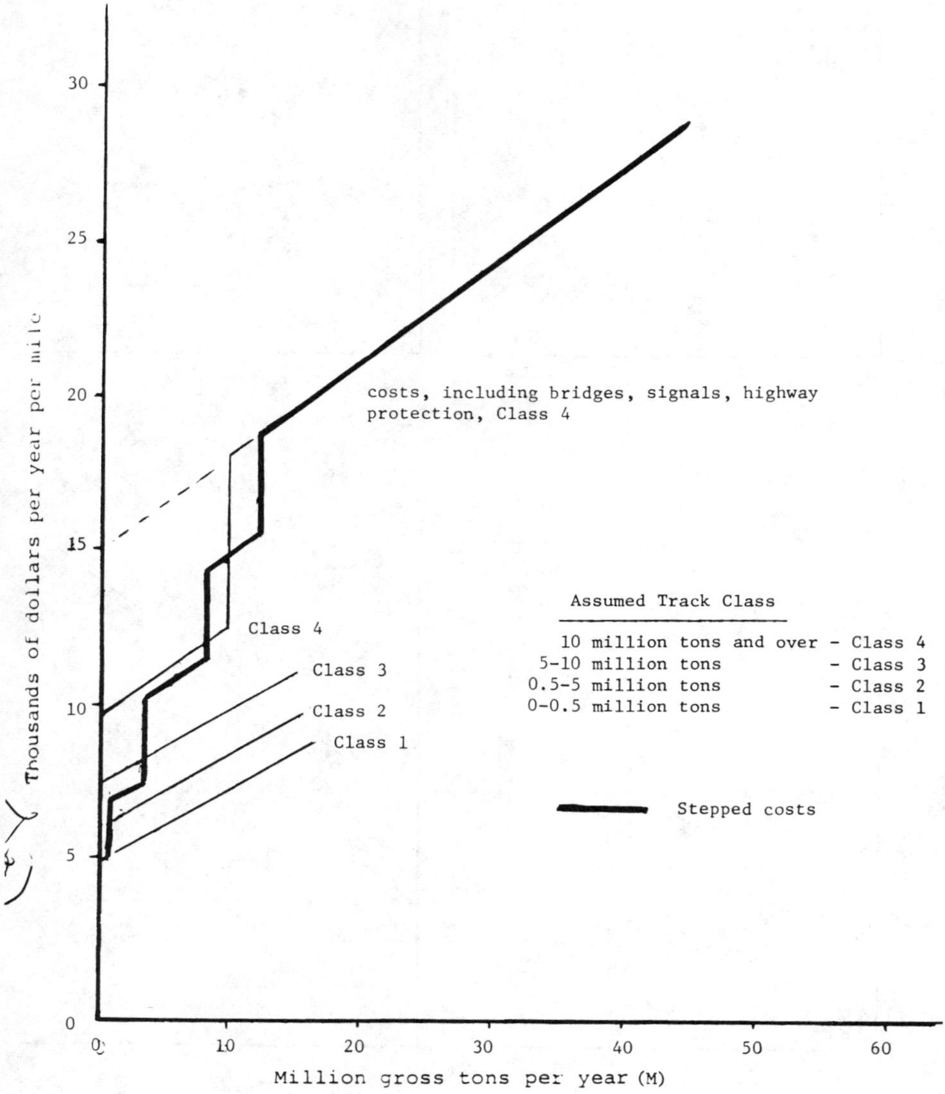

Figure 5. Composite Line Segment Cost Function

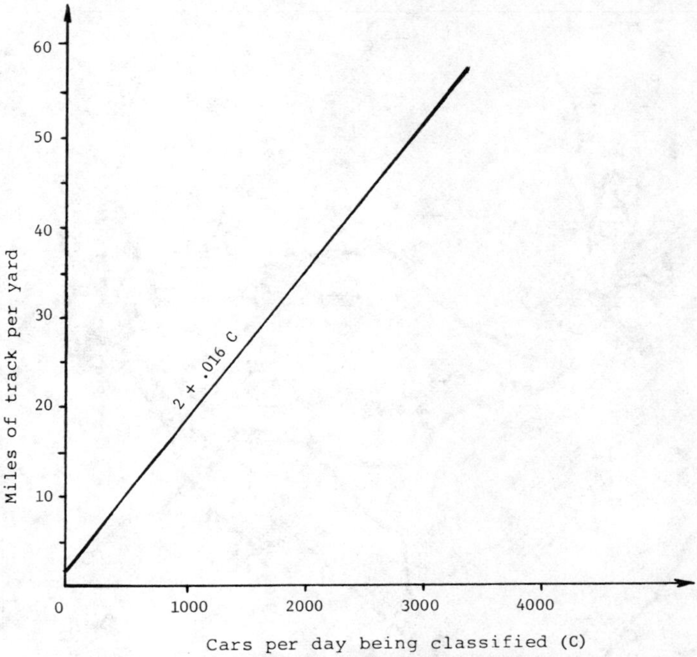

Figure 6. Yard "cost" function expressed in terms of miles of track

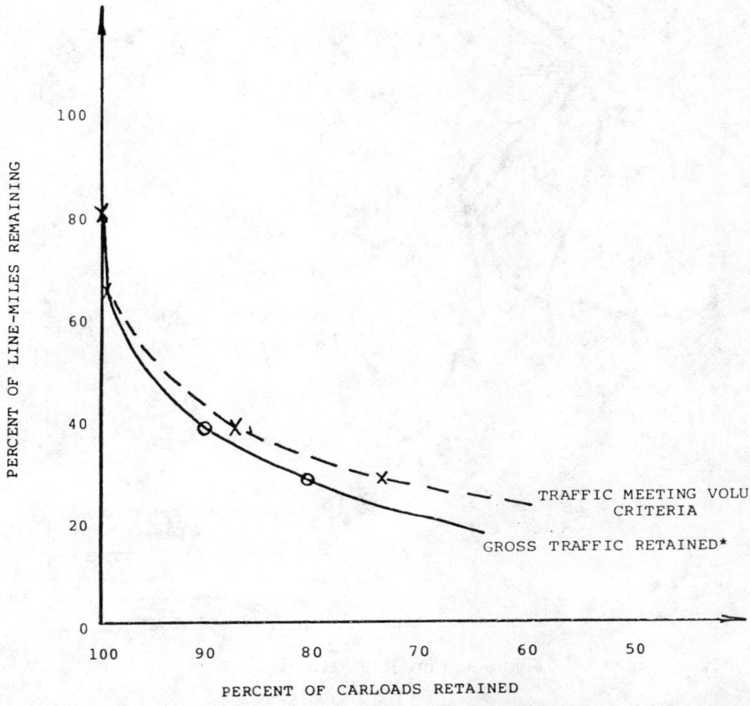

Figure 7. Percent of Line-Miles Retained as a function of the Percent of traffic retained

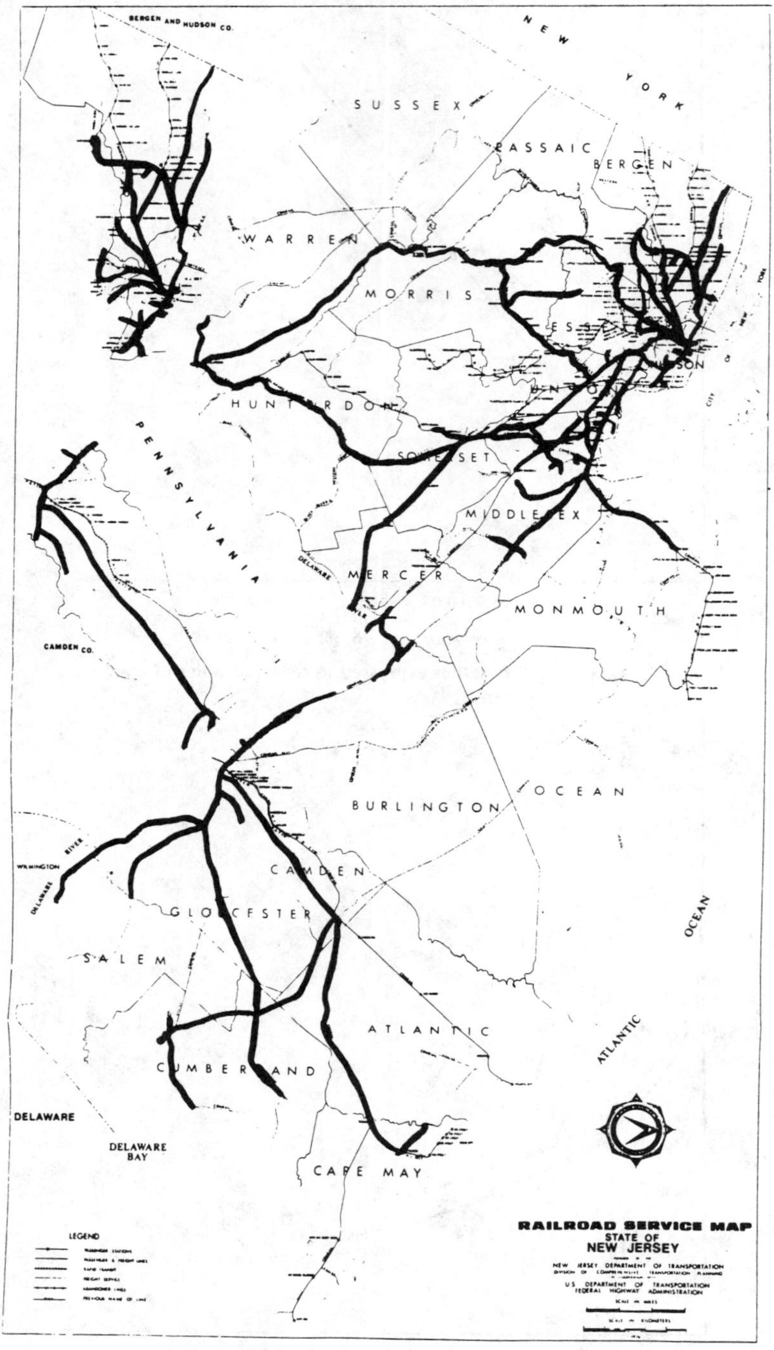

Figure 8. Core Rail Network

Analysis of the Transportation Elements of Alternative Logistics Concepts for Disposal of Spent Nuclear Fuel

by Michael S. Bronzini, David P. Middendorf,** and Robert E. Stammer, Jr.****

ABSTRACT

The U.S. Department of Energy (DOE) has proposed to site in Tennessee a Monitored Retrievable Storage (MRS) facility for handling spent nuclear fuel. The MRS would gather spent fuel from all (or only Eastern) reactors, consolidate and repackage it, and send it on to the first nuclear waste permanent repository, which presumably will be located at a site in the West. The DOE studies supporting the MRS place considerable emphasis on the transportation advantages which can be gained by consolidating the shipments of spent fuel to permit efficient rail transport in large casks to the Western repository. This paper estimates the transportation impacts of moving spent nuclear fuel, both with and without the proposed MRS facility. The principal finding is that measures can be taken, both with and without MRS, to reduce transport requirements. With reduced transportation requirements, the differences in impacts between the alternative systems are relatively small. Thus, the choice between alternative nuclear waste disposal systems should *not* rest on transportation and handling cost or risk.

I. INTRODUCTION

A. Background

The U.S. Department of Energy (DOE) has prepared a proposal to Congress (*1*,*2*) to site in Tennessee a Monitored Retrievable Storage (MRS) facility for handling spent nuclear fuel. The MRS would gather spent fuel from all (or only Eastern) reactors, consolidate and repackage it, and send it on to the first nuclear waste permanent repository, which presumably will be located at a site in the West. Three repository sites have been recommended for detailed characterization studies—Deaf Smith Canyon, Texas; Yucca Mountain, Nevada; and Hanford, Washington. Sites recommended for the MRS are in Oak Ridge and Hartsville, Tennessee (*3*).

The DOE studies supporting the MRS place considerable emphasis on the transportation advantages which can be gained by consolidating the shipments of spent fuel to permit efficient rail transport in large casks to the Western repository. In the DOE analysis, the MRS alternative is compared with direct shipment from the reactors to the MRS in existing casks, which results in numerous long-haul truck shipments.

B. Objectives

The purpose of this study was to estimate the transportation impacts of alternative systems for moving spent nuclear fuel from reactors to a repository, both with and without the proposed MRS facility. In order to make a fair comparison, the systems considered included measures to reduce transportation requirements, again with and without the MRS. The purpose of making these impact estimates was to clarify the comparisons between alternatives, rather than to determine their precise impacts or recommend any one of them. Both national and state impacts were of concern.

C. Procedures

Basic inputs to the study consisted of forecasts of spent nuclear fuel generation at each reactor, present and anticipated future shipping cask characteristics, and preliminary or representative routes by truck and rail between reactors, repositories, and the MRS. These data were used to estimate cask-miles, shipment-miles, and ton-miles, for both Tennessee and the United States, for a number of alternatives. Existing transport models and data were used to estimate unit cost and risk factors for rail and truck transport. Finally, the unit factors were applied to the estimated transportation requirements to produce total cost and risk estimates for the alternatives. These procedures are discussed in greater detail in succeeding sections.

II. SPENT FUEL FORECASTS

Two different forecasts of the amount of spent fuel generated at each operating reactor in the United States were used in the study. The first was that made by DOE for their MRS studies (*1*). The second was prepared for this study (*4*) by the Nuclear Assurance Corporation (NAC), following and expanding on some work done by the Tennessee Valley Authority. DOE forecasts a cumulative waste discharge of 130,000 metric tons of uranium (MIU) by the year 2020, whereas the study team's figure for the same year is 86,000 MIU.

For most of the transportation work, it was desirable and meaningful to estimate average annual impacts. Typical annual transportation patterns were based on the total waste which would be accumulated five years prior to reaching steady state operation, defined by DOE as receiving 3,000 MIU of waste per year. The five-year lag allows fresh spent fuel to age sufficiently for safer and easier handling. Since DOE presently plans to have the waste disposal system in steady state operation by 2003, the spent fuel on hand as of 1998 was the basis for most of the analysis. Note that since the impact estimates are based on the average annual transportation activity for handling 3,000 MIU of nuclear waste, the results are not particularly sensitive to the forecast of the total quantity of waste. Only the duration of the impacts is affected.

III. TRANSPORTATION ALTERNATIVES

The transportation alternatives considered included variation of the following factors:

- at-reactor fuel consolidation;
- cask design;
- transport mode; and
- MRS functions.

Table 1 summarizes the cases considered.

Cases A through C use the DOE spent fuel forecast. Cases AA, B, D, and E exclude MRS from the

TABLE 1
Transportation Alternatives Analyzed

Case	Spent Fuel Forecast	At-Reactor Consolidation	MRS(a)	Transport Scheme(b)
A	DOE	No	C/P/S	DOE
A'	DOE	No	C/P/S	DOE/ST
AA*	DOE	No	C/P/S	DOE-2(W)
AA	DOE	No	No	DOE-2
B	DOE	Yes	No	Team
BB	DOE	No	No	Team
C	DOE	No	C/P/S	Team
C'	DOE	No	C/P/S	Team/ST
D	Team	Yes	No	Team
E	Team	No	No	Team
F	Team	No	C/P	Team
F*	Team	No	C/P	Team(W)
G'	Team	Yes	S	Team/ST
H'	Team	No	S	Team/ST
I	Team	No	C/P/S	Team
I'	Team	No	C/P/S	Team/ST

(a) C = consolidation, P = packaging, S = storage

(b) DOE indicates preliminary DOE transportation plan, and DOE-2 indicates revised DOE plan featuring larger casks. ST indicates use of dual purpose storage/transport casks. Team indicates use of the research team's improved transportation scheme. (W) indicates western reactors ship directly to the repository.

nuclear waste disposal system, and B, D, and G′ feature a readily achievable amount of at-reactor consolidation of the spent fuel. Cases A and A′ use the preliminary DOE transportation scheme, which places the fuel in current generation truck and rail casks and assigns spent fuel shipments to transport modes according to the present split between reactors with rail and truck-only access (70 percent and 30 percent, respectively.) Case AA is a revised DOE plan which assumes the use of future casks with approximately twice the capacity of the current cask models. The remaining cases use an improved transportation scheme (*4*), which is designed to maximize use of large casks and rail transport. Unlike the DOE transport schemes, in which rail shipments of spent nuclear fuel from the reactors are limited to one cask per train, the study team's improved transport scheme uses dedicated trains carrying up to five rail casks per train. A "prime" in the case designation (A′, C′, etc.) indicates that dual purpose storage and transport casks are used, rather than separate casks for each function, to minimize fuel handling at the MRS facility. An asterisk (*) indicates that the reactors located in Arizona, California, Oregon, and Washington would ship their waste directly to the repository, rather than to the MRS, to eliminate extensive backhauling. Finally, most cases with MRS use it as a full service facility which provides the three functions of consolidation, packaging, and storage. The exceptions are case F, which excludes storage, and cases G′ and H′, which provide storage only.

Given the sizable number of variables, it is very easy to generate additional cases of interest, and question the selection of the 16 defined above for detailed study. However, these 16 cases include the most likely ones and bracket the range of expected results. Also, there was not sufficient time available to consider additional alternatives.

IV. TRANSPORTATION REQUIREMENTS

A. Introduction

Transportation requirements were expressed in terms of cask-miles, ton-miles, and shipment-miles of spent nuclear fuel in Tennessee and in the United States. Ton-miles and shipment-miles (the latter sometimes referred to as trip-miles) are commonly used and reported measures of the quantity of transportation service. Shipment-miles are related to cask-miles by the average number of casks per shipment. For example, five casks of spent nuclear fuel shipped 500 miles is the same as 2,500 cask-miles. If each cask is a separate shipment, this is also 2,500 shipment-miles. On the other hand, a five-cask shipment of spent nuclear fuel by dedicated train over a distance of 500 miles is equivalent to 500 shipment-miles. Ton-miles are useful in estimating transportation costs. Cask-miles and shipment-miles are important for estimating both the radiological and nonradiological risks of transporting spent nuclear fuel.

B. Analytical Procedures

The following steps were taken to estimate the transportation requirements of a spent nuclear fuel disposal system under the 16 different cases or scenarios described earlier:

1. determine the number of tons, casks, and shipments of spent nuclear fuel to be shipped from each reactor;
2. determine the number of cask-miles, ton-miles, and shipment-miles of spent nuclear fuel between each reactor and an MRS facility or between each reactor and a geological repository;
3. for each scenario involving an MRS facility, determine the number of cask-miles, ton-miles, and shipment-miles of spent nuclear fuel between the MRS facility and the geological repository; and
4. combine the results of the reactors-to-MRS and MRS-to-repository legs of the spent nuclear fuel movements and calculate the average annual number of cask-miles, ton-miles, and shipment-miles.

In each of these steps, a spreadsheet model was created and used. These models enabled the study team to analyze the 16 scenarios fairly quickly on a microcomputer. In fact, without them, the analysis could not have been completed within the time available. These procedures are described in detail in (*5*). Some brief comments about the analysis follow.

Three of the 16 scenarios allowed for consolidation of spent fuel rods at the reactors. In each of these cases, both intact fuel assemblies and consolidated fuel packages would be shipped from the reactors to either an MRS facility or a geological repository. NAC provided information on the extent to which each reactor would consolidate spent fuel as well as factors for converting numbres of intact assemblies into equivalent numbers of consolidated fuel packages.

Three transportation schemes were analyzed. One of them corresponded with the scheme used by DOE in its preliminary analysis of the need for and feasibility of an MRS facility. DOE considered only current cask models: the NLI ½ for truck shipments and the IF300 for rail movements. The truck casks hold one PWR (pressurized water reactor) and two BWR (boiling water reactor) intact fuel assemblies and weigh 23 tons fully loaded. The rail casks hold 7 PWR and 18 BWR intact assemblies and weigh 68 tons when fully loaded with PWR assemblies and 70 tons when fully loaded with BWR assemblies. DOE assumed that a reactor would ship its spent nuclear fuel by rail only if the plant was currently able to receive and load casks by that mode. Otherwise, trucks would be used. Information on the current cask-handling and modal capabilities of nuclear power plants was taken from (*6*). The initial DOE transport scheme was used in only two of the 16 scenarios. A revised DOE transport scheme, as specified in the final submission to Congress (*2*), was used in two scenarios. This revised scheme features truck and rail casks with approximately twice the capacities specified above, and is identical to the initial plan in all other respects.

The third transport scheme used in this study was one developed by the study team. It included a new generation of casks: 25- and 40-ton truck casks, and 70-, 100-, and 125-ton rail casks. Details of this transport scheme are presented (*4*). Basically, the scheme assumes that each reactor will use the largest cask that it is capable of handling and will ship by the corresponding mode. The study team's transportation scheme was used in 12 of the 16 scenarios.

The study team assumed that rail shipments of spent nuclear fuel from the reactors would go by dedicated or special-purpose trains rather than by regular commercial rail. Dedicated rail shipments are much faster and avoid many of the yard delays associated with general rail freight. It is also unlikely that the railroads will want to mix casks of spent nuclear fuel with other freight in a train. Each dedicated train was assumed to carry five rail casks regardless of cask size. Truck shipments, on the other hand, were assumed to consist of only one cask per truck.

The second and third steps in the analysis required information on highway and rail miles between the reactors and the MRS facility or repository. These data were provided by the Oak Ridge National Laboratory (ORNL). The ORNL data provided a state-by-state breakdown of highway and rail miles between each reactor and each candidate MRS and repository site, and a further breakdown of rail and highway route miles by rural, suburban, and urban population density groups. The latter information was needed for the transportation rail analysis. The route and mileage data came from ORNL's highway and rail transportation routing models, which have been used to route shipments of hazardous materials over a computerized representation of the U.S. highway and rail networks (*7,8*).

A series of spreadsheet models was created using the ORNL data. One set of spreadsheets was developed for rail transport and another set for truck. Each set contained five worksheets—one for each of the two candidate MRS sites and one for each of the three candidate repository locations. Each spreadsheet multiplied the number of casks, tons, or shipments from each reactor by the various route miles to produce the following statistics:

- rail or truck cask-miles, ton-miles, or shipment-miles in Tennessee (total, rural, suburban, and urban); and
- rail or truck cask-miles, ton-miles, or shipment-miles for the United States (total, rural, suburban, and urban).

These spreadsheets had to be rerun for each of the 16 scenarios. Similar spreadsheets for the MRS-to-repository movements were created and run for each case involving the MRS facility.

In the final step of the analysis, the results of all of the reactor-to-MRS and MRS-to-repository spreadsheets were combined into one spreadsheet. Up to this point, all calculations had been based on cumulative quantities of spent nuclear fuel on hand as of the end of 1998. The summary spreadsheet divided the cumulative cask-miles, ton-miles, and shipment-miles by the approximate number of years it would take to dispose of this amount of spent fuel at the rate of 3,000 MIU per year. This was 14 years for cases involving the DOE's spent fuel forecasts and 12 years for those involving the study team's own set of forecasts. The result was an estimate of the average annual transportation requirements of a spent nuclear fuel disposal system for cases either with or without an MRS facility.

C. Findings

The numerical results of this analysis for the 16 cases studied are given in (*5*). Some representative results are given in Table 2.

TABLE 2
Average Annual Transportation Requirements (a)

Case	MRS	Transport Scheme	Cask-Mi. (000)		Ship-Mi. (000)	
			US	TN	US	TN
A,A'	MRS	DOE	3,309	529	2,584	464
AA*	MRS	DOE-2(W)	1,412	175	1,152	167
AA	No MRS	DOE-2	3,572	165	3,572	165
B	No MRS	Team	857	26	523	21
BB	No MRS	Team	1,060	34	657	27
C,C'	MRS	Team	770	64	305	34
D	No MRS	Team	905	27	554	21
E	No MRS	Team	1,122	35	696	28
F	MRS	Team	792	67	317	37
F*	MRS	Team(W)	671	55	271	30
G'	MRS	Team	782	59	283	30
H'	MRS	Team	961	72	351	38
I,I'	MRS	Team	792	67	317	37

(a) Nevada repository, Oak Ridge MRS

With or without an MRS facility, the transportation requirements of a spent nuclear fuel disposal system can be greatly reduced by consolidating the movement of the spent fuel as much as possible. This means placing the spent fuel in large casks and shipping as much of it as possible by rail in dedicated trains.

This conclusion is supported by Figure 1 which compares average annual cask-miles in the United States for several representative scenarios with an MRS (if any) in Oak Ridge and a geological repository in Nevada. For example, the number of cask-miles in the AA case (DOE-2 transport plan, no-MRS) is about the same as the number in the A scenario (DOE transport, MRS). In the latter case, DOE based its analysis on a transport scheme utilizing current cask models and current rail access. In its revised analysis, DOE considered a new generation of casks with twice the capacity of the current generation, although it assumed no further improvements in rail access to nuclear power plants.

The study team's improved transport scheme would reduce transportation requirements even further. Average annual cask-miles under Case BB, a no-MRS scenario, are 70 percent less than under the DOE-2 no-MRS case (AA) and about 25 percent less than under the DOE-2 MRS scenario (AA*).

Given a transport scheme that takes advantage of each reactor's largest cask-handling capability and uses rail transport as much as possible, other meas-

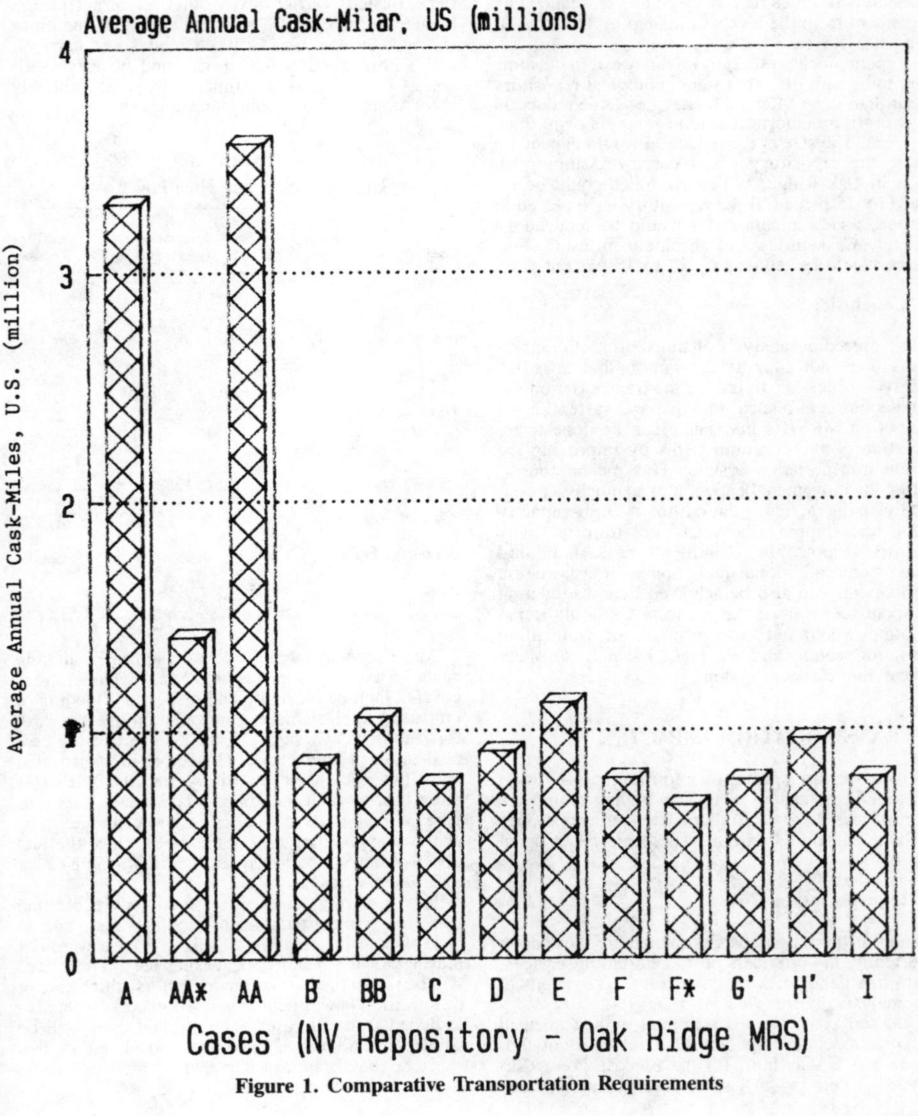

Figure 1. Comparative Transportation Requirements

ures, such as the presence of an MRS facility or fuel rod consolidation at the reactors, have a much smaller effect on transportation. This is evident in Figure 1, which shows that the average annual cask-miles for the United States are similar for scenarios B through I, all involving the study team's improved transport scheme.

Cask-miles, ton-miles, and shipment-miles of spent nuclear fuel in Tennessee will obviously be higher if there is an MRS facility in the state because all of the spent fuel will then be shipped into and out of the state. However, as the results in Table 2 show, the amount of transportation of spent fuel through Tennessee can be substantially reduced by implementing the study team's improved transport scheme. Even with an MRS in the state, cask-miles, and shipment miles in Tennessee can be reduced by 90 percent from the levels estimated by DOE for its original MRS case (case A).

If spent nuclear fuel from the western reactors were to be sent directly to the geological repository rather than to an MRS in Tennessee, some reduction in overall transportation requirements would be achieved. The size of the reduction would depend on where the repository was located. Assuming an MRS in Oak Ridge, U.S. cask-miles could be reduced by 15 percent if the repository were placed in Nevada, and shipment-miles would be reduced 14 percent. Cask-miles and shipment-miles in Tennessee would be reduced by 17 to 18 percent.

C. Conclusion

The preceding analysis of transportation requirements does not support the notion that an MRS facility is necessary to reduce the transportation demands of a spent nuclear fuel disposal system. Even without an MRS facility, much can be done to reduce transportation requirements by improving the current transportation system. This means consolidating the movement of spent fuel as much as possible by using a new generation of high-capacity casks, and shipping the spent fuel from as many reactors as possible in multiple-cask dedicated trains. Some additional reduction in transportation requirements can also be achieved by consolidating the spent fuel rods at the reactors. Certainly transportation would not be a primary or compelling reason for needing an MRS facility as part of a spent nuclear fuel disposal system.

V. TRANSPORTATION IMPACTS

Given the transportation requirements as estimated in the preceding section, two types of impacts were estimated for each alternative—transportation cost and transportation risk. These are taken up in turn below.

A. Transportation Cost

The transportation costs estimated are those costs incurred by the providers of transportation services, the railroad and trucking companies. The details of this analysis are reported in (*5*).

Unit rail costs were determined with a linehaul rail cost model developed by The University of Tenenssee and CACI, Inc., for the Electric Power Research Institute (*9*). All spent fuel was assumed to move in unit trains consisting of two locomotives, five flatcars carrying casks, six spacer cars, and a messenger car carrying inspection, security, and supervisory personnel. Truck costs were estimated with a truck load cost model originally developed for the U.S. Department of Transportation (*10*) and adapted for use in this study. In both cases, the cost of a return trip with empty casks was computed and added to headhaul cost, to obtain total round trip cost.

The unit transport costs obtained are given in Table 3. It is important to recognize that these are the *costs* incurred by the carriers, not the *rates* which might be charged to the shippers. As with any commodity movement, the rates will be determined through negotiation between the shippers and carriers. Besides carrier costs, other important factors include service to be provided, annual quantities shipped, competition between modes and carriers, and a host of other operational and business variables. Thus the costs estimated here are probably lower bounds on the transportation expense.

TABLE 3

Unit Transport and Handling Costs

Cask Type	Transport $/Ton-Mile	Handling $/Cask
Truck		
DOE 25 ton	0.152	2,000
Team 25 ton	0.152	2,000
Team 40 ton	0.095	2,500
Rail		
DOE 70 ton	0.1285	4,000
Team 70 ton	0.1285	4,500
Team 100 ton	0.0932	5,500
Team 125 ton	0.0768	6,000

Also shown in Table 3 are estimated cask handling costs based on estimates in (*11*). The operations covered include receiving and inspection, loading or unloading fuel elements, and readying of the cask for shipment. These costs are assessed in the cost analysis each time a cask is loaded or unloaded, that is, at the reactor, the repository, and the MRS (for those cases which include it). The use of dual purpose storage/transport casks in some of the cases avoids one handling cost at the MRS, since the fuel does not have to be transloaded from a storage cask to a transport cask.

Two important categories of cost were omitted from the transportation and handling cost, due to lack of both information and time. These were demurrage charges and cask leasing fees. These fees will be set on the basis of many factors which cannot be specified now. Fuel consolidation costs were also excluded from all cases. The omitted costs will be similar for all the alternatives considered in this study, so the differences between alternatives will be largely unaffected.

Table 4 presents a summary of the average annual transportation and handling costs estimated for the 16 cases, for the Nevada repository and the Oak Ridge MRS site. To help clarify the choices between alternatives, Table 4 ranks them by annual transport and handling costs, from lowest to highest.

Based on the estimated transport and handling costs, the following findings are offered.

1. Improving the transportation system has the most dramatic impact on cost, with or without MRS. Once this step is taken, all of the alternatives are similar in cost, varying by only about $4.5 million per year. The relatively modest transportation improvement proposed by DOE reduces annual cost by over $16 million. An additional $17 million is saved (compare AA and BB) by implementing the study team's transportation scheme.
2. At-reactor consolidation of spent fuel produces cost savings of about $2 million per year.
3. The use of dual purpose casks has the potential of saving about $1 million per year, primarily because of reduced handling cost.

TABLE 4

Transport/Handling Cost Ranking

Case	Spent Fuel Forecast	At-Reactor Consolidation	MRS	Transport Scheme	Annual Cost(a) ($ Million)
B	DOE	Yes	No	Team	8.56
D	Team	Yes	No	Team	9.03
BB	DOE	No	No	Team	10.46
G'	Team	Yes	S	Team/ST	10.71
C'	DOE	No	C/P/S	Team/ST	10.72
F*	Team	No	C/P	Team(W)	10.76(b)
E	Team	No	No	Team	11.07
I'	Team	No	C/P/S	Team/ST	11.08
C	DOE	No	C/P/S	Team	11.69
F	Team	No	C/P	Team	12.05
I	Team	No	C/P/S	Team	12.05
H'	Team	No	S	Team/ST	13.11
AA*	DOE	No	C/P/S	DOE-2(W)	20.47
AA	DOE	No	No	DOE-2	28.07
A'	DOE	No	C/P/S	DOE/ST	37.00
A	DOE	No	C/P/S	DOE	37.97

(a) Costs are for Nevada repository, Oak Ridge MRS.

(b) Estimated by reducing Case F transport cost in proportion to the reduction in cask-miles.

4. Spent fuel disposal systems without MRS tend to be slightly lower in overall cost than those systems with MRS.
5. Handling costs are nearly as large as transport costs. Thus further work to develop better estimats of handling cost is warranted.
6. Since only 16 cases could be considered here, additional work to define and analyze preferred transportation options for the spent fuel disposal system is needed. This additional work should include input from prospective transportation carriers.

B. Transportation Risk

The risks associated with various waste management schemes were estimated in terms of the expected fatalities for each case. After reviewing a number of potential risk assessment models, recent risk factors generated by RADTRAN III computers runs at Sandia National Laboratories were deemed to be the most appropriate for use with the transportation requirements (cask-miles, shipment-miles, and ton-miles) generated as described earlier.

The RADTRAN III risk factors consider both the consequence of an event and the likelihood of its occurrence. The risk factors are stated in terms of radiological and nonradiological risks, expressed as fatalities and injuries per shipment-kilometer, for both incident-free (normal) and accident operating conditions. Separate factors are provided for each mode (truck, rail) and population density grouping (urban, suburban, rural).

Radiological risks are attributable to the spent nuclear fuel being transported, while nonradiological risks are essentially independent of the commodity transported and are associated with the actual transport operations rather than the payload being transported. Incident-free or normal operating conditions describe risks associated with radiological emissions as a vehicle travels along a roadway or via rail, as well as nonradiological risks such as vehicle emissions or pollutants that cause health hazards. Accident operating conditions involve those situations where a truck or railcar is involved in a transportation mishap. Occupational risks consider the transport operators and crew who are at risk in various situations. Nonoccupational risks consider the likelihood of fatalities to residents, pedestrians, and other individuals (i.e., the general public).

The complete results of applying the unit risk factors to the various transportation scenarios are presented in (*5*). Risks to both Tennesseans and to all citizens are stated. Table 5 summarizes the fatalities that might reasonably be expected for an MRS facility in Oak Ridge and a final repository in Nevada. Table 6 shows the range of values and converts the average fatalities into the expected time intervals between fatalities.

The major findings of the transportation research team concerning risks are that risks are extremely small, and nonradiological risks are much higher than radiological risks. Other types of transport (e.g., movement of gasoline, chemicals, coal, general freight, and even automobile occupants) incur greater relative risks than spent nuclear fuel transport. Thus transportation risks should not be a major determining factor in the final MRS decision. Furthermore, in view of the extremely low risks, it is not worth expending much additional effort to develop refined unit risk factors. These findings assume that current regulatory and design standards will remain in effect. Any change in these standards will produce different risks.

TABLE 5
Expected Average Annual Fatalities

Case	Residents	Radiological Fatalities	Nonradiological Fatalities
A & A'	TN	.029	.183
	US	.159	1.630
AA*	TN	.009	.038
	US	.060	.594
AA (no MRS)	TN	.010	.031
	US	.176	.607
B (no MRS)	TN	.001	.014
	US	.028	.686
BB (no MRS)	TN	.002	.017
	US	.035	.815
C & C'	TN	.002	.054
	US	.015	.818
D (no MRS)	TN	.001	.015
	US	.030	.729
E (no MRS)	TN	.002	.018
	US	.038	.874
F, I, & I'	TN	.002	.057
	US	.015	.841
F*(a)	TN	.002	.046
	US	.013	.719
G'	TN	.002	.053
	US	.013	.874
H	TN	.025	.064
	US	.016	1.063

(a) Estimated from Case F, based on the reduction in shipment-miles.

TABLE 6
Range of Likely Radiological and Nonradiological Fatalities

	Low Value	High Value
Radiological Fatalities		
TN	1 per 1000 years (B & D)	1 per 34.5 years (A & A')
US	1 per 76.9 years (F* & G')	1 per 5.7 years (AA)
Nonradiological Fatalities		
TN	1 per 71.4 years (B)	1 per 5.5 years (A & A')
US	1 per 1.7 years (AA*)	1 per 7.4 months (A & A')

Note: Associated transportation case shown in parentheses.

VI. CONCLUSIONS

The principal findings of this study:

1. Measures can be taken, both with and without MRS, to improve the nuclear waste disposal transportation system and reduce transport requirements.
2. With improved transportation, the differences in transportation requirements and impacts of the alternative systems are comparatively small.
3. The radiological risk of all alternatives studied is extremely small.
4. Nonradiological risk exceeds radiological risk, but is still very small, and much smaller than the risk of most other types of transportation.
5. The choice between alternative nuclear waste disposal systems should *not* rest on transportation and handling cost or risk.
6. Additional work is needed to develop good estimates of spent fuel handling costs, and to define and analyze in detail the preferred transportation systems for spent fuel disposal.

This research was accomplished over a very short period of time (13 weeks) in order to provide results which could be used by the State of Tennessee in formulating a position on the DOE proposal. Thus there was no time available to try alternative methodologies and assumptions. In most respects, a more refined analysis would use the same basic approach, but with greater attention to developing better input data. It is the writers' considered opinion that the limited analysis which was conducted treats the alternatives fairly, and that a refined analysis would not produce major changes in the results and conclusions.

ACKNOWLEDGEMENTS

This research was sponsored by the State of Tennessee Safe Growth Team, under a grant provided by the U.S. Department of Energy. Mr. Ben Smith and Dr. Ruth Neff monitored the study for the State. Technical guidance was received from a Nuclear Waste Technical Advisory Panel organized by the State. Dr. E. William Colglazier, Directory of the Energy, Environment, and Resources Center, University of Tennessee, was the coordinator of the overall research.

REFERENCES

1. U.S. Department of Energy. *The Need for and Feasibility of Monitored Retrievable Storage—A Preliminary Analysis*. DOE/RW-0022, Washington, D.C., 1985.
2. U.S. Department of Energy. *Monitored Retrievable Storage Submission to Congress—Volume 2*. Washington, D.C., January 1986 (Preliminary Information).
3. U.S. Department of Energy. *Screening and Identification of Sites for a Proposed Monitored Retrievable Storage Facility*. DOE/RW-0023, Washington, D.C., 1985.
4. Energy, Environment, and Resources Center, The University of Tennessee. *Evaluation of the Need, Feasibility, and Siting of the MRS in Tennessee*. Knoxville, TN, December 1985.
5. Transportation Center, The University of Tennessee and Department of Civil and Environmental Engineering, Vanderbilt University. *Monitored Retrievable Storage of Spent Nuclear Fuel: Transportation Studies*. Transportation Center, Knoxville, TN, December 1985 (Revised October 1986).
6. Pacific Northwest Laboratory. Spent Nuclear Fuel Shipping Cask Handling Capabilities of Commercial Light Water Reactors. PNL-5384, Richland, WA, April 1985.
7. Joy, D. S. and Johnson, P. E. *Highway, A Transportation Routing Model: Program Description and Revised User's Manual*. ORNL/TM-8759, Oak Ridge National Laboratory, Oak Ridge, TN, October 1983.
8. Peterson, B.E. *Interline, A Railroad Routing Model: Program Description and User's Manual*. ORNL/TM-8944, Oak Ridge National Laboratory, Oak Ridge, TN, November 1983 (Draft).
9. Bronzini, M. S. and Sherman, D. M. *Railroad Routing and Costing*. EA-4716, Electric Power Research Institute, Palo Alto, CA, August 1986.
10. Bronzini, M. S. and Miller, R. C. *Freight Transportation Energy Use: Volume II-Methodology and Program Documentation*. DOT-TSC-OST-79-1, II, U.S. Department of Transportation, Transportation Systems Center, Cambridge, MA, July 1979.
11. Daling, P.M. and Engel, R. L. *Analysis of Near-Term Spent Fuel Transportation Hardware Requirements and Transportation Costs*. PNL-4574/UC-85, Pacific Northwest Laboratory, Richland, WA, January 1983.

END NOTES

* Department of Civil Engineering, Penn State University, University Park, PA.

** Department of Computer Science, Purdue University, West Lafayette, IN.

*** Department of Civil and Environmental Engineering, Vanderbilt University, Nashville, TN.

Assessing the Impact of Consolidation on Inbound Vendor Traffic

by Charles F. Myers, Joseph L. Fanelli*, and Dan C. Boger***

ABSTRACT

The Defense Logistics Agency annually receives at its six major supply depots approximately 650,000 shipments representing some 250 million pounds. This paper proposes a set on consolidation points which will collect these inbound shipments, build truckload lots, and ship them to the depots under lower, government-negotiated rates. The consolidation points are selected by minimizing the volume-weighted distances from the vendors to the points. Cost savings from the net reduction in transportation costs are derived, and advantages and disadvantages are discussed.

I. INTRODUCTION

As the integrated logistics concept has become more widely used by organizations, one area which has come under increasing scrutiny is that of the potential cost savings generated from the consolidation of inbound traffic from vendors. By consolidating a large number of small shipments into fewer large shipments, which are then forwarded at lower cost to the ultimate destination, potentially large cost savings may be realized. The savings may be computed as the difference between (1) the cost of shipping the freight directly to the ultimate destination under the higher, small shipment (less-than-load) rates and (2) the cost of shipping the freight to the consolidation point with transshipment on to the ultimate destination under the lower, large shipment (fullload) rates.

The Defense Logistics Agency (DLA) annually receives at its six major supply depots approximately 650,000 small shipments representing some 250 million pounds. Currently, all of these shipments are sent by the vendors directly to the depots via small parcel or surface carrier using less-than-truckload (LTL) or truckload (TL) rates. This procedure is relatively costly since the vendors cannot use the government-negotiated rates available to DLA but must use class or discounted class rates.

Using an analytic model, this paper obtains a set of consolidation points located strategically throughout the continental United States which will collect the LTL shipments, build truckload lots, and ship them to the six depots under government-negotiated TL rates. The consolidation points are selected by minimizing the volume-weighted distances from the vendors to the consolidation points. Cost savings from the net reduction in transportation costs are derived, and advantages and potential problems of the consolidation approach are discussed.

II. THE FACILITY LOCATION PROBLEM

Determining the location of the consolidation points discussed above is one of a large class of problems generally known as facility location problems. Such problems, e.g., plant or warehouse site selection, occur when a facility with known demands and supplies must be sited to optimize some objective function. Revelle, Marks, and Liebman (1970) and Francis and White (1974) provide overviews of many of these problems, discuss alternative site selection criteria, and present numerous analytic models for obtaining solutions.

The first formulation of the facility location problem was provided by Weber (see Hall (1985)). Given a set of customer locations and customer demand quantities, Weber selected a single facility site by minimizing the transportation distance, weighted by demand quantities, from the customer locations to the potential site. While more recent formulations have minimized the transportation cost, again weighted by demand quantities, Weber's formulation assumes implicitly that transportation distance is a reasonable proxy for transportation cost. Hall (1985) provides a discussion of the reasonableness of this proxy.

The selection of multiple facilities considerably complicates the problem since the evaluation of alternative combinations of facilities grows combinatorically. To simplify the selection process, a finite-set approach has been proposed and used. This approach selects M facilities from a set of N possible sites, which avoids searching over a continuum of possible sites. This restriction of the continuum is meaningful since many potential sites in the continuum may not be feasible due to prior use, topography, etc. Additionally, Hakimi (1964) showed that if the M facility problem is formulated on a network, one of the optimal solutions to the problem has the facilities located on the nodes of the network. Therefore, one need search only over the nodes and not the continuum. Efroymson and Ray (1966), among others, have developed a branch-and-bound algorithm for the problem in this M-from-N form.

While formulated as an optimization problem, the multisite location problem is of a sufficiently complicated structure that finding the optimum is not a straightforward process. Hence, heuristics have been developed to find a solution which is near the optimum. Kuehn and Hamburger (1963) have developed a bumpand-shift heuristic, and Hall (1985) has provided a heuristic, general set-covering approach to solving the above form of the problem.

All of the above formulations of the problem use a decision criterion of either transportation distance or transportation cost. Since in most cases cost is clearly the operant criterion, distance is used only as a proxy for cost. The model described here uses both cost and distance as criteria in a two-stage heuristic. The first stage selects a set of candidate consolidation sites using the criterion of minimum demand-weighted distance. The second stage evaluates transportation costs, both direct and consolidated, for the candidate sites and chooses the final set of cost-minimizing sites.

This approach is simpler than prior approaches using the criterion of transportation cost since it uses the minimum distance criterion to limit the size of the set of candidate sites which must be costed. Although this approach may not yield the true optimum (hence, its heuristic nature), it should yield a reasonable approximation, because Cooper (1964) and Larson and Stevenson (1972) showed that transportation cost is relatively insensitive to changes in locations near the optimum.

III. SELECTION OF CONSOLIDATION POINTS

Based upon the above discussion of the general problem, consolidation points are to be selected by minimizing weighted transportation distances from the vendors to a finite set of potential consolidation sites. The distances, based on actual highway mileages, are weighted by the volume, measured in pounds, of shipments from the vendor that are received by any of the six depots.

A. Vendor/Origin Point Determination

In order to reduce the vendor sites to a manageable number, all vendors in the continental United States are grouped into 723 separate origins. This is done by aggregating vendor locations to the three-digit zip code level for those shipments which occurred in fiscal year 1985. The zip codes for vendor locations are stored for use in distance calculations.

B. Potential Consolidation Site Determination

Thirty candidate consolidation locations are picked based upon the following criteria: accessibility by major highways, the availability of motor carrier transportation services, the concentration of vendor weight, and location at or near the six major DLA supply depots. The first two criteria are assessed using a recommended list of points from the Pooled Distributors of America, Inc. (PDA). PDA also provided other information to assist in the selection of candidate sites.

The criterion of concentration of vendor weight is checked via a graphical analysis of vendor weights by originating states which assisted in eliminating a number of potential sites. The last criterion is used because it is necessary for the six DLA depots to be active participants in the consolidation program and because consolidation at the depots, the ultimate destinations of the vendor shipments, would eliminate some portion of the transshipments. The three-digit zip codes of the candidate sites are stored for use in distance calculations. Table 1 shows the thirty candidate sites.

C. Shipment Data

Shipment weights are obtained from automated historical information on vendor contracts which were filled during fiscal year 1985. Shipment data are not directly available, so the contract data are used. Some small purchase contract information was lost prior to being collected in the contract data with an estimated loss of approximately 20 percent of the small purchases. This does not materially affect the analysis, since this represents less than 2.2 percent of total weight.

TABLE 1
POTENTIAL CONSOLIDATION POINTS

Zip	Location	Zip	Location
021	Boston, MA	606	Chicago, IL
100	New York, NY	641	Kansas City, MO
132	Syracuse, NY	701	New Orleans, LA
171	*Mechanicsburg, PA	722	Little Rock, AR
191	Philadelphia, PA	731	Oklahoma City, OK
212	Baltimore, MD	752	Dallas, TX
232	*Richmond, VA	772	Houston, TX
303	Atlanta, GA	802	Denver, CO
322	Jacksonville, FL	844	*Ogden, UT
336	Tampa, FL	850	Phoenix, AZ
352	Birmingham, AL	900	Los Angeles, CA
381	*Memphis, TN	941	San Francisco, CA
432	*Columbus, OH	953	*Tracy, CA
482	Detroit, MI	972	Portland, OR
554	Minneapolis, MN	981	Seattle, WA

* indicates DLA Supply Depot location.

Additionally, only unit weight, and not shipping weight, is available from the contract data. The unit weight is multiplied by the quantity shipped to obtain a shipment weight, and a 10 percent packing factor, obtained from standard commercial practices, is added.

A weight of 10,000 pounds is the standard weight break for LTL versus TL traffic in all DLA shipments. Therefore, all shipments which are between 70 and 10,000 pounds are put into the shipment file for analysis. This approach limits the analysis to freight shipments, therefore eliminating the below 70 pound parcel post shipments from further consideration.

D. Procedure for Consolidation Site Selection

Using the three-digit zip codes of the 723 origins and the 30 candidate destinations, a matrix is constructed with 723 rows and 30 columns. The entry in each cell of the matrix is the actual highway mileage between the two points calculated using the RandMcNally distance based on three-digit zip codes. This mileage is weighted by the number of pounds shipped from the origin during 1985.

Following the structure of the general finite-set approach facility location problem discussed above, an algorithm which used a simple greedy approach based on weighted-mile savings was developed using Kuehn and Hamburger (1963) as a basic model. Prior analysis indicated that nine to twelve potential consolidation sites would yield the minimum cost set. The model was then used to select the consolidation sites in a stepwise fashion beginning with nine sites and incrementing the solution set by one site until a total of twelve potential sites were selected. The algorithm performed the following steps:

1. The point with the least overall weighted-miles is selected assuming all vendors feed a single point.
2. Then each other point is evaluated to determine the weighted-mile savings that could be achieved by reassigning vendors with less weighted-miles to that point.
3. The point with the greatest weighted-mile savings is then admitted to the solution.
4. These weighted-mile savings calculations are continued until the desired number of points are achieved.

This approach resulted in the selection of four solution sets containing nine, ten, eleven, and twelve potential consolidation sites. The four solution sets produced by this algorithm are shown in Table 2. The points are listed in the order in which they were admitted to the solution set. Once the results are obtained, each solution set is run through the cost model discussed below, beginning with the nine set solution, until the marginal contribution to cost savings of the last site added to the solution set is at or near zero dollars. This occurs at the tenth site, Portland, OR. The ten sites are then plotted against the aggregated vendor tonnage and visually checked to insure adequate coverage. This is shown in Figure 1.

TABLE 2
FOUR SOLUTION SETS OF POTENTIAL CONSOLIDATION POINTS

<u>Solution Set No. 1</u>

1. *Columbus, OH
2. Los Angeles, CA
3. New York, NY
4. Birmingham, AL
5. Chicago, IL
6. Dallas, TX
7. *Richmond, VA
8. Philadelphia, PA
9. Jacksonville, FL

<u>Solution Set No. 2</u>

1. *Columbus, OH
2. Los Angeles, CA
3. New York, NY
4. Birmingham, AL
5. Chicago, IL
6. Dallas, TX
7. *Richmond, VA
8. Philadelphia, PA
9. Jacksonville, FL
10. Portland, OR

<u>Solution Set No. 3</u>

1. *Columbus, OH
2. Los Angeles, CA
3. New York, NY
4. Birmingham, AL
5. Chicago, IL
6. Dallas, TX
7. *Richmond, VA
8. Philadelphia, PA
9. Jacksonville, FL
10. Portland, OR
11. Boston, MA

<u>Solution Set No. 4</u>

1. *Columbus, OH
2. Los Angeles, CA
3. New York, NY
4. Birmingham, AL
5. Chicago, IL
6. Dallas, TX
7. *Richmond, VA
8. Philadelphia, PA
9. Jacksonville, FL
10. Portland, OR
11. Boston, MA
12. Kansas City, MO

*indicates DLA Supply Depot locations.

Figure 1

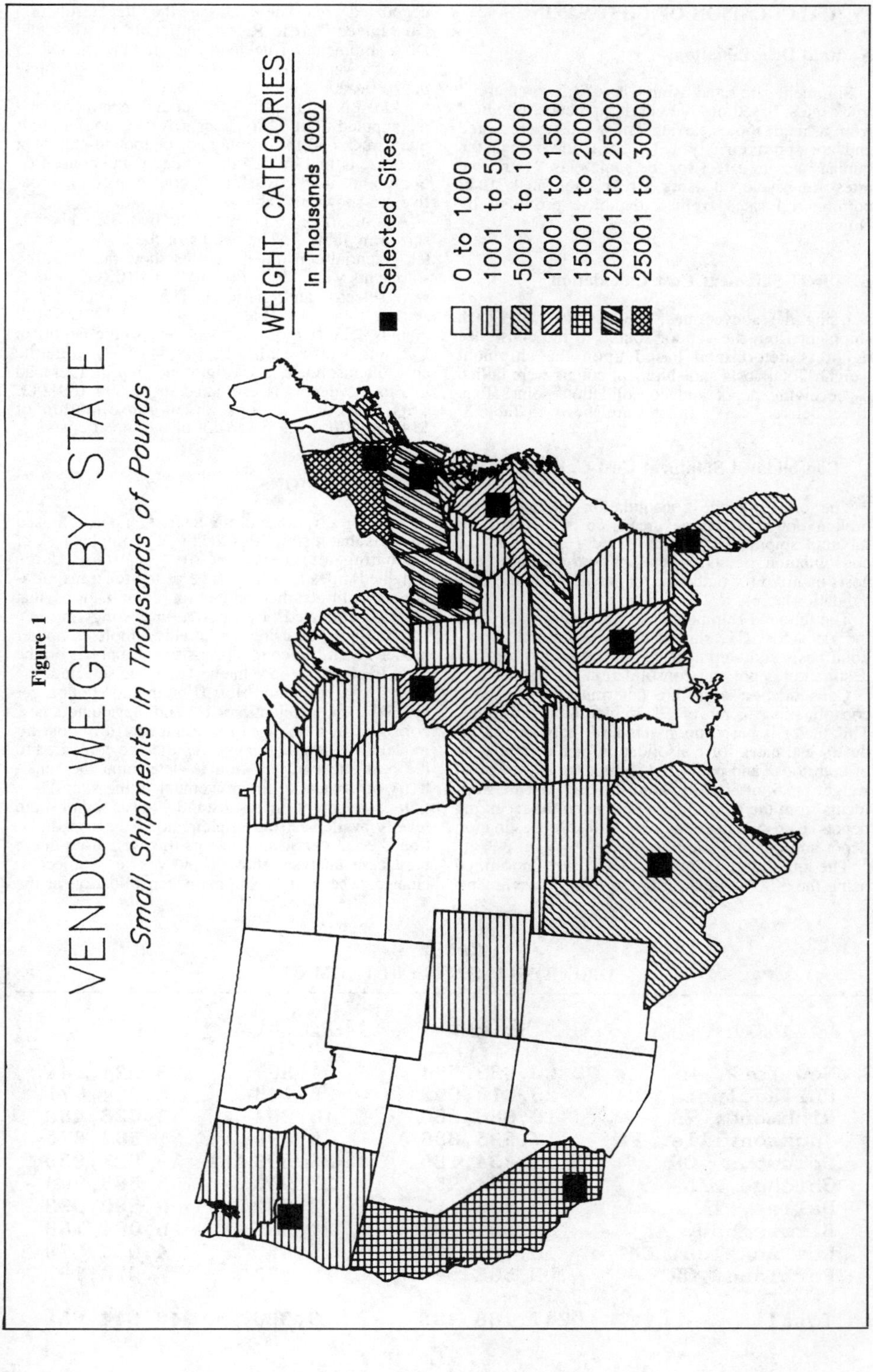
VENDOR WEIGHT BY STATE
Small Shipments In Thousands of Pounds
WEIGHT CATEGORIES
In Thousands (000)
Selected Sites
0 to 1000
1001 to 5000
5001 to 10000
10001 to 15000
15001 to 20000
20001 to 25000
25001 to 30000

IV. CALCULATION OF COST SAVINGS

A. Rate Determination

Shipments are rated using Class 77.5 commercial truck rates, based on three-digit zip codes, obtained from a major motor carrier. These rates were tested and are approximately 1.3 percent lower than the bureau rates in effect for the same class. Intrastate rates are developed using a regression analysis of commercial rates from a database provided by Numerax.

B. Direct Shipment Cost Calculation

Using the above rates, the cost to make each shipment from the vendor source to the consignee depot is determined based upon the shipment weight. Total costs and shipment counts were tallied by receiving depot and consolidation point. Total direct delivery cost estimates are shown in Table 3.

C. Consolidated Shipment Cost Calculation

The costs for the consolidation alternative are broken down into three areas, costs incurred for inbound shipments from the vendor origins to the consolidation points, costs for consolidation, and costs incurred for outbound shipments from the consolidation points to the depots.

The inbound shipment costs are determined using the rates for LTL shipments as discussed above. Total costs and shipment counts are tallied by final destination depot and consolidation point.

Consolidation costs are determined by using a consolidation factor of $1.35 per hundredweight. This factor is based upon standard, commercial industry estimates for consolidation of a combination of both loose and palletized shipments. A truckload weight break of 30,000 pounds is used for the shipments from the consolidation points to the receiving depots. Consolidation costs are tallied by destination depot and consolidation point.

The outbound shipment costs are determined using the 30,000 pound minimum and government-negotiated rates. These rates are the First Destination Guaranteed Traffic Rates applicable to truckload DLA shipments. Truckload lots are apportioned on at least a weekly basis. In the event that less volume than 30,000 pounds is available, the weight per truckload is reduced in 5,000 pound increments and recomputed until 52 loads or a 10,000 pound minimum truckload lot is obtained. Outbound shipment costs are tallied by destination depot and consolidation point. Total costs for the consolidation alternative are shown in Table 4.

A comparison of costs for the two alternatives is shown in Table 5. The results of the analysis using ten potential consolidation sites show that 213,388 shipments weighing between 70 and 10,000 pounds were selected and analyzed. The shipments represent a total aggregate weight of 247,188,498 pounds. The cost of direct delivery to the depots in LTL lots is estimated to be $48,544,652, while the cost to consolidate near origin and ship in truckload lots to the depots is estimated to be $34,189,282. This reflects a savings under consolidation of $14,355,370 or 29.6 percent of the direct costs.

V. CONCLUSIONS

A major advantage of the heuristic approach used above is that it simplifies the highly complex task of evaluating alternative facility locations. Using weighted miles as a first-stage proxy for transportation cost limits the number of combinations that must be evaluated. Transportation costing can then be accomplished using a relatively simple computer model which can conduct sensitivity analysis on the data. In addition, once the best sites are selected, the analyst can bump-and-shift (Keuhn and Hamburger (1963)) the consolidation sites and reevaluate a new configuration using the transportation costing model. The new alternative can then be compared to the best site configuration to determine the sensitivity of savings to the movement of the site(s).

It is apparent that substantial cost savings are readily available from implementing the consolidation alternative. Several factors that were used in the base case analysis above, however, are subject to change. The cost of shipment consolidation at the

TABLE 3
DIRECT SHIPMENT COST ESTIMATES

Location	Weight	Shipments	Cost
New York, NY	66,869,594	51,907	$13,924,644
Philadelphia, PA	28,615,902	28,143	5,824,071
Richmond, VA	19,686,560	18,264	3,626,462
Jacksonville, FL	7,595,886	5,895	1,567,875
Columbus, OH	31,434,185	23,506	5,627,939
Chicago, IL	28,624,952	27,943	5,583,908
Dallas, TX	12,035,157	13,382	2,590,092
Birmingham, AL	33,391,096	21,537	5,084,786
Los Angeles, CA	16,426,620	19,988	4,036,678
Portland, OR	2,508,546	2,823	678,197
Total	247,188,498	213,388	$48,544,652

TABLE 4
CONSOLIDATION COST ESTIMATES

Location	Weight	Shipments	Consolidation Cost
New York, NY	66,869,594	51,907	$9,278,511
Philadelphia, PA	28,615,902	28,143	4,325,910
Richmond, VA	19,686,560	18,264	2,606,942
Jacksonville, FL	7,595,886	5,895	1,055,492
Columbus, OH	31,434,185	23,506	4,177,628
Chicago, IL	28,624,952	27,943	4,310,574
Dallas, TX	12,035,157	13,382	1,617,129
Birmingham, AL	33,391,096	21,537	3,935,135
Los Angeles, CA	16,426,620	19,988	2,194,763
Portland, OR	2,508,546	2,823	687,199
Total	247,188,498	213,388	$34,189,282

consolidation sites appears to be one area where the above costs may be overstated. Negotiations with potential consolidators should decrease substantially the consolidation costs shown above. Additionally, there are other advantages as well as potential problems for the consolidation alternative.

The first advantage of consolidation is that it is expected to increase competition among vendors. Under consolidation, vendors will bid to the consolidation point of their choosing, usually the one which produces the lowest overall cost to the vendor. Since there will be a number of consolidation points located throughout the U.S. to choose from, this should reduce the amount of transportation-related costs included in the bid price of the item. Hence, this should increase the opportunity for vendors to compete who may otherwise have had to ship long distances at greatly increased transportation costs. This could make the difference in whether or not they obtain future contracts.

Another advantage of shipment consolidation is that it will encourage more small-dollar contractors to bid F.O.B. destination. This is a significant advantage to the reviewer of the bids since F.O.B. origin bids must be researched for the additional per item transportation costs necessary to place origin and destination bids on the same basis. This can sometimes require that the per item rate be extended to seven or more decimal places. This extension does not ensure that the estimated rate is what is actually paid when the item is shipped.

A final advantage of consolidation is that it provides a means to increase inventory line item visibility and management. Currently, advance notice of the arrival of an item at a depot is usually not given by the vendor. Under consolidation, however, during the process of building a truckload, a manifest listing the items by stock number and quantity will be constructed and transmitted to the depot in advance of the shipment. This will allow the receiving depot an opportunity to schedule these receipts and would allow the diversion of mission critical items directly to the customer.

Some potential problems resulting from consolidation are also apparent. Perhaps the most serious is the hold time attributable to the consolidation process. It is essential that inbound freight not be unduly delayed; therefore, the maximum time that a

TABLE 5
COST COMPARISON

Location	Consolidation Cost	Direct Cost	C-D Cost
New York, NY	$9,278,511	$13,924,644	($4,646,133)
Philadelphia, PA	4,325,910	5,824,071	(1,498,161)
Richmond, VA	2,606,942	3,626,462	(1,019,520)
Jacksonville, FL	1,055,492	1,567,875	(512,383)
Columbus, OH	4,177,628	5,627,939	(1,450,311)
Chicago, IL	4,310,574	5,583,908	(1,273,334)
Dallas, TX	1,617,129	2,590,092	(972,963)
Birmingham, AL	3,953,135	5,084,786	(1,149,651)
Los Angeles, CA	2,194,763	4,036,678	(1,841,915)
Portland, OR	687,199	678,197	9,002
Total	$34,189,282	$48,544,652	($14,355,370)

() indicate savings to DLA

shipment will remain at a consolidation point must be established.

Another potential problem is the time lag in bringing the consolidation points up to capacity once the decision to implement them has been made. Unless managed carefully, the traffic flow into the consolidation points may be very slow at first, with considerable direct traffic still flowing. If this should happen, the adverse costs of both alternatives will be incurred.

Since the consolidation points will be processing large numbers of shipments each day, some type of automated information handling system will be required. Compatibility between the systems at the points and the systems currently in use at the depots must be assured.

Also, rules and procedures must be considered to cover loss and damage of freight at the consolidation points. This is essential because the contractor at the consolidation point will be acting as an agent for DLA when receiving freight from vendors.

All in all, it appears that the consolidation alternative is advantageous, given that the potential pitfalls listed above are managed carefully.

REFERENCES

Cooper, L. (1964), "Hueristic Methods for Location-Allocation Problems," *SIAM Review*, Vol. 6, 37-53.

Efroymson, M.A. and T.L. Ray (1966), "A Branch-bound Algorithm for Plant Location," *Operations Research*, Vol. 14, 361-368.

Francis, R.L. and J.A. White (1974), *Facility Layout and Location,* Englewood Cliffs, NJ: Prentice-Hall, Inc.

Hakimi, S.L. (1964), "Optimal Locations of Switching Centers in a Communications Network and Some Related Graph-Theoretic Problems," *Operations Research,* Vol. 12, 450-459.

Hall, R.W. (1985), "Heuristics For Selecting Facility Locations," *Logistics and Transportation Review,* Vol. 21, No. 4, 353-373.

Kuehn, A.A. and M.J. Hamburger (1963), "A Heuristic Program for Locating Warehouses," *Management Science,* Vol. 9, 643-666.

Larson, R.C. and K.A. Stevenson (1972), "On Insensitivities in Urban Redestricting and Facility Location," *Operations Research, Vol. 20, 595-612.*

Revelle, C., D. Marks, and J.C. Liebman (1970), "An Analysis of Public and Private Location Models," *Management Science,* Vol. 16, 692-707.

ENDNOTES

* Transportation Analysis Group, Operations Research and Economic Analysis Office, Defense Logistics Agency.

** Associate Professor of Economics, Departments of Administrative Sciences and of Operations Research, Naval Postgraduate School.

Railroad Cost Structure—Revisited

by Anthony Barbera, Curtis M. Grimm,** Kent A. Phillips,*** and Leslie J. Selzer****

ABSTRACT

Dating back to the landmark work of Meyer, Peck, Stenason and Zwick (1959), economists such as Borts (1960), Healy (1962), Griliches (1972), Keeler (1974), and Harris (1977), Harmatuck (1979), Spady and Friedlaender (1981), Breautigam, Daughety, and Turnquist (1982) and Caves, Christensen, Tretheway and Windle (1985) have attemped to estimate railroad cost structure. However, there are a number of data limitations common to most of these studies. First, accounting data is often used which does not correspond to economic reality. For example, betterment accounting techniques historically treated maintenance expenses as current expenses rather than additions to capital stock. Thus, the economic value of capital is not accurately reflected, and this may well result in biased estimates of economies of density. Second, these studies have relied on older data (no more recent than 1975). Consequently, potential impacts on cost structure from regulatory reform have not been previously estimated.

This paper presents new estimates of railroad structure, improving upon previous estimates by using more recent data and superior estimates of capital costs. A data set consisting of Class I railroads for the years 1979 through 1983 is used. Major findings are strong economies of density and length of haul, but no economies of firm size. Implications for both public policy and management strategy are discussed.

This paper presents new estimates of railroad structure, improving upon previous estimates by using more recent data and superior estimates of capital costs. A data set consisting of Class I railroads for the years 1979 through 1983 is used. Major findings are strong economies of density and length of haul, but no economies of firm size. Implications for both public policy and management strategy are discussed.

I. INTRODUCTION

The cost structure of the railroad industry has long been of interest to transportation economists and railroad managers. An important question, both for regulatory policy and strategic management, is whether railroads can lower their average costs by expanding their operations. This question is complicated by the fact that railroads can expand their operations (i.e., increase their ton-miles of output) in three distinct ways. First, firms can seek to achieve **economies of density** by adding more traffic to an existing route structure. Average costs are reduced by spreading the fixed cost per mile of route over more units of output, and by using more efficient methods of production as increasing volume warrants correspondingly higher capital invest ment on the route. Economies of density can also be derived from reduced operating costs, such as lower classification and switch ing charges, as traffic density on an existing route is increased.

Second, firms can boost output by increasing their length of haul. **Economies of length of haul** imply that unit costs decrease as length of haul increases. These economies are achieved mainly by spreading origination and termination costs over more output.

Third, firms can expand output by merging with other firms or purchasing additional lines. The cost advantages of size apart from economies of density or length of haul are referred to in the railroad literature as **economies of scale.** To estimate scale economies one must measure average costs associated with various firm sizes while holding traffic density and average length of haul constant. Economies of scale are more likely to be associated with the administrative rather than operating functions of the firm, although there may be opportunities for more efficient utilization of plant, equipment and crews with respect to total firm size.

Theoretically, any or all of these economies could exist in the rail industry. However, it is crucial to distinguish between these three types of economies when formulating business or public policy. The existence of economies of density may provide rationale for natural monopoly regulation of railroads, for differential pricing of rail services, and for parallel mergers to increase traffic density. The presence of economies of length of haul may support a policy of end-to-end mergers and the focus of marketing efforts on longer haul traffic. Economies of scale, on the other hand, may provide a rationale for mergers of unrelated rail lines.

Dating back to the landmark work of Meyer, Peck, Stenason and Zwick (1959), economists have attempted to estimate the railroad industry's cost structure. The state-of-the-art was advanced by other economists, such as Borts (1960), Healy (1962), Griliches (1972), Keeler (1974), and Harris (1977), who offered alternative equation specifications and new insights into the applicable economic theory.

More recently, Friedlaender and Spady (1981), Harmatuck (1979), Braeutigam, Daughety, and Turnquist (1982) and Caves, Christensen, Tretheway and Windle (1985) have employed a translog functional form. This equation specification is more flexible than the simpler ones used in earlier studies. More recent studies have also included factor prices as independent variables.

In spite of the differences in econometric techniques and data, these studies have shown strong and somewhat consistent evidence of economies of density, virtually no evidence of economies of scale, and mixed results on economies of length of haul. The overall results are summarized in more detail in Jara-Diaz (1982), Keeler (1983), Winston (1985), and Daughety (1985).

There are, however, a number of limitations or weaknesses in the studies cited above. First, previous research has relied on older data—none more recent than 1975. Consequently, potential impacts on cost structure from regulatory reform brought about by the 4R Act of 1976 and the Staggers Act of 1980 have not been previously estimated. Furthermore, although this pre-1975 data was audited by both the railroads and the ICC, it has been shown to be inconsistent over time and between carriers. In recent years, however, the ICC has perfected its automatic data processing audits of railroad data. Thus, recent data, particularly post-Staggers, is measurably better than the data previously used to estimate railroad cost functions.

In addition, most of the capital data in previous studies were derived from annual reports filed with the ICC and reflect accounting expenses rather than economic costs. Because of the accounting conventions employed, the reported expenses may not correspond to the underlying economic costs. Most importantly, until 1983, the railroads utilized a special railroad accounting technique for roadway property, known as Retirement Replacement Betterment Accounting (RRB). Under this accounting convention programed roadway maintenance expenses are treated as current expenses rather than as additions to capital stock. Under conventional accounting systems such programed maintenance expenditures are capitalized and depreciated. It is generally agreed that depreciation accounting more closely links expenditures to the physical consumption of the assets. Since major capital expenditures are treated as expenses under RRB accounting, the economic value of roadway consumption is not accurately reflected.

The purpose of this paper is to improve and update estimates of railroad cost structure, utilizing more recent and refined data vis-a-vis previous studies. Data is drawn from the years 1979-1983. Moreover, recent ICC studies are utilized to more accurately assess capital costs on a rational economic basis.

This paper is organized as follows. Section II provides details on the methodology and data used in the study, while section III provides the results of a translog cost function estimation. Revisiting railroad cost structure is warranted by the ever greater importance for both public and business policy in the post-Staggers Act environment. Section IV explores the implications of the findings for regulators and rail managers.

II. METHODOLOGY AND DATA

The methodology used to estimate railroad cost structure is the translog cost function. Details of the procedure are provided in the Technical Appendix.

The data used in this study were developed for Class I railroads for the years 1979 through 1983. Table 1 lists all of the variables used to construct the data set along with the sources from which they are abstracted. This data can be divided into four broad classes: size variables, operating expenses, capital costs and factor data.

A. Size variables

Operating statistics quantify the physical characteristics of the rail system, including the size of the system and the level of output. The operating statistics used in this study are taken from the Annual Report of Class 1 Railroads (R-1) to the Interstate Commerce Commission (ICC). These data items include: net ton-miles (NTM), net freight tons (NFT), and miles of road (MR).

B. Operating costs

The dependent variable in the translog regression is the total costs for each railroad, obtained from summing operating and capital costs. Operating expenses reflect the reported expenditures for variable inputs. This data, derived from the annual reports, include: maintenance-of-way and structures (MWS), maintenance of equipment (ME), transportation expense (TE) and traffic expense and general expense (TFEGE).

C. Capital costs

Capital costs are defined to include the return to physical capital and the amortization of the investment over its expected life. The computation of capital costs is more complex than that for operating expenses because the lives of capital assets in the railroad industry are extremely long. A major improvement of our study over most previous estimates of rail costs is an assessment of capital costs based on current economic value in lieu of accounting valuation. Accordingly, the derivation of capital expenditures used in the study will be explained in detail.

Current capital costs (CURKC) measures the market value of capital assets in the current year, i.e., the current net book value of road property and equipment. This figure is computed by the Depreciation Branch (DB) of the ICC and is derived using the following procedure. The historic (book) value of each asset or group of assets is brought to current levels using a price index peculiar to that category of assets. This process yields an estimate of gross current value. The DB also has developed estimates of current depreciation (CURDP), reflecting consumed life and/or economic obsolescence of the assets. By deducting the accumulated depreciation from the current gross investment, CURKC is computed. Current land values (LAN63) were calculated by using a DB study conducted in 1963, which estimated land values, and indexing them to a current level using asset specific indices developed by the Department of Agriculture.

The return to capital is computed by applying the real pre-tax cost of capital to the railroad net investment base. A nominal cost of capital is first computed and then converted to a real value. The nominal cost of capital used in this study differs from that used in prior studies in that differences in the cost of equity capital are recognized for the major rail systems. Virtually all prior studies employ a uniform weighted cost of capital. Differences in the cost of capital, however, may affect investment levels and therefore should be explicitly recognized. Since the

TABLE 1
Sources of Underlying Data

Variable	Description	Source
Size Variables		
NTM	Net ton-miles	R-1 Sch 755 L 128 Col. B (78-82), L 114 (83)
NFT	Net freight tons	R-1 Sch 755 L 121 Col. B (78-82), L 107 (83)
MR	Miles of road	R-1 Sch 700 L 57 Col. D
Operating Costs		
MWS	Maintenance of ways and structures	R-1 Sch 410 L 151 Col. F
ME	Maintenance of equipment	R-1 Sch 410 L 324 Col. F
TE	Transportation expense R-1	R-1 Sch 410 L 528 Col. F
TFEGE	Traffic and general expense	R-1 Sch 410 L 619 Col. F
Capital Costs		
REAL	Real cost of capital	ICC Financial Analysis
CURKC	Current value of track, road property & equipment, under depreciation accounting	ICC Depreciation Branch
CURDP	Accumulated depreciation on current valuation of track, road property and equipment	ICC Depreciation Branch
LAN63	Current value of land	Indexed 1963 special study
Factor Prices and Shares		
FG	Fuel gallons	R-1 Sch 750 L 4 Col. B
FE	Fuel expense	R-1 Sch 750 L 5 Col. B
FP	Fuel price per gallon	Calculated
LH	Labor Hours	ICC Wage Statistics Form A L 909 Col 7
LE	Labor expense	Form A L 909 Col 11
LP	Labor price	Calculated
RIMS	Regional Index	AAR's Railroad Materials-Supplies Cost Recovery Index
RIPS	Regional Index	AAR's Railroad Purchased Services Cost Recovery Index

cost of equity is more likely to reflect carrier differences, this study utilizes individually computed equity costs for thirteen major rail carriers. These carriers are: Burlington Northern, CSX, MoPAC, Norfolk Western, Denver and Rio Grande Western, Atchison Topeka and Santa Fe, Southern, Soo, Kansas City Southern, Southern Pacific, Seaboard Coastline, Illinois Central Gulf and Union Pacific. The cost of equity for each of the above carriers was also extended to its subsidiaries.

The cost of equity capital for the carriers listed above was based on the discounted cash flow (DCF) methodology adopted by the ICC for use in its annual revenue adequacy determinations for the Class I railroads. Under the DCF methodology the cost of equity is computed by solving the following equation:

$$K = \{(Do/Po)\ (1+G/2)\} + G$$

WHERE;

K = pre-tax nominal cost of equity
Do/Po = dividend yield per share
G = projected growth rate in earnings

The DCF methodology has been applied in Commission revenue adequacy proceedings for a number of Class I railroads. The cost of equity for these observation was taken directly from the record in the individual Interstate Commerce Commission proceedings. For those carriers not covered by the record and for years in which a record did not exist, the equation was estimated using year end data from Moody's Stock Guide. The growth rate "G" was estimated using the earnings per share for the prior six years. The weighted nominal cost of capital was then computed using the appropriate debt/equity capital structure for each year and the industry average current cost of debt. The cost of capital for other than the thirteen carriers identified was based on the annual weighted system average cost of capital.

The nominal cost of capital was restated to a real level using the following formula:

$$r = [(1+n)/(1+i)]-1$$

Where;

r = real cost of debt or equity or debt
n = nominal cost of equity or debt
i = inflation as measured by the GNP deflator

D. Factor prices and shares

Four factors were used in the study: capital, fuel, labor, and materials/purchased services. Gallons of fuel purchased (FG) and fuel expenses (FE) are derived from the R-1 annual reports. Statistics relating to labor input levels (LH) and wage compensation (LE) are derived from the year ending Report of Employees, Service and Compensation, I.C.C. Wage Statistics, Form A. Several factor price variables are calculated from data abstracted from ICC sources. These are price per gallon of fuel (FP) and hourly labor wage (LW) for each of the catagories captured. Finally, the Association of American Railroads' (AAR) "Railroad Cost Recovery Index" is used to obtain regional indices on materials and supplies (RIMS) and purchased services (RIPS). The factor share for materials and supplies/purchased services is calculated by subtracting labor and fuel expenses from total operating costs.

III. RESULTS

Parameter estimates for the translog cost function are provided in Table 2. The R-square of the equation (.96) indicates a very strong fit. All first order terms are of the expected sign and statistically significant. The elasticities of substitution for all pairs of inputs are provided in Table 3, with details of elasticity derivation given in the Technical Appendix.

The normalization of data employed in the study allows the first order terms to be interpreted as the mean elasticity of cost with respect to the ith coefficient, as detailed in the Technical Appendix. The coefficient on net ton-miles provides evidence on economies of length of haul. Given that net freight tons is also included in the equation, the parameter measures the impact of changing ton-miles while holding tons constant, thereby also altering length of haul. Specifically, the first order coefficient on net ton-miles indicates that a one percent increase in the

TABLE 2
Regression Coefficients
(Standard Errors in Parentheses)

First Order Terms	Coefficient	T-Statistic			
Constant	-.364 (.057)	-6.417	Labor x Labor	.037 (.018)	2.005
NTM	.416 (.077)	5.399	Labor x Capital	-.094 (.010)	-9.553
NFT	.224 (.068)	3.303	Labor x Materials	.058 (.018)	3.175
MR	.390 (.075)	5.216	Capital x Capital	.222 (.010)	22.277
Energy Price	.072 (.002)	46.226	Capital x Materials	-.099 (.011)	-8.758
Labor Price	.320 (.007)	46.059	Materials x Materials	.074 (.023)	3.275
Capital Price	.177 (.008)	23.372	NTM x Energy	.047 (.003)	13.949
Materials Price	.431 (.006)	76.634	NTM x Labor	-.049 (.015)	-3.249
			NTM x Capital	-.059 (.011)	-4.921
Interaction terms			NTM x Materials	.061 (.016)	3.730
NTM x NTM	.495 (.195)	2.535	NFT x Energy	-.027 (.003)	-9.090
NFT x NFT	.088 (.144)	.607	NFT x Labor	.060 (.014)	4.433
MR x MR	-.011 (.099)	-0.114	NFT x Capital	.042 (.010)	3.977
NTM x NFT	-.385 (.154)	-2.492	NFT x Materials	-.075 (.015)	-5.159
NTM x MR	-.187 (.129)	-1.449	MR x Energy	-.025 (.003)	-9.558
NFT x MR	.348 (.103)	3.368	MR x Labor	.012 (.011)	0.964
Energy x Energy	.063 (.006)	9.832	MR x Capital	.026 (.009)	2.800
Energy x Labor	-.001 (.004)	-0.280	MR x Materials	-.013 (.013)	-0.990
Energy x Capital	-.029 (.003)	-11.602	Time Trend	.070 (.013)	5.597
Energy x Materials	-.033 (.008)	-4.268			

TABLE 3
Elasticities of Substitution Between Factors

	Energy	**Materials**	**Labor**	**Capital**
Energy	-0.94	-1.74	0.95	2.05
Materials		-0.95	1.95	-0.33
Labor			-1.76	-1.67
Capital				-0.13

average length of haul increases total costs by only .416 percent. Thus, evidence of economies of length of haul is provided.

The attributes of rail cost structure of paramount interest are returns to density and scale. Following the procedure outlined by Caves, Christensen, Tretheway and Windle (1985) to calculate returns to scale and density, the present results show strong economies of density but no economies (or diseconomies) of scale. The Technical Appendix reviews the methodology employed, while Table 4 summarizes the empirical evidence on returns to density and scale from the present and previous studies of rail cost structure. As discussed earlier, previous studies have relied on data no more recent than 1975 and have employed a wide range of specifications, econometric techniques and data. It is then noteworthy that the key results regarding economies of scale and density are largely consistent with the previous work. More specifically, the current study confirms the lack of scale economies found across all of the previous studies. Note that the coefficient capturing returns to scale varies between a narrow range of 0.93 - 1.03 across the seven studies, where a value less than 1 denotes decreasing returns to scale while a value greater than 1 denotes increasing returns. In no case is the value significantly different from 1, so that this study in concurrence with all of the previous ones points to constant returns to scale. This implies that there is no cost advantage from size independent from the effects of density and length of haul.

However, evidence regarding returns to density is perhaps more indicative of railroad cost structure in that these economies influence the optimal number of carriers in a given market. Increasing returns to density up to and beyond the level of demand in a given market would imply that a natural monopoly exists. Referring again to Table 4, the earlier studies of Keeler, Harris and Harmatuck all revealed strong evidence of returns to density. However, the next two studies, by Caves, Christensen and Swanson, and Friedlaender and Spady, found little evidence of

TABLE 4
Returns to Scale and Density: Present and Previous Studies

	Returns to Density	**Returns to Scale**
Keeler (1974)	1.79	1.01
Harris (1977)	1.72	1.03
Harmatuck (1979)	1.92	0.93
Caves, Christensen and Swanson (1981)	1.00	1.01
Friedlaender & Spady (1981)	1.16	1.02
Caves, Christensen, Tretheway and Windle (1985)	1.76	0.98
[Source: Caves, Christensen, Tretheway and Windle (1985)]		
Present Study	1.59	0.98

Returns to Density = 1/(.416 + .224) = 1.59

Test if returns are significantly different from 1: T-Statistic = 4.749 - Reject null hypothesis of constant returns in favor of increasing returns to density.

Returns to Scale = 1/(.416 + .224 + .390) = 0.98

Test if returns are significantly different from 1: T-Statistic = 1.047 - Cannot reject null hypothesis of constant returns to scale.

returns to density1. A later study by Caves, Christensen, Tretheway and Windle, again found strong economies of density. It is then significant that the present study finds strong evidence of returns to density, confirming the result found in the majority of earlier studies.2

In summary, it is notable that technological and regulatory changes in the rail industry since 1975 (the last data point of previous rail cost studies) have not substantially affected railroad cost structure. To clarify, while railroads have reduced costs through measures such as abandonments, the fundamental relationship between average costs, density and scale of operation has not been measurably altered. To confirm this, specifically with regard to the impacts of the Staggers Act of 1980, a dummy variable was employed to obtain separate estimates of scale and density economies for the periods 1979-1980 and 1981-1983. The results over the two periods were not significantly different.

IV. POLICY IMPLICATIONS

The results of this study have important policy implications for both regulators and railroad management. The presence of economies of density and length of haul are particularly meaningful because of the recent changes in regulation brought about by the Staggers Rail Act of 1980. These changes resulted in new rules and guidelines which are intended to permit greater ratemaking freedom in today's competitive environment. With this increased pricing latitude, a solid understanding of the railroad industry's cost structure is necessary.

Clearly regulators have relied on prior research in formulating merger policies and maximum rate rules. However, as was recognized in the introduction to this paper, all prior research of industry cost structures was based on pre-Staggers data. The findings in the present study indicate that, from a technological perspective, the cost structure for the industry has not changed materially as a result of regulatory reform. Thus, it appears that the ICC's assessment of the cost structure of the railroads, based on prior research, is consistent with current technology.

Because of the increased reliance on the marketplace to set rates, the results of this study have perhaps their most important implications for railroad management. Managers must be fully cognizant of the impact density and length of haul have on their companies' cost structures. The presence of economies of density and length of haul influence decision making in the areas of operations, rate making, abandonments and mergers. From an operations perspective, the presence of significant economies of density suggest that management should attempt to maximize line segment densities wherever possible. Economies of length of haul further suggest that carriers should focus on longer haul movements. From a ratemaking perspective, the study results indicate that the railroads should price traffic so as to maximize the volume of compensatory traffic using its lines. Contract rates aid the railroads in this area, since they permit carriers to adjust operations for maximum efficiency.

Finally, railroad managers should conduct further cost studies such as this one, utilizing where possible line or corridor specific expense and output data not publicly available. Results from studies of this type would provide managers with greater insight into their infrastructure, i.e., the cost of providing service over specific lines, profitability of current traffic and the cost impact of soliciting additional traffic. Cost and density data on a line specific basis would thus permit greater precision in pricing competitive traffic and allow managers to determine the optimal mix of inputs for the most cost efficient system. Given that prices for the majority of rail services are set in the competitive marketplace, information on optimal operational networks and attendant cost structures is of the utmost importance to railroad executives.

ENDNOTES

* Data Resources, Inc.
** College of Business and Management, University of Maryland, College Park, MD.
*** Interstate Commerce Commission.

[1] Keeler (1983) argues that, properly interpreted, the studies are consistent with the earlier work.

[2] Work in progress by Tenpao Lee and C. Philip Baumel finds significant returns to density weaker than the present study and constant returns to scale.

BIBLIOGRAPHY

Borts, G. H., "The Estimation of Rail Cost Functions," *Econometrica* 28, January 1960, pp. 108-31.

Braeutigam, Ronald R., Andrew F. Daughety, and Mark A. Turnquist, "The Estimation of a Hybrid Cost Function for a Railroad Firm," *Rev. Econ. Statist.*, Aug. 1982, 64, pp. 394-404.

Braeutigam, Ronald R., Andrew F. Daughety, and Mark A. Turnquist, "A Firm Specific Analysis of Economies of Density in the U.S. Railroad Industry," *Journal of Industrial Economics*, Vol. 33, No.1, Sept. 1984, pp. 3-20.

Caves, Douglas, Laurits Christensen, and J.A. Swanson, "Productivity Growth, Scale Economies, and Capacity Utilization in U.S. Railroads, 1955-1974," *American Economic Review*, Vol. 71 Dec. 1981, pp. 994-1002.

Caves, Douglas, Laurits Christensen, Michael Tretheway, and Robert Windle, "Network Effects and the Measurement of Returns to Scale and Density for U.S. Railroads," in *Analytical Studies in Transport Economics*, A. Daughety, Ed., Cambridge University Press, New York, 1985.

Daughety, Andrew F., "Transportation Research on Pricing and Regulation: Overview and Suggestions for Future Research," Presented at the NSF Conference on Transportation Research, Evanston, Illinois, March 1985.

Friedlaender, Ann F. and Richard H. Spady, *Equity, Efficiency and Resource Rationalization in the Rail and Regulated Trucking Industries*, MIT Press, Cambridge, 1981.

Griliches, Z., "Cost Allocation in Railroad Regulation," *Bell Journal of Economics* 3, Spring 1972, pp. 26-41.

Harris, Robert G., "Economies of Traffic Density in the Rail Freight Industry," *Bell Journal of Economics* 8 (2), Autumn 1977.

Harmatuck, Donald J., "A Policy-Sensitive Railway Cost Function," *Logistics and Transport Review,* Vol. 15, No. 2, 1979.

Healy, Kent T., *The Effects of Scale in the Railroad Industry,* Committee on Transportation, Yale University, New Haven, Connecticut, 1961.

Jara-Diaz, Sergio, "The Estimation of Transport *Transport Review, Vol. 2, No. 3, 1982, pp. 257-278. Jara-Diaz, Sergio and Clifford Winston, "Multiproduct Transporta tion Cost Functions: Scale and Scope in Railway Operations," in Eighth European Association for Research in Industial Economics,* Vol. 1, Nicklaus Blattner, et. al., eds., U. of Basel, 1981. Keeler, Theodore E., "Railroad Costs, Returns to Scale, and Excess Capacity," *Review of Economics and Statistics* 56, May 1974.

Keeler, Theodore E., *Railroads, Freight and Public Policy,* Brookings Institution, Washington, D.C., 1983.

Meyer, John R., Merton J. Peck, John Stenason, and Charles Zwick, *The Economics of Competition in the Transportation Industries,* Harvard University Press, Cambridge, MA, 1959.

Sammon, John P., "Returns to Traffic Flow Concentration in the Railroad Industry," *Transportation Research Forum Proceedings XIX,* 1978.

Winston, Clifford, "Conceptual Developments in the Economics of Transportation: An Interpretive Survey," *Journal of Economic Literature,* Vol. XXIII, March 1985.

Winston, Clifford, "Conceptual Developments in the Economics of Transportation: An Interpretive Survey," *Journal of Economic Literature,* Vol. XXIII, March 1985.

TECHNICAL APPENDIX

The measures of returns to scale, returns to density and elasticities of substitution presented in the paper were computed with the parameter estimates of a four input, three output translog cost function (TLC) with Hicks neutral technical change. If we define Q_i as the ith normalized output and P_i as the ith normalized input price where the normalizing factors are the sample means, then the cost function which was estimated for this paper is written as follows:

$$(1) \quad \ln C = \alpha_0 + \sum_i^4 \alpha_i \ln P_i + \sum_i^3 \beta_i \ln Q_i + \alpha_t t + \sum_i^4 \sum_j^4 \theta_{ij} \ln P_i \ln P_j + \sum_i^3 \sum_j^3 \delta_{ij} \ln Q_i \ln Q_j + \sum_i^4 \sum_j^3 \phi_{ij} \ln P_i \ln Q_j$$

where ln is the natural log and C is normalized total factor cost.

The appeal of (1) to many researchers is that it has potentially very desirable theoretical properties while at the same time it provides a functional form which may be relatively easily estimated. Theory tells us that (1) can be viewed as derivable from a concave, twice differentiable, continuous, multi-output transformation function under cost minimizing behavior if certain restrictions hold. Specifically, the cost function itself must be linearly homogenous in prices and must itself be concave. Linear homogeneity was imposed parameterically wtih the linear restrictions:

$$(2a) \quad \sum_i \alpha_i = 1$$

$$(2b) \quad \sum_i \theta_{ij} = \sum_j \theta_{ij} = 0$$

$$(2c) \quad \sum_i \phi_{ij} = 0 \text{ for } j=1..3.$$

Concavity must be checked once the parameter estimates have been derived. A necessary condition for concavity is that the own price elasticities must be nonpositive. As the estimates of the own price elasticities of must be nonpositive. As the estimates of the own price elasticities of substitution in the paper illustrate, this condition is satisfied.

For estimation purposes, we augment (1) and (2) with cost share equations. Taking the partial derivative of (1) with respect to the (log of) the ith input price and applying Shepard's Lemma gives

$$(3) \quad S_i = \alpha_i + \sum_j \theta_{ij} \ln P_j + \sum_j \phi_{ij} \ln Q_j \qquad i=1...4$$

where S_i is the ith cost share. Thus, we are able to introduce more degrees of freedom by estimating the cost function (1) jointly with the share equations (3) subject to the restrictions in (2). Note that the share equations involve no new parameters.

To estimate the model, we must eliminate one of the share equations for, otherwise, the error matrix would be singular. We then use an iterative version of Zellner's seemingly unrelated regression technique, Zellner (1982), which produces maximum likelihood estimates and assures that the parameter estimates will be invariant with respect to the share equation which is excluded from the system.

The most useful theoretical property of TLC is that the elasticities of substitution are estimated directly and may be different for different pairs of inputs. In contrast, the Cobb Douglas production/cost function (CD) requires that the elasticity of substitution must be unity for all pairs of inputs while the Constant Elasticity of Substitution production/cost function (CES) requires that the elasticity of substitution must be constant and identical among all pairs of inputs (although it may differ from unity). Thus, while the CES is a generalization of the CD, the TLC is more general than both and provides a far richer set of potential technology estimates.

The elasticities of substitution which are presented in the paper where derived from the following formulas:

$$(4a) \quad \sigma_{ij} = (\theta_{ij} + S_i S_j)/S_i S_j \quad i \neq j,$$

$$(4b) \quad \sigma_{ii} = (\theta_{ii} + S_i^2 - S_i)/ S_i^2.$$

Following Caves et al (1984), returns to density is given by

$$RTD = 1 / (E_{c1} + E_{c2})$$

where E_{c1} and E_{c2} are the elasticities of cost with respect to Q_1 and Q_2. In the paper, Q_1 is net freight tons (NFT), Q_2 is net ton miles (NTM) and Q_3 is miles of road (MR). Returns to scale is then given by

$$RTS = 1 / (E_{c1} + E_{c2} + E_{c3}).$$

Therefore, RTS and RTD depend upon the estimated output elasticities.

Taking the partial derivative of (1) with respect to the (log of) the ith output gives

$$(5) \quad E_{ci} = \beta_i + \sum_j \phi_{ij} \ln P_j + \sum_j \delta_{ij} \ln Q_j.$$

In the paper, we present the mean output elasticities which are derived by simply inserting mean price and output into (5). Since all prices and outputs are normalized by their sample means, all terms in (5) vanish except for β_1. By estimating the model with normalized data, therefore, the first order term, β_1, is the direct estimate of the mean elasticity of cost with respect to the ith output.

The Cost Structure of the U.S. Railroad Industry Under Deregulation*

*by Tenpao Lee*** and C. Phillip Baumel****

ABSTRACT

The concept of revenue adequacy of the railroad industry has been a national goal since 1976. The Interstate Commerce Commission "Constrained Market Pricing" guidelines emphasize increased rates on so-called captive traffic to achieve railroad revenue adequacy. The purpose of the paper is to estimate the potential contribution of cost-saving opportunities to the goal of railroad revenue adequacy.

A simultaneous system of equations of cost and demand functions is estimated by using the jointly generalized least squares method. Economies of traffic density, firm size, and shipment characteristics are estimated and alternative public policies are suggested to help achieve the goal of revenue adequacy by restructuring the railroad industry.

I. INTRODUCTION

Prior to and after World War II, the rate of return on net investment in the railroad industry was very low (Association of American Railroads, *Railroad Facts*). Since 1960, several major railroad companies declared bankruptcy and the Rock Island Railroad was liquidated. Federal regulation was believed to be partly responsible for the low earnings, bankruptcies and liquidation because the regulations made it difficult for railroads to shed unprofitable operations and to adjust quickly to meet shifting demands, intermodal competition and rising costs.

In 1976, the Railroad Revitalization and Regulatory Reform Act (4R Act) introduced the concept of revenue adequacy into railroad rate-making. Revenue adequacy was defined as a level of earnings sufficient to enable a carrier to meet all expenses, retire a reasonable amount of debt, cover plant depreciation and obsolescence, and earn a return on investment adequate to attract new capital. The Staggers Rail Act of 1980, which substantially deregulated the railroad industry, retained the concept of revenue adequacy. In August, 1985, the ICC adopted "Constrained Market Pricing (CMP) to serve as guidelines in determining the reasonableness of rail captive coal rates" to achieve the objective of revenue adequacy (Interstate Commerce Commission, 1985). The ICC has begun to draft rate reasonableness guidelines for non-coal traffic (Traffic World, May 5, 1986).

CMP emphasizes increased freight rates based on the inelastic demand for rail transport of so-called captive traffic. CMP does address the cost side by stating that captive coal shippers should not be required to pay more than is necessary for efficient service where inefficiencies can arise from operating, structural and pricing policies. However, most of the public debate has focused on the price implications of CMP. There has been little research and public debate on the potential contribution of adjustments in the railroad industry to cost reduction and, hence, to railroad revenue adequacy through non-price alternatives.

Several railroad cost functions have been estimated using historical data. Borts conducted a statistical cross-sectional analysis of the variance of freight costs for Class I railroads based on 1950 data. Keeler estimated a Cobb-Douglas multiproduct cost function based on 1968-70 railroad data. Harris estimated a linear cost function using 1972-73 railroad data and included average length of haul as an explanatory variable. Sidhu, Charney, and Due used a linear model to estimate long-run average cost functions for Class II railroads based on the 1968 and 1973 Class II railroads based on the 1968 and 1973 Class II railroad cross-sectional data. Harmatuck argued that inflexible work rules and the standardization of certain railroad operating procedures make it more appropriate to estimate cost functions based on activities. He estimated a joint translog cost function using 1968-70 railroad data. Caves, Christensen, and Swanson (1980) estimated the growth in productivity in the rail industry by fitting a translog cost model to 1951-74 railroad data to estimate the elasticities of total cost with respect to output and factor prices. Friedlaender and Spady estimated a translog cost function of Class I railroads based on 1968-70 cross-sectional data. Caves, Christensen, and Swanson (1981) estimated a generalized translog cost model of Class I railroads based on 1955-74 cross-sectional data. Braeutigam, Daughety, and Turnquist estimated a hybrid cost function for a single railroad firm by fitting a flexible translog model to 1969-77 time series data. They called their model a hybrid because they incorporated engineering information (speed of services) to improve model specification.

Caves, Christensen, Tretheway and Windle argue that the earlier rail cost studies have yielded biased estimates of returns to density. We believe that the earlier studies are of limited value for evaluating alternative policies to deal with the revenue adequacy problems of the railroad industry and deriving the implications of alternative policies on agriculture for the following reasons:

1. All previous studies are based on old data ranging from 1955 to 1974 with 1968–1970 the most popular study years. The structure of today's railroad industry is significantly different from the railroad industry that was analyzed by previous studies.

2. Since the underlying true cost function can not be identified, the earlier studies can only be interpreted for the railroad industry in the time period studied. No conclusions can be drawn outside the data range.
3. Model specification of earlier studies usually includes both capital price and firm size which violates the domain of either a short-run or a long-run cost function. A short-run cost function should not include a capital price variable while a long-run cost function should not include a firm size variable. Moreover, only two of the previous studies use more advanced theoretical specification, more modern functional form and more thorough cost accounting (Caves, Christensen, Tretheway and Windle).
4. Under deregulation, output price and quantity are determined simultaneously and endogeneously. Theoretically, it is more meaningful to estimate cost and demand functions simultaneously.

The purposes of this paper are to estimate a simultaneous system of cost and demand functions of the railroad industry under a deregulated environment, to identify structural changes on the railroad industry that could reduce railroad costs, to formulate public policies that could help achieve revenue adequacy of the railroad industry from a lower cost approach rather than from a higher price approach, and to identify the direction of future research opportunities.

II. A SIMULTANEOUS SYSTEM OF COST AND DEMAND

Since firms in general, and railroads in particular, typically cannot adjust all factor inputs in an optimal fashion in the short run, it is likely that they are operating along an efficient long-run cost function. Thus, estimates of long-run cost functions are often subject to bias because firms are not in long-run equilibrium. Therefore, we specify and estimate the following short-run cost and demand functions:

$$SRTC = f(r_i\text{'s}, Q, S, SC_i\text{'s}, D) \quad (1)$$
$$P = LRMC \quad (2)$$
$$Q^d = f(P, MKTG, D) \quad (3)$$

where:

SRTC = short-run total freight costs,
r_i's = input prices of labor and fuel,
Q = output measured as revenue ton-miles,
S = firm size measured as road miles,
SC_i's = shipment characteristics include average length of haul in miles, load factor measured as net ton-miles divided by gross ton-miles, percent of unit-train operation measured as total unit-train miles divided by total freight train miles,
D = dummy variable for year,
P = price level of rail services measured as average revenue per ton-mile,
LRMC = long-run marginal costs,
Q^d = demand for rail services measured as revenue ton-miles,
MKTG = total marketing expenses measured as the summation of marketing expenses and advertising costs.

A. Model Specification

In the short-run, capital expenditures are fixed and their impact on the short-run cost function depends on the amount of capital rather than the price of the capital. Hence, capital price is not included in our short-run cost model. Firm size in route miles is included as a proxy variable to represent the fixed amount of capital and to reflect the existence of economies of scale as well. Shipment characteristics included average length of haul, unit train miles as a percent of total miles, and load factor. The concept of rate tapering implies that average costs are inversely related to length of haul. Low cost unit-train operations reduce average freight cost. The load factor reflects the mix of commodities as well as the percent of empty car miles. The year dummy variable reflects the intercept differences of the 1983 and 1984 cost functions. The limited number of degrees of freedom prohibits testing for slope differences between 1983 and 1984 and the intercept and slope differences among railroads.

The demand for freight transportation is largely a function of rates. Shipment characteristics, such as length of haul, load factor, and percent of unit-train operation also affect demand indirectly through costs and prices. Therefore, these variables were excluded from the demand function. Competitive truck and barge rates faced by each individual railroad are not available. Marketing expenditures are included in the demand function to identify it from a supply function. A dummy variable is used to reflect the differences in demand among different years.

The linkage between a cost function and a demand function is the firms' pricing behavior. Friedlaender and Spady found that long-run marginal cost pricing without subsidies is both feasible and desirable on higher-density mainline rail routes. They argue that long-run marginal cost pricing for all commodities in the railroad industry is, in principle, feasible but they doubt that either the real world marketplaces or political considerations would actually allow it. Under deregulation, railroads have substantial pricing freedom; hence, since October, 1980, prices have become endogenous variables. Moreover, the actual or potential competition to bring prices down to long run marginal costs is more likely to occur under deregulation and, in fact, rail grain rates have fallen sharply since 1980. As our study is based only on 1983-84 data from deregulated Class I railroads with relatively high traffic density, we assume that pricing behavior is based on long-run marginal costs. This assumption could be very restrictive. In a competitive industry, this will provide a stable equilibrium. A natural monopoly will always result in losses so long as there are too many firms in the industry (Keeler, 1983).

B. Functional Form

There are many functional forms that can meet economic regularity conditions and reflect the characteristics of demand and cost functions of the railroad industry (Berndt and Khaled). Theoretically, one is not likely to identify the best functional form because the underlying true functions are unknown. We selected the translog function as our cost model because the translog function is a second order approximation of Taylor's expansion of an arbitrary function, and also because it was used in almost al previous rail cost estimations, and thus our results can be more readily compared with those of previous studies.

Rail transport services are inputs into other industries. Therefore, the demand for rail services are derived demands from the using industries. Assuming that the cost functions of other industries can be approximated by a second order approximation of Taylor's expansion, the input factor demands for rail services are first order approximations of Taylor's expansions. Hence, we select a Cobb-Douglas model to estimate the demand function, as the Cobb-Douglas model is a first order approximation of Taylor's expansion of an arbitrary function.

Based on Shepherd's lemma, input share functions can be derived from a translog function by taking a partial derivative with respect to the input price.

Based on the envelop theorem, the long-run marginal cost function can be derived from the short-run cost function by taking a partial derivative of the short-run cost function with respect to firm size to find the optimal firm size and inserting the optimal firm size into the short-run marginal cost function to derive a long-run marginal cost function. Hence, the whole system based on a translog cost function is as follows:

$$
\begin{aligned}
SRTC = {} & \alpha_o + \Sigma\alpha_i \ln(r_i) + \alpha_Q \ln(Q) + \alpha_S \ln(S) \\
& + \Sigma\alpha_{SCi} \ln(SC_i) + \tfrac{1}{2}\Sigma\alpha_{ii} \ln(r_i)\ln(r_i) \\
& + \tfrac{1}{2}\alpha_{QQ} \ln(Q)\ln(Q) + \tfrac{1}{2}\alpha_{SS} \ln(S)\ln(S) \\
& + \tfrac{1}{2}\Sigma\beta_{ii} \ln(SC_i)\ln(SC_i) + \beta_{ij} \ln(r_i)\ln(r_j) \\
& + \Sigma\beta_{iQ} \ln(r_i)\ln(Q) + \Sigma\beta_{iS} \ln(r_i)\ln(S) \\
& + \Sigma\Sigma\Theta_{ij} \ln(r_i)\ln(SC_j) + \Theta_{QS} \ln(Q)\ln(S) \\
& + \Sigma\Theta_{Qi} \ln(Q)\ln(SC_i) + \Sigma\delta_{Si} \ln(S)\ln(SC_i) \\
& + \Sigma\Sigma\delta_{ij} \ln(SC_i)\ln(SC_j) + \tau D
\end{aligned}
$$

$$
\begin{aligned}
Share_i = {} & \alpha_i + \alpha_{ii} \ln(r_i) + \beta_{ij} \ln(r_j) + \beta_{iQ} \ln(Q) \\
& + \beta_{iS} \ln(S) + \Sigma_j \Theta_{ij} \ln(SC_j)
\end{aligned}
$$

$$
\begin{aligned}
P * Q/SRTC = {} & \alpha_Q - \Theta_{QS}\alpha_S/\alpha_{SS} + (\alpha_{QQ} - \Theta_{QS}\Theta_{QS}/\alpha_{SS}) \ln(Q) \\
& + \Sigma(\beta_{iQ} - \beta_{iS}\Theta_{QS}/\alpha_{SS}) \ln(r_i) \\
& + \Sigma(\Theta_{Qi} - \delta_{Si}\Theta_{QS}/\alpha_{SS}) \ln(SC_i)
\end{aligned}
$$

$$
\ln(Q) = \lambda_0 + \lambda_1 \ln(P) + \lambda_2 \ln(MKTG) + \lambda_3 D
$$

III. ESTIMATION PROCEDURE AND HYPOTHESIS TESTING

The seemingly unrelated regression technique (Zellner) was used to estimate the cost function, share functions, pricing identity, and demand function as a single multivariate system. The seemingly unrelated regression technique will increase the efficiency of the estimation as long as restrictions across equations are imposed and covariances among equations do not equal zero.

Each variable has been rescaled by its geometric mean which permits us to interpret the results as an approximation to the true underlying cost function in the neighborhood of this point. Prediction outside the data range is, however, of limited meaning. Total cost and demand were deflated by output level and firm size, respectively, to obtain average costs and demand per route mile. Total way and structures costs were deducted from total costs. Hence, we are implicitly assuming that railroad firms are in disequilibrium with respect to the stock of way and structures. Moreover, restrictions across equations were imposed to satisfy the economic requirements of linear homogeneity in input prices and pricing behavior assumption.

The Glejser's test of heteroscedasticity is performed against output level and firm size, respectively. The validity of the model is tested by comparing the predicted values with the actual values of the cost and demand functions.

A cost function corresponds to a homothetic production structure if, and only if, the function can be written as a separable function in its output level and factor prices. Hence, homotheticity of the cost function is determined by testing $\beta_{iQ} = 0$ for all input i. A homothetric cost function can be a homogeneous function if, and only if, the elasticity of cost with respect to output is constant. Hence, homogeneity is determined by testing $\alpha_{QQ} = \beta_{iQ} = \theta_{QS} = \theta_{Qi} = 0$. If all parameters of the second order terms are equal to zero, the translog will be reduced to a Cobb-Douglas. Hence, the fit of a Cobb-Douglas function to the railroad data is determined by testing whether all parameters of the second order terms equal zero.

The elasticities of substitution and input demand elasticities in terms of cost function developed by Uzawa are estimated based on the following equations:

$$
\begin{aligned}
e_{ij} &= (CC_{ij})/C_iC_j) = b_{ij}/(S_iS_j) + 1 \\
e_i &= (b_{ii} + S_i(S_i - 1))/S_i
\end{aligned}
$$

Economies of firm sizes are estimated by taking a partial derivative of the short-run average cost function with respect to firm size while holding the traffic density constant and then adjusted by the way and structures effects. Economies of traffic density and shipment characteristics are estimated by taking a partial derivative of the short-run average cost function with respect to output and shipment characteristics, respectively, and then adjusted by the way and structures effects.

IV. DATA

The data in this analysis were taken from 1983 and 1984 railroad statistics (Association of American Railroads, January 1983 and June 1983). Beginning in 1983, the railroad industry changed from a retirement, replacement, betterment accounting system to a depreciation accounting system; thus, 1983 and 1984 data were not comparable with previous data. Moreover, the Staggers Act was enacted in October, 1980, and assuming the railroad companies needed time to adjust to the deregulated environment, the 1983 and 1984 data are the latest available comparable data which reflect the impacts of railroad deregulation.

Total short-run costs are defined as total annual freight operating costs. Way—the track—and structures costs are the sum of way and structures depreciation and interest. Output level is measured by total revenue ton-miles. Firm size is measured by route miles in service. Percent of unit-train operation is estimated by dividing total unit-train miles. Unit-train miles were collected by the Association of American Railroads (AAR) according to the ICC definition. Marketing expenditures include marketing, advertising, and public relations costs. Price (freight rate level) is measured by revenue per ton-mile. All data were available from the AAR, except marketing expenditures, which were available from the AAR, except marketing expenditures, which were obtained from ICC reports (U.S. Interstate Commerce Commission 1980 and 1981).

There were 30 and 28 Class I railroads in 1983 and 1984, respectively. The Western Maryland Railway Company was included in the reports of the Baltimore & Ohio Railroad Company in 2983. The Elgin, Joliet and Eastern Railway Company was excluded because it is primarily a switching railroad serving the Chicago area. Hence, we have 55 observations for these two years.

V. THE RESULTS

Table 1 presents the results of structure and goodness of fit tests of the model. The tests for homothetic and homogeneous procution structures were accepted. The test for functional form indicates that the translog model fit the railroad data better than the Cobb-Douglas model. The heteroscedasticity test indicates that heteroscedasticity did not exist in the railroad data with respect to either output levels or firm size.

The validity of the whole system is tested by comparing the actual cost and demand values with the predicted cost and demand values. The results indicated that the differences between actual cost and predicted cost, and actual demand and predicted demand were both insignificant.

Table 2 presents the coefficients of the cost and demand models. All first order variables in the short run average cost model were significant at the five percent level. The signs indicate that short run average costs increase as labor and fuel prices and size of firm increase. Short run average costs decline as quantity, length of haul, unit trains and load factor increase. Input share equations can be derived directly from the cost function. The only significant variables in the input share equations are the intercepts and length of haul. Fuel share increases directly and labor share varies inversely with length of haul.

Price—freight rate—was the major variable that significantly explained the variance among railroads' demands. As a large portion of the commodities carried by railroads are relatively low value, high volume commodities, it seems reasonable that shippers' selection of transport mode is based heavily on freight rates.

Table 3 presents the estimated elasticities at the mean values. The elasticity of substitution between labor and fuel and the own factor demand elasticities were insignificant. The estimated elasticities of traffic density and shipment characteristics indicate that average cost declined significantly with increases in traffic density, average length of haul, unit train operations and load factor. Load factor was the most

TABLE 1

Test results of goodness of fit and structure of the model.

Hypothesis	F value
1. Short-run homothetic production structure	0.014
2. Short-run homogeneous production structure	0.764
3. Cobb-Douglas vs. the translog	2.645*
4. Heteroscedasticity	
a. output	0.48
b. firm size	0.32
5. Validity of the model	
a. cost function	0.38
b. demand function	0.23

*Significant at 5 percent level.

TABLE 2
Estimated coefficients of the cost and demand models.

Model	Variable	Parameter estimate	Standard error
Cost	Intercept	-0.338	0.027*
	r_1 (labor price)	0.863	0.034*
	r_2 (fuel price)	0.137	0.034*
	Q (output in ton-miles)	-0.157	0.022*
	S (size in road miles)	0.163	0.030*
	SC_1 (length of haul)	-0.217	0.067*
	SC_2 (percent of unit-train)	-0.112	0.017*
	SC_3 (load factor)	-1.445	0.298*
	Q x Q	0.028	0.011*
	Q x S	0.020	0.026
	Q x SC_1	0.020	0.038
	Q x SC_2	-0.002	0.005
	Q x SC_3	-0.185	0.141
	Q x r_1	0.014	0.056
	Q x r_2	-0.014	0.056
	S x S	-0.019	0.036
	S x SC_1	0.146	0.060*
	S x SC_2	0.016	0.006*
	S x SC_3	-0.313	0.207
	S x r_1	0.035	0.063
	S x r_2	-0.035	0.063
	SC_1 x SC_1	-0.182	0.078*
	SC_1 x SC_2	-0.316	0.011*
	SC_1 x SC_3	-0.388	0.456
	SC_1 x r_1	-0.291	0.057*
	SC_1 x r_2	0.291	0.057*
	SC_3 x SC_2	0.198	0.086*
	SC_3 x SC_3	-0.530	0.785
	SC_3 x r_1	-0.055	0.215
	SC_3 x r_2	0.055	0.215

TABLE 2
Estimated coefficients of the cost and demand models. (Continued)

Model	Variable	Parameter estimate	Standard error
	$SC_2 \times SC_2$	-0.019	0.003*
	$SC_2 \times r_1$	0.001	0.009
	$SC_2 \times r_2$	-0.001	0.009
	$r_1 \times r_1$	0.143	0.142
	$r_2 \times r_1$	-0.143	0.142
	$r_2 \times r_2$	0.143	0.142
	D_1 (year)	0.057	0.026*
Demand	Intercept	-0.358	0.233
	P	-0.592	0.186*
	MKTG	0.029	0.040
	D	-0.057	0.097

*Significant at 5 percent level.

TABLE 3
Estimated elasticities at mean value.

Elasticity	Coefficient	Standard error
1. Elasticity of substitution between		
labor and fuel	-2.18	3.16
2. Factor demand elasticity		
a. labor	-0.21	0.28
b. fuel	-0.68	1.58
3. Elasticity of traffic density	-0.19	0.03*
Elasticity of firm size	0.01	0.04
Elasticity of length of haul	-0.26	0.08*
Elasticity of unit trains	-0.13	0.02*
Elasticity of load factor	-1.73	0.36*

*Significant at 5 percent level.

TABLE 4
Comparison of returns to scale and density from previous studies.

	Economies of density	Economies of firm size	Data base years
Friedlaender and Spady (1981)	-0.14	-0.02	1968-70
Caves, Christensen, and Swanson (1981)	0	-0.01	1955-74
Harmatuck (1979)	-0.48	0.08	1968-70
Harris (1977)	-0.42	-0.03	1972-73
Keeler (1974)	-0.44	-0.01	1968-70
Caves, Christensen, Tretheway, and Windle (1984)	-0.43	0.02	1951-75
Lee and Baumel (1986)	-0.19	0.01	1983-84

important variable affecting average variable costs. The estimated elasticity of firm size were not significant. Although we did not estimate the minimum efficient traffic density, one might conclude that at current traffic levels, the economies of traffic density were not exhausted in 1983 and 1984. The minimum traffic density level where economies of density are exhausted should be higher than current traffic levels in those years.

VI. COMPARISON WITH OTHER STUDIES

Table 4 compares the estimated elasticities of density and firm size of seven studies along with the base data years. The estimated returns to firm size vary little among the seven studies. Four of the six pre-Staggers data studies found substantial returns to density while Caves, Christensen and Swanson (1981) found no returns to density and Friedlaender and Spacy found modest returns to density. Our results show higher returns to density than Friedlaender and Spady but lower than Caves, Tretheway and Windle, Harmatuck, Harris, and Keeler. Moreover, our results confirm that returns to density have not been exhausted under deregulation and returns to firm size are not significant.

VII. CONCLUSIONS AND POLICY IMPLICATIONS

The 4-R Act and the Staggers Rail Act of 1980 established a national goal of railroad revenue adequacy. The ICC has established Constrained Market Pricing guidelines on coal and is developing similar guidelines on non-coal products. These guidelines would mainly achieve revenue adequacy through rate increases on commodities with relatively inelastic demands for rail transport. There is less emphasis on the role of cost reduction in achieving railroad revenue adequacy.

The purposes of this paper are to estimate the cost reduction potential of further structural adjustments in the railroad industry, to identify alternative public policies that could be pursued to help the railroad industry achieve revenue adequacy through lower costs rather than from higher rates, and to identify additional research needs to help achieve these objectives. The reason for the analysis is that previous studies are based on single equation models and out-of-date data that do not reflect the deregulation of and the major structural changes in the railroad industry in the 1970s.

Our results indicate that, on the average, a one percent increase in traffic density will decrease short-run costs by 0.19 percent. These results suggest that deregulation and the massive structural adjustments in the railroad industry from 1950 to 1980 have not exhausted the economies of density. Historically, railroads have attempted to increase traffic density by abandoning rural branch lines. Shippers, in general, have vigorously resisted rail line abandonment. More recently, railroad companies have attempted to increase traffic density by selling low density lines to short line or regional railroads that presumably have lower cost structures. Increasing density through abandonment or sale of low density rail lines has been attractive to railroad management because these methods not only reduce costs and increase profits, but also reduce net investment which further increases return on investment and hence revenue adequacy.

Short-line railroads operating in rural areas have generally not been profitable. However, the Staggers Rail Act of 1980 permits newly formed short line and regional railroads to operate without labor contracts and without honoring prior labor agreements. While these newly formed railroad companies have lower labor costs, they generally face high debt loads and declining agricultural and manufacturing traffic levels since 1980. If successful, these railroads will fill the need of serving lower traffic rural lines and still allow the Class I railroads to lower their average costs by shedding these lines to increase traffic density. If these short line and regional railroads fail, returns to density will likely drive the railroads to further rail abandonment.

Our results show no railroad economies of size. At the same time, the economies of length of haul were significant and substantial. These findings suggest that mergers to increase size have little impact on costs but mergers could have a significant impact on costs if they result in increased length of haul and density.

The most substantial cost reducing option in the model is the load factor. A one percent increase in load factor results in a 1.73 percent decrease in average short-run costs. There are numerous opportunities to increase the load factor. Examples include loading grain and fertilizer cars to capacity, utilizing back hauls such as shipping meat in empty refrigerator cars, shipping fertilizer in empty grain cars, and placing leased rail cars in railroad car pools to reduce the empty mileage. One railroad company recently reported that a sample of 759 randomly selected loaded trailers moving on railroad flat cars on the average used only 60 percent of the trailers' cubic capacity and only 68 percent of the weight capacity (Traffic World, June 10, 1986). While some shippers utilize a limited amount of trailer on flat car shipments, fuller utilization of the capacity would reduce average costs and allow the railroad industry to achieve revenue adequacy at lower rate levels. Moreover, the railroad industry could contribute to the load factor by using incentive and contract rates to encourage fuller utilization of rail car and trailer capacity and by increasing computer-to-computer data exchange among railroads, between railroads and truckers, warehouses and shippers to fully utilize rail car capacity and to reduce empty miles. More computer-to-computer data exchange could also increase length of haul and density as a partial substitute for mergers and rail abandonment.

The economies of unit trains are significant but less substantial than density, length of haul and load factor. Major progress has been made in unit train utilization. One railroad firm had over 50 percent of its total car miles in unit trains.

In summary, the railroad industry has substantial opportunities to reduce costs. Most of the opportunity is in increasing load factor and length of haul. Lesser opportunities are in increased density and further expansion of unit trains. Captive shippers can lower their concerns about railroad pricing powers by helping the railroad industry realize these cost savings to increase profits and revenue adequacy.

VIII. OPPORTUNITIES FOR FURTHER RESEARCH

Economists have substantial research opportunities in transportation even though transportation problems are less severe than they were in the 1970 decade. First, little is known about the cost structure of newly formed short line and the rapidly growing regional railroads. Cost studies of these railroads would provide information on the probability of these railroads successfully filling the vacuum created in rural areas as Class I railroads continue to shrink in size to take advantage of economies of density.

Secondly, no research has been conducted on the impacts of alternative shippers' production and price policies on the structure of the Class I railroads, regional and short line railroads and other transportation industries. These policies could have substantial impacts on the structure and costs of the railroad industry.

Third, as more post deregulation and post depreciation accounting data become available, future research on the cost structure of the railroad industry should emphasize the problem of pooling time series and cross sectional data. Moreover, as additional years of data become available, the additional data will provide the degrees of freedom needed to experiment with models to deal with what Caves, Christensen, Tretheway and Windle refer to as unobserved network effects.

REFERENCES

Association of American Railroads, *Analysis of Class I Railroads 1983*, Washington, D.C., 1984.

Association of American Railroads, *Analysis of Class I Railroads 1984*, Washington, D.C., 1985.

Association of American Railroads, *Railroad Facts*, Washington, D.C., various issues.

Berndt, E. R. and U. S. Khaled, "Parametric Productivity Measurement and Choice Among Flexible Functional Forms," *Journal of Political Economy* 87, (December 1979): 1220-45.

Borts, G., "The Estimation of Rail Cost Functions," *Econometricia* 28, (January 1960): 108-31.

Braeutigam, R., A. F. Daughety, and M. A. Turnquist, "The Estimation of a Hybrid Cost Function for a Railroad Firm," *Review of Economics and Statistics* LXIV (August 1982): 394-404.

Caves, D. W., L. R. Christensen, and J. A. Swanson, "Productivity in U.S. Railroads, 1951-1974," *Bell Journal of Economics* 11, (Spring 1980): 166-81.

Caves, D. W., L. R. Christensen, and J. A. Swanson, "Productivity Growth, Scale Economies, and Capacity Utilization in U.S. Railraods, 1955-74," *American Economics Review* 71 (December 1981): 994-1002.

Caves, D. W., L. R. Christensen, M. W. Tretheway, and R. J. Windle, "Network Effects and the Measurement of Returns to Scale and Density for U.S. Railroads," Working Paper No. 1014, Faculty of Commerce and Business Administration, University of British Columbia, Vancouver, March 1984.

Friedlaender, A. F., and R. H. Spady, *Freight Transportation Regulation,* Cambridge, Massachusetts: The MIT Press, 1980.

Harmatuck, D. J., "A Policy Sensitive Railway Cost Function," *The Logistics and Transportation Review* 15 (1979): 277-315.

Harris, R. G. "Economies of Traffic Density in the Rail Freight Industry," *Bell Journal of Economiccs* 8 (Autumn 1977): 467-82.

Keeler, Theodore E., *Railroads, Freight, and Public Policy,* The Brookings Institution, Washington, D.C., 1983.

Keeler, Theodore E., "Railroad Costs, Returns to Scale, and Excess Capacity," *Review of Economics and Statistics* 56 (May 1974): 201-8.

Sidhu, N. D., Alberta Charney, and John Due, "Cost Functions of Class II Railroads and the Viability of Light Traffic Density Railway Lines," *Quarterly Review of Economics and Business* 17 (Autumn 1977): 7-24.

Traffic World, "ICC Says It Plans Rule for Rail Rate Guidelines on Non-Coal Commodities," No. 5, Vol. 206, Whole Number 4123, Washington, D.C., May 5, 1986.

Traffic World, "ICG Head Warns of 'Phantom 5 Feet' Threatening Rail Intermodal Gains," No. 10, Vol. 206, Whole Number 4128, Washington, D.C., June 10, 1986.

U.S. Interstate Commerce Commission, *Transpotation Statistics for the U.S.,* Washington, D.C.: U.S. Government Printing Office, 1983-84.

Uzawa, H., "Duality Principles in the Theory of Cost and Production," *International Economic Review* 5 (1962): 261-91.

Zellner, A., "An Efficient Method of Estimating Seemingly Unrelated Regressions and Tests for Aggregation Bias," *Journal of the American Statistical Association* 57 (1962): 348-68.

ENDNOTES

* Journal Paper J-11476 of the Iowa Agriculture and Home Economics Experiment Station, Ames; Project 2439.

** Niagara University, NY.

*** Iowa State University, Ames, IA.

The UP-UKT Merger: Operating Plan and Operating Savings

*by Michael B. Godfrey**

ABSTRACT

The application for control of MKT by UP was filed with the ICC on November 14, 1986, and accepted in a decision served on December 16, 1986. Formal proceedings are expected to conclude in December 1987, and a decision is expected early in 1988. This paper summarizes the Operating Plan for the merged railroads. The Operating Plan describes how the two railroads will be integrated and includes transportation, engineering, mechanical, communications, information systems, and staff coordination. It also describes the consequent impacts on service, traffic density, terminal switching, labor force, and costs.

The consolidation will result in service improvements in several corridors, due to shorter, faster routes, new trains, added through blocks, and reduced switching delays. These service improvements will be supported by major investments in main line and yard upgrading, plus the construction of numerous UP-MKT connections.

Normal year postmerger operating and administrative savings are expected to be $61 million, nearly half of which will result from terminal savings ($17 million) and administrative consolidations ($13 million). MKT and UP terminal operations will be consolidated at St. Louis, Kansas City, Wichita, Muskogee, Tulsa, Dallas, Ft. Worth, Waco, San Antonio, and Houston. Abandonment applications have been filed for twelve line segments, totaling 326 miles. Operating and administrative consolidations will result in the elimination of approximately 1,100 positions.

I. INTRODUCTION

Railroad merger applications to the Interstate Commerce Commission normally include an operating plan which describes how the merged railroads will be operated as a single entity after merger. An operating plan addresses such functions as transportation, maintenance of way, maintenance of equipment, communications, information systems, and staff support. This paper summarizes the development and contents of the operating plan for the proposed merger of UP and MKT, and the savings resulting therefrom. The paper concludes with some observations on the role of economies of scale in management planning for mergers, and the realization of any such potential scale economies in actual practice.

In an Agreement dated May 22, 1985, Union Pacific Corporation and Missouri Pacific Railroad contracted with Katy Industries, Inc., and the Missouri-Kansas-Texas Railroad Company to acquire control of MKT. MPRR is a wholly owned subsidiary of UPC, and MKT is 98% owned by Katy Industries. The Purchase and Consolidation Agreement provided for consolidating MKT into MPRR.

The Agreement was contingent on MKT's acquiring at least 60% of some 667,000 Registered Certificates Representing a Charge on Income, dated January 1, 1958, having a total face value of approximately $73 million. MKT held approximately 19% of the Certificates. These Certificates provided that no dividends could be paid on MKT stock until at least 60% of them had been retired; any company acquiring MKT was expected to be subject to the same restriction. When two successive tender offers for the needed 41% of the Certificates were unsuccessful, the Agreement was terminated in August 1985. The Agreement was subsequently reinstated on June 3, 1986, and MKT's third tender offer was successful.

MKT operates 3,147 route miles in six states, while UP operates 21,416 route miles in 21 states. Figure 1 shows the MPRR, UPRR, and MKT routes in their common territory. In terms of freight revenue and revenue ton-miles, MKT is approximately 6% of the size of UP. Table 1 shows these and other comparisons. Of MKT's route miles, 942 miles (30%) consist of trackage rights on other railroads, including 683 miles on UP. In contrast, only 2,101, or 10% of UP's route miles are operated under trackage rights, of which 420 miles are on MKT.

From an operating perspective, the proposed consolidation seeks to accomplish two major goals:

(1) To ensure the long-term viability of MKT's physical plant, a significant part of which, because of UP's trackage rights on MKT, is an integral part of UP's network;

(2) To realize significant operating savings and service improvements through shorter routes, consolidation of facilities and operations, and consolidation of staff, administrative, and management functions.

II. THE PLANNING PROCESS

Section 1180 of Title 49 of the Code of Federal Regulations requires that applications for consolidation or control involving Class I railroads include an operating plan. Section 1180.8 stipulates that an operating plan must describe the resulting patterns of service, impacts on passenger and commuter services, equipment requirements and utilization, plans to eliminate any deferred maintenance or delayed capital investments, gains in service, operating efficiencies, and any other merger benefits.

Thus, an Operating Plan Group was organized, consisting of the following task forces:

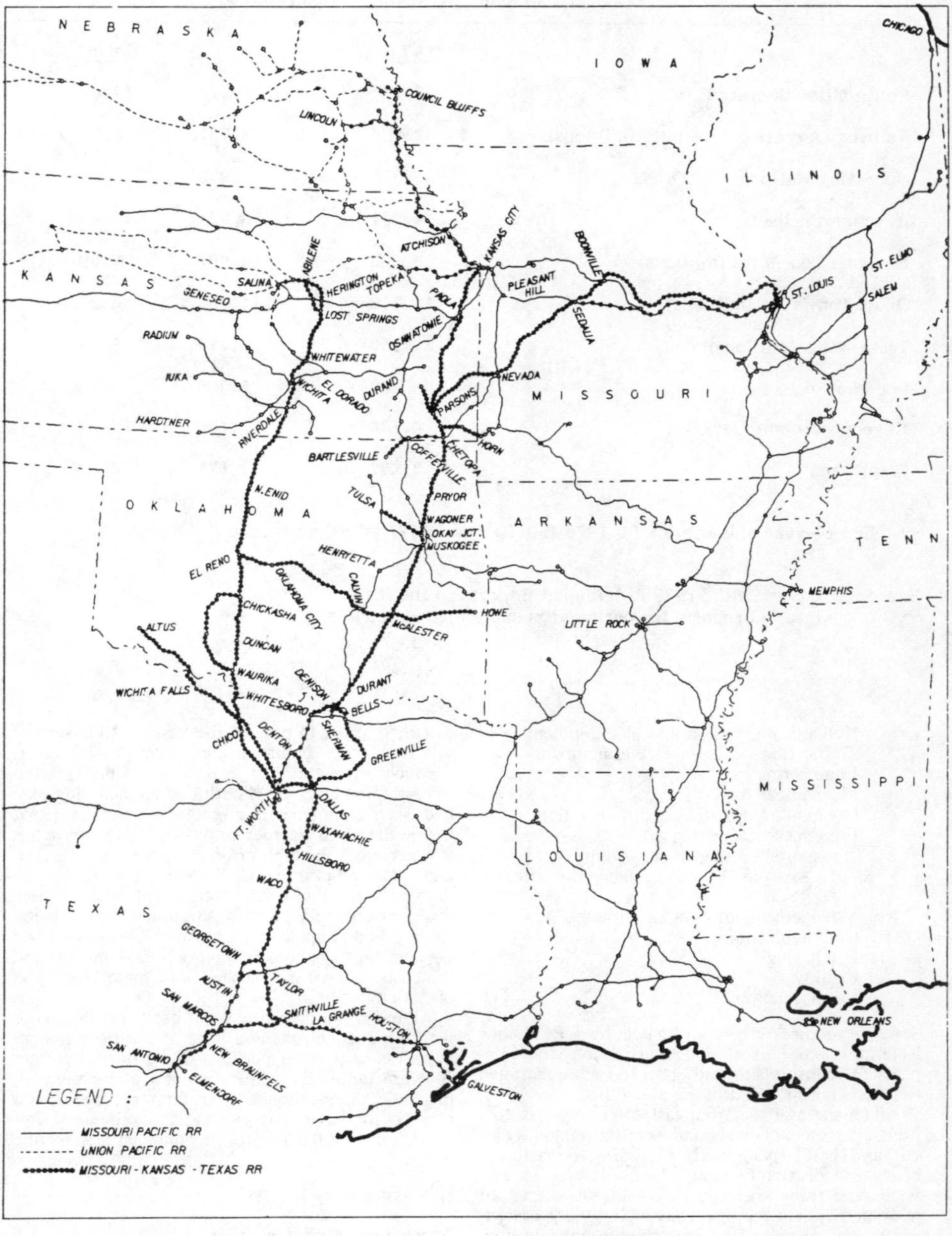

Figure 1. MP, UP, and MKT Routes.

TABLE 1
MKT and UP Summary Statistics (1986)

	MKT	UP	Ratio
Route Miles Operated	3,147 a/	21,416	14.8%
Railway Operating Revenues (millions)	$ 232.4	$ 3,767	6.2%
Operating Ratio	75.5%	79.8%	
Revenue Carloads	354,251	3,286,021	10.8%
Revenue Ton-Miles (millions)	8,100	136,097	6.0%
Gross Ton-Miles (millions)	13,315	271,568	4.9%
Train-Miles (millions)	3.37	37.14	9.1%
Locomotives	262	2,705	9.7%
Revenue Freight Cars	4,136	75,406	5.5%
Employees	2,697	32,675	8.3%

a/ End-of-year value in MKT's 1986 R-1 Report (3,377 miles) corrected for changes to 6-30-87.

Sources: UP and MKT 1986 R-1 Annual Reports to the ICC
Internal reports for carloadings and employee counts

Network integration and abandonments
Traffic flows and network assignment
Engineering
Mechanical
Operating Department administration
Equipment utilization
Transportation area study groups
1. Missouri-Kansas-Oklahoma-Nebraska
2. Texas
Management information systems
Communications
Purchasing
Costing
Labor relations

Separate studies were coordinated by UP's Labor Relations and Personnel Department to develop plans for consolidating all staff and administrative functions not included in the above list.

All Operating Plan Group task forces reported to a full-time project manager and were staffed jointly by UP and MKT personnel, including department heads and district/division officers as appropriate. Each task force was charged with inspecting all relevant physical facilities, reviewing work assignments and force levels, and submitting a written report describing current operations and all recommendations for postmerger operations. Approximately forty individuals participated (part time) in the Operating Plan Group.

The overall objective in planning the consolidated operation of UP and MKT to integrate all functions, personnel, and facilities to the maximum feasible extent, in order to provide the best possible service to the shipping public at the lowest possible cost.

Much of the work of inspecting facilities, determining force levels, and outlining an operating plan had been completed during the summer of 1985, before the initial merger study was halted. When it was resumed, the high-priority tasks were to review and update all prior work, to identify potential abandonments, to acquire current train and car movement data, to resume the numerous terminal consolidation studies, and to update the unit costs used to estimate the costs of track construction or abandonment, and the costs or savings of changes in force levels and operations.

Although the operating plan[1] itself was the major work product, results and information from it were in turn used in other portions of the overall merger study, including the preparation of abandonment applications, applications for terminal trackage rights, analyses of labor, energy, and environmental impacts, and for the pro forma financial statements.

III. METHODOLOGY

A. Definition of Base Year

To quantify merger impacts, such as operating savings, requires that a test, or base period be designated. Operations and performance during this base period are then examined and compared: (1) without the merger, and (2) with the merger. It is important to understand that all analyses of postmerger operations use base-year data, adjusted to reflect *only* merger-related impacts. The same is true for the pro forma financial statements. Thus, for example, no forecasts of traffic changes are used, other than merger-related diversions, and debt and interest payments likewise reflect only base year conditions and anticipated merger impacts.

The base year employed in evaluating the UP-MKT consolidation was 1985. Accordingly, economic costs and benefits were calculated using 1985 cost, wage, salary, and benefits levels. Revenue movement data from UP and MKT's 1985 revenue waybill records were used to analyze the impacts of the Operating Plan on traffic flows and tonnage densities. In order to represent the impacts of route and terminal consolidations more accurately, 1985 revenue movement data were adjusted to reflect all approved and pending abandonment applications, the recent closure of UP's Muskogee Tulsa line, changes in scheduled train routes, and changes in major unit coal train movements since January 1, 1985. After these updating adjustments, the 1985 revenue movement data became the "base-year" revenue movement data. The impacts of all operating changes were measured against this movement data.

Similarly, in order to develop a more accurate Operating Plan and to avoid including efficiencies that had already been achieved through changes in operations since 1985, the most recently available information on actual operations was employed. Accordingly, force levels, train schedules, train densities, blocking plans, job assignments, and equipment fleet sizes reflect June/July 1986 conditions. (MKT train densities reflect January-September 1986.)

Finally, since the traffic diversion study had already been completed in 1985, using 1984 revenue movement data, it was indexed to reflect 1985 traffic conditions and revenue levels. Traffic diversions added 80 car or vanloads per day.

B. Determining Car Flows and Traffic Densities

The base year car-flow data described above were supplemented with diverted revenue movements identified in the traffic diversion study.[2] An empty return factor was applied in the direction opposite the movement of all loads, generating the number and routing of empty car movements. Factors were derived from 1985 MKT and MPRR data for each of fifteen car types. The empty return ratio for trailers and containers was assumed to be 0.162. Together, the loaded movements and the empty returns constituted the total post-consolidation car flows.

A network assignment analysis was employed to determine changes in routings and traffic densities due to the addition of diverted traffic, deletion of traffic lost to abandonments, and rerouting of traffic to more efficient routes. The resulting link and terminal counts were used to check average train sizes and switch engine workloads.

Gross tons (used for computing gross ton-miles) for carload traffic were developed by adding to the net tons involved in each loaded movement (1) the tare weight extracted from the Universal Machine Language Equipment Register (UMLER file), and (2) the same tare weight, multiplied by the appropriate empty return factor to represent the tonnage for the empty return. Gross tons for intermodal traffic were computed assuming that 1.72 trailers or containers would move per flat car.

C. Timing of Traffic Gains and Consolidation Benefits

The Operating Plan considers the impact of phased gains in diverted traffic. Diverted traffic is expected to be gained at the rate of 70% in the first year of unified operation and 100% in the second and succeeding years.

The Operating Plan also considers the impact of phased gains in operating efficiency. Due to the time required to complete planned track and terminal upgrades, to construct needed connections and other improvements, and to negotiate labor agreements, complete coordination and consolidation of operations will not occur immediately upon consummation of the merger. Thus, 40% of the recurring operating savings are expected to be realized in the first year of unified operation, 80% in the second year, and 100% in the third year and thereafter.

IV. NETWORK CONSOLIDATION

The consolidated UP-MKT network will maintain service to all currently-served major market areas, to all major shippers and receivers, and between all significant market pairs. Traffic will be rerouted in some corridors to utilize shorter routes, to improve service, to achieve consolidation economies (including line abandonments), and to avoid significant track upgrading costs that could not be recovered through operating savings. Duplicative terminal operations will be consolidated wherever possible.

The major UP and MKT routes and terminals will be consolidated as follows.

St. Louis-Dallas/Ft. Worth

All MKT traffic using the St. Louis gateway will be rerouted. MKT Dallas, Ft. Worth, and south Texas traffic will be diverted to UP's St. Louis-North Little Rock-Ft. Worth or Houston routes. Some Conrail-MKT, and CSX-MKT Texas traffic currently moving via TRRA and the St. Louis gateway will be diverted to UP's St. Elmo and Salem, Illinois, gateways, respectively, thereby avoiding St. Louis altogether. The balance of MKT's St. Louis traffic—primarily traffic to Kansas and Oklahoma points—will be handled via Kansas City on existing UP trains. This will allow portions of MKT's Sedalia-Parsons line be abandoned, while preserving service to the Kansas City Power and Light Company's Montrose, Missouri, generating station, located 64 miles northeast of Ft. Scott, Kansas. MKT's St. Louis operations will be consolidated into existing UP facilities and into the Alton & Southern Railroad, a terminal railroad in East St. Louis.

Kansas City-Dallas/Ft. Worth

UP and MKT cross at grade at Paola, Kansas, approximately 45 miles southwest of Kansas City, Missouri. Between Kansas City and Paola, MKT does not have its own track, but instead uses trackage rights over Burlington Northern. South of Paola,

MKT and UP diverge, crossing again at Wagoner, Oklahoma. Trains now using MKT's Glen Park Yard in Kansas City will instead originate and terminate at UP's Neff Yard and will operate over the UP route. Two daily manifest trains that bypass UP's Neff Yard will be routed over the shorter MKT (BN) route, along with selected unit coal and grain trains. At Paola, connections will be constructed between the UP and MKT lines so that all route combinations may be used. Glen Park Yard will continue to be used, but only for industry switching and auto loading.

Between Paola and Wagoner, the major route will be the UP line, because it has higher speed limits, longer sidings, better rail condition, and easier grades than the MKT route. Train density will increase by 1.2-2.5 trains per day on this line, depending on location. Unit coal trains for the Grand River Dam Authority generating station at Pryor, Oklahoma[3], and four manifest trains will operate over the MKT line. Parsons Yard will be closed and retired, except for limited industry switching and several receiving tracks for holding or storing trains.

Between Muskogee and Durant, most of the UP line will be abandoned and all traffic consolidated on the MKT line. The MKT route is 30.1 miles shorter and has adequate siding capacity to handle the consolidated traffic. The existing jointed rail will be relaid with new 133 lb. continuous welded rail to accommodate the added tonnage of UP traffic and allow maximum speeds to be raised from 40 or 50 mph to 60 mph. A new siding will be built in MKT's Muskogee Yard.

All traffic, including MKT Dallas traffic currently routed via MKT's Greenville line, will move over the joint MKT/UP route between Durant and Ft. Worth via Whitesboro. Dallas traffic will be handled between Ft. Worth and Dallas on UP's double-track line. This will allow the abandonment of two segments of MKT's Denison-Dallas line, which is laid with 90 lb. rail and has maximum speeds between 25 and 40 mph.

Salina-Ft. Worth

UP and MKT train operations between Salina and Ft. Worth will be consolidated onto a single route, using the following line segments: Salina, Kansas-UP-McPherson, Kansas-UP-Whitewater, Kansas-MKT-Wichita, Kansas-MKT-Enid, Oklahoma-MKT-Ft. Worth, Texas. The consolidated route is 468 miles long and is 25 miles shorter than the current MKT route; 118 miles shorter than the UP route via Durand, Coffeyville, and the MKT trackage rights between Wagoner and Durant, Oklahoma; and 148 miles shorter than the UP route via Durand, Coffeyville, and the UP line to Durant. Six sidings will be extended on the MKT line between Wichita and Ft. Worth.

Ft. Worth-Houston

Between Ft. Worth and Waco, UP has trackage rights on MKT. Thus, the two railroads already share a common route for 85 miles. South of Waco, the two railroads diverge. The MKT route to Houston is 61 miles longer to UP's Settegast Yard (ten miles northeast of downtown Houston), where most of MKT's Houston switching will be relocated after consolidation. Due to UP's shorter mileage, better accessibility to Settegast Yard, higher speed, and availability of excess capacity, all trains will use the UP route south of Waco.

Ft. Worth-San Antonio

Between Ft. Worth and Taylor, both railroads operate on MKT for 162 miles; between Taylor and San Marcos, both railroads operate on UP for 64 miles. Between San Marcos and San Antonio, a distance of 51 miles, the two railroads are located on approximately parallel alignments. A connection at Ogden, 24 miles north of San Antonio, was completed in June 1986 to enable UP unit coal trains destined to the J. T. Deely generating station in Elmendorf, Texas, to use MKT trackage between Ogden and San Antonio. The MKT route must be retained south of Ogden, both for the Elmendorf coal trains and for service to on-line industries (by two through trains and a local).

All San Antonio Yard operations will be consolidated in UP's SoSan Yard, and MKT's Sloan Yard will be closed and retired, except for access to local industries. Most road train operations will also be consolidated on UP between San Marcos and San Antonio, allowing 14 miles of MKT track to be abandoned north of Ogden. Access to MKT industries in New Braunfels will be preserved by retaining MKT trackage in New Braunfels. Terminal trackage rights over SPT have been requested in San Antonio to gain a direct route into SoSan Yard for trains which must utilize the MKT track.

San Antonio-Houston

MKT and UP both operate a daily train pair between San Antonio and Houston. UP also operates a tri-weekly unit train that hauls aggregates to various points in the Houston area from quarries located north of San Antonio. After consolidation, all San Antonio-Houston traffic will be handled via the MKT route, which is 67 miles shorter between SoSan Yard and Settegast Yard. Terminal trackage rights over SPT have been requested in Houston in order to gain access to a direct route into Settegast Yard for trains utilizing MKT. Portions of the MKT route will be upgraded in order to remove existing 10 mph speed restrictions, and two long sidings will be constructed.

The various route consolidations will allow the abandonment of 326 route miles, of which three miles (in New Braunfels, Texas) will be reclassified as industry track. Also, UP plans to spend $73 million on rail relay (342 miles), bridge upgrading, yard rehabilitation, constructing twelve new connections, extending or constructing nine sidings, and miscellaneous other projects.

A. Post-Consolidation Through Train Service

Following consolidation, the changes described below will be made in through freight train operations.

- MKT's extremely infrequent trains from Kansas City to Topeka and its approximately weekly trains to Atchison/Omaha/Council Bluffs will be discontinued. MKT traffic will be handled by daily UP trains.
- MKT's St. Louis—Dallas and St. Louis—Parsons trains will be abolished. Two new trains, operating five days per week, will be

established to provide through-train service between St. Louis (UP Dupo Yard) and Dallas (MKT Mockingbird Yard) via Little Rock, saving nearly a day in comparison to base-period MKT schedules. Other MKT St. Louis traffic will be carried to Kansas and Oklahoma points via Kansas City, or in existing trains to Texas points.

— MKT's Kansas City to Houston and Waco to San Antonio trains, which provide connecting service from Kansas City to San Antonio, will be replaced by a new train operating directly from Kansas City to San Antonio. UP's Ft. Worth to Houston train, which operates approximately once a week, will operate daily. UP's existing Kansas City-Houston train pair will continue to be operated.

— MKT's San Antonio to Waco train will be replaced by a five-day-a-week train from San Antonio to Ft. Worth. UP's existing Ft. Worth-San Antonio train pair will continue to operate daily.

— MKT's Houston to Kansas City train will be replaced by increasing UP's Houston to Ft. Worth train's frequency to daily and adding a second connecting train from Ft. Worth to Kansas City. Normally, this train will be consolidated at Coffeyville with UP's North Little Rock to Kansas City train.

— MKT's Kansas City-Texas train pair, which handles Oklahoma traffic to/from Pryor, Muskogee, and McAlester, will be replaced by a train pair between Kansas City and Oklahoma City operating via MKT's Parsons line six days per week. This train pair will also replace a tri-weekly local which operates between McAlester and Oklahoma City.

— MKT's Texas Specials will be retained, but not as pure TOFC trains. Instead, they will carry all Kansas City-Dallas traffic, on an expedited schedule. They will also carry Tulsa TOFC traffic, which will make an expedited connection at Muskogee, and Ft. Worth TOFC traffic, which will be set out in Ft. Worth, providing much faster handling for this traffic.

— MKT and UP both operate a daily train pair between Houston and San Antonio. These will be retained, with changes in routing (noted in Section 4) and blocking. Aggregates traffic from Georgetown to the Houston area will connect with these trains at Smithville.

— MKT's Herington-Ft. Worth trains will operate daily instead of five days per week between Wichita and Ft. Worth to serve OKT shippers. When traffic levels justify it, this service will be supplemented by locals between Wichita and Chickasha. The Wichita-Herington line will be served as needed from Wichita.

B. Changes in Blocking and Other Service Improvements

The most significant blocking and service improvements are the following:

— North Platte and Kansas City will make a San Antonio block, thus avoiding classification and delay at Ft. Worth.

— Kansas City will make Tulsa, Pryor, and Oklahoma City blocks, avoiding classification at Parsons, Muskogee, and/or McAlester.

— UP Dallas carload traffic from Kansas City, North Little Rock, New Orleans, and points beyond will move direct to Dallas, instead of first being reclassified in Centennial Yard in Ft. Worth.

— Outbound traffic from Dallas to North Little Rock and points beyond will move direct to North Little Rock, instead of being backhauled to Centennial Yard.

— Coffeyville and Ft. Worth will make the Kansas City interchange blocks now made at Parsons, but these blocks will be handled daily, whereas MKT moved only the CNW block daily. Also, MKT northbound traffic to Kansas City from Ft. Worth and San Antonio that was classified in Parsons will move directly to Kansas City.

— Settegast Yard in Houston will make TOFC blocks for San Antonio and Kansas City, thereby expediting handling at these two locations.

— Service to Oklahoma City, now provided three days a week, will be increased to six days a week.

— Service to most OKT shippers will be more frequent, due to increased train frequencies south of Wichita.

— Cars currently interchanged between UP and MKT will spend less time in terminals due to the elimination of interchange delays and the establishment of new through blocks and better connections.

The various service and routing improvements will result in savings of 4.5 million car-miles and 42,000 car-days per year for system cars, equivalent to an ownership of 123 freight cars.

C. Intermodal Facilities

Both railroads have intermodal facilities at St. Louis, Kansas City, Dallas, Houston, and San Antonio. In fact, MKT has two intermodal facilities in Dallas, one of which also handles Ft. Worth TOFC.

At each of these locations, all operations will be consolidated into the UP facility. These consolidations will preserve existing service and utilize existing capacity more efficiently, thus enhancing the consolidated system's ability to compete in the intermodal market. For example, relocating MKT Ft. Worth TOFC traffic from Mockingbird Yard in Dallas to Centennial Yard in Ft. Worth will both improve TOFC service to Ft. Worth and eliminate the cost of maintaining duplicative facilities.

Four of the UP facilities are equipped with lift-on, lift-off equipment, which allows trailers to be handled more rapidly and reduces operating costs. Only one of MKT's ramps is so equipped. Also, all UP facilities can handle containers, while the only MKT ramp that can do so is at Kansas City.

In addition to the locations listed above, MKT has circus ramps at Parsons and Wichita, Kansas; Tulsa and Oklahoma City, Oklahoma; and Denison and Waco, Texas. In order to preserve service to current TOFC shippers and to remain competitive in local markets, these facilities will be kept in operation as long as traffic volumes warrant. Intermodal facility consolidations will result in a reduction of 18 jobs.

D. Terminal Consolidations

MKT's yards are in Kansas City, Parsons, Enid, Denison, Ft. Worth, Dallas, Waco, and Houston. MKT also has auto ramps in Kansas City and San Antonio. UP's major midwest and southwest yards are in North Platte (Nebraska), Kansas City, North Little Rock, Ft. Worth, San Antonio, and Houston. At Mesquite, eight miles east of Dallas, UP has a major intermodal and auto unloading facility. Most UP classification in St. Louis is handled by the Alton and Southern.

MKT and UP terminal functions will be consolidated at St. Louis, Kansas City, Wichita, Muskogee, Tulsa, Dallas, Ft. Worth, Waco, San Antonio, and Houston. MKT facilities at Muskogee, Tulsa, Dallas, and Waco will become the principal yards in those locations, and MKT's Muskogee Yard and Mockingbird Yard in Dallas will be upgraded. Elsewhere, the UP facility will become the principal yard. Other MKT and UP yards that are retained in these locations will be used primarily for industry support. Operations at MKT's yards in Parsons and Denison will be greatly reduced.

The various terminal consolidations will result in the elimination of 17 seven-day yardmaster positions and the equivalent of 47 five-day yard switch-engine jobs. Thirty-three fewer yard locomotives will be needed.

In Houston and San Antonio, UP and MKT are requesting terminal trackage rights on SPT, since in both locations there are no direct connections between UP and MKT that avoid having to "run around" trains when operating between MKT and UP trackage.

V. CLERICAL FUNCTIONS

UP has greatly improved the productivity of its yard office employees by continually adding to the capabilities of its Transportation Control System (TCS). Substantial amounts of clerical work have been automated or eliminated. The combination of TCS and recent advances in telecommunications technology have allowed UP to establish a National Customer Service Center in St. Louis for the entire railroad. This National CSC provides all services to shippers and receivers that were formerly provided by field agencies and local CSC's, including tracing, billing, ordering empty cars, diverting cars, issuing switching instructions, updating demurrage records, and releasing cars. The merger will make possible a net savings of 103 clerical and agent positions due to the combined impacts of yard consolidations, TCS, and bringing MKT customer service functions into the NCSC.

VI. MECHANICAL FACILITIES

MKT has centralized its locomotive heavy repair and component rebuild functions at the Parsons, Kansas, locomotive facility, which was constructed in 1908. Likewise, UP has consolidated comparable work functions for the southern portion of the railroad at North Little Rock, Arkansas. The complex at North Little Rock includes a system locomotive and freight car air brake shop, an automated freight car wheel shop, a semi-automated locomotive wheel shop, and a modern locomotive repair facility that was opened in July 1984. This shop has sufficient capacity without additional investment in equipment or facilities to perform heavy repairs and preventive maintenance for the consolidated UP and MKT locomotive fleets.

In addition, similar locomotive components are rebuilt at both the UP and MKT facilities. This work will also be consolidated at North Little Rock. A central material warehouse at North Little Rock inventories material required for locomotive repairs, and a distribution network is in place to transfer repaired components to outlying running repair facilities.

MKT has centralized its program freight car repairs at Denison in a car shop that was constructed in 1910. UP has two system freight car repair shops to cover the southern portion of the railroad, one at DeSoto, Missouri, and the other at Palestine, Texas. Significant capital expenditures have been made in the last ten years to modernize the DeSoto facility and to purchase new machinery and shop equipment.

Since the car types that were worked during the base study year by MKT and are programmed to be worked over the next three years are similar to those worked in the DeSoto shop, the MKT Denison facility will be closed and the programs transferred to the more modern and efficient DeSoto facility. MKT's freight car wheel work will also be transferred from Denison to the UP's wheel shop in North Little Rock.

Facility and terminal consolidations, plus a reduction of 52 locomotives, will allow 224 mechanical positions to be eliminated.

VII. MAINTENANCE OF WAY

All engineering administration, planning, design, and construction functions now performed at MKT's Denison headquarters will be consolidated either into the UP Omaha office or into UP's Southern and Central Region engineering offices. Signal and maintenance-of-way equipment shop facilities will also be consolidated. A total reduction of 115 engineering positions is planned as a consequence of eliminating duplicative staff and supervisory positions and performing more track maintenance on MKT lines with high-production track maintenance gangs and machines. There will be an annual savings of $1.5 million from using common tie and ballast sources and a consolidated rail-welding contrast, plus a one-time inventory reduction savings of $1.4 million.

VIII. STAFF AND ADMINISTRATION

All MKT departments will be fully integrated into the existing UP organization. MKT's field operating organization will be integrated into UP's existing divisional structure, with MKT territory being added to UP's Illinois, Kansas, Texas, and Gulf Divisions. This will allow a reduction of 33 positions.

An additional 261 jobs will be eliminated by consolidating MIS, communications, accounting/finance, marketing, sales, purchasing, and industrial development.

IX. LABOR IMPACT

The total impact of the merger on the labor forces of the two railroads is a net job loss of 1,120 positions and a transfer of 622 positions. Job losses will result from buyouts, furloughs, and attrition. Of the loss, 859 are Operating Department (including intermodal) positions. Approximately one-half of the transfers are train crew personnel, two-thirds of whom are being transferred due to changes in crew districts and home terminal locations. Many of the details of the labor impact, such as which railroad's employees are affected, the locations to and from which employees will transfer, and the actual number of employee transfers are difficult or impossible to predict, because they will depend on the provisions of the implementing labor agreements, the decisions of individual employees to accept buyouts, and the exercise of employee seniority.

All active nonagreement MKT employees will be offered a voluntary severance program. Those electing not to participate will be offered UP employment, which in most instances will require a change in residence.

For agreement employees, UP and MKT are required by law to apply the labor protective conditions prescribed in the ICC's decision in Finance Docket No. 28250, *New York Dock Railway-Control-Brooklyn Eastern District Terminal*, 360 ICC 60 (1979). These conditions protect displaced and dismissed employees against loss of earnings for up to six years. In lieu of receiving continuing income payments, employees who are furloughed due to the merger can elect to receive a lump sum separation allowance of up to 17 months pay.

The labor impact analysis is stated in terms of full-time position equivalencies, rather than in terms of an employee count. For example, clerical and yard switch engine crew positions are based on working five shifts per week, while agent positions are based on six shifts per week. Road train crew positions are based on the mileage limits used to regulate the number of pool train and engine crews. In some cases, the actual number of employees affected may be greater, due to employees who, for a variety of reasons, may work fewer hours (or miles) than the full-time position would permit.

Average crew sizes for yard switch engines, locals, and through-freight trains were computed from 1985 ICC Wage Statistics reports for MPRR and MKT:yard-4.18, local-4.31, through-freight-4.37. These were used to calculate the labor impacts resulting from changes in yard switch engine shifts and in freight train miles.

The labor impact includes scheduled relief jobs and, where applicable, the number of extra-board positions needed to fill in for vacations, sickness, personal leave, etc. (The train and engine crew extra-board additive was 16%.)

X. OPERATING SAVINGS

Total normal year operating savings after consolidation will be $60,866,000. Of this total, $48,168,000 is accounted for by the Operating Plan, and the remainder by Administrative savings. The Operating Plan total of $48.2 million includes the following amounts:

- — $17.1 million in terminal operations
- — $7.0 million in line-haul operations, including abandonments
- — $10.9 million in equipment maintenance coordination
- — $7.1 million in maintenance-of-way coordination
- — $2.5 million in operating organization coordination
- — $2.5 million in MIS coordination
- — $1.1 million in communications coordination.

Table 2 presents a complete summary of the expected merger benefits.

XI. RAILROAD MERGERS AND ECONOMIES OF SCALE

Having participated in developing the operating plan for the UP-MKT merger, plus having experienced firsthand the UP-MP-WP merger, has been most instructive. The following comments on railroad mergers may be of interest.

A. Why merge?

From a brief review of the railroad literature, one would think that the major economic issue in rail mergers is the presence or absence of scale economies. See, for example, Kneafsey[7], Keeler[8], and Caves, et. al.[9]. While this concern is certainly valid from a national rail policy perspective, it is, at best, a severely oversimplified model of railroad management decisionmaking. For example, both the UP-MP-WP and the UP-MKT merger applications cite several other major purposes that these mergers sought to accomplish, including expanded single system service, the development of new markets for rail services, the enhancement of intra and intermodal competition, and, in the case of UP-MKT, the preservation of MKT's service and route structure. While scale economies are important as a source of service improvements and operating savings, they are demonstrably not the only factor affecting merger decisions.

B. Some comments on scale economies.

First, the extent to which a merger is parallel vs. end-to-end is a major factor in determining the magnitude of any achievable scale economies (e.g., operating savings). The UP-MP-WP merger was primarily an end-to-end merger. As such, economies were realized at a few common points, such as Kansas City, Omaha/Council Bluffs, and Salina. There were, of course, other savings, including reductions in the size of staff departments, equipment reductions, and facility consolidations (such as roadway machine repair). UP and MKT, in contrast, have far more common points and, hence, far greater operating savings, proportionately, than did UP-MP-WP.

Second, regardless of the extent of parallelism, a "core" scale economy savings should result from the consolidation of staff, administrative, and other centralizable functions. Thus, scale economy savings can be categorized into "core" savings, associated with centralized functions, and "parallelism" savings, associated with consolidations at common

TABLE 2
Summary of Benefits
$000

	Year 1 Annual	Year 1 One Time	Year 2 Annual	Year 2 One Time	Year 3 Annual	Year 3 One Time	Normal Year
Revenue Gains:							
Revenue from Diversions	12,764		18,234		18,234		18,234
Revenue from Abandonments	(304)		(434)		(434)		(434)
Total Increased Revenue:	12,460		17,800		17,800		17,800
Operating Benefits: (a)							
Terminal Operations	6,826	12,380	13,653	12,742	17,066	11,595	17,066
Line Haul Operations	2,797	(28,741)	5,594	(23,764)	6,993	2,658	6,993
Equipment Utilization	0	1,139	0	1,139	0	569	0
M of E Coordination	4,345	1,003	8,690	370	10,863	185	10,863
M of W Coordination	2,850	1,481	5,701	38	7,126	19	7,126
Operating Dept. Coordination	1,021	(180)	2,042		2,553		2,553
MIS Coordination	(48)	0	1,966		2,458		2,458
Communication Coordination	444	(404)	887		1,109		1,109
Administration Coordination	5,079	17,232	10,158	384	12,698	192	12,698
Total Operating Benefits:	23,314	3,910	48,691	(9,091)	60,866	15,218	60,866
Labor Protection	(23,749)		(28,207)		(17,687)		0
Total Benefits:	12,025	3,910	38,284	(9,091)	60,979	15,218	78,666

(a) Net of additional cost of handling diverted traffic.

points. While further distinctions could be made, I simply want to demonstrate that analyses of railroad mergers should look deeper than plain vanilla economies of scale.

Third, I cannot emphasize too strongly the role of management in achieving potential scale economies. It is easy to conclude from a distance that, for example, the merging railroads' accounting, purchasing, planning, and public relations functions can be consolidated with major savings in force levels. But management has to take the initiative to translate this potential into hard reality, as UP's management has been able to do since consummation of the UP-MP-WP merger. UP has implemented all administrative and departmental consolidations specified in the UP-MP-WP merger application. Economies of scale do not occur spontaneously; a company is not a "black box" which automatically follows an economic equation of state. Rather, the savings resulting from a merger depend critically on the ability of management to achieve them.

Fourth, there is the time dimension. Economies of scale take time to achieve, even with the requisite management skills. The UP-MKT merger operating plan recognizes this, as explained in Section 3.3.

Finally, technology also plays a part. A portion of the operating savings projected for the UP-MKT merger results from implementing UP's more advanced computer and communication systems on MKT. This nicely illustrates Kneafsey's[7] conclusion that larger railroads tend to have more highly developed technology.

ACKNOWLEDGEMENT

Developing the Operating Plan required the concentrated efforts of numerous individuals at MKT and UP, too many to mention here. Their work is hereby acknowledged. Special thanks to Dale Salzman, who was the Project Manager for the initial development effort in 1985.

ENDNOTES

* Union Pacific Railroad.

1. The Operating Plan appears in the merger application as Appendix E of Volume 1 (MP/KT-9), ICC Finance Docket No. 30800.
2. For a description of a traffic diversion study, see Randall R. Reyff and Bengt Muten, "Estimating Revenue and Cost Changes Resulting from the Southern Pacific and Santa Fe Merger," TRF Proceedings, Vol. XXVII, No. 1, pp. 86-91.
3. GRDA has recently signed a transportation contract with BN; however, MKT is contesting its legalty.
4. The Lower Colorado River Authority and the City of Austin own a generating station at La Grange, Texas, served by MKT.
5. Oklahoma Gas and Electric.
6. Public Service of Oklahoma.
7. J. T. Kneafsey, "Mergers, Technical Change, and Returns to Scale in the Railroad Industry," Proceedings of the Thirteenth Annual Meeting of the Transpn. Res. Forum, Volume XIII, No. 1, pp. 439-457, 1972.
8. T. H. Keeler, "Railroad Costs, Returns to Scale, and Excess Capacity," Review of Econ. & Stat., Vol. LVI, No. 2, pp. 201-208, 1974.
9. D. W. Caves, L. W. Christensen, and J. A. Swanson, "Productivity Growth, Scale Economies, and Capacity Utilization in U.S. Railroads, 1955-1974," Amer. Econ. Rev., Vol. 71, No. 5, pp. 994-1002, 1981.

A Branch and Bound Procedure for Optimal Train Dispatching

*by Edwin R. Kraft**

ABSTRACT

This paper presents a deterministic train dispatching algorithm for minimizing delays. The algorithm is suitable for use in real-time applications in train dispatching centers, as well as for planning use as part of railroad line capacity simulators.

Simulation test results indicate the procedure can reduce train delay by as much as 55% under heavy traffic conditions. At more moderate traffic levels a savings of 15% represents a more realistic expectation. Performance is relatively insensitive to random train speed fluctuations expected to be encountered in a real-world operating environment.

The paper outlines major design goals, limitations and capabilities of dispatcher assist systems and the simulation techniques upon which they are based. A probability model of train delay is derived to show how dispatching decisions can be made on a continuous, sliding scale basis to minimize the weighted average of train delays. Finally, a combinatorial train dispatching algorithm is presented, and its performance is compared with a previously existing "local optimization" technique.

I. INTRODUCTION

This paper will present a deterministic train dispatching algorithm for minimizing delays. The algorithm is intended for real-time applications in train dispatching centers, as well as for planning use as part of railroad line capacity simulators. Although test runs were performed using a CSX in-house model, the procedure is adaptable to any deterministic train dispatching simulator. It can be used with either conventional TCS technology, or with any of the new Advanced Train Control System technologies.

II. GOALS FOR DISPATCHER ASSIST SYSTEMS

To succeed, any train dispatcher assist program must be both cost effective and user-friendly. It should be designed with at least the following in mind:

- *Safety.* In evaluating the safety of any given system, one must examine how the system is used by dispatchers on a daily basis. Convenient means must be provided to accomplish all routine operational functions. Otherwise, dispatchers may use "tricks" to manipulate the computer. System functions may be used in a potentially unsafe manner, not anticipated by the system designers.
- *Response Time.* Probably the most common (and also most easily avoided) cause of failure of dispatching projects is an overworked central computer. Some systems have required several minutes simply to clear a train from a siding. In one case, it has been faster to maintain manual records rather than key the information into the computer. This results in the deactivation of any CPU-intensive programs and reversion to manual methods.

  Considering the declining cost of computer hardware, and the fact that the computer comprises only a few percent of total project cost, this is the last place one should attempt to economize. **Fast response time** will be especially critical to the success of any system making extensive use of graphics and optimization techniques.
- *Accuracy.* A high percentage of the decisions the system proposes must be "correct" and satisfactory to the dispatcher. If the system's proposals do not make sense, the decision making logic will simply be turned off and the whole effort will be wasted.

  Even in the worst case circumstance, the system should not "lock up" the railroad. A lock up occurs when two or more trains are facing each other on single track, with no moves possible until one or more trains backs out of the way. Such an occurrence is particularly embarrassing when management has claimed that a given system is capable of operating the railroad "hands off." A severe loss of both management and system credibility is the result. The system must be designed to interact with, not replace human intelligence. It is most important to eliminate the "surprise factor". The system should be capable of simulating far enough in advance that it will help the dispatcher anticipate congestion **prior to its occurrence**. After presenting possible solutions, it should allow ready interaction to permit the plans to be changed. Once an operating strategy has been selected by the dispatcher, it should be automatically implemented without exception. The system should monitor compliance with the plan and alert the dispatcher to review the situation if a significant deviation occurs.

III. LIMITATIONS AND CAPABILITIES OF DISPATCHER ASSIST SYSTEMS

The train dispatching problem has traditionally been modeled using one of three methods:

- *Event Oriented Simulation.* This process yields a detailed simulation based on track configuration, signal characteristics and train performance capabilities. The results are often plotted on a time-distance diagram. Most operating personnel can readily understand and relate to the analysis.

Normally at least two runs, and frequently more, are performed in order to determine how delay varies as a function of volume or track configuration. Sometimes a formal statistical analysis is performed on simulation results. One must be certain to simulate a long enough period to assure statistical convergence of results. Otherwise, a comparison between two simulations "Before" and "After" may be misleading. Simulation is the preferred technique for evaluating major operating changes where a definite impact can be detected within a reasonably short simulated time period.

- *Statistical Analysis*. Simulated or actual train movement data can be analyzed using statistical tools to reveal functional relationships among the underlying parameters. For example, one might want to perform a multiple regression analysis to determine the relationship among train delay, train volume and train length.

  One must always be careful when drawing conclusions from a statistical analysis. Simple correlation does not necessarily imply a cause-and-effect relationship.

- *Probability Models*. Probability theory can be used to construct mathematical models of train delay. Such models establish a direct cause-and-effect relationship through rigorous mathematical derivation.

  Probability models can be quite useful as quick screening tools. They are also the preferred tool for the evaluation of smaller investment proposals: The reason is that statistical sampling error in a simulation can easily mask the true benefits of a small investment. Although a probability model may contain some simplifying assumptions, the analysis is completely free from sampling error.

  The branch and bound technique is not a modeling procedure in itself, but it can be used in conjunction with event-oriented simulation to optimize train meeting patterns to minimize delay.

  In "locally optimized" simulation, delay minimization is accomplished on a one-at-a-time individual train meet basis, subject to the constraints imposed by various priority schemes. By viewing the problem in the classical mathematical programming framework, train dispatching constraints can be handled in a more eloquent, exact way.

  The branch and bound technique defines an efficient framework for the development of a combinatorial train dispatching algorithm. Using the branch and bound technique, the train dispatching problem can be solved as a constrained optimization problem having the following characteristics:

Objective function:
Minimize (weighted) sum of train delays.

Subject to:
- All trains meet their schedules.
- No crew exceeds 12 hours of service.
- All industries or mines are served.

Limitations: A "feasible" solution, meeting all constraints, may not exist for every dispatching problem. For example, it may be impossible for a local switcher to serve all industries and still make it back to the yard within the 12-hours of service limitation. Or, the dispatcher may be faced with a choice between delaying a piggyback train, causing it to fall behind schedule, or delaying a coal train, causing it to exceed its 12 hours of service limitation.

A computer can simulate each option and clearly present the options to the dispatcher. By simulating far enough in advance, the computer might identify the possibility of outlaw and precise location so that a relief crew could be standing by when the event actually occurs. Or it might allow the dispatcher sufficient time to "negotiate" with the next division, to see if it would be possible for the piggyback train to make up the lost time.

Human decision making also is critical in the handling of local switching items and maintenance of way activities. These are "big ticket" items where errors can be very costly. Many local switching crews have agendas which are purposely designed to approach 12 hours of service. Accomplishing all scheduled activities within a 12-hour time frame may require a higher train priority than what might otherwise be indicated. Dispatching decisions might be referred to humans based on the following criterion:

- Any time a dispatching constraint must be violated.
- Work items on single track.
- Work items on multiple track exceeding a specified duration.
- Any projected delays to trains with extensive switching agendas.

Capabilities: Although some limitations exist, requiring a continued human presence in dispatching centers, many factors **can** be considered in automated analysis of routine conflicts. These factors include:

- *Train fleeting*—Existing "local optimization" algorithms tend to bunch trains into fleets, whereas in validation comparisons, human dispatchers often try to maintain spacing between following trains. Bunching of trains on a single track line can lead to a large increase in total delay as shown in the following figure.

  Once the fleets have formed, the local optimization algorithms do a poor job of efficiently meeting the fleets to minimize overall train delay. Human dispatchers tend to minimize the complexity of their problems by preventing the bunching in the first place.

  As compared to local optimization, the branch and bound procedure will do a much better job of dealing with the natural bunching of trains which occurs on high density single track lines. Once bunching occurs, the program can efficiently resolve complex multiple train interactions. Or, if future congestion is projected, the

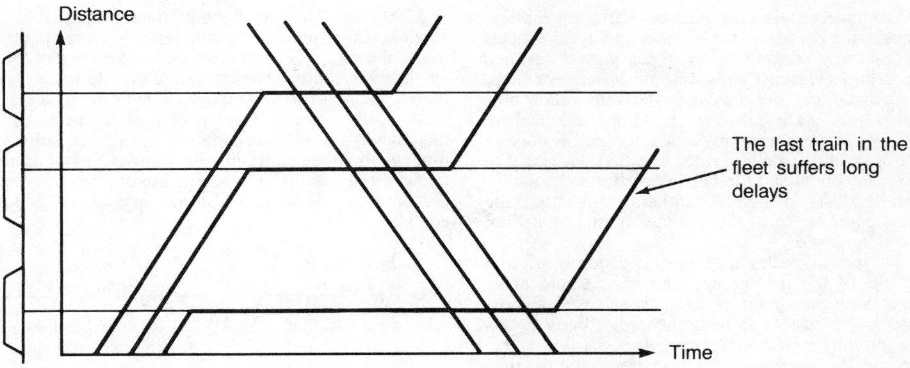

Figure 1. Destruction of a Fleet

system may decide to keep trains spaced rather than allow bunching to occur.

- *Yard Capacity*—One component of train delay is time loss waiting to be received at a destination rail yard. If two or more trains follow each other with insufficient spacing, the yard may not be able to receive all the trains as they arrive. Dispatchers, aware of the limited capacity of yard facilities, try to space train arrivals into destination yards.

  Some rail lines act as "conveyor belts" between two rail yards: all trains originate and terminate at the same yards. These lines are strongly impacted by destination yard capacity. Traffic flow must be uniform with minimal train fleeting on such lines. Division management endeavors to dispatch trains at uniform intervals, and the dispatchers must try to break up any fleets which develop.

  Other lines handle trains originating and terminating at many different yards. These lines are less impacted by destination yard capacity, but may experience higher peak loadings due to uncoordinated train originations.

  The branch-and-bound dispatching procedure can be calibrated for either type of rail line by including minimum train spacing requirements in the train delay calculation.

- *Lock-ups and loops*—The train dispatching problem can become quite intricate, particularly when trains having multiple combinations of restrictive work items and schedule constraints interact. Various model structures and complex decision making rules have been devised in an attempt to produce highly reliable dispatching logic. However, most of these models continue to be plagued with unfortunate and frustrating lock-up and logical looping conditions.

  The branch and bound algorithm promises to offer greatly increased reliability of model operation. Should an inadvertent loop or lock-up condition develop in one of the branches, a feasible non-looping solution is still likely to be developed out of one of the others.

  In practice, lock ups and loops generally result from complex dispatching situations, the resolution of which, at best, usually generates substantial train delay. Therefore, the probability of discarding an optimal solution from a looping condition is quite low.

- *Train schedules*—Train schedules, 12 hour time constraints, and various trains' time values can be considered exactly and explicitly. If the program generates a meet decision which results in violation of any constraint, this branch is immediately ruled out. Once an infeasibility occurs, further simulation is terminated.

  By restricting the number of feasible outcomes, added constraints actually simplify the problem to be solved and reduce program execution time.

  In this context, it is important to differentiate the impacts of schedule constraints from those of train priorities. Priorities, as defined here, arise out of the different values of time of various trains. (A longer train, or one with more locomotives, should receive a heavier weighting. Weightings may also appear if the traffic carried by one train is more time sensitive than that carried by another.) Because of the limitations of local optimization dispatching models, it has been necessary to model schedule constraints indirectly using artificially constructed absolute priority schemes. Explicit treatment of schedule and other constraints eliminates the need for such approximations. Train delays can be traded off on a time-value basis to minimize total cost.

IV. TRADING OFF TRAIN DELAYS

When all schedule constraints are satisfied, the branch and bound procedure automatically minimizes the weighted average of train delays. These tradeoffs are made on a continuous, "sliding scale" basis rather than according to a discrete, absolute priority scheme.

The tradeoff takes into account differing running times of the opposing trains, acceleration time losses for restarting and pulling out of passing sidings, and the values of each train's time in dollars per hour.

An analytical formula can be derived to show how such a tradeoff is made in the simple case where only two trains are involved in a conflict. To derive this optimal dispatching rule, it is first necessary to derive an analytically tractable probability model of train delay. Then, by differentiating and setting the result equal to zero, the optimal train dispatching rule is obtained.

One way to reach a train meet decision is through the use of the first come, first served rule. This dispatching policy serves as a "base case" against which other plans can be compared. However, this rule minimizes weighted train delay *only under certain specific conditions*: for example, when the running times of the opposing trains are equal, when the acceleration/pullout penalties are equal, and when each train has an identical value of time in dollars per hour. If these conditions are not met, it may be desirable to introduce a "time adjustment" parameter to clear the second train instead.

The arrival times of two trains can be compared by subtracting one from the other. If the difference is positive, train A will arrive first; if negative, train B will arrive first. **The proper train to clear first can be chosen by comparing this computational difference with a (nonzero) threshold value.** If the difference is greater than the threshold, train A is cleared; if less, or negative, train B is cleared. Specification of a nonzero threshold, called "time advantage," appropriately generalizes the rule for the effects of different single track running times, acceleration penalties and time values.

A nonzero threshold gives one train an "advantage" over the other, because the favored train can arrive second but still be cleared for first movement. For example, suppose Train A has a 5 minute "time advantage" over Train B. Then, train A may be expected to arrive up to 5 minutes later than B, but A would still be cleared onto the single track. If the threshold is zero, then the first come, first served rule is applied.

The appropriate time advantage for local resolution of any conflict is determined from the equation given below:

$$^{A}OPT = \frac{W_B\,(R_A + D_{min_B}) - W_A\,(R_B + D_{min_A})}{(W_A + W_B)}$$

where:

^{A}OPT	=	Optimal time advantage for train A, in minutes.
W_A	=	Time value weighting of train A
W_B	=	Time value weighting of train B
R_A	=	Single track run time of train A, in minutes.
R_B	=	Single track run time of train B, in minutes.
D_{min_A}	=	Acceleration Penalty to A if train A is delayed, in minutes.
D_{min_B}	=	Acceleration Penalty to B if train B is delayed, in minutes.

Example

An eastbound merchandise train valued at $600/hour is in conflict with a westbound coal train valued at $200/hour. The merchandise train will cross the single track segment in 40 minutes, while the opposing coal train requires an hour. At the west end of the single track section is high speed double track, so the merchandise train would be able to begin acceleration to full speed immediately after the meet. The merchandise train's acceleration penalty is only 3 minutes, but the coal train would have to pull out of a low speed passing siding. The coal train's penalty is 20 minutes. What is the proper time advantage to apply?

Merchandise Train A	Coal Train B
$W_A = 600$	$W_B = 200$
$R_A = 40$ min	$R_B = 60$ min
$D_{min_A} = 3$ min	$D_{min_B} = 20$ min

$$^{A}opt = \frac{200\,(40+20) - 600\,(60+3)}{600 + 200}$$

$$= -32.25 \text{ minutes}$$

Suppose the merchandise train arrives exactly 32.25 minutes later. Then the dispatcher should be indifferent to which train is cleared first: the delay costs should be equal. **Let's test this.** Suppose the merchandise train is cleared:

Coal train delay = 32.25 + 40 + 20 = 92.25 min
92.25 ($200)/60 = $307.50 cost of delay

If the coal train were cleared:

Merchandise train delay = -32.25 + 60 + 3 = 30.75 min
30.75 ($600)/60 = $307.50 cost of delay

Using the formula, the optimal time advantage can be computed to minimize delay, simultaneously considering differing values of train time, different running speeds and acceleration penalties. For those who are mathematically inclined, the derivation of this rule is contained in the Appendix to this paper.

V. BRANCHING PROCEDURE

For any set of conflicts, all train meeting possibilities can be defined by a set of sequential train meeting instructions. For four conflicts, the following sixteen instruction lists define all possibilities: EEEE, EEEW, EEWE, EEWW, EWEE, EWEW, EWWE, EWWW, WEEE, WEEW, WEWE, WEWW, WWEE, WWEW, WWWW. For example, the third list, "EEWE", says: resolve the first two conflicts in favor of the eastbound train, the third conflict in favor of the westbound, and the fourth in favor of the eastbound. In theory, the optimal strategy could be found by simulating all sixteen possibilities completely through from beginning to end. In practice, this approach would not be very efficient. The branch and bound procedure reduces the required computer time by saving intermediate results and by terminating the examination of alternatives as soon as it becomes obvious they cannot be optimal.

When a conflict is recognized, the model status is archived, creating a new branch. Once the original data has been saved, the meet is first resolved in a locally optimum manner. Then the original data is reloaded so that the conflict can be resolved in favor of the other train. As the simulation proceeds, a tally of total weighted train delay is maintained for each active alternative branch. Also, at each step, the model status can be checked against the predefined schedule and hours of service constraints. The most promising active alternative is then selected for further examination. Figures 2 and 3 graphically display this process.

As the number of meets occurring within the defined planning horizon increases, the number of possible train meeting strategies increases **exponentially**. Two sidings, one on either side of the section of single track where the conflict occurs, are potential train meeting locations. Because there are two possible resolutions for each conflict, there are 2^n possible meeting strategies, where n is the number of meets to be planned.

The number of train **meets**, in turn, depends upon the number of **active trains** and upon their **relative positions** on the rail line. The number of train meets, **n**, is not the same as the number of trains.

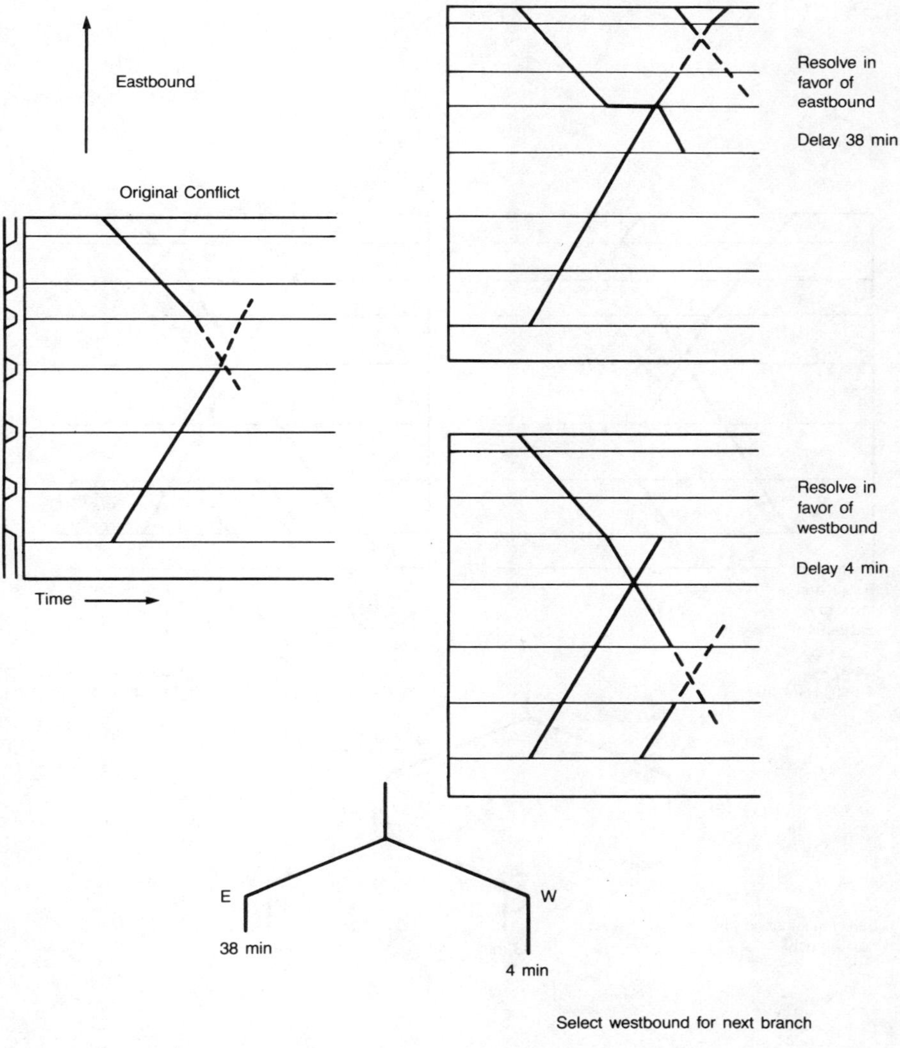

Figure 2. Branching Procedure for Train Meet Optimization

For example, if trains A and B eastbound are to meet train C westbound, there are three trains but only two meets: A meeting C, and B meeting C.

While in theory a large number of train meeting combinations are possible, in practice it is not necessary to simulate all the possibilities to obtain good results. **Since the algorithm is built upon local optimization, the optimal solution is almost always found within the first 20-30 branches.** By limiting the number of branches to 30, substantial savings in computer time are possible with negligible degradation in performance.

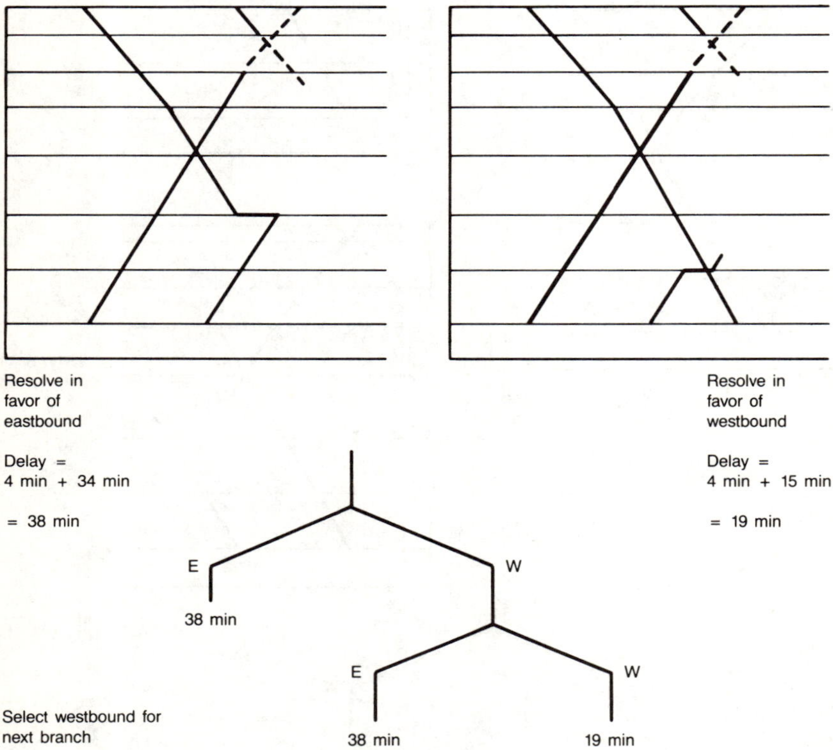

Figure 2. Branching Procedure (continued)

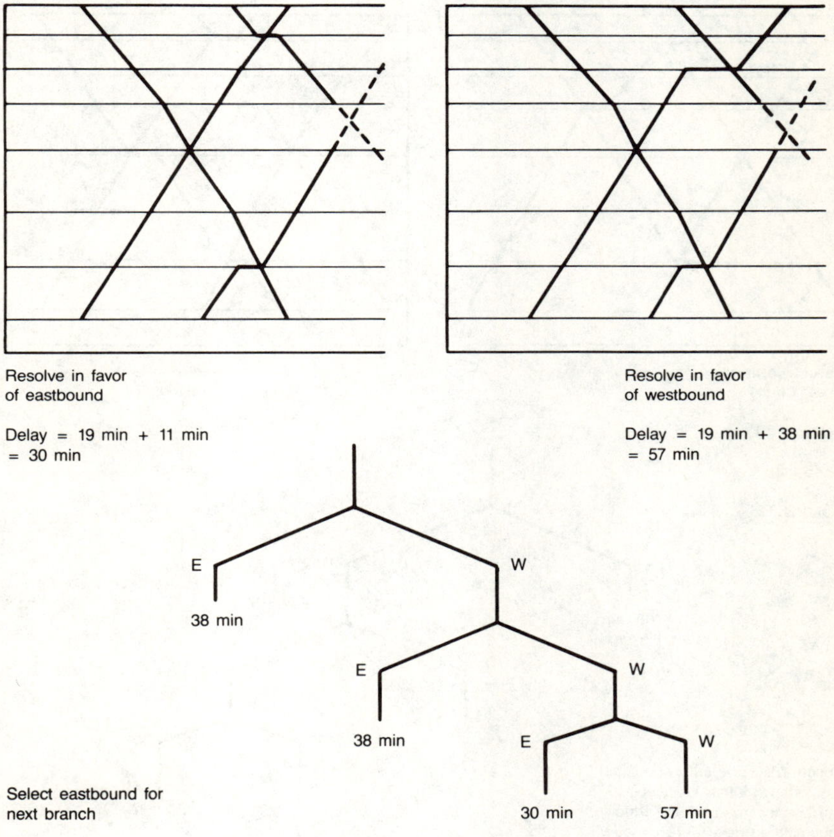

Figure 2. Branching Procedure (continued)

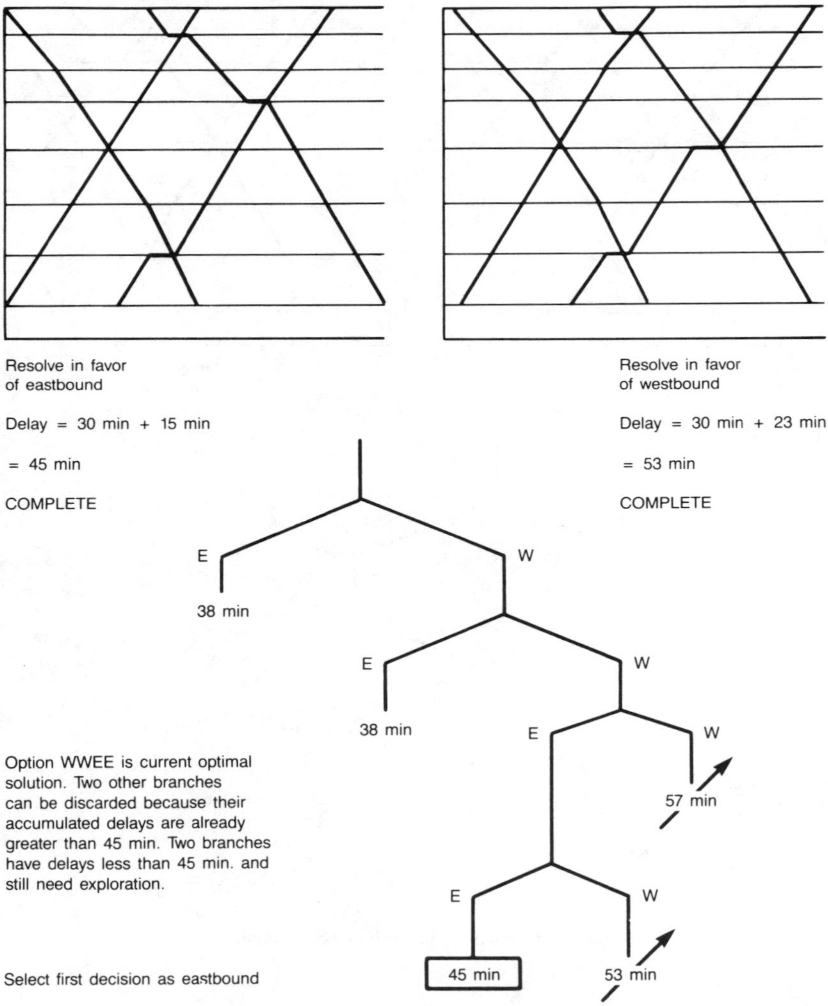

Figure 2. Branching Procedure (continued)

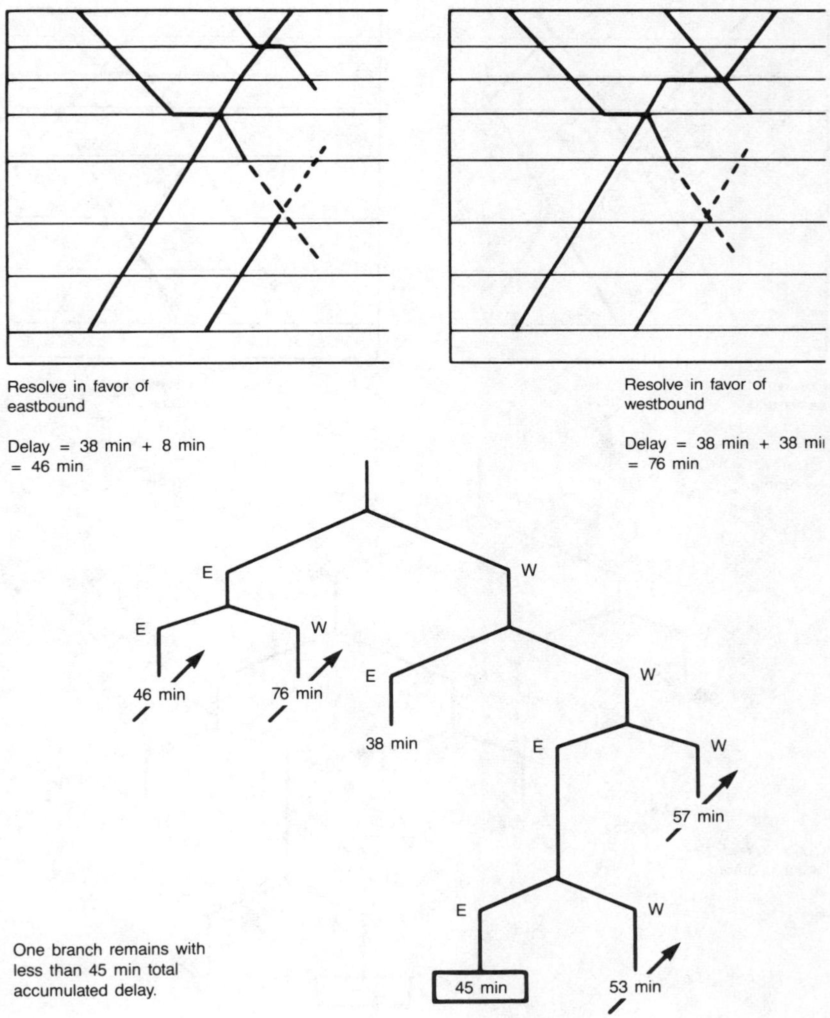

Figure 2. Branching Procedure (continued)

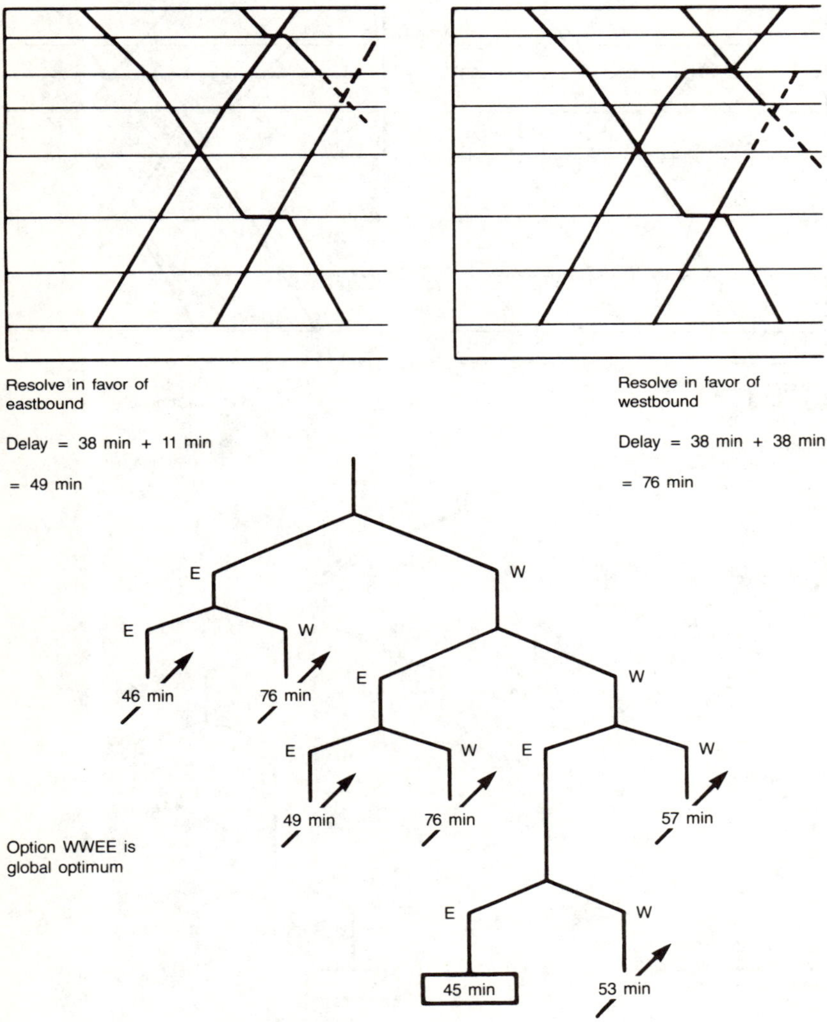

Figure 2. Branching Procedure (continued)

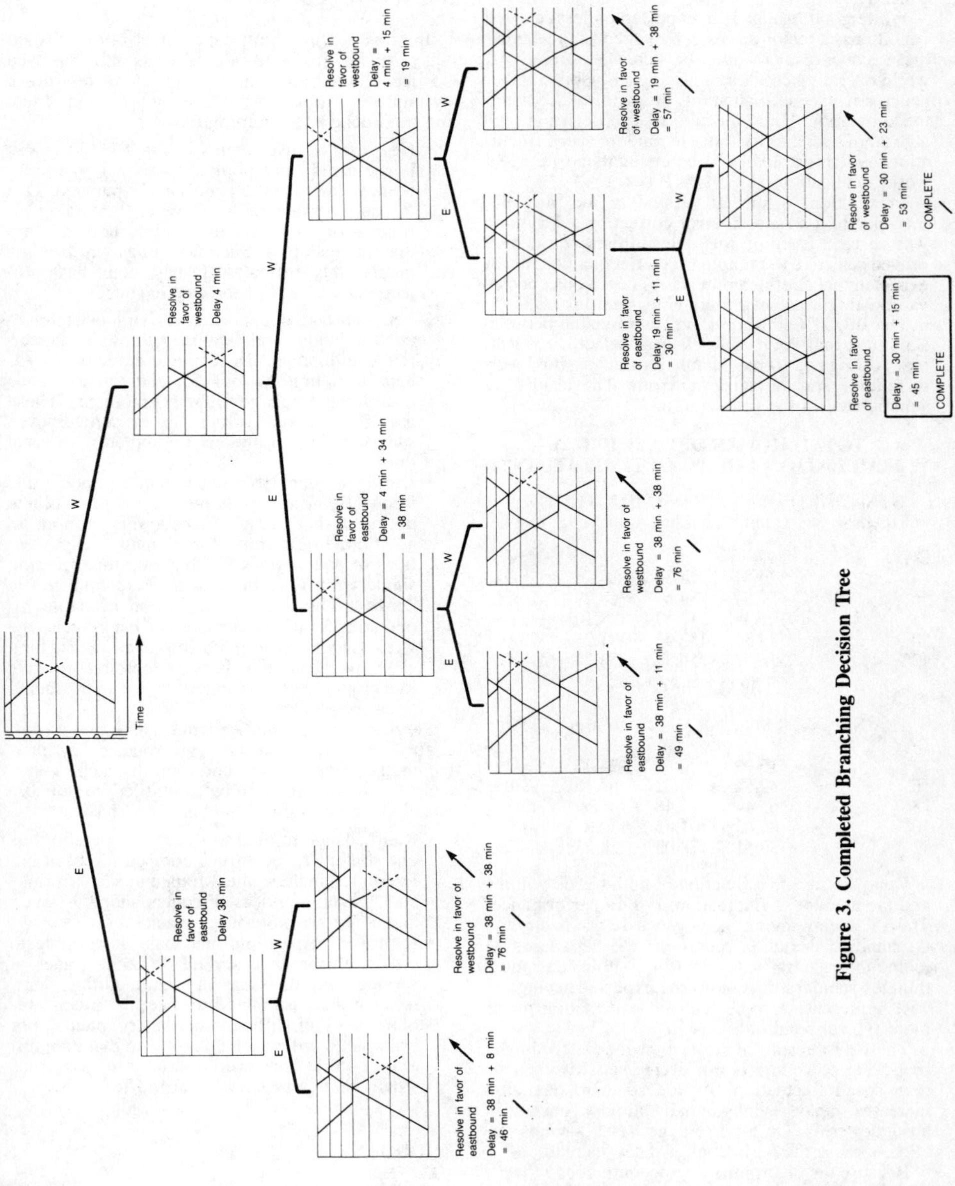

Figure 3. Completed Branching Decision Tree

VI. SIMULATION TEST RESULTS

To estimate the performance of the branching procedure relative to a local optimization technique, the procedure was implemented in FORTRAN, and a simulation model to test the program was constructed.

While good results were expected under idealized conditions, a major question remained as to whether these same results could be achieved under real world operating conditions. Using a random number generator, speed fluctuations were introduced into the simulation. The objective was to determine the algorithm's sensitivity to these random speed fluctuations which are likely to be encountered in a real world application.

In addition, a special version of the program, incorporating a random error correction subroutine, was tested. Even if the deterministic procedure proved sensitive to random speed fluctuations, it was expected that a refined procedure, based on expected values, might achieve more consistent results.

The B&O "Old Main Line" subdivision between Baltimore and Point of Rocks was selected for testing. A series of 10-day simulations at varying levels of volume and variance were run. The results are summarized in the following table:

TOTAL HOURS DELAY: 10 DAY SIMULATION LOCAL OPTIMIZATION

	STANDARD DEVIATION (MINUTES/HR.)			
# TRAINS/ DAY	0 Min.	9 Min.	16 Min.	21 Min.
4	20.54	24.39	25.76	28.94
6	54.84	59.67	68.29	74.51
8	101.38	117.57	128.10	142.08
10	133.51	152.23	171.23	178.61
12	312.07	350.31	353.76	410.16

BRANCHING

# TRAINS/ DAY	0 Min.	9 Min.	16 Min.	21 Min.
4	16.79	20.67	24.47	27.25
6	42.52	50.52	62.10	65.03
8	65.44	84.46	101.59	112.13
10	85.65	101.43	132.81	147.04
12	139.51	178.05	228.66	258.29

Delay savings is a function of both traffic volume and the amount of fluctuation in train performance. The 55% delay savings achieved at 12 trains/day and no standard deviation represents the best case. At more moderate traffic levels, (e.g. 6 trains/day and 9 minutes standard deviation) the expected savings of 15% represents a more realistic expectation under more typical conditions.

These test results indicate that the deterministic branching procedure **is not overly sensitive** to random speed fluctuation. As the standard deviation increases, delay savings remains almost constant in absolute terms. On a percentage basis, savings decline because the base level of delay increases.

Because the deterministic procedure does so well, a sophisticated error correction procedure, based on expected values, **is unnecessary**. After considerable effort, we have so far succeeded in reducing expected delay by only an additional 2% through use of an expected-value based procedure.

VII. CONCLUSIONS

Improved dispatching capabilities offer substantial prospects for both reduced costs and improved service. Capital productivity and rate of return can be improved by increasing line capacity and reducing train delays to a minimum.

- New capital investments in fixed facilities can be deferred, and plant rationalization opportunities can be more aggressively pursued. Due to past abandonments, however, remaining rerouting options tend to involve more circuity than in years past. Such reroutings will become increasingly difficult to justify, as the easy projects have already been implemented.

  Yet, substantial opportunities still exist to reduce redundant rail facilities through intercarrier coordination. To facilitate agreement, perhaps it is time to look for alternatives to the traditional trackage rights agreement. These may include joint ownership, or perhaps even sale of joint facilities to a completely external third party.
- During summer track maintenance periods, it is frequently necessary to move 24 hours' volume during a short 12 to 16 hour time period. In addition, Engineering forces may occupy sections of double track for days at a time, forcing single track operations in areas of high traffic density. The track configuration available for operating trains is constantly changing as the work forces move. With improved dispatching capabilities, it may become possible to give longer windows to Engineering forces, reducing track maintenance costs.

Service can be improved through more precise control over individual train performance. By interfacing classification yard and train dispatching systems, train arrivals can be controlled to reliably make connections on a "just in time" basis.

- Railroads are turning to short/fast operations as one strategy to meet truck competition head-on. Examples include the "Expediters", "Sprint" and "Laser" services. Frequent short/fast operations require more line capacity.
- Improved dispatching can make it possible to effectively produce several different grades of service using the same physical facilities. Services can be customized to meet the precise requirements of different market segments. For example, it will now be possible to maintain fast and reliable intermodal schedules on lines which also carry heavy coal traffic.

ENDNOTE

*CSX Transportation.

APPENDIX

Derivation of the Probability Model and Dispatching Rule

This train dispatching rule is derived for the case of light traffic volume. The sum of delay to each train multiplied by that train's value of time is minimized. Delay is the total time loss relative to passing through an area at full speed. Delay includes both waiting time for an opposing train to clear single track and acceleration time loss after the meet has been completed.

The expected value of delay to a single train, the "train of interest," arriving at the entrance to a single track section, is determined first. Then, by summing the expected delays of each individual train, total train delay is estimated.

Define:

N_{opp} = Number of trains/day entering the single track segment in the direction opposite that of the train of interest.

D_{max} = The maximum time the train of interest will be held waiting for an opposing train to clear, which is the sum of the single track running time and the time advantage applied by the dispatcher. The unit of time of D_{max} must be the same as used in N_{opp}; if N_{opp} is a daily volume, then D_{max} is measured in days.

D_{min} = Minimum time lost while pulling out of the passing siding and reaccelerating to full speed.

P(delay) = The probability of delay to the train of interest.

Z = The expected number of opposing train arrivals during time period D_{max}.

AWAIT = Given that delay occurs, the average time the train of interest must wait for an opposing train to clear the single track segment ahead.

ADV = Time advantage, the threshold difference in train arrival times used by the dispatcher to determine which train will be cleared across the single track segment first.

R = Running time of the single track segment at full speed, without stops at the sidings on either end.

If opposing train arrivals are uniformly distributed throughout the day, then the expected number of arrivals during any shorter period D_{max} is proportional to the length of the time period D_{max}, giving:

$$Z = \text{Expected number of arrivals during time period } D_{max}$$
$$= N_{opp} \times D_{max} \qquad (1)$$

If train arrivals are assumed to follow a Poisson process, the probability of exactly k arrivals during D_{max} is:

$$P(k) = \frac{e^{-Z} Z^{k}}{k!} \qquad (2)$$

Delay will occur if one or more arrivals occurs in period D_{max}.

$$P(\text{delay}) = 1 - P(k=0)$$
$$= 1 - e^{-Z}$$

Taking a Taylor series expansion of e^{-Z}:

$$e^{-Z} = 1 + (-Z) + (-Z)^2/2! + (-Z)^3/3! \; . \; . \; . \; .$$
$$= 1 - Z + Z^2/2! - Z^3/3! \; . \; . \; .$$
$$P(\text{delay}) = 1 - (1 - Z + Z^2/2! - Z^3/3! \; . \; . \; .)$$
$$= Z - Z^2/2! + Z^2/3! - Z^4/4! \; . \; . \; .$$

If traffic density is light, then the value of Z will be small and the Taylor series can be approximated by dropping the Z^2, Z^3 terms, yielding:

$$P(\text{delay}) = Z$$
$$P(\text{delay}) = Z = N_{opp} \times D_{max} \qquad (3)$$

The maximum wait D_{max} at a location (for meeting one train) is the running time of the opposing train minus any time advantage. In the example cited in the body of this paper, the coal train had a maximum wait of 40 + 32.25 = 72.25 minutes. The merchandise train had a maximum wait of 60 − 32.25 = 27.75 minutes.

If opposing train arrivals are uniformly distributed, the average wait to trains which are delayed is half the maximum wait, plus a fixed acceleration penalty, which occurs regardless of the length of the wait for opposing traffic. Call this average wait time AWAIT.

$$AWAIT = D_{max}/2 + D_{min} \qquad (4)$$

One of the initial assumptions was that the probability of more than one arrival in D_{max} is negligible. Equation 4 holds true as long as D_{max} is not increased to the point where this initial assumption is violated.

The expected value of delay to the train of interest, and also to each train arriving at the entrance to a single track section is the probability of delay to that train, times the average delay given that delay occurs:

$$\text{Expected delay to each train} = P(\text{delay}) \times AWAIT \qquad (5)$$

For trains in direction A:

$$D_{max} = R_B + ADV$$
$$P(\text{delay}) = N_B({}_BR + ADV)/24$$
$$AWAIT = (R_B + ADV)/2 + D_{min_A}$$

Likewise, for trains in direction B:

$$D_{max} = R_A - ADV$$
$$P(\text{delay}) = N_A(R_A - ADV)/24$$
$$AWAIT = (R_a - ADV)/2 + D_{min_B}$$

Using equation 5, compute the expected value of delay to each train in each direction. Each term is premultiplied by W, the cost or priority weighting factor, and by N_A and N_B, the number of trains in directions A and B respectively.

$$D_{tot} = W_A N_A \left(N_B (R_B + ADV)/24 \left((R_B + ADV)/2 + D_{min_A}\right)\right)$$

$$+ W_B N_B \left(N_A (R_A - ADV)/24 \left((R_A - ADV)/2 + D_{min_B}\right)\right) \quad (6)$$

Combining terms:

$$D_{tot} = N_A N_B/24 \left(W_A/2 \, (R_B + ADV)^2 + W_A (R_B + ADV) D_{min_A}\right.$$

$$\left. + W_B/2 \, (R_A - ADV)^2 + W_B (R_A - ADV) D_{min_B}\right) \quad (7)$$

Now, to optimize, differentiate with respect to ADV and set the result equal to zero:

$$dD_{tot}/dADV = N_A N_B/24 \left(W_A (R_B + ADV) + W_A D_{min_A}\right.$$

$$\left. - W_B (R_A - ADV) - W_B D_{min_B}\right)$$

$$0 = W_A R_B + W_A ADV + W_A D_{min_A} - W_B R_A + W_B ADV - W_B W_{min_B}$$

$$0 = W_A (R_B + D_{min_A}) - W_B (R_A + D_{min_A}) + ADV (W_A + W_B)$$

Gathering terms yields the expected result:

$$A_{OPT} = \frac{W_B (R_A + D_{min_B}) - W_A (R_B + D_{min_A})}{(W_A + W_B)} \quad (8)$$

Short Line Operation of Light Density Rail Networks: Economics and Public Policy

*by Denver D. Tolliver, Frank J. Dooley and Daniel L. Zink**

ABSTRACT

Short line railroad operation of light density Class I rail networks is changing the organizational structure of American railroads. The accelerated rate at which Class I rail lines are being spun off to short line operators suggests that substantial economies may be gained through this restructuring. However, little information exists regarding the economic considerations of this industrial reorganization. In addition, no clear public policy has been developed regarding light density track sales, as the federal and state governments deal with the question of their appropriate role in the matter.

The objective of this paper is to develop a method to compare the relative economics of light density rail line operation by Class I and short line operators. System costs and revenues are structured under two different scenarios for a network of Burlington Northern branch lines in North Dakota. Results support the contention that substantial economies are gained through restructuring under short line operation, particularly from reductions in labor costs.

I. INTRODUCTION

The industrial organization of American railroads is currently in a state of transition. Since 1980, over 150 short line and regional railroads have been organized (Wastler and Keeney). Many of these new railroads have been organized to operate lightdensity line segments abandoned or spun-off by Class I carriers. As Class I carriers attempt to sell most if not all of their light density trackage to independent short line operators, a new pattern of railroad organization is emerging. In the future, the railroad industry might be characterized as a "short line/trunk line" system. Under such a pattern, short line operators will provide local service over light-density portions of the network, while trunk line carriers will provide through freight service over high-density main lines and principal feeder lines.

The economic and public policy ramifications of a short line-trunk line system upon railroads, labor, and shippers are not well understood. Although the Interstate Commerce Commission has routinely approved the sale of light-density lines by Class I carriers, it has not adopted a formal policy towards the formation of short lines. The lack of a clear policy and the absence of any investigative procedure of branch line sales has been questioned. Legislation has been introduced in the Senate (S. 1026 and S. 1040) which would allow any party to protest a proposed short line sale to the ICC (Wastler and Keeney). In addition, several states have enacted or considered legislation allowing local review of any short line sale.

Ideally, appropriate public policy decisions should be made with complete economic information available to the policy makers. However, little information currently exists regarding the economics of short line/trunk line systems.[1] The objective of this paper is to provide a method for comparing the profitability of a current Class I network to a proposed short line/ trunk line system. The product of this exercise will be a tool which will provide policy makers with greater information as to the positive and negative economic consequences of the formation of short line railroads upon shippers, carriers, rail labor, and the general public.

II. RESEARCH DESIGN

The research which underlies this paper consisted of two principal steps. First, methodologies for estimating both light-density Class I carrier costs and short line railroad costs were developed and translated into computer models. Second, the models were applied to a designated set of light-density line segments in North Dakota which were recently sold by the Burlington Northern to a short line railroad. The results of the case study were then interpreted and implications for public policy developed. Before detailing the estimation of system costs and revenues, the general method is overviewed.

A. General Approach

The model employs a systems approach or perspective. System costs and revenues have been estimated for a designated set of lines under two different scenarios. Under the first scenario, the set of lines was analyzed as a light-density subsystem of a Class I carrier's network. Under the second scenario, the same set of lines was analyzed as a short line component of a short line/trunk line network.

System costs, as defined in this study, consist of two basic components: "on-line" costs and "off-line" costs. On-line costs comprise the operating, capital, and opportunity costs associated with serving and maintaining the set of light-density lines. Online costs consist of both variable and fixed components. Offline costs reflect the variable expense associated with moving the traffic generated from the light-density network over other parts or subsystems of the Class I carrier's network.

Under Scenario 1, on-line costs have been calculated so as to reflect the train operating and work-rule environment and fixed cost characteristics of a Class I carrier. Under Scenario 2, on-line costs have been estimated so as to reflect the potential labor economies and scale of operation of a short line railroad. Under both scenarios, system costs are the sum of the on-line fixed and variable costs and the off-line variable costs.

System revenues represent the revenues generated by the traffic which originates or terminates on the network of lines. System revenues have been determined under each scenario by multiplying the Class I carrier's tariff rates for each commodity, market, and service level by the volume of traffic. System revenues are compared to system costs under each alternative to determine the operating profit or loss.

B. On-Line Railroad Costs

The estimation of on-line costs for the Burlington Northern Railroad, the Class I carrier in this analysis, consists of a three-step process. First, a series of unit costs were calculated from the accounting and operating data contained in BN's 1986 R-1 report. Second, annual levels of activity or "service units" were estimated for the network of lines through the use of an operating model. Third, the annual service units were multiplied by the unit costs to yield on-line expenses.

1. Calculation of Unit Costs

The on-line unit costs used in this study closely follow the cost categories and definitions prescribed by the ICC in Ex Parte 274 Sub. No. 11, *Abandonment Regulations–Costing*. Most of the operating unit costs were calculated from BN's R-1 report using Schedule 755 (Operating Statistics), Schedule 415 (Equipment Expenses), and Schedule 410 (Railway Operating Expenses). In each case, an expense item (e.g. locomotive fuel) was divided by an output measure (e.g. locomotive hours) to produce a unit cost (e.g., fuel per locomotive hour).

In addition to the R-1 unit costs, a normalized maintenance of way (MOW) and net liquidation value (NLV) were calculated per mile of track. These unit costs reflect the light-density nature of the set of line segments and regional considerations such as the value of land.

Table 1 lists the principal on-line unit costs which were used in the study. Column (b) of Table 1 shows the 1986 Burlington Northern estimates for each item. Column (d) shows the associated service unit. Column (e) lists the source of the expense and/or output measures. A more detailed explanation of the calculation of the R-1 unit costs and the derivation of NLV and MOW values may be found in Tolliver (1987) and Zink, respectively.

2. Operating Model

The basic on-line service units included: (1) train-miles, (2) train-hours, (3) road locomotive hours, (4) road locomotive miles, (5) locomotive gross ton

TABLE 1.
On-Line Unit Costs for the Burlington Northern and Short Line

	Unit Cost			
(a) Cost Item	(b) Burlington Northern	(c) Short Line	(d) Service Unit	(e) Source
Locomotive Administration	4.9388	a	loc. hour	R-1
Locomotive Repairs	0.0049	b	loc. GTM[c]	R-1
Locomotive Depreciation, Rent, Lease	2.645	b	loc. hour	R-1
Locomotive ROI	4.169	b	loc. hour	R-1
Locomotive Machinery	0.000086	0.000065	loc. GTM[c]	R-1
Locomotive Fuel	31.25	31.25	loc. hour	R-1
Locomotive Servicing	0.21142	0.50695	loc. mile	R-1
Train Crew Wages	20.72	8.50	trn/engine man-hour	Qrtly Wage Stats
Freight Car Day Cost	10.52	10.52	car-day	BN
Freight Car Mile Cost	0.55	0.55	car-mile	BN
Train Inspection	5.6712	3.303	train-mile	R-1
Dispatching	0.23	a	train-mile	R-1
Crossing Protection	0.0178	a	train-mile	R-1
Signals/Interlockers	0.1037	a	train-mile	R-1
Maintenance of Way	3500.00	2980.00	track-mile	UGPTI[d]
Opportunity Cost on NLV	3505.00	2245.00	track-mile	UGPTI[d]
Administrative Overhead	0.0827	0.2985	car-mile	R-1

[a]Accounted for under other short line cost categories, primarily administrative overhead.
[b]Not applicable under full service locomotive lease.
[c]Gross-ton-miles.
[d]Upper Great Plains Transportation Institute.

miles, (6) car-days, and (7) car-miles. Train and locomotive hours involve two components, running time and time spent switching at stations.

In calculating the BN service units, the network of lines was organized into way train routes based on existing classification points and BN timetables. Discussions with short line management provided similar information for the short line. Mileages from each station to the classification yard, as well as round trip way train miles were estimated from timetables and distance tariffs. Two types of way train service were modeled: (1) assigned or scheduled way train service, and (2) unassigned or on-demand service. Large multiple-car and trainload consignments were assumed to receive unassigned way train service. Single car and small multiple car shipments were assumed to be spotted and pulled during assigned way train service. The frequency of scheduled way train service for each route was obtained from Burlington Northern trainmasters.[2]

Using North Dakota Public Service Commission data, the number of 26- and 52-car trains consigned at each station was calculated. This number was assumed to constitute the annual number of unassigned locals for each route. Once the number of assigned and unassigned locals was determined the number of train-miles and train-hours on the network were estimated. Train hours running were calculated by dividing the round trip distance for assigned locals by the maximum allowable running speed. A similar calculation was performed for multiple-car shippers based on the distance from the station to the division point.

Switching times at each station were estimated by multiplying BN's average switching time per car by the number of cars switched. Adjustments were made to switching times to account for the efficiencies of multiple car and trainload shipments.

Car days on the network were modeled using operating estimates of pickup, delivery, and loading/unloading times. Assigned locals were assumed to drop the empties along a given route on the first day of a cycle. Loading was assumed finished by the second day, and on the third day the cars were pulled to the classification yard. Multiple-car shippers were restricted to 24-hour loading times. Thus, only the running time from classification yard-to-station and back again was considered in the pickup and delivery time for unassigned locals.

C. Short Line Railroad Costs

The general methodology for developing short line operating costs was to adjust Class I operating costs on a unit cost basis to reflect the expected differences in labor costs and operating characteristics of a smaller, more flexible railroad. Information was provided by short line management and verified through personal interviews with Burlington Northern officials, rail labor representatives, and John Gohman, a short line consultant. Column (c) of Table 1 depicts short line unit costs expressed on a service unit basis which is comparable to Class I railroads. A general discussion of the unit costs and some of the reasons for the differences between the two systems follows.

Several of the unit costs shown in Table 1 are identical for the BN and short line systems. It can be expected, for example, that locomotive fuel costs, car-day costs, and car-mile costs will not vary appreciably between the two systems. Several of the cost components do vary, however, primarily due to differences in labor costs, capital cost assumptions and the proportion of administrative overhead contained in each system.

Four of the Class I unit costs in Table 1 were not directly applicable to a short line operation. Locomotive administration and train dispatching were assumed to be handled by management and central office clerical staff and are reflected in administrative overhead. Crossing protection and signal/interlocker operations are assumed to be administered by management, train crews, and maintenance of way personnel.

The train crew wages for the Burlington Northern in Table 1 were estimated from 1986 ICC Quarterly Wage Statistics. The figure of $20.72 per man-hour includes salaries, constructive allowances, and time paid-for but not worked. A fringe benefit factor of 38% was allocated to the Class I wage rate. The hourly train crew wage for the short line is $8.50. A fringe benefit rate of 30 percent was allocated to the short line train crew wage rate. The crew size was adjusted down from the 4 man crew used by BN to a 2.5 man short line crew. The smaller crew size reflects the labor flexibility and work-rule environment which characterizes short line railroad operations.

The BN's cost of capital was assumed to equal the ICC's prescribed replacement cost of capital rate of 15.8 percent. The short line has a lower capital cost, 10 percent, because it obtains its capital from other lending sources.

Administrative overhead for the short line railroad in Column (c) of Table 1 represents an annual network cost. This value is not directly comparable to the Burlington Northern's unit cost which consists of a car-mile allocation of general and administrative expenses such as accounting, marketing, sales, legal and secretarial, research and development, etc. The short line's administrative overhead includes managerial and clerical salaries and wages, office space, equipment, and utilities. In addition, it was assumed that some of the other short line tasks (locomotive administration, dispatching, crossings, signals/interlockers) are borne by the administrative staff and are therefore included in the administrative overhead unit cost. The larger traffic base of the Class I carrier may allow it to reduce this fixed cost item far below that of the short line. Thus, while the short line may have a much lower absolute administrative overhead cost, on a per unit of output basis it may experience much higher fixed costs.

Locomotive costs were also computed on a somewhat different basis. Locomotives for the short line operator were assumed to be covered under a full service lease on a cost per day basis. This lease includes locomotive repairs, and also reflects costs similar to the Class I locomotive depreciation, rent, lease and return on investment.

D. Off-Line Cost

Off-line costs include all variable costs of moving the traffic from the division point or classification yard to the final destination, or in the case of inbound traffic, from the point of origin to the classification point for delivery. Offline costs have been calculated in this study using Burlington Northern's Rail Form A (RFA).

A series of adjustments have been made to RFA to more closely reflect the realities of the situation. First, an intertrain or classification yard switch has been substituted for an industry switch. This models the interface between way train operations which were accounted for in the on-line calculations, and through train operations which are accounted for in the offline costs. Second, the actual car-day and car-mile unit costs for BN covered hopper cars, as shown in Table 1, have been used in lieu of RFA's system average. Third, a range of adjustments have been made to account for the efficiencies associated with multiple-car and trainload shipments. These include adjustments to terminal switching costs, car ownership expenses, clerical costs, circuity, and intertrain switching. A more detailed explanation of these adjustments may be found in Tolliver (1982).

E. Revenue and Traffic

Traffic volumes for six grains and oilseeds to four destinations were obtained from actual shipments reported monthly by individual North Dakota grain elevators. Data for the crop year July 1, 1985 - June 30, 1986 were used in the analysis. Data for other commodities and nongrain shippers were not available. Freight rates for shipments to primary markets were obtained from the Burlington Northern Tariff 4022-E. The BN tariff rates remained in effect after the short line sale. Rates included single, 26-car and 52-car rates, and were allocated by station to the appropriate service level. Rates for shipments to miscellaneous markets were not readily available and were estimated via an averaging procedure. Rates for the same commodities from North Dakota to all other destinations were taken from the 1986 ICC waybill sample and averaged to reflect costs of shipments to miscellaneous markets.

III. CASE STUDY

The branch line network chosen for analysis consists of 670 miles of primarily light density Burlington Northern branch lines located in southeast and central North Dakota (Figure 1). The system includes 9 branch lines and one secondary main line, and constitutes approximately 15% of the entire North Dakota rail network. Except for the secondary main line between Breckenridge and Casselton, all of the line segments in the network are classified as light-density lines. In 1986, the average gross ton miles (GTM) per segment mile for the regional network was 754,209. Excluding the Breckenridge to Casselton segment, the average traffic density was 220,606 GTM per mile. By comparison, the entire BN system averaged 13.2 million GTM per segment mile in 1986. A total of 13,452 carloads of grains and oilseeds originated on the network in the 1985-86 crop year.

The proposed short line network would be served by three classification yards, Jamestown, New Rockford, and Breckenridge. All traffic from the proposed short line railroad would interline with the Burlington Northern at one of these points. Eight way train routes were identified on the system and would be served by scheduled and unscheduled way train service from one of the three classification yards.

The rail network analyzed in the model serves 49 grain shippers, of which 29 are single car shippers and 20 are multicar or trainload shippers. In the 1985-86 crop year 68.1% of all grain shipments originating on the network were shipped by rail. Primary destinations for these shipments include the

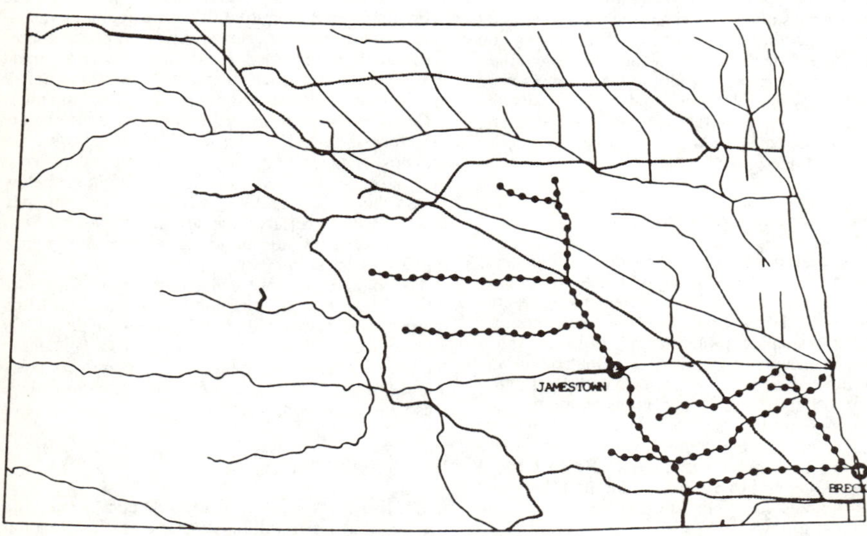

Figure 1. Burlington Northern Branch Lines in Southeast and Central South Dakota.

terminal markets of Minneapolis/St. Paul, Duluth/Superior and the Pacific Northwest A growing proportion of grain and oilseeds from North Dakota, however, are being shipped to other markets in the southwest and central United States, and to Gulf port destinations.

A. Profitability Analysis

Results of the model applied to the above described network of lines are presented in Table 2. Cost components of the analysis were aggregated and compared for the Class I and short line systems. Substantial differences in absolute and percentage terms were noted for several of the major cost categories.

The largest decrease in costs came from train crew savings of the short line railroad. Cost reductions of $1.68 million, or 75.5%, were realized from crew size reductions and a lower rail labor wage rate. Maintenance of way costs were reduced by $0.35 million (15%), while transportation costs (other than crew costs) dropped by $0.25 million (42%). Administrative costs, however, increased sharply by over 300 percent. This higher overhead expense proportion makes intuitive sense given the higher fixed cost nature of short line operations. The larger traffic base characteristic of the Class I carrier is simply not present on the short line network.

Total on-branch system costs for the short line operator were $2.5 million less than for the Class I carrier, a reduction of 24.7 percent. Total off-branch costs and total costs are identical for both systems. From a profitability perspective, therefore, the Class I carrier is barely able to recover total on- and off-branch costs at existing traffic levels and specified costs of capital. For the same traffic base, the short line operator earned a profit of $2.7 million.

IV. CONCLUSIONS

Congress and several state legislatures are currently in the process of developing short line railroad policy. However, little information presently exists concerning the relative economics of short line/trunk line systems. This paper has presented a methodology which can provide policy makers valuable information concerning the potential economies to be gained from operating a set of lines as a short line railroad as opposed to maintaining the same set of lines as a light-density subsystem of a Class I carrier's network. The methodology involves calculating a series of on-line unit costs from the Class I carrier's R-1 report, and adjusting the labor component to reflect the potential labor economies and scale of operation of a short line railroad. An operations model is used to estimate annual service units for the set of lines, and adjustments are made to Rail Form A to account for the efficiencies of multiple-car and trainload shipments during the off-line cost calculations.

The model has illustrated that major cost differences exist between short line and Class I operations in the areas of train crew wages, locomotive overhead and servicing, and general administration and overhead. The flexibility of short line work rules and lower wages, the lack of stratification in short line labor categories, and high Class I equipment overhead are the major factors which result in differences in on-line costs.

TABLE 2.
Cost Comparison of Burlington Northern Railroad and Short Line Railroad by Component Cost of Operation.[a]

Cost Component	Burlington Northern Railroad	Short Line Railroad	Percent Difference
Train Crew Costs[b]	2,221,350	543,716	-75.5
Locomotive Costs[c]	1,187,996	1,369,562	-15.3
Car Costs[d]	1,087,633	993,906	-8.6
Transportation Costs[e]	582,635	336,289	-42.3
Maintenance of Way Costs	2,338,000	1,990,640	-14.9
Administrative Costs	104,203	434,000	+316.5
Property Taxes	303,270	303,270	-
Opportunity Cost on NLV	2,337,905	1,479,687	-36.7
Other Costs	61,799	250,000[f]	-
Total On-branch Costs	10,224,792	7,701,072	-24.7
Total Off-branch Costs	12,305,560	12,305,560	-
Total Revenue	22,741,824	22,741,824	-

[a]Costs were estimated assuming a 15.8% cost of capital for freight cars, for locomotives, and fixed roadway expenses for the BN and a 10.0% cost of capital for the short line.
[b]Includes wages, fringe benefits, alimony and crew overnight costs.
[c]Includes locomotive repairs, depreciation/rentals/leases, return on investment, servicing, fuel, overhead and machinery.
[d]Includes car-day and car-mile costs.
[e]Includes train inspection/lubrication, dispatching, crossing protection and signal/interlockers costs.
[f]Includes opportunity cost on working capital and insurance costs.

A. Policy Implications

The paper has highlighted several important implications for federal and state policy makers. First, the Class I carrier's rate of return on line-segments which are short line candidates may be so low that the lines may be abandoned in the future if they are not sold. The sale of light-density lines to short line operators may thus help to preserve branch line trackage, thereby provided continued rail service to the shippers on these lines. Second, the development of short line railroads may enhance the economic viability of railroad operators. Because of lower operating costs, short line operators may be able to earn a higher return on investment than Class I carriers on the same lightdensity subsystem network. At the same time, the Class I carriers earnings should be enhanced as it is able to spin off marginally profitable rail line segments. Third, the higher profitability of short line operators may mean that more dollars are available and allocated to maintaining the track, roadway, and structures. As a result, short line development may thus serve to lessen or eliminate deferred maintenance on light-density lines, which could enhance the economic viability of the branch line network.

B. Future Short Line Research

The development and application of the short line model has produced useful results and policy implications. However, much remains to be gained through sensitivity analysis and simulation. The model provides the capability to simulate the effects of increased modal share, increased frequency of trainload and multiple-car shipments, and more efficient way train operations on the profitability of the network. Future research efforts should assess the impacts on these changes and evaluate the effects on profitability of varying key parameters.

The impacts of short line railroad development on related areas must be addressed as well. Effects on labor and shipper service levels are only two of possibly many other issues which should be addressed.

REFERENCES

Charney, Alberta H., Nancy D. Sidhu, and John F. Due. "Short Run Cost Functions for Class II Railroads." *The Logistics and Transp. Review*, 13(4):345-359, 1977.

Due, John F. "New Railroad Companies Formed to Take Over Abandoned or Spun Off Lines." *Transp. J.*, 24(1):30-50, 1984.

Hackney, Tom and Jean Arndt. "North Dakota Short Line Seminar." Fargo, North Dakota, February 11, 1987. Interstate Commerce Commission, Ex Parte 274 (Sub-No. 11). *Abandonment Regulations - Costing*, January 6, 1987.

Interstate Commerce Commission Rail Services Planning Office. *Guidelines for Evaluating the Feasability of Short Line Operations*. Washington, 1982.

Mittleider, John F., Denver D. Tolliver, and Harvey G. Vreugdenhil. *North Dakota Line Segment Analytical Model— A Technical Description.* UGPTI Rep. 50; Agr. Econ Rep. 173, Upper Great Plains Transp. Institute, North Dakota State Univ., Fargo, 1983.

R. L. Banks and Associates, Inc. *Short Line Techniques to Improve Financial Viability of Light Density Lines - Major Railroads*. Report to the FRA and DOT, Washington, DC., 1974.

Sidhu, Nancy, D., Alberta H. Charney, and John F. Due. "Cost Functions of Class II Railroads and the Viability of Light Traffic Railway Lines." *Q. Review of Econ. and Business*, 17(3):7-24, 1977.

Tolliver, Denver D. *Multiple Car Rail Costs in North Dakota*. UGPTI Staff Series No. 22, Upper Great Plains Transp. Institute, North Dakota State Univ., Fargo, 1982.

Tolliver, Denver D. *A Rail Line-Segment Costing Model*. Unpublished manuscript, Upper Great Plains Transportation Institute, North Dakota State Univ., Fargo, 1987.

Wastler, Allen R. and Kathleen R. Keeney. "BN Line Sale Talks Spark State, Federal Measures." *Traffic World*, April 27, 1987, pp 33-34.

Zink, Daniel L. *Analysis of Short Line Railroad Development in North Dakota*. UGPTI Rep. 52; Agr. Econ. Rep. 181, Upper Great Plains Transp. Institute, North Dakota State Univ., Fargo, 1984.

ENDNOTES

* Upper Great Plains Transportation Institute, North Dakota State University, Fargo, ND.

1. The limited information available in this area is dated and primarily concerned with the operation of Class I or Class II light density lines, not short line operations per se. See Banks; Charney, Sidhu, and Due; Sidhu, Charney, and Due.
2. The frequency of service was assumed to be the same for both the Class I and short line operation. Further research is being conducted to analyze the cost effects of increasing the frequency of service.
3. Further research is underway to determine whether differences in the quality of service may increase traffic for the short line operator.

Mandatory Trackage Rights: Economic Consequences[1]

*by Allen R. Ferguson**

ABSTRACT

Legislation is pending in Congress once more that would so amend existing law as, in effect, to require railroads to provide trackage rights over much of the nation's railroad network. Regulatory reform proposals in Canada incorporate similar provisions for up to 500 miles of track. The purpose of such legislation is to increase competition in situations where markets are served by a single railroad.

Relatively little analysis has been performed on the economic consequences of mandatory trackage rights, the conditions under which they would, increase competition, enhance efficiency or reduce inequities, and the conditions under which the prospective gains from trackage rights could be achieved voluntarily or could be attained only through mandatory action. These are the questions to be addressed in the proposed paper. The four main portions of the paper are as follows:

First, the proposition that all the major economic consequences hinge upon the rental paid for access to the owner's line is established.

Second, the characteristics of the rental that could be expected to emerge from voluntary arrangements are deduced from standard economic theory. In doing so the determinants of owning road's asking price and the prospective tenant's offer price are specified and the predictable characteristics of a final bargain are described.

Third, using the voluntary analysis as a point of departure the nature of mandatory rights that could be imposed in the absence of appropriate inter-carrier agreement is examined. It is established that both the nature of such rights and their economic consequences depend upon whether, in the *status quo ante,* the owning road was behaving as a monopolist (as well as having sole railroad access to the market in question).

Finally, in the concluding section the conditions under which mandatory trackage rights could be expected to serve the economic interest of the shippers and the public at large through increased economic efficiency and equity are specified.

I. THE PROBLEM

In light of charges that railroads often fail to offer adequate interline service at competitive rates, Congress is considering proposals that would force any railroad that provided the only rail service in a market to grant trackage rights to other railroads. Deregulatory proposals that have been considered in the Canadian House of Commons would impose similar requirements for distances up to 500 miles.[2]

Were mandatory trackage rights (MTR) imposed on all sole providers of rail service in markets lacking inter-modal competition, they might conceivably become virtually ubiquitous in the American railroad network, with financial, physical and organizational impacts permeating the entire system.

The goal of the bills is to eliminate railroad monopoly and to reduce rates. The present author takes as suitable social objectives, attaining competitive rates and services and preserving as much of the railroad system *as shippers are willing to pay for.* The purpose of this paper is to explore whether widespread mandatory trackage rights (MTR) are compatible with that broader policy goal.

Five basic facts largely determine the impacts of MTR and delineate the conditions under which they could be constructive.

First: Until recently, the industry was continuously deteriorating, physically and financially, and many important railroads are still financially weak.

Second: With economies of density and scope and with competition in most of its markets, a railroad can cover its large fixed, joint and common costs only if, in markets where it does have market power, it charges rates substantially above costs; failing that, it will be unable to provide all the service its shippers would be willing to pay for.

Third: The presence of only one supplier does not, in the railroad case and in many others, necessarily give rise to monopolistic practices. Whether an anti-competitive situation exists depends solely on whether the seller charges greater than competitive rates.[3] Regulation or sometimes the mere threat of competition, from any source, can keep rates below monopoly levels.[4]

Fourth: The extent of economies of scale (density) and scope, the relative operating costs of the tenant and owning road and whether traffic on the line is increased or merely redistributed, all, influence the value of the rights to both parties and the appropriate rental and, therefore, the impact on the industry, shippers and consumers.

Fifth: Efficiency requires that the railroad system be integrated. The Interstate Commerce Act requires inter-change, and authorizes the Commission to compel carriers (1) to establish joint rates (2) to provide reciprocal switching and trackage rights around terminal areas; and (3) to grant trackage rights as a condition of mergers. Consequently, there are no absolute barriers to movement of freight and cars among railroads; the degree of cooperation and coordination between carriers, and the apportionment of the gains from interchange activities are subject to price negotiation between the participants.

In Section II, the conditions under which voluntary trackage rights (VTR) could be expected to be created and their consequences are described. In Section III, the potential contribution of rational MTR and the realistic prospects of such beneficial trackage rights are identified. In Section IV, the consequences of individual and, more important, of general MTR are discussed. Section V is "Summary and Conclusions."

II. VOLUNTARY TRACKAGE

A. The Voluntary Agreement

The discussion is facilitated by reference to the Diagram, in which the landlord railroad, RR_I, owns rail link A-B over which RR_{II} [defined to include shippers and others who are potential railroad operators] wants trackage rights.[5]

Diagram

```
                                  RRII

                         2 = 2 = 2 = 2
                        =             =
A   1 = 1 = 1 = 1   B                  C
                        =             =
            RRI          1 = 1 = 1 = 1
```

The following simplifying conditions are assumed:

- there is no change in the volume of traffic over A-B or A-C, only to diversion from RR_I;
- neither economies of density nor scope are significant;
- transportation between A-B is the only issue, no broader strategic considerations are involved;
- we are interested only in solutions that are sustainable in the long run.

VTR can exist—if managements of both railroads are seeking to enhance their firms' going-concern value—only when a price, i.e. trackage rental, can be agreed upon. Obviously, this requires that RR_I's asking price not exceed RR_{II}'s offer price. The maximum price RR_{II} could be expected to offer for access to A-B under any circumstances is slightly less than its total expected increase in going-concern value. The minimum RR_I should accept voluntarily is an amount that would at least slightly exceed the costs and losses it would expect to incur as a consequence of RR_{II}'s access.[6]

B. RR_{II}'s Bid Price

The potential benefit to RR_{II} and, potentially to shippers and the public, depends on the rate RR_I charged in the *status quo ante*.

"Low" Initial Rate: If RR_I's rate over A-C were no greater than RR_{II}'s incremental cost of serving A-C, access to A- B would be of no value to the latter. It could at best cover its costs of the new service, and, hence, its bid price should be precisely zero. No voluntary arrangement could be expected.

RR_{II}'s incremental costs, are those costs (other than the rental) that it would incur only if it obtained and exercised access to A-B. Even in the long run, these are less than total costs, because they exclude all capital charges on A-B and any of RR_I's joint and common system costs. Those costs are solely the responsibility of RR_I. If they are not covered, line A-B must eventually deteriorate. RR_{II} could then simply withdraw from serving A-B with little (if any) capital loss; RR_I would have lost some, or all, of its investment in A-B. "High" Initial Rate: If RR_I had been charging freight rates that produced old revenue greater than RR_{II}'s incremental cost, RR_{II} could gain by serving A-C and charging the same rates as RR_I.[7] If necessary, RR_I could afford a rental approaching the total difference between its incremental cost and RR_I's prior revenues.

If variable costs of the two railroads were equal and RR_I were charging competitive rates, the difference between revenue and RR_{II}'s variable cost would equal RR_I's fixed cost. However, if RR_{II}'s operating cost were less than RR_I's or if RR_I were charging monopolistic prices, the difference between revenues and RR_{II}'s operating cost would be greater than RR_I's fixed cost. RR_{II} could therefore, pay something more than RR_I's variable costs, contributing to fixed costs. A necessary condition for a voluntary agreement is, then that RR_{II}'s costs be less than RR_I's.[8]

C. RR_I's Asking Price

The costs that RR_I must cover include the incremental costs it would incur solely because of RR_{II}'s use of A-B, such as extra signalling, terminal and switching costs and any congestion costs, as well as RR_I's contribution to maintenance costs. In addition, diversion of traffic to RR_{II} could give rise to severance pay to workers laid off. Probably more important, RR_I would be deprived of its margin on the diverted traffic. For RR_I to be able to sustain A-B in the long run, that margin must cover a contribution to fixed costs including replacement cost plus the current real market rate of return on the capital investment in A-B, and, to be able to sustain its system, the margin must also cover a portion of the joint and common costs of RR_I's network, such as administration and communication.

If the rental were too low to cover all those costs, A-B would deteriorate. Further, since components of a railroad's network are typically mutually supportive, deterioration of A-B would impose costs on users of other parts of RR_I.[9]

RR_I might face another kind of loss. If it is a monopolist, that is, charging a rate above unit costs[10] on A-B, and if RR_{II} could enter the market, the latter could, in the absence of a high rental, set rates below RR_I's rates, reducing or eliminating the owning road's profit. In addition, if RR_{II}'s operating costs were less than RR_I's, continued exclusion of RR_{II} from A-B might protect RR_I from the competitive consequences of its own inefficiency. [RR_{II} should, of course, be willing to pay some premium, approaching the cost difference, for access to A-B.]

It follows that, RR_{II}, if it obtained rent-free access to A-B, could earn a net return above its costs at any transportation rate [charged by both carriers, since they would be serving a single market] that was above its own short-run incremental costs. In contrast, RR_I must receive rates plus any rentals that cover all long-run costs if it is to sustain line A-B in its existing condition indefinitely. In addition, it would insist on covering any monopoly profits or on some offset to its inefficiencies. Consequently, there are many circumstances in which no agreement on rental would be possible.

D. Conclusions on Voluntary Rights

A necessary condition for VTR is that the tenant railroad's operating costs be less than those of the owning road. In that case, substituting operations by RR_{II} would provide an increment of net revenue to be divided between the parties or, conceivably, to be passed on to shippers. The apportionment of that new surplus would depend on the bargaining power of the parties.

In contrast, no VTR is possible when the two railroads have equal variable cost, because the trackage rights would create no saving to motivate agreement. In that case if RR_{II} were to pay any rental greater than RR_I's incremental cost and a suitable contribution too fixed cost, it could not expect to survive in that market. If it offered less than RR_I's incremental plus fixed cost RR_I would, of course, refuse the offer; further, if RRI were making a profit, it would refuse any offer that did not compensate for reduction in that profit.[12]

E. Consequences

Any voluntary trackage rights agreement could be expected with confidence to benefit the public: it would increase efficiency without an offsetting decrease in equity, and the going-concern value and hence the viability of at least one railroad would to rise. VTR would normally not maximize economic welfare, however; the motive for agreement is to increase and to share profits, not to reduce them.

If, after the agreement, the railroads' earnings exceeded the market rate of return, rates would be greater than and traffic less than the efficient levels. If, in contrast, returns to both railroads remained inadequate to sustain their services at a level shippers were willing to pay for, economic efficiency would be served by their retaining the surplus created by the trackage rights agreement.[13] Only if competition over A-B-C were to emerge, would one expect shippers to benefit through any substantial rate reduction.[14] Mere duopolistic rivalry between RR_I and RR_{II} would not be likely to produce competitive rates.

III. MANDATORY TRACKAGE RIGHTS

The discussion to this point provides the basis for determining both the conditions under which any MTR could serve the public interest and the prospects that actual MTR would do so.

A. The Potential Role of Mandatory Trackage Rights

For mandatory trackage rights (MTR) to serve the public interest, all of the following conditions must be fulfilled: (1) the prospective tenant must have lower operating costs than the owning railroad, or one of the two railroads must be charging rates above the competitive level;[15] (2) the rental imposed in mandating the trackage rights must be economically valid; (3) a superior regulatory remedy must not be available.

Condition (1) has been discussed; if it is not fulfilled, there is no room for any increase in economic efficiency. To recapitulate, if RR_{II}'s costs are less than RR_I's, the carriers have incentive to enter into efficient MTRs. However, if costs are equal [or if RR_{II}'s are the greater], there is no incentive for such agreement, but, if RR_I is charging monopoly rates, even where the carriers' costs are equal, there is an opportunity for MTR to enhance efficiency. However, MTR, with rentals less than would be voluntarily agreed upon, cannot enhance efficiency if the owning railroad is earning less than the market rate of return on its system.

Condition (3) is obvious, and rate regulation is likely to have considerable advantages over mandatory trackage rights.[16] In this Section, we focus on (2), the required characteristics of a suitable mandatory rental payment.

Even where condition (1) is fulfilled, in order for mandatory trackage rights actually to be beneficial to the public and to shippers in the long run, the mandated rental payment must fulfill four requirements.

First: it must cover at least RR_I's costs as discussed above. Otherwise, the viability of RR_I at its current size and quality would be jeopardized.

Second: MTR must reduce the resource costs of the total operation of routes A-B or A-B-C, or reduce the extraction of monopoly profits, or both. Any decrease in real cost increases efficiency *per se* by saving resources. Reduction of pre-existing monopoly profits would also increase efficiency.

Third: the increase in efficiency must be passed forward to shippers or, if one or both railroads incurred losses in the *status quo ante,* must be used to reduce them. If both roads were at least covering costs before MTR, simply dividing the gains between the carriers would yield no benefits to shippers. On the other hand, if either or both carriers had been making inadequate returns, permitting at least one of them to reduce losses would benefit shippers and the public—viability of a useful railroad facility would be enhanced.

Fourth: the rental must be affordable to the prospective tenant or it would not take advantage of its power to exercise trackage rights.

Thus, the conditions under which MTR can produce public benefits are limited. However, where "non-economic" factors preclude voluntary agreement, MTR may have a productive role to play. Many such problems could preclude efficient VTRs. For example, the parties may estimate or value various of the relevant parameters differently; there may be inadequate communication between them; there might be parochial jealousy of proprietary prerogatives. In such cases, some form of mediation or compelling an owning railroad to grant trackage rights to a tenant might be economically desirable.

B. Realistic Prospects for Mandatory Trackage Rentals

These required characteristics of mandated rentals must be compared with the kinds of rentals one could expect from the regulatory or judicial process.

Mandatory trackage rights imposed by the ICC have emerged largely as conditions of mergers, where their purpose has been to prevent erosion of competition. The ICC's announced intent has been ". . . to set compensation for the trackage rights which will put the tenant in the same position as the owning carrier."[17] In contrast, the purpose of the proposed legislation is not just to maintain but to increase competition. Therefore one could expect rentals less favorable to owning roads than those the ICC has established in merger cases. Thus, the ICC approach, largely as reflected in the August 30, 1984, decision in the UP/MP acquisition case, is taken as a "favorable" bound of prospective realistic mandatory rentals and, as such, are compared with rentals required for economic efficiency.

A typical ICC rental consists of (1) a variable cost contribution to compensate RR_I for the direct incremental costs it incurs because of RRII's use of A-B and (2) an "interest rental" to contribute to ownership costs, including return on capital. Both components are proportional to the tenants use of A-B, measured in tons or ton-miles.

The "interest rental", which poses several problems, equals the tenant's fractional share of use of A-B multiplied by the product of (a) the estimated value of A-B and (b) a computed rate of return. The value of A-B is estimated to approximate its previous value to the owning road. The rate of return is calculated to equal RR_I's past system-wide return on capital. Both the calculated value of A-B and the rate of return are apparently fixed for the duration of the trackage rights contract.

IV. EVALUATION OF MANDATORY TRACKAGE RIGHTS

We can now evaluate prospective realistic mandatory trackage rentals. In fact, we find that it is unrealistic to expect efficient levels of compensation, both in the case of isolated grants of trackage rights and in the more important case of virtually universal MTR.

A. Individual Mandatory Trackage Rights

It is unlikely that any standard procedures could incorporate the necessary principles established in Section III. The ICC approach may be as good as is practical but it contains serious flaws.

"Interest Rental":

The most complex and, perhaps, the most serious problems arise in connection with the capital charge, which the Commission calls "interest rental".

First: In all cases where, as ICC has typically found, railroad earnings are inadequate, interest rental on recent capitalized earnings would *assure* that line A-B *could not be sustained or replaced* without subsidy.[18] In the opposite case, if RR_I were earning monopoly profit—the situation the legislation is intended to alleviate— an interest rental based on the capitalized value of those earnings would tend to perpetuate that monopoly profit, as well as to preclude RR_{II}'s cutting rates to the competitive level in most circumstances, thus defeating the purpose of the proposed legislation, and yielding no significant rate benefit to shippers.

Second: The procedure introduces excessive rigidity. RR_I's existing rate of return and the value of A-B are frozen into the interest rental for the foreseeable future.

Third: It is superficially plausible that the tenant's rental should be proportional to use. Unfortunately, that would increase RR_{II}'s marginal cost with each additional carload of traffic by an amount equal to the increment in the rental. In contrast, a fixed rental charge would, by definition, not affect RR_{II}'s marginal cost. Consequently, a variable interest rental tends toward higher rates and less service than would a fixed one.

Fourth: In many actual cases there would remain, in addition to these problems, a fundamental flaw in setting a rental such that RRI's future earnings were held to no more than competitive returns. As long as returns on RR_I's system [efficiently operated] lay below both the market cost of capital and the return that shippers would be willing to pay rather than to forego the services, contributions from all sources to fixed, joint and common costs should be calculated on principles analogous to Ramsey pricing. Thus, not only should rates charged to shippers on each segment of the system, be set in accordance with an inverse elasticity rule, but also each segment of the system should yield revenues reflecting its value to shippers. Thus, RR_{II} should make a greater than average contribution, per unit of traffic, to system costs if its demand for use of A-B is relatively inelastic.[19]

The ICC sometimes appears to be aware of at least one manifestation of this problem in that it has on occasion allowed a rental high enough to reflect a particular advantage to the tenant, such as reduced circuity.

Rental and Variable Cost:

ICC's calculation of the variable-cost component of the rental is also inadequate; it omits congestion and contingency costs.

On a route where traffic is light, congestion cost tends to be negligible, which may explain the Commission's routine ignoring of it. However, on heavily used routes subject to the pending legislation it could be important.

The owner faces a number of contingencies. Its capital costs, state and local taxes and some maintenance of way costs do not vary with usage, but RR_{II}'s contribution to those costs does. Also, it is more difficult for the owner to abandon a line than for the tenant to reduce or discontinue its service. Thus, RR_I receives variable contributions toward fixed obligations. Further, RR_I runs the risk, for example, that cyclical or secular decline in demand in the affected market or increased intermodal competition could leave it with unliquidated investment in A-B. Under the Commission's procedures, RR_{II} bears no such risk, but obtains basically riskless access to RR_I's investment in A-B.

Thus, some inequity inheres in the ICC approach. Further, contingency and congestion costs are just as valid economic costs as any others.

Small Railroads:

Short-line and new regional railroads typically provide the only rail service in their markets. Consequently, they would be particularly vulnerable to the inequities and inefficiencies inherent in the standardized system of determining rental payments. New railroads would be subject to the risk that as soon as they became profitable, they would be vulnerable to (in effect) take-over by another railroad or a large shipper through imposition of trackage rights. The ICC approach, by freezing recent past earnings and valuations into the rental payment, could foreclose their ever earning economic profits.

Since creating new enterprises is always risky, the threat in the proposed statutes of denying unusually high profits on the grounds of eliminating monopoly, would decrease investor's expected future returns and, therefore, reduce the supply and increase the cost of capital to new railroad ventures.[20]

B. Ubiquitous Mandatory Trackage Rights

Economic Impacts:

If inefficient mandatory rentals were to be imposed only infrequently, as in the past, the damage would be limited. In contrast, the damage of ubiquitous trackage rights is potentially enormous.

Under current conditions, pervasive Ramsey pricing is necessary for a vigorous and efficient railroad industry. High-end discriminatory prices, which must recover costs that cannot be recouped in competitive markets, can be successfully charged only where railroads have substantial market power, precisely where MTR would be imposed under pending legislation.

One would expect that rentals established in such markets to benefit successful plaintiffs would yield revenues to the owning road lower than the preexisting, allegedly monopoly, returns. If competition within the new duopoly were to induce rate reductions, ubiquitous MTRs could, segment by segment, threaten virtually every rate at the high end of the railroad price structure.[21] This small fraction of all rates contribute disportionately to funds for investment and heavy maintenance.[22] According to the AAR, in 1983 restricting all rates to 180% of variable cost would eliminate two-thirds of the railroad net operating revenues.[23]

Organizational Impacts:

The organizational effects of ubiquitous mandatory trackage rights might be as significant as the financial impacts.

By providing opportunities for low-cost operators to enter the most lucrative railroad markets, universal MTR would threaten rail unions. Also, to forestall the destruction of its pricing structure, the railroad industry could be expected to strive to reestablish some form of cartelization.

In addition, if MTR were applied in all cases where only one railroad served a market, two physically interlocked railroad industries could conceivably be created: first the owning railroads constituting a "fixed-cost industry", characterized by very high levels of fixed, joint and common costs, and, second, tenants, using the capital-intensive infrastructure but being a "variable-cost industry". Over an extended period, more fundamental change could emerge as well. Widespread use of privately owned railroad facilities by many "foreign" carriers (or shippers) could lead toward a nation-wide "CONFAC", that is, public ownership of the infrastructure with private ownership of rollingstock. The CONFAC idea was advanced in the mid-1970s as a possible solution to the problems posed by the Penn-Central bankruptcy. In such a system, decisions as to the level of investment would be made in the political arena for the major fixed facilities and in the market arena only for rollingstock.

V. SUMMARY AND CONCLUSIONS

Criticism of the railroads' alleged failure to provide adequate interline service at competitive rates has led to some legislative proposals to subject all rail lines —main lines and short lines— to potential mandatory trackage rights as a means of reducing railroad rates and monopoly profits. The risks to the public, the shippers and the industry in such a policy are very large. Mandatory trackage rights would endanger the entire industry by precluding virtually all high-end rates; that would jeopardize the recent progress toward improved financial and physical condition of the railroad system.

The potentially adverse impacts of ubiquitous mandatory trackage rights would arise from the cumulative effects of individual MTRs, which in most actual cases could be expected to compensate owners inadequately and to perpetuate existing deficiencies in railroad revenues; where any monopolistic profits did exist they would tend not to be eliminated but perpetuated.

Although actual prospects for productive mandatory trackage rights are poor, there are some stringent conditions under which they could theoretically provide public benefits:

- The prospective tenant had lower operating costs than the owning railroad —that differential provides a potential saving to be realized by carriers or, shippers and the public;

or

- At least one of the railroads was making monopoly profit, whose elimination could reduce shippers' costs, with no reduction in the viability of the profitable railroad; thus so long as the affected railroads are in fact revenue inadequate, MTR cannot serve the public interest;

and

- Mandatory rentals are economically sound;
- No superior regulatory remedy is available.

Fulfillment of these conditions is necessary but not sufficient to assure maximization of shippers' and consumers' welfare through MTR. It is also necessary that duopolistic behavior by the owning and tenant railroads not be permitted to raise or maintain rates above the competitive level.

Thus, the fact that only one railroad serves a particular market does not constitute even a presumption that MTR could increase competition, economic efficiency or equity. The prospects that mandatory rentals would actually meet the criteria stated above are small. The ICC's current approach to setting mandatory trackage rentals is inefficient and inequitable but may be the best that realistically can be expected.

ENDNOTES

*AFE, Inc.

1. This paper is based on research sponsored by the Association of American Railroads. The views expressed are solely those of the author

and do not necessarily reflect the position of opinions of the AAR or any of its members. The author is greatly indebted to one of the anonymous referees for a very meticulous review that pointed out significant errors and to other reviewers who identified some lack of clarity in the initial version of the paper. The paper discusses none of the legal issues, such as the compatibility of the proposed legislation with the antitrust laws, as addressed, for example, by the FTC in its letter to the House Subcommittee on Commerce, Transportation, and Tourism, dated June 9, 1986.

2. Mandatory trackage rights are also imposed, of course, in merger cases and were proposed for the SFSP merger.
3. For simplicity, "rate" [or price] is used throughout to encompass the value of the quality of service; alternatively, one might consider that rate comparisons are made assuming service quality is unaffected.
4. For a discussion of the possible applicability of the theory of contestable markets to the railroad industry see, for example, the author's "Mega-Railroads and the Theory of Contestable Markets," TRF: *Proceedings,* Vol. 27, Number 1, Twenty-Seventh Annual Meeting, September 22–24, 1986, Seattle, Washington, 72–79.
5. I assume that there is no inter-modal, source or product competition.
6. Thus, if managements are profit maximizers, no VTR can be expected that does not promise some net gain to be divided between the participants.
7. If, subsequently, RR_I were to lower its rate, that would reduce or eliminate RR_{II}'s gain.
8. The discussion to this point would indicate that RR_I's earning profits might also lead to VTR, but, as shown below, further analysis reveals that the existence of RR_I profits are irrelevant in this context.
9. This discussion of costs largely parallels Baumol, W.J., *Statement #1,* Finance Docket Nos. 30,000 and 30,000 (Sub-No. 1), *Trackage Rights Compensation*.
10. RR_{II}'s total cost, being a cost at which A-B could be served by a potential competitor becomes the upper limit of RR_I's simulated competitive cost—stand-alone-cost. Therefore any rate above RR_{II}'s cost is a monopoly rate.
11. That is, costs "above the rail," including relevant marketing, sales, administrative costs, direct train operating costs, replacement of rolling stock and the market return on capital invested therein, but excluding any fixed capital or maintenance-of-way costs.
12. Because of the simple examples used here, this suggest that all traffic would shift to RR-II. Probably, in most realistic situations, only a portion of it would be diverted.
13. If, after implementation of the agreement, one carrier had greater than adequate revenues and the other did not, the analysis would, of course, be slightly more complicated.
14. However, even under monopoly, the reduction in long-run marginal costs that is a precondition for durable voluntary traffic rights would tend to cause some downward adjustment in rates, unless the change led to a reduction in the elasticity of demand for service over the route.
15. I have not discussed the case where RR_{II} earned greater than competitive profit, but it should be clear that if it were both more efficient than RR_I and charging monopolistic prices, there would be ample room for MTR to provide public benefits. For a discussion of the requirements that profits be excessive before constraints on rates or earnings can serve the public interest, cf, eg, Willig, Robert D. *Verified Statement,* Docket 39176, *The Chlorine Institute, Inc. v. The Atcheson, Topeka and Santa Fe Railway,* pp 2 and 3.
16. Monopoly profits in any railroad market, if the carrier's system revenues are adequate, indicate a failure of regulation under the Staggers Act, which empowers the ICC to disapprove, in markets where railroads have market dominance, rates it considers unreasonable.
17. Decision, Finance Docket No. 30,000 (Sub-Nos. 16, 18 and 25) [the UP-MP-WP consolidation], Decided August 20, 1984, 10.
18. Also, the Commission has sometimes accepted replacement cost new less depreciation (RCNLD) as a basis for determining the value of RR_I's assets. In fact, only a return equal to the real market cost of capital on the net replacement cost of assets, adjusted for technological change, provides financing adequate for maintaining the quality and quantity of facilities.
19. For an analogous discussion, cf Baumol William J., *Statement No. 1,* Verified Statements of Applicants in Response to *Applications for Trackage Rights,* Finance Docket Nos. 30000 and 30000 (Sub-No. 1) (Trackage Rights Compensation), p3.
20. Also it may well be impossible for the judicial process to meet the demands of economic efficiency. At several points, we have shown that it is unsound to incorporate into a formula for future payments values based rigidly on past performance. Mandating a payment that produces greater than past returns to the owning road may be inconsistent with legal precedent.
21. As pointed out above, if, as is likely, competition within the duopoly did not emerge any rate benefit to shippers would be small.
22. It is to be remembered that a key difference between VTR and MTR rental is that the former would include and the latter would exclude (or diminish the preexisting profits of RR_I.
23. Source: letter of William H. Dempsey, dated March 14, 1986.

An Assessment of the Rail Competitive Motor Carrier Industry (1977-1986)

*by K. Eric Wolfe, Vijaya Krishna, and William F. Huneke**

ABSTRACT

During the last decade, significant demographic changes in the rail competitive motor carrier industry have occurred. These changes have been brought about by not only motor carrier deregulation (Motor Carrier Act '80, and the Surface Transportation Assistance Act '82) but also by many exogenous factors. Other factors which have influenced these changes are: the overall economy, fluctuating energy prices, altered tax structures (Economic Recovery Act '80, Tax Overhaul '86), and the rate of assimilation at which these new opportunities have been implemented by different sectors of the motor carrier industry. Examples of these changes include:

1. Energy price increases (during the late 1970's) in combination withhistorically high (real and nominal) interest rates only to significantly decline six years later;
2. Interstate highway weight limits were increased to a uniform nationwide level of 80,000 lbs;
3. Motor carrier entry into new markets was eased by reduction/removal of entry restrictions and motor carriers were allowed greater pricing freedom; and,
4. Larger vehicles were allowed (e.g. single trailer lengths up to 53', abolition of overall vehicle length laws, 108" trailers, nationwide 28' doubles).

The purpose of this paper is to focus on the intercity truckload sector of the motor carrier industry, (the sector which offers the greatest direct competition to the railroad industry) in order to identify, qualify, and quantify the impact of these changes on the motor carrier industry. The analysis will examine changes in tractor and trailer characteristics, annual mileages, wages, commodities, and other salient competitive factors.

The analysis concludes by identifying certain important trends in the truckload sector: the vast increase in non-union company drivers, longer trailers, increased productivity, and changes in commodity mix. All of these trends underlie, and help predict, future increases in the competitiveness of the truckload motor carrier sector.

I. INTRODUCTION

This paper presents and analyzes major demographic changes which have occurred over the last decade in the rail-competitive sector of the motor carrier industry. The term "rail-competitive sector" applies to those motor carriers which are engaged in long haul truck-load (TL) transportation under single bills-of-lading. Costs and operational methodologies of TL firms differ significantly from those associated with the movement of less-than-truckload (LTL) quantities, as LTL operations include terminal and overhead costs associated with the consolidation of small shipments before transit. Traditionally, rail-competitive TL motor carriers were categorized as: Irregular Route Common Carriers, Contract Carriers, Agricultural and Exempt Carriers, Private Carriers, and Owner Operators.[1]

Before deregulation, extensive federal and state motor carrier regulation served to maintain distinct boundaries between the different types of motor carrier operations. As a relatively narrow set of limits on the size and weight of equipment prevailed, (e.g. total vehicle length limitations effectively limiting trailer length to 45 feet or less), analysis of the rail-competitive motor carrier industry was relatively easy. Published tariffs and operating rights defined market opportunities and the rates charges in those markets. Extensive use of unionized labor allowed the analyst use of the National Motor Freight Agreement as a bench mark for labor costs. Motor carriers were required to file extensive annual reports with the Interstate Commerce Commission while the Federal Government undertook wide-range surveys of truck inventory and use, and manufacturers at regular 5 year intervals.[2] Other public sector companies also produced extensive data bases which allowed monitoring of the essentially static motor carrier industry.[3]

Since deregulation, the boundaries which once set these motor carrier categories distinctly apart, have largely dissolved. For example, private carriers can now obtain common carrier back-haul authority, owner operators can obtain their own operating authority, contract carriers are no longer arbitrarily limited to the number of firms they may serve.[4] Additionally, motor carrier deregulation coupled with a decreased Federal presence in data collection significantly curtailed the ability to examine the now rapidly evolving rail-competitive motor industry. Although much has been written concerning the new opportunities under deregulation, relatively little analysis on this specific sector of the total motor carrier industry has been performed to monitor the actual adoption and incorporation processes of these new market and operational opportunities. The effects of exogenous factors such as energy prices, and interest rates on this adoption process have also been largely ignored. Of course, much of the reason for this absence of research has been an absence of reliable and consistent data from the pre- and post-regulatory period.[5] Even from a post-regulatory perspective only, many current motor carrier analysts are stymied by an information shortage.[6]

This study will be using a proprietary longitudinal data base, specifically designed to evaluate on an on-going basis this market segment. Changes in rail-competitive motor carrier operational characteristics will be evaluated with respect to both the regulatory and economic environment which existed during the study period 1977 to 1986.

II. EXOGENOUS EVENTS AFFECTING MOTOR CARRIERS

A number of exogenous events have buffeted the motor carrier industry during the study period. These include deregulation, changes in truck size and weight limits, variations in fuel prices, insurance rate increases, and rail deregulation.

A. Deregulation

The Motor Carrier Act of 1980 substantially deregulated trucking by reducing entry requirements for new carriers and allowing new and existing carriers to serve larger market areas. The granting of 48 state operating authority is now routine and route restrictions have been eliminated. Carriers also have far greater pricing freedom to meet changing market conditions.

Deregulation also spawned a far more competitive and dynamic motor carrier industry. Observation of firm entry and exit rates show just how dynamic trucking has become. From 1980 to 1985, 15,000 new firms have entered the trucking industry while about 7,500 firms left the industry.[7] In spite of the continuing economic recovery, the number of motor carrier failures continues to increase because of intense competition in the marketplace. (Refer to Table 1)

Passage of the Stagger's Act in 1980, which deregulated the railroad industry, has also affected the status of the rail-truck competition. Under deregulation, railroads could now vary their rates (within a zone of rate flexibility) and service levels, and rationalize their plant size to meet changing market conditions. Additionally, deregulation also permitted the formulation of contracts between shippers and railroads.

If markets change (and trackage becomes unprofitable) it used to take years before the railroads were allowed to abandon that trackage. Under the Railroad Revitalization and Regulatory Reform Act (4-R Act enacted in 1976), unprotested trackage abandonments take effect in 75 days while protested cases must be settled within 225 days. Deregulation (starting with the 4-R Act) has seen the need for revenue adequacy as a basis for future railroad growth and stability. Instruments such as the Rail Cost Adjustment Factor (RCAF) allow the railroads to recoup inflationary cost increases and obviate the traditionally lengthy and expensive general rate increase procedure. Due to the highly-competitive environment, however, the rail industry has declined to take the full amount of the inflation-recoupment increases allowed under the RCAF. This is understandable as both inter-modal and intra-modal competition is intense.[8]

However, the trends after rail and truck deregulation have usually favored trucking in that railroads have generally have only been able to use their deregulatory freedoms to try to retain market share rather than increasing market penetration. This phenomenon is not limited to the higher-valued manufactured commodities as several bulk commodities have also been affected.[9] For example, railroads have seen their market shares fall for coal and grain since deregulation.

COMMODITY GROUP	1980 RAIL MARKET SHARE	1985 RAIL MARKET SHARE
Coal Shipments	62.9%	60.7%[10]
Grain Shipments	40.9	25.6[11]
Total Intercity Freight	37.5	36.9[12]

B. Changes in Truck Size and Weight Limits

The Surface Transportation Assistance Act of 1982 (STAA) mandated nationwide weight limits to 80,000 pounds, allowed 102" wide trailers and forbade overall length limits. In addition the Act mandated that the maximum trailer length limit (set by the states) could not be less than 48 feet. Congress began the move to the 80,000 pound limit in the Federal-Aid Highway Amendments of 1974 by changing the allowable gross weight limit from 73,280 to 80,000 pounds. However, until the STAA, Illinois, Arkansas and Missouri held to the lower weight limit. These states thus created a barrier to nationwide legal operation of 80,000 pound vehicles before 1983.

Several states, particularly in the midwest, have recently enacted legislation to increase trailer lengths to 53 feet. There are now 26 states that

TABLE 1
Motor Carrier Failure Rates Have Increased Since Deregulation

YEAR	FIRM FAILURES	TOTAL NUMBER OF FIRMS	FAILURE RATE (PER 10,000 FIRMS)
1977	274	Not Available	Not Available
1978	162	67,038	24.2
1979	186	68,451	27.2
1980	382	72,146	52.9
1981	610	75,167	81.2
1982	960	79,715	120.4
1983	1,228	83,262	147.5
1984	1,409	78,078	180.5
1985	1,533	80,621	190.1
1986	1,548	85,008	182.1

Source: American Trucking Association

permit trailer lengths greater than 48 feet. (Refer to Table 2) Such length increases would have a more dramatic effect on rail-truck competition were the 80,000 pound limit to be increased.

C. Fuel Cost Changes

During the 1977 to 1986 study period, fuel prices have risen and fallen dramatically, especially when adjusted for inflation. Fuel costs represent 17.5 percent of the costs of a rail-competitive truck while they only make up 10 percent of railroad costs.[13] Thus, increases in fuel prices are a competitive advantage for railroads while decreases benefit trucks.

Table 3 displays the variation in real fuel prices. The oil price rise brought by the Iranian Revolution increased real diesel fuel prices by 70 percent from 1978 to 1981. However, Organization of Petroleum Exporting Country's (OPEC) continued inability to control oil production and prices has allowed oil prices to recede. Adjusted for inflation, fuel prices are now less than they were in 1977. With modest decreases in energy prices forecast for the near future, the motor carrier industry will continue to benefit relative to the railroad industry.[14]

D. Insurance

Trucking industry sources claim that their insurance premiums rose from 50 to 600 percent from mid 1984 to late 1986.[15] This equates to an average increase in truck costs (on a loaded mile basis) of about four percent. Property damage and liability insurance now makes up about 5 percent of truck costs or 5.4 cents per loaded mile.

Unable to pool their risk to as great an extent as the larger motor carrier firms, Owner-Operators (O-O) have been harder hit by insurance cost increases. It is estimated that 1985 insurance costs for O-Os are 22 percent above those for company operations.[16]

In general, the insurance cost increases have hit the trucking industry a little harder than railroads because railroads are essentially self insured. As the railroads are such a large exposure base, they essentially pay their own property and casualty losses over the long-run.[17] Since the implicit cost for railroad insurance must also reflect the on-going liability crisis, rail insurance cost advantages are thought to be small.

E. Tax Reform

Tax reform has mixed results for rail-truck competition. The 1981 legislation allowed railroads to make their tax accounting similar to other industries like trucking by allowing railroads to change the method for depreciating track assets. The result was a large and extraordinary write-off of track assets—assets that previously had to be kept on the books without being depreciated, some for as long as 99 years.[18]

TABLE 2

State Maximum Semi-Trailer Length Regulations
(as of January 1987)

	Maximum Semi-Trailer Lengths			
	48′	48′1″ to 52′11″	53′ or more	Total Truck Lengths
Number of States:	25	4	19	3*
States:	Alabama	Arizona	Delaware	Colorado
	Alaska	Louisiana	Idaho	Nevada
	Arkansas	Mississippi	Illinois	North Dakota
	California	Tennessee	Indiana	
	Connecticut		Iowa	
	D.C.		Kansas	
	Florida		Kentucky	
	Georgia		Michigan	
	Hawaii		Missouri	
	Maine		Nebraska	
	Maryland		New Mexico	
	Massachusetts		Ohio	
	Minnesota		Oklahoma	
	Montana		Oregon	
	New Hampshire		Pennsylvania	
	New Jersey		South Dakota	
	New York		Texas	
	North Carolina		Wisconsin	
	Rhode Island		Wyoming	
	South Carolina			
	Utah			
	Vermont			
	Virginia			
	Washington			
	West Virginia			

* These three states do not have maximum semi-trailer length limitations but do have overall tractor and semi-trailer length restrictions. Each of three states allow sufficient overall length to permit the use of a 53′ semi-trailer, without the need for a special permit.

Source: Telephone interviews with transportation agency representatives in the various states.

TABLE 3
Average Real Diesel Fuel Price

YEAR	SIX MONTH PERIOD	PRICE (Cents/Gallon)
1977	2	51.9
1978	1	48.6
1978	2	50.1
1979	1	65.9
1979	2	79.4
1980	1	79.4
1980	2	82.4
1981	1	85.8
1981	2	84.7
1982	1	78.7
1982	2	76.8
1983	1	63.7
1983	2	66.9
1984	1	64.0
1984	2	66.6
1985	1	61.4
1985	2	68.1
1986	1	38.0

Sources: National Energy Information Center & Highway Statistics. Fuel price is deflated by using the GNP deflator reported in **The Survey of Current Business.**

However, the 1986 reform favors less capital intensive industries like trucking rather than railroads. Under tax reform the motor carrier industry on the whole will benefit due to lower corporate tax rates and acceleration of depreciation for motor carrier assets.[19] Although both the truck and rail industry lost tax benefits (due to repeal the of the investment tax credit), the alternative minimum tax will have a larger adverse effect on railroads. With the minimum tax, if railroad earnings continue to fall, railroads could pay as much as $1 billion more in taxes over the next six years. The repeal of the investment tax credit could cost the railroad industry $4 billion over the next six years.

F. General Economic Conditions

The domestic economy has also undergone many changes which impact on the level and type of transportation services demanded. Historically high real interest rates and intense foreign competition have forced domestic manufacturers to reduce inventory levels and adjust the size of orders to match production needs. Shipments of large quantities have been replaced with smaller shipment sizes. This shift in transportation demand is illustrated (in the rail industry) by the decline in boxcar loadings in favor of trailer or container loadings.

RAIL CARLOADINGS ORIGINATED
(Percent of Total)

Year	Total Boxcar*	TOFC/COFC
1980	19.4%	9.1%
1981	17.1	9.6
1982	15.3	11.9
1983	14.1	13.9
1984	12.7	14.7
1985	12.4	15.8
1986	11.7	16.6

*Includes equipped, unequipped and refrigerated boxcars
Source: Form CS 54, Transportation Division, AAR

The continued shift from a production to service economy has also accelerated. In 1950, goods as opposed to services, represented 56.3 percent of the Gross National Product (GNP). By 1980 this figure had slipped to 48 percent, while 1986 figures indicate goods represent only about 40 percent of total GNP.

Downsizing of products in response to the energy crisis has also curtailed the size and scope of many traditional transportation markets. Smaller automobiles, housing units with lesser square footage (thereby requiring less lumber, concrete, roofing singles, etc.) illustrate the effects of such downsizing.

Although high real interest rates adversely affected the motor carrier industry by increasing equipment carrying costs, especially among owner operators, continued excess capacity in the motor vehicle manufacturing industry depressed equipment prices.[20] As a result, equipment carrying charges did not rise as sharply as interest rates alone would have dictated. Volume purchasers, those able to buy in lots of 100 or more pieces of equipment, were actually able to save significantly on their equipment acquisition costs.

Higher interest rates tend to favor motor carriers over rail in that trucks generally carry smaller shipments, have shorter transit times, and allow the shipper to hold smaller inventories. With high real interest rates, these factors become more important, and thus tilt shippers toward motor carriers.

III. PREVIOUS RESEARCH

Recent studies using the National Motor Transport Data Base (NMTDB) have focused on the rail-competitive segment of the motor carrier industry. In examining changes during the 1977 to 1982 period, Larkin and Wolfe (1983) reported that significant increases in efficiency had occurred and were continuing to occur, although the full effects of the Surface Transportation Act had yet to take effect. Among several examples of cost-cutting they cited were: owner-operators were increasingly entering into lease arrangements which served to guarantee them traffic while in effect replacing the use of union-company drivers as union wages continued to out pace non-union wages (in nominal and real terms); the "standard" 40 foot long trailer was rapidly being replaced with 45 foot equipment; although larger engines (greater than 400 horsepower) were being increasingly employed, fuel efficiency was also increasing.[21] The derived demand sensitivity of the motor carrier industry to overall economy was also illustrated as the average length of haul and annual miles travelled by motor carriers rose and fell in step with the overall economy. Corsi, et.al. (1985) comparing the O-O segment of rail-competitive motor carriers between 1978 and 1984 concluded that the O-O segment had not improved their efficiency since the passage of the Motor Carrier Act in 1980. This finding was determined by examining two efficiency measures: the portion of total loaded to total empty trips, and the portion of trips were deadhauling occurred.[22] As neither of these ratios improved during their study period, it was concluded that no efficiency gains had taken place. O-O's were seen to be in a deteriorating position as the average age of their equipment was rising while gains in revenue were only just keeping pace with increases in cost thus endangering their future ability to recapitalize their equipment. Corsi concluded by stating that although the then recent ICC decision to allow direct leasing of O-Os by private carriers might reveal new opportunities, it was too early to tell especially in light of the "decline in the overall share of trips carried by private carriers."[23]

Although there would appear to be major differences between the conclusions reached in the two studies using essentially the same data from the NMTDB, these differences can be readily explained. The declining share of private motor carriers in the data base can be explained by the method under which the data is collected. When a private motor carrier is transporting its own traffic it is properly reported as a private carrier. However, when that same private carrier exercises its common carrier backhaul authority, it may be instead reported as an irregular route common carrier. Although the O-Os did not reduce their empty to loaded ratio of trips, this is not the only measure of efficiency. During this time frame, the O-Os did slightly reduce their portion of total empty miles. (Refer to Figure 1) In effect they were deadheading more to obtain new opportunities, i.e. backhauls. While it is true that O-Os (referring to Figure 1) have apparently only held their own (in terms of group efficiency) during the longer period of 1977 to 1986, increased productivity and operational efficiency (e.g. longer, wider trailers, heavier loads permitted, and more fuel efficient tractors) has allowed them to remain a force in an increasingly competitive marketplace. The fact that their margins did not increase is the result of rate compression due to increased intra- and inter-modal competition. It is generally recognized that the O-O portion of the motor carrier industry is very volatile. Although a large portion of the individual O-Os may leave the industry due to economic downturns, inefficient management or operational procedures, other individuals continue to enter the market to replace them.[24] As the O-O share of all movements in the data base has not seriously declined, it appears that productivity gains have allowed the O-Os to remain a significant factor in an every more competitive marketplace.

Finally, the adoption and diffusion process of any

Figure 1.
Owner Operator Empty to Total Mileage Ratios Have Remained Essentially Stable

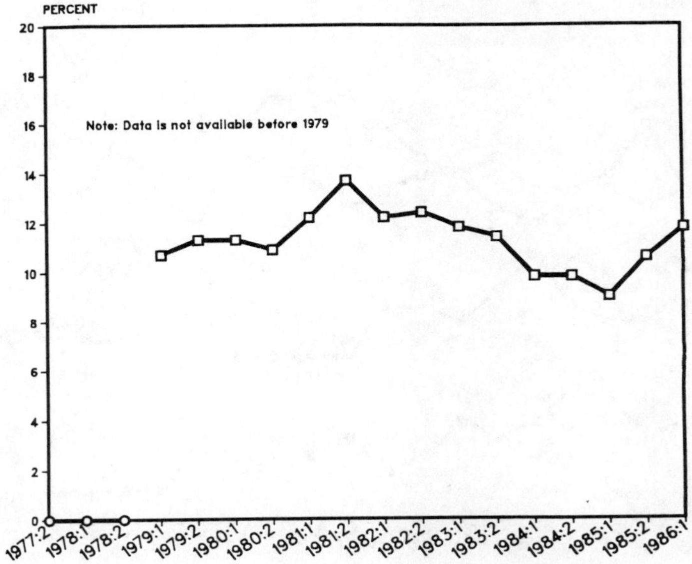

new idea or technology takes time. Only now, after several years of deregulation and an expanding economy (to provide and finance these new market opportunities) may the enhanced competitive results of deregulation be better observed.

IV. DATA BASE DESCRIPTION

The primary source of data employed in this paper is from the National Motor Transport Data Base (NMTDB). This data base was begun in 1977 under a joint effort by the Economics and Finance Department of the Association of American Railroads (AAR) and Transportation Research and Marketing (TRAM). Since 1983, TRAM has continued to interview motor carriers at an average of eighteen strategically located truck stops along the interstate highway system. Except for the first half of 1977, when the original questionnaire and enumeration process were being developed, an average of over 16,000 questionnaires are collected annually. As each questionnaire can record two loaded and empty movement pairs (current and previous movements), over 300,000 movements were recorded during the 1977 to 1986 study period. In addition to information regarding loaded and empty movements, information on driver, equipment, and operational characteristics are also collected.

In the actual analysis of data, no statistical significance tests were performed owing to the light density of the sample.[25] Finally, although the Regular Route Common Carriers (RRCC) aré not considered to be directly rail-competitive, they are occasionally interviewed. In this paper all observations recording the use of RRCC were excluded from the analysis as were team driver operations. The latter make up a very small portion of the sample.[26] Data from the top 50 TL carriers (in terms of revenues) as compiled by Data Resources Incorporated was also employed to provide perspective for many of these changes.[27]

V. CHANGING RAIL-COMPETITIVE MOTOR CARRIER CHARACTERISTICS

To evaluate the new opportunities afforded the motor carriers industry since deregulation, three demographic areas were investigated. These included: equipment, driver and commodity characteristics.

A. Equipment

One key development during the 1977-1986 period has been improved equipment utilization. Annual tractor mileages have increased from 108,000 to 117,000 while the number of loaded trailers reported fully loaded (from either a cubical or weight perspective) has increased from 55 to 93 percent. During the DRI study period of 1980 through 1985, TL drivers were seen to travel an additional 13,000 miles per year.[28]

During the period 1977 to 1984, the relative mix of trailer types remained constant. Flatbed trailers posted a slight decline in percent of the fleet reflecting the downsizing of the domestic steel industry. Dry van and the more versatile and costly refrigerated van trailers competed for dominance with approximately equal shares. (Refer to Figure 2) However, since 1984, the less costly dry van has been employed in ever increasing numbers.

Figure 2. Trailer Distribution (1977-1986)

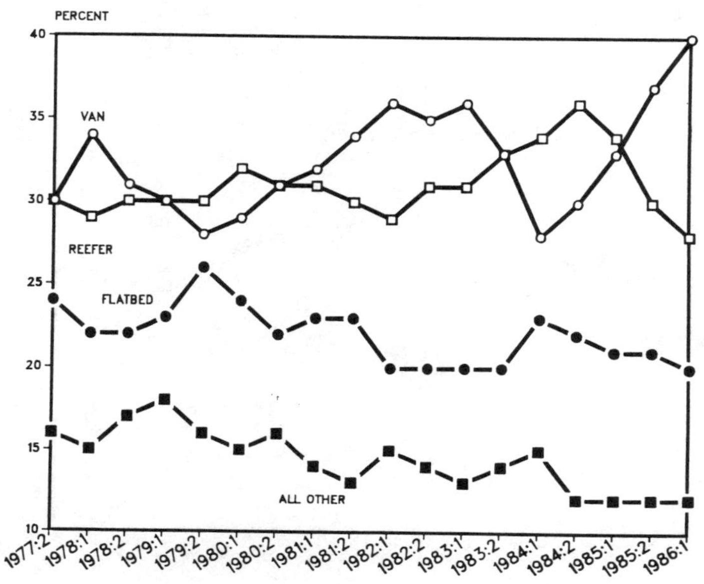

Perhaps the single largest change observed among the equipment related data is the market penetration of larger trailers. During the 1977 to 1983 time frame, the 45 foot trailer made steady progress in domination of the fleet. (Refer to Figure 3) Nearly 60 percent of the rail-competitive intercity truck trailer fleet in 1983 was 45 feet in length. Although the STAA had allowed 48 foot trailers in 1982, it was not until 1983 that the shift away from the 45 foot standard became apparent. Within just two years, the longer trailers went from less than 2 percent to over 60 percent of the market, while the 45 foot trailer fell back to 1977 levels. The one year delay between the STAA and market response can be explained in several ways. First, between July 1981 and November 1982, the economy was in recession. Existing equipment was able to handle demand and little excess funds were available for new capital equipment. In addition, many carriers also took a "wait and see attitude" before making the several year commitment that the new equipment would encumber.[29] Overall, DRI reported that equipment expenditures doubled between 1980 and 1985 from 900 million to 1.8 billion dollars.

The increased use of higher horsepower engines which had been the trend since 1977 leveled off at the beginning of 1984. Tractors with power plants of less than 350 horsepower actually rose slightly since 1983. (Refer to Figure 4) Unless the Pinwheel Amendment or other similar legislation is passed, this leveling in the use of highly powered tractors may continue as those carriers who employed them in markets requiring heavier loads and high speeds (e.g. Western doubles) have already done so.[30] Despite the use of more powerful and larger engines, and self-reported transit speeds that averaged more than 63 miles per hour, the data indicates that fuel efficiency has also increased. (Refer to Figure 5)

In addition to the adoption of this more efficient technology is the more productive use of these motor carrier assets. While motor carrier equipment utilization is generally a mirror of the overall economy, motor carriers are increasingly filling their larger equipment. (Refer to Figure 6)

From 1977 to 1986, the more efficient use of larger, more fuel efficient equipment has enabled the motor carrier industry to reduce costs (in real terms) by 10 percent.[31]

B. Driver

As large gains in equipment efficiency have taken place, so to have cost reductions occurred in human capital. During the later 1970's as many as one in four intercity drivers was a union member. With a brief hiatus during 1981, unionization among these drivers has fallen to where less than one in twenty is union member. (Refer to Figure 7) Non-union members earning 23 cents per loaded mile are far more attractive than union members drawing 28 cents per loaded mile. (Refer to Figure 8) In either case, motor carrier real wages have actually declined since 1977 by 22 to 24 percent. During the 1981 to 1985 time frame, DRI reported annual wages decline 7 percent in real terms.

The ability for owner operators to lease their services to motor carrier firms under trip or permanent lease arrangements helps guarantee them work. Currently, more than 90 percent of all O-Os report that they were involved in some sort of permanent lease arrangement. (Refer to Figure 9) Although it was

Figure 3.
Since The STAA Was Passed, Longer, More Productive Equipment is Being Used (All Trailer Types)

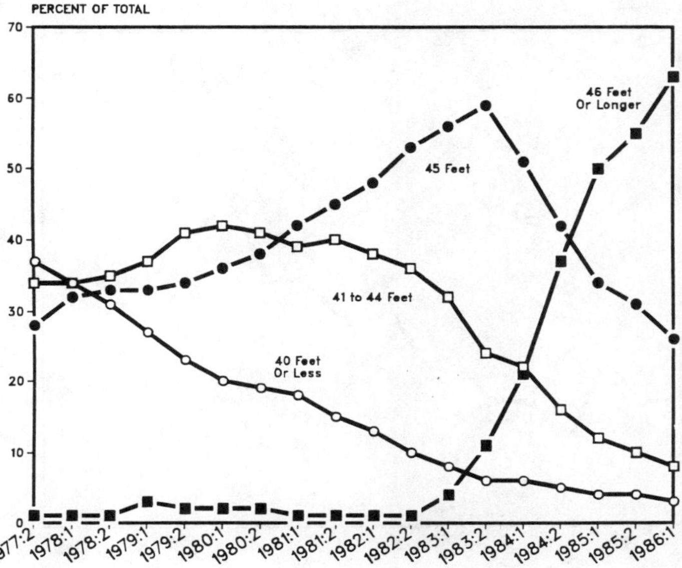

Figure 4.
Larger Engines are Being Used Allowing Drivers to Haul Larger Loads and Maintain Higher Highway Speeds

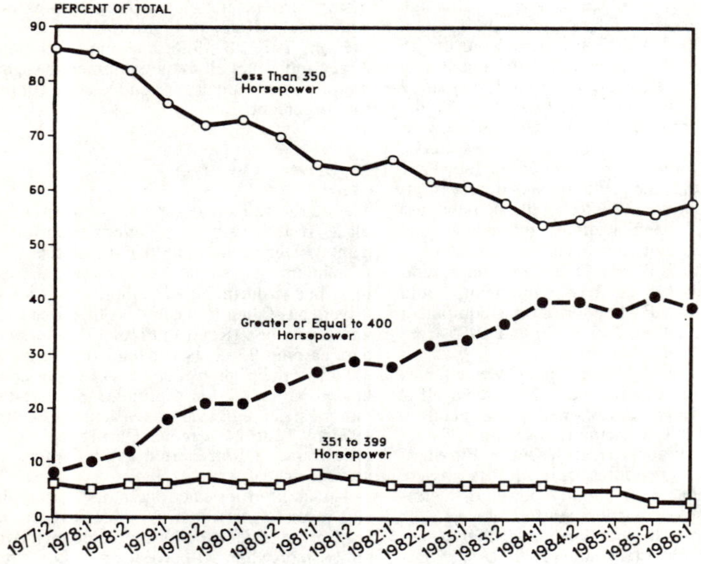

Figure 5.
Average Fuel Efficiency Has Increased (All Trailer Types)

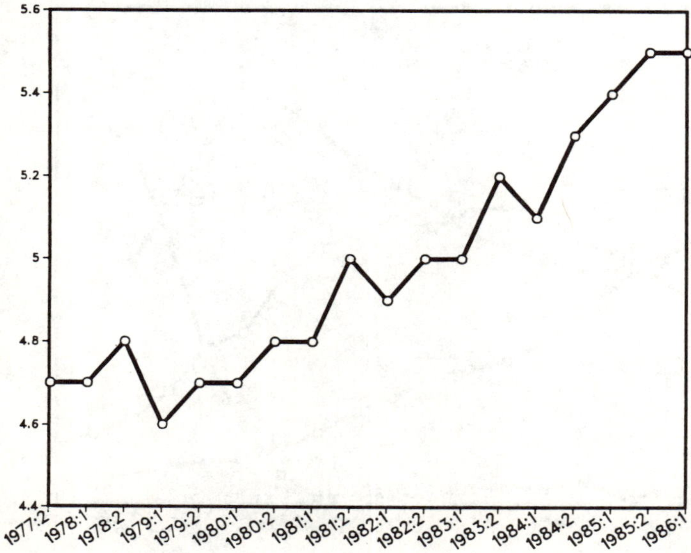

Figure 6.
Motor Carriers Have Increased Utilization (Percent of Trucks Reported to be Fully Loaded)

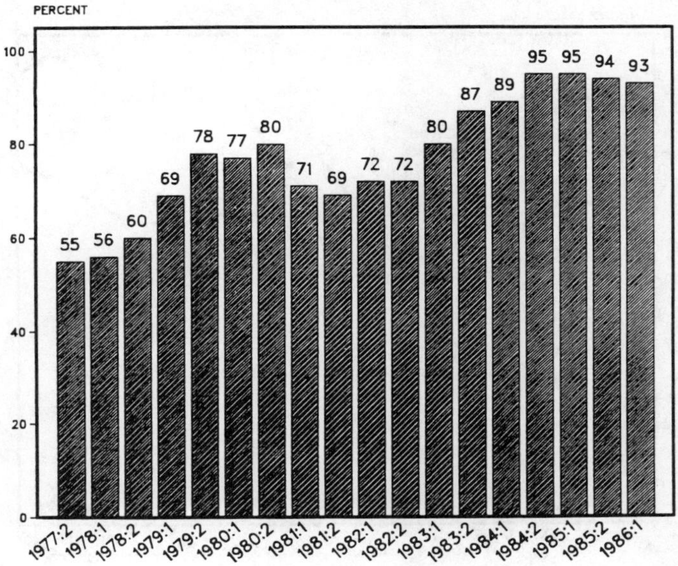

Figure 7.
Motor Carrier Union Membership Has Declined (1977-1986)

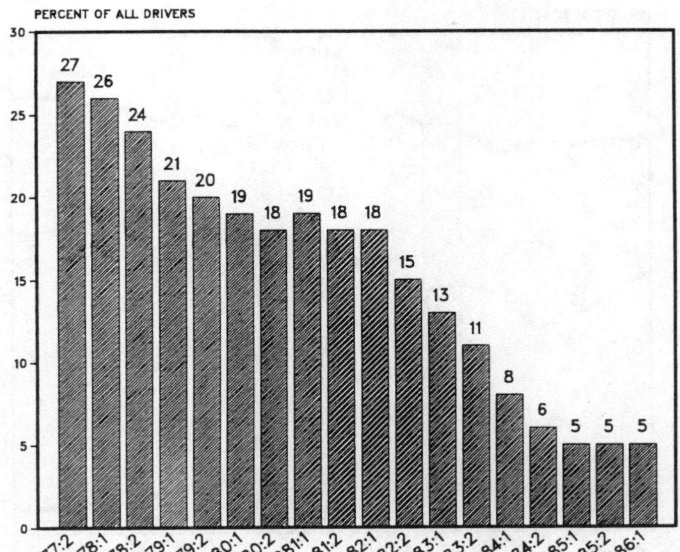

Figure 8.
Union vs. Non-Union Wages
Company Drivers Only
(1977-1986)

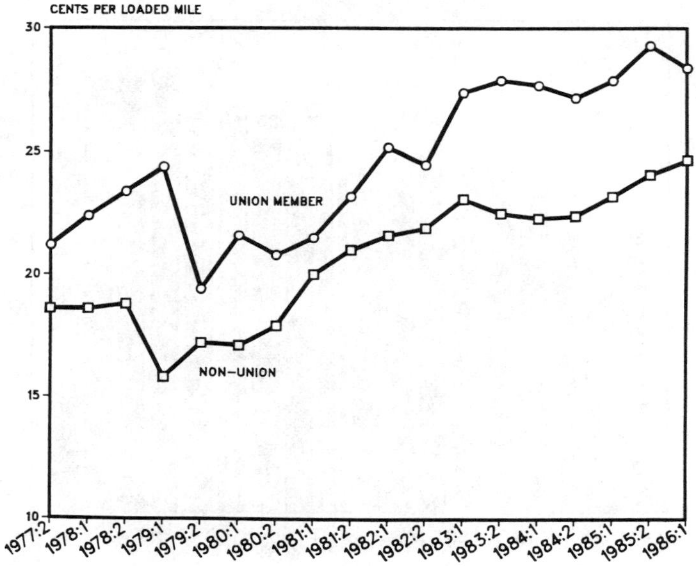

Figure 9.
Owner-Operators Increasingly Enter Into Leasing Arrangements
(Recessionary Periods Are Highlighted in Shaded Areas)

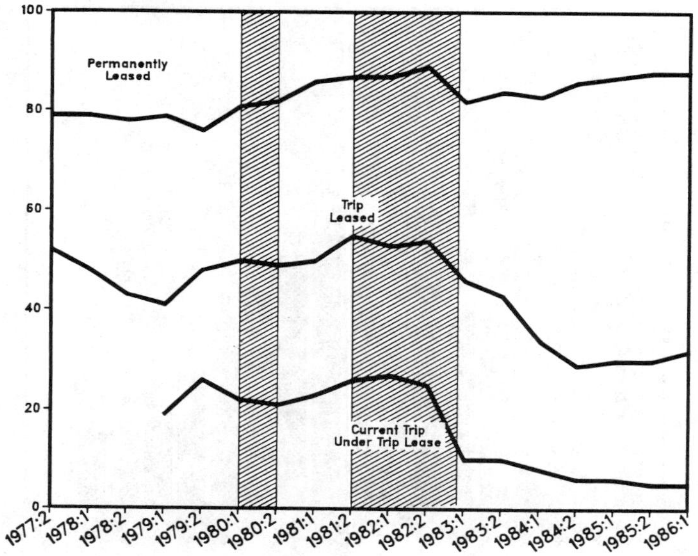

Figure 10.
Food and Farm Products Movements Dominate Intercity Rail-Competitive Motor Carrier Traffic Mixes
(1977-1986 Average)

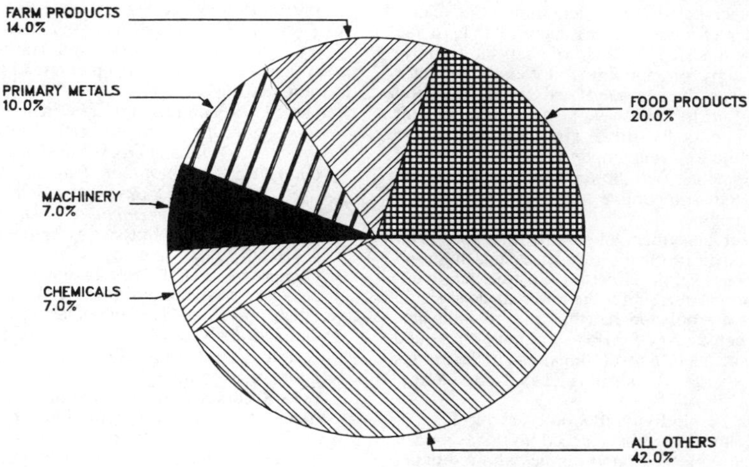

once common to envision the O-O sector of the market as true independents, due to increased market competition and increased insurance costs, more and more O-Os have sought the protection of lease arrangements.

While further erosion in the union's share of the market would have little additional impact on cost reductions (due to their small presence in the market), relative cost differences between union and non-union drivers may continue to fall as unions accept wage reductions or only nominal increases to maintain employment of their members.

C. Commodity

The broad shifts in the driver sector are not mirrored in commodity shares. Food and farm products accounted for about one-third of total movements throughout the 1977-1986 period. On the average, primary metals had the next largest share accounting for one in ten loaded movements. Machinery and chemicals each maintained a 7 percent share. (Refer to Figure 10)

A few shifts in commodity shares did occur. Primary metals' share fell 42 percent, and machinery's share fell 25 percent. This no doubt reflects the declining importance of the industrial sector to the U.S. economy. On the other hand, chemicals increased their share by 60 percent. (Refer to Table 4).

Rail competitive trucking's reliance on chemicals and food and farm products could increase in the future. Research at the AAR's Intermodal Policy Division indicates that railroads will find it increasingly difficult to hold onto these commodities to avoid diversion to truck competition.[32]

VI. CONCLUSIONS AND IMPLICATIONS

The rail-competitive motor carrier industry has become more efficient and more effective. Due to a variety of reasons, the motor carrier industry did not immediately undergo significant changes in operational methodologies allowed them by deregulation. Given recessionary pressures of low market demand, high interest rates and energy costs, the time required for existing firms to reposition themselves to

TABLE 4
Portion of All Truck Movements

YEAR	FOOD PRODUCTS	FARM PRODUCTS	PRIMARY METALS	MACHINERY	CHEMICALS
1977	19%	14%	12%	8%	5%
1986	19	11	7	6	8

Source: National Motor Transport Data Base

compete with new entrants in the new "open" transportation market, many carriers were unable to make significant changes during the 1980 to 1983 period. However, since 1983 with an expanding economy, relatively low energy and interest costs, significant changes in the rail competitive motor carrier sector have occurred. These changes built upon and exacerbated a consistent trend of increased motor carrier efficiency dating back to at least the beginning of this study's period of analysis. By relying increasingly on non-union drivers, the motor carrier industry has lowered real wages rates by almost one-fourth. They have also lowered per unit equipment costs by increasing annual tractor mileages while lowering empty mileages, and use of larger equipment. With lower costs in labor and equipment, rail-competitive trucking has become a formidable competitor.

The recent enactment of the 65 mile per hour (MPH) speed limit on the rural interstate highway system will have little effect on rail-truck competition.[33] Motor carriers prefer the lower speed since 55 MPH saves not only on fuel but also maintenance and tire expense. At 55 MPH "engines, tires, and other components turn revolutions, and brake linings absorb less energy. The result is less wear and longer service life."[34]

Although the individual motor carrier firm or owner operator may not succeed in today's marketplace, the increased opportunities allowed under deregulation have certainly benefited the trucking industry as a whole. Incremental cost reduction trends and increased competitiveness will continue as the newer technologies and operational methodologies diffuse to a greater extent in the motor carrier industry. Temple, Barker & Sloan (TBS) in its analysis of the rail industry's condition pointed out that the competitive "environment has changed radically and will continue to change". This analysis highlighted reductions in truck costs and growing shipper emphasis on total logistics and higher quality service. These trends will put more pressure on the rail industry especially if the proposed new size and weight limits (as exemplified by the Pinwheel Amendment) are enacted.[35]

ENDNOTES

*Association of American Railroads. The views expressed in this paper are solely those of the authors and may not reflect the views of the Association of American Railroads or its members.

1. Regular route common carriers (RRCC) are usually engaged in relatively short-haul-less-than-truckload (LTL) service. Although the railroad industry does carry LTL freight, it has traditionally represented only a small portion of total railroad tonnage. For example, STCC 47 tonnage originated represented only 0.007 percent of total railroad tonnage originated in 1986. Source; *1986 Freight Commodity Statistics*. While additional LTL traffic moves under other STCCs (e.g. 44 and 45), due to the relatively short lengths of haul involved, it has not been traditionally considered directly competitive with LTL trucking. Additionally, in the collection of the NMTDB, RRCC data is not actively sought. Therefore, while certain LTL movements by RRCC might be rail competitive, and some partial RRCC information is contained within the NMTDB, to provide consistency over the time frame studied, this category of motor carriers has been excluded from the analysis.
2. Examples include the Department of Commerce Commodity Transportation Survey and Truck Inventory and Use Survey.
3. TRNC's "The Blue Book of the Trucking Industry," and the Transportation Association of America once provided a plethora of transportation statistics.
4. Until 1979, contract carriers were not allowed to serve more than eight separate shipper.
5. Wyckoff, D. D. and D. H. Maister, *The Owner-Operator: Independent Trucker,* Lexington Mass, Lexington Books, D. C. Heath, 1975; D. H. Maister, *Management of Owner-Operator Fleets,* Lexington Mass, Lexington Books, D. C. Heath, 1980; and M. H. Agar "Toward an Owner-Operator Theory of the Carrier," *Proceedings the Transportation Research Forum,* 1984, pp 410-414, among others have presented detailed analyses of the rail-competitive motor carrier market from either a pre- or post-regulatory viewpoint.
6. Data Resources Incorporated in a recent assessment of the motor carrier industry after deregulation states "Most portrayals of the motor carrier industry rely on data that come from the handful of carriers (typically less-than-truckload) that are either publicly held or large enough to both filing accurate reports with the ICC." They go on to state "Although lack of reliable data makes a bottom-up analysis of TL activity impossible, individual carrier reports as well as strong circumstantial evidence support the assertion that truckload carriage was the fastest-growing segment in the motor carrier industry during the first half of the 1980's. Source: Data Resources U.S. Review, p. 23-24, March 1987.
7. Glaskowsky, Nicholas A., *Effects of Deregulation on Motor Carriers,* Eno Foundation, Westport, Conn., p. 8, 1986.
8. Over capacity in the inland waterway industry has also added to competition in the area of bulk commodities.
9. The 1977 Commodity Transportation Survey (CTS) indicated that the rail industry transported 24.2 percent of all manufacturing tonnage shipped. (Manufacturing tonnage was defined as those commodities falling within the Standard Industrial Classification (SIC) codes of 20 through 39). By contrast, the 1983 CTS indicated that total rail market share had fallen to 19.6 percent. Source: Bureau of the Census, Business Division, Washington, D.C.
10. Association of American Railroads, *The Coal Book,* 1986, pp. 18, 22, and 24.
11. Association of American Railroads, *The Grain Book,* 1986, pp. 8 and 14.
12. Transportation Policy Associates, *Transportation in America,* 4th Edition, 1986, and Supplement, November 1986.)
13. Lane, L. Lee, *"Innovation in Trucking: Advanced Truckload Firms,"* Transportation Research Board Record, forthcoming; *1986 Yearbook of Railroad Facts,* Association of American Railroads.

14. Data Resources Incorporated (DRI) forecasts a 3.4 percent decline in the Wholesale Price Index for gas fuels (WPI053) during the 1986 to 1989 period.
15. *Commercial Carrier Journal,* "The Insurance Dilemna," July 15, 1985, p. 143. Thomas Donohue, American Trucking Association President statement to the U.S. Senate Committee of Commerce, Science and Transportation, September 27, 1985.
16. Telephone conversations with several insurance industry agents.
17. The financial vehicle employed to fund these losses is a confirmation of self-insurance of predictable losses and the purchase of insurance to cover catastrophic losses. Refer to Wolfe, *"An Examination of Risk Costs Associated with the Movement of Hazardous Materials,"* Proceedings of the Transportation Research Forum, pp. 228-229, 1984.
18. Under depreciation accounting the initial track installation is added to the property accounts as an asset. The cost of the property (less salvage value) is prorated and annually written off over the life of the asset. Under retirement-replacement-betterment (RRB) accounting the asset was never written down. Track replacements were written off in the year they occurred as expenses. If improvements were made, the incremental investment would be added to the "frozen" asset base. Only when the line was abandoned would the original investment and capital improvements be written off as expenses. Salvage was treated as an asset. Refer to *"Alternative Methods of Accounting for Railroad Track Structures,"* ICC, Docket No. 36988, Decided January 26, 1983.
19. For example, under the old tax plan a $75,000 tractor would be written off in three years ($18,750, $28,500, and 27,750 in annual depreciation). The new tax law accelerated this to ($25,000, $33,332, and 16,668). Source: Ernst & Whinney, *"Equipment Replacement Strategies of Truckload Carriers and the Tax Reform Act of 1986,"* special study prepared for the AAR, p. 13, January, 1987.
20. "Excess production capacity has existed in the tractor manufacturing industry for the past five years. This trend is likely to continue for the next five years, depressing prices for new equipment." Source: Ibid #19, p. 7.
21. Larkin, Frances M., and K. Eric Wolfe, "Rail-Competitive Truck Characteristics," *Proceedings of the Transportation Research Forum,* Volume 2, pp. 145-155, 1983.
22. At the conclusion of a loaded trip, commodities for a return load are often not available at the same location. The process of driving with no load to another location (in order to secure a load for the return trip, the backhaul) is considered to be the deadhaul.
23. Corsi, Thomas M., Michael H. Agar, and Merrill J. Roberts, *"Dynamics of Owner-Operator Behavior and Profitability: 1978 Versus 1984,"* Proceedings of the Transportation Research Forum, pp. 123-136, 1985.
24. When quizzed as to what alternative occupation might be followed if the respondent could not be an 0-0, most 0-0s indicated that they would be farmers, mechanics or join the armed forces. Although 0-0 income has declined in real terms, nominal annual earnings of $22,000, the "psychic income" derived from being "your own boss" continue to draw new entrants into the ranks of the 0-0s. Source: NMTDB.
25. During the study period, the average length of haul reported for solo drivers in the data base was roughly 1,400 miles; the average miles traveled on all annual basis averaged 113,000. Thus, the average truck is estimated to make 81 loaded trips per year. As estimates of the total number of trucks in intercity service range from six to eight hundred thousand, the total loaded truck movements might range between 49 and 65 million. Since the database collects 32,000 or so loaded movements per year, it is sampling all intercity movements at a rate of less than 0.1 percent.
26. In addition to RRCC, observations involving household goods movers, cargo weights of less than 5,000 pounds or lengths of haul less than 250 miles were excluded from this analysis. These excluded observations made up less than 5 percent of the total database.
27. The DRI total motor carrier financial database includes as many as 250 financial and operating statistics for over 2,200 Class I and II motor carriers. For selection of those carriers thought to be the most rail-competitive, only data from the top 50 firms was used. Source: ICC and ATA.
28. Data Resources Incorporated *"Review of the U.S. Economy,"* March 1987, p. 28.
29. Uncertainty as to what extent twin 28 foot and single 48 foot trailers would be allowed to operate in all states also delayed the employment of larger trailers. For example, the Federal Highway Administration had to take the state of Connecticut to court to force acceptance of larger equipment.
30. This plan would allow seventeen western states to raise vehicle weight limits to Federal bridge formula levels. This would mean the end of the 80,000 pound limit in the west. Turnpike doubles (paired 48′ trailers) could be operated at 134,000 pounds and would be especially rail-competitive due to their lower operational cost structures.
31. This estimate (based in constant 1986 dollars) is from the AAR truck cost model based on the following assumptions: 45′ trailer in 1977 versus 48′ in 1986, 4.8 miles-per-gallon in 1977 versus 5.5 in 1986, and 106,000 annual miles in 1977 versus 120,000 in 1986.
32. AAR, Intermodal Policy Division unpublished paper, *"Advanced Truckload Competition,"* 1986.
33. Of the 43,291 miles of the U.S. interstate highway system, over 78 percent (33,910 miles) could be designated for 65 MPH travel. Source: Associated Press/Senator Steve Symms (R-Idaho).
34. Standley, Gerald F., "The Economics of 55," *Commercial Carrier Journal,* December 1986, p. 9. See also *Saving Truck Fuel: 2 Case Histories,* Trucksawe, Ontario Ministry of Energy.
35. Temple, Barker & Sloane, Inc., *"The Changing Rail Industry,"* Washington Representatives Briefing, January 7, 1987.

Mobility Barriers in the Motor Carrier Industry

*by Thomas M. Corsi and Curtis M. Grimm**

I. INTRODUCTION

The Motor Carrier Act of 1980 (MCA) substantially reduced regulation in the motor carrier industry. The post-MCA environment gives motor carriers far greater freedom in deciding which markets to enter and what commodities to carry. The purpose of this paper is to investigate a particular class of motor carrier strategic decisions—commodity specialization. With the advent of deregulation, there has been increasing attention directed toward the strategic management literature and its application to trucking strategies.

Organizational strategy deals with major decisions by firms as they attempt to meet the challenges of the marketplace (Hambrick 1980). Dimensions of strategy include pricing policies, customer service, advertising and marketing efforts, financial control measures, and many other management actions. It is generally agreed that a firm's performance is at least partially determined by management decisions regarding approaches to executing the basic functions of the firm.

Research to date on firm strategies in the motor carrier industry (McGee, 1982; Tye, 1983; Walters, 1985) has been largely descriptive, with a focus on interviews and case studies. Both emphasize expansion of existing services and diversification into new services, appropriate reactions to the environmental changes brought by the MCA of 1980. However, there are disagreements among analysts regarding specific motor carrier strategies. For example, Tye (1983) foresees a decrease in selective service as rates on previously unprofitable traffic rise, while McGee (1982) envisions "greater concentration of carriers' assets to meet traffic movements." Such statements reflect a degree of uncertainty over what strategies are being practiced and point to the need for further strategy research.

Two empirical studies of motor carrier strategies have been performed. Booz, Allen, and Hamilton (1984) identified management and operating characteristics of carriers that performed above and below average. They found that successful firms focused more on marketing, customer service, and firm diversification than did their less successful counterparts. The study used discriminant analysis to separate the firms and then analyzed the groups to ascertain the most important differences. Not surprisingly, since firm strategy has been demonstrated to greatly affect performance in many industries, strategic factors such as marketing, service quality, and diversification were found to explain much of the difference in performance between firms.

The second empirical study was performed by Corsi, Grimm, and Smith (1986). The authors used Porter's (1980) framework regarding generic strategic groups to derive and operationalize two key dimensions for motor carrier strategies: low cost and product differentiation. Then, motor carrier annual report data for 1977 and 1984 was presented to test the hypothesis that on average, carriers increased their emphasis on these strategic dimensions with deregulation. Support was found for this hypothesis, with strongest evidence on a greater emphasis on product differentiation in 1984 versus 1977.

These previous studies have provided insight into actual strategic factors employed in the industry. They also support other theoretical and empirical findings that strategy affects performance. However, there has been no analysis of a key motor carrier strategic variable, choice of product focus. Motor carrier firms can choose to carry a variety of commodities: general freight (less-than-truckload), general freight (truckload), household goods, refrigerated products, petroleum products, etc. Given the freedoms granted carriers to change their focus, it is of interest to determine the degree to which firms have shifted commodity focus and the types of shifts which have occurred most frequently. In addition, this paper will examine the performance implications of a change in focus. The investigation will be guided by strategic management and economics literature on mobility barriers and the experience curve. The following section will review this literature and relate it to motor carrier commodity focus. Then, evidence on commodity shifts will be presented and analyzed.

II. EXPERIENCE CURVE AS A MOBILITY BARRIER

The decision by a motor carrier to switch commodity focus is analogous to a major change in strategy or strategic group. As discussed by Caves and Porter (1977), mobility barriers are elements which deter firms from duplicating another firm's strategy. In the process of strategic interaction, firms attempt to build and maintain mobility barriers. Other firms may try to overcome existing barriers by switching strategies; however, they should anticipate reactions from firms whose strategic territory is threatened.

Perhaps the most important mobility barrier is the cost and productivity advantage acquired by firms experienced with a particular strategy. The concept of an experience curve, popularized by the Boston Consulting Group (1972), posits a relationship between costs and accumulated volume of production. As discussed by Hax and Majluf (1982), the concept has been most commonly applied to manufacturing processes such as production of integrated circuits, air conditioners, or primary magnesium. Empirical studies performed for all of the above products reveal a substantial cost advantage for firms with larger cumulative production over time. The cost advantage derives from factors such as employee learning, specialization and redesign of labor tasks, product and process improvements, and system rationalization. The benefits of experience are often summarized under the label "know-how," which represents a rich knowledge of operational, managerial, and technological factors affecting costs and productivity in the firm.

The concept of the experience curve can be applied to the motor carrier's choice of commodities. It would appear likely that firms switching commodity focus would be at an initial disadvantage versus carriers with significant experience hauling that commodity. Thus, we would expect firms that changed their primary commodity between 1977 and 1985 to have higher costs, lower productivity, and lower profits than firms which carried the same commodity in both 1977 and 1985. Furthermore, we would anticipate firms which shifted commodities may perform worse than those firms which carried the same commodity in 1977 and 1985.

Our purpose then is to investigate the extent of mobility barriers, particularly the experience curve, in the motor carrier industry. We will first investigate the proclivity of strategic change regarding commodity focus and identify the most common commodity alterations. To the extent that strong mobility barriers exist, we would expect few firms to change their commodity strategy between 1977 and 1985. Then, we will examine more carefully the performance impacts of this change. Specifically, 1985 performance of firms changing commodity focus will be compared with firms specializing in the same commodity who did not switch from their 1977 focus. To the extent that the experience curve is applicable, we would expect firms inexperienced with handling a particular commodity to have higher costs, lower productivity, and lower profits. To gain further insight on the ramifications of switching commodity focus, we will compare profitability of firms who changed focus with that of the firms with the same primary commodity in 1977 who did not change.

III. DATA AND METHODS

The authors obtained 1977 and 1985 data tapes from the American Trucking Associations. These tapes were compiled from the Annual Reports of Class I and II motor carriers filed with the Interstate Commerce Commission (ICC) in those respective years. In its annual report, each carrier denotes its primary commodity. The seventeen commodity classifications, used in the annual reports correspond to the categories in the ICC's commodity classification scheme (See list in Table 1).

Each carrier's primary commodity designation in 1977 was compared to its 1985 designation. The data across segments and time were summarized in tabular form to indicate the following: (a) percentage of firms by segment with the same primary commodity designation in 1977 and 1985 and (b) percentage of firms by segment shifting commodity focus between 1977 and 1985. The data in these tables document the extent to which carriers have changed their commodity focus during the transition from a highly regulated environment to a less regulated one. In addition, the analysis shows which segments in 1985 have the highest percentage of carriers who have entered from other segments between 1977 and 1985.

After documenting the extent to which firms changed their primary commodity during the transition, the paper will focus on investigating the impacts of this major strategic change. Specifically, the annual report data will be used to compare the 1985 costs, productivity, and profits of firms that changed their primary commodity between 1977 and 1985 with those of the firms specializing in the same commodity who made no such change between 1977 and 1985. The difference of means test will be employed by segment to indicate the statistical significance of the observed mean difference in costs, productivity, and profits between the two groups of carriers. The cost measures used are the following: (1) operating expenses per mile (ton-mile); (2) line haul expenses per mile (ton-mile); (3) pick-up and delivery expenses per ton; (4) general and administrative expenses per mile (ton-mile); (5) insurance and safety expenses per mile; (6) traffic and sales expenses per mile; and (7) maintenance expenses per

TABLE 1
Stability in Commodity Focus by Industry Segment, 1977-1985

Industry Segment	% of Firms with Same Primary Commodity 1977 and 1985	# of Firms
General Freight (LTL)	94	227
General Freight (Truckload)	50	105
Heavy Machinery	75	28
Petroleum Products	85	92
Refrigerated Solids	79	58
Agricultural Commodities	52	50
Building Materials	79	61
"Other" Specialized Carriers	75	316
Overall	77	937

mile. The productivity measures compared are: (1) vehicles miles per truck and (2) ton-miles per vehicle mile. The profit measures used are: (1) operating ratios and (2) net income/total assets.

In addition, the 1985 profits of the carriers who shifted their primary commodity between 1977 and 1985 will be compared to the 1985 profits of firms with the same primary commodity in 1977 who did not make such a strategic change. Again, observed differences between the two groups will be tested for statistically significant differences through the use of a difference of means test.

A few comments regarding data limitations are in order. The Annual Report data used in this study is limited by the extent to which carriers failed to file reports in either 1977 and 1985. However, the large sample size (937) ensures a rich representation of motor carrier behavior. Moreover, for the purposes of this analysis the failure by a limited number of firms to report does not bias study results unless under- reporting is significantly different for carriers who shifted commodity focus as opposed to those who made no such shifts. However, there is no basis for such a bias.

Secondly, a more complete study of commodity behavior is precluded in that only the primary commodity is indicated on the annual reports. To the extent that carriers are newly handling traffic in other segments without changing their primary focus, some of a firm's commodity diversification activities are masked. The paper then is limited to analysis of the major strategic change of altering one's primary commodity.

The remaining sections of this paper summarize the extent to which firms made major strategic changes by shifting their commodity focus during the transition years, evaluate the impact of this type of major change, provide insights into the existence of mobility barriers in the industry, and assess the implications of the empirical results for the industry managers.

IV. SHIFTS IN COMMODITY FOCUS

Table 1 substantiates by industry segment the extent to which there was stability among motor carriers in their primary commodity focus between 1977 and 1985. The data in Table 1 are based upon the primary commodity identification supplied by 937 Class I and II motor carriers in both 1977 and 1985.

Overall 77 percent of the firms made no change in their primary commodity focus during the study period. However, the overall indication averages a substantial variation in shifts among the individual industry segments. Those with the highest levels of strategic shifts in commodity focus are: the general freight (truckload) and agricultural commodity segments. Those with very low levels of strategic shifts are: the petroleum products and general freight (LTL) carriers.

Among the 105 firms with a general freight (truckload) focus in 1977, only 50 percent had that same designation in 1985. Of the 53 firms who changed their commodity focus, 34 shifted into the general freight (LTL) designation, while 13 moved into the "other" specialized commodity category. Among the 50 firms with agricultural commodities as a primary focus in 1977, only 52 percent kept that focus during the transition period. Of the 24 firms who shifted their commodity designation, 8 moved to the refrigerated solids segment, 7 switched to "other" specialized commodities, and 5 went to the general freight (truckload) segment.

In sharp contrast, 85 percent of the 92 petroleum products carriers in 1977 also had the same commodity focus in 1985. Similarly, 94 percent of the 227 general freight (LTL) carriers in 1977 also had the same commodity focus in 1985.

The remaining four industry segments (heavy machinery, refrigerated solids, building materials, and "other" specialized carriers) all had between 70 and 79 percent of the firms in 1977 keep the same commodity focus throughout the study period. Among the 78 "other" specialized carriers who shifted their commodity focus between 1977 and 1985, 50 moved into the general freight (truckload) segment.

Table 2 analyzes the extent of commodity shifting by industry segment. It shows that overall 23 percent of the motor carriers in 1985 shifted their commodity focus between 1977 and 1985. Again, this overall average masks some important variations in

TABLE 2
Extent of Commodity Shifting by Industry Segment, 1977-1985

Industry Segment	% of Firms who Shifted into the Segment between 1977-1985	# of Firms
General Freight (LTL)	15	251
General Freight (Truckload)	62	136
Heavy Machinery	32	31
Petroleum Products	6	83
Refrigerated Solids	22	59
Agricultural Commodities	19	32
Building Materials	19	59
"Other" Specialized Carriers	17	286
Overall	23	937

shifting patterns among the industry segments. For example, among the 136 general freight (truckload) carriers in 1985, 62 percent shifted into this segment between 1977 and 1985 from another segment. Among the 31 heavy machinery carriers in 1985, 32 percent changed their commodity focus between 1977 and 1985. In contrast, among the 83 petroleum products carriers in 1985, only 6 percent moved into this segment between 1977 and 1985.

As posited earlier, existence of strong mobility barriers in the motor carrier industry would be substantiated by the finding of little shifting in commodity focus by the firms between 1977 and 1985. Results in Tables 1 and 2 substantiate that some of the industry's segments have much higher mobility barriers than do others.

Some segments, such as general freight (LTL) and petroleum products carriers, appear to have characteristics which makes shifting into another segment or out of the segment difficult. The general freight (LTL) carriers have extensive terminal networks or a heavy capital investment compared to carriers in other industry segments. Furthermore, such carriers require extensive customer support systems and marketing expertise that transcend the requirements in other industry segments. The petroleum products carriers have highly specialized, expensive trailers requiring high maintenance. Furthermore, these assets are not transferable to other industry segments. Thus, in these two segments the required infrastructure development and equipment specialization would increase the likelihood that firms could gain significant cost and productivity advantages through experience. Hence, this preliminary evidence points to the existence of existence mobility barriers in these two industry segments.

In contrast, other segments, such as general freight (truckload), have a much lower percentage of their firms staying in the same segment between 1977 and 1985. Furthermore, a high percentage of firms in these segments in 1985 have shifted into that segment between 1977 and 1985. The general freight (truckload) segment does not have specialized equipment needs nor does it require extensive capital expenditures for terminals. Certainly, evidence in this segment points to a less complicated operating environment, fewer advantages to be gained from experience, and less likelihood of mobility barriers.

Having documented the extent of strategic shifts in product focus and discussed the implications regarding the existence of mobility barriers in the industry, it is critical to examine the performance of the carriers who shifted their commodity focus with those who made no shifts as a way of confirming the strength of mobility barriers in the industry. It is obviously important to focus the performance evaluation on some of the industry segments (general freight—LTL) identified in a preliminary way as either having mobility barriers or not (general freight—truckload).

V. PERFORMANCE IMPACTS OF SHIFTS IN COMMODITY FOCUS

The objective of this section is to explore the existence of mobility barriers due to experience advantages by examining the performance impacts of a shift in commodity focus. To the extent that experience advantages exist in the industry, we anticipate that firms who changed their commodity focus between 1977 and 1985 would have higher costs, lower productivity, and lower profits than would firms who did not switch during the study period. Further insight on the impact of shifting commodity focus is gained by examining the profitability of firms who changed focus with that of the firms with the same primary commodity in 1977 who did not change.

There are three commodity groups with sufficient number of firms changing to their focus to allow for formal statistical comparisons. The performance impacts of these shifts in commodity focus are displayed in Table 3. First, carriers who switched their focus between 1977 and 1985 to the "other" specialized commodity segment outperformed "other" specialized carriers who held that focus throughout the study period. The firms for which this focus was new had statistically significantly lower costs (operating expenses per mile/ton-mile; line haul expenses per mile/ton-mile; and general and administrative expenses per mile/ton-mile), higher productivity (ton-miles/vehicle miles), and lower operating ratio and higher profits (net income/total assets). Thus, in the "other" specialized commodity sector, there appear to be no significant experience curve advantages. On the contrary, the firms with shifting focus are able to significantly outperform the stable firms across a wide range of key dimensions.

Similarly, carriers moving into the general freight (truckload) segment during the transition outperformed those who stayed in that segment throughout on key cost and productivity measures. Specifically, the carriers who did not shift focus had statistically significantly higher averages than did the "shifters" on numerous cost measures (operating expenses per mile/ton-mile; line haul expenses per mile/ton-mile; general and administrative expenses per mile/ton-mile; traffic and sales expenses per mile). The carriers with a strategic change in focus had statistically significantly higher performance than did the non-shifters in the following productivity categories: vehicle miles/vehicles and ton-miles/vehicle miles. Interestingly, however, despite the cost and productivity advantages, carriers with the new commodity focus did not have statistically significantly higher profits than did the non-shifters. The explanation for this finding may lie in the highly competitive rate environment in this particular industry segment. Carriers with the new strategic focus may have passed along cost and productivity gains to shippers in the form of major rate discounts, perhaps, in an effort to establish market presence. Nonetheless, the results clearly substantiate the earlier indication of low mobility barriers in the general freight (truckload) segment. In addition to a high percentage of firms shifting out of and into the segment, the entering carriers have cost and productivity advantages over the non-shifters.

Finally, turning to the third commodity, Table 3 reveals that there are no statistically significant performance differences between carriers who switched their strategic focus to general freight (LTL) carriers during the transition and those who stayed in this category throughout. These results though not as strong as the findings in the other two segments, also point to an absence of mobility barriers. If there were clear learning curve advantages from experience, the expectation is that the new entrants would have statistically significantly higher costs, lower

TABLE 3
Performance Impact of Shifts in Commodity Focus

1985 Measures	Industry Segments		
	New "Other" Specialized	Stable "Other" Specialized	Sig. of Diff
Cost			
Operating Expense			
Per Mile/Ton-Mile		+	***
Line Haul Expense			
Per Mile/Ton-Mile		+	**
General & Admin. Expense			
Per Mile/Ton-Mile		+	*
Insurance & Safety Expense			
Per Mile/Ton-Mile			
Traffic & Sales Expense			
Per Mile			
Maintenance Expense			
Per Mile			
Productivity			
Miles Per Truck			
Ton-Miles/Vehicle Miles	+		*
Profit			
Operating Ratio		+	*
Net Income/Total Assets	+		**

	New GF (LTL)	Stable GF (LTL)	Sig. of Diff
Cost			
Operating Expense			
Per Mile/Ton-Mile			
Line Haul Expense			
Per Mile/Ton-Mile			
General & Admin. Expense			
Per Mile/Ton-Mile			
Insurance & Safety Expense			
Per Mile/Ton-Mile			
Traffic & Sales Expense			
Per Mile			
Maintenance Expense			
Per Mile			
Productivity			
Miles Per Truck			
Ton-Miles/Vehicle Miles			
Profit			
Operating Ratio			
Net Income/Total Assets			

*** Statistically significant difference of means at .01 level of confidence
** Statistically significant difference of means at .05 level of confidence
* Statistically significant difference of means at .10 level of confidence

The mean values and t-statistics are available on request from the authors.

TABLE 3 (continued)
Performance Impact of Shifts in Commodity Focus

1985 Measures	Industry Segments		
	New GF (Truckload)	Stable GF (Truckload)	Sig. of Diff
Cost			
Operating Expense Per Mile/Ton-Mile		+	***
Line Haul Expense Per Mile/Ton-Mile		+	**
General & Admin. Expense Per Mile/Ton-Mile		+	***
Insurance & Safety Expense Per Mile/Ton-Mile			
Traffic & Sales Expense Per Mile		+	***
Maintenance Expense Per Mile			
Productivity			
Miles Per Truck	+		***
Ton-Miles/Vehicle Miles	+		***
Profit			
Operating Ratio			
Net Income/Total Assets			

*** Statistically significant difference of means at .01 level of confidence
** Statistically significant difference of means at .05 level of confidence
* Statistically significant difference of means at .10 level of confidence

The mean values and t-statistics are available on request from the authors.

productivity, and smaller profits. These findings, however, indicate that carriers shifting their focus to the general freight (LTL) designation do not perform differently from the established, non-shifting carriers. Thus, the low levels of shifting out of and into the general freight (LTL) segment suggesting the existence of mobility barriers stemming from learning curve advantages are not substantiated by the performance results.

In summary, a switch in commodity focus does not appear to disadvantage a carrier, contrary to expectations based on the mobility barrier literature. In fact, the carriers who shifted out of the "other" specialized commodity segment between 1977 and 1984 had statistically significantly higher profits than did the carriers who remained in that segment throughout. There were no statistically significant profit differences between the carriers who changed their strategic focus from general freight (LTL) between 77 and 1985 and those who kept the same focus throughout. Likewise, there were no profit differences between the carriers who moved from the general freight (truckload) segment during the transition and those who made no such change. The Table 4 results are consistent with the findings outlined in the discussion of Table 3, there appear to be no statistically significant cost, productivity, and profit advantages to be gained from experience. In all segments, carriers with a new strategic focus did no worse (and in two segments did significantly better) than did those who remained in the segment. The evidence found in this empirical investigation does nor indicate mobility barriers across all segments in the motor carrier industry.

VI. SUMMARY AND IMPLICATIONS

There do appear to be barriers associated with certain segments, such as petroleum products, as evidenced by the small number of firms moving into or out of this commodity. However, where sufficient movement has occurred to conduct formal statistical tests, the results do not substantiate significant cost,

TABLE 4
Profitability Impacts of Shifts in Commodity Focus

1985 Measures	Industry Segment		
	Shifters from "Other" Specialized	Stable "Other" Specialized	Sig of Diff
Operating Ratio		+	**
Net Income/Total Assets	+		***
	Shifters from Gen Frt (LTL)	Stable Gen Frt (LTL)	Sig of Diff
Operating Ratio			
Net Income/Total Assets			
	Shifters from Gen Frt (Truckload)	Stable Gen Frt (Truckload)	Sig of Diff
Operating Ratio			
Net Income/Total Assets			

*** Statistically significant difference of means at .01 level of confidence
** Statistically significant difference of means at .05 level of confidence
* Statistically significant difference of means at .10 level of confidence

The mean values and t-statistics are available on request from the authors.

productivity, and profit advantages to firms based on their accumulated "know how" or experience. In fact, the findings indicate that the decision by carriers to shift their product focus does not disadvantage them in any systematic way. Evidence developed in the paper shows that in several industry segments firms shifting product focus outperform both the carriers who had the same original product focus but did not make a product shift as well as the carriers who already had the product focus that the "shifters" selected in the latter period.

The industry's violative environment may explain why there is no consistent evidence of mobility barriers based on the experience curve in the motor carrier industry. Industry procedures and practices have changed radically in conjunction with the sharp reduction in regulation. It is perhaps not surprising in such an environment that past experience under regulation is not particularly advantageous for the stable motor carriers. The environment seems to have favored carriers who were bold enough to shift their strategic focus to take advantage of market opportunities in segments where the established carriers failed to alter practices to adapt to the new environment.

This paper has identified the extent to which motor carriers have changed their commodity focus during the transition years and assessed the impacts of such changes. This research could be extended through examination of firms' rationale in switching commodity focus. Also specific decisions that enabled these aggressive firms to outperform carriers who made no changes in product focus should be evaluated.

Furthermore, the authors believe that a change in commodity focus is only one of many major strategic variables that should be evaluated. The authors intend to investigate other major strategic decisions such as geographic scope of operations. We believe a systematic assessment of key strategic variables will yield much information about strategic management decisions in a dynamic regulatory environment.

REFERENCES

Booz, Allen & Hamilton, "The Motor Carrier Executive Survey Report: Strategies for Success in the Motor Carrier Industry," June, 1984.

Boston Consulting Group, *Perspectives on Experience*, Boston Consulting Group, Inc., Boston, MA 1972.

Caves, Richard E., and Porter, Michael, "From Entry Barriers to Mobility Barriers," *Quarterly Journal of Economics*, 91, 1977.

Corsi, Thomas M., Grimm, Curtis M., and Smith, Raymond D., "Motor Carrier Strategies in a Changing Environment: An Empirical Analysis," *Transportation Research Forum Proceedings*, 1986.

Hambrick, Donald C., "Operationalizing the Concept of Business Level Strategy in Research," *Academy of Management Review*, 5, 1980, pp. 567-575.

Hax, Arnoldo C., and Majluf, Nicolas S., "Competitive Cost Dynamics: The Experience Curve," *Interfaces*, Vol. 12, No. 5, October 1982, pp. 50-61.

McGee, Michael, "Changing Market Structure for the For-Hire Motor Carrier," *Transportation Research Forum Proceedings*, 1980, pp. 12-17.

Porter, Michael, *Competitive Strategy*, New York, Free Press, 1980.

Tye, William B. "Fundamental Elements of a Marketing Audit for a More Competitive Motor Carrier Industry," *Transportation Journal*, Spring 1983, pp. 5-22.

Walters, Timothy C., "Market Oriented Strategic Management for Motor Carriers," *Transportation Research Forum Proceedings*, 1985, pp. 554-558.

ENDNOTE

* Transportation, Business and Public Policy, College of Business and Management, University of Maryland, College Park, MD.

Sales Force Management in a Deregulated Environment: General Freight Carriers

by Paul R. Murphy, Jr., and Thomas M. Corsi***

ABSTRACT

This study investigates the sales force management practices of general freight motor carriers in the post-MCA competitive environment. The results support the contention that carriers have already changed or intend to change significant aspects of their sales force management activities in response to their new environment.

The significantly greater reliance by large carriers than by smaller ones on advertisements, customers, and universities as recruiting sources suggests that the larger firms are making a concerted effort to recruit salespeople who have advanced technical skills as well as those with a more thorough knowledge of customers based on direct experience.

The results show that a significant portion of the carriers intend to add various psychological tests to their selection criteria in an effort to be more scientific and systematic in obtaining salespeople whose skills are highly compatible with the stressful competitive environment in which the carriers now find themselves.

Finally, the results denote that the larger carriers devote a significantly higher share of their sales force training programs to pricing issues than do the smaller carriers. It is reasonable to expect that the larger carriers have a wide range of customers with specific price/service needs and, as a consequence, a need for salespeople with much more extensive training in pricing.

I. INTRODUCTION

The examination of sales force management in the motor carrier industry is of particular relevance for several reasons. First, the significant differences between the marketing of a service (such as truck transportation) and the marketing of a good place primary emphasis on achieving an efficient, productive sales force in a service industry. Second, the Motor Carrier Act of 1980 (MCA) created a more competitive environment requiring motor carriers to distinguish their services (especially on the price dimension) from those of their competitors. Such differentiation dictates that the industry's sales force acquire additional skills that were of significantly less importance in a highly regulated environment.

Given the critical importance of the sales force in the motor carrier industry, it is surprising that no systematic, empirical investigation of the industry's sales force management practices have been conducted to date. The objective of this paper is to provide such an investigation of the practices of one industry segment—namely, the general freight carriers.

A. Differences Between Goods and Services Marketing

While the marketing of services has much in common with the marketing of goods, important differences do exist. One distinguishing characteristic is that services are more intangible than tangible and, as a consequence, cannot be easily displayed.[1] For instance, a computer salesperson can bring portable computers to potential clients, but a motor carrier salesperson cannot produce such a tangible item to demonstrate freight services offered by the company.

Second, unlike goods marketing, services marketing involves simultaneous production and consumption of the individual service.[2] This particular quality suggests that services must be provided in the right way. As such, discourteous service personnel can have significant ramifications for future business; inconsiderate and rude sales personnel could lose customers permanently for a motor carrier.

Third, services may be far less standardized than are goods.[3] One would expect little variability from items produced by the same machine, as the operating environment remains relatively unchanged during the production process. However, with services, the operating environment may differ tremendously from production to production, thus causing variability in the particular service. For instance, transportation on-time performance is generally better in good weather than it is in bad weather.

Finally, goods and services differ in that services cannot be inventoried.[4] That is, tangible items can be stored to meet uncertainties in either supply or demand. However, the truck capacity that goes unused on a particular trip cannot be saved for use at a later date.

These distinct differences between the marketing of goods and the marketing of services result in differences in the types of individuals and the skills needed for selling services as opposed to selling goods. For example, personal selling may play a greater role in services marketing than it does in goods marketing. Zeithaml points out that one strategy for dealing with the intangibility of services is for personal sources to be used more often than nonpersonal sources.[5] A study by Kelly and George found the relationship between the buyer and the salesperson is more important in services marketing than it is in goods marketing. Moreover, they discovered that customers need more assistance when buying services, primarily because services cannot sell themselves.[6] With respect to transportation, Hynes finds that the motor carrier user may have difficulty in distinguishing between different carriers' services and that carrier choice "*may be influenced by the quality and training of the carrier's salespeople. . . .*"

B. The New Competitive Environment

Sales force management in the motor carrier industry has taken on new importance as a result of the new competitive environment ushered in with the passage of the MCA of 1980. It is logical to expect that carriers, forced for the first time to compete with each other on a price dimension, would have to devote additional resources to marketing and, in particular, the development of new skills in their sales force. It is not surprising, then, that a recent study by Cross and Beier found marketing planning to be a relatively new activity in many motor carrier companies. Moreover, they discovered *personal selling* to be an extremely important component of a carrier's marketing program.[8]

Sales force management in the new environment can provide a carrier with a noticeable competitive edge over rival firms. *Distribution Magazine* conducted a survey (August 1986) of transportation users, asking them to rate particular carriers on four marketing dimensions (service, convenience, price, and sales). The respondents rated the industry as being weakest on the sales dimension, which included knowledge by the sales force of services and customers' needs.[9] Increased attention to sales force personnel could allow carriers to cultivate an effective, efficient sales force that is willing to work with users to satisfy their freight needs. Such a sales force could provide a carrier with a noticeable competitive advantage in the post-MCA environment.

C. Research Objectives

Having established the importance of effective sales force management in the motor carrier industry, particularly in the post-MCA environment, it is all the more critical that a systematic investigation be made of its sales force management techniques. The current research provides an investigation of the sales force management practices of one large industry segment most directly impacted by the MCA—i.e., the general freight segment. The main research objective is to develop evidence to answer the following questions.

- What sources are used to recruit motor carrier salespeople?
- How do the sources used differ among carriers based on their size, their profitability, and the geographic scope of their operations?
- What techniques are used to select motor carrier salespeople?
- How do the selection techniques differ among carriers based on their size, their profitability, and the geographic scope of their operations?
- What is the content of sales force training programs used in the motor carrier industry?
- How does the content differ among carriers based on their size, their profitability, and the geographic scope of their operations?

It is expected that the answers to these questions will be of general use to motor carriers in all industry segments as a basis for comparison with their own current sales force management techniques. Furthermore, the results will identify whether large and/or more profitable motor carriers in a particular industry segment exhibit any distinguishing recruitment, selection, and/or training practices which other carriers may want to model in modifying their current programs. Finally, the results concerning sales force management practices in a service industry undergoing vast changes in its regulatory environment will serve as a basis for comparison with the experiences of other service industries and/or other industries undergoing rapid environmental changes due to the changes in regulations.

II. DATA AND METHODS OF ANALYSIS

A. Data

The authors collected information on sales force management practices by means of a mail survey sent to all Class I and II motor carriers of general freight. Specifically, the authors obtained from the Interstate Commerce Commission the computer list of all ICC-regulated motor carriers. From that list, all Class I and II motor carriers classified as Instruction 27 carriers—i.e., general freight carriers with at least 75 percent of their revenues derived from hauling general freight—were selected for sampling. The 1985 Annual Reports of all Class I and II motor carriers were also checked to verify that all carriers meeting the criteria of Instruction 27 carriers were included in the universe. All carriers thus identified were sent questionnaires. Excluding firms who had discontinued operations since the end of 1985, 347 firms were identified as meeting the criteria cited above.

A concerted effort was made to send the survey to the highest ranking sales or marketing executive. If such a person could not be identified, then the survey was addressed to the company President. Each of the firms received an initial mailing of the survey. Non-respondents were sent a follow-up survey approximately one month after the initial mailing. These two mailings resulted in 134 usable surveys being returned—a response rate of 39 percent.

The 347 firms constituting the universe of general freight carriers with at least 75 percent of their revenues from general freight (i.e. Instruction 27 carriers) were divided into twelve strata based on their 1985 operating revenues for the purpose of developing expansion factors. As shown in Table 1, the response rate in each stratum is at or above 21 percent with the rate exceeding 50 percent in five strata. On the basis of the response rate in each stratum, expansion factors were developed for each stratum through the application of the following formula: $1 / r$, where r is the response rate in a particular stratum.

In the remaining sections of the paper, results are reported using the expansion factors developed in Table 1. Thus, tables reporting the number of firms answering a particular question are based on an expansion of survey results to the universe. Furthermore, tables reporting percentage of firms answering a particular way are also based on the application of the expansion factors to each stratum.

TABLE 1
Expansion Factors for Revenue Groups General Freight, LTL Carriers

Revenue Groups* ($000's)	Response Rate	Expansion Factor
402-1,967	0.25	4.00
2,031-2,993	0.21	4.77
3,081-4,894	0.29	3.42
5,131-7,399	0.37	2.73
7,882-12,235	0.53	1.89
12,548-17,488	0.44	2.23
17,877-24,911	0.52	1.93
25,223-48,651	0.52	1.93
53,439-99,663	0.65	1.53
103,390-244,752	0.44	2.29
302,221-673,676	0.57	1.75
1,329,571-1,428,750	0.33	3.00

* Based on 1985 Total Operating Revenue

B. Analysis Methods

In discussing the recruitment and selection techniques of the general freight carriers, frequency distributions are presented showing the expanded number of motor carriers who currently use, plan to use, and never use a particular technique. Logit analysis is employed to assess the combined influence of firm size (annual revenues), profitability (operating ratio), and geographic scope of operations (number of states) on the decision to use each selection and recruitment technique explored.
The information on sales force training content was continuous in nature (i.e., the questionnaire included information on the percent of total training effort devoted to each content item covered in the course). The content of the training programs currently in use is given by showing the expanded number of firms falling into a particular percentage group (e.g. the expanded number of firms that spend less than 10 percent of their training program on operations). The continuous nature of the training variables enables the use of multiple regression to discover relationships between training content and the combined influence of the carrier attribute variables (firm size, operating ratio, and the geographic scope of operations).

The remainder of this paper is devoted to the presentation of empirical results regarding the sales force management practices of general freight carriers in the areas of recruitment, selection, and training of new salespeople. A final section summarizes empirical results and draws implications regarding future studies of sales force management in the motor carrier industry.

III. RECRUITMENT TECHNIQUES

A. Univariate Analysis

An initial component in sales force management concerns sources of recruitment for new salespeople. Information on the recruitment tactics of motor carriers, presented in Table 2, indicates that the most popular source of recruitment is from people cur-

TABLE 2
Sources of Recruitment

Source	Currently Use	Plan to Use	Never Use
People in Company	290* (90)**	27 (8)	5 (2)
Customers	218 (70)	12 (4)	82 (26)
Advertisements	209 (65)	22 (7)	89 (28)
Competitors	197 (64)	29 (10)	80 (26)
Universities	98 (32)	69 (23)	134 (45)
Employment Agencies	50 (17)	16 (5)	233 (78)

Totals may not add to universe of 347 due to non-responses

* Number of Firms
** % of Firms

rently within the company, while the least popular recruitment source is employment agencies. A likely explanation for these findings is cost-based: i.e., in-house recruiting is characterized by low out-of-pocket costs, while employment agencies require high out-of-pocket expenditures. Moreover, recruitment from within may also mean that the acculturation process for new sales personnel is kept to a minimum, which should reduce the level of organizational disfunction.

Moreover, recruitment from educational institutions (colleges and universities) is the source most likely to be added in the future by motor carriers. In fact, 23 percent of the carriers plan to use universities as a recruiting source. No other recruiting source is planned to be added by more than 10 percent of the carriers. One explanation for this is the changing nature of the industry, in that advanced formal education is seen as essential in adapting to the new marketplace, which is characterized by change. College educated personnel are often viewed as being better able to accept change and to adapt to change than are non-college educated workers. Furthermore, college educated workers may possess some of the technical skills (e.g.. exposure to personal computers and particular software packages) that are now viewed as critical in the new competitive environment.

B. Multivariate Analysis

Table 3 displays the results of the logit analysis indicating the contribution of the firm attribute variables (size, profitability, and scope of operations) in combination to explain the decision to use each of the recruitment techniques examined.

The logit results presented in Table 3 demonstrate that firm size, profitability, and scope of operations, taken together, significantly influence the decision to use a particular recruitment technique. For example, profitability and scope of operations contribute significantly (at the .10 level) towards an explanation of the carrier decision to recruit by means of advertisements. Specifically, the positive coefficient for the operating ratio variable suggests that firms with higher operating ratios are more likely to use advertisements than are those with lower operating ratios. Additionally, the positive coefficient for scope of operations suggests that firms operating in a larger number of states are more likely to use advertisements than are those with more limited operations.

Concerning the influence of operating ratios on use of advertisements, it is observed that while 82 percent of the motor carriers with an operating ratio of 1.01 or more use advertisements as a recruiting tool, only 57 percent of those with an operating ratio

TABLE 3

Influence of Firm Attributes on Recruitment Sources: Logit Results

Source	Firm Variables Revenues(1)	# of States	Operating Ratio	Chi-Square Test(2)
Advertisements	.129(3) (1.08)(4)	.047 (1.90)***	5.32 (1.90)***	11.7*(5)
Customers	.578 (4.21)*	-.139 (0.65)	8.07 (2.77)*	27.3*
Universities	.364 (3.15)*	.010 (0.56)	-4.14 (1.57)	22.0*
Competitors	.032 (0.28)	.044 (1.96)**	5.67 (2.12)**	9.6**

* significant at the .01 level
** significant at the .05 level
*** significant at the .10 level

(1) Log transformation performed.
(2) Chi-square statistic on the log-likelihood ratio to test the hypothesis that all coefficients except the constant are equal to zero. Chi-square statistics exceeding tabled values indicate the coefficients are statistically significantly different from zero at the shown levels of significance.
(3) Positive coefficients indicate an increase in the likelihood that a particular source will be used, while negative coefficients indicate a decrease in the probability that a source will be used.
(4) T-statistic.

less than 0.97 employ advertisements. A possible explanation for this finding is that less profitable firms, looking for an infusion of new talent that might improve their financial performance, are attracted to advertisements since they provide the most cost effective way for them to reach out to a group of individuals not directly targeted by other recruiting sources—such as from within the company or from customers. The less profitable firms may regard advertisements as the best way to attract new salespeople who are unconnected with the company's past experience as a basis for striking out along new lines.

In explaining the relationship between scope of operations and the use of advertisements, it should be noted that approximately 80 percent of the motor carriers operating in ten or more states have advertisements as part of their recruiting package, while about 58 percent of the companies operating in one or two states recruit new salespeople through advertisements.

Logit results show that firm size and profitability contribute significantly (at the .01 level) to an explanation of the decision to use customers for recruiting purposes. The positive coefficient for firm size indicates that larger firms are more likely than are smaller ones to use their customers as a source for new salespeople. The positive coefficient for operating ratio demonstrates that increases in operating ratio are associated with increased likelihood that customers will be used in recruiting.

In terms of the influence of firm size, 44 percent of motor carriers with annual revenues of less than $2.7 million utilize customers as a recruiting source, while 75 percent of those with revenues in excess of that amount recruit new salespeople from existing customers. Thus, larger companies are more apt to use customers as a source of sales force recruits. From a customer's perspective, the small trucking companies may be unattractive potential employers. For example, firms with less than $2.7 million in annual revenue may not have pension plans or competitive medical benefits.

The logit results show that firm size adds significantly (at the .01 level) to an understanding of the use of colleges/ universities as recruiting sources. The positive coefficient suggests that larger firms are more likely than are smaller ones to bring their recruiters to colleges/universities. Specifically, 30 percent of the truckers with under $8 million in revenues recruit at schools, compared to 55 percent of those with greater revenues. Recruitment at colleges/universities is often a costly and time-consuming practice; smaller motor carriers may have neither the resources nor manpower to devote to large-scale university recruitment.

The final recruiting source analyzed by a logit model is the use of competitors. Two variables (scope of operations and profitability) helped explain the carrier decision to use competitors for recruiting. Specifically, the model's positive coefficient for the profitability variable denotes a greater likelihood to use competitors among firms with higher operating ratios. Furthermore, the positive coefficient on the geographic scope variable shows that firms with more extensive operations rely more heavily on competitors than do firms with a more limited operation. These results suggest an effort by the less profitable motor carriers to reverse their financial plight by "raiding" talented salespeople (with possible personal connections to major accounts) from their competitors.

In short, the results regarding recruitment sources indicate that the larger general freight carriers are significantly more likely than are smaller firms to use advertisements, customers, competitors and universities as sources of recruitment. The combination of these results suggest that the larger general freight carriers have been making a concerted effort to recruit salespeople who have greater technical skills as well as those with more thorough customer knowledge. This trend is consistent with the new competitive environment which, as suggested earlier, requires carriers to put into the field salespeople who have the ability to grasp complex pricing issues as well as to relate to the needs and concerns of the customer. Furthermore, the logit results indicate that firms with high operating ratios are more likely than are firms with lower operating ratios to use a wide range of recruiting sources in an effort to reverse their financial situation.

IV. SELECTION TECHNIQUES

A. Univariate Analysis

Once a company has attracted a pool of sales force applicants, it must then select new salespeople from these applicants. The information in Table 4 indicates that three techniques—personal interviews, application blanks, and reference checks—enjoy the most popularity as sales force selection techniques in the motor carrier industry. These results are in concert with selection practices in other industries. For example, Stanton and Buskirk reported that application blanks and personal interviews are the two most commonly used selection tools by businesses.

Application blanks can provide a relatively cheap and quick method for increasing a company's knowledge about potential salespeople. Interviews, on the other hand, provide a way to see how effective an applicant is in a social situation. Social adroitness is certainly an important attribute for any salesperson to possess. The importance of reference checks as a selection tool in the motor carrier industry reflects the industry's concern for honest, trustworthy employees. Given that motor carriers generally transport higher-valued freight, suspectable to theft, reference checks may be helpful in determining whether a potential salesperson has ever been accused, caught, and/or convicted of pilferage.

Moreover, the data in Table 4 suggest that three items— personality tests, intelligence tests, and aptitude tests—are going to be more prevalent in the selection of future motor carrier salespeople. Each of these tools represents one type of psychological testing and is an indication that motor carriers are striving to improve the effectiveness of their selection process. This quest for efficiency is consistent with the carriers' need in the new competitive environment to minimize the costs associated with high sales force turnover and to maximize the likelihood of hiring salespeople who will make an effective contribution. Psychological tests are justified on the grounds that they can measure dimensions that may not be fully explored by other selection techniques.

Concerning selection techniques, perhaps the most interesting finding is the indication that a size-

TABLE 4
Selection Techniques

Technique	Currently Use	Plan to Use	Never Use
Personal Interviews	317* (98)**	5 (2)	0 (0)
Application Blanks	307 (96)	0 (0)	16 (4)
Reference Checks	302 (94)	11 (4)	7 (2)
Physical Exams	162 (52)	24 (8)	127 (40)
Personality Tests	101 (33)	64 (21)	138 (46)
Aptitude Tests	63 (22)	56 (19)	170 (59)
Intelligence Tests	56 (20)	52 (18)	178 (62)

Totals may not add to universe of 347 due to non-responses

*Number of Firms
** % of Firms

able portion of the carriers intend to add psychological testing to their set of selection criteria. The addition of these tests clearly seems to indicate an attempt by the carriers to upgrade the quality of their sales force in the new competitive environment. The planned use of these tests suggests that carriers recognize the need to identify certain characteristics associated with effective selling and then select individuals possessing those characteristics. There is a clear recognition that the new competitive environment has placed new demands on their salespeople who must, as noted in the introduction, be more effective individual communicators with the customers.

B. Multivariate Analysis

Logit results in Table 5 show that there are several statistically significant firm attribute variables that

TABLE 5
Influence of Firm Attributes on Selection Sources: Logit Results

	Firm Variables			
Source	Revenues(1)	# of States	Operating Ratio	Chi-Square Test(2)
Physical Exams	.200(3) (1.90)***	.014 (0.80)	6.217 (2.49)**	11.6*
Personality Tests	-.055 (0.51)	.050 (2.72)*	1.915 (0.77)	9.5**
Intelligence Tests	.282 (2.05)**	.031 (1.65)	1.78 (0.59)	14.6*

----
* significant at the .01 level
** significant at the .05 level
*** significant at the .10 level

(1) Log transformation performed.
(2) Chi-square statistic on the log-likelihood ratio to test the hypothesis that all coefficients except the constant are equal to zero. Chi-square statistics exceeding tabled values indicate the coefficients are statistically significantly different from zero at the shown levels of significance.
(3) Positive coefficients indicate an increase in the likelihood that a particular source will be used, while negative coefficients indicate a decrease in the probability that a source will be used.
(4) T-statistic.

explain company decisions to employ various selection techniques. For example, firm profitability and size contribute significantly to an understanding of carrier decisions to use physical exams in the selection process. Approximately 50 percent of the motor carriers with operating ratios less than or equal to 1.00 use physical exams as a selection tool, while nearly 80 percent of the carriers with operating ratios of 1.01 or more use them. Therefore, less profitable motor carriers make greater use of physical as a selection technique than do more profitable firms. The positive coefficient for firm size shows that firms with higher revenues are more likely than smaller ones to use physical exams in their selection process.

Logit analysis also revealed that larger firms are more likely than are smaller ones to employ the various psychological tests in selection. For example, almost 50% of the carriers with operations in ten or more states make use of aptitude tests, compared with 14% of the firms operating in fewer than ten states. Similarly, 52% of the firms operating in ten or more states employ personality tests as a selection tool; 30% of the carriers with more limited operations make use of personality tests.

V. TRAINING CONTENT

A. Univariate Analysis

Once new salespeople have been selected, it is crucial that they be carefully trained. As pointed out earlier in this paper, the quality and training of motor carrier sales personnel may be the key factor in the overall effectiveness of a carrier's sales force management program. Table 6 contains information on the content of training programs for new motor carrier salespeople. Three items—selling, operations, and pricing—dominate the training programs for new sales personnel. Not surprisingly, selling information represents the most popular component of a sales force training program. To insure that sales personnel are well-versed in selling tactics, approximately 200 motor carriers devote more than 20 percent of their training programs to selling techniques.

Operations information appears to be the next most popular component of a salesperson's training program. Approximately 145 motor carriers spend upwards of 20 percent of their training content on operations information. Salespeople must be knowledgeable about the capabilities of their company, including such matters as terminal locations and transit times, in order to construct price/service packages for customers.

Pricing practices represent the third major area of content in the training programs of motor carrier salespeople. Approximately 130 motor carriers devote more than 20 percent of their training programs to pricing issues. The changing nature of motor carrier pricing practices provides support for the importance of pricing as a component of sales force training. That is, pricing freedoms granted the industry since the MCA of 1980 allow salespeople to develop rate packages for particular shippers. Therefore, it is essential for salespeople to understand the costing and pricing practices of their company so that they have maximum flexibility in procuring business for their respective companies.

B. Multivariate Analysis

Multiple regression results, presented in Table 7, show relationships between training content and firm attribute variables (size, profitability, and scope of operations). For example, firm variations in the pricing content of their training program are significantly related to scope of operations. The positive regression coefficient indicates that as the number of states operated in increases so too does the portion of training devoted to pricing issues. Motor carriers

TABLE 6

Training Content

Number of Firms Reporting a Particular Item as Percentage of Formal Training Program

Item	0-9%	10-19%	20-29%	30-49%	>50%
Selling	32	75	91	68	37
Operations	44	113	76	48	20
Pricing	48	126	103	24	2
Company Philosophy	116	116	60	6	4
Competitor Information	134	118	44	5	2
Transportation Information	159	110	22	10	0
Claims	223	70	4	5	0
Government Regulations	266	29	6	0	0

Totals may not add to universe of 347 due to non-responses

TABLE 7
Influence of Firm Attributes on Training Content: Multiple Regression Results

	Firm Variables			
Item	Revenues	# of States	Operating Ratio	F-Statistic
Selling(1)	.00000006(2) (4.20)*(3)	-.148 (1.18)	37.45 (2.20)*	7.05*
Operations	.00000005 (2.57)*	.051 (.046)	-24.81 (1.66)**	3.01**
Pricing	.000000004 (0.37)	.177 (2.33)**	15.97 (1.55)	3.50**

----
*Significant at the .01 level
**Significant at the .05 level

(1) Dependent variable is the percent of training content devoted to a particular item.
(2) Positive coefficients indicate increases in independent variable are associated with increases in training content devoted to a particular item, while negative coefficients indicate the opposite.
(3) T-statistic.

operating in one or two states spend 12 percent of their training programs on pricing issues; companies operating in three or more states report that pricing information accounts for slightly more than 17 percent of their sales force training programs. Larger carriers, with more customers than the smaller carriers, are likely to have pricing policies that vary by customer (e.g., a high volume user may be entitled to a greater discount than is a low volume user). Given this scenario, it is imperative that salespeople be knowledgeable about rates so that customers are charged the correct prices.

Table 7 also indicates that there are significant relationships between the amount of selling information included in a training program and firm size, and operating ratio. The regression coefficients for both these variables are positive indicating that increasing firm size and higher firm operating ratios are significantly (at the .05 level) related to an increasing amount of selling information in the training program. It appears that the relationship between selling information and firm size is best explained by looking at the age of sales force recruits. That is, carriers with revenues of $23 million or more spend about 30 percent of their training programs discussing selling issues. This carrier group also has the highest proportion (29 percent) of new sales personnel between 20 and 29 years of age. These younger people are less likely to have previous sales experience and must therefore be trained more thoroughly with respect to selling techniques.

The multiple regression results denote that amount of training content devoted to operations is significantly related to firm size and operating ratio. The negative firm size coefficient indicates that larger firms spend less time on operations than do the smaller firms. The negative coefficient for operating ratio establishes that firms with higher operating ratios spend less time on operations than do firms with lower operating ratios. This finding may be an indication that the less profitable firms need to spend additional time on operations in their training programs in order to improve their service reliability.

The results concerning the content of sales force training programs currently used by general freight carriers is not surprising given the new competitive environment. The training programs emphasize three components (selling techniques, operations, and pricing practices) that are critical to successful selling in the new environment. Salespeople must use effective selling techniques and develop pricing alternatives that are plausible and profitable given known potential constraints.

V. SUMMARY

This paper has investigated the sales force management activities of general freight carriers by examining their recruitment, selection, and training practices. The results provide significant support for the belief that carriers have already changed or intend to change significant aspects of their sales force management activities as a response to the new post-MCA competitive environment.

Indeed, the significantly greater reliance by large carriers than by small carriers on advertisements, customers, and universities as recruiting sources suggests that the larger firms are making a concerted effort to recruit salespeople who have advanced technical skills as well as those with more thorough knowledge of customers based on direct experience. Both of these skills are critical for the development of an effective sales force in a competitive environment.

Additionally, the results show that a significant portion of the carriers intend to add various psychological tests to their selection criteria in an effort to be more scientific and systematic in obtaining salespeople whose skills are highly compatible with the stressful competitive environment in which the carriers now find themselves.

Furthermore, the results indicated that the larger, more geographically dispersed carriers devote a significantly larger share of their sales force training program to pricing than do the smaller more geographically limited carriers. It is reasonable to expect the larger carriers with extensive operations to have a wide range of customers with specific price/service needs and, as a consequence, a need for salespeople with much more extensive training in pricing.

While these results support the notion that carriers are modifying sales force management activities in response to the new competitive environment, they clearly show that the larger carriers with extensive geographic operations are adapting significantly more quickly than are the smaller carriers with less extensive operations.

The recruitment and selection information indicates that carriers with high operating ratio are more likely than are carriers with lower operating ratio to experiment with a variety of the available techniques. While the willingness to change in the face of difficulties is positive, it may also indicate a lack of strategic focus on the part of these less profitable carriers. This lack of strategic focus, as pointed out by Porter, generally leads to less profitable operations than firms with a defined strategic focus.

There are other additional lines of investigation that follow from this initial inquiry. Further research could focus on compensation and motivation issues. In addition, the effectiveness of particular sales force management techniques could be investigated. Given that personal selling is recognized to be an important element of the marketing mix in a service industry, knowledge about sales force management issues should allow motor carriers to develop a more comprehensive and cohesive marketing program. Improved personal selling will permit motor carriers to increase their customer base, which should go a long way to insuring survival in the turbulent post-MCA environment.

ENDNOTES

*School of Business, John Carroll University, University Heights, OH.
**College of Business and Management, University of Maryland, College Park, MD.

1. Rathmell, John. (1974), *Marketing in the Services Sector,* Cambridge, Mass.: Winthrop.
2. Berry, Leonard. (1980), "Service Marketing is Different," Business, (May-June), pp. 24-28.
3. Berry, (1980).
4. Lovelock, Christopher. (1980), "Why Marketing Management Needs to be Different for Services," in *Marketing of Services,* J. Donnelly and W. George, eds., Chicago: American Marketing Association, pp. 5-9.
5. Zeithaml, Valerie, A. Parsuraman, and Leonard Berry. (1985), "Problems and Strategies in Services Marketing," *Journal of Marketing,* (Spring), pp. 33-46.
6. Kelly, J. Patrick and William George. (1980), "Perceptions of the Personal Selling Function in Service Marketing: A Field Study," in *Marketing in the 1980s: Changes and Challenges,* Chicago: American Marketing Association, pp. 244-247.
7. Hynes, Cecil. (1986), "Marketing, Strategy, and Practice in the Motor Carrier Industry," *Transportation Practitioners' Journal,* (Spring), p. 307.
8. Cross, James and Frederic Beier. (1986), "Market Planning, Status and Need in Small and Medium Class I Motor Carriers," *Transportation Journal,* (Spring), pp. 12-19.
9. "Quest for Quality," *Distribution,* (August, 1986), pp. 10-24.
10. In the analysis, the "plan to use" and "never use" responses are combined into one category in order to sharpen the contrast with the firms who show current use. Thus, firms are divided into two categories for the analysis: (1) firms who currently use a particular recruitment or selection technique and (2) firms who plan to use or never use such a technique.
11. Becker, M. H. (1970), "Sociometric Location and Innovativeness: Reformulation and Extension of the Diffusion Model," *American Sociology Review,* Vol. 35, 267-304.
12. Stanton, William and Richard Buskirk. (1974), *Management of the Sales Force,* 4th Edition, Irwin.
13. Wotruba, Thomas. (1981), *Sales Management: Concepts, Practices, and Cases,* Goodyear Publishing.
14. Stanton and Buskirk (1974).
15. Porter, Michael. (1980), *Competitive Strategy,* New York: Free Press.

The Definition and Measurement of Financial Fitness in the Trucking Industry

by Garland Chow, Mark Vasina*, Bahram Adrangi**, and Richard D. Gritta***

ABSTRACT

Canadian trucking industry is moving towards a more deregulated environment in which the market is expected to govern the decisions and subsequent penalties and rewards of its members. The United States regulatory environment has already been substantially deregulated with respect to entry and rate control. In such an environment, shippers, industry suppliers, carriers and public officials are more concerned about the financial fitness and condition of motor carriers. Greater competition may results in greater rates of entry and exit. Shippers want to minimize the risk of having freight delayed or lost by carriers declaring bankruptcy. Suppliers want to minimize their dealings with potentially bankrupt carriers which may not be able to pay their bills. Motor carriers can better plan their marketing and operating strategies given knowledge of impending exit from the industry. Finally, the fitness test is the remaining element of entry regulation un U.S. regulatory reform and potentially under Canada's version of trucking re-regulation. Public officials would like new entrants to be fit enough to actually provide the service that they promised the public when applying for new authority.

This paper describes the C-Score Model which combines financial statement analysis with statistical techniques to develop an index of financial distress. The C-Score Model was determined to be superior to existing models that are publicly available. The C-Score model was used to examine the relative stability of the general freight segment of U.S. trucking since 1981, compare the financial condition of U.S. versus Canadian trucking, and compare the relative strength of different segments of the general freight industries in Canada and the United States.

I. INTRODUCTION

The financial fitness of motor carriers has always been a concern of shippers, industry suppliers, carriers, and public officials. This concern will be heightened by the deregulation of the Canadian trucking industry. The U.S. experience with deregulation has already shown that greater competition resulted in greater rates of entry and exit and greater uncertainty about the quality and availability of service. Bankruptcy since 1980 has become commonplace in the U.S. trucking industry. Dun & Bradstreet report that 4207 trucking companies have gone bankrupt in the years 1981 through 1984, and the rate of failures in trucking since 1978 has risen faster than the failure rate of business in general.[1]

This paper describes an approach to measuring and evaluating financial distress which may be used to evaluate the financial condition of individual carriers or segments of the motor carrier industry. The application of financial models is demonstrated using Canadian and U.S. data.

II. THE IMPORTANCE OF FINANCIAL CONDITION

The financial condition of a carrier is important for many reasons, although the reasons may vary for different participants in the trucking services market.

> Shippers want to minimize the risk of having freight delayed or lost by carriers declaring bankruptcy. Shippers may prefer long term shipper-carrier relationships to insure service availability. Carriers in financial distress may also reduce service quality which is not perceptible until damage or delays occur.
>
> Suppliers want to minimize their dealings with potentially bankrupt carriers which may not be able to meet their obligations. These include suppliers of capital, fuel, insurance and equipment. An equipment dealer, for example, may extend credit to carriers for up to three years to finance the purchase of line haul equipment.
>
> Carrier knowledge of impending bankruptcy or financial strength of its competitors is essential in strategic planning. Strategic pricing and marketing decisions take into account the strength of competition. A carrier should be prepared to take advantage of additional freight volume that may no longer be shipped by carriers going out of business. Carriers considering acquisition and merger should evaluate the financial condition of prospective carriers as with any investment.
>
> Public officials would like new entrants to be fit enough to actually provide the service that they promise the public when applying for new authority. The fitness test is the cornerstone of fair entry regulation under Canada's proposed extra-provincial trucking deregulation. More importantly, public officials are still charged with the responsibility to insure that the public has access to efficient but safe transportation. Financially distressed carriers must be a concern of regulators if they have less regard for public safety on the highways.

These important concerns by a wide spectrum of participants in the trucking industry indicate a strong need for an objective but timely approach to the evaluation of financial conditions in the trucking industry or of individual carriers.

III. EVALUATING FINANCIAL DISTRESS

Financial analysts have long searched for more accurate methods of gauging financial health and forecasting insolvency. Single financial ratios used individually are not adequate measures of the financial condition of a firm. A carrier with negative earnings may have enough liquidity to stay in business long enough to return to profitabilty. A carrier with positive earnings may not have the cash flow to meet current obligations. Most lenders and credit analysts will look at a variety of financial indicators to make a balanced assessment of one's credit worthiness. These indicators are often grouped into separate catagories, each of which measures a particular aspect of financial strength. The following is one such breakdown:

Profitability Ratios—Ratios in this group gauge the rate of profit relative to some base. That base can be sales, equity, or assets resulting in profit margin, return on equity and return on assets respectively. Common ratios used in the trucking industry are operating ratio and return on transportation investment. Higher profitability ratios are favorable and decrease the risk of bankruptcy.

Turnover Ratios—These statistics indicate the efficiency of assets, capital and labor. The widely used ratio in this group is the capital turnover ratio. Productivity measures are variants of this type of performance measure. However productivity information is not always in standard income and balance sheet statements. Greater efficiency is correlated with lower bankruptcy risk.

Leverage Ratios—This catagory of ratios indicate the extent to which a firm used debt to finance its asset base. The excessive use of debt can create significant profitability and liquidity problems should revenues not reach breakeven volumes. Interest charges are fixed charges which are unavoidable. Excessive debt is thus a direct cause of bankruptcy risk.

Liquidity Ratios—These ratios measure the ability of the firm to pay its current debts (current liabilities) as they become due. The less liquid the firm, the greater the danger that creditors will force the firm into involuntary cessation of operations. Liquidity ratios include the current ratio, debt coverage and cash flow. A variation of cash flow used in trucking is the cash throw ratio.

All of these types of ratios (except certain productivity ratios as noted above) can be derived from basic financial statements of motor carriers. The challenge is to determine what importance should be placed on each dimension of financial performance. The use of predictive models which combine financial statement analysis (ratio analysis) with statistical techniques is one approach to overcoming this difficulty.

IV. A MODEL OF FINANCIAL DISTRESS FOR TRUCKING

Bankruptcy models have been applied to a variety of industries. The most publicized are the Altman Z models.[2] Chow and Gritta demonstrated that the Z" version provides reasonable estimates of the financial condition of selected segments of the trucking industry.[3] The Z" version was found to be superior to the Z or Z' versions since they did not rely on the market value of equity and did not include an asset turnover ratio. Most trucking companies are not publicly held firms and trucking companies have fewer assets than manufacturing firms. Thus the Z" version better reflects the economic and ownership characteristics of the trucking industry.

The most current version of the Z model, the ZETA model is claimed to be applicable to a wide range of industries since it was developed from a cross section of bankruptcies of publicly held firms. In 1982, the ZETA model was used to suggest that Carolina Freight Carriers was a candidate for bankruptcy.[4] Hindsight verifies what most industry observers believed at the time; Carolina Freight was in fact one of the outstanding carriers in the trucking industry in terms of financial strength and future profitability. A model that is estimated from the data of carriers in the trucking industry may have made a better prediction for the same reasons that the Z" model was superior to the Z or Z' models. Furthermore when the economic and regulatory environment changes, the factors that best explain financial distress may change. Chow, Gritta, Adrangi and Ebelt have shown that the financial indicators that best predicted financial distress in the U.S. trucking industry before deregulation are different from those after deregulation in 1980.[5]

A model of financial distress which we shall call the "C" model was estimated. Actually a set of models were developed, where each model was specific to different segments of the U.S. trucking industry. Motor carriers were classified as financially distressed if they had gone bankrupt or were technically insolvent. The financial distress model was developed using Multiple Discriminant Analysis (MDA) where the dependent variable was a 0 - 1 variable indicating whether the carrier was financially distressed (1) or not financially distressed (0). Dependent variables were financial ratios as discussed above. MDA provided statistically determined weights for financial ratios which significantly characterized firms in financial distress.

One of the early versions of the C model was calibrated from a sample of U.S. general freight carriers in financial distress between 1982 and 1986 and has the following form:

$$C = a + b_1X_1 - b_2X_2 - b_3X_3 - b_4X_4 - b_5X_5$$

Where:

C — Financial distress score
X1 — Retained Earnings/Total Tangible Assets
X2 — Total Liabilities/Total Tangible Assets
X3 — Shareholder Equity/Total Tangible Assets
X4 — Current Liabilities/Total Tangible Assets

X5 — Working Capital/Total Tangible Assets

The financial data was obtained from the carrier annual report statements submitted to the Interstate Commerce Commission (ICC) as reported by the American Trucking Associations, Inc.[6] Financial data was collected for the four years prior to the year of bankruptcy or classification of technical insolvency for carriers in financial distress. Financial data for 1979 to 1983 was used for nonbankrupt carriers.

The C model outperforms the best publicly available model of its kind in use, Altman's Z" model. A comparison of the two models reveals the superiority of the C model. The Z" model's classification rules lead to a correct classification rate of 76 percent. When we apply similar rules with the C model, we correctly classify 87 percent of the population of general freight carriers.

The C model estimates were tested for the accuracy of the model over longer periods. It was found that overall accuracy was reduced by 5 percent when we tried to predict four years in advance. Theoretically the accuracy will continue to decrease for periods past four years. However we were unable to validate this hypothesis since we believe that the model is not applicable to the industry under the regulated environment that ended in 1980.

The simplest interpretation of the C model is that a positive C score corresponds to a probability of bankruptcy *less* than 0.50 while a negative C score corresponds to a probability of bankruptcy *greater* than 0.50. Viewed in this manner, a C score of 0 corresponds to a 0.50 probability of bankruptcy, and the higher the C score the lower the associated probability of bankruptcy. This "probability of bankruptcy" is based on the degree to which a firm "resembles" a firm that is known to have failed or be in a prolonged state of technical insolvency.

The C scores and the probabilities of bankruptcy have a one-to-one correspondence and the information content of one rating system is identical to the other. Figure 1 is a depiction of this informational relationship. The symmetry of the systems and the relative ease of computing C scores lead us to choose C scores over probabilities as the preferred method of presenting results and using the model.

Figure 1

C SCORES AND PROBABILITIES OF BANKRUPTCY

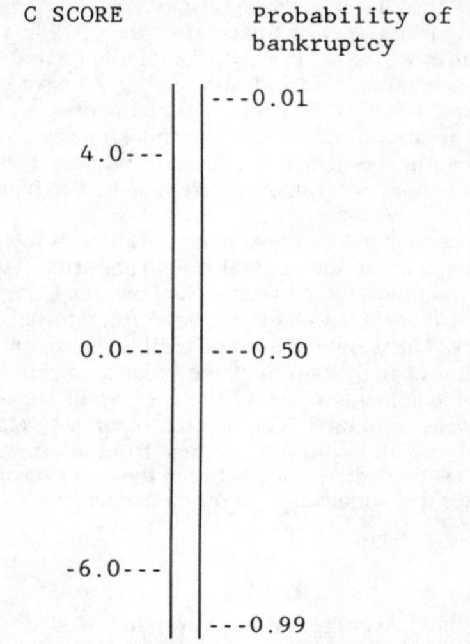

V. TRENDS IN THE FINANCIAL CONDITION OF THE U.S. GENERAL FREIGHT TRUCKING INDUSTRY

The C model was used to estimate the number and percentage of carriers in our U.S. general freight carrier data base which are candidates for financial failure from 1981 to 1985. Only Instruction 27 carriers, carriers who earned 75 percent or more of their revenues from intercity movement of general freight, with average shipment weights below 5000 pounds were selected to concentrate on specialists in this segment of the industry. All of these firms will not necessarily go out of business but they do have the financial characteristics of carriers who have gone out of business in the past or are classified as technically insolvent. The estimates are shown in Table 1. Eighteen percent of the Class I and II

TABLE 1

Financial Condition of the General Freight/LTL Segment of the Trucking Industry (1981-1985)

YEAR	NUMBER OF FIRMS WITH LOW RISK OF BANKRUPTCY	NUMBER OF FIRMS WITH HIGH RISK OF BANKRUPTCY	FIRMS FOR WHICH NO PREDICTION IS MADE	TOTAL NUMBER OF FIRMS*
1985	138 (59%)	43 (18%)	52 (22%)	233 (99%)
1984	174 (63%)	43 (16%)	59 (21%)	276 (100%)
1983	171 (62%)	53 (19%)	53 (19%)	277 (100%)
1982	188 (61%)	59 (19%)	59 (19%)	306 (99%)
1981	215 (67%)	47 (14%)	61 (19%)	323 (100%)

* Sum may not equal 100 due to rounding.

carriers specializing in the transport of general freight were in financial distress in 1985. The financial condition of the industry has improved marginally since 1982 when approximately 19 percent of the industry was in financial distress. 1982 was the year in which the effects of deregulation and the sluggish economy on industry viability peaked. It represented the last year of continual decline in the overall financial wellness of the industry since pre-deregulation. As can be seen, the industry is just now regaining the financial strength that it had in 1981.

The aggregated C scores may mask trends in specific segments of the general freight industry. Table 2 displays the weighted average and median C scores for selected segments of the general freight trucking industry. The weighted averages reflect the financial condition of the group as if the balance sheets and income statements of all of the carriers in the segment were combined. The median separates exactly half of the individual C scores from highest to lowest. A large difference between the two measures indicates that some very important carriers are either doing extremely well or extremely poor. In all groups there will be carriers performing well above and well below these two measures of the average carrier.

TABLE 2

Weighted Average (WTD-AVG) and Median C Scores for Selected Segments of the U.S. General Freight/LTL Trucking Industry

SEGMENT	YEAR	WTD-AVG C-SCORE	C-SCORE MEDIAN
ALL:			
	1985	.75	.65
	1984	.74	.68
	1983	.68	.68
	1982	.64	.64
	1981	.78	.68
LONG-HAUL:			
	1985	.79	.51
	1984	.79	.40
	1983	.71	.40
	1982	.63	.29
	1981	.71	.34
SHORT-HAUL:			
	1985	.69	.70
	1984	.63	.72
	1983	.61	.71
	1982	.65	.71
	1981	.84	.75
REVENUES < =$25M:			
	1985	.71	.67
	1984	.70	.78
	1983	.71	.71
	1982	.76	.67
	1981	.93	.71
REVENUES =$25-50M:			
	1985	.58	.65
	1984	.61	.56
	1983	.74	.93
	1982	.69	.92
	1981	.90	1.02
REVENUES =$50-100M:			
	1985	.85	.76
	1984	.70	.81
	1983	.68	.67
	1982	.52	.72
	1981	.67	.64
REVENUES =>$100M:			
	1985	.75	.57
	1984	.75	.43
	1983	.65	.38
	1982	.63	.29
	1981	.74	.37

The financial condition of carriers grouped according to revenue size is shown in Table 2. Overall the strongest group of carriers in 1985 were those carriers in the $50 to $100 million annual revenue class. The very large and the very small carriers followed and the carriers in the $25 to less than $50 million revenue bracket were the most financially weak. There is of course variability in the financial strength within each group. This is particularly true for the large carrier group where the average C score is 18 points above the median C score. This suggests that some very large carriers accounted for the bulk of the financial viability in this group. The financial strength of both small carrier groups (less than $50 million) have declined over the 5-year period while the larger carriers have increased in strength.

The financial condition of carriers grouped according to type of geographic market (lenght of haul) is also shown in Table 2. Long haul carriers are defined as those carriers with an average length of haul exceeding 500 miles. There is a sharp contrast in the fortunes of the carriers in each group over time. The long haul carrier group showed increasing financial strength since 1981 while the short haul carriers moved in the opposite direction. This certainly reflects the economics of the two markets. The long haul sector has become increasingly concentrated as the least efficient have left the market. The remaining carriers consolidated and strengthened their positions as there are some economies of scale and capital requirements in this sector. Entry is substantially easier in the short haul market preventing a similar retrenchment by the carriers in that competitive arena. The short haul market is also characterized as one in which non union competition is strong.

VI. FINANCIAL CONDITION OF CANADIAN TRUCKING

Firm specific data for Canadian carriers were not available for this research. Thus estimates comparable to those in Table 1 cannot be made (the percentage of the Canadian trucking industry which is in financial distress). However, aggregate statistics of selected segments of the Canadian trucking industry are available from Statistics Canada publications and weighted average C scores can be computed that are similar to the scores shown in Table 2.[7] The C score model has not been validated on a Canadian carrier sample. Further, the financial accounts used by Statistics Canada are not completely matched with the financial accounts used by U.S. carriers in their reports to the ICC. However the inherent financial structure and basic balance sheet accounting practices should not differ simply because one trucking company is domiciled in one country or the other. The C score computations are thus useful for identifying relative financial performance and caution must be used in interpreting Canadian - U.S. comparisons.

The results of these computations are shown in Tables 3 and 4. Statistics Canada provides detailed statistics on intercity general freight carriers for

TABLE 3
Weighted Average C-Scores General Freight
Canadian and U.S. Trucking Industry
(1984)

Population	WTD-AVG C Score
Canadian Carriers	
All General Freight	.21
Group 1 - Extensive Utilization of Owner Operators.	.14
Group 2 - Moderate Utilization of Owner Operators.	.15
Group 3 - Company Drivers	.54
U.S. General Freight	.65

1984. These carriers were defined as for-hire carriers earning $2 million or more annually and generating more than 50 percent of their revenues from general freight shipments carried over distances of 25 kilometres and more by Statistics Canada. These carriers were subdivided into three groups:

Group One Carriers spend 30 percent or more of their operating expenses to purchase the services of broker operators.

Group Two Carriers spend less than 30 percent of their operating expenses to purchase the services of broker operators but do utilize them.

Group Three Carriers rely completely on company drivers and do not purchase any services of broker operators.

The C scores shown in Table 3 indicate that group three general freight carriers, those which do not rely on broker operators at all, are in the best financial condition. The figures indicate that each segment is not financially distressed as a whole but of course, individual carriers may be well above or well below the average. However the Canadian general freight sector's financial condition is well below that of the U.S. general freight carrier segment as indicated by a weighted C score which is three times higher. However the group three Canadian general freight carriers perform relatively well compared to their U.S. counterparts.

TABLE 4
Weighted Average C-Scores
Canadian and U.S. Trucking Industry
by Geographic Segments (1984)

Population	WTD-AVG C Score
Canadian	
All Canada	.19
Newfoundland	.36
P.E.I	.43
Nova Scotia	.04
New Brunswick	.42
Quebec	.25
Ontario	.18
Manitoba	.21
Saskatchewan	.05
Alberta	.22
British Columbia	.11
Yukon	-.47
N.W.T.	.17
U.S.	
All U.S.	.63
Class I	.63
Class II	.62

The C scores of carriers by geographic domicile were computed for all Canadian motor carriers regardless of principal commodity transported. Again C scores were computed for U.S. carriers for comparison and these scores are displayed in Table 4.

The overall C score for all Canadian carriers is heavily weighted to the performance by Ontario domiciled carriers who earned over 40 percent of industry revenues. The financial condition of carriers in the Yukon, Nova Scotia and Saskatchewan was quite poor. With the exception of Nova Scotia, the Maritime provinces are the domicile of the most financially strong carriers in the country. Even Quebec's trucking industry exhibits an above average financial index. However none of the provinces or the all Canadian average is close to the financial strength of the U.S. carrier industry as measured by an average C score of .63 which is more than three times the Canadian average and 50 percent better than the best provincial segment domiciled in P.E.I. On the whole, the performance of carriers in the eastern portions of Canada were above the all Canadian average while the performance of carriers in the western and northern portions of the country were below.

VII. CONTINUING RESEARCH AND POTENTIAL APPLICATION OF THE C SCORE MODEL

In this paper an early version of the C score model was used to monitor trends in the trucking industry or for specific industry segments. The relative stability of the industry may be assessed by examining the distribution of C scores as was demonstrated by the analysis of the General Freight segment of U.S. trucking. Comparisons between the financial condition scores of the U.S. and Canadian trucking industries suggest that Canadian motor carriers are in worse financial condition than its neighbors (and potentially competitors) to the south. Comparisons within Canada suggest that carriers domiciled in the eastern provinces are generally better off financially than carriers in the west and far north. Finally, within the general freight segment of Canadian trucking, it appears that carriers which rely on company drivers as opposed to broker operators are the most financially stable.

Validation tests (not reported here) indicate that the model predicts two to four years into the future with a slight decline in accuracy. Thus it can be used as an early warning system in assessing the future financial condition of the trucking industry over a medium time horizon. Obviously predictions about the financial prospects of individual carriers can be made but this is best accomplished in a comparative and historical context. A carrier's financial performance is best evaluated relative to its peer group and what it has accomplished before.

The long term goal of the research that developed the C model is to identify and validate the linkages between deregulation, financial condition and safety. Academic and industry literature suggests that the linkage between motor carrier deregulation and motor carrier safety is the financial condition of the motor carrier. Decreased profitability of the industry and individual carriers result in poor financial conditions which provide an incentive for carriers to reduce expenditures on maintenance, to defer the replacement of unsafe equipment, reduce safety programs or encourage unsafe driving practices. The objective measurement of carrier financial distress is the first step in this process. The next steps are to test the hypothesized relationships between safety and financial condition, and between regulation and financial condition. The effect of deregulation can be deduced from these two results.

ENDNOTES

*Faculty of Commerce and Business Administration, University of British Columbia, Vancouver, B.C.

**School of Business Administration, University of Portland, OR.

1. American Trucking Associations, Inc., *American Trucking Trends: 1985*, Alexandria, Va. (1985).
2. Altman, E.I., *Corporate Financial Distress: A Complete Guide To Predicting, Avoiding, and Dealing with Bankruptcy*, New York: John Wiley, (1983).
3. Chow, G. and R.D. Gritta, "Motor Carrier Bankruptcy in an Uncertain Environment," *Transportation Law Journal*, XIV (Winter 1985), 39-57.
4. "Companies That Face Financial Strain," *Business Week*, (May 17, 1982).
5. Chow, G., R.D. Gritta, B. Adrangi and R. Ebelt, "The Financial Status of U.S. General Freight Motor Carriers before and after the Motor Carrier Act of 1980" *Proceedings of the 1986 Annual Meeting of the Decision Sciences Institute* (Nov. 1986), pp. 267-271.
6. American Trucking Associations, Inc., *1984 Motor Carrier Annual Report*, Alexandria, Va. (1985).
7. Statistics Canada, *1984 Trucking in Canada*, Catalogue 53-222, Ottawa, (1986).

Transport Markets and Firm Behavior: The Backhaul Problem

*by Wesley W. Wilson**

ABSTRACT

In this paper, the behavior of "exempt" motor carriers is examined both theoretically and empirically. These firms, in traveling back and forth between two locations, jointly produce commodity movements in two markets. In outgoing or "fronthaul" movements, firms haul exempt from regulation commodities (e.g., agricultural products) from an origin to a destination. On return or "backhaul" movements, firms are randomly loaded with commodities subject to Interstate Commerce Commission regulation.

In general terms, it is maintained that price and entry regulation by the ICC affect firm behavior through the probability the firm will obtain a backhaul load on return trips. It is argued the backhaul probabilities vary across firms and markets, producing different levels of economics rents for different firms.

The empirical analysis consists of two components. First, the hypotheses that backhaul probabilities vary across both firms and markets differentiated by space and time are tested. The second empirical analysis formalizes a set of firm level optimization principles and tests a market level proposition that prices adjust such that expected marginal profits are equated across markets. A second hypothesis tested addressed the question of whether or not prices adjust to the marginal firm (i.e., the firm not obtaining backhaul loads).

The results suggest that backhaul probabilities do vary across both firms and markets, that expected profits are equated across markets, and fronthaul rates adjust to the marginal firm. The marginal firm in this study is inherently inefficient in the sense that one-half of its miles traveled are empty and the market adjusts to this inefficiency. Partial deregulation of motor carriage eases entry requirements and if marginal firms seek to increase their margin by gaining entry to backhaul markets, fronthaul prices would be expected to fall under deregulation. Thus, the results suggest that enhancing the efficiency of these marginal carriers can only serve to decrease the prices paid by the fronthaul shippers.

I. INTRODUCTION

Commodities are transported between fixed locations, and the backhaul problem is a term used to describe the situation where traffic levels in one direction do not match traffic levels in the opposite direction [Pederson et al., (1979)]. The general purpose of this paper is to examine transport markets and firm behavior in the context of this so-called backhaul problem. In this regard there are two specific issues addressed—the variability of backhaul probabilities across firms and markets, and the adjustment of unregulated (exempt) fronthaul prices to varying backhaul probabilities.

First, previous studies have provided evidence that the backhaul problem affects firms asymmetrically.[1] That is, there exists firm specific attributes which give rise to differing levels of backhauls (loaded return trips) across firms. Second, firms often serve multiple markets, and it has been maintained elsewhere that the extent to which firms travel empty on return trips varies across markets.[2] This study forms both market and firm specific backhaul probability estimates, and tests for differences between firms and markets.

Given that the backhaul problem affects firms asymmetrically and varies across markets, a second issue addressed in this study pertains to the determination of fronthaul prices. Nicholson (1958) suggests that, in lieu of regulation, backhaul prices adjust downward to a relatively small separable marginal cost, if fronthaul prices adjust such that the firm covers round trip costs when a backhaul load is not obtained. Miklius and DeLoach (1965) suggest that "[Nicholson's] argument confuses expected with realized revenues." That is, the decision to make the trip is based on expected round trip profit and as such there is no reason to expect that fronthaul revenues cover round trip costs when the firm is not loaded on the return trip. In such a framework, it follows that fronthaul prices adjust downward in response to greater backhaul probabilities.

There is some limited existing evidence that fronthaul prices adjust to differences in backhaul probabilities across firms. Beilock and Kilmer (1986) suggest circumstances under which rates tend toward the costs of the marginal supplier. In their model, the marginal supplier is the firm for which a load is not waiting at the destination of the movement in question (i.e., the uncommitted carrier). In such cases, a positive impact of movement distance on the probability a firm is loaded was posited and found significant. In this paper, following the tradition of Nicholson and Miklius and DeLoach, the analysis presented here is based on round trips. Hence, in the model discussed below the carrier is taken as committed to round trips and the marginal supplier is the firm without backhaul traffic on the returning leg of the round trip.

The remainder of the this paper is organized as follows. Section II provides the analytical framework underlying the backhaul schedules. Section III provides a market model with the salient features of both Nicholson and Miklius and DeLoach addressed. Finally, a summary and concluding remarks are provided in Section IV.

II. BACKHAUL PROBABILITY SCHEDULES

Most previous research surrounding the backhaul problem has focused primarily on price determination and/or firm pricing strategies.[3] Recently, Beilock and Kilmer (1986) have provided a model generating differing backhaul statistics across firms and/or markets. In this section their model is briefly summarized, then the empirical methodology employed in this study to obtain firm and market specific backhaul probability estimates is presented.

Beilock and Kilmer provide a discrete choice framework leading to an empirical specification that a firm is loaded between two locations as a function of distance, market conditions, and carrier specific characteristics. In their model, market conditions are represented by month specific dummy variables representing distinct market conditions. Carrier specific characteristics in this model are represented by a set of dummy variables, and interactions thereof, including whether the firm has ICC authority, whether the firm is an owner-operator or part of a fleet, and the firm's base location. A further firm attribute is whether the firm has a prior commitment at the destination of the movement and is represented by a dummy variable.

The data employed in this analysis represent trucking firms traveling from origins in North Dakota to two primary destinations in Minnesota (Duluth and Minneapolis), surveyed at two points in time (1976 and 1979).[4] These carriers haul primarily agricultural commodities (grain) and regularly serve the same elevator(s). As such, they operate as firms committed to providing round trips, with the trip originating in North Dakota. Thus, return loads from Minneapolis and Duluth can be interpreted as backhaul traffic as viewed by the firms.[5]

The proportion of trips each firm takes to each destination Duluth (d) and Minneapolis (m = 1 − d), the total proportion of loaded backhauls to total trips (b), and various firm characteristics including the number of years the firm has been in business (A) are observed in the data. Age in business is a proxy variable taken to represent human capital (marketing expertise) and/or a tendency (ability) of such carriers to obtain ICC authority over time.[6]

Let b_D and b_M represent the proportion of backhaul loads from Duluth (D) and Minneapolis (M) to total trips taken to Duluth and Minneapolis, respectively. The equation estimated is based on the following identity

$$(1) \qquad b = b_D d + b_M (1 - d)$$

Conditioning (1) over d and parameterizing $E(b_i \mid i)$ as a function of time in business yields the equation estimated[7]

$$(2) \quad b = Prb(b_D \mid A)\, d + Prb(b_M \mid A)(1 - d) + \varepsilon$$

$$= \left[\frac{e^{ax}}{1 + e^{ax}}\right] d + \left[\frac{e^{bx}}{1 + e^{bx}}\right](1 - d) + \varepsilon$$

where: $ax = a_o + a_1 D + a_2 A + a_3 DA$;
$bx = b_o + b_1 D + b_2 A + b_3 DA$;
a,b = parameter vectors to be estimated;
D = a dummy taking a value of 1 if 1979 data and a value of 0 if 1976 data; and
A = years in business.

In the backhaul equation, differences in firms in obtaining backhauls from a given location is reflected in the years in business variable (A), while differences in backhauls across markets is reflected in differences in the parameter vectors (a and b). Finally, the extent relative market conditions change through time is reflected in the dummy terms.

Equation (2) was estimated by nonlinear estimated generalized least squares.[8] The results are reported in Table 1. As expected, age in business has a positive impact on the probability of a backhaul

TABLE 1

Estimated Generalized Nonlinear Least Squares Estimates.

Parameter	Estimate[a]	Standard Error
a_o	-5.015*	1.578
a_1	2.707	1.950
a_2	.1591**	.0816
a_3	-.1359	.1480
b_o	-6.449*	2.215
b_1	4.292**	2.586
b_2	2.399*	.8471
b_3	-2.108	.8565

correlation (predicted, actual) = .66[b]

N = 104

[a] A (*) and (**) denote significance at the 5 and 10 percent levels, respectively.

[b] The correlation was based on the predicted backhaul probability and the actual proportion of backhaul trips rounded to total trips. The correlation of the corresponding weighted variables was 96.

from each market. Also as expected, the impact of age in business has a relatively larger impact on the probability of a backhaul from Minneapolis than from Duluth. That is, Minneapolis is a major distribution center for the northern central plains while little westward bound traffic originates in Duluth. These facts are suggestive of greater demand for backhaul transport services which, when coupled with the observation that fewer trips are taken to Minneapolis, is consistent with the hypothesis that firms are loaded more often from Minneapolis. The empirical result that age in business has a larger effect in Minneapolis than Duluth simply implies that older firms (with ICC authority and/or more human capital) have greater chances in obtaining a backhaul in a "high" backhaul market than in a "low" backhaul market.

Three central hypothesis tests were performed. These include the hypotheses that the backhaul probabilities are the same across firms, markets, and time. The results are reported in Table 2, and the null hypothesis in each case could not be accepted. Thus, the results suggest that the backhaul probabilities vary not only across firms, but also markets and through time.

The estimated probability schedules are graphically portrayed in Figures 1 and 2 for 1976 and 1979 respectively. The evidence suggests quite clearly backhaul probabilities vary across firms and markets differentiated by time and space. Estimated probabilities for the sample evaluated at different time in business values are shown in Table 3.

Differences between 1976 and 1979 are observed in the backhaul statistics reported in Table 3. The average estimated backhaul probability increased in Duluth between the two years, while falling considerably in Minneapolis. This finding suggests only that relative fronthaul-backhaul market conditions vary through time across markets. Comparing 1976 with 1979, it is noted that a greater percentage of trips were taken to Minneapolis in 1979 than in 1976. Hence there is greater competition in Minneapolis for backhaul loads than in Duluth. However, the differences in proportions of trips taken is quite small. Another explanation for which no data exist would be that the number of available backhaul loads decreased from 1976 to 1979 in Minneapolis and increased in Duluth.[9]

The values provided in Table 3 are suggestive of substantial differences in estimated backhaul probabilities across markets and across firms in the Minneapolis market. But, two issues arise. First, despite higher estimated backhaul probabilities and potentially higher returns for Minneapolis than for Duluth round trip movements, a greater proportion of truck movements flow to Duluth. Second, differences in estimated backhaul statistics also exist across firms. The two issues investigated in the next section pertain to the level of fronthaul prices. That is, do fronthaul prices adjust to differences in backhaul statistics across markets and firms?

III. PRICE ADJUSTMENT IN FRONTHAUL MARKETS

The above evidence was provided suggesting that backhaul probabilities vary across both firms and markets. In this section, a model depicting fronthaul price adjustment is posited allowing direct tests of the hypotheses that fronthaul prices adjust such that expected profit maximizing firms are indifferent between markets served and that fronthaul prices adjust to the marginal firm.

All firms are taken to face a set of offers consisting of a destination h and a fronthaul price (p_h^f). Randomness enters as associated with each destination is a backhaul probability (ρ_h) and a distribution of backhaul prices (P_h^b). Each firm, when confronted with a set of offers, is assumed to choose that offer yielding the highest expected profit per round trip mile (D_h). Observed in the data are firms serving both Duluth and Minneapolis leading to the hypothesis that fronthaul prices adjust such that expected profits per round trip mile are equal across markets resulting in expected profit maximizing firm being indifferent between markets served. Note that this

TABLE 2
Wald Statistics

Test	Wald Statistic	Prob > W
Intertemporal Differences H_o: $a_1 = a_3 = b_1 = b_3 = 0$	8.5342	.07386
Market Asymmetries H_o: $a_o = b_o$, $a_1 = b_1$, $a_2 = b_2$, $a_3 = b_3$	12.8074	.01226
Firm Asymmetries H_o: $a_2 = a_3 = b_2 = b_3 = 0$	17.7611	.00137

Figure 1. Conditional Backhaul Probability Estimates (1976)

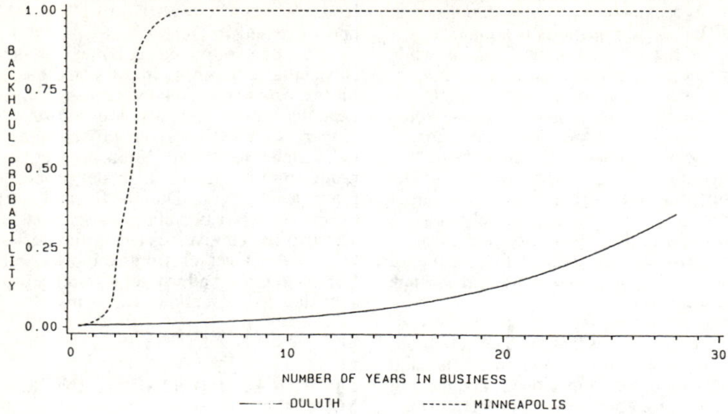

Figure 2. Conditional Backhaul Probability Estimates (1979)

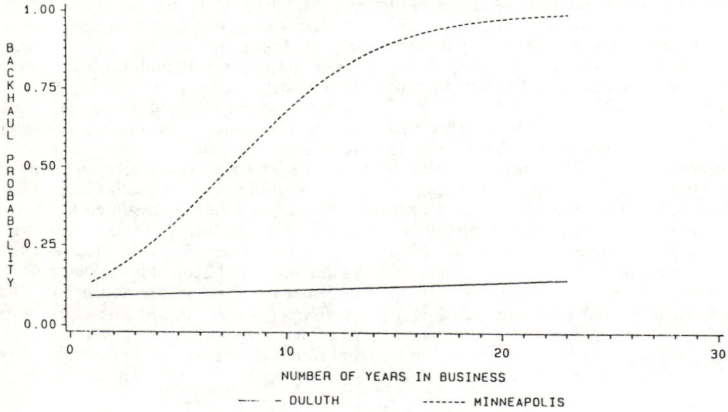

TABLE 3
Estimated Backhaul Probabilities.

Years in Business	Duluth 1976	Duluth 1979	Minneapolis 1976	Minneapolis 1979
1	.00772	.09230	.01715	.1342
5	.01449	.10039	.99613	.3322
10	.03153	.11131	1	.6813
20	.13775	.13630	1	.9753
Average	.0453	.1076	.7801	.4985
Proportion of Trips	.754	.73	.246	.27

condition holds for all firms including the marginal firm which in these markets is the no backhaul firm.

The second issue addressed rests with adjustment of fronthaul prices across firms, rather than markets. First, entry and exit into agricultural transportation by truck is generally considered to be relatively easy. Thus, by the usual argument, fronthaul prices cannot yield economic profits. However, the salient feature differentiating Nicholson (1958) from Miklius and DeLoach (1965) rests with downward adjustment of fronthaul prices with respect to increasing backhaul probabilities. That is, firms successful in obtaining backhaul loads may use any profits associated with those backhaul loads to drive down fronthaul prices to attract additional loads. Here, this issue is treated as an empirical issue. That is, do prices adjust to the marginal or no backhaul firm?

Somewhat more formally, these equations are represented by

$$(3)\quad \begin{aligned} &EM\Pi_i = EM\Pi_j \quad \text{for all destinations i, j} \\ &E\Pi = ER - EC \end{aligned}$$

where $EM\Pi_i$ = expected marginal profits per round trip mile to destination i;
$E\Pi$ = expected total profits over the time period;
ER = expected total revenues over the time period;
EC = expected total costs over the time period.

For the marginal firm, the expected marginal profit and profit terms reduce to $M\Pi^f$ and Π^f, where f is a superscript representing fronthaul.

As stated previously, the data were made available by the Upper Great Plains Transportation Institute and cover two years (1976 and 1979). The data generally provide cost and operating statistics of firms hauling agricultural commodities from North Dakota origins to Minneapolis and Duluth, Minnesota. Limited fronthaul price data were observed in the 1976 data set, and no other price data were observed. These limited data were combined with available rail data and used to construct rate indices for 1976 and 1979.[10]

Cost data were available in both the 1976 and 1979 data sets. In modelling costs, a distinction was made between directly assignable costs and overhead costs. Directly assignable costs include labor, fuel, usage-based depreciation, tire, and management costs which were directly assigned to a given trip. Overhead costs consisted of repair, license, insurance, and interest costs. In defining expected marginal profits per mile, repair costs which vary with miles traveled were assumed equivalent on a marginal basis across markets. Remaining overhead costs were taken to be fixed, given the firm's equipment decision.

The equations estimated are the following

$$(4)\quad \begin{aligned} &P_h^f - MC_h = M\Pi_h^f + \rho_h \gamma_h \quad \text{for h = Duluth, Minneapolis} \\ &R^f - C = \Pi^f + \Sigma_h \beta_h \rho_h M_h \end{aligned}$$

where $P_h^f - MC_h$ = the fronthaul price associated with the h destination net of directly assignable costs per round trip mile;
$R^f - C$ = total fronthaul revenue net of total costs;
$M\Pi_h^f$ = mean fronthaul profit for the average firm associated with the h destination per round trip mile;
Π^f = mean total fronthaul profit for the average firm over the time period; and
γ_h, β_h = parameters to be estimated, these parameters capture any covariation in the dependent variables emanating from backhaul probabilities.

Since there were no backhaul data available, beyond that discussed in Section III, coefficient interpretations and tests are confined to the marginal or no backhaul firm. The equations estimated are given by[11]

$$(5)\quad \begin{aligned} NP_D &= \gamma_o + \gamma_1 D + \{\gamma_2 + \gamma_3 D\}\rho_D + e_D \\ NP_M &= \beta_o + \beta_1 D + \{\beta_2 + \beta_3 D\}\rho_M + e_M \\ R^f - C &= \psi_o + \psi_1 D + \{\psi_2 + \psi_3 D\}\rho_D M_D \\ &\quad + (\psi_4 + \psi_5 D)\rho_M M_M + \varepsilon_\Pi \end{aligned}$$

where: NP_i = Fronthaul price net of directly assignable costs per round trip mile to destination i (i = Duluth, Minneapolis).
$R^f - C$ = Annual fronthaul revenues net of costs
D = 1 if 1979 data, = 0 if otherwise
ρ_i = The probability of a backhaul from destination i as predicted in the last section.
M_i = Miles traveled to destination i. (i = Duluth, Minneapolis).

In (5), γ_o and $\gamma_o + \gamma_1$ represent the mean markup of fronthaul price over directly assignable costs per round trip mile for 1976 and 1979 for the average firm, while β_o and $\beta_0 + \beta_1$ represent the same for Minneapolis. Tests for the equality of marginal fronthaul profits per round trip mile for the marginal firm in 1976 and 1979 are then given by

$$(6)\quad \begin{aligned} &H_o: \gamma_o = \beta_o \qquad H_o: \gamma_o + \gamma_1 = \beta_o + \beta_1 \\ &H_A: \gamma_o \neq \beta_o \qquad H_A: \gamma_o + \gamma_1 \neq \beta_o + \beta_1 \end{aligned}$$

In addition, ψ_o and $\psi_o + \psi_1$ represent the mean level of accounting fronthaul profits for the average firm in 1976 and 1979. Tests for compensatory fronthaul profits in an accounting sense for 1976 and 1979 are then given by

$$(7)\quad \begin{aligned} &H_o: \psi_o = 0 \qquad H_o: \psi_o + \psi_1 = 0 \\ &H_A: \psi_o \neq 0 \qquad H_A: \psi_o + \psi_1 \neq 0 \end{aligned}$$

Specifically, if fronthaul prices adjust such that the marginal or no backhaul firm is compensated, then $\psi_o \geq 0$ in 1976 and in 1979 $\psi_o + \psi_1 \geq 0$.

Estimation of (5) is conducted using two estimators—seemingly unrelated regressions (SUR) and a modified two step estimator (MTS). Under the maintained hypothesis that the probability functions reported in the last section yield consistent probability estimates, each of these estimators yield consistent parameter estimates. However, as estimated probabilities form "generated regressors" on the right-hand side of (5) the SUR covariance structure estimates are inconsistent and any hypothesis performed are suspect when applied to this case.[12] The MTS has been shown elsewhere to be consistent in this regard.[13]

Heuristically, the MTS estimator consists of two stages. First, consistent estimates of the probability schedule parameters and the economic system parameters in (5) are obtained as discussed in Section II and OLS respectively. Equation (5) is then approximated by a first order Taylor's series expansion around those estimates, and SUR is performed on the linearized system.

Coefficient estimates using both the SUR and MTS estimators are reported in Table 4. With regard to the coefficients of interest (the intercept terms), both estimators yield similar results. In 1976 esti-

TABLE 4

SUR and MTS Estimation Results

Dependent Variable	Coefficient	Independent Variable	SUR	MTS[b]
NP_D	γ_o	1	.1319*	.1261*
			(.0100)	(.0123)
	γ_1	D	-.0686	-.1574
			(.0694)	(.1093)
	γ_2	ρ_D	.067275	2.6709*
			(.0843)	(1.1915)
	γ_3	$D\rho_D$	-.3579	-20.8178
			(.6528)	(34.9487)
NP_M	β_o	1	.0999*	.1098*
			(.0176)	(.0248)
	β_1	D	-.0670*	-.0496
			(.0258)	(.0354)
	β_2	ρ_M	-.0045	-.01295
			(.0183)	(.0272)
	β_3	$D\rho_M$	-.0552	-.0738
			(.0404)	(.0532)
$R^f - C$	ψ_o	1	1184.48	773.529
			(2970.67)	(4150.38)
	ψ_1	D	-249.846	5335.359
			(4333.24)	(6096.266)
	ψ_2	$\rho_D M_D$	.49298*	10.803*
			(.05634)	(5.0133)
	ψ_3	$D\rho_D M_D$	-2.3362*	53.207
			(.18865)	(131.175)
	ψ_4	$\rho_M M_M$	-.0580	-.06552
			(.06645)	(.07009)
	ψ_5	$D\rho_M M_M$	-.21104*	-.24796
			(.0863)	(.15951)

[a]A (*) indicates significance at the five percent level.

[b]The MTS estimator requires the backhaul parameters in Table 1 to be estimated jointly with the economic system parameters. As these are "nuisance" parameters, they are not reported but are available from the author upon request.

mated per-round trip mile fronthaul marginal profits (not including variable repair costs) were about 13 and 10 cents for Duluth and Minneapolis, respectively. In 1979, these figures depend somewhat upon the estimation procedure used, but range from about −3 to 6 cents per-mile. With the SUR estimator again marginal fronthaul profits in Duluth are larger than in Minneapolis (6 cents versus 3 cents). Using the MTS, firms lose about 3 cents per round trip mile to Duluth and earn about 6 cents per round trip mile to Minneapolis.

Inspection of these results and those of Section II tends to support the hypothesis that fronthaul prices adjust downward in response to higher backhaul probabilities in the sense that marginal fronthaul profits per round trip mile are lower in Minneapolis (the high backhaul market) than in Duluth (the low backhaul market). However, as reported in Table 5, the differences are generally not significant. Specifically, only one test performed suggests marginal fronthaul profits are significantly different (1976-SUR). But, from the discussion above, this test was performed using an inconsistent covariance estimator.

The second issue investigated pertains to fronthaul price adjustment to firms with differing backhaul probabilities. In 1976 the mean accounting fronthaul profit of the average firm was estimated as about $1200 using SUR and/or $800 using MTS, while in 1979 SUR results suggest profits of about $900, and MTS results suggest losses of about $4600. Again this evidence provides some support for the hypothesis that fronthaul prices adjust to the marginal or no backhaul firm (profits ≥ 0)—fronthaul prices cover round trip costs when not loaded on the backhaul. As reported in Table 5, regardless of the estimation procedure or year, the estimated values are insignificantly different from zero, adding further support for the hypothesis.

IV. SUMMARY AND CONCLUSIONS

This paper investigated firm behavior and price adjustment in transport markets which were characterized in terms of the backhaul problem. Data reflected trucking firms that generally traveled from one origin to one of two destinations and back to the original origin. On the outgoing movement, they haul primarily agricultural products for which the transport price adjusts freely. On the return trip, firms commonly but not always travel empty. The evidence presented suggests the degree to which firms travel empty depends on firm specific characteristics but equally, on the market served.

Given that backhaul probabilities vary across both firms and markets, the manner in which fronthaul prices adjust to such differences was addressed through a market model. The evidence suggests that fronthaul prices adjust across markets such that the marginal (in this case the no backhaul) firm is indifferent in an expected profit maximizing sense to the actual market served. This result helps to explain why the low backhaul market (Duluth) receives a higher proportion of truck movements than the high backhaul market (Minneapolis).

However, backhaul probabilities also vary across firms. This issue, emanating from Nicholson (1958) and Miklius and DeLoach (1965) was cast in a market model and treated as an empirical issue. That is, do fronthaul prices adjust to the no backhaul firm (Nicholson) or do fronthaul prices not compensate the no backhaul firm (Miklius and DeLoach)? The results for the particular market examined tend to support Nicholson's maintained hypothesis that fronthaul prices support the no backhaul firm.

In one sense the marginal or no backhaul firm in this case is the least efficient firm in the market. One-half of such carriers' miles traveled are traveled empty. When fronthaul prices adjust to such inefficiencies, a necessary cost of shippers demanding services from such carriers requires the shipper to "pay" for the empty return movement.

On one hand, the evidence provided here tends to support Nicholson's assumption that fronthaul prices cover round trip costs. However, as backhaul prices and associated costs are not observed, the profitability of backhaul traffic and/or the level of backhaul prices cannot be addressed. Either backhaul profits are retained by those truckers receiving backhauls or alternatively those profits are captured in intermediate markets (e.g., brokerage markets). However, since the hypothesis is that prices adjust to the marginal firm, from the agricultural perspective the resulting policy prescription is to raise the level of the marginal firm. There are a variety of mechanisms to achieving this result, depending on the source of the backhaul problem. First, to the extent that regulation

TABLE 5
Hypothesis Tests.

Test:	F - Statistic[a] SUR	MTS
Equality of Marginal Profits		
1976	4.4948*	.5215
1979	.2806	.9016
Compensatory Fronthaul Rates		
1976	.1680	.0412
1979	.0928	2.2192

[a]A (*) indicate significance at the five percent level.

(i.e., operating authority) is the source of the backhaul problem, allowing agricultural carriers to haul regulated commodities on the backhaul is an obvious policy prescription. The Motor Carrier Act of 1980 goes far in this regard through fitness authorities and expansion of the agricultural cooperative exemption. Alternatively, an extension effort promoting firms attaining operating authority may well be warranted. Second, finding backhaul traffic can be costly both in terms of search costs and in terms of opportunity costs. If the source of the backhaul problem pertains to matching up shippers with truckers, another policy prescription pertains to transforming search activities from truckers to other entities. For example, added use of brokers can be promoted. Alternatively, searching for backhaul traffic may include a traffic manager. Since most of the carriers are small, such an investment may not be warranted. However, elevators dealing with multiple truckers may find such an investment coupled with a round trip pricing strategy profitable.

Finally, this research points to a wide variety of potential extensions. First, in one regard the markets anlayzed here pertained to carriers operating symmetrically between market pairs and traveling relatively short distances (200—600 miles). As such, one extension might be to consider the robustness of these findings in other markets. As shown by Beilock and Kilmer (1986), as distance between markets increases, expected profit-maximizing firms in some cases have a motive to search more intensively for backhaul loads. Second, the data employed are somewhat dated. Thus, a second extension would point toward an investigation analogous to that of this paper using more current data. Specifically, this analysis was conducted with pre- Motor Carrier Act of 1980 data. As entry into regulated areas is simplified under the new legislation, an analysis conducted with more current data seems appropriate. It is noted, however, that discussions with other industry analysts suggest that at least in the markets analyzed here truckers generally are not thought to have ICC authority either in the past or in the present.

REFERENCES

Baseman, R., Daughety, A., and Inaba, F. 1977. "Complementarity in Production and Empty Backhauls in Transportation." Paper presented at Western Economics Association annual meeting, 1977.

Beilock, R. and Kilmer, R. L. 1986. "The Determinants of Full-Empty Truck Movements." *American Journal of Agricultural Economics* 68:67-76.

Cosgriff, J. G. "The Costs and Operations of Exempt Motor Carriers in North Dakota," Upper Great Plains Transportation Institute Report No. 33, North Dakota State University (1978).

Duncan, G. M. 1987. "A Simplified Approach to M-Estimation with Application to Two-State Estimators." Forthcoming *Journal of Econometrics*.

Harmatuck, D. J. 1985. "Backhaul Pricing: Mathematical Programming and Game Theoretic Approaches to Allocating Joint Costs." *Transportation Research Forum, Proceedings*. 26:474-482.

Harvey, A. C. 1977. "Estimating Regression Models with Multiplicative Heteroscedasticity." *Econometrica* 44:461-465.

Miklius, Walter, Kenneth L. Casavant, and James Cornelius. 1975. "Motor Carriers of Exempt Agricultural Commodities: Characteristics and Operating Practices." Report Prepared for U.S. Department of Agriculture. Washington State University.

Miklius, W. and DeLoach, D. B. 1965. "A Further Case for Unregulated Truck Transportation." *Journal of Farm Economics* 47:933-947.

Nicholson, H. W. 1958. "Motor Carrier Costs and Minimum Rate Regulation." *Quarterly Journal of Economics* 72:139-152.

Pagan, A. 1984. "Econometric Issues in the Analysis of Regressions with Generated Regressors." *International Economic Review* 25:221-247.

Pederson, L. E., Mittelhammer, R. C., and Casavant, K. L. 1979. "Factors Affecting Interstate Backhauling of Exempt Agricultural Commodities by Regulated Motor Carriers: A First Look." *Transportation Journal* 18:46-52.

Wilson W. 1986. *Transport Markets and Firm Behavior: The Backhaul Problem*. Unpublished dissertation, Washington State University.

Wilson, W. W., Griffin, G. C., and Casavant, K. L. 1982. "Costs and Characteristics of Operating Interstate Motor Carriers of Grain in North Dakota." Upper Great Plains Transportation Institute Report No. 46, North Dakota State University.

Wilson, Wesley W. 1987. "The Exempt Trucker: Rent-Seeking in Backhaul Markets." Mimeo, Department of Agricultural Economics, Washington State University.

ENDNOTES

* Department of Agricultural Economics, Washington State University, Pullman WA. Comments from Greg Duncan, Fred Inaba, William Hallagan, Ken Casavant, and Rick Beilock are gratefully acknowledged. Also, the author wishes to express his thanks to Gene Griffin and the staff of the Upper Great Plains Transportation Institute for providing the data used.

1. See Wilson et al. (1987), Beilock and Kilmer (1986), and Miklius et al. (1965).
2. For example, see Miklius and DeLoach (1965).
3. In addition, to those cited earlier, see also Harmatuck (1985), Wilson et al. (1982), Baesemann et al. (1977) for further reference.
4. The data are fully described in Wilson (1986), Cosgriff (1978), and Wilson et al. (1982).
5. In Beilock and Kilmer, the analog to these firms is that these firms have prior commitments in North Dakota and as such have opportunity costs associated with delays to search for and handle loads [Beilock and Kilmer (1986), p. 71].
6. See testimonial evidence "Regulatory Problems of the Independent Owner-Operator in the Nation's Trucking Industry (Part 1)," in Hearings before the Subcommittee on the Activities of Regulatory Issues of the Committee of Small Business, House of Representatives 94th Congress.

7. It is noted that the firm is assumed to take backhaul probability schedules as given in the sense that the proportion of backhaul loads to total loads from a given location is unaffected by the proportion of trips taken to that location, i.e., $E(b_D/d,A) = E(b_D/A)$ and $E(bM/M, A) = E(bM/A)$.
8. A multiplicative heteroscedastic model [Harvey (1979)] was employed using a quadratic expansion of the x's in (1) to form weights. First round non-linear ordinary least squares results and the heteroscedastic model results are reported in Wilson (1986).
9. Other explanations can be found in Wilson (1987).
10. The procedure used is fully described in Wilson (1986). In short, 1976 rates were constructed by using available rates to estimate a rate schedule as a function of distance and destination. Since 1979 prices were not available, the 1976 prices were regressed on constructed rail rates in 1976 and these parameter estimates were used with available 1979 rail rates to construct 1979 truck prices for available 1979 rail rates to construct 1979 truck prices for each destination.
11. There are no other variables included in the model, the hypothesis being that all other variables explaining differences in profit margins are included in the intercept terms which are then interpreted as mean total fronthaul profits and mean marginal fronthaul profits for the average firm.
12. See Pagan (1984) and Duncan (1987) for a general discussion of the "generated regressors" problem.
13. The proof is available from the author upon request, or see Wilson (1986).

Highway User Taxes and Infrastructure Improvements: The Question of Benefits

by Edward A. Morash and Charles R. Enis***

ABSTRACT

The imposition of user taxes is a subject of considerable importance and contemporary concern for carrier management, government investment and promotional policy, and efficient resource allocation. Although infrastructural improvements are frequently based on intended benefits and beneficiaries, the actual measurement of the policy impacts is rarely undertaken. The present study provides a market-based methodology to empirically investigate the net-cost benefit of a major legislative initiative to upgrade and rehabilitate the nation's highway and bridge system—the Surface Transportation Assistance Act of 1982. Utilizing step-wise regression and ANOVA pro cedures, the study analyzes the economic impact of the user tax package on the motor carrier industry, on types of carriers, and on interactive effects amongst carrier characteristics. Implications for public policy and carrier management are then outlined.

I. INTRODUCTION

Public policy actions can have a significant impact on an industry as well as on the competitive position of individual firms within the industry. In the case of highway investment and user taxation, infrastructural improvements are often justified based on particular benefits and beneficiaries. However, both the actual measurement of the purported net benefits and an identification of the actual beneficiaries are rarely undertaken in any systematic fashion. Furthermore, from a carrier management viewpoint, to the extent that carriers can alter their operations and target markets, such public policy actions can create important "messages" and incentives for firm behavior.

The present study provides a methodology for determining whether a significant public policy action, the Surface Transportation Assistance Act of 1982 (STAA), is a net positive economic benefit for the motor carrier industry as well as for specific types of motor carriers. The STAA was selected as the focus of analysis since it substantially increased user taxes on the motor carrier industry yet at the same time provided substantial benefits in the form of improved infrastructure and reduced operating restrictions for potential carrier productivity gains. The specific objective of the present research is to obtain an understanding of whether the benefits received from an improved infrastructure and relaxed operating restrictions are worth the increased taxation sustained by the trucking industry as a specific user group.

The next section of this paper will provide a brief overview of the user tax and benefit provisions of the STAA. After considering various agency predictions as to the potential net benefits and beneficiaries of this legislation, Section II will outline the market-based methodology employed in this study to ascertain the real economic impact of the STAA. The results of the step-wise regression and ANOVA analyses will then be presented followed by conclusions and implications for highway financing, user taxation, and carrier management.

A. STAA User Taxes and Benefits

For the first time since 1966, the STAA substantially increased user taxes on the motor carrier industry (Taff 1986, p. 59). The major increases were for the vehicle fuel tax (a variable tax) and for the heavy vehicle use tax (a fixed component).[1] The fuel tax was increased from 4 cents per gallon to 9 cents per gallon, and eventually to 15 cents per gallon for diesel fuel.[2] In turn, the heavy vehicle use tax was increased from $162 to $1600 per vehicle for the heaviest trucks but ultimately reduced to $550 per vehicle.[3] These heavy trucks were intended as a particular focus of the STAA tax package reflecting the "costs occasioned" approach of earlier governmental studies (see, e.g., U.S. D.O.T., "Final Report on the Federal Highway Cost Allocation Study" 1982).

Although overall truck taxes were significantly increased, the STAA also offered potential productivity benefits for the motor carrier industry. Authorized funds included 58 billion dollars over four years to rehabilitate and restore the nation's deteriorating highway and bridge infrastructure, to complete the interstate system, and for other related highway projects (U.S. Congress 1983). Clearly, improved infrastructure would mean more efficient motor carrier operations, decreased repairs and accidents, reduced claims for shipment damage, and better highway transit times.

The STAA also provided for relaxed operating restrictions on carriers which would translate into additional productivity gains for some types of carriers. Specifically, an 80,000 pound nationwide standard would preempt lower state limits on the interstate system. Single 48 foot trailers and double trailers of 28 feet each were also authorized on all interstate and certain other designated highways, while permissible trailer widths were increased from 96 inches to 102 inches (U.S. Congress 1983). To the extent that carriers can improve their operations and increase their payloads with these larger and heavier vehicles, then these STAA provisions should represent additional productivity gains for affected carriers. In total, the STAA offered the trucking industry many benefits in the form of an improved infrastructure and fewer operating restrictions, but at a substantial cost in the form of significant increases in user taxes.

B. Agency Predictions for Motor Carrier Benefits and Costs

At the time of the STAA, no studies attempted to assess the economic benefits of improved highway infrastructure, although two studies did attempt to estimate the aggregate net benefit of the relaxed size and weight limitations and the increased user taxation. Both the U.S. Department of Transportation (DOT) and the American Trucking Associations (ATA) predicted that the Act would be a net positive economic benefit for the trucking industry as a whole, although the ATA's estimate (0.8 billion annually) was much more conservative than the DOT's (3.2 billion annually).[4] Presumably, a consideration of the additional benefits from improved highways would raise these estimates further.

Several agencies and interest groups also made apriori predictions as to the economic impact of the STAA on different types of motor carriers.[5] In general, those carrier types expected to be negatively disadvantaged by the STAA were heavy vehicle operators (variable x_1 in this study), long-haul carriers (x_3), rail competitive carriers (x_4), owner-operators (x_7), and specialized carriers (x_9). In contrast, those carrier types expected to be positively advantaged by the STAA were less-than-truckload (LTL) carriers (x2), carriers with high power unit utilization (x_5), large firms (x_6 and x_8), and carriers operating primarily in the more congested East (x_{10}). In essence, the benefits and burdens of the STAA were not expected to be distributed evenly amongst motor carriers, although little attempt was made to empirically evaluate these hypotheses by the various agencies making the predictions.

II. METHODOLOGY AND DATA SOURCES

The primary research question addressed in this study is whether a user tax package such as the STAA is a net positive economic benefit or cost to the motor carrier industry as well as for specific types of motor carriers. Although a long period of time is required to fully evaluate the economic impact of public policy actions, the capital market has been shown to be an efficient processor of information (Fama 1970; Schwert 1981). According to the rational-expectations/efficient-markets framework, the stock market instantaneously impounds all publicly available relevant information into security prices to reflect actual economic values. Furthermore, there is a great deal of empirical support for this theoretical perspective.[6]

To answer the primary research question, the current research proceeded in three phases. In the first phase, market models were constructed for each of 39 motor carriers over a 235 day trading horizon. These market models utilized daily share price data and represent the long-term "systematic" relationship between a firm's securities and an overall market index. Appendix A provides a more technical explanation of the market models. The 235 day time frame was chosen to symmetrically encompass the 23 day STAA period. This STAA period embraced the eight critical event days that represented the legislative history of the STAA (Congressional Record, Vol. 128, 1982) and included:

Event Day$_d$	Critical Event
1	Bill was considered and passed by House of Representatives
2	Senate Finance Committee approved a tax package
3	Bill called to Senate floor
4	Senate floor debate during which trucking industry vigorously opposed bill
5	Senate passed amended bill with lower heavy truck taxes
6	House agreed to conference report
7	Senate agreed to conference report; final Congressional approval
8	Presidential signature

Abnormal security returns were then computed for each day of the STAA period including the eight critical event days. At the risk of oversimplication, the market models represent a standard or benchmark while the abnormal returns represent cost-benefit deviations, either favorable (positive) or unfavorable (negative), from that standard (see Schwert 1981; Beaver and Landsman 1983). Appendix A also provides a detailed explanation of abnormal returns. The following null hypothesis was statistically tested to determine whether motor carriers as a group had received a net positive gain or loss from the user tax package:

> H_0 (1): The mean abnormal returns associated with the specific critical event days do not significantly differ from zero.

A rejection of this hypothesis would indicate that the STAA tax package was a net economic benefit (cost) to the motor carrier industry, while an acceptance of this hypothesis would indicate that the STAA was relatively neutral for the industry.

The second phase of the research utilized a step-wise linear regression model to test whether the economic costs and benefits of the STAA were differentially distributed amongst various types of motor carriers. Cumulative abnormal returns (CAR_i) served as the dependent variable in the model. The CAR_i for each of the 39 trucking firms was the sum of the abnormal returns for each firm over the eight critical event days (see Appendix A).[7] Ten firm-specific variables $x_1 \dots x_{10}$ derived from the literature review and discussed in the Introduction (x_k variables) served as the independent variables in the step-wise regression procedure. This regression model represented the best "explanatory" model after evaluating all possible combinations of the x_k variables while minimizing the loss in "degrees of freedom" (see, e.g., Bolch and Huang 1974). Ten null hypotheses were then tested for the x_k variables as derived from the transportation literature:

$H_0(k+1)$: Variable x_k is not associated with the net economic effects of the STAA on the trucking industry.

Where:

k = 1 ... 10 (see Table 2).

A variable appearing in the step-wise regression model represents a rejection of null hypothesis (k+1) while an x_k variable not appearing in the model means an acceptance of the (k+1) null hypothesis for that variable.

In the third phase of the research, ANOVA tests were performed to identify any interactive effects between any two x_k variables. The CAR_i for the trucking industry served as the dependent variable. The independent variables in each model were two specified x_k variables and their two-way interactive effect. Ten variables taken two at a time required the formulation of 45 ANOVA models.

The firm-specific data used in the step-wise regression and ANOVA procedures were obtained from the 1983 edition of the *TRINC's Blue Book of the Trucking Industry*. The daily returns for the market and for firm securities graded on the major exchanges were obtained from the *CRSP* tapes. The daily returns for securities traded over-the-counter were compiled from share price data listed in the *Standard and Poors Daily Price Record*. The sample size consisted of 39 motor carriers which represented the entire universe of all publicly traded trucking firms. These motor carriers account for over 40 percent of industry revenues, so that the study results would have relevance for a major portion of the trucking industry.

III. RESULTS

A. Economic Impact on Trucking Industry

To test whether the market perceived the STAA to be a net positive benefit or cost for the motor carrier industry as a whole, Table 1 shows the mean abnormal returns for all of the motor carriers for each day of the STAA period. The eight critical event days during the STAA period are indicated in column (3) and the mean abnormal returns in column (4).

Several observations should be noted from Table 1. First, the mean abnormal returns for four of the critical event days were significantly different from zero and negative, while six of the eight critical event days were negative. The two event days which were positive were not statistically significant. One

TABLE 1
Mean Abnormal Returns for Motor Carriers
During the STAA Period

(1) Date	(2) Estimation Period Day_t	(3) Event Day_d	(4) Mean Abnormal Returns for all Motor Carriers
12/06/82	106	1	-.0045
12/07/82	107		-.0050
12/08/82	108		-.0013
12/09/82	109	2	-.0058*
12/10/82	110	3	-.0085*
12/13/82	111		-.0033
12/14/82	112		-.0005
12/15/82	113	4	-.0172*
12/16/82	114	5	.0087
12/17/82	115		-.0075
12/20/82	116		-.0047
12/21/82	117	6	-.0046
12/22/82	118		.0012
12/23/82	119	7	-.0092*
12/27/82	120		-.0087*
12/28/82	121		-.0012
12/29/82	122		-.0048
12/30/82	123		.0015
12/31/82	124		.0101*
01/03/83	125		.0016
01/04/83	126		-.0044
01/05/83	127		.0043
01/06/83	128	8	.0042

*Significant from zero at the .05 level (t-test).

was a "good news" day for the trucking industry which was expected to be positive since the heavy truck taxes were reduced on that day (event day 5, see Table 1 and the Methodology section); while the other event day was the pro forma Presidential signing of the Act into law (day 8). Finally, four of the five days that the trucking abnormal returns were significantly below zero were critical event days (Table 1). Consequently, $H_0(1)$ is rejected. In total, these findings suggest that the market perceived the STAA to have an unfavorable impact on the motor carrier industry as a whole.

B. Types of Motor Carriers

To test whether the costs and benefits of the STAA were differentially distributed amongst different types of motor carriers [hypotheses (2). . . (11)], a step-wise regression model was constructed utilizing cumulative abnormal returns (CAR_i) as the dependent variable and the motor-carrier specific variables $x_1 \ldots x_{10}$ as the independent variables (x_k variables; see top portion of Table 2). The expected signs of the regression coefficients associated with each independent variable were as discussed in the Introduc-

TABLE 2
Step-Wise Regression Model:
CAR_i with Motor-Carrier Specific Variables

(A) Motor-Carrier Specific Variables (x_k) and Their Measures

x_1 Average Load in Tons, a measure of heavy vehicle operators

x_2 LTL Percentage (percentage of LTL shipments)

x_3 Average Haul, Distance

x_4 Rail Competitiveness (average load weighted by percent of TL shipments)

x_5 Power Unit Utilization (total miles divided by total power units)

x_6 Total Miles (all vehicles), a measure of firm size

x_7 Owner-Operator Percentage (percent of power units using owner operators)

x_8 Firm Size (number of shares)

x_9 Specialized (coded 1) vs. General Commodities Carrier (coded 0)

x_{10} Eastern States Carrier (coded 1) (0 otherwise)

(B) Best Model: Dependent Variable CAR_i

x_k Variable	$\hat{\alpha}$, $\hat{\beta}_k$ Parameter Estimate	Standardized Parameter Estimate	t-value	Prob. >|t|	Model d.f.	Model R^2	Model Sig.
Intercept	-.1852	.000	3.27	.00			
x_3 Average Haul, Distance	.0003	.572	2.90	.01*			
x_5 Power Unit Utilization	1.8245	.303	1.96	.05*			
x_6 Total Miles	-.0008	-.476	2.67	.01*			
x_7 Owner-Operator Percentage	-.1957	-.490	2.95	.01*	4,34	.354	.004

*H_0 (k + 1) is rejected at the .05 level.

tion, where a favorable impact of the STAA would be indicated by a positive coefficient and an unfavorable impact by a negative parameter estimate.

The results of testing hypotheses (k + 1) are reported in Table 2. Hypotheses (4), (6), (7), and (8) which are associated with variables x_3, x_5, x_6, and x_7 respectively are rejected. Consistent with the expectations discussed in the Introduction, power unit utilization is positively associated with CAR_i, and reliance upon owner operators is inversely associated with CAR_i. On the other hand, contrary to expectations, the market perceived the STAA to have a favorable impact on long distance carriers, and an unfavorable impact on large carriers relative to other motor carriers. According to the results presented in Table 2, the profile of a motor carrier that would be adversely affected by the STAA would be a large short haul carrier with idle power units that relies heavily upon owner-operators.

C. Interactive Effects between Motor-Carrier Specific Variables

ANOVA models were constructed for various combinations of two of the ten x_k variables for a total of 45 such models. The purpose of this analysis was to identify possible interactive effects between any two of the x_k variables. Five of the 45 models had two-way effects that were statistically significant. These models are shown as panels (A) through (E) in Table 3. In a separate analysis, the x_k variables for these five models were also broken down into high and low (or dichotomous) motor carrier groupings in order to ascertain the nature of the interaction effects.

The interactive effects reported in Table 3 have interesting implications and interpretations. For example, according to Table 2, rail competitiveness (x_4) is not a significant variable, although its negative correlation did suggest a possible expected unfavorable impact of the STAA onrail-competitive firms. However, the interactive effect between x_4 and x_{10} in panel (A) of Table 3 indicates that the negative impact of the STAA onrail-competitive motor carriers is much more unfavorable for those truckers not operating in the East. Similar results are achieved in panel (B) where the interactive effect indicates that rail competitive carriers which haul primarily special commodities are much more adversely affected than general commodities carriers. Furthermore, according to panel (C) of Table 3, specialized carriers

TABLE 3

ANOVA Models Containing Significant Interactive Effects Between Motor-Carrier Specific Variables

Dependent Variable: CAR_i

		Source	d.f.	Mean Square	F	Probability > F
(A)	X_4	Rail Competitiveness	1	.0181	4.03	.05**
	X_{10}	Eastern States Carrier	1	.0012	0.27	.61
		X_4 by X_{10}	1	.0192	4.28	.04**
		Error	29	.0045		
(B)	X_4	Rail Competitiveness	1	.0181	3.89	.05**
	X_9	Spec. vs. General Commod.	1	.0004	0.08	.77
		X_4 by X_9	1	.0154	3.30	.08*
		Error	29	.0047		
(C)	X_2	LTL Percentage	1	.0011	0.04	.85
	X_9	Spec. vs. General Commod.	1	.0492	1.70	.20
		X_2 by X_9	1	.0872	3.02	.09*
		Error	30	.0289		
(D)	X_2	LTL Percentage	1	.0058	0.19	.67
	X_7	Owner Operator Percentage	1	.0594	1.89	.18
		X_2 by X_7	1	.0989	3.15	.08*
		Error	26	.0314		
(E)	X_1	Average Load	1	.0293	1.08	.31
	X_5	Power Unit Utilization	1	.1372	5.05	.03**
		X_1 by X_5	1	.0910	3.35	.07*
		Error	26	.0272		

**Significant at the .05 Level.
* Significant at the .10 level.

should also feel an especially unfavorable impact of the STAA if their percentage of shipments that are LTL is relatively low (i.e.,truckload carriers). Similarly, panel (D) shows that truckload carriers (i.e., low LTL) that make extensive use of owner-operators will also be adversely impacted by the STAA. In contrast, panel (E) implies a favorable market reaction to motor carriers that have a relatively high degree of power unit utilization and relatively low average loads. In total, although variables x_1, x_2, x_4, x_9, and x_{10} are not statistically significant in the stepwise regression model in Table 2, these factors in tandem with other variables provide some interesting insights.

IV. CONCLUSIONS

The economic consequences of changes in tax policy involving transportation infrastructure are important public policy and carrier management considerations. The substantial public investment that is required to provide a national highway system that serves the needs of commercial motor transportation, national defense, and the general public requires that the purported beneficiaries and benefits of these public goods be scrutinized and justified. Furthermore, from a carrier management perspective, properly understood user tax changes can create important firm incentives for adapting carrier practices and market strategies toward obtaining public policy benefits and avoiding economic costs.

The specific focus of the present research was to gain an understanding of the net marginal cost or benefit to the motor carrier industry that is associated with the combination of a user tax increase, an improved right-of-way system, and reduced operating restrictions. The results of the analyses indicate that the STAA tax and benefit package was a net economic loss to the motor carrier industry as a whole. However, the costs and benefits of the legislation were not evenly distributed amongst different types of motorcarriers. Those types of motor carriers benefiting from the user tax package included long-haul carriers and carriers with high power unit utilization. Both of these carrier types would be able to spread the fixed heavy vehicle use tax over more miles and to achieve greater absolute productivity gains from infrastructural improvements and relaxed size and weight limitations, all of which are per-mile benefits. In contrast, those firms disadvantaged by the STAA include carriers which rely heavily on owner-operators and larger firms. Similarly, the interactive ANOVA tests show the profile of a disadvantaged motor carrier as a specialized, truckload carrier of rail competitive traffic that extensively utilizes owner-operators and who does not operate in the East. These latter interactive results provide interesting insights and are in general agreement with the "main effects" expectations discussed in the transportation literature. In total, the results suggest that from a carrier management viewpoint, firms might wish to alter those variables under their direct control; e.g., power unit utilization, intensity of owner-operator use, and traffic composition to properly adapt to the new user tax environment.

Since public policy actions frequently have unintended consequences, it is important to monitor the actual cost-benefits and beneficiaries of these actions. For example, the finding that the STAA user tax package was a net economic loss to the motor carrier industry should be of concern to policymakers and carrier management. Furthermore, a major focus of the STAA was to shift highway cost allocation to heavy vehicles under the "costs occasioned" theory. However, no direct evidence was found in this study that heavy vehicles were disadvantaged by the legislation, while subsequent legislation reduced their tax burden further. In contrast, owner-operators and the carriers which utilize them were found to be adversely impacted by the STAA. If these consequences are considered inappropriate public policy impacts, then corrective policy modifications would be warranted.

REFERENCES

American Trucking Associations, *The Effect of Increased Highway Taxes on Motor Carrier Operating Expenses and Profitability,* (Statistical Analysis Dept., ATA, April 1983), pp. 1-26.

Beaver, W. and W. Landsman, *Incremental Information Content of Statement 33 Disclosures,* (Stamford, Conn.: Financial Accounting Standards Board, 1983).

Bolch, Ben W. and Cliff J. Huang, *Multivariate Statistical Methods for Business and Economics,* (Englewood Cliffs, NJ: Prentice-Hall, Inc., 1974).

Davisson, John F., "Tax Reform's Impact on Transportation," *Handling and Shipping Management,* February 1987, pp. 30-32.

Enis, Charles R. and Edward A. Morash, "Market Protection, Deregulation, and the Question of Industry Losses," *Transportation Law Journal,* Vol. 15, No. 1, December 1986, pp. 89-104.

Fama, Eugene, "Efficient Capital Markets: A Review of Theory and Empirical Work," *Journal of Finance, Vol. 25, No. 2, May 1970, pp. 383-417.*

Hoffer, George E., "Estimating the Impact of Changes in Recent Cost Allocation Methodology and Highway User Fees on the Allocated Cost and Revenue Responsibility of Motor Carriers," Transportation Journal, Vol. 23, No. 1,Fall 1983, p. 32.

Malkiel, B.G., "Equity Yields, Growth, and the Structure of Share Prices," American Economic Review, Vol. 53, No. 4, December 1963, pp. 1,004-1,031.

Nelson, James C., "The Pricing of Highway, Water, and Airway Facilities," *The American Economic Review,* Vol. 52, No. 2, May 1962, pp. 426-435.

Rao, Kant and Thomas D. Larson, "Capital Investment, Performance, and Pricing in Highways," *Transportation Journal,* Vol. 21, No. 3, Spring 1982, pp.22-33.

Schwert, G. William, "Using Financial Data to Measure Effects of Regulation," *Journal of Law and Economics,* Vol. 24, No. 1, April 1981, pp. 121-158.

Taff, Charles A., *Commercial Motor Transportaiton,* Seventh Edition, (Centreville, Md.: Cornell Maritime Press, Inc., 1986).

U.S. Comptroller General, *The Surface Transportation Assistance Act of 1982: Comparative Economic Effects on the Trucking Industry* (General Accounting Office, April 6, 1984).

U.S. Congress, *Surface Transportation Assistance Act of 1982* (P.L. 97-424), 97th Cong. 2d Sess. (January 6, 1983).

U.S. Department of Transportation, *Final Report on the Federal Highway Cost Allocation Study,* Report of the Secretary of Transportation to the U.S. Congress (U.S. Government Printing Office, May 1982).

VI. ENDNOTES

*Graduate School of Business, Michigan State University, East Lansing, MI

**College of Business Administration, Pennsylvania State University, University Park, PA.

1. The tax on truck sales was also increased from 10 to 12 percent for truck chassis and bodies over 33,000 lbs. gross vehicle weight (GVW) and for truck trailer and semi-trailer chassis and bodies over 26,000 lbs. GVW. Taxes were eliminated on truck parts, lubricating oils, and tires under 40 lbs. (U.S. Congress 1983).
2. The Tax Reform Act of 1984 (TRA 84) further increased the variable component to 15 cents per gallon of diesel fuel but reduced the fixed component to $550 per vehicle for the heaviest trucks.
3. Vehement trucking opposition to the heavy vehicle use tax was successful in shifting some of the tax reponsibility from the fixed component to the variable component in TRA 84 (see previous note). The present analysis focuses on the STAA because TRA 84 provided no significant benefits, did not change the overall level of taxation, and merely shifted some tax responsibility from heavy vehicles to high mileage vehicles. Based on any study findings, it would be possible to predict the direction of the change; for example cost impact would be shifted from heavy trucks to high mileage trucks.
4. For a summary of these studies, see: U.S. Comptroller General 1984, p. 25.
5. For a discussion and summary of these various agency predictions, see: U.S. Comptroller General 1984. See also American Trucking Associations 1983.
6. See: Beaver and Landsman 1983, p. 2; Enis and Morash 1986, pp. 95-96; Schwert 1981, pp. 121-158; and Malkiel 1963, p. 1,009.
7. For additional technical and in-depth discussion of the CAR_i, see Schwert 1981; and Beaver and Landsman 1983.

APPENDIX A

MARKET MODELS AND ABNORMAL RETURNS

Market models were constructed for each of 39 motor carriers over a 235 day trading period. The parameters for the market models used to compute abnormal returns associated with the critical event days were estimated using the ordinary least squares regression model in equation (1).
For each carrier security i ... n:

$$R_{it} = \hat{\alpha}_i + \hat{\beta}_{1i} M_{t-1} + \hat{\beta}_{2i} M_t + \hat{\beta}_{3i} M_{t+1} + \hat{E}_{it} \quad (1)$$

where

R_{it} = return on carrier security i on day t.

M_{t-1}, M_t, M_{t+1} = returns on the CRSP value-weighted market portfolio (including dividends) on days $t-1$, t, and $t+1$, respectively.

$\hat{\alpha}_i$ = estimated intercept for security i.

$\hat{\beta}_{1i}$, $\hat{\beta}_{2i}$, $\hat{\beta}_{3i}$ = estimated regression coefficients for lagged, contemporaneous, and lead market returns, respectively, for carrier security i.

$\hat{E}_{it}$ = estimated residuals.

t = estimation period days 1 ... 235.

i ... n = security 1 ... 39.

Conceptually, these market models are "benchmarks" or standards, and represent the long-term systematic relationship between a firm's securities and the overall market index (e.g., CRSP portfolio).

Abnormal returns for each carrier's common stock were then computed for each of the eight critical event days using equation (2). For each carrier security i ... n:

$$AR_{id} = R_{id} - (\hat{\alpha}_i + \hat{\beta}_{1i} M_{d-1} + \hat{\beta}_{2i} M_d + \hat{\beta}_{3i} M_{d+1}) \quad (2)$$

where:

AR_{id} = abnormal return for carrier security i on critical event day d of the STAA period.

R_{id} = return on security i on critical event day d of the STAA period.

M_{d-1}, M_d, M_{d+1} = returns on the CRSP value-weighted market portfolio on days d-1, d, and d+1 of the STAA period regarding event day d, respectively.

d = critical event days 1 ... 8.

Mean abnormal returns for the entire sample of 39 motor carriers were determined for each critical event day of the STAA period. For the step-wise regression and ANOVA analyses, cumulative abnormal returns (CARi) were also calculated by summing the abnormal returns for each firm over the eight critical event days. The CAR_i served as dependent variables in the step-wise regression and ANOVA models which were used to test the relationships between the STAA economic impact and specific motor carrier characteristics as independent variables.

Abnormal returns are a measure of the net economic impact of the STAAon individual motor carriers. Conceptually, they represent abnormal deviations from the market model of equation (1). For example, if investors expect the increased user taxes in conjunction with an improved infrastructure and less restrictive size and weight limitations to be a net economic benefit (cost) to a given trucking firm, then the enactment of the STAA will be perceived as a favorable (unfavorable) event by the market, and such will be reflected in share price behavior so that the abnormal returns for the given carrier would be positive (negative). Thus, positive abnormal returns for the firm from equation (2) on STAA critical event days indicate a net economic benefit while negative abnormal returns from equation (2) indicate anet economic loss to the carrier.

Separate Freight Rates for Weight and Cube

*By Miles H. Sonstegaard**

ABSTRACT

Separate flat prices for the carrying of weight and the accommodation of cargo volume are proposed. Adoption of this pricing method can be viewed as a refinement of the venerable principle that the charge per unit of weight should rise as density decreases. Advantages of this refinement include (a) a particularly simple relationship between ideal price and marginal cost and (b) true revelation of customer valuations of proposed transportation services. In the small-numbers case, such revelation also requires that prices be based on a submarginal bid. A history of separate prices might provide additional guidance for transportation-system design.

I. INTRODUCTION

The cost of owning and operating line haul vehicles is, ceteris paribus, an increasing function of both weight-carrying capability and volumetric capacity. A vehicle having a given weight capability and a relatively small volumetric capacity can be redesigned for additional cube at relatively low marginal cost. But as volume is increased with weight capability held constant, the marginal cost of adding volume tends to increase. An analogous statement applies where volume is constant and weight capability increases. There is a long tradition of adjusting rates to weight density, or charging by volume instead of weight when density drops below a specified level.[1] What is investigated here is the imposition of separate charges for carrying the weight of a shipment and for providing vehicular space for it. Ideally, each of these services would be priced at its own marginal cost. In the time frame of the individual trip, market-period analysis would appear to be applicable. At the other extreme, long-run analysis would apply to the design of vehicles, roadways and terminals.

Marginal cost pricing for weight and cube could encourage better utilization of existing equipment. The mixing of high-density and low-density shipments within vehicles might avoid some of the current waste of potential transportation service where some vehicles run half full but loaded to weight capability while others run stuffed but lightly burdened. And the existence of a history of separate rates might constitute some improvement on the information now available to the designers of vehicles and of route and terminal facilities.

Marginal costs are discussed briefly in Section II, a method of dual pricing is presented in Section III, possible advantages of such pricing and problems related to it are discussed in Section IV. The main argument is summarized and conclusions are drawn in Section V.

II. MARGINAL COSTS

It is for rather complex reasons that the incremental cost of increasing the cubic capacity of a vehicle rises, and tends to rise more and more steeply with successive increments of cube. As height increases, the strength requirements of the freight compartment become more demanding. Loads imposed by gusting cross-winds increase, not only because of increased area but also because the wind blows harder at greater heights. Increasing the transverse section of the freight compartment tends to increase air resistance. And greater vehicle height trends to raise the center of gravity and thereby to decrease roll stability. But the sharper increases in marginal cost would seem to come, in many cases, when overall height limits were reached, so that further increases in cube would require more compact design of undercarriage, power unit, and so on. Wheels can be made smaller and more numerous in order to permit the cargo compartment to extend downward closer to the roadway, but construction and maintenance costs rise sharply enough to give the overall marginal cost of cube an increasingly steeper slope as cube is added. The implied context here has been highway or rail. If the cubic capacity of inland barges were greatly increased, then on some routes overhead clearances might be insufficient for the passage of empties, and this would necessitate costly provisions for making otherwise empty backhauls in ballast.

Similarly, where the volumetric capacity of a vehicle is fixed and weight-carrying capability is incremented, one would expect successive tons of added capacity to cost more and more. Where the constraint is a load limit on a roadway (or a draft limit for some barge of standard length and width), increases in net carrying capacity come at the price of reducing tare weight, which requires costly shifts to high-strength-to-weight designs and materials.

III. PRICING

Where n identical units of a good are to be sold simultaneously on the basis of sealed bids, a method of minimizing the tendency to underbid is to charge a flat price and to set that price equal to the highest submarginal bid. (Vickrey, 1961, pp. 26, 27). Suppose for example that three identical units of a good were available and the following five bids, in dollars per unit, were received: 15, 12, 10, 9 and 7. (Here 10 is regarded as the *marginal bid* and 9 as the *highest submarginal* bid.) The three highest bids would each take a unit and each unit would sell for 9 dollars. Under this method each bidder knows that his bid can affect the price charged only if it is unsuccessful. Because no successful bid will be used to set the price, the bidder suffers no disadvantage in revealing his true valuation via his bid. But he could injure himself by understating, for he might thereby forgo the purchase of a unit at a price that he was actually willing to pay. It is presupposed that the reservation price for each available unit is zero. This method of pricing will be referred to hereinafter as the *submarginal-price method.*[2]

Where weight and cube are to be priced separately, the foregoing pricing method may be extended appropriately. An extension that is here termed the *submarginal isoexpense-line method* will now be described. Each would-be shipper from a given origin to a given destination for a specified departure time places a bid with the carrier. This bid specifies the weight and cube of the shipment and the maximum amount that the bidder is willing to pay for the service. When the bidding deadline arrives, some feasible subset of the set of bids is accepted. A feasible subset is one that is consistent with the weight and cube limitations placed on the carrying unit. (This unit might be a truck, semitrailer, rail car, train, barge, barge tow, airplane, van cargo container, or some other means of transportation.) If there is more than one feasible subset, the one chosen must maximize the total declared customer benefit, where the customer's offer to pay is regarded as his declared benefit. If there is a tie, one of the maximal subsets is chosen via a fair random process.

Corresponding to each bid there exists an *isoexpense line,* such as that shown in the graph in Figure 1. The horizontal axis measures the price of weight p_w, which might be expressed in cents per pound. The vertical axis measures the price of cube p_v which might be expressed in cents per cubic foot. If a particular bid (call it *A*) specified 30 pounds, 10 cubic feet and $3.00, this would imply the shipper's willingness to pay up to 10 cents per pound if cube was free or 30 cents per cubic foot if weight was free. The p_w intercept of the isoexpense line is therefore at 10 cents per pound; the p_v intercept is at 30 cents per pound. The bidder has declared, in effect, that he will accept any pair of prices (for weight and cube) that does not lie above his isoexpense line.

Figure 1. Isoexpense Line

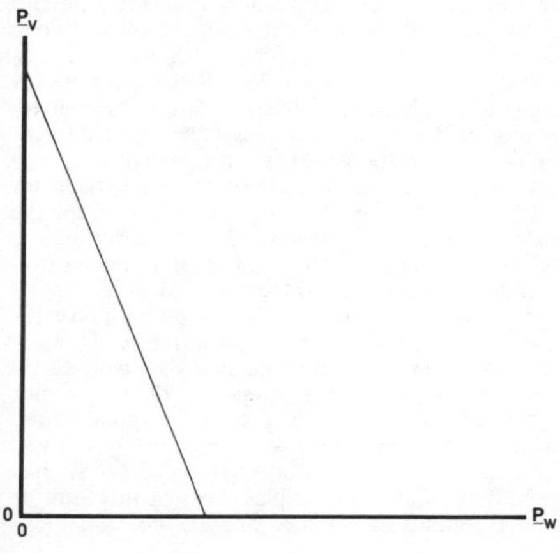

The isoexpense line for bid *A* is shown in Figure 2, along with the respective isoexpense lines for bids *B* and *C*. Let *A, B* and *C* be the only accepted bids. Then the point that graphically represents the pair of prices actually charged must not lie above any of these isoexpense lines, for if it did, the terms of some accepted bid would be violated. In other words, the price pair actually selected by the carrier must not lie outside the heavy boundary (shown in Figure 3), which is composed of portions of the isoexpense lines for bids *A, B* and *C*. This boundary will be termed the *contractual boundary.* The ray out of the origin is drawn such that its slope is equal to the actual density of the load (comprising the shipments associated with bids *A,B* and *C*) it is termed the *density ray.* Portions of the isoexpense lines for bids *D* and *E* also appear in Figure 3; neither bid *D* nor bid *E* was accepted. The *marginal bid* is that accepted bid whose isoexpense line touches the density ray closer to the origin than does the isoexpense

Figure 2. Isoexpense Lines for Bids *A, B* and *C*.

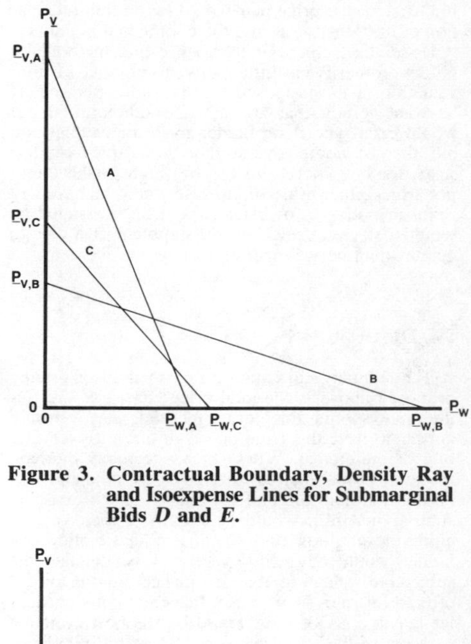

Figure 3. Contractual Boundary, Density Ray and Isoexpense Lines for Submarginal Bids *D* and *E*.

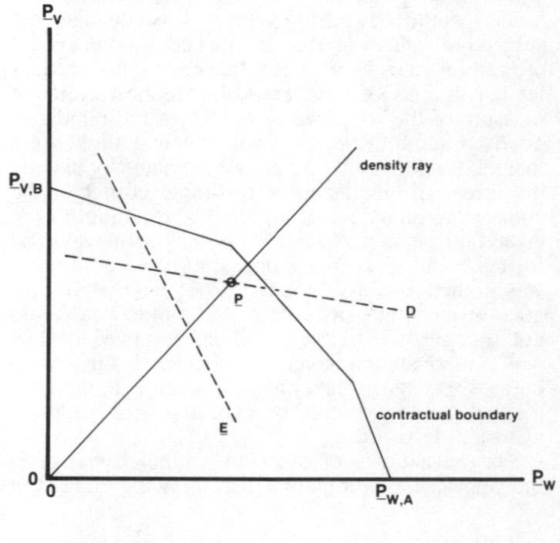

line of any other accepted bid. A *submarginal bid* is any rejected bid whose isoexpense line touches the density ray at a point below the contractual boundary. The submarginal bid whose isoexpense line intersects the density ray further from the origin than does the isoexpense line of any other submarginal bid is here defined as the *highest submarginal bid*.

A flat price for weight and a flat price for cube are set by the carrier according to the following rules:

*R*1. The prices must reflect the density of the actual load.

*R*2. If a submarginal bid exists, the prices must be consistent with some point on the isoexpense line of the highest submarginal bid.

*R*3. If no submarginal bid exists, the prices must be consistent with some point on the isoexpense line of the marginal bid.

R1 requires that the carrier's price pair lie on the density ray. R2 requires that, if a submarginal bid exists, the price pair must lie on the isoexpense line of the highest submarginal bid. Thus Figure 3 shows the price pair *p* at the intersection of the density ray and the isoexpense line of bid *D*. If no submarginal bid existed, the price pair would lie on the intersection of the density ray and the contractual boundary.

Under the submarginal expense-line method the bidder ordinarily has little motive to misstate his true valuation of the proposed transportation service. If he were certain that at least one submarginal bid would exist, and if no bidder made more than one bid, then he would have no rational motive to understate, for any successful bid he might make could not affect either price. In no case would he have any rational motive to overstate, for if he overstated he would risk acceptance of his shipment at a charge greater than he was willing to pay.

IV. DISCUSSION

The submarginal expense-line method of pricing proposed above is the dual-price analog of the single-price submarginal price method, and it would appear to have the same power to elicit true revelation of preference. Where there are many bidders, the joint probability (a) that any particular bid will be the marginal bid and (b) that there will exist no submarginal bid, would be low. And therefore, in either the single-price or the dual-price situation, the bidder would have little motive to understate the subjective valuation that he placed on the object offered for sale. In any case, the sum of any particular feasible set of bids would be the best available measure of the social value of the overall bundle of goods put up for sale. On a naive view it might seem that the carrier could obtain more revenue by placing his price pair in the most favorable corner of the contractual boundary of Figure 3. Or he might even go so far as to depart from flat pricing for weight and for cube and to charge more, simply because a bid was higher. But as Vickrey (1961) has so ably argued in the single-price case, these practices would not necessarily increase the seller's revenue, for they would motivate the buyer to understate his true willingness to pay. In the context of Figure 3, the contractual boundary could be expected to shift downward and leftward.

The advantages of the submarginal isoexpense-line approach might include the following: better use of vehicle capabilities for the individual trip; more efficient scheduling of shipments; better geographic distribution of the existing fleet of vehicles; and more appropriate design of vehicles, route facilities, and terminal facilities. Available vehicle capacities (for handling weight and cube) are utilized efficiently in the sense that the combination of shipments carried will represent a total value of service at least as large as that of any other feasible combination. Consider a sample example in which five shipments *a, b, c, d* and *e* are tendered, the respective relative weights (in arbitrary units) being 6, 7, 8, 9 and 10 and the respective relative volumes 10, 9, 8,7 and 6. The respective bids, in dollars, are 20, 19, 18, 17 and 16. The available vehicle has a weight capability of 21 units and a cubic capacity of 27 units. The feasible combination that maximizes the sum of the associated bids is *a, b* and *c*, with a total declared value for the transportation service of \$57.00. The highest submarginal bid is *d*. The load density is 0.777 weight unit per volume unit (= 21 weight units ÷ 27 volume units). The point on the price graph where the density ray intersects the isoexpense line of bid *d* is where p_w = \$1.18 per unit weight and p_v = \$0.92 per unit volume. These prices, applied respectively to the actual weight and actual volume of the vehicle load, yield a total revenue of \$49.62. Where the transportation was supplied under competitive conditions, presumably no public subsidy would be needed. In some instances, however, the situation might be naturally monopolistic; that is, the market might be served best by having just one supplier. There a subsidy that made up the difference between the total declared customer benefit and the sales revenue might be appropriate. In the foregoing example the subsidy would be \$7.38 (= \$57.00-\$49.62). The purpose of such a subsidy would be to motivate the carrier to act so as to maximize total declared customer benefit rather than total sales revenue; for example, to distribute its fleet so as to serve its customers as well as possible.[3].

The scheduling of shipments may be more efficient under dual pricing because the prices guide the shipper in the timing of shipments of various densities destined to go out in various directions. Suppose for example that in the next twenty days several high-density shipments need to move east and several low-density shipments need to move west. If prices are reasonably predictable in either a precise or a stochastic sense, then the shipper may be able to ship east when the price of weight is low and west when the price of volume is low. Separate prices for weight and cube can also guide the carrier in redistributing high-cube and low-cube vehicles geographically in response to shifting demands.

In the longer run, long-term average prices for weight and for cube could be useful guides for the optimal design of vehicles. If, for example, the prices of cube tended to exceed the marginal cost of adding cubic capacity and the prices of weight to fall short of the marginal cost of adding weight-carry capability, then a shift toward lower ratios of weight to cube would be indicated. Such a shift would tend to bring the mean price of weight and the mean price of cube closer to their respective marginal costs. And if optimal vehicle design persistently clashed with existing route and terminal capabilities, then the benefits of making adjustments in the limiting facilities should be weighed against the costs.

In the section on pricing, only the major features of a scheme are presented. The main objective is to motivate disclosure of the valuation that the customer places on the proposed transportation service. This is done by charging flat prices, that is, by avoiding price discrimination. In practice, various refinements might be found to be worth while; presumably, the optimal combination of refinements would depend on the mode of transportation and on other aspects of the situation. Suppose for example that the carrier, instead of simply maximizing the sum of the accepted bids, maximized the excess of this sum over the incremental cost of carrying weight on the trip in question. This would be worth while if this incremental cost was significant. And in some cases this cost could be negative, as where a ship's captain needed ballast for a backhaul. Where the incremental cost of carrying weight was thought to be close to zero, the foregoing refinement might not be worth bothering with. Futhermore, practical considerations might sometime dictate departures from the major features of the proposed scheme. Suppose that a linehaul highway rig is scheduled to leave shortly but has enough unfilled space to accommodate one-half of the cargo just arriving at the terminal on board a local pickup and delivery truck. The only practical way to fill out the linehaul load before departure time might be to unload from the local vehicle on a last-in, first-out basis and pay no attention to the bids. And there are doubtless many other instances in which the proposed scheme would be modified in practice.

It would be easy to overstress the importance of separate charges for weight and cube. Information on densities can be recorded to serve as a general guide for the scheduling of vehicles and shipments and the design of transportation plant. It is particularly straightforward, however, to compare weight and cube prices with estimates of the corresponding marginal costs. But the greatest significance of flat dual pricing would seem to be as a means of eliminating price discrimination and thereby eliciting true revelation of the valuations that the customers place on transportation services. None of this is to deny the importance of other refinements in pricing, such as the imposition of a flat charge per shipment to cover the carrier's transactions costs, the substitution of constructive volume for actual volume where package size or shape violates standards, penalties for low strength-to-density ratios, and so on. Because of rapid reductions in the costs of information handling, it would be surprising to find anything less than a very sizable backlog of unexploited opportunities in the areas of standards for and pricing of transportation services.

V. SUMMARY AND CONCLUSIONS

Under dual pricing, the charge for transporting an indivisible unit of cargo includes one charge for carrying the weight and another charge for accommodating the volume. An additional flat per-shipment charge covering order processing would not be inconsistent with dual pricing. The charge for weight is actual shipment weight times the price of weight p_w; the charge for cube is the assessed volume times the price of volume p_v; p_w and p_v are flat prices with respect to any particular origin-destination-vehicle-load combination. (Assessed volume may exceed measured volume where size or shape standards are violated.) Although a pair of flat prices p_w, p_v necessarily implies some price schedule p_w = f(d), where d is the density of the shipment, a density-based price schedule does not necessarily imply, for any given load density, a pair of flat prices applicable to all of the shipments (of various densities) that comprise the load. To avoid discrimination in the general case, it is necessary to apply flat prices. To elicit the true valuations that customers place on particular shipments, it is necessary to avoid discrimination, which amounts to a rewarding of understatement. And to find which of the feasible combinations of proposed shipments should actually be put on board in order to maximize social benefit, it is necessary to elicit true valuations. The basing of prices on the highest submarginal bid is also necessary to true revelation. A carrier following this latter practice may seem to forgo revenue that, according to the actual bids, appears to be available. But if he abandons the practice, then he can reasonably expect the bids to shrink as bidders shift from honest revelation to defensive understatement.

REFERENCES

McPherson, Logan G. 1909. Railroad freight rates. New York: Henry Holt and Company.

Sonstegaard, Miles H. 1986. "Economic Evaluation for U.S. Railroads." *Transportation Research-A*. 20A (1):25-31.

Vickrey, William. 1961. "Counterspeculation, auctions, and competitive sealed tenders. *Journal of Finance*. 16:8-37.

ENDNOTES

*Department of Economics, University of Arkansas, Fayetteville, AR.

1. McPherson (1909, p. 153) mentions "the original simple distinction between articles light in weight and heavy in weight." He presents a quotation, from the Eleventh Annual Report of the Interstate Commerce Commission, that lists approximately a score of more or less distinct factors used in making railroad rates; one of these factors is "the space occupied and weight."
2. Vickrey (1961), who proposed this method, refers to it as "'first-rejected-bid' pricing."
3. This would be similar to what Sonstegaard (1986, pp. 28-30) proposes, in a single-price context, for regulation of tolls charged by companies operating railroad routes but neither owning nor operating any rail vehicles. There the subsidy motivates the tollway operator to reveal his true reservation price.

Petroleum Pipeline Deregulation: A Critical Analysis of the Primary Analytic Justifications for Deregulation

*by Leonard L. Coburn**

ABSTRACT

Recently, two significant reports were issued that call for petroleum pipeline deregulation. The Department of Justice issued its report in May 1986 recommending deregulation of all crude oil petroleum pipelines in the lower 48 states. It also recommended deregulation for most product pipelines, except for eleven pipeline systems. In November 1986, the Association of Oil Pipe Lines issued a report prepared by Dr. David J. Teece. Dr. Teece's study recommended that the petroleum pipeline industry is sufficiently competitive that the rates of all pipelines can be deregulated.

The paper analyzes the market definitions used by the Department of Justice and Dr. Teece. Justice relied upon a hypothetical monopolist model to develop its markets. While this model appears to lead to a good definition of product markets, transportation services divided into four segments at either end of the crude and product pipelines, the geographic markets derived from this model, 181 areas based upon Commerce's Bureau of Economic Analysis areas, may not comport with reality. Dr. Teece appears to use the hypothetical monopolist model to define his product market, the delivered commodity, which appears to be a contrived market in order to develop the geographic markets based upon shipment data. The paper develops four models for market definition, places the reports within the context of these models, and critically analyzes the application of the models to the petroleum pipeline industry. The paper concludes that the author is more comfortable with the inherent conservatism of the Justice approach rather than the sweeping findings made by Dr. Teece.

I. INTRODUCTION

Petroleum pipeline deregulation is now being considered by the Reagan Administration as one of many ways to decrease federal regulation and improve the nation's competitiveness. In addition to being included in the overall legislative packages submitted in response to the competitiveness initiative, petroleum pipeline deregulation is part of a more limited bill dealing just with energy initiatives.

This is the first time any Administration formally proposed the deregulation of petroleum pipelines, although proposals to deregulate the industry have been submitted in recent years in response to industry initiatives. With petroleum pipeline deregulation advancing on the legislative agenda, analytic justifications of the concept play increasingly important roles. Recently, two major studies were published that provide the primary analytic justification for petroleum pipeline deregulation. In May 1986, the Department of Justice issued its study, *Oil Pipeline Deregulation*, supporting the overall idea of deregulation, but holding back on complete deregulation of all pipelines. [18] Later in 1986, the Association of Oil Pipe Lines (AOPL) issued a report prepared by Dr. David J. Teece, *Oil Pipeline Rate Deregulation*, supporting complete deregulation of rates but not deregulation of other service aspects of pipeline regulation. [16]

These two studies provide the principal analytic justifications for the two different legislative approaches to petroleum pipeline deregulation. The Reagan Administration's approach to deregulation is to deregulate entirely all pipelines that operate in competitive markets but maintain complete regulation over pipelines that exhibit undue market power. In the Department of Justice's study, all crude oil pipelines (except the Trans Alaska Pipeline System) would be deregulated immediately and all but eleven product pipelines would be deregulated immediately. The eleven product pipelines would be subjected to additional analysis in an administrative forum to determine whether they should be deregulated. The AOPL-industry approach to deregulation would be to deregulate the rates of all petroleum pipelines immediately and maintain regulation over the service aspects of petroleum pipelines.

Both approaches depend upon there being sufficient competition in the petroleum pipeline industry to justify deregulation. Both studies analyze competition in the industry and both studies reach the conclusion that there is sufficient competition to support the two different legislative approaches to deregulation.

The purpose of this paper is to discuss each of these studies and to discuss the market definition methodologies used by each to arrive at their conclusions pointing out the strengths and weaknesses of each methodology.

II. THE JUSTICE AND AOPL STUDIES

A. *The Department of Justice Study*

The final version of the Department of Justice's study on oil pipeline deregulation was issued in May 1986. A preliminary version, *Competition in the Oil Pipeline Industry: A Preliminary Report*, was issued in May 1984. [17] Taken together, the two reports represent the Department of Justice's analytic justification for petroleum pipeline deregulation.

Justice relies on its Merger Guidelines for its market definition. A separate intellectual analysis of market methodology is missing from the report. The report merely indicates what the product market is and what the geographic markets are. The product market is "transportation services" broken down into four types of product markets—crude origin, crude destination, product origin, and product destination. The geographic markets are broken down into the 181 Department of Commerce Bureau of Economic Analysis areas or BEAs for short.

Based upon these markets and using a Herfindahl-Hirschman Index (HHI) cut-off level of 2500 as well as a host of other assumptions about market power, the Department of Justice applies the data it published in its preliminary report (supplemented with additional data) to determine which pipeline systems are competitive and which possess substantial market power to justify continued regulation. After examining all pipelines in the lower 48 states, the Department concludes that no crude oil pipeline warrants continued regulation. Turning to product pipelines, however, the Department finds that five pipelines exhibit substantial market power that they should not be deregulated until they undergo additional administrative procedures, while six others raise enough questions that more analysis is necessary (Justice calls them close calls).

B. The AOPL-Teece Study

The study issued by the AOPL prepared by Dr. David Teece takes a different approach. It uses totally different product and geographic market definitions. For its product market, Teece rejects Justice's reliance on transportation services and uses instead a product market composed of the delivered commodity itself. For the geographic market, Teece relies on shipment data, and especially the approach developed by Elzinga and Hogarty, [4,5,11] to determine the extent of his geographic markets. Based upon this market analysis, Teece concludes that all pipelines operate in sufficiently large markets that no pipeline can exert substantial market power.

While Dr. Teece is comfortable with his conclusion, he also addresses the eleven pipelines posed by Justice as sufficiently troublesome to warrant more scrutiny. Teece's pipeline-by-pipeline analysis draws him to the conclusion that even using Justice's approach, no pipeline in the lower 48 states has substantial market power to warrant continued regulation.

The two principal reports arrive at similar conclusions using very different methodologies. Justice's product and geographic markets are based upon the intellectual analysis of its Merger Guidelines, which, in turn, are based upon a hypothetical monopolist being able "to impose a small but significant and nontransitory increase in price above prevailing or likely future levels" in order to define product and geographic markets. Dr. Teece, on the other hand, relies on a different product market although he appears to use the hypothetical monopolist theory to arrive at his definition. Dr. Teece's geographic markets are determined by shipment analysis, quite different from Justice's hypothetical monopolist approach.

The following section discusses market analysis in more depth, developing various models used for market determination and discussing both the Justice and Teece reports in light of the models.

III. MARKET ANALYSIS

The definition of the market is perhaps the most important part of competition analysis. The existence of a competitive industry or competitive structure must be found within definable limits. To say that some industry is competitive is virtually meaningless unless the product and geographic scope of the industry are provided. The following models provide a framework within which to assess the Justice and Teece reports.

Four market definition models describe the various approaches used to define a market. [1] A market can be viewed as a set of products and a geographic area

1. within which cross-elasticity of demand is high or
2. that has been insulated historically so that little flows in and little flows out or
3. within which prices tend toward equality or
4. within which a hypothetical monopolist would be able to raise prices profitably.

No matter which market model is used, there must be an independent assessment of how much market power will be tolerated, over what time period the market power will be tolerated, or during which time period market forces will be allowed to operate to dissipate the market power, and the determination of a baseline price from which market power will be measured. [1]

A. The Cross-Elasticity Model

The cross-elasticity model posits that the market for products within a geographic area is determined by high cross-elasticity of demand. Supply cross-elasticity often is treated separately as supply substitution in all models. Some, however, would include the supply here, rather than a separate issue applicable to all models. Cross-elasticity of demand means the percentage change in the quantity demanded of one product given some percentage change in the price of another product. If the cross-elasticity is high, that is, if a change in the price of one product causes a large change in the quantity demanded of the other product, then both products would be treated as being within a single market. The geographic development follows along similar lines, that is, if the change in price in a product in one territory causes a large change in the quantity demanded of the other product in another territory, then both territories would be considered as being in the same market.

The cross-elasticity model has been used extensively in antitrust analysis of product markets, but finds lesser use in defining geographic markets. The idea of substitutability has broad appeal in defining which products compete with each other and courts have used this test, although often lacking in specificity, to decide whether products actually do compete with each other. Rarely, however, have courts used the model to define the geographic areas within which products compete, usually for the practical reason that it is very difficult to determine cross-elasticity measurements with a substantial degree of reliability. Despite its shortcomings, the cross-elasticity model has many advocates in both the legal and economic professions.

B. The Historical Insularity Model

The historical insularity model defines a market as the products and geographic area that were historically insulated over a specified period of time so that little has flowed out and little has flowed in. The model developed primarily in the geographic market context and examined shipment flows in the context of two tests—Little Out From Inside (LOFI) and Little In From Outside (LIFO). The LOFI test looks at the supply side of the equation and asks what is the smallest region that accounts for a large proportion of the shipments from a producing area. The LIFO test examines the demand side of the equation and asks similarly what is the smallest region that accounts for a large proportion of the shipments that originate from within the region. The LOFI test makes sure that exports to other regions are taken into account while the LIFO test makes sure that imports are accounted for from other regions. The market keeps expanding until the two tests are satisfied simultaneously. The historical insularity test can be used for product market definition as well. The LOFI analogue would ask what is the smallest group of products required to account for nearly all of the shipments of a given product. The LIFO analogue would ask does a product consist of most of the purchases of the product identified by the LOFI test.

The historical insularity test finds its greatest application in determining geographic markets. Many antitrust commentators find the LIFO-LOFI test easier to work with than most other tests; however, many indicate that its prime use should be for evidentiary purposes and not as a way to decide whether the market exists in the first place. In other words, once the product and geographic markets have been determined, historical shipments are useful to determine whether there is evidence to support the theory.

C. The Price Equality Model

The third model looks at the market in terms of the products and geographic area within which prices tend toward equality. The important inquiry is the closeness of price relationships. Close relationships signify the existence of a single market; relationships that are not close signify the existence of separate markets.

The price relationship model probably finds its greatest support among economists and it certainly was the basis of Marshall's market analysis as early as 1920. While the model may find considerable support in the economic literature, the antitrust application of it has been less enthusiastic. Most commentators find that the drawbacks to the model outweigh its usefulness.

D. The Hypothetical Monopolist Model

The final model for market determination is the hypothetical monopolist model. In this model, products and geographic area are determined by looking at whether a hypothetical monopolist could raise prices profitably. In the product market, the examination looks at whether the monopolist could raise its price by a given percentage for a given time period. If many buyers would shift to alternatives so that the monopolist's price increase would not be profitable, then the products included in the market are too narrow. The market would be expanded until the monopolist's price increase is profitable. The geographic market would follow a similar development, expanding the size of the market until the monopolist's price increase is profitable.

This model finds its greatest support from the Department of Justice, since it was adopted in Justice's 1982 and 1984 Merger Guidelines. It is an extension of the focus on market power as the primary criterion for merger and antitrust analysis. The inquiry is on whether market power can be exercised within the defined market in light of the intramarket competitive conditions. The inquiry focuses on the criteria used to evaluate whether competitive conditions are adequate to thwart the exercise of market power in the relevant market if a prior determination was made that a hypothetical monopolist could exercise market power unless competitive conditions existed to stop it.

The next sections critically examine the Justice and Teece studies in the context of these models.

E. The Department of Justice Report

The Department of Justice relies upon the hypothetical monopolist model for the determination of product and geographic markets. This model finds its strongest support from the Department of Justice and is the basis for its Merger Guidelines.

The Department of Justice defines the product market as transportation services in four types of product markets, crude origin, crude destination, product origin, and product destination. Justice examines the markets at either end of the pipeline to define the product market where a hypothetical monopolist could increase prices at a profit (or decrease prices in the case of a monopsonist). In the crude origin market, the crude may be consumed in the area or transported out of it. Thus, a pipeline transporting crude from an area competes with other facilities transporting the crude out of the area or with local refineries consuming the crude in the area. In the crude destination market, pipelines transport crude into an area, along with other transportation facilities. In addition, the transportation modes compete with local crude production. In product origin markets, pipelines must compete with other transportation modes carrying the product from refineries. Also, the product can be consumed in the product origin market so that local consumption must be considered a competitor. Finally, in product destination markets, pipelines compete with other transportation facilities carrying the product into the market. Plus, local production, that is, refining, must be considered.

Justice then indicates how the exercise of market power can result in competitive problems. Justice indicates that "the exercise of market power can result in (1) the reduction of crude production in crude origin markets, (2) the reduction of refinery utilization in crude destination and product origin markets, and (3) the reduction of petroleum consumption in product destination markets." Thus, it is to be assumed from what Justice states that a hypothetical monopolist would be able to exert its market power by a small but significant and nontransitory price increase and remain profitable within the confines of the four types of product markets for transportation services.

The greatest problem with the Justice report is its lack of explanation of how it arrived at these product markets and the general assumption that the reader can apply the theoretical aspects of the Merger Guidelines without more detailed explanation. All it says about the four-market approach is that it is similar to approaches taken in other reports. It is here, however, that the clue to Justice's derivation of the four-market approach can be found.

John Hansen in his doctoral dissertation at Yale and his subsequent book, *U. S. Oil Pipeline Markets*, developed the four-market approach to product definition. [8] Hansen offers a better description of how he arrived at his definition. After surveying the literature and finding that previous analysts used markets as large as the entire U. S. or as small as Petroleum Administration for Defense Districts (five multistate regions developed during World War II for petroleum defense planning), Hansen determined that these markets were much too large, since pipelines in different parts of the areas did not compete with each other. Moreover, he argued that pipelines did not only compete with other pipelines over similar routes. He maintained that pipelines competed for the right to carry product out of an area, no matter where its destination, as well as the right to carry product into an area, no matter where its origin. Thus, he developed the idea that pipelines competed to provide a transportation service either from or to a specific area and he broke these types of services down into the four product markets used by Justice.

Justice can be held somewhat remiss for not developing the argument in the detail provided by Hansen. Certainly, the product market development would have been more explicable if this type of analysis were included rather than forcing the interested reader to develop this information on his or her own.

Another antitrust agency recently considered the issue of petroleum pipeline product markets. The Federal Trade Commission in its report, *Mergers in the Petroleum Industry*, agreed with the Department of Justice's approach to product market definition. [6] The FTC concluded that "the relevant product market for the transportation of crude oil or petroleum products for purposes of merger analysis should embrace economically substitutable petroleum transportation services." The FTC also broke down the relevant product market into four types of markets indicating that "the aim is to assess the degree of competition faced by petroleum pipelines from other modes of transportation at each stage of transportation. . . ."

For the geographic component of market definition, Justice again uses its hypothetical monopolist theory without detailed explanation of its application. Justice derives 181 geographic market areas. These 181 geographic areas are the Department of Commerce's Bureau of Economic Analysis market areas, or BEAs. Justice states that these are reasonable approximations of market activity because they are the maximum area that can be served by the highly competitive trucking sector. The radius of the average BEA is 72 miles, while Justice says that most trips by tank trucks take place within a 50-mile radius of their origin. Moreover, the BEAs are centered around major cities and Justice argues that both consumption of oil products and refining capacity are concentrated in metropolitan areas. Justice thinks that the BEAs are reasonable approximations for crude destination, product origin, and product destination markets. Justice finds some difficulty applying the BEA analysis to crude origin markets and indicates that it will make adjustments as necessary with respect to crude origin markets.

While Justice is to be applauded for assembling its data in such discrete geographic markets, Justice's explanation for its geographic market approach appears rather flimsy. It cites only one source for its proposition that trucks usually ship within 50 miles of origin. No actual experience appears to be taken into account, even after many commentors on Justice's report indicated that truck transportation can compete over much longer distances. Justice may violate the spirit of its own Merger Guidelines by totally discounting the actual marketplace experience and continuing with its assertion that competitive trucking is limited in the distance it competes. Furthermore, the assertion that BEAs may be appropriate merely because they contain a major city means very little if consumption patterns are quite different from the area denoted by the BEA. It does not appear that Justice tested the BEAs to see if real market experience approximated these areas. As a result, Justice's approach to the geographic market area may be quite arbitrary. BEAs may be used for convenience rather than replicating reality. While administrative feasibility is an important part of market definition, it is equally important that the geographic area developed has some relationship with the real world. It is not clear that this is the case in Justice's approach.

The major drawback that many commentators find with the hypothetical monopolist model is its administrative feasibility. Some have called it "completely nonoperational" or "simply impracticable" or "devoid of any commercial reality." [1] Its reliance on a hypothetical price increase presents the practical problem of identifying the price at which transactions occur. Others indicate that it may lead to markets either being overinclusive or underinclusive. Finally, the theory rests upon "neoclassical price theory, which views buyers and sellers as rational, optimizing actors, with more or less perfect information about prices and product attributes." [1] But most economic transactions do not occur under such perfect conditions and therefore the theory may not be an accurate predictor of economic reality.

These general criticisms of the model show up in its application to the petroleum pipeline industry and most clearly with respect to geographic markets. It is not at all clear that the geographic markets developed by Justice have any relationship with economic reality. It may be that Justice is correct in its geographic market development, but the Justice study does not present a convincing case. A great deal more work needs to be done to justify the application of the hypothetical monopolist model to the derivation of the geographic markets in the petroleum pipeline industry.

The FTC also appears to adopt the hypothetical monopolist model in its analysis of petroleum pipeline geographic markets even though it does not explicitly say so. The FTC focuses on the question of whether other pipelines would respond to an attempt by a competing pipeline shipping out of the area, or a pipeline shipping into an area, to raise its price for transportation. The FTC agrees with Justice that market power must be examined at either end of the pipeline. The FTC, however, does not attempt to quantify its pipeline geographic areas preferring to adopt a case-by-case approach to its merger analysis.

The FTC adopts a shipments approach to its geographic market definition of crude oil and refined products markets. It may be inferred that it would use a shipments analysis in its pipeline geographic market approach as well; however, the FTC explicitly states that many other factors such as ownership structure, regulation, and entry conditions would enter into its evaluation.

F. The AOPL-Teece Study

The AOPL-Teece study approaches market definition from an entirely different perspective. Teece is much less explicit in how he arrives at his definition of a product market; however, it appears he is applying some variant of the hypothetical monopolist model. Teece rejects Justice's conclusion that transportation services is the appropriate product market; rather, he states that the product market is the market for the delivered commodity. In arriving at this definition, Teece says, "The point is that if a pipeline cannot exert control over the price of the commodity which it is shipping, it cannot exert monopoly power over the transportation services it is providing." [16] This approaches the hypothetical monopolist model most closely, since Teece is concerned about the exertion of market power in the market.

Teece basically merges the transportation of the commodity into the delivery of the commodity and says that the two are inseparable for product market definition. It is not so obvious why this merger of transportation and delivery is necessary or is valid. For example, in every other transportation industry, the focus is on transportation services and not the delivery of commodities. It is possible to separate out transportation from delivery in other industries and, in fact, it is most appropriate to do so. Why is it not possible to separate out transportation services from the delivered commodity in the petroleum pipeline setting?

If the analysis focused on the water transportation of crude and products, it would not be necessary to merge the transportation and the delivery of the commodities to determine whether water transportation was competitive or not. An independent analysis could be conducted to examine the structure of the waterborne petroleum transportation industry without having to introduce the delivery of the commodities to come to conclusions about the nature of the competitive process in water transportation. The same inquiry could be made into the truck transportation of petroleum. Therefore, it is not clear why pipelines have to be singled out as being different.

One basic difference between petroleum pipelines and other modes is that petroleum pipelines carry only petroleum. This difference should not lead to Teece's result. A single product carrier can exert market power over the transportation of the product. While the effects may be felt in other segments of the industry, that is not the same thing as stating that it is essential to move downstream or upstream to other segments in order to measure competitive effects. There is no doubt that the price effects will show up in upstream or downstream markets, but transportation costs do have an independent character. Transportation costs are a component of the upstream or downstream price and can be affected independently of other components of the commodity price depending upon the competition faced in the marketplace. For example, Colonial Pipeline, the largest products pipeline in the nation operating from the Gulf Coast to New York, designs its tariff depending upon whether it competes with water transportation, with other pipelines, or with no one else. Colonial's tariffs to inland destinations are higher on a barrel-mile basis where there is limited or no competition than its tariffs to cities served by water. Colonial may be exerting some limited market power in some markets even though it has no control of the commodities in the markets where it accepts products or in the markets to which it delivers.

Another distinction in the petroleum pipeline industry is the high degree of vertical integration. Most petroleum pipelines are owned by major petroleum companies that have extensive crude production, large refining capacity, and widely dispersed product marketing. Vertical integration may affect the ability to exert market power in any particular segment (Justice discusses this in some detail), but it does not imply that market power cannot exist only in transportation segments. There does not appear to be any valid reason why transportation services cannot exist as a separate product market. If there is a reason, Teece is not convincing.

Perhaps Teece is muddling the product market definition so that he can apply the historical insularity test to the determination of the geographic markets. The LIFO-LOFI test works quite well in determining geographic markets for the delivered commodity, but does not work particularly well in determining geographic markets for transportation services. It appears that Teece's overall goal is to determine markets so that he can make use of his shipments data to come up with much broader geographic markets to suit the purpose of finding extensive competition in both crude and product markets. Based on his particular market definition, he reaches the conclusion that there is sufficient competition in the pipeline industry to warrant rate deregulation.

For example, in the crude analysis, Teece finds that there is only one principal crude market, while some states that neither produce nor consume crude are not in any market. There also are separate markets that rely on imported foreign crude, namely a market in Pennsylvania/Delaware/West Virginia, and markets in New York, New Jersey, Maryland, and Virginia. Teece concludes that while conceptually these are separate markets, they all are dependent on the same foreign market and therefore they really can be considered as one foreign crude market. In the product area, Teece's analysis leads him to conclude there is really only one market for refined products. Teece utilized a 90 percent LOFI-LIFO criterion to arrive at these conclusions. That is, 90 percent of the products produced in the area had to be consumed in the area, while no more than 10 percent of the products consumed in the area could come from outside the area. Teece then used a computer model to determine at the state level how much flowed in and out of the state and applied the 90 percent criterion to these flows. Teece found that for both crude oil and refined products no state qualifies as a separate market because no state produces at least 90 percent of its consumption and consumes at least 90 percent of its production. Teece then applied the model just to crude, with the results indicated above, and just to products with the results also indicated above. Since the separate analysis for crude and products led to similar results for the combined crude-product market, Teece concludes that this is strong evidence that the two should be combined into one market.

A potential problem with Teece's shipment model is that he starts with states. It may be possible that these geographic areas are too large to begin with so that only by constantly expanding the LIFO-LOFI tests can both be met simultaneously. The LIFO-LOFI tests may be met in SMSAs, or BEAs, or something smaller than a state. Since no data are available to test this hypothesis, it is not possible to test its validity.

There is no doubt that the shipments analysis, or historical insularity model, has support in the literature and in the courts as an acceptable approach to market determination. But many critics indicate that it should be used only for evidentiary purposes and not for theoretical analysis. Critics charge that this model says little about the market's susceptibility to the exercise of market power, since it does not measure responses to price changes within the market. The model is a static picture of markets and is not particularly good in characterizing the effects of dynamic actions.

The historical insularity model is particularly susceptible to the problem posed when one market already is subjected to pricing by a monopolist and the second market has competitive prices. Shipment data may indicate that this is one broad market, since there are cross shipments due to the excessively high prices charged by the monopolist in the first market. Without these high prices, cross shipments would be reduced measurably and two markets would exist. The model will not indicate existence of this problem (the problem being that the first market already is being subjected to market power) but will indicate that the broader market already is competitive, when in reality market power is being exerted and is undetected.

The FTC in its *Mergers in the Petroleum Industry* Report attempted to construct geographic markets for crude oil and refined products using shipment data. [6] For crude oil, the FTC concluded that crude oil trades in a world market, but for merger analysis there may be instances when smaller markets need to be analyzed. The FTC indicated that domestic crude markets may be appropriate. In this regard the FTC stated that a market made up of Alaska North Slope (ANS) type oil may be a relevant product submarket and that the Petroleum Administration for Defense District (PADD) V (comprised of Alaska, Arizona, California, Hawaii, Nevada, Oregon, and Washington) may be a relevant geographic submarket.

In its analysis of refined petroleum markets, the FTC focused on shipment data to determine relevant geographic markets with emphasis on gasoline and middle distillates. The FTC examined inter-PADD movements (there are five PADDs in the United States) and concluded that for merger analysis PADD III (Gulf Coast states) was a relevant market despite substantial flows of product out of the area. Since there were virtually no flows into the region, the FTC concluded that there were no supply alternatives within PADD III from outside and competitive effects should be limited to PADD III. Similar analysis for other PADDs led the FTC to conclude that PADD IV and PADD V may be distinct geographic markets for gasoline and middle distillates.

While the FTC placed many caveats on its analysis, it indicated that refined product markets most likely were smaller than the entire United States and most likely approximated some of the regions defined as PADDs. The FTC analysis is quite distinct from Dr. Teece's analysis and indicates that there is more than one way to examine shipment data with differing results.

The Teece and Justice market definitions lead to different ways to determine whether market power exists in the industry. Teece merely lists all the pipelines in the country with their market shares and finds that for the crude market the Herfindahl-Hirschman Index (HHI) is 486 and for the product market the HHI is 669. Both of these are quite low and indicate a competitive market structure.

Justice examines the HHI in all of its markets using a 2500 HHI level to test for market power. This takes much longer since there are so many product and geographic markets to consider, as defined by the Justice Department. Ultimately Justice concludes that there is no crude oil pipeline in the lower 48 states that can exert market power. But on the product side, Justice finds that five pipelines definitely are troublesome and could exert market power in the various markets they operate in, while six product pipelines are close calls and require additional analysis based on better evidence.

IV. NEED FOR FURTHER RESEARCH

The conclusions reached in this paper point to a need for additional research in the area of petroleum pipeline competition. In the first instance, it may be worthwhile for a new analyst to take a fresh look at the question of product market definition. Most analysts writing in the area start with the definition proposed by Justice—transportation services. Dr. Teece's analysis challenges this approach. Another look at the question now appears warranted.

The question of geographic markets definitely needs more work. Justice's approach requires the application of its theory to actual experience. While its theory may be correct, it would be on sounder footing if additional research could demonstrate the link between theory and reality. Dr. Teece's approach also could benefit from more research. For example, the FTC developed its geographic markets using shipment data and found different markets. A more rigorous examination of the shipments methodology with particular emphasis on the approaches to data analysis would aid in the understanding of the application of the model to actual experience.

V. CONCLUSION

There are other inherent problems in each of the studies; however, for the purposes of this analysis, only market definition is examined in depth. Each study has its strength and weaknesses. Justice has the better product market definition, while its geographic market definition is quite problematical.

Teece, on the other hand, has the more problematical product market definition, while the geographic market definition has more support in the literature despite its drawbacks.

In the view of this writer, Teece overstates the level of competition in the industry because of his market definition. His focus on the delivery of the commodity rather than the transportation of the commodity virtually assures that his analysis will lead to the largest market definition possible. Justice, on the other hand, has the better argument concerning product definition even though Justice did not provide much analysis in its studies. Justice's geographic markets, the 181 BEAs rest on rather flimsy and scanty evidence. Teece's geographic markets, dependent solely on shipments data, while having more concrete support, may not be appropriate if transportation services were the relevant market and not commodity deliveries.

We are left with two excellent attempts to provide analytic support for petroleum pipeline deregulation. Both are flawed efforts, but the best to date. While they are adequate for generalizations about the degree of competition in pipeline markets, neither would hold up if it had to be utilized in specific antitrust litigation. Additional evidentiary support would be required before one could feel confident of the result.

In the final analysis, this writer is more comfortable with the inherent conservatism of the Justice approach as opposed to the more broad-brush treatment of Teece. It always is easier to decide to deregulate a pipeline once additional evidence is proposed than to try to reregulate or apply antitrust remedies after the fact. Given the uncertainties in the petroleum pipeline marketplace, it is better to take a step-by-step approach rather than deregulate everything all at once only to find out later that the analysis did not comport with reality.

REFERENCES

1. ABA Antitrust Section, Monograph No. 12, *Horizontal Mergers: Law and Policy*, American Bar Association, Chicago, IL, 1986.
2. Dunfee, Thomas W., Louis W. Stern, and Federick D. Sturdivant, "Bounding Markets in Merger Cases: Identifying Relevant Competitors," *Northwestern University Law Review*. Vol. 78, 1983.
3. Elzinga, Kenneth G. and Thomas F. Hogarty, "The Problem of Geographic Market Delineation in Antimerger Suits," *The Antitrust Bulletin*. Vol. 18, 1973.
4. Elzinga, Kenneth G. and Thomas F. Hogarty, "The Problem of Geographic Market Delineation Revisited: The Case of Coal," *The Antitrust Bulletin*. Vol. 23, 1978.
5. Elzinga, Kenneth G., "Defining Geographic Market Boundaries," *The Antitrust Bulletin*. Vol. 26, No. 4, 1981.
6. Federal Trade Commission, *Mergers in the Petroleum Industry*, Federal Trade Commission, Washington, D. C., 1982.
7. Griese, N. L., *Current Trends in Defining Oil Pipeline Markets for the Purpose of Measuring Competitiveness*, Vol. XXVI Proceedings of Transportation Research Forum, Washington, D. C., 1985.
8. Hansen, John A., *U. S. Oil Pipeline Markets*, The MIT Press, Cambridge, MA, 1983.
9. Harris, Robert G. and Thomas M. Jorde, "Market Definition in the Merger Guidelines: Implications for Antitrust Enforcement," *California Law Review*. Vol. 71, 1983.
10. Harris, Robert G. and Thomas M. Jorde, "Antitrust Market Definition: An Integrated Approach," *California Law Review*. Vol. 72, 1984.
11. "Relevant Markets in Antitrust," *The Journal of Reprints for Antitrust Law and Economics*. Vol. 14, No. 2, 1984.
12. Scherer, Frederick, *Industrial Market Structure and Economic Performance*, 2d ed. Rand McNally, Chicago, IL, 1980.
13. Stigler, George J. and Robert A. Sherwin, "The Extent of the Market," *Journal of Law and Economics*. Vol. 28, 1985.
14. Teece, David J., *The Economic Regulation of Oil Pipelines: The Implications of Competition for Regulatory Policy*, Center for Research in Management, University of California, Berkeley, CA, 1985.
15. Teece, David J., *U. S. Petroleum Pipelines and Competition: The Rationale for Deregulation*, Center for Research in Management, University of California, Berkeley, CA, 1985.
16. Teece, David J., *Oil Pipeline Rate Deregulation*, Association of Oil Pipe Lines, Washington, D. C., 1986.
17. U. S. Department of Justice, *Competition in the Oil Pipeline Industry: A Preliminary Report*, Department of Justice, Washington, D. C., 1984.
18. U. S. Department of Justice, *Oil Pipeline Deregulation*, Department of Justice, Washington, D. C., 1986.
19. Werden, Gregory J., "The Use and Misuse of Shipments Data in Defining Geographic Markets," *The Antitrust Bulletin*. Vol. 26, No. 4, 1981.
20. Werden, Gregory J., "Market Delineation and the Justice Department's Merger Guidelines," *Duke Law Journal*. Vol. 1983, 1983.
21. Werden, Gregory J., "A Closer Analysis of Antitrust Markets," *Washington University Law Quarterly*. Vol. 62, 1985.

ENDNOTE

* Division of Competition, U.S. Department of Energy.

JOURNAL OF THE TRANSPORTATION RESEARCH FORUM

Student Award Papers

1987

Improving Truck Transportation of Agricultural Commodities: A Case Study of Fresh Produce Shipped from Florida Farmers' Markets

*by N. Devadoss**

1st Prize—Best Graduate Student Paper

ABSTRACT

In the truck transportation of fresh fruits and vegetables most hauls involve in multiple load pickups while collecting a full truckload order. Multiple load pickups invariably increases unproductive mileage, operating costs, and the overall turnaround time of trucks. This paper investigates the inefficiencies of multiple-stop pickups and savings of shuttle services between pickup points in an effort to marshal complete loads in one location. The study uses Florida fresh fruits and vegetables from Florida Farmers' markets as a case study.

A computer simulation model using "SLAM" was developed to demonstrate the savings that can be realized through the institution of a shuttle service. Implementation aspects of the shuttle service and a set of recommendations needed for the effective use of shuttle service are also given. This paper also recommends the use of shuttle service at the growing regions to improve the overall efficiency of the operation of truck service.

I. INTRODUCTION

Agriculture is considered one of the most important sectors in U.S. economy. Within the area of agricultural produce, sales of fresh fruits and vegetables contribute well over half of the agricultural revenue. Hence, fresh fruits and vegetables are an important item of agricultural produce (Manalytics, 1977).

In being moved from grower to consumer, perishable fresh produce may take a variety of channels through various middle men, transportation modes, and transfer points. In recent decades, the fresh produce industry has moved to almost total reliance on trucking for its domestic shipment, with the exception of some piggyback operations from the west coast (Manalytics, 1977; USDA, 1982). Moreover, due to the seasonality of the produce, the demand placed on the trucks also vary accordingly.

Transportation cost of fresh fruits and vegetables (also commonly referred as produce types) is an important component of the overall retail cost (approximately 15-20%). Past studies have shown that the transportation service is mainly provided by the owner-operator trucks, who are normally required to make multiple pickup stops depending on the demand and supply of trucks, and the availability of produce types (Beilock and Fletcher, 1983; Pavlovic et al., 1980; Maze, 1982). The extra services provided in the form of multiple pickup stops are generally at little or no cost to the buyers of transportation (Pavlovic et al., 1980). The extra cost is normally borne by the truckers thus leading to the tranportation inefficiency and consequently leading to resource misallocation.

In a case study of the transportation of Florida fresh fruits and vegetables, it was found that a shuttle service between markets would yield a substantial amount of savings in the transportation cost (Devadoss, 1985). A computer model using "Simulation Language for Alternate Modeling (SLAM)," has proved that by avoiding the multiple pickup stops through a shuttle service can lead to substantial cost savings in waiting, operating and social costs. In this paper, the ineffficency of the present operation, strategies for implementing shuttle service, and the cost savings to the individual truckers and to the society are presented (Devadoss, 1985).

II. BACKGROUND OF PRODUCE TRANSPORTATION IN FLORIDA

The shipment of produce from Florida varies depending on the season, month, and even day of production period (Figure 1). In order to help Florida farmers to get a fair price for their produce, state farmer's markets were established as early as 1939 in Sanford, Florida. All of the present fifteen markets serve as points of sale and transfer of produce from grower to buyer (Figure 2). Approximately one-third of all vegetables are handled at the state farmers' markets (FDACS, 1982a). The remaining produce types are directly shipped from the packing houses directly to the buyers (FDACS, 1982b).

Normally, the grower-brokers bargain with buyer-brokers and then buyer-brokers will purchase the individual portions of the produce. Usually buyer-brokers would want their produce loaded onto trucks after negotiating with truck-brokers for truck rates.

Based on the type of demand (quantity and produce type) from buyers, a truck load shipment may include only one type of produce, such as a truck load of tomatoes (commonly known as a "straight load"), or more than one type of produce, such as one-third is sweet corn, one-third is celery, and one-third is lettuce (commonly known as a "mixed load"). Most buyers prefer a mixed load rather than a straight load because at the retail level, consumer normally want a variety of fresh fruits and vegetables.

Because all the produce types are not available at the same markets, trucks are often directed to make multiple pickup stops to gather the required load. Sometimes, even though the produce may be available at a particular market, due to advantageous

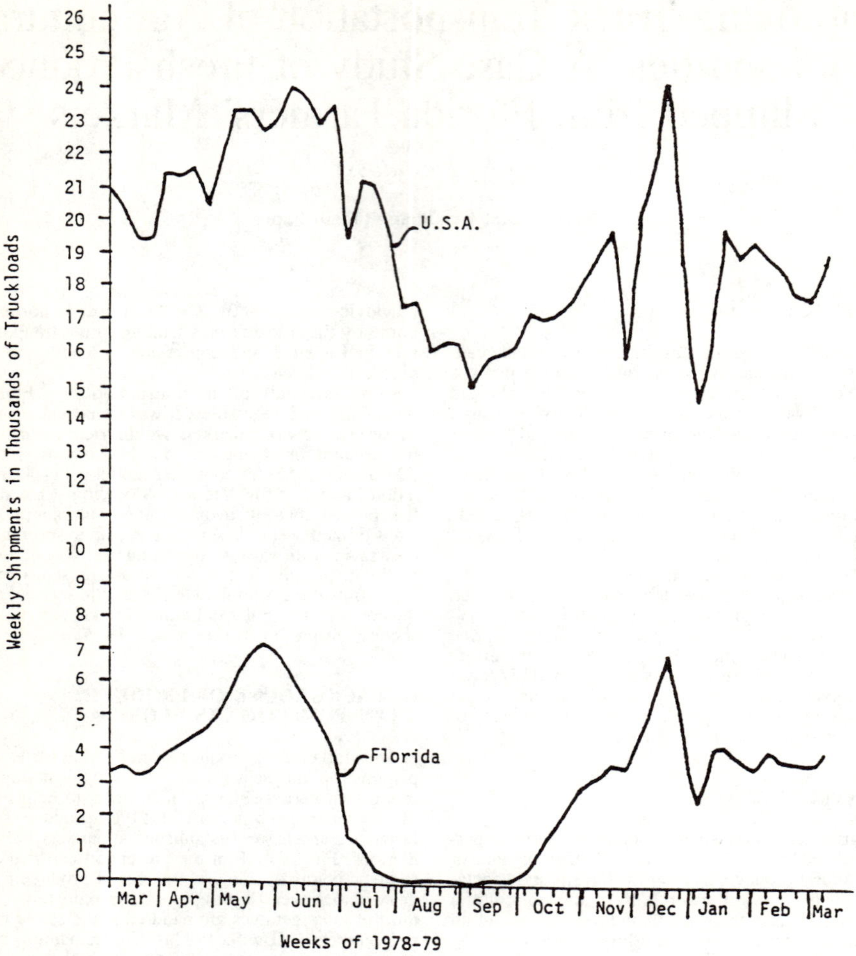

Source: Compiled from U.S.D.A. statistics

Figure 1. Weekly Produce Shipments by Truck

selling price of produce at other markets trucks are directed to pick up produce at these places. From the results of a limited survey, the percentage of pickup stops with respect to the load type may be tabulated as shown in Table 1.

From Table 1, it may be inferred that just because a truckload shipment is a straight load, it is not necessarily all picked up at the same market. Nor is it always the case that mixed loads require more than one stops. However, the proportion of straight load trucks that require multiple pickup stops is comparatively less. The average number of load pickup stops for a typical truck was estimated at 2.32 (Pavlovic et al., 1980).

Multiple pickup stops may entail driving hundreds of miles out of the way (off of the final destination route), waiting for the produce to be packed in some instances, and waiting to be loaded, thereby incurring additional unnecessary costs. Apart from the above costs which are frequently borne by the truck operators, society also bears considerable costs. Social costs result from the extra consumption of valuable petroleum resources, wear and tear of roads and bridges, externalities such as traffic congestion and air pollution due to the extra driving caused by the multiple pickups, and roadway safety is diminished by increased truck traffic. Hence, it is desirable to eliminate multiple pickups or to reduce them as much as possible by providing the required produce at the demanded loading places, such as Farmers' markets.

A typical scenario considering two Farmers' Markets highlights the problems associated with multiple pickups. Consider the case of ten independent

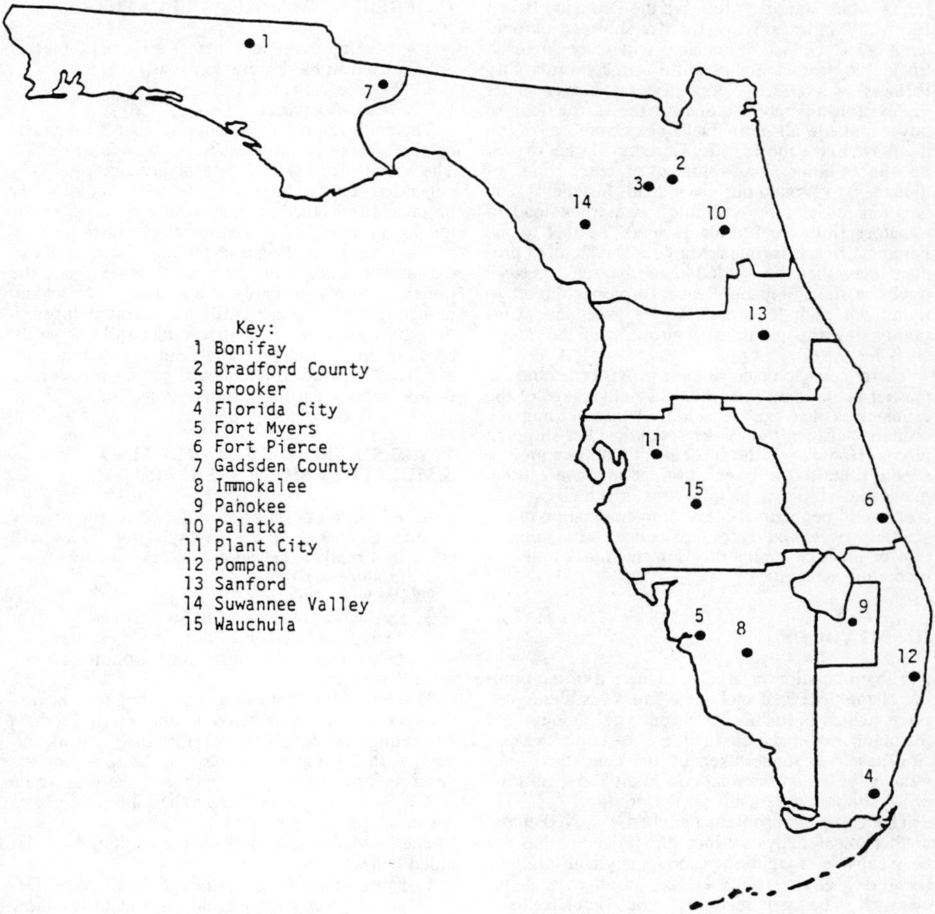

Figure 2. Florida State Farmers' Produce Markets

TABLE 1
Shipments by Load Type and Number of Pickup Stops

Pickup Stops Required	Load Type		
	Straight Load	Mixed Load	Total[a]
Single Pickup Stop	35%	10%	45%
Multiple Pickup Stops	13%	42%	55%
Total (All Pickups)	48%	52%	100%

Source: Pavlovic et al., 1980
[a] Sample Size: 131

trucks which are ready to leave the Pompano Beach Farmers' Market, except for the shortage of one-tenth of a truck load of tomatoes with respect to each truck. If tomatoes are plentiful at the Florida City Farmers' Market at a competitive price, then under the existing operation based on the instructions of buyer-brokers, all individual trucks would move on their own to the Florida City Farmers' Market to pick up the balance (one-tenth of a truck load of tomatoes) to "round out" each load. Instead, it may be more efficient to "shuttle" one truck load of tomatoes from the Florida Farmers' Market to the Pompano Beach Farmers' Market. The shuttled produce may then be loaded onto the ten "needy" trucks at the Pompano Beach Farmers' Market to round out each load thus saving time and effort involved with operating, waiting, and the social cost.

There seems to be no suitable provision within the present operational system to take advantage of the efficiencies that can be achieved by shuttling the produce between the markets and/or other shipping places. However, if the subsequent load pickups are on the route to the buyer destination, then a truck may make additional pickup stops, in which case the cost would be relatively less. Implementation of the shuttle service would require a considerable integration of produce availability information within the marketing system.

III. METHODS

The elimination or at least reduction of the number of multiple load pickups of trucks is a complex issue, which is mainly based on type, amount and the number of unsatisfied requests. A complex decision has to be undertaken by the concerned individuals, based on the total cost and the overall time of satisfying the demand for the produce.

The complex problem cannot be solved using mathematical analysis, since this procedure has not been capable of yielding general analytical solutions (Bowersox et al., 1972). Hence, an alternate methods such as heuristic methods, simulation have to be used (Pritsker, 1979). Here, a computer simulation model using SLAM is developed to analyze the complex problem of multiple load pickups.

IV. PRESENT MARKET OPERATIONS

The existing main activities at a Farmers' market are depicted in the Figure 3, namely

1. Produce arrival
2. Produce departure

The net inventory of produce at a market depends on the amount of produce brought by farm trucks. The actual quantity and type of produce may vary depending on the harvesting season. Occasionally produce from other markets would be brought for temporary storage for subsequent transactions.

The quantity and type of produce shipped out of the market depends on the buyers' demand and the initial inventory available at a market. The buyers' demand, which is generally represented through their brokers, can be satisfied either fully or partly based on the inventory available at a particular market. The "unsold" produce will be put into storage for the subsequent day's transactions.

V. FRESH PRODUCE MARKETING SIMULATION MODEL (FRMAS)

Based on the description outlined above, it may be inferred that there are five types of events will occur at a market (Devadoss, 1985), namely

1. Produce arrival
2. Produce departure
3. Evaluation of the unsatisfied demand
4. Arrival and update of the shuttle produce
5. Departure of trucks and update of the inventory.

These five types of events (including the existing two events) are considered in the Fresh Produce Marketing Simulation Model (FRMAS). It is developed with 24 hours (one day) as its time horizon because, depending upon the time of the year, some of the Farmers' Markets and other shipping places are in operation 24 hours a day.

The FRMAS model follows the methodology detailed below:

1. Farm trucks bring produce to the markets. Depending on the type and amount of produce, the net inventory is updated.
2. Demand for produce is created by buyers and they are arranged by their brokers. Buyer-bro-

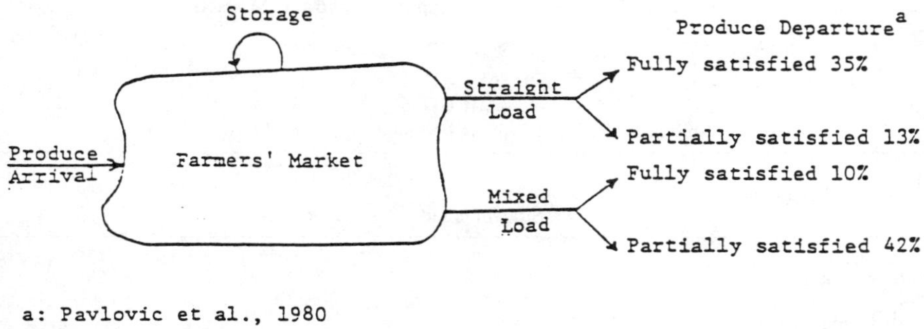

Figure 3. Existing Flow of Produce at a Typical Farmers' Market

kers would try to satisfy the demand as much as possible with the locally available inventory.
3. The unsatisfied demand for produce, with respect to a truckload order, will be reported by buyer-brokers to the proposed central authority at the market.
4. The evaluation of the excess demand will be done for each one of the markets at specific intervals of time for each type of produce demanded.
5. The total unsatisfied demand for each type of produce is checked with the available supply at other markets. Then the efficient way of satisfying the excess demand must be analyzed. Bringing the required "shuttle" load from a single market or multiple markets or allow the "needy" trucks to pickup the unsatisfied demand on their own.
6. Various "Measures of Effectiveness (MOE's) due to the shuttle service such as, savings in waiting cost, operating cost, social cost, and fuel cost will be calculated.

VI. DATA

The level of detail needed is "microscopic" in nature, detailing the shipping activities on a daily basis. Unfortunately, the published sources of information available provide some information are of "macroscopic" level (USDA, 1982; FDACS, 1982a). Therefore, secondary sources of information, personal interviews, telephone conversation were used to obtain the required data (James, 1981; Pavlovic et al., 1980).

The basic data needed for the analysis are
1. Types of Tractor-Trailer Trucks and Farm Trucks
2. Type of truck load carried (Straight load/Mixed load)
3. Proportion of produce in a mixed load truck
4. Operational details of markets/shipping places
5. Cost of time, operational, social cost of trucks

VII. ANALYSIS

In the case study four Farmers' Markets have been chosen, namely Florida City Market, Immokalee Market, Plant City Market, and Pompano Beach Market. These markets account for 63% of the produce shipped from the entire marketing system. Moreover, these markets are located within a reasonable driving distance (60-160 miles) of one another, making them attractive as alternative potential supply markets for satisfying excess demand. From the multitude of produce shipped out of Farmers' Market system, nine major types of produce have been selected for analysis, based on the annual sales of produce from these outlets. The selected major type of produce are: 1. beans, 2. cabbage, 3. cucumbers, 4. egg-plant, 5. okra, 6. peppers, 7. potatoes, 8. squash, and 9. tomatoes.

Based on the net inventory available at other markets, a minimum cost of satisfying the excess demand of the produce has to be chosen from one of the following:
1. Shuttle option, and
2. Independent pickup option (status quo operation).

A. Shuttle option

The whole number of truckloads that needs to be shuttled is analyzed with respect to the cost, which consists of the waiting cost of "needy" trucks, the driving cost of the shuttle truck, and social cost due to wear and tear of highway structures. However, for decision-making on a daily basis only the first two costs have been considered. This is because truckers are more concerned with the overall profit and out-of-pocket costs than with the indirect social costs.

B. Independent pickup option

In order to satisfy the unfulfilled demand of a truckload order, a truck may have to visit a single market or group of markets before proceeding to the final destination. The excess distance of miles due to pickup may vary depending on the location of pickup markets. However, if the pickup market is on the enroute to the final destination then there may not be any excess miles. Even then a "needy" truck may have to wait for the loading of the produce.

Comparing the net costs involved with the above options, the decision maker has to decide the best means of satisfying the demand.

VIII. RESULTS

The simulation model has been developed to demonstrate the potential of shuttle service and the savings that can be realized by buyer-brokers, and operators of trucks with excess unsatisfied demand. The various measures of effectiveness (MOE's) of interest to analyst were calculated using the model. The MOE's represent savings (Table 2) in
1. time (waiting) cost
2. operating cost
3. fuel cost, and
4. social cost

From the MOE's, it can be inferred that there are definite savings to the individual truckers, and to the society.

TABLE 2
Measures of Effectiveness

MOE's	Daily Independent Pickup Cost	Daily Shuttle Service Cost	Daily Net Savings due to shuttle	Annual Net Savings due to shuttle
Waiting Cost ($)	46,473	28,186	18,287	181,216
Operating Cost ($)	70,874	42,240	28,634	3,525,700
Fuel Cost ($)	30,213	20,440	9,773	1,206,782
Social Cost ($)	30,192	25,472	5,720	825,480

IX. IMPLEMENTATION

In a deregulatory environment, it is difficult to answer the question of how the concept of shuttling would be implemented. The best approach would be to demonstrate, at least on a small scale, the benefits and profits that can be derived from the shuttle service. When there is positive gain to all of the individuals concerned, the shuttle idea can sell itself. To begin with, the concept of shuttling produce can be adopted at the Farmers' Market level during peak harvesting season. Shuttling can be implemented in one of the following ways:

1. The personnel in-charge of the market, such as the Manager or Assistant Manager of the market, could be made responsible for coordinating the shuttling activity and can act as the central shuttling authority..
2. The second approach would be to authorize an individual or a firm to be the authority for the shuttle service at each individual market.

The second approach might be implemented with relative ease, especially in a free market system such as that of the fresh fruits and vegetables industry. Whatever approach is adopted, the implementation of the proposed model demands a "dynamic management information system," with respect to each market. This can be accomplished at a modest cost by providing remote computer terminals/personal computer at each one of market/shipping places.

While implementing the shuttle idea, the following strategies can be tried:

1. Bringing in a shuttle load in response to the amount of demand created for produce at the market (risk aversion approach). This approach has been used in the FRMAS model.
2. Bringing in a shuttle load in advance, in anticipation of the demand for produce. Considering overall waiting cost of the "needy" trucks, this strategy may be cheaper.
3. The above two strategies may be combined so that when there is an unutilized capacity in a shuttle truck after loading the required produce, extra produce might be loaded in anticipation of a demand.

One of the other important questions that needs to be addressed under the implementation of the shuttle service is how to persuade truckers to haul a shuttle load, which is basically a short-haul trip, especially when there is shortage of trucks. Normally, truckers prefer to haul a long distance shipments due to the high revenue. In order to attract truckers to haul a shuttle load, some kind of incentives are needed, such as the following:

1. Shuttle truckers could be paid for their service based on the time involved with the shuttling, and not on distance basis, as majority of the time is spent in loading and unloading of shuttle trucks compared to the time involved with line haul.
2. Shuttle truckers who bring in produce could be given high priority to take the next available long line-haul load, or a long distance load might be arranged in advance.

X. CONCLUSIONS

Multiple pickups occur due to variety of complex factors, the main reason being that required produce types are not available at all markets at a competitive price. Hence, the basic solution is to try to provide the required produce at the market at a competitive price by some suitable means such as shuttling, as indicated in the previous section.

Initially, the shuttle service could be adopted at the Farmers' Markets on an experimental basis. After studying the operation for a few harvesting seasons, shuttle service could be extended to all other shipping places within the state.

The benefits derived from the shuttle service may be divided among all participants in the industry. Some of the benefits are:

1. As the transportation cost decreases, there is a good possibility that farmers will get a better price for their produce.
2. The truckers can make more revenue-generating trips on an annual and seasonal basis, which may act as a catalyst to attract more trucks.
3. Quick and expeditious transportation means more business to the state in general, which may stand to have a competitive edge over other producing regions.
4. Lower transportation cost means lower unit retail price, a reduction which ultimately benefits the consumers.
5. Reduced wear and tear of bridges and roadway structures results in a savings to society.

The normal days' total savings as estimated by the simulation model for a group of four markets and nine types of produce, add up to $62,410. The estimated annual savings due to the shuttle service is $5.74 million, which is quite substantial.

ACKNOWLEDGEMENTS

The research described in this paper was carried out as part of the requirement for Engineers degree while the author was at the University of Florida. The author wishes to acknowledge the valuable comments of his committee members during the course of this research.

REFERENCES

Beilock, R., and G. Fletcher, *The Structure and Characteristics of the Florida Exempt Perishables Trucking Industry*, IFAS, University of Florida, Economic Report 109, December 1983.

Bowersox, D. J., O. K. Helferich, P. Gilmour, F. W. Morgan Jr., E. J. Marien, M. L. Lawrence, R. T. Rogers, *Dynamic Simulation of Physical Distribution System*, Graduate School of Business Administration, Michigan State University, East Lansing, 1972.

Devadoss, N., *Improving Truck Transportation of Fresh Fruits and Vegetables at the Florida Farmers' Markets*, Unpublished Engineers Report, University of Florida, May 1985.

Florida Department of Agriculture and Consumer Services (FDACS), *Florida State Farmers' Markets Forty-Eighth Annual Report, 1981-82*, 1982a.

Florida Department of Agriculture and Consumer Services (FDACS), *Division of Fruit and Vege-*

table Inspection Annual Report, 1981-82, 1982b.

James, M. L., *Analysis of Market Layout Alternatives for Optimum Produce Transfer*, Unpublished Masters Thesis, University of Florida, 1981.

Pavlovic, K., G. Long, D. P. Reaves, and T. H. Maze, *Domestic Transportation for Florida Perishable Produce*, Transportation Research Center, University of Florida, 1980.

Pritsker, A. A. B., and C. D. Pegden *Introduction to Simulation and SLAM*, System Publishing Corporation, West Lafayette, Indiana, 1979.

Manalytics, Inc., *A Long-Term Study of Produce Transportation*, Prepared for the Federal Railroad Administration, U.S. Department of Transportation, Washington, D.C., 1977.

Maze, T. H., "Price Response of Truck-Service Suppliers in an Unregulated Market," *Transportation Reserach Record 889*, 1982, pp. 26-32.

United States Department of Agriculture (USDA), *Fresh Fruits and Vegetable Shipments*, Agricultural Marketing Service, Annual Report, 1980-82, 1982.

ENDNOTE

* Graduate Student, Department of Civil Engineering and Environmental Sciences, University of Oklahoma, Norman, OK.

Towards a Level-of-Service Measure for Mass Transit

*by Jonathan C. Levine**

2nd Prize—Best Graduate Student Paper

I. INTRODUCTION

Public transit systems may be properly evaluated for policy purposes in a number of ways. Among the more common tools for evaluation are measures of a transit system's performance from a fiscal point of view, such as passengers or revenues per operating mile, or farebox recovery ratios. Less developed are evaluation techniques that adopt the perspective of the transit passenger and attempt to measure the quality of the service received and facilitate service comparison over time or between transit systems. The concept may be referred as an extension of "level of service" (LOS) measures commonly used in street and highway planning to the realm of public transit.[1] This paper will examine some of the attempts to develop transit LOS measures, suggest criteria for the evaluation of such measures particularly as they are influenced by land use patterns, and offer proposals towards the development of a standardized LOS measure for transit operations.

II. PURPOSES OF A TRANSIT LOS

The need for a transit LOS measure stems from transit's shift over the past few decades from private to public ownership. When profit was the motive for running streetcars, profit would also be the criterion by which changes in service were evaluated.[2] Current transit decisions are less singular and require multifaceted evaluation criteria. When considered within its specific land use context (that is, avoiding comparisons between systems operating under widely differing density conditions) a transit LOS may serve as such an evaluation measure. Botzow[3] proposes LOS for the purposes of rating existing modes and specifying service goals to be sought. Allen and DiCesare[4] extend the goals to include policy formulation and monitoring, enabling the linkage of subsidy to performance, facilitating transit agency management and generating public information and involvement. Flyvbjerg[5] emphasizes the public involvement aspect, seeing transit LOS primarily as a tool to elicit effective citizen participation in transit planning.

Flyvbjerg also raises the question of the appropriateness of a performance measure (typically a tool of a technocratic planning style) to a more open system of planning of which citizen participation is an integral part. He shows that there is no conflict between the two as long as the measure itself is inexpensively produced, directed at evaluating specific problems (and not just macro or region-wide issues), and most importantly, understandable. To Flyvbjerg a measure is understandable to the lay person if it is scaled in real terms; for example travel time in his scheme is measured in minutes and is not combined with other service elements to generate an arbitrary "A" thorough "F" measure.

This approach gives rise to several difficulties in a tradeoff between the values "understandable" and "usable." Specifically, unless a LOS measure develops a weighting system to render service elements comeasureable, the lay person will be unable to use it to decide whether a service change that, say, increased frequency while decreasing speed but holding reliability constant, leaves the community better or worse off than before. Moreover, without such a measure, a statement such as "our city gets 'X' level service while their city gets 'Y'" would be impossible, and the value of the measure for comparison purposes would be sharply cut. In addition, certain service variables (notably comfort) may be readily measurable but even the untransformed units of measurement (acceleration per second per second, cubic meters of ventilated air per second) are utterly meaningless to the lay person. For these reasons this paper will focus on LOS determination efforts that attempt to achieve a single measure reflecting transit's many aspects. They may thus sacrifice a measure comprehensibility but are more than compensated by their increase in usefulness.

III. DEFINITION AND MEASUREMENT OF SERVICE VARIABLES

LOS measures such as these will contain three chief elements; a definition of relevant service variables, an approach to their measurement, and a scheme to weight the outcomes to reach a single measure of service level.[6] Mundy[7] criticizes LOS measures that take this form on the grounds that they focus on traditional transit to the exclusion of the broad spectrum of alternatives between transit and the private automobile. But comparing service levels between a range of modes is a daunting task, in part due to the fact that certain service attributes (such as headways between transit vehicles) are mode-specific. For this reason this paper will focus on measures of the type described above.

In keeping with the goals of LOS measurement, the following criteria are proposed to guide the acceptance or rejection of proposed LOS measures:

The measures should be user-oriented and service-oriented. Thus the measurements themselves will avoid judgments regarding what the transit operator may reasonably be expected to provide, focusing instead on what the patron actually receives. The fare the passenger pays must not be included in a level of service measure. Indeed LOS may fuel debates for fare changes based on better or poorer service; including the fare itself within the measure would confuse the issue.

The number of variables must be small enough for the measure to be usable. Thus the approach of Allen and DiCesare[8] who specified at least twenty seven variables (including such marginal items as average curb-to-vehicle step height) to be included in a LOS measure is to be avoided.

The number of variables must be great enough to reflect different significant attributes of public transit. Polus and Shefer[9] based their measure on speed alone, Bakker on seat availability,[10] while Flyvbjerg[11] based his on travel time and cost. These extremely limited approaches, though enjoying the advantage of simplicity, sacrifice too much information about other trip aspects to be useful LOS measures.

No significant aspect of the transit trip should be a priori excluded from the analysis. Alter[12] arbitrarily excludes "hygiene" variables from his LOS measure, asserting that a lack of cleanliness, comfort or safety can discourage a rider but their presence can never motivate a potential rider. The fallacy of this statement is readily apparent; the rider who was previously repelled by poor cleanliness, comfort or safety may well be wooed back by improvements in these areas. Moreover, surveys have shown safety and comfort to be important determinants of mode choice.[13]

The above criteria suggest the definition of a handful of service variables that can describe the major aspects of the transit trip. Following is a description of the major areas to be considered.

A. In Vehicle Travel Time

Botzow[14] measures trip time in miles per hour, specifying goals together with his measures: 35 miles per hour in line haul operations, 6 to 15 miles per hour in CBD areas, based on a maximum door-to-door travel time of one hour. Alter[15] avoids the complexity of such arbitrary definitions by basing his travel time variable on the ratio between transit trip time and automobile trip time. Subsequent authors[16] have followed Alter's approach, with the exception of Flyvbjerg[17] who uses minutes spent in travel as the relevant measure of travel time.

Flyvbjerg's approach also combines in vehicle trip time with out of vehicle trip time, thus ignoring the significantly higher psychological cost of the latter[18]. The approach makes two important contributions, however. First, Flyvbjerg bases his analysis on a survey of the population served that uncovers the desired trips and trip times, matching them with timetables to derive actual trip times. This approach enables him to measure another variable he calls "hidden waiting time," being the difference between the time a person actually arrives at a destination and the time he or she needed to arrive. Thus a worker who has to be on the job at eight in the morning but is forced to arrive at 7:45 due to bus scheduling is enduring fifteen minutes of "hidden waiting time."

A combination of the approaches suggested by Alter and Flyvbjerg may prove particularly useful. In recognition of the fact that in the American context the chief competitor of public transit is the private automobile, Alter's relative speed measure should be used. But in order to ensure that the relative speed is calculated for a typical trip rather than an ideal one (and to measure service for potential users as well as actual), Flyvbjerg's survey approach may be employed.

The survey approach will also generate a "hidden waiting time" variable, though this information should not be automatically lumped into travel time; it is reasonable to expect it to have a different cost to the passenger from travel time itself. On the other hand, the "hidden waiting time" effect is at least partly captured by the frequency of service variable described below; empirical analysis would likely show the two variables to be describing the same kinds of costs to the passenger. I am proceeding on the assumption that service frequency captures much of "hidden waiting time" but more adequately describes the disutility of fixed schedules to the customer.

B. Service Frequency

The frequency of service is generally accepted as an important determinant of transit LOS, and attempts to measure this service characteristic within an LOS context have taken two approaches. The first approach (which was adopted by Botzow[19]) utilizes the common approach of adding one-half the average headway to the travel time. This technique leads to several difficulties, however. First, as Botzow himself points out, when headways rise to much above ten minutes passengers begin to use timetables and the implicit presumption that the average wait equals half the average headway no longer holds.[20] (For this reason Botzow proposes adding half-headway times to travel time to a maximum of five minutes, thus placing a system with thirty minute headways on the same level as one with ten minutes separating its vehicles.)

The second reason the "half average headway" method is problematic is the implicit presumption that waiting time is viewed by the passenger equally as "costly" as in vehicle travel time, a presumption disproven by numerous studies.[21] An approach separating the two and allowing them to carry differing weights (as discussed in Section IV below) will better reflect their impact on the service levels passengers perceive.

Such an approach is advocated by Alter[22] who uses average headways directly as a measure of service. Though this may be the best measure it should not be applied blindly; for example a transit operator running very frequent commuter service during peak hours but little service the rest of the day would not be evaluated on an all day basis. Rather, two LOS measures would need to be computed for such an operator: one for commute hours, the other for the off-peak.

Another potential pitfall of this measure would be an assumption that the relationship between frequency and passenger-perceived service is linear. In fact it is markedly nonlinear over the relevant range; a change from 5 to 20 minute headways is significantly worse than a change from 30 to 45 minute headways, from the passengers' point of view.

C. Accessibility and Out of Vehicle Travel Time (OVTT)

The range of activities a passenger needs to engage in just to ride a transit vehicle is great. Horowitz[23] writes of bus travel: "In contrast to competing modes, a relatively small percentage of the effort expended in bus transit travel actually results in phyiscal progress toward a destination. Transit users spend substantial amounts of time planning their trips, coping with unfamiliar areas, dealing with crowds, contending with the weather, walking, waiting, transferring, or otherwise satisfying the basic constraints of the bus mode." This range of OVTT activities and the varying costs passengers attach to each of them makes accessibility a particularly difficult variable to measure in an LOS context. The difficulty stems not so much from an inability to define or measure the range of these activities as a need to keep LOS measures simple and inexpensive to produce.

An approach adopted by Allen and DiCesare[24] calculates that the ratio of route-kilometers to catchment area-kilometers to determine the degree of coverage the transit operator is offering. The difficulty with this approach is in its inherent assumptions regarding land use patterns throughout the catchment area. If both population and employment are relatively concentrated along certain corridors, the region may be quite well served by a skeletal network. (Such may be the case in cities with radial street and highway patterns, for example). If the inherent assumption of uniform distribution of population and employment throughout the catchment area is a rough approximation of reality, this measure may be appropriate as long as it is used in conjunction with a measure of route directness (such a number of transfers) and service frequency.

Perhaps more all-purpose measures are suggested by Bullard[25], who used simple OVTT or Alter[26], who used distance to and from the trip ends[27]. An advantage of OVTT over distance as a measure of accessibility is that OVTT, unlike distance, can measure time spent in transferring[28]. On the other hand, OVTT can underestimate the difficulties of access to transit since it values identically a walking trip and a driving trip of equal time to a bus stop. Hence, if OVTT is selected as the measure, an auto access dummy variable should be added, while if distance to the trip ends is used to indicate accessibility, time spent in transferring will have to be measured separately.

In either case, the number of transfers required should be counted separately to allow for the likelihood that the very fact of transferring imposes a psychological cost on a passenger over and above the time spent in transferring.

D. Service Reliability

Reliability can be roughly defined as the passenger's degree of certainty of arriving according to schedule. Hence Botzow's variable of average minutes of delay per trip[29] is probably inadequate as it says nothing of the likelihood of experiencing a significant delay. Regular delays of 5 to 10 minutes are likely to be perceived quite differently from the occasional long delay due to traffic jams or equipment failure.

Alter[30] measures reliability so as to capture this aspect; he proposes to compute the percentage of vehicles that run between one minute early and three minutes late.[31] Reliability of this sort must be measured relative to service frequency; the effects of lateness on a 12 vehicle per hour line are much less severe than on a 1 vehicle per hour line.

E. Comfort and Security

Comfort and security, dismissed by Alter as "hygiene" factors, may in many cases determine mode choice. But they are particularly difficult aspects to measure, since they are the most subjectively determined of all LOS elements.

For example, security may be measured on the basis of transit crimes or accidents per million vehicle kilometers[32] (security from crime is likely to be the dominant concern). But a number of intangible or uncontrollable factors (such as the systems safety "image" or the fear of crime while travelling to and from the system) may shape people's preferences more than actual crimes.

The measurement of comfort poses difficulties of a different nature. First, comfort in transit is composed of a large number of elements, including seat availability and quality, crowdedness, ventilation, temperature, cleanliness, odors, and for standees, acceleration, jerk and sway. Secondly, these elements are perceived differently by different individuals, and perhaps most importantly, they are perceived as a bundle; most passengers do not consciously consider each of these elements, but evaluate the entire trip as "comfortable" or "uncomfortable." Perhaps the most comprehensive approach to the development of a "comfort" variable would be an experiment in which passengers were asked to rank various transit rides on a "comfortable" "uncomfortable" scale; the results could then be regressed on the comfort elements mentioned above.

If this comprehensiveness proves unfeasible, an acceptable simplification may be to adopt a single variable (seat availability would be a likely candidate) as a proxy for all others. The strength this admittedly simplified approach is that the availability of a seat enables a passenger to withstand many other affronts to his or her comfort.

IV. WEIGHTING OF SERVICE VARIABLES

As discussed above, the combination of the diverse service elements comprising the transit trip experience in a single LOS measure demands a scheme that would assign varying weights to the different elements. Most authors proposing multivariate LOS measures have in fact used weighting systems, but the weights used, even if reflective of the authors' knowledge of transportation demand characteristics, were remarkable arbitrary.

For example, Botzow[33] based his weights on a survey of Washington D.C. transit riders, calculating them from the frequency of criticism reported regarding each service element. The following weights are thus produced: Speed, 30 points; density (i.e., crowding) 25 points; temperature, 15 points; delay and acceleration (i.e., jerk), 10 points each; and ventilation and noise, 5 points each.

The weaknesses of this approach are several. First, it is based on what people say they care about and not their observed behavior. Thus the results are most likely skewed toward the most noticeable shortcomings of the D.C. system. Furthermore, the survey questioned current system riders only. By neglecting potential riders the results were probably biased toward certain transit characteristics. The precise degree and direction of bias cannot be determined without further empirical analysis but 60 points out of 100 devoted to comfort related variables (density, acceleration, temperature, ventilation and noise) is probably an overstatement of comfort's relative importance.

In a similar fashion, Bullard[34] developing a weighting system based on ridership surveys, while Alter[35] created a weighting system he identified as "arbitrary." "Ridership," he writes "is a response to an offered LOS. As such it (ridership) is an important performance indicator but in no way directly measures LOS[36]." The problem with this approach is that it gives rise to the arbitrary determinations that Alter made, which in turn carry the danger of identifying "improvements" in LOS that drive riders away and "deterioration" that attracts them. Logically a change in the bundle of attributes that constitute LOS is an improvement if and only if it has the **potential** to attract a larger ridership.

Such a view would lead the LOS measure developer to models of modal split to examine variable coefficients as potential bases for service element weights. Existing modal split models might also provide some answers on the validity of the more arbitrary LOS weights. Unfortunately, a direct comparison of existing LOS measures with existing multinomial modal split measures is hampered in two significant ways. First, crucial LOS elements, such as comfort, are missing from most modal split models. Secondly, variables are frequently measured differently in modal split models and LOS measures. For example, travel time tends to be measured in absolute terms in modal split models but as a ratio (transit to auto) in LOS measures, thus rendering the coefficients incomparable.

But limited comparisons are possible through hypothetical scenario testing. Using the comparable elements of a single logit modal split model and an single set of LOS weights one can compare the assessment each scheme would give to a change in transit service. The LOS weights and the logit model coefficients used for this comparison are presented in Table 1. It should be noted that both the logit model and the LOS system contained a number of other variables; only those that were directly comparable were utilized.

Two scenarios are presented in Table 2; using a reference automobile trip of 20 minutes IVTT and 3 minutes OVTT, two hypothetical bus trips are compared under Bullard's LOS weights and Train's logit coefficients. The results demonstrate the inconsistency of the LOS measure with the modal split model; the second bus trip will attract a smaller mode share than the first, yet its LOS "increased" from something under a "C" to a flat "C." The "improvement" in service levels drove away passengers!

Valid LOS measures thus must justify their weighting schemes. The methodology for doing so is well known; the maximum likelihood estimation used in multinomial logit modal split analysis is

TABLE 1

LOS Weights and Logit Modal Split Model Coefficients

Service Element	Logit Coefficients[37]	LOS Weights[38]
IVTT	−.0201	10
Walk Time	−.0464	10
Number of Transfers	−.0656	10

TABLE 2

Comparison of Hypothetical Transit Trips Under LOS Weights and Logit Model Coefficients

Trip #1:

Element	Bus	Auto	Utility (Bus)	Utility (Auto)	LOS (Bus)
IVTT	30	20	−0.603	−0.402	D
Walk Time	3	3	−0.139	−0.139	B
Transfers	2	0	−0.131	0	D
Total			−0.873	−0.541	

Bus' market share under this scenario: 41.8%
Overall Level of Service Ranking: C−

Trip #2:

Element	Bus	Auto	Utility (Bus)	Utility (Auto)	LOS (Bus)
IVTT	35	20	−0.703	−0.402	E
Walk Time	6	3	−0.278	−0.139	C
Transfers	0	0	0	0	A
Total			−0.981	−0.541	

Bus' market share under this scenario: 39.2%
Overall Level of Service Ranking: C

entirely appropriate for generating LOS weights as well. Such studies would have to be tailored to LOS purposes in their variable specification. But the relative stability of the coefficients of modal split models over time and across geographic regions offers hope for similar stability in LOS weights.

V. LAND USE PATTERNS AS A DETERMINANT OF TRANSIT LOS

It is easy to show how land use patterns are a direct determinant of LOS; higher density, for example, will almost always allow a greater frequency of service. Traffic congestion will constrain bus speed. Even the layout of the street network determines service characteristics; the grid system enables quite direct travel to all points[39], while a radial system may allow for more frequent service along the "spokes."

But more important to the design of LOS measures is the question of whether and how LOS should be measured differently according to land use conditions. For example, if fifteen minute headways are considered level "D" for urban areas, might they be considered level of service "A" or "B" for suburban districts? Several authors[40] have proposed such more lenient standards for less dense areas. However no author has considered explicitly the justifications for varying the measure to suit the area.

Such justification needs to be carefully examined. If LOS is intended to be a **unit of measure,** it should be independent of any **standard** that is to be advocated or imposed. Arbitrarily varying the measure to suit local conditions is akin to defining 70 degrees fahrenheit to be "the desired ambient temperature" whether that temperature is measured in an office building, an old age home, or a food storage locker.

Two possible justifications come to mind when one is considering varying LOS according to land use conditions. Two of them have no proper bearing on LOS measurement.

A. *Transit operators cannot reasonably be expected to provide the same level of service in suburban areas as in urban.* While this statement is true it irrelevant to LOS measurement. As opposed to measures of financial performance, LOS is intended to be user-oriented; that which the operator may be expected to provide bears no relationship with what the user in fact receives.

B. *Residents of suburban communities want less transit than those of urban areas.* Again the statement is probably true, but need have no bearing on LOS measurement. One of the chief purposes of an LOS measure is the stimulation of public involvement in transit planning, and such involvement almost always entails differences of opinion between citizens. There will be ardent transit advocates even in the suburbs; to artificially inflate the LOS they are reported to receive is to deprive them of an important tool for influencing debate.

C. *LOS is perceived differently under varying conditions.* This is the sole proper justification for varying LOS measurement according to local conditions. It does not justify adjusting the vehicle headway categories in areas of sparser development[41], but it may have influence in other areas.

For example, as discussed above, the importance of on-time reliability grows with the headways between vehicles; a once-an-hour bus must be on time, while delays on a very frequently served line would be barely noticed. Thus it would be entirely legitimate to grant a "D" ranking to a suburban half hourly service with an 85 percent on-time record, but an "A" ranking to an 8 minute headway urban service with the same record[42].

Another example of LOS perceptions varying according to local conditions might be in the area of comfort. On a very short CBD ride, for example, the lack of a seat might be of little importance (many people would find it barely worthwhile to take off their coat and set down their parcels anyway); such a ride might be ranked "B" in comfort terms. But in a 45 minute line haul from the suburbs, the lack of a seat would certainly be ranked "F" in its discomfort producing potential.

In a similar fashion, comfort variables such as acceleration, deceleration, jerk and sway may be measured relative to seat availability, as they are most important to standees. Systems where virtually all passengers have a seat (as is typical for suburban systems) may be permitted a fairly rough ride and still be granted a high comfort grade. Systems in which standing is common will have to ride more gently to be given a high ranking.

V. CONCLUSION

For a transit level of service measure to fulfil its potential in transit evaluation, planning and citizen participation, the concept would need to become widely standardized and accepted. Despite the efforts of a number of authors, thus far nothing approaching a standardized measure has been developed. This is largely attributable to the difficulty of generating general acceptance of a measure that includes so many subjectives. If there is a way to overcome this obstacle it is through more careful specification of LOS models and greater justification of the systems developed to rate the relative importance of each element of the LOS measure.

ENDNOTES

* Ph.D. Student, City and Regional Planning; M.S. Student, Transportation Engineering, University of California, Berkeley, CA.

1. Botzow, H. "Level of Service Concept for Evaluating Public Transport." Transportation Research Record 519, p. 73-84.
2. Bullard, Diane L. and Christiansen, Dennis L. "Level-of-Service Concept Applied to Public Transportation." Texas Transportation Institute, The Texas A&M University System, College Station [1981]. Technical Report 1067-1F, Study 2-10-81-1067, p. 1.
3. Ibid, p. 73.
4. Allen, William G., and DiCesare, Frank D. "Transit Service Evaluation: Preliminary Identification of Variables Characterizing Level of Service." Transportation Research Record 606 [1976], p. 43.
5. Flyvbjerg, Bent, et al. "Evaluation of Public Transport: Method for Application in Open

Planning." Transportation 13 [1986], pp. 23-52.
6. The LOS measure may be computed for the transit system as a whole, a geographical or temporal subunit, or a single line.
7. Mundy, Ray A. "Mass Transit Guidelines Versus a Consumer Orientation in Public Transit Systems." Transportation Research Record 625 [1977], p. 34.
8. Allen and DiCesare, op. cit.
9. Polus, Abishai, and Shefer, Daniel. "Evaluation of a Public Transportation Level of Service Concept." Journal of Advanced Transportation 18:2 [1984], pp. 135-144.
10. Bakker, J. J. "Transit Operating Strategies and Level of Service. Transportation Research Record 606 [1976], p. 5.
11. Flyvbjerg, op. cit.
12. Alter, Colin H. "Evaluation of Public Transit Services: The Level of Service Concept." Transportation Research Record 606 [1976] p. 40.
13. Mundy, op. cit., p. 35.
14. Botzow, op. cit., p. 74.
15. Alter, op. cit., p. 38.
16. For example, Bullard and Christiansen, op. cit.
17. Flyvbjerg, op. cit. pp. 27-31.
18. Horowitz, Alan J. "Subjective Time in Bus Trnasit Travel." Transportation 10 [1981] 149-164.
19. Bozkow, op. cit., p. 75.
20. On the other hand, half average headway may be a reasonable proxy for waiting time plus "hidden waiting time."
21. For example, Bakker, op. cit., p. 2.
22. Alter, op. cit., pp. 38-39.
23. Horowitz, op. cit., p. 149.
24. Allen and DiCesare, op. cit., p. 43.
25. Bullard, op. cit., p. 2.
26. Alter, op. cit., p. 38.
27. Similarly to in vehicle travel time, these measures are most appropriately based on survey data.
28. Though time spent walking to a transit vehicle will have a different "cost" from that spent transferring from vehicle to vehicle, the cost of each will be higher than that of in-vehicle travel time; I am combining the two for simplicity.
29. Botzow, op. cit., p. 75.
30. Alter, op. cit., p. 38.
31. Whether a vehicle leaving one minute early is to be considered "reliable" may be debated; Bullard (op. cit., p. 21) uses "zero minutes early" as her standard.
32. Allen and DiCesare, op. cit., p. 43.
33. Botzow, op. cit., p. 82.
34. Bullard, op. cit., p. 29.
35. Alter, op. cit., p. 39.
36. Ibid.
37. Train, Kenneth. "Work Trip Mode Split Models: An Empirical Exploration of Estimate Sensitivity." Working Paper #7602, Urban Travel Demand Forecasting Project, Institute of Transportation Studies, University of California, Berkeley, 1976, p. 14.
38. Bullard, op. cit., p. 30.
39. Bakker, op. cit., p. 2.
40. For example, Alter, op. cit., and Dawson, W. R. "An Approach to Urban Transit Service Standards in a Regional Context for Small and Medium-Sized Communities." Transportation Forum Vol. 1-3 [1984], pp. 5-12.
41. If a single scale of headway categories is to be used to describe service under all land use conditions, the scale must be finer grained than the six-point A through F scale. It would be inadequate to lump all service with headways above 30 minutes into the "F" category.
42. This approach was used by Alter, op. cit., p. 39.

TL-PIGGY, The Intermodal Decision Assistant

*by Paul C. Heymann**

Best Undergraduate Student Paper

OVERVIEW

This paper develops and applies a model that helps a small domestic shipper decide between two similar modes. Since deregulation, commercial truck and trailer-on-flatcar (TOFC) have emerged as major competitors for long distance high-rated commodities. Each has its own cost and service implications for the shipper and receiver. The increased flexibility made possible by deregulation has made the decision even more complicated. Intermodal loadings are at record levels—4,605,099 in 1985.[1] Can the small shipper use piggyback, and if so, how?

This project involved three stages. First, a decision flowchart was developed to formalize the mode selectionn process. The associated text aids in understanding the complex issues involved. Implementing software flowchart might use a rule-based decision support system, but writing such a program was beyond the scope of this project. Commercial "Expert Systems", such as those writte in Prolog, could be used as a starting point.

The second phase was building a spreadsheet to quantify the decision factors. The spreadsheet is dual choice (truckload carrier vs piggyback), but could be adapted to include other options. These two stages comprise the decision model.

The third stage applies the model to Charles Jacquin et Cie, a medium-sized Philadelphia distiller, blender and distributor of liqueurs and cordials. The objective was to obtain "real-world" input, and to provide the company with a working decision model. Details are omitted at the manufacturer's request.

This report draws several conclusions about TOFC service in general, and about the shipping decision in particular. Piggyback is a valid option, but the small shipper cannot use it randomly to improve his logistics. The complete conclusions are at the end of the paper.

I. DEVELOPING THE DECISION PROCESS FLOWCHART

The flowchart serves as a decision aid for the shipper, and helps develop spreadsheet inputs. It is also a beginning point for designing related software. A review of logistics software indicates a gap in commercial software that helps in common carrier vs piggyback decision.[2] Large corporations will have a sophisticated logistics strategy that assists with modal choice, and may have developed inhouse software.

For the small shipper, however, the decision is less formalized. Typically the traffic manager chooses based on tradeoffs between customer preference, available rates, carrier relationships, shipment particulars, and personal experience. While a satisfactory decision is probably made most of the time, this method has at least two drawbacks. First, without specific calculations to account for all of a shipment's costs, the traffic manager and the firm cannot be sure that the shipping decisions are actually optimal. Second, relying on the traffic manager's experience and knowledge leaves the small firm vulnerable should he leave. Therefore, formalizing the decision process will make the results more justifiable and consistent.

Decisions made with the flowchart will follow through to actual calculations made with the spreadsheet. Thus, the flowchart helps with the qualitative and strategic factors, while the spreadsheet provides answers to some of the quantitative questions.

A. FLOWCHART INPUTS

The following discussion elaborates on the flowchart inputs, to help the shipper make the "Yes/No" judgments required. It follows the flowchart's order, at Exhibit 1.

—Is The Shipment/Commodity Containerizable?

The commodity being shipped must be one that can be shipped by truck, since piggyback is a substitute for truck service. Piggyback may be less suitable for refrigerated cargo because the trailers are left unattended for long periods.

—Is Mode Specified By Receiver?

In many cases the receiver specifies the mode or carrier. This will apply whether FOB shipper or FOB receiver. However, customer preference should not end the search, as the shipper may find a better deal for the consignee elsewhere. It benefits both parties to lower shipping costs, regardless of who pays. The shipper should continue to seek rates that lower the delivered cost and increase the market coverage of his products. He must maintain a rapport with the receiver and advise him of these opportunities. However, negotiating a particular rate may depend on which party controls the traffic and/or pays the freight charges.

—Is Shipment Of Normal (Non-Urgent) Priority?

Clearly, if the shipment is an emergency resupply or is needed to keep a production line going, direct truck service has the edge. In normal moves where cost is a factor, the choice is broader.

—Is Shipment A Trailerload, By Volume Or Weight?

One disadvantage of TOFC is inflexibility: no stop-offs are allowed enroute, although the drayman may accomodate split shipments within the same city. Also, reconsignment enroute is difficult or impossible. Thus, the shipment should only be between two points, such as distribution centers. Piggyback rates are on a per-trailer basis, regardless of weight, so better results are obtained with a full trailer. (LTL shipments do move economically by piggyback, once consolidated by the carrier, freight forwarder, or shippers association).

Exhibit 1:
TL-PIGGY Flowchart

This flowchart documents the process involved in a modal choice shipping decision. In this case, the decision is between for-hire truck (motor carrier or owner operator) and piggyback (TOFC) for the linehaul. The flowchart's dovelopment and use is explained in Part I of the text.

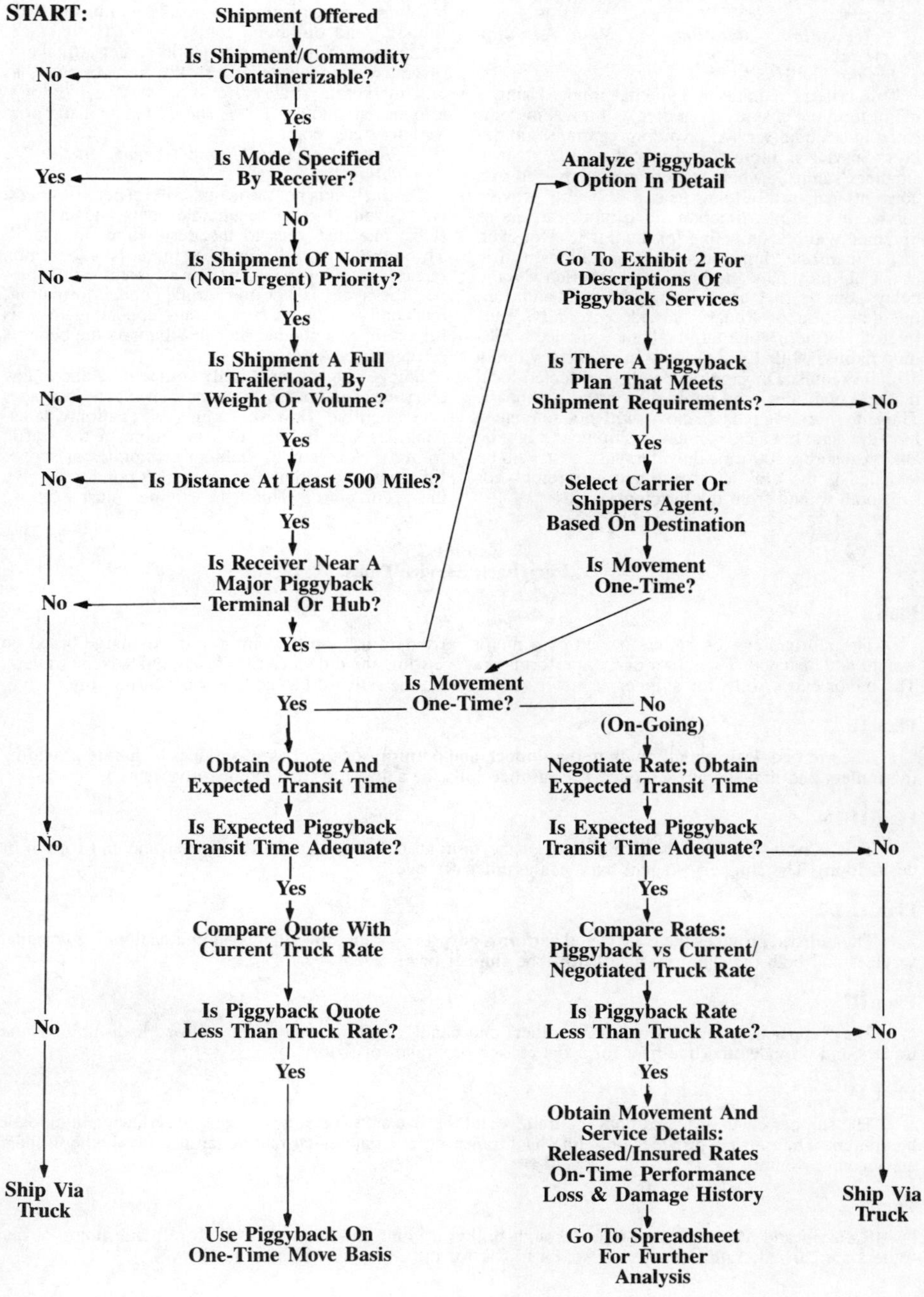

—Is Distance At Least 500 Miles?

Here, the cutoff is arbitrary. The point of indifference varies with the specific traffic lane. For example, Soo Line piggyback trains are competing in the Chicago-Twin Cities market (420 miles). Other sources such as Booz, Allen & Hamilton indicate 600 miles may be more appropriate. The shipper should modify the 500 mile guideline as necessary.

—Are Shipper and Receiver Near A Major Piggyback

Terminal or Hub?

This criteria requires judgmental input. Using a major terminal ensures the trailer will travel in dedicated intermodal service, avoiding yardings and delays. Service is more frequent and reliable than at smaller "ramps", where trailers may be placed in a conventional train. Being near a hub will provide service in multiple directions, and more carriers or draymen will be competing for the traffic. However, just as in airline "hub and spoke" operations, traffic at the big terminals and hubs may experience more delays, due to truck and rail congestion. In addition, the definition of "near" varies regionally. Burlington Northern's hub terminals may service a 250-mile radius, while Conrail looks at markets within a 50-mile radius. Drayage rates will be affected local market conditions, and are higher in the Northeast. Thus, a short piggyback move will not tolerate a long drayage, but a cross-country shipment might be able to absorb a 200-mile dray. Drayage costs will be lessened if the drayman hauls a trailer (empty or full) both to and from the terminal.

—Analyze Piggyback Option In Detail.

If the above criteria have been met, piggyback still may or may not be satisfactory. Exhibit 2 describes the basic features of the common piggyback services. The key issue is whether piggyback is compatible with the company's overall logistics strategy. (If not, perhaps the strategy needs reevaluation). Numerous factors will come into play here, involving input from many people within the firm. Inbound and outbound logistics, traffic balance, trailer ownership, and production strategy (make to inventory or make to order) will be important, as will marketing strategy (cost vs service). A long-term decision to use TOFC should not be made from a narrow viewpoint.

—Select Carrier/Shippers Agent, Based On Destination.

Destination is not the actual criteria here. Rather it is: "Given a set of acceptable agets and carriers, select one that goes to the destination involved". Throughout this discussion, there is a common thread: the shippers agent. For a typical small shipper, the agent is key to getting good information, rates and service. A competent shippers agent is as important as a good carrier in achieving the benefits of piggyback service.

Shippers agents are highly competitive due to low entry costs and because one of their inputs (drayage) is competitive. However, agents with nationwide capabilities will be able to serve more of the traffic from a given source, yielding economies of scope. Information asymmetries on market conditions give the agent power. Thus, the shipper must exercise

Exhibit 2:
Piggyback Service Plans

Plan I

The railroad carries trailers owned by a motor carrier from ramp to ramp at a fixed charge based on weight and distance. The motor carrier performs drayage while the rail carrier loads and unloads the trailers. The motor carrier bills the shipper at motor carrier rates; the railroad has no contact with the shipper.

Plan II

The railroad deals directly with to the shipper, and performs or supplies all services including providing the trailers and drayage at each end. The railroad bills the shipper at truck-competitive rates.

Plan II 1/4

This plan is the same as Plan II, except the railroad only handles the drayage at one end (origin or destination). The shipper or agent handles the other drayage.

Plan II 1/2

The railroad provides the trailers and performs only the loading, unloading and line-haul. Other terminal services and both drayages are handled by the shipper or his agent.

Plan III

The shipper uses his own or leased trailers and handles both drayages. The railroad loads/unloads the trailers and provides the line-haul for a flat charge per trailer or flatcar.

Plan IV

The shipper owns or leases both the trailers and the flatcars. The shipper loads, ties down and unloads his trailers. The railroad provides only line-haul transport, at a flat per-car rate, regardless of whether trailers are loaded or empty.

Plan V

Railroad and Motor Carrier offer through billing at joint rail-truck rates. Both rail and motor carrier trailers are carried. Either party may solicit traffic for the other, regardless of territory.

skill in selecting a competent agent, and should monitor his performance, service and cost.

The selection of railroads will be limited, although drayage may allow the shipper to economically access several competing carriers. In many cases, the shippers agent will choose the carrier and handle all direct transactions.

—Is Movement One-Time (or On-Going)?

Both one-time and continuig moves require rate quotes from the agent or carrier. However, the one-time or occasional move will usually hinge on cost, assuming that the expected service is adequate. The shipper will not want to devote much time to this relatively minor decision, and may prefer to stick with the present choice of carrier. A full-blown analysis of the other costs (inventory, loss & damage, etc) is unnecessary for these moves. Continuing moves, however, will require detailed analysis (using the spreadsheet), as they represent large blocks of traffic to major customers. These decisions have long-term strategic implications. Modal choice will affect not just rates, but also inventory, buffer, and L&D costs.

—Obtain Rates and Expected Transit Times: Is Expected Piggyback Transit Time Adequate?

This may be easier said than done. Rates for one-time moves can be furnished quickly by a shippers agent, often is one hour or less. Negotiating rates for long-term volume moves is a different, more involved situation. Factors such as balance (drayage and line-haul), commodity, volume/frequency, density of flow on each route, number of destinations, and minimum traffic guarantees or percentages will be important, as will levels of service required. As mentioned above, TOFC is less flexible than truck. However, the market is more stable, in that carrier routes are fixed, rates fluctuate less and multi-year contracts work better.

When negotiating rates, the shipper should be aware that an agent may purchase a block of train capacity (wholesale) for retail to various shippers. To the agent, the "marginal cost" of an extra trailer during slack periods will be less than his normal rates. Conversely, during busy seasons train space is at a premium, and rates may be higher.

A competent agent can accurately estimate expected transit time, but consistency of service must also be taken into account. While the carriers compile these statistics, the small shipper is at a disadvantage in finding them out. All service aspects are important, but transit time is more easily obtained and the most comparable. Thus, expected transit time will be used here as a proxy for service. If the TOFC shipment is not fash enough, there is no need to look further for on-time performance history.

—Compare Rates: Piggyback vs Truck

Here, the shipper can use the rate provided by his present carrier, or substitute other choices, such as owner operators or a private fleet. The decision for occasional moves relies mostly on rates alone. For long-term decisions, the other factors mentioned earlier come into play. The flowchart at this point refers the shipper to the spreadsheet; where these other costs are compared.

—Obtain Movement And Carrier Service Details.

Compared to transit time, these other inputs for the spreadsheet may be more difficult for shippers to obtain. Depending on the commodity's sensitivity to service quality and L&D, they will vary in importance. Overall statistics are available, but actual damage levels vary by route and commodity. Experience may prove to be the best predictor. The shipper must therefore recognize the spreadsheet's limitations.

II. DEVELOPING THE SPREADSHEET

Lotus 1-2-3 was selected for calculating the "numbers" aspects of the decision. Lotus enjoys widespread acceptance among small businesses, and is easily modified for individual situations.

The spreadsheet gives an estimate of the tangible and intangible costs of a shipping decision, focusing on those aspects of the shipment where costs differ between modes. Cost aspects that would be the same are not included. For example, since lot size (trailerload) is generally the same for both, this consideration is omitted. However, it might affect heavy shipments, since piggyback trailers are at a slight weight disadvantage to highway trailers.

Unfortunately, the spreadsheet does not indicate how well the modal choice fits into the firm's overall logistics strategy. The long-term decision must consider customer service requirements, in/outbound flow balance, warehousing, and must be compatible with the manufacturing operations. These strategic issues require the judgmental input of the marketing, operations, and traffic managers. Software packages such as Expert Choice can help evaluate each mode's effectiveness in meeting these concerns.

TL-PIGGY Went to Market . . .[3]

The selected variables include rates, inventory costs, loss & damage, and allowances for other related costs. An applied example of the spreadsheet is at Exhibit 3. This section helps the user select good values for the variables. The spreadsheet has two parts. The User Section takes rates (A) and other inputs. Parts B, C, and D of the Calculations Section apply the user inputs to formulas and display the results back in the User Section.

A. Direct Costs (Rates)

The truckload shipper usually pays one bill, either directly to the carrier, or to a shippers agent. Most small piggyback shippers will use shippers agents or associations, which do business on commission or on the difference between trainload rates (purchased from the railroad), and single trailer rates (sold to the shippers). Many shippers agents also offer the convenience of extra services such as a single invoice, shipment tracing, ad drayage management at each end. Some rail carriers such as Conrail deal primarily with shippers agents and other "volume" shippers. On the other hand, CSX Transportation uses its own sales force to contact the shippers, and can use its CMX truck line to move the trailers. In general, trailerload motor carriers also "retail" their service directly to the shipper.

Regardless of the carrier's marketing strategy, the user must obtain a rate quote. Rate inputs are not based on a general distance or modal cost formula. While this function could be added, many cost factors are not related to mileage, causing actual rates to vary significantly from actual rates. Shippers can often use modems or floppy disks to access their carrier's current rates. Shippers agents also provide

Exhibit 3:
T-L PIGGY: The Intermodal Decision Maker
See Part II of text for discussion and explanation of variables.

USER SECTION: Enter Variables Below "VVV"
**Means Do Not Enter A Value

	Motor Carrier Values VVV	Piggyback Values VVV
Enter externally-generated variables:		
A. QUOTED RATES:		
Either: Origin Drayage	(n/a)	0
Line Haul Rate	0	0
Destin Drayage	(n/a)	0
Or: Rate (From Carrier/Shippers Agent)	1175	1074
** Total Quoted Rates	1175 **	1074 **
TRANSIT TIME:		
Times From Carrier/Agent	2 days	4 days
SERVICE FACTORS:		
Maximum Expected Delay	1 days	3 days
Frequency Early/Late (.##)	5 %	7 %
Loss Frequency (.##)	5 %	3 %
Loss Percent (.##)	4 %	8 %
Enter company-generated variables:		
Average Shipment Value (Per	60000	
No. Trlrs Shipped Yearly Customer)	6 trlrs	
Extra Variable Costs (Dunnage, etc)	0	100
Company Discount Rate (.##)	0.12	
Inventory Holding Cost (.##)	0.25	
Retailer Markup (.##)	0.75	
Labor Overtime Premium ($/hr)	5.5 per hr	
Man Hrs to Offload Trailer	4	
Truck-Warehouse Space Factor:	2	
Warehouse Annual Cost ($/SqFt)	1.35	
Claim Delay (Filing to Receipt)	45 days	45 days
Clerical Hours Per Claim	3 hrs	3 hrs
Hourly Cost, Claims Clerk	12	
TOTALS: DIRECT COSTS (RATES):	$1,175	$1,074
INTANGIBLE COSTS:	172	504
TOTAL ESTIMATED VARIABLE COSTS:	$1,347	$1,578

CALCULATIONS SECTION:

B-1. INVENTORY COSTS (Resulting From Differences In Shipping Time)

Formula: No. Days Dock-to-Dock * Daily Holding Cost * Shipment Value

Inv Holding Cost 0.25
Avg Shipment Value 60000

Total Shipping Time Inventory Cost: 82.19 164.38

B-2. COST OF SHIPPING TIME VARIABILITY (Difference Between Modes)

— Cost of handling late shipments

Formula: # Trlrs Yrly * Freq Late Arrival * Premium * Man-Hours

	Motor Carrier	Piggyback
Overtime Premium	5.5	5.5
Man Hrs to Unload	4	4
Freq of Late Shipmts	0.05	0.07
# Trlrs Used Yearly	6	6
Avg Extra Labor Cost/Shipmt:	6.6	9.24

Exhibit 3: Cont'd

—Carrying Costs on Inventory Buffer

Formula: Avg Shipment Value * Daily Inv Holding Cost * Days Max Delay

Buffer = (# Trlrs Yrly/365) * Maximum Expected Delay

Buffer (calculated)	0.016 trlr	0.049 trlr
Buffer Cost, Per Shipment	41.10	123.29

—Extra Warehouse Cap'y (Opportunity Cost on Space, Investment)

Formula: Extra Space Required * Cost Per Sq Ft / No. trlrs Yrly
Extra Space Req = Trlrs In Buffer * 45′ * 8′ * Space Factor

Buffer = (# Trlrs Yrly / 365) * Maximum Expected Delay

Buffer (calculated)	0.016	trlr	0.049	trlr
Trk-Warehse Space Factor:	2		2	
Extra Space Required	11.8	Sq. Ft.	35.5	Sq. Ft.
Yrly Warehouse Cost	1.35	Per Sq. Ft.	1.35	Sq. Ft.
No. Trlrs Yrly	6		6	
Space Costs Per Shipment:	2.66		7.99	
Total Costs, Time Variation:	50.36		140.52	

C. LOSS & DAMAGE COSTS

	Motor Carrier		Piggyback	
Loss Frequency	0.02		0.03	
Percentage Lost	0.04		0.07	
Avg Shipment Value	60000		60000	
Average Loss	48		126	
File/Obtain Claim	36	Avg Cost	36	A.C.
Claim Processing	45	days	45	days
Holding Cost, Daily	0.000684	percent	0.000684	pct
Retailer Markup	75%		75%	

—Total L&D Costs, per shipment:

Formula: Admin Costs = Loss Freq * Hrs Per Claim * Clerical Wage Rate
Int Costs = Avg Loss * Days to Receive Claim * Holding Cost
Profit Impact = Retailer Markup * Average Loss

Administrative Costs	1.80	1.08
Interest Costs	1.48	3.88
Profit Impact	36.00	94.50
Total:	39.28	99.46

D. ADDITIONAL COSTS

	Motor Carrier	Piggyback
— Extra Dunnage Costs For TOFC	0	100

V. TOTAL: Average Per Shipment Variable Costs:

Direct Rates	1175.00	1074.00
In-Transit Inventory	82.19	164.38
Shipping Time Variation	50.36	140.52
L&D, Insurance	39.28	99.46
Other Costs (dunnage)	0.00	100.00
DIRECT COSTS (RATES):	1175.00	1074.00
INTANGIBLE & OTHER COSTS:	171.83	504.36
TOTAL COSTS:	1346.83	1578.36

Notes: 1. Direct Costs (Rates) and dunnage apply to one-time decisions. For longer term decisions, include all intangible costs. Long term capital investments are not considered.

2. The above costs are borne as follows:

	FOB Origin		FOB Destination	
	Shipper	Receiver	Shipper	Receiver
Direct Rates		*	*	
In-Transit Inventory		*		*
Shipping Time Variation		*		*
L&D	both		both	
Additional Costs	*		*	

current information on rates and service. Exhibit 4 gives some sample piggyback rates.

Quoted rates are affected by many non-mileage variables. While rate negotiation is a separate subject altogether, it plays a large part in the deregulated marketplace. To get the best possible rate for a given volume, the traffic manager must consider these variables, among others:

1. Traffic Direction. Carriers often experience different overall volumes on the two directions of a particular lane. For example, volumes to the Northeast are generally greater than volumes originating there, because the region is a net consumer. Thus, rates can be expected to be lower for Philadelphia originations that terminate, for example, in Florida, Southern Canada, or the West Coast. Rates also vary seasonally, for example in cycle with perishables traffic.

2. Related to direction is balance. If the individual shipper offers a backhaul of raw materials to balance his headhaul of finished goods, he can negotiate a lower rate. This applies to drayage and to linehaul costs. Further synergy may be possible if other nearby shippers participate in the load balancing.

3. Volume and consistency of traffic. Shippers who can guarantee steady business to the carrier will be in a better position to negotiate contracts, which should produce better service at a lower rate. Even if only major flows to primary customers are placed under contract, savings will result.

4. Relative sizes of shipper, receiver and carrier. Clearly, the party with the largest volume will be in a better position to negotiate a favorable rate. Thus, a small shipper may relinquish control over carrier selection to new or existing large customers. They in turn will then have a larger traffic block to leverage a lower rate from the carrier with, whether the carrier is big or small. This lowers the shipper's delivered price to his customer.

5. "Piggyback Plan" used. Five basic plans, plus variants, are employed by rail carriers to describe the different intermodal services available. While somewhat complicated, each has its cost and service characteristics, allowing the shipper to select a service that meets his needs. Most small shippers use Plan II ½ or a variant. This plan allows him to move loaded trailes to the terminal to guarantee that shipments make the train, while leaving the destination drayage to the agent or carrier. Using the carrier's trailers also frees the shipper from worries about empty backhauls or trailer ownership.

6. Liability for loss and damage. So-called released rates, with a flat rate per-trailer and a limit on declared value, will be lower than fully insured rates. However, this has considerable risks, especially for high value cargo subject to theft or damage.

B. Inventory Costs

Inventory costs can be separated into two components. First, there are carrying costs on the goods while they are in the shipping "pipeline" to the receiver. Second, the receiver must maintain an extra "buffer", to minimize the impact of transit time variation while the shipment is in the pipeline.

Significant differences exist between the service quality offered by rail and truck carriers, in terms of

Exhibit 4:
Sample Piggyback Rates

These rates are representative for typical medium distance piggyback moves. Rates are dollars per 40 or 45 foot trailer, regardless of contents or weight.

Plan II 1/2: Rail-owned trailers hauled ramp-to-ramp by carrier. Shipper or agent performs drayage at each end of move.

*NOTE: Drayage costs are not included in these rates. Thus, the total cost for a move from Philadelphia to New Orleans would be about $1,074 because of the drayage to Potomac Yard. (See Exhibit 3).

Origin Ramp: Alexandria, Virginia (CSX)

Destination:	One Trailer	Two Trailers
Atlanta	502	445
Augusta	436	386
Birmingham	650	577
Charleston SC	440	390
Charlotte NC	326	289
Chattanooga	533	472
Ft. Lauderdale	921	817
Miami	926	822
New Orleans*	896	795
Raleigh	327	290
Tampa	809	718
Wilmington, NC	392	348

Source: Seaboard System Intermodal Pricing Directory #2, published June 8, 1984; effective July 1, 1984.

transit time and variability therein. For this reason, pure boxcar service has lost market share in the last 40 years, and is not even considered as an option for our typical small shipper. However, while truck generally outperforms TOFC, the difference is smaller, and in some longer markets piggyback has the edge.

1. Inventory Costs Due To Different Transit Times. Motor carriers usually offer lower transit times, especially for short and medium distances. The indifference point varies, depending on the relative quality of the carriers' routes. The truck industry's average over-the-road speed is about 40 mph, compared to the piggyback average speed of perhaps 30 or 35 mph. Exhibit 5 shows some typical intermodal services from Philadelphia.

Piggyback service is primarily slower than truck because of terminal delays and drayage. An all-truckload move travels directly from dock-to-dock, with a minimum of delay as drivers and tractors are switched. The piggyback trailer must be taken (sometimes 50 or more miles) to the intermodal terminal, which may be congested or located inconveniently with respect to the highways. At the yard, trailers must be checked in and be loaded one at a

Exhibit 5:
Typical Piggyback Services From Philadelphia

Origin City, Rail-road		Destin	Arrive Time	Hours	Approx Miles	Avg MPH
		To Southeast				
Phila, CSX		Richmond	1015	15.3	245	16
	Lv 1700	Raleigh	1700	24	450	19
		Charlotte	0100	32	515	16
		Atlanta	0800	39	770	20
		Birmingham	0130	33.5	940	28
		Jax, Fla	0700	38	910	24
		Miami	1400	45	1310	29
		Tampa	1800	49	1150	23
Alexandria VA, NS		Atlanta	0025	16.7	640	38
	Lv 0745	Birmingham	0515	21.5	800	37
	*	New Orleans	1600	32.3	1160	36
		To Midwest				
Phila, CR		Toledo	0120	21.8	580	27
	Lv 0600	Chicago	0545	24.8	830	33
		CNW Connection		42.8	840	
	Lv 0810	Indianapolis	0850	24.5	730	30
		St. Louis	1235	29.4	960	33
		SP Connection		34.9	960	
To Southwest/West Coast			Arrive Time	Overall Hrs	Miles	Avg Speed
St. Louis, SP		Kansas City	0310	44.1	1240	28
	Lv 1800	Dallas	2400	65.9	1670	25
		Houston	0300	68.9	1780	26
		Los Angeles	2400	110.9	3000	27
Chicago, CNW/UP		Denver	2200	65.9	1880	29
	Lv 2350	Los Angeles	2340	91.9	3155	34
		Oakland	2305	92.1	3275	36
		Seattle	1130	104.4	3530	34

Sources: Official Railway Guide, Nov-Dec 1986; Conrail schedules. Rand McNally Rail Mileage Charts.

Notes:

1. *Rail transit time to New Orleans is 32 hours. Because of rail loading delays and drayage time, dock-to-dock time is estimated by the shipper's agent at 4 days. (See example at Exhibit 3).
2. Trailers must arrive at terminal at least 3 hours before departure. Trailers are generally available for pickup 4 hours after arrival.
3. Schedules are for demonstration of typical piggyback performance and general information only.

Exhibit 5, Cont'd

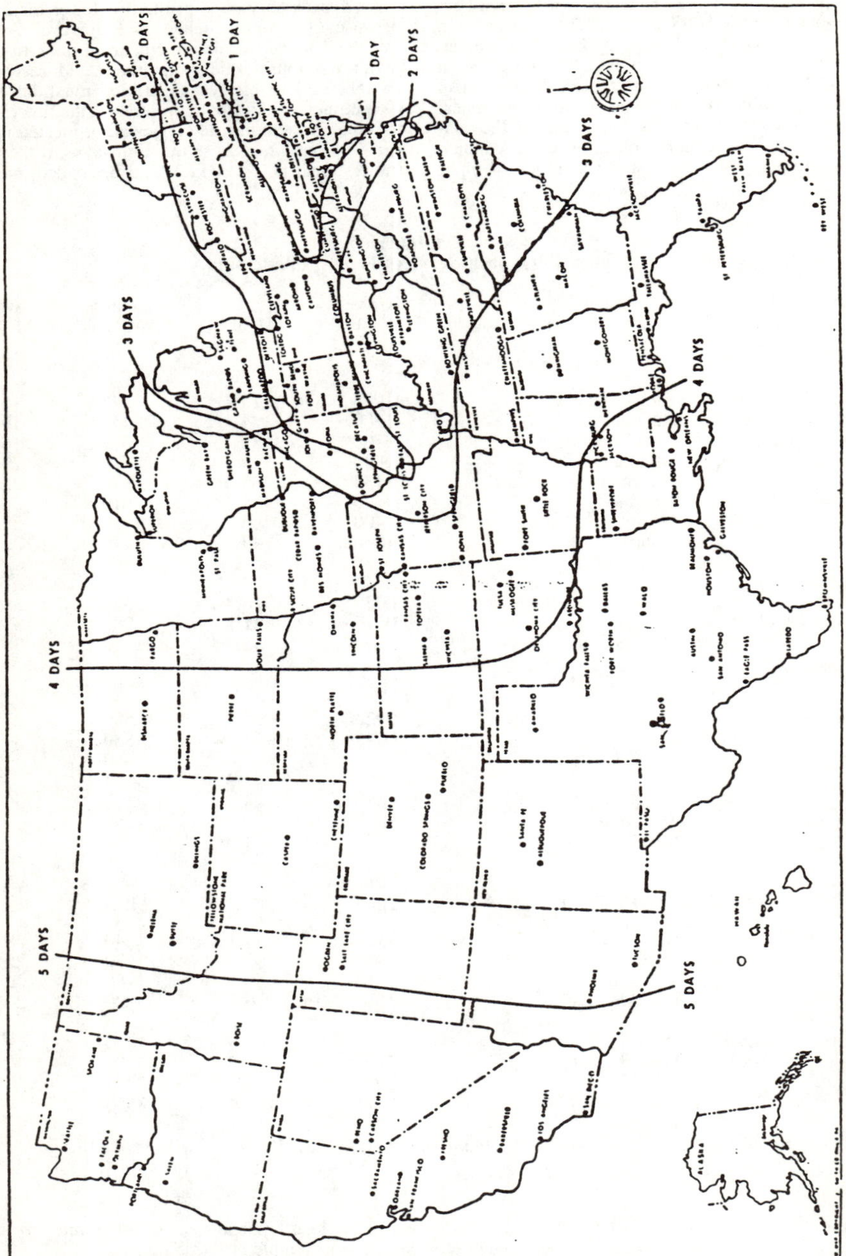

time onto the train. Thus, rail carriers require that trailers arrive 2 or 3 hours before departure time; similar time lags are found at the other end. Shippers must consider travel time to and waiting time at the terminal. However, waiting for drayage is not included in the spreadsheet, as the shipper would also wait for the motor carrier to spot and pick up his trailer.

The conclusion is that the shipper must analyze rail service on a route-by-route basis before estimating transit times, while truck transit times can be estimtaed fairly accurately using average speeds. Exhibit 6 shows typical truck times that can be expected to various parts of the country, from Philadelphia. A good shippers agent is also familiar with current data, and can give accurate estimates.

Once the total transit time has been estimated in days, it is a simple matter to calculate the opportunity cost of the shipment's value while in transit, based on the company's inventory holding cost. It is important to note the difference between the company's cost of capital and the (higher) holding cost of inventory. Inventory may be perishable, lose value if late, or be subject to loss, damage, and theft. The shipper should apply a higher cost to holding inventory than the overall company cost of capital.

2. Inventory Costs Due To Transit Time Variation. These costs may be more significant in the shipping decision. While the receiver may accept a moderate lag between ordering and receiving his goods, a well-run organization cannot tolerate shipments that usually show up on Wednesday, but sometimes on Thursday or even on Saturday. Much has been written on the growing importance of "just-in-time" delivery. Just as rail piggyback generally takes longer than truck, the time enroute is also more variable. Delays are likely to occur at each intermediate handling, and TOFC simply has more handlings: drayage, loading on railcars, train assembly, switching the cars to other trains, and interchanging with connecting carriers. In addition, the trains are less flexible in moving past other trains or delays, compared to trucks. Rail carriers have worked to minimize the impact of these delays. Considerable slack is built into schedules, so that trailers are usually delivered or ready for pickup at the advertised time, regardless of when the train actually arrived. Truckload motor carriers, on the other hand, are able to achieve a tighter delivery window because delays are less likely to occur on a typical single-line shipment. Nevertheless, small shippers (especially) must take sales promises, advertisements and published performance statistics with a grain of salt. Various surveys have been published to compare shipper satisfaction with the various modes.[5]

These costs fall upon the receiver, regardless of who makes the shipping decision. The shipper must take these costs into account, as they affect the true delivered cost to his customers. This cost is significant because using TOFC on a regular basis means the buffer must be carried year-round, even if the stock itself is rotated. By using modal choice to provide better delivery, the producer may be able to expand his market area, or at least protect it from incursions by competitors. If the piggyback option costs less even after these receiver-borne costs are accounted for, then it merits serious consideration.

Developing specific spreadsheet inputs may be difficult. The objective is determining a buffer size that ensures no stockouts or production stoppages occur due to delayed shipments. (Some firms may tolerate those stockouts, however). The buffer size is determined by the quantity of goods used and the maximum expected delay for the carrier or mode. Of course, piggyback delivery times are more variable and have the largest expected delay. While the receiver's actual buffer may differ from the calculated buffer, the spreadsheet will estimate the incremental buffer that a piggyback choice would imply.

3. Two other costs must be considered. One is the extra space required for the buffer (either the warehouse space could be smaller or a greater variety of goods could be stocked in the given space). Another is the extra wages needed to unload and store late or bunched up shipments.

The shipper must estimate how many square feet of warehouse are needed to store a square foot of trailer capacity (the "truck-warehouse factor). Also needed is the annual cost per square foot of warehouse space. The spreadsheet then calculates the buffer cost based on the number of trailers shipped yearly to the receiver.

Labor overtime costs may seem fictitious, because the trailer could always wait to be unloaded. However, assuming that trailers must be unloaded quickly to minimize demurrage and get the inventory flowing, extra labor will be needed to deal with a congested dock situation or to get trailers unloaded after hours. This is one of the costs of an uneven supply operation. In fact, if the emptied inbound trailers are being reloaded outbound, these problems and costs are at least doubled. Therefore, shipper estimates of how often the trailers will arrive early, late or after hours are important, albeit best obtained from experience.

C. Costs of Insurance and Loss & Damage

Loss & damage histories depend on the mode's vulnerability to accidents, thefts or equipment failures. The spread between TOFC and truck is less than that between conventional boxcar and truck. The shipper must enter estimates of variables for which specific values may be hard to verify. Three methods of obtaining them are possible.

For a particular motor carrier, the shipper presumably already has historical data indicating the frequency and severity of losses. This suggests experience as one source of data.

Second, shippers agents will have information on expected losses, because they handle the actual claim filing and settlement with the carriers.

A third method of estimating would be using the AAR's average Payout/Revenue ratio for a particular commodity or for TOFC (FAK). The shipper could use that ratio to calculate estimated damage as a percent of the rate paid. AAR statistics are broken out by commodity and cause. Overall payout to revenue ratios declined from 1.97% in 1970 to .75% in 1981. The shipper could adjust this by estimating his cargo's vulnerability to the damage causes as listed in the AAR statistics.[6]

One assumption is that shipments will be fully insured, and that reimbursements will cover the value of the goods lost. However, other costs NOT covered by insurance must be considered. These include the cost of filing the insurance claim, the opportunity cost of the claim's value between loss and reimbursement, and the profit impact.

Exhibit 6

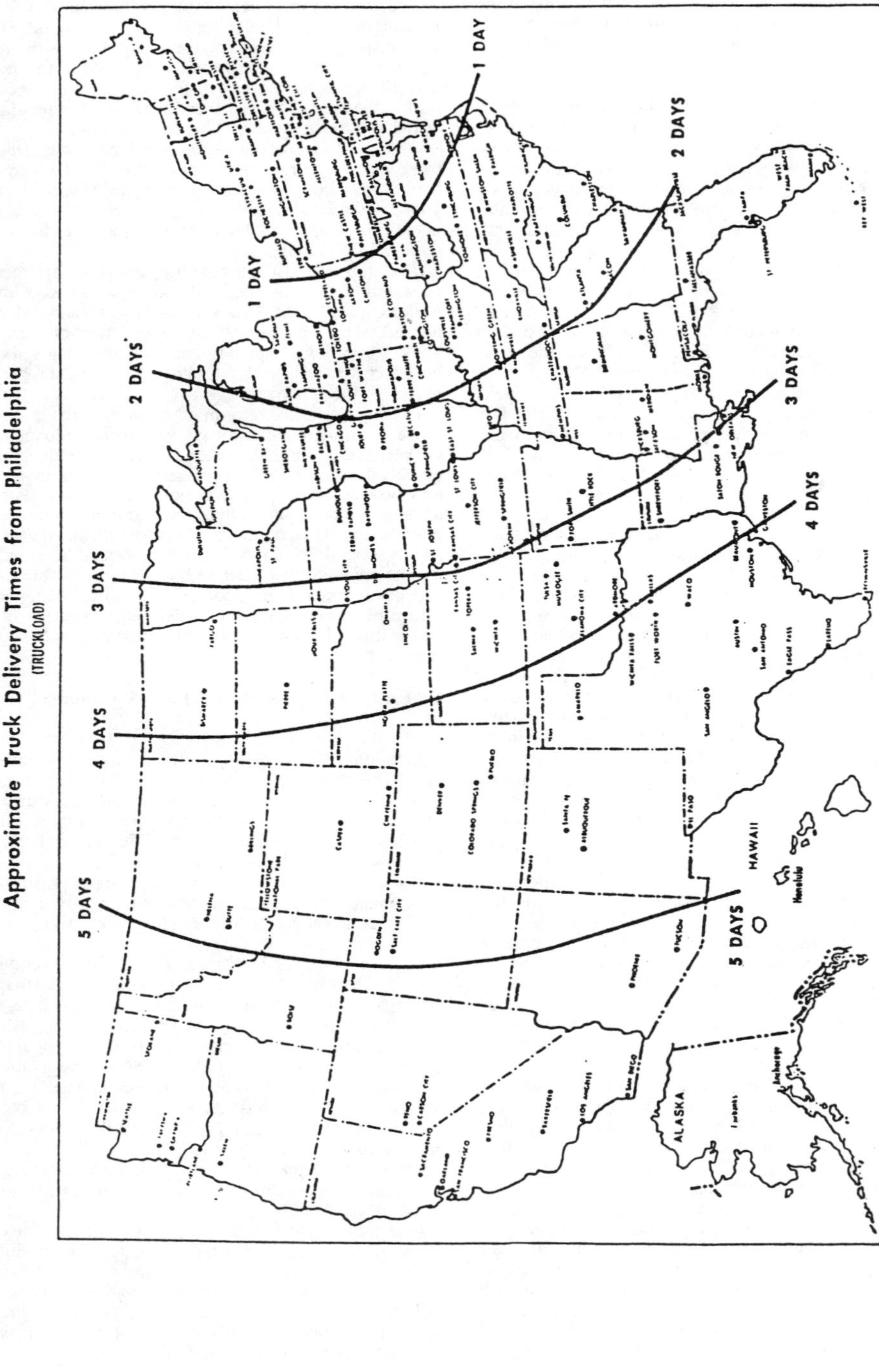

Source: 1977 Delaware Valley Guide to Transportation

A problem in estimating these non-reimbursed costs is the frequency that they occur. Because the spreadsheet estimates costs on a per-shipment basis, the shipper must be able to input loss frequency and the average amount of loss per incident, for each mode. L&D may vary by carrier, by region and by commodity, and the loss may be partial or total. Piggyback L&D will differ from the rail industry in general. Applying percentages to an average shipment will be an approximation, but it must be done for long-term decisions.

Several other factors may or may not cost the receiver money. If the lost shipment caused a stock-out and subsequent loss of sales, there will be a negative profit impact on the receiver. An extra order may be necessary to replace the goods, possibly with an expedited (more expensive) shipment. Finally, inconveniences caused by the loss and damage will have an adverse effect on customer goodwill, both at the wholesale and retail level. Not all of these costs will happen every time a shipment experiences loss or damage. However, the model takes the liberty of estimating an average loss. The foregone retail markup on the lost/damaged goods is used on the basis that it follows the normal cost of doing business, plus a fair return.

D. Additional Costs

The shipper will incur extra costs when changing from one mode to another. In the case of piggyback, the costs are generally small. The shipping container (trailer) and materials handling equipment are the same. Loading facilities do not need modification as they would for boxcar.

Some potential costs should be monitored, however. Trailers handled by piggyback experience all the forces of a normal highway move, and are subjected to higher shock and rocking forces. Some fragile commodities require extra dunnage, especially if the trailer is not "cubed out" and has room for the load to shift. The model ignores the one-time costs that would be incurred to train personnel to load trailers differently.

Piggyback trailers are built more sturdily to withstand the forces of being lifted on/off the railcar, and to guard against internal damage from shifting loads. However, rail carriers supply these trailers to shippers who use Plan II or its variants; thus this cost may be ignored. Indeed, switching to piggyback would allow a shipper who owns or leases his trailers to reduce the size of this fleet. On the other hand, a shipper desiring to use his own trailers for piggyback would face a cost hurdle if they were not suitable for TOFC service. These savings or costs would be part of a long-term modal choice and capital investment decision, and to be weighed against reductions in per-shipment transportation costs.

III. APPLICATION AND CONCLUSIONS

The model was used to analyze a typical shipment from Philadelphia to New Orleans. Specific inputs were developed from available data and from the company, as shown in Exhibit 3. The total cost (rates plus intangibles) was then calculated using the spreadsheet formuale. While TOFC does offer a lower rate, the added inventory, L&D, and dunnage requirements raise the true cost above that of truck.

Two conclusions can be drawn about TOFC in general, and another four about shipping decisions in particular. First, piggyback service is not a generic commodity, but must be analyzed on a by-corridor basis. For example, southbound traffic suffers a lengthy delay if ramped in Philadelphia, as shown in Exhibit 5. Competitive service to the South requires a long drayage to Alexandria, which eats up much of TOFC's long-haul cost advantage. Westbound shipping times are better, as rail facilities are less congested. Conrail may also be more concerned about making its long-haul western connections than its southern ones.

Second, shippers agents play a key role in the small shipper's logistics strategy. The agent is the shipper's link to the carrier and to the drayman, providing extra "value-added" services incidental to the move. Information asymmetries in the marketplace make "best choices" difficult. Such things as schedules, transit times, and service quality are not readily at hand for comparison. Here, the shippers agent serves as a source of marketplace information and power, counterbalancing those asymmetries.

With respect to the shipping decision, the first conclusion is that the receiver is often the decision-maker, not the shipper. In the liqueur industry, distillers, distributors and retailers are more concerned about meeting their respective sales orders than in saving a few dollars (1-2 cents per bottle) on shipping costs.

Second, the marketplace offers a wide variety of different quality services. The shipper will limit his search once he has found a satisfactory arranagement. This leads to a strong tendency against switching carriers or modes "experimentally". In other words, don't fix something that's not broken.

Third, while the transportation rate is the biggest part of the tangible cost, other costs such as blocking and bracing can alter the balance. For shipment-by-shipment decisions, only direct costs and adequacy of service need be considered. For long-term decisions over specific routes, intangibles such as inventory cost and loss & damage must be considered. Both cost types are included in TL-PIGGY, but the shipper should be aware of the distinction and also which party bears the difference costs. In addition, strategic decisions to use piggyback will affect the size of capital investment in a shipper's private fleet.

Fourth, the shipper may have the option to limit declared value and use released rates. While these rates are lower, the savings must be weighted against the substantial risk of absorbing L&D costs. Insurance should be considered part of the basic rate, not something added on. This is even more true for high-value commodities that are vulnerable to theft and damage enroute.

ACKNOWLEDGEMENTS

The assistance of Professors W. Bruce Allen and Paul R. Kleindorfer in developing the flowchart and spreadsheet is gratefully acknowledged. Mr. Joseph Toporczyk of Jacquin's and Mr. Jeffery Saunders of M&M Mars are thanked for their time and effort in critiquing the results of this study.

GENERAL REFERENCES

W. Bruce Allen, Mohamed M. Mahmoud and Douglas McNeil, "The Importance of Time in Transit and Reliability of Transit Time for Shippers, Receivers, and Carriers," Transportation Research, Volume 19B, 447-456, 1985.

Richard D. Armstrong, "The Relative Importance of Linehaul and Terminal Costs for Individual Motor Carrier Shipments," Transportation Research Forum, 1985, p. 459-461.

Booz, Allen and Hamilton, "Piggyback, The Efficient Alternative for the 80's," (New York, Transamerica Interway, March 1980).

Charles River Associates: "Competition Between Rail and Truck in Intercity Freight Transportation" (Washington DC: US Department of Transportation, Contract DOT-OS-A9-060, December 1969).

Garland Chow, "Microcomputers in Logistics," Logistics and Transportation Review (Vol 20/ Number 4): 471-501.

Council of Logistics Management, "Official Railway Guide," November/December 1986.

Council of Logistics Management, "Official Intermodal Guide," Fall/Winter 1986.

John H. Mahoney, "Intermodal Freight Transportation" (Westport, CT: Eno Foundation for Transportation 1985).

Philadelphia Chamber of Commerce, "Business Firms Directory of the Delaware Valley", 1987 Edition.

Philadelphia Chamber of Commerce, "Delaware Valley Guide to Transportation," 1977 Edition.

US Department of Commerce: "Census of Transportation" (Washington DC: 1977 & 1982).

ENDNOTES

*The Wharton School of Business.

1. Student Association of American Railroads, "Railroads Facts 1986", page 26.
2. Arthur Andersen & Company, "Survey of Software for Physical Distribution—1986".
3. The decision model's nickname, TL-PIGGY, is derived from TrailerLoad and Piggyback, and from the old rhyme where This Little Pig went to market, this one went to town, etc.
4. Consolidated Rail Corporation Boxcar/Intermodal Department List of Qualified Volume Shippers, dated August 8, 1986.
5. For example, see the "Summary and Highlights of DOT's Industrial Shipper Survey (Plant Level)", December 1975, published by Lana Batts of the American Trucking Association.
6. Gus Welty, "Loss & Damage Prevention, There's Cause For Celebration," Railway Age, May 31 1982, pp 17-19.

PROCEEDINGS—
TWENTY-SECOND
ANNUAL MEETING

Canadian
Transportation Research Forum

St. John's, Newfoundland

June 1987

An Alternative Mechanism For Payment of the Crow Subsidy: The Western Grain Stabilization Programme

*by Paul D. Earl and Edward W. Tyrchniewicz**

I. INTRODUCTION

The case for subsidizing freight rates for Western grain and grain products is a complex one, resting more compellingly on historic factors than on economic ones. These historic factors go back to before the turn of the century when farmers in the West felt themselves discriminated against by the railways, and most notably, the CPR. This discrimination was alleged to be of several varieties: rates in Western Canada which were higher than for similar distances in the East; the so-called "monopoly clause" which protected the CPR from the construction of competing lines; preferential treatment to elevators over loading platforms in the allocation of boxcars; and finally, the abiding (albeit arguable) belief of Western farmers that they are the only shippers who are compelled "to pay the freight both ways": on the grain moving out, and on the manufactured goods moving in. These grievances of course do not constitute grounds for subsidization of freight rates; but they were instrumental in creating the Crow's Nest Pass Agreement which is the institutional and legal starting point for the situation we are now in.

The Crow's Nest Pass Rates have developed a rich and complex web of myth and misinformation since their inception in the Agreement. But myth (leaving aside the question of misinformation) is the stuff from which political realities are spun. It is widely believed by Western grain producers that the freight rate provisions of the *Western Grain Transportation Act* protect them from burdensome transportation charges. This belief that the producer benefits from subsidized rates has received acceptance outside the agricultural community. The Turgeon Royal Commission of 1949 spoke of three great pillars of national transportation policy: the Maritime Freight Rates Act for the East, water competition for the Central region, and the Crow Rates for the West. Of course, economists can show that whatever benefit the rates might have provided is now capitalized into land values. But the historic and hence political justification for grain transportation subsidies means that they are here to stay, and that being so, the debate in recent times has turned on the issue of how they ought best be paid. As the focus of this paper is the question of the method of payment, we will first review the current state of this debate.

II. THE METHOD OF PAYMENT DEBATE

Two competing methods occupy the field: firstly, that favored by the three Prairie Wheat Pools and the Provinces of Quebec and Manitoba and now incorporated into the WGTA: to pay the subsidy to the railways; secondly, that favored by United Grain Growers, livestock interests, the Western Wheat Growers Association, various commodity groups, the Province of Alberta and in the reports from three major investigations into the matter (Gilson and the two reviews carried out under Section 62 of the *Western Grain Transportation Act*): to pay the subsidy to farmers. The pros and cons of these two competing methods should be understood because they form the evaluative basis for the third way to be explored in this paper. *The Report of the Commitee of Inquiry on the Crow Benefit Payment* contains a thorough summary of the matter (pp. 7-9), of which the following is a precis.

The advantages of the "pay the railways" approach are:

a) it is administratively simple;
b) the benefit accrues to the actual shippers of grains;
c) when combined with the inflation protection and "safely net" provisions of the *WGTA*, it moderates freight rate increases; and,
d) it provides direct accountability to the government for railway performance.

The disadvantages are:

a) it perpetuates incentives for the production of statutory grains, rather than encouraging other, possibly better, uses of prairie agricultural resources;
b) it perpetuates disadvantages against western livestock production and processing (this is the so-called "livestock hurt" issue);
c) it creates a competitive disadvantage for modes of transportation other than rail; and
d) it masks real transportation costs to the producer, and reduces incentives for efficiency.

The "pay the producer" approach has the following advantages:

a) it would be "resource neutral";
b) it would remove farm gate grain price distortions discouraging livestock production and processing;
c) it would make full transportation costs visible to producers, thus creating market accountability and incentives for efficiency in the transportation system.

Paying producers on an acreage basis would have the following disadvantages:

a) it would dilute the benefit to actual shippers because the payments would be spread over all land capable of producing statutory grain;
b) it would not recognize individual farm management practices;

c) it could work to the disadvantage of agriculture in other parts of Canada by providing payments, which would be available only in the West, on crops other than grain;
d) it was seen as administratively difficult; and
e) it would be difficult for producers to understand.

The Committee was careful to ascribe the alleged disadvantages of paying the subsidy to the producer specifically to an acreage based system. However, the disadvantages which they identify would, to a greater or lesser degree, apply to any "pay the producer" scheme.

The Committee's criteria for evaluating different method of payment proposals were as follows:

a) conduciveness to agricultural development in Canada;
b) resource neutrality;
c) provision of equity;
d) encouragement of efficiency both in grain handling and transportation and in allocation of agricultural resources in general; and
e) administrative feasibility.

Criterion a) is contained in the *Western Grain Transportation Act* (WGTA); criteria b) through d) were provided to the Committee by the then Minister of Transport, Don Mazankowski.

The Grain Transportation Agency (GTA) was charged with conducting the more general review of the WGTA provided for under Section 62 of the Act. This required a further examination of the method of payment issue. Section 62 required that specific attention be paid to the inflation sharing provisions of the *Act*, having regard to the relationship between the price of grain and the producers' transportation costs. The Minister emphasized the importance of this section, suggesting that the producers' "ability to pay" was to be considered the most important criterion for recommendations relating to the method of payment.

Both the Committee of Inquiry and the GTA recommended variations on the "pay the producer" option. The Committee's recommendation was for a Grain Transportation Refund to be paid to producers on the basis of their net sales of eligible grain. The Agency's recommendation was for a one time "payout" of the net present value of the government commitment direct to producers.

Neither of these two systems has proven acceptable to the federal government. Although the government has never made clear precisely what its objection to either system is, no doubt the continued opposition of the Prairie Wheat Pools and the Province of Quebec have been important factors. Whether the proposal put forward in this paper would fare any better remains to be seen. It does, however, have some inherent advantages over these two systems.

Let us first look at the conceptual framework for what is proposed.

III. THE CONCEPTUAL FRAMEWORK

A. Definition of Terms

It is necessary to have a clear understanding of some of the terms that will be used in this paper. These are as follows:

Eligible cost is defined in the WGTA and means the total cost in dollars of moving statutory grain by rail in Western Canada. The *average eligible cost per tonne* is this total divided by the total tonnage moved. In this paper, when the term *full freight rate* is used, it may be taken for practical purposes to be the same as the average eligible cost per tonne. In 1986/87, this was approximatley equal to \$31. On some 30 million tonnes, the total eligible cost is close to \$1.0 billion.

The *producer share* of the freight rate is that portion of the full freight rate (or the average eligible cost per tonne) paid by the grain farmer. In 1986/87, this was approximately equal to \$6 per tonne, or about 19% of the total. The *government share* is the remainder, about 81% of the total, or \$25 per tonne.

The *Crow Benefit* is a term defined in the WGTA, and is approximately equal to the railway losses on grain in 1981/82. It is a fixed dollar amount established in the WGTA at \$658.6 million.

The WGTA provides for increases in the cost of moving grain by rail to be shared between the producer and the government. The term *cumulative share of the cost increase* is defined in the WGTA and is the government's accumulated portion of these cost increases. It is a per tonne figure, and its current value is approximately \$2.30 per tonne.

The *government commitment* is the total amount of the subsidy which the government will provide for grain transportation in any one year. It is equal to the sum of the Crow Benefit plus the cumulative share of the cost increase multiplied by the total tonnes which move in that year. For example, if 34 million tonnes of grain move by rail to the designated destinations in the crop year 1986/87, the government commitment will be approximately:

$$658.6 + (2.30) \times (34) = \$736.8 \text{ million} \quad (1)$$

The *Freight rate multiplier* is approximately the factor by which the old Crow Rate (or what is called in the *Act* the base rate scale) must be multiplied in order to calculate the full freight rate. If the average level of the old Crow Rate is taken to be \$4.89 per tonne, then the multiplier for 1986/87 is approximately \$31 divided by \$4.89, or 6.3.

It will be important to understand these terms clearly when we come to describing how the subsidy might be administered through the Western Grain Stabilization Programme.

B. The Proposal

The proposal is as follows:

1. Grain producers would pay the *full freight rate* as calculated under the WGTA. This would include both the producer share and the government share of the rate.
2. The federal government would continue to calculate the *government commitment* to Western grain transportation in the same way that it now does.
3. Each year, the *government commitment* would be placed into the Western Grain Stabilization Fund, and would be used for stabilization payments as required under the *Western Grain Stabilization Act*. (The Fund, and the operation of the Stabilization Programme are described below).

Since this would represent an enrichment of the Stabilization Fund, some changes would have to be made to the payout provisions of the Stabilization Programme for producers to receive the full benefit. In order to describe these changes, it is first necessary to understand how the Programme works, and how the operation of the Fund would be altered by a change in the freight rate.

C. The Operation of the Stabilization Programme

Table 1 illustrates, using a simplified model with hypothetical numbers, how the Stabilization Programme works. Basically, it is intended to smooth out the grain producers' net cash flow (NCF). NCF is approximately equal to producers' gross returns from grain sales less the costs of production. The programme is designed to ensure that NCF in any one year does not fall below the average NCF in the preceding five years. When the payments are triggered out of the Stabilization Fund to augment farmers' incomes. This is illustrated in the first five columns of Table 1. In years 1 to 3, and in years 9 to 11, NCF exceeded the preceding five year average. Farmers would not receive a stabilization payment in these years. However, in the other years (4 to 8, and 12) farmers would receive payments equal to the deficits shown in Column (5)). (It should be noted that in actual practice the farmers' income is not fully stabilized as this simple model indicates, and accordingly the payments would not cover the full deficit.)

This Stabilization Fund is sustained from two sources: a check-off from producers' gross receipts, and a government payment approximately equal to double the producers' check-off. To be actuarially sound, the amounts paid into the Fund must be equal to the amounts paid out. In the example shown in Table 1, a total of $6,960 million was paid out of the Fund over the twelve years. If this were representative, approximately $580 million would have had to be paid into the Fund each year, approximately $190 million from producers, and $390 million from government. (In fact, the check-off is approximately 2% of producers' gross receipts, and the government levy is 2% over the farmers', or about 4%. These levels were chosen after extensive analysis as those which would make the Fund actuarially sound.)

The only other thing to note at this point is that over the long term, assuming there is no long term trend up or down in NCF, the sum of the surpluses (Column (4)) must equal the sum of the deficits (Column (5)).

Now let us look at the impact of the producer paying higher freight rates.

D. The Impact of the Farmer Paying Higher Rates

It is assumed in Table 1 that the increase in producer freight rates is an even $1000 million annually (Column (6)). The producers' "new" net cash flow is shown in Column (7) as $1000 million less than Column (2). Now it is clear that, in this simple example, a corresponding "new" five year average column (not shown) would also be $1000 million less than Column (3) and the surpluses and deficits would be equal to those shown in Columns 4 and 5. Hence increasing the freight rate to the farmer (ignoring the phasing in of the increase) would have little impact on the Stabilization Programme.

E. Impact of Paying the Government Commitment into the Stabilization Fund

It was noted above that the Stabilization Fund is considered to be actuarially sound under the present arrangements. Thus the payments out of the Fund are sustained by the payments in. Therefore, if the payments into the Fund were increased by the amount of the government commitment, unless there were a corresponding increase in payments out, the Fund would grow over time at an average annual rate equal to the amount of the government commitment. Accordingly, if the producres are to be paid what is owing to them, the payout provisions of the Stabilization Programme would have to be made more generous.

How then ought the payout provisions to be altered so that the producers receive the full amount of the freight rate subsidy? At first glance, it might be supposed that this could be accomplished by stabilizing producers' income at the "old" five year average: i.e. by bringing producers' incomes up to the level of the "new" five year average plus the government commitment. This is illustrated in Columns 8 and 9 of Table 1. These columns represent the difference between the "new" NCF (Column (7)) and the old five year average (Column (3)). The deficits are more frequent and larger, and the payout from the Fund are greater.

Columns 5 and 6 show what is owed to the producers. The total to be paid out over the twelve years is $6960 million in stabilization monies, plus $12,000 million in government commitment. However, the sum paid out in Column (9) is only $14,560 million. The reason for the difference can be seen by comparing year 1 with year 4. In year 4, the producers are owed the government commitment of $1000 million plus the stabilization and that is indeed what they get. But in year 1, although they are owed $1000 million, they only get $400 million. Thus it can be seen that in years such as year 1, the payout is reduced by the amount of the surplus.

Hence, in order to pay out to producers the full amount owing to them, their NCF would have to be stabilized at a slightly higher level than the "old" five year average. How much higher? By an amount approximately equal to the sum of the surpluses. However we noted two things earlier in this paper: first, that the sum of the surpluses is equal to the sum of the deficits; second, that the sum of the deficits is equal to the sum of the levies. Accordingly, if producers' NCF were to be stabilized at the level approximately equal to:

a) the five year average NCF as it would be under subsidized rates;
b) *plus* the amount of the levies (which in the simple example illustrated here was calculated to be about $590 million annually).

Then the producers would receive the full amounts due to them under the two programmes. This is illustrated in Columns 10 to 12 of Table 1.

This then is the conceptual framework for paying the grain transportation subsidy—legally known as

TABLE 1

Illustrating Stabilization Programme and Proposal Using Hypothetical Figures

(1) Year	(2) NCF	(3) 5 Yr. Ava.	(4) Difference Surplus	(5) Difference Deficit	(6) Inc. in Freight	(7) "NEW" NCF	(8) Difference Surplus	(9) Difference Deficit	(10) "Stab." Level	(11) Difference Surplus	(12) Difference Deficit
	(					$ millions					)
1	3,800	3,200	600	—	1,000	2,800	—	400	3,780	—	990
2	3,800	3,240	560	—	1,000	2,800	—	440	3,820	—	1,020
3	5,400	3,360	2,040	—	1,000	4,400	1,040	—	3,940	460	1,020
4	3,400	3,800	—	400	1,000	2,400	—	1,400	4,380	—	1,980
5	3,600	3,840	—	240	1,000	2,600	—	1,240	4,420	—	1,820
6	1,800	4,000	—	2,200	1,000	800	—	3,200	4,580	—	3,780
7	2,000	3,600	—	1,600	1,000	1,000	—	2,600	4,180	—	3,180
8	1,000	3,240	—	2,240	1,000	—	—	3,240	3,820	—	3,820
9	2,600	2,360	240	—	1,000	1,600	—	760	2,940	—	1,340
10	3,600	2,200	1,400	—	1,000	2,600	400	—	2,780	—	180
11	3,200	2,200	1,000	—	1,000	2,200	—	—	2,780	—	580
12	2,200	2,480	—	280	1,000	1,200	—	1,280	3,060	—	1,860
			5,840	6,960	12,000		1,440	14,560		460	26,550

the government commitment—through the Stabilization Programme. Let us now turn to the practicalities of implementing it.

IV. THE PRACTICAL ASPECTS OF USING THE STABILIZATION PROGRAMME

A. In Respect of the WGTA

The proposal seems to present no problems from the point of view of the *Western Grain Transportation Act*. The calculations which are now done to determine the cost of moving grain, the cost increases, the apportionment of cost increases between the producer and the government, could all be done as they now are. The differences in procedure would arise only in the last few steps. The Freight Rate Multiplier would be calculated as it now is, but the farmer would pay the full freight rate thus calculated instead of only a portion. The government commitment would be paid into the Stabilization Fund instead of to the railways. (This is not quite true. The CN adjustment portion of the government commitment would presumably be paid directly to CN as it now is. The remainder of the government commitment would then go into the Fund).

The WGTA rate setting process would in fact be simplified somewhat. The most troublesome aspect of the present rate setting procedure lies in the need for a forecast to be made of the volume to be moved in the upcoming year. Since the Crow Benefit portion of the government commitment is a fixed dollar sum, its value per tonne varies with volume, and since the volume cannot be forecast with precision, the government portion of the rate can never be set correctly before the beginning of the crop year. The result is that the government always either overpays or underpays its commitment and some provision must be made for subsequent correction. Under this proposal however, there would be no need to determine the government commitment prior to the beginning of the crop year. It could be done after the end of the year when the volume is known and it is time to calculate the payout under Stabilization.

B. In Respect of the Stabilization Programme

From the point of view of the mechanics of the Stabilization Programme, there would appear to be no difficulty. The Western Grain Stabilization Administration would have to calculate the new level to which farmers' NCF would be stabilized. As analyzed above, this new level would be equal to actual NCF plus the average of the previous five years of government commitments plus a further augmentation approximately equal to the levies or premiums paid into the Stabilization Fund. This would be a straightforward calculation.

There are however four minor and two major difficulties. The minor ones are as follows:

First, some phase-in provisions would have to be worked out. In the simple model illustrated in Table 1, NCF under subsidized rates was compared with NCF under compensatory rates. But in fact, there would be a transition procedure for moving from Column (2) to Column (7) of Table 1. This however should not pose a major difficulty, and would be a one-time calculation. It should be noted that the Stabilization Administration's practice for introducing major changes such as this one is to recalculate previous years' NCF as if the change had been in place.

Second, the Stabilization Programme is a voluntary one, and not all farmers are covered by it. Farmers may have any number of reasons for staying out of Stabilization at the present, but it is unlikely they would wish to stay out if their Crow subsidy were at stake. There would have to be provisions made for those who are currently out of the Programme to get in without penalty.

Third, in moving from Column (2) to Column (7), there would be a decrease in farmers' gross receipts, but not much change in the stabilization payments. This may require an increase in the check-off levels in order to keep the Fund actuarially sound.

Fourth, and the most significant of the minor problems, would be the need to calculate the added differential above the "old" five year average to which the producers' income would have to be stabilized. The analytical approach shown above, suggesting that this differential is approximately equal to the sum of the levies, gives only a starting point, and to be certain that the plan would be actuarially sound, and that farmers would be getting their due, a thorough analysis, probably entailing the development of a simulation model, would have to be undertaken.

The two major problems are of course much more severe. The first, and probably the lesser, is that the Stabilization Programme places an upper limit on the level of income which can be stabilized. Gross receipts over $60,000 p.a. are not subject to a levy, and no stabilization payments are paid in respect of them. If this provision were left unchanged under this proposal, the producer would not receive his share of the Crow subsidy on grain generating gross income over the $60,000 level. This excludes about 20 to 30% of receipts.

Procedurally, the easiest way to remedy this problem would be to remove the ceiling on eligible receipts. This would, however, involve changing a major feature of the Programme. A procedurally more difficult approach would be to divide the payout into two portions: one related to Stabilization alone, and one related to the Crow subsidy portion. This, however, would seem to defeat one of the major benefits of this proposal, namely its administrative simplicity (see the discussion below under pros and cons).

The second, and more serious major practical difficulty is that the Stabilization Programme covers different grains than does the Crow subsidy. This has two parts to it. First, some grain eligible for a stabilization payment is not eligible for the statutory freight rate. Feed barley moving to the lower mainland area of British Columbia is an example. Second, certain grains eligible for the statutory rate are not part of the Stabilization Programme. Corn, Alfalfa and sunflower seeds are probably the most important of these. The first part of this problem is the more general one of dilution (see the Committee of Inquiry's discussion of disadvantages of "pay the producer" methods of payment, above); the second part is unique to this proposal. One element of the solution to the second part of this problem could be to bring all primary products now eligible for statutory rates within the Stabilization Programme. For the most part this could be done under existing legislation because the Stabilization Act allows for grains under the *Canadian Grain Act* to be included. However, Commodities like alfalfa which are not grains would pose a problem.

These then are the practical problems of greater or lesser severity which would be encountered in implementing the proposal. Now let us turn to the pros and cons of the proposal.

V. PROS AND CONS

This proposal can be evaluated under various headings, most of which are implicit in the findings of the Committee of Inquiry and in the instructions given to the Committee and the GTA by the Minister.

A. Resource Neutrality

Under this heading can be included the issues of the "livestock hurt," equity, and agricultural production across Canada. The livestock hurt arises fro the artificial increase in grain prices on the Prairies caused by the low freight rate on grain. As the producer, under this proposal, would pay the full freight rate, the livestock hurt issue would be resolved.

The Committee of Inquiry defined the equity criterion in terms of the livestock problem, and in terms of identifying all losses and benefits arising from policy changes, and providing compensation for losses. In relation to their own proposal they suggested that the equity criterion would be met because of the benefits to the livestock sector, and because the GTR would keep the transportation subsidy within the grain sector. Finally, they suggested that the criterion relative to agriculture across Canada would be best met by removing "constraints to non-grain agricultural production and marketing activities such as livestock production and other forms of processing in the western region of Canada." As the proposal made in this paper has some similarities to the GTR in its bottom line effects on the producers, these conclusions can be extended to the present proposal.

B. Dilution

As noted above, the Stabilization Programme covers grains which do not qualify for the statutory rates. This means that, if paid out through Stabilization, the government commitment would be spread among a larger number of recipients. This effect has become known as "dilution." This proposal therefore, in common with other "pay the producer" schemes, dilutes the subsidy benefits for the current beneficiaries.

C. Administrative Simplicity

This proposal is administratively the simplest proposal advanced to date. As discussed above, virtually no substantive changes would be required in the administrative procedures under either the WGTA or the Stabilization Programme to put this proposal into effect. The price of this administrative simplicity is the problems referred to earlier.

D. Comprehensibility to Producers

Although the conceptual framework for making this proposal work takes some careful thought, the basic idea is straightforward. The producers would pay the full freight rate, and the government would pay its share into the Stabilization Fund. The payout provisions would be altered to that the producers get the full transportation subsidy.

One of the criticisms has been levelled at various proposals relative to the statutory rates has been that they are too complex for producers to understand. Since this present proposal involves only a modification to a programme which is familiar to producers, it should be acceptable under this criterion.

E. Efficiency

The essence of the efficiency argument is that because the present arrangements mask the full costs of transportation, a less than optimal allocation of resources in the transport sector is achieved. Only if producers are aware of the full freight costs, it is argued, will they seek to maximize efficiency in the grain handling and transportation system. As this proposal would have the producers pay the full rate, the efficiency issue would be dealt with.

F. Ability to Pay

An examination of Table 1 will show that one effect of this proposal is to provide a further stabilization of producers' NCF. Comparing Column (5) with Column (12), it can be seen that producers would receive payout far more frequently under this proposal than they do under the present Stabilization Programme. This means that producers' final income for the year (NCF plus deficit payment from Stabilization) would more frequently follow a relatively more stable five year average line than a relatively more erratic annual NCF line.

It follows that the proposal in this paper better satisfies the criterion of ability to pay. The producers would receive the largest payout from the funds in those periods of lowest NCF, and their income would be stabilized at a higher level than it would be with just the Stabilization Programme alone. The price of this higher level of stabilization of course is lower peak NCFs.

G. Freight Rate Stability

A matter closely related to the ability to pay issue is that of freight rate stability. One of the more contentious issues in the freight setting, process under the WGTA is the fact that the producers' share of the freight rate varies erratically from year to year. The reason for this is that the Crow Benefit portion of the government commitment does not vary with volume. Its per tonne value therefore varies up and down, causing the producers' share of the rate to fluctuate accordingly. This has an inherently destabilizing impact on producers' income.

Under this proposal, the per tonne variation of the Crow Benefit does not affect either the freight rate or the Stabilization payout. This is a further stabilizing effect of the proposal on producer income.

H. Railway Accountability

This proposal as outlined above, in common with all other pay the producer schemes, would not provide any specific mechanism for ensuring railway accountability. However, a simple modification to the proposal would allow for this. Instead of having the producers pay the full rate, they could pay some amount less than the full rate. For sake of illustration, this might be in the order of $2 per tonne. At the end of the year when the government commitment is calculated, the deficit in railway revenues (which, at the $2 level would amount to some $60 million) could be paid as a lump sum to the railways on the condition that their performance had been satisfactory.

I. The Stabilization Proposal Compared with the GTR and the GTA Payout

The proposal in this paper has a strong similarity to the GTR proposal made by the Committee of Inquiry. Hence, it has most of the strengths of that scheme. It is however superior to the GTR in two ways:

a) it is administratively much simpler;
b) with the modification suggested above, it could easily be made to provide for railway accountability.

It is not as good as the GTR in paying the Crow subsidy to the "right" producers. This is because of the second major problem identified above, that the Stabilization Programme and WGTA cover different grains. In particular, it is not as good as the GTR in respect of the dilution factor. The Committee of Inquiry recommended that the GTR be paid to an individual producer on his net grain sales (sales less purchases) thus inhibiting any tendency for a livestock producer to sell his grain into the market and buy back his feed requirements. This proposal does not have the same netting out provision, and grain could be forced into the market in order to qualify for the stabilization payout.

The GTA payout proposal is widely held by economists to be the most desirable way to pay the Crow subsidy. It would pay the subsidy out once and for all, and would remove all distortions from the transportation market. Unfortunately, it has two major drawbacks that were not addressed by the GTA report:

a) it provides no simple way to provide for inflation sharing; any effective inflation sharing formula would likely entail maintaining all the administrative machinery now in place for the rate setting procedure, and having an ongoing program of subsidization which would tend to offset the benefits achieved by the payout;
b) it provides no straightforward way of ensuring railway performance; again, some modifications of the payout idea, tending to offset some of its inherent advantages, would be necessary to implement penalties for poor performance.

The proposal presented here resolves the inflation sharing problem, and could be modified to resolve the railway accountability problem, as described above.

J. Shift from Volume Base to Value Base

One of the effects of this proposal which would require careful further examination is its potential impact on individual payout. At present, the producer receives his individual proportion of the government commitment on the basis of the volume of grain he ships. Under this proposal, the individual payment would be influenced by the value of the grain, and the kind of margins which exist between production costs and prices. No work has been done on this particular matter.

K. International Trade Issues

One final matter that would also have to be addressed is the implication of this proposal under the

GATT. To date, the Stabilization Programme has not been considered a violation of the GATT; however, the United States has continually claimed that the subsidization of freight rates is. The impact of this proposal could either be a pro or a con, depending how the matter might come to be viewed internationally. It could either jeopardize the Stabilization Programme or neutralize the grain freight rate subsidy. At this point, the issue is simply noted. Further analysis would have to be done to determine the implications for these trade arrangements.

VI. ILLUSTRATION USING ACTUAL FIGURES

Let us finally take a look at how the actual Stabilization figure would look under this proposal. Table 2 shows actual net cash flow figures as determined by the Western Grain Stabilization Administration. The second column, which contains the actual NCF figures for the years shown, corresponds to Column (2) on Table 1. Column (3) is the preceding five year average, corresponding to Column (3) in

TABLE 2
Actual Net Cash Flow Figures:
1966/67–1984/85

(1) Year	(2) NCF	(3) 5 Year Avg. NCF	(4) Diff.	(5) Actual Payments
	(	$'000,000		)
1966/67	868	611	+ 257	
1967/68	670	714	− 44	
1968/69	229	736	− 507	
1969/70	273	634	− 361	
1970/71	539	548	− 9	
1971/72	849	516	+ 333	
1972/73	1225	512	+ 713	
1973/74	1848	623	+2225	
1974/75	2345	1147	+1198	
1975/76	2282	1561	+ 721	
1976/77	1551	1910	− 359	115 (1977)
1977/78	1875	2050	− 175	253 (1978)
1978/79	1949	2180	− 231	
1979/80	2802	2000	+ 802	
1980/81	4613	2092	+2521	
1981/82	3322	2558	+ 764	
1982/83	3633	2912	+ 721*	
1983/84	3177	3264	− 87*	223 (Jan-Jul/84)
1984/85	2530	3509	− 739*	522
1985/86	2002	3455	−1453*	Not Available

NOTES:
1. Column (2) corresponds to Column (2) on Table 1
2. Column (3) corresponds to Column (3) on Table 1
3. Column (4) equals Column (2) - Column (1)

SOURCE:
Figures for Columns (1), (2) and (3) provided by Western Grain Stabilization Administration, and quoted in the Review of the Western Grain Transportation Act (except 1985/86 provided directly from WGSA); figures of Column (5) from the WGSA Annual Report, 1984/85.

* COMPARE WITH COLUMN (7) TABLE 4.

Table 1. The fourth column is the difference between the actual NCF and the preceding five year average. When this column shows a negative number, a payout could be triggered from the Stabilization Fund. However, unlike in the simple model described above, the full deficit is not made good because of the eligibility restriction which sets the maximum grain sale proceeds at $60,000 annually. Column (5) shows the actual payouts from the Fund, and these can be seen to be substantially less than the calculated shortfall in Column (4). It should be noted that the Stabilization Programme used to operate on a calendar year rather than crop year, and this makes comparisons between actual payout and NCF shortfalls calculated on a crop year basis different. Table 3 shows approximately how the NCF figures would alter if the producers were to pay full statutory rates. The calculations are shown in the notes to the Table, and basically involve deducting from the annual NCF figures shown in Table 2 the approximate amount of additional freight which producers would pay. The "new" NCF figures are shown in Column (7), and the "new" preceding five years average NCF is shown in Column (8). Column (7) corresponds to Column (7) in Table 1.

The calculations in Table 3 start with the NCF figures as calculated by the WGS Administration and work to the "new" NCF figures in Column (7). In fact, if the proposal contained in this paper were implemented, the Administration would have Column (7) from its own data, and would have to have Column (1) reconstructed from CTC data on freight rates and volumes moved.

In Table 4 the new level of NCF from which the stabilization payments must be determined is calculated. As discussed above, the "old" five year average NCF must be augmented by the amount of the premiums paid into the Fund in order to determine the level to which the producers' income must be stabilized. Column (5) in Table 4 corresponds to Column (10) in Table 1. Column (7) is the difference between:

a) the Producers' actual "new" NCF as it would be under compensatory freight rates, and,
b) the sum of: (i) the "old" preceding five year average NCF as it would be under the subsi-

TABLE 3
Net Cash Flow Figures At Compensatory Freight Rates
1981/82–1984/85

(1)	(2)	(3)	(4)	(5)	(6)	(7)	(8)
Year	NCF	Volume Moved	Total Rate	Prod's Rate	Dec. in NCF	"New" NCF	"New" 5yr.av.
	$ m.	m.t.	$/t	$/t	$ m.	$ m.	$ m.
81/82	3322	30.4	22.05	4.89	522	2800	2200
82/83	3633	33.1	23.08	5.66	577	3056	2560
83/84	3177	33.9	25.00	5.72	653	2523	2880
84/85	2530	27.2	25.90	7.70	495	2053	2990
85/86	2002	27.5	28.42	6.00	605	1397	2919

SOURCES AND CALCULATIONS:
1. Column (2) is from Table 2.
2. Column (3) is from the CWB Annual Report, except for the 1985-86 figure which was provided by the GTA.
3. Columns (4) and (5) are re-constructed from CTC data.
4. Column (6) equals Column (3) x [Column (4) - Column (5)].
5. Column (7) equals Column (2) - Column (6), and corresponds to Column (7) in Table 1.
6. To calculate the five year averages in Column (8), the total freight rate in Column (4) was decreased by 5 percent per annum from 1981/82 back to 1975/76. Column (7) for these years was calculated using these approximations for the total rate, CWB data for Column (3), $4.89 for the producers' rate in Column (5) and calculations described for Columns (2) and (7).

TABLE 4
Augmented Five Year Average NCF To Which Producers' NCF is Stabilized

(1) Year	(2) NCF	(3) Levies	(4) 5 Year Av.NCF	(5) Stab. Level	(6) "New" NCF	(7) Difference
	($'000,000)					
81/82	3322	168	2558	2726	2800	+ 74
82/83	3633	184	2912	3096	3056	- 40*
83/84	3177	174	3264	3438	2523	- 915*
84/85	2530	152	3509	3661	2053	-1628*
85/86	2002	120	3455	3575	1397	-2218*

SOURCES:
1. Column (2) from Table 2.
2. Column (3) from WGSA Annual Report, 1984/85. Figure for 1985/86 provided directly by WGSA.
3. Column (4) from Table 2.
4. Column (5) = Column (3) + Column (4); this column corresponds to Column (10) on Table 1.
5. Column (6) from Table 3.
6. Column (7) = Column (6) - Column (5).

* COMPARE WITH COLUMN (4) TABLE 2.

dized rate, and (ii) the levies paid into the Stabilization Fund.

The key figures to compare are the last three figures in Column (7) of Table 4 and the last three figures of Column (4) of Table 2. As expected, the Table 4 figures are larger, and are larger by more than the additional freight charges paid by the producers as shown in Column (6) of Table 3.

VII. CONCLUDING COMMENTS

The payment of the Crow subsidy through the Stabilization Programme has been shown to have certain advantages over the existing system of "pay the railway" contained in the WGTA and over the previously recommended methods of paying the producer *viz.*, the GTR recommended by the Committee of Inquiry and the "payout" recommended by the Grain Transportation Agency. It also has some problems which these other methods do not posses. The proposal seems to be administratively feasible, and a rough calculation of its effects using actual WGSA data does not indicate any problems in its application.

This paper is put forward in the hope that the idea might receive some critical public copmment. Should the idea be found acceptable, further work would have to be done to determine whether the difficulties described earlier could be resolved and to develop an implementation plan.

ENDNOTES

* University of Manitoba Transport Institute. Mr. Earl is currently on educational leave from the Grain Transportation Agency.

Productivity Implications of Vehicle Weight and Dimension Regulations in the Transportation of Petroleum Products

by Gordon A. Sparks and Russell D. Neudorf*

I. INTRODUCTION

There has been considerable interest in the issue of vehicle weight and dimension regulations in Canada as well as in other parts of the world in recent years. This interest is primarily motivated by the increased productivity/efficiency of larger trucks. While it is intuitively obvious that there are potential cost savings associated with the use of larger trucks in any particular haul, it is not obvious as to the magnitude of these potential savings. It is therefore of interest to analyze the costs associated with operating trucks of different sizes. This is of particular interest in the context of variable (and changing) regulations governing the size and weight of trucks permitted to operate on Canadian roads. Understanding the implications of changing vehicle size and permitted weights is an obvious first step in determining the potential benefits associated with changing vehicle weight and dimension regulations.

II. MEASURE OF PRODUCTIVITY

Productivity of a transportation system can be measured in a number of different ways. One of the most common and probably the most useful when considering the motor carrier industry is in terms of costs per unit of useful output. This is usually measured in terms of cents/tonne km of payload or for a particular vehicle, in terms of $/loaded km or $/running km. (If the vehicle cubes-out, a more useful measure of output may be cents per cubic metre km. This involves essentially "re-scaling" the measure of useful output by the relative density of the commodity involved. For example, if the density of the commodity is such that the vehicle can haul only ½ of its weight capacity, then it costs twice as much per tonne km as it would if weight governed rather than cube.)

III. COST MODEL

In order to determine the productivity of a particular haul situation or compare the productivity of one method versus another, it is necessary to have an analysis framework or methodology. Numerous models have been developed with most falling into what would be considered a general category of costing models. Most such models are deterministic in nature and draw heavily upon methods of engineering economics. One such model is illustrated in Figure 1. (1)

As illustrated in Figure 1, the productivity (as measured in cents/tonne km) is a function of a large number of inputs and/or assumptions about a relatively large number of variables. This results in a situation wherein the productivity as determined by the model can be highly variable depending upon the value of inputs and/or basic assumptions made. This is indeed a reasonable representation of the real world. This high variability in results does however present a difficult situation for those attempting to make general comments about productivity. The usual approach has been to determine the productivity (cents/tonne km) for a number of typical situations to provide a "feel for" the various situations. This approach is clearly helpful however it does have its limitations because of the large number of variables and the large range in potential values for many of these parameters. (This point will be illustrated in the examples later in the paper.)

There is another approach which is possible which can contribute to a more complete understanding of the relationships between the various parameters and productivity. This is a three stage approach which involves the use of a deterministic model to carry out a deterministic sensitivity analysis as the first stage, the development of a probabilistic model as the second stage and a stochastic sensitivity analysis as a third stage. This approach is described and illustrated in the following sections.

IV. DETERMINISTIC SENSITIVITY

In order to determine the relative importance of various variables in determining productivity as measured by cents/tonne km, a sensitivity analysis is undertaken using the deterministic model illustrated in Figure 1.

This is done as illustrated in Table 1 for a 5 axle tanker unit by specifying three values for each parameter. These values represent a minimum value, a maximum value and a most likely value for each parameter. (Observed values would be expected to fall inside the range defined by the minimum and maximum values about 98% of the time, i.e. ± 2 standard deviations from the mean.) The deterministic model is then used to evaluate the sensitivity of productivity to individual parameters by setting the value of all parameters equal to their most likely value except for one which is evaluated at its maximum and minimum value. The results of such a sensitivity analysis for a 5 axle tanker using the deterministic costing model illustrated in Figure 1 are summarized in Table 1 and illustrated in Figure 2. (Note: No allowance has been made for adminis-

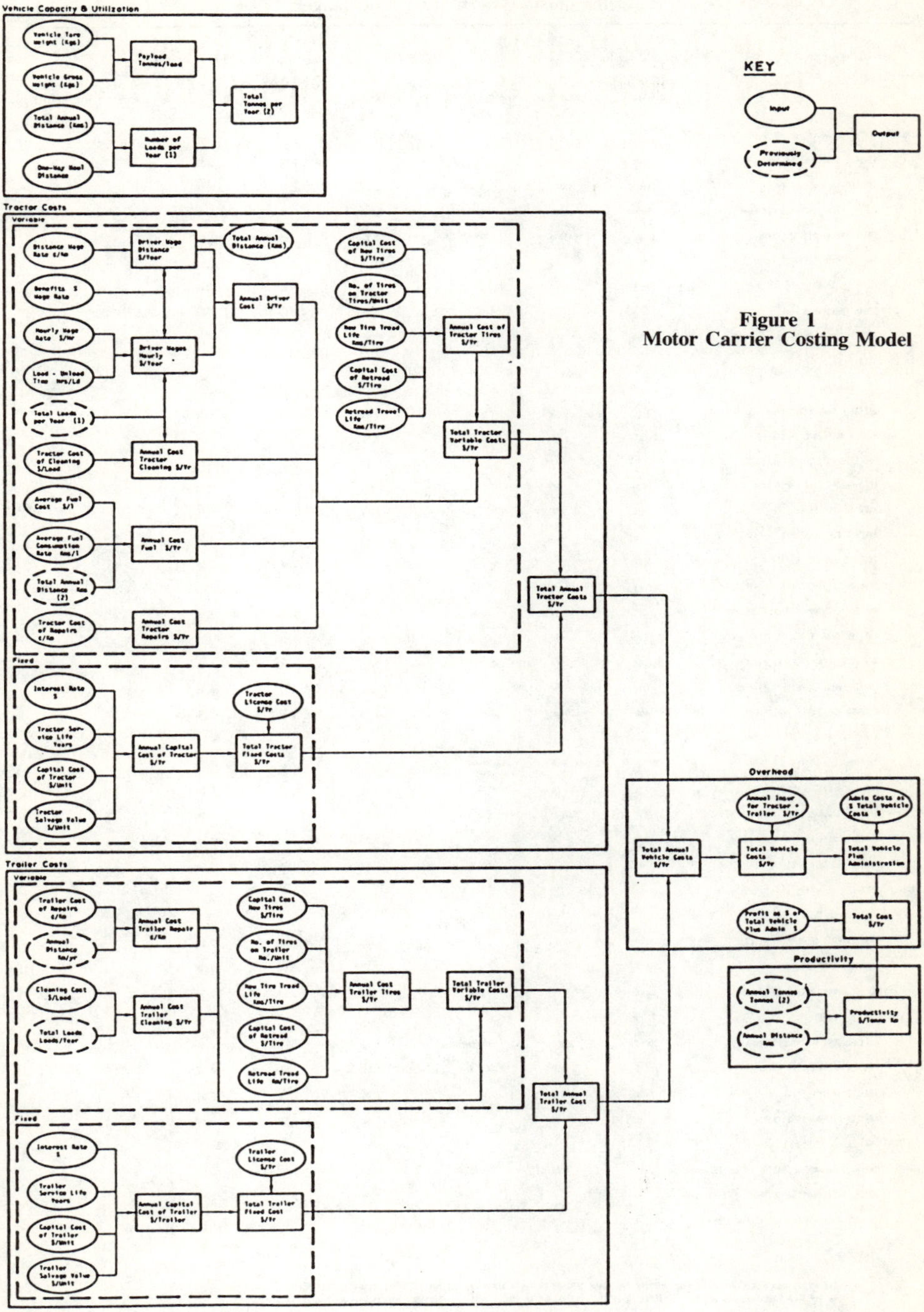

Figure 1
Motor Carrier Costing Model

TABLE 1
Deterministic Sensitivity 5 Axle-Tanker

INPUT PARAMETER	TYPICAL VALUE - INPUTS			ASSOCIATED PRODUCTIVITY (CENTS/TONNE KM)				
	MINI(1)	MAXI(1)	MOST LIKELY	MINI	MAXI	MOST LIKELY	% VARIATION	SENSITIVITY
VEHICLE CAPACITY AND UTILIZATION								
VEHICLE GROSS WEIGHT (KGS)	37500	37500	37500	2.58	2.58	2.58	0	
VEHICLE TARE WEIGHT (KGS)	11500	14500	13000	2.43	2.75	2.58	12	**
PAYLOAD (KGS PER LOAD)	23000	26000	24500	NA	NA	NA		
TOTAL ANNUAL DISTANCE (KMS)	90000	240000	170000	2.42	3.08	2.58	26	**
ONE WAY HAUL DISTANCE (KMS)	500	900	700	2.54	2.65	2.58	4	
TOTAL LOADS	188.9	340.0	242.9	NA	NA	NA		
TOTAL TONNES PER YEAR	4344.4	8840.0	5950.0	NA	NA	NA		
AVERAGE SPEED (KMS/HR)	80	100	90	2.58	2.58	2.58	0	
TRACTOR COSTS								
VARIABLES								
DISTANCE WAGE RATE ($/KM) (2)	0.12	0.18	0.15	2.42	2.74	2.58	12	**
HOURLY WAGE RATE ($/HR) (2)	6.5	12.5	9.75	2.53	2.62	2.58	3	
BENEFITS AS % OF WAGE RATE (%)	25	40	32	2.53	2.64	2.58	4	
LOAD-UNLOAD TIME (HRS/LOAD)	0.5	1.5	1	2.5	2.55	2.58	2	
AVERAGE FUEL PRICE (C/L)	25	35	30	2.49	2.67	2.58	7	
AVERAGE FUEL CONSUMPTION RATE (KMS/L)	2.6	1.8	2.2	2.49	2.7	2.58	8	
TRACTOR COSTS OF REPAIR ($/KM)	0.02	0.06	0.04	2.5	2.66	2.58	6	
TRACTOR COST OF CLEANING ($/LOAD)	1	5	3	2.57	2.59	2.58	1	
NO. OF TIRES ON TRACTOR (TIRES/UNIT)	10	10	10	2.58	2.58	2.58	0	
CAPITAL COST OF NEW TIRES ($/TIRE)	260	460	360	2.57	2.59	2.58	1	
NEW TIRE TREAD LIFE (KMS/TIRE)	100000	320000	200000	2.56	2.6	2.58	2	
CAPITAL COST OF RETREAD ($/TIRE)	240	330	280	2.57	2.59	2.58	1	
RETREAD TREAD LIFE (KMS/TIRE)	160000	600000	360000	2.55	2.63	2.58	3	
FIXED								
CAPITAL COST OF TRACTOR ($/UNIT)	80000	110000	90000	2.54	2.67	2.58	5	
TRACTOR SERVICE LIFE (YRS)	7	10	8.5	2.55	2.62	2.58	3	
TRACTOR SALVAGE VALUE ($/UNIT)	20000	33000	27000	2.55	2.61	2.58	2	
INTEREST RATE (%)	8	14	10	2.55	2.64	2.58	3	
TRACTOR LICENSE COST ($/YR)	1500	2200	2000	2.57	2.58	2.58	0	
TRAILER COSTS								
VARIABLE								
TRAILER COST OF REPAIRS ($/KM)	0.04	0.07	0.055	2.53	2.62	2.58	3	
CLEANING COST ($/LOAD)	1	5	3	2.57	2.59	2.58	1	
NO. OF TIRES ON TRAILER (NO./UNIT)	8	8	8	2.58	2.58	2.58	0	
FIXED								
CAPITAL COST OF TRAILER ($/UNIT)	30000	50000	40000	2.53	2.62	2.58	3	
TRAILER SERVICE LIFE (YRS)	6	10	8	2.56	2.61	2.58	2	
TRAILER SALVAGE VALUE ($/UNIT)	5000	13000	10000	2.56	2.59	2.58	1	
TRAILER LICENSE COST ($/YR)	22	22	22	2.58	2.58	2.58	0	
OVERHEAD								
ANNUAL INSURANCE - TRACTOR & TRAILER ($/YR)	3300	4400	3900	2.56	2.59	2.58	1	
ADMIN. COSTS % OF TOTAL VEHICLE COSTS (%)	0	0	0	2.58	2.58	2.58	0	
PROFITS AS % TOTAL VEHICLE PLUS ADMIN. (%)	0	0	0	2.58	2.58	2.58	0	

(1) VALUES WOULD BE EXPECTED TO FALL OUTSIDE THE RANGE DEFINED BY THE MINIMUM AND MAXIMUM 2% OF THE TIME.
(2) IN SOME CASES DRIVERS ARE PAID A MILEAGE RATE FOR DRIVING AND AN HOURLY RATE FOR LOADING / UNLOADING / WAITING.

tration costs and profits. The analysis therefore allows for only vehicle operating costs as opposed to total operating costs. This is appropriate for our purposes herein because we are primarily concerned with the impact of changes in VWD regulations on vehicle productivity.)

From Figure 2 it is noted that the various parameters can be generally categorized relative to the sensitivity of productivity to changes in values of their individual parameters (i.e. % variation column in Table 1). They might be generally classified as relatively insensitive, moderately sensitive and very sensitive. (The parameters to which productivity is highly sensitive are identified in Table 1 by **).

Given the results of the sensitivity analysis it is possible to conclude that issues related to productivity will be primarily determined by the values of parameters to which productivity is very sensitive, to a lesser extent to those to which productivity is moderately sensitive and largely unaffected by those to which productivity is relatively insensitive.

For the example illustrated in Table 1 and Figure 2 it is possible to conclude that productivity is:

1) highly sensitive to:
- vehicle capacity and utilization as measured by vehicle tare weight and total annual distance
- tractor cost as measured by distance wage rate

2) moderately sensitive to:
- benefits as % of wage rate
- fuel price
- fuel consumption rate
- tractor repair costs, and
- capital cost of tractor

3) relatively insensitive to all other parameters.

From the above, it can be seen that the critical parameters in terms of productivity/economic viability are the relative magnitudes of vehicle capacity, utilization and labor costs.

It is important to note that this analysis represents a particular operating scenario. This scenario is described by the values assigned to the costing variables. The costs (productivity) would be substantially different if the scenario changed. To illustrate, it has been assumed that a new tractor costing between $80,000—$110,000 with a service life of 7-10 years and a salvage value of $20,000—$33,000 was used on the haul. If a used tractor with a capital cost of $25,000 were used the cost incurred and therefore the productivity would be substantially different. It is for these reasons that it is very difficult to generalize when dealing with costing and therefore productivity measures.

Figure 2. Deterministic Sensitivity 5 Axle Tanker

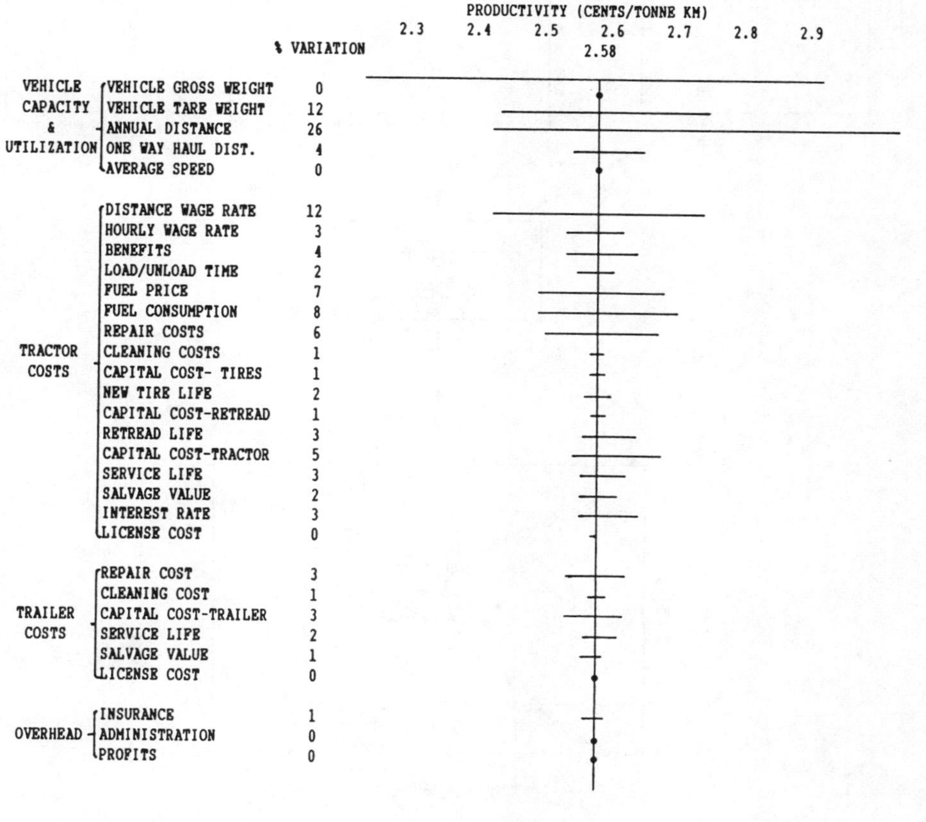

V. PROBABILISTIC MODEL

The deterministic model detailed in the previous section indicated that the productivity (vehicle costs) of a "typical" 5 axle tanker operation is in the order of 2.58 cents per tonne km (given the assumptions used). It was further illustrated how these costs could vary up to 2-26% by varying any one of a number of parameters to which productivity is sensitive.

The deterministic sensitivity analysis summarized in the previous section is very informative in terms of identifying individual important parameters. It would also be of considerable interest to determine the combined effect of varying a number of sensitive parameters at one time as opposed to varying them one at a time as was done in the previous section. This type of analysis is possible if the previous deterministic model is extended to include a probabilistic structure.

Such a probabilistic model is conceptually illustrated in Figure 3 for those parameters for which the deterministic model was determined to be very sensitive. (2) In effect, the probability distributions for the highly sensitive parameters are combined (as per the costing model) to provide a probability distribution for productivity as measured by the cost per tonne km. The expected value as provided by the probabilistic model is equivalent to the value provided by the deterministic model when the most likely values for all parameters are used. The probability distribution for productivity as provided by the probabilistic model gives a measure of uncertainty in this measure of effectiveness (as measured by the variance of the probability distribution). This uncertainty in the measure of productivity is the result of the (combined) uncertainties of the input parameters. As noted from Figure 3, for the given case (i.e. 5 axle tanker), the expected cost would be 2.58

Figure 3. Probabilistic Model
5 Axle Tanker

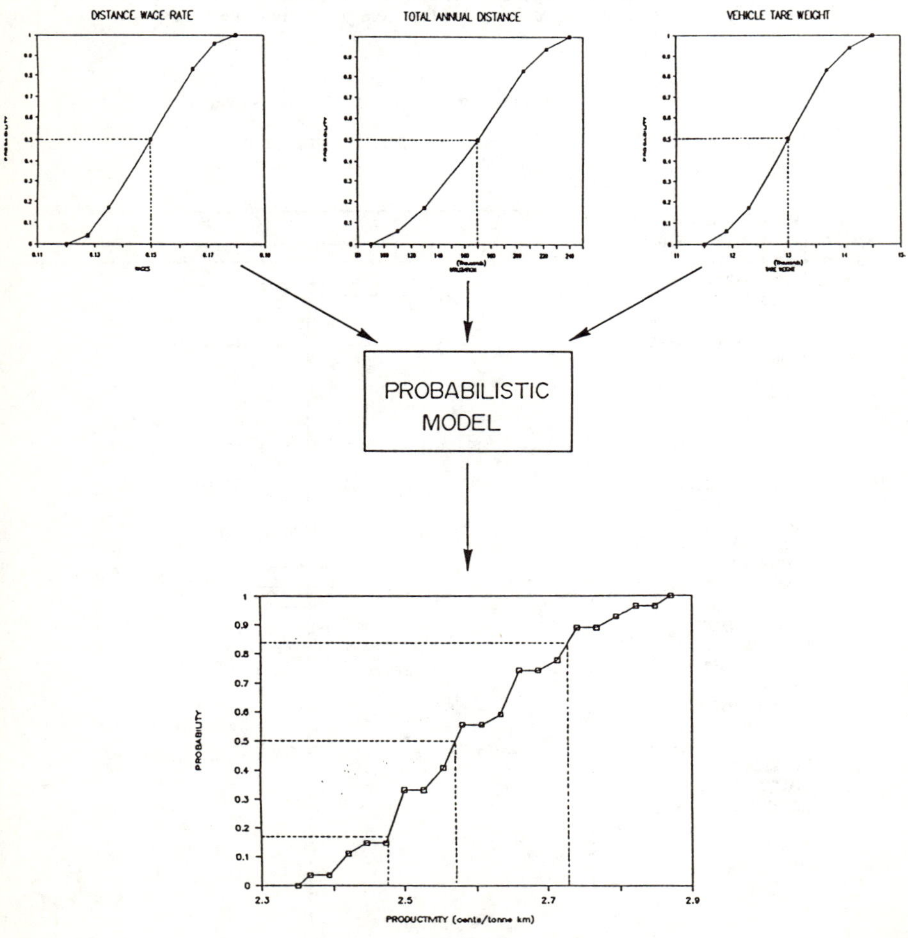

cents per tonne km and the range (i.e. ± one standard deviation) would be 2.5–2.7 cents per km. (Note: There are some minor discrepancies between the deterministic model and the stochastic one because of descretness of the data input into the stochastic model.)

VI. STOCHASTIC SENSITIVITY

Another issue that is of potential interest is the question of how does the probability distribution of productivity vary as the probability distribution for a given parameter varies. This is in effect a measure of the stochastic sensitivity. The result of such an analysis for the 5 axle tanker is illustrated in Figure 4.

From Figure 4, it can be noted that the distribution of productivity is not very sensitive to changes in the distribution of any one of the parameters analyzed. From this, one would conclude that little is to be gained from improved information about these parameters.

VII. MODEL APPLICATIONS

The previous sections have presented the details of two models (one deterministic and one stochastic) which illustrate the functional relationship between the parameters which determine motor carrier productivity.

These models and the associated analysis result in a good understanding of productivity and the parameters that affect it. The primary issue of concern here is however how do changes in vehicle weight and dimension regulations affect productivity. The use of

Figure 4. Stochastic Sensitivity 5 Axle Tanker

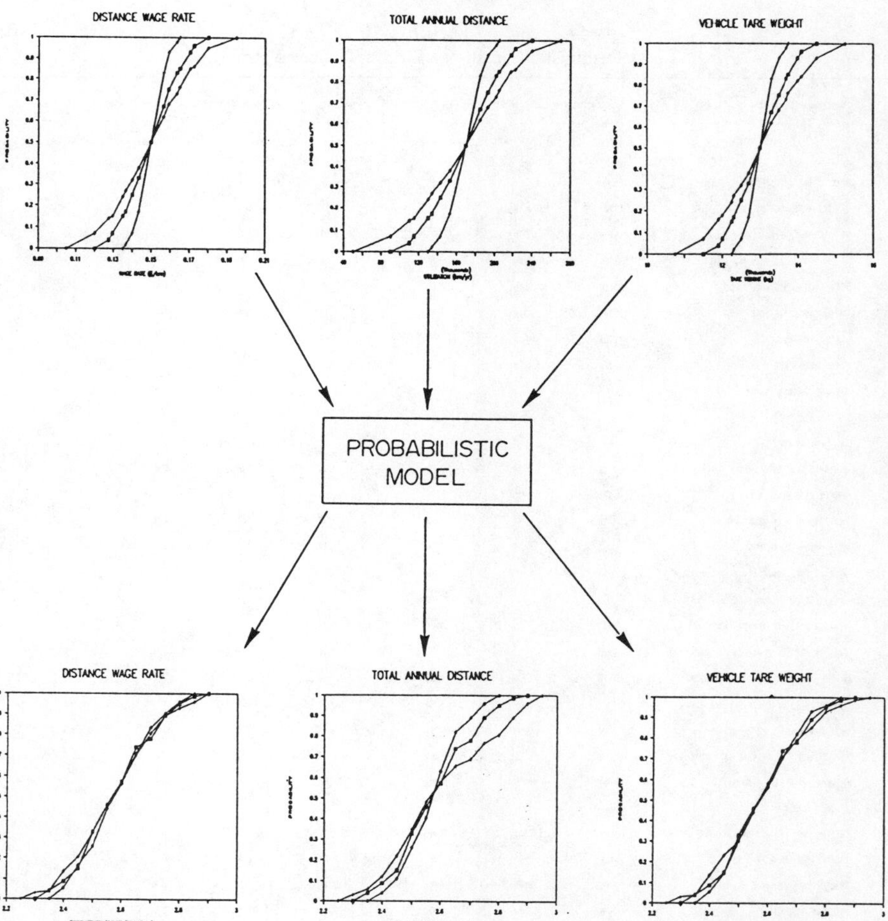

the previous models in analyzing these types of questions might be illustrated by the following example.

Suppose current regulations (and other considerations) dictate that a particular haul be undertaken using a 5 axle semi-trailer tanker at a maximum gross vehicle weight of 37,500 kgs. (The operation is described by the data in Table 1). Suppose a regulator is considering changing the VWD regulations so that the haul under consideration could be carried out using a 7 axle A-Train at 53,500 kgs. (The operation being described by the data in Table 2.) One might carry-out a deterministic sensitivity analysis similar to that illustrated in Table 2. One could also illustrate the results as per Figure 2, develop a stochastic model similar to that in Figure 3 and carry-out a stochastic sensitivity as per Figure 4. (Space does not permit so doing here.)

(Note that the three most sensitive parameters are now, annual distance, distance wage rate, and fuel consumption. That is fuel consumption has replaced vehicle tare weight in moving from the 5 axle semi to the 7 axle unit.) Similar calculations could be performed for 7 axle B-Trains, and 8 axle A and B-Trains. (See (3) for details.)

The implications of the proposed change in VWD regulations on productivity for each type of vehicle are illustrated in Figure 5. From Figure 5 it is noted that the expected productivity for the 5 axle unit is approximately 2.6 cents per tonne km with a standard deviation of about .1 cent/tonne km and for the 7 axle A-Train unit it is about 2.2 cents per tonne km, also with a standard deviation of about .1 cent/tonne km. Put another way, the productivity of the 5 axle unit would be expected to be approximately 2.6 cents/tonne km and lay in the range of 2.5 and 2.7 cents per tonne km 68% of the time (i.e. expected value plus or minus one standard deviation). Similarly the productivity of the 7 axle A-Train unit would be expected to be about 2.2 cents/tonne km and lay in the range of 2.1–2.3 cents/tonne km 68% of the time.

In short, the change in regulations would be expected to result in a change in vehicle productivity that would be detectable (i.e. the probability distributions for the 5 axle unit and the 7 axle A-Train in

TABLE 2
Deterministic Sensitivity 7 Axle-Tanker (A Train)

INPUT PARAMETER	TYPICAL VALUE - INPUTS			ASSOCIATED PRODUCTIVITY (CENTS/TONNE KM)				
	MINI(1)	MAXI(1)	MOST LIKELY	MINI	MAXI	MOST LIKELY	% VARIATION	SENSITIVITY
VEHICLE CAPACITY AND UTILIZATION								
VEHICLE GROSS WEIGHT (KGS)	53500	53500	53500	2.17	2.17	2.17	0	
VEHICLE TARE WEIGHT (KGS)	16300	19300	17800	2.08	2.26	2.17	8	**
PAYLOAD (KGS PER LOAD)	34200	37200	35700	NA	NA	NA		
TOTAL ANNUAL DISTANCE (KMS)	90000	240000	170000	2	2.66	2.17	30	**
ONE WAY HAUL DISTANCE (KMS)	500	900	700	2.14	2.22	2.17	4	
TOTAL LOADS	188.9	340.0	242.9	NA	NA	NA		
TOTAL TONNES PER YEAR	6460.0	12648.0	8670.0	NA	NA	NA		
AVERAGE SPEED (KMS/HR)	80	100	90	2.17	2.17	2.17	0	
TRACTOR COSTS								
VARIABLES								
DISTANCE WAGE RATE ($/KM) (2)	0.13	0.19	0.16	2.05	2.28	2.17	11	**
HOURLY WAGE RATE ($/HR) (2)	6.5	12.5	9.75	2.14	2.19	2.17	2	
BENEFITS AS % OF WAGE RATE (%)	25	40	32	2.13	2.21	2.17	4	
LOAD-UNLOAD TIME (HRS/LOAD)	0.5	1.5	1	2.12	2.22	2.17	5	
AVERAGE FUEL PRICE (C/L)	25	35	30	2.1	2.25	2.17	7	
AVERAGE FUEL CONSUMPTION RATE (KMS/L)	2.3	1.3	1.8	2.07	2.35	2.17	13	**
TRACTOR COSTS OF REPAIR ($/KM)	0.02	0.06	0.04	2.12	2.23	2.17	5	
TRACTOR COST OF CLEANING ($/LOAD)	1	5	3	2.16	2.18	2.17	1	
NO. OF TIRES ON TRACTOR (TIRES/UNIT)	10	10	10	2.17	2.17	2.17	0	
CAPITAL COST OF NEW TIRES ($/TIRE)	260	460	360	2.16	2.19	2.17	1	
NEW TIRE TREAD LIFE (KMS/TIRE)	100000	320000	200000	2.16	2.19	2.17	1	
CAPITAL COST OF RETREAD ($/TIRE)	240	330	280	2.17	2.18	2.17	0	
RETREAD TREAD LIFE (KMS/TIRE)	160000	600000	360000	2.22	2.15	2.17	3	
FIXED								
CAPITAL COST OF TRACTOR ($/UNIT)	85000	125000	105000	2.11	2.23	2.17	6	
TRACTOR SERVICE LIFE (YRS)	6	10	8	2.14	2.23	2.17	4	
TRACTOR SALVAGE VALUE ($/UNIT)	22000	38000	30000	2.15	2.2	2.17	2	
INTEREST RATE (%)	8	14	10	2.14	2.24	2.17	5	
TRACTOR LICENSE COST ($/YR)	2500	3200	2900	2.17	2.17	2.17	0	

TABLE 2 (continued)
Deterministic Sensitivity 7 Axle-Tanker (A Train)

INPUT PARAMETER	TYPICAL VALUE-INPUTS			ASSOCIATED PRODUCTIVITY (CENTS/TONNE KM)				
	MINI(1)	MAXI(1)	MOST LIKELY	MINI	MAXI	MOST LIKELY	% VARIATION	SENSITIVITY
TRAILER COSTS								
VARIABLE								
TRAILER COST OF REPAIRS ($/KM)	0.06	0.105	0.08	2.12	2.24	2.17	6	
CLEANING COST ($/LOAD)	1	5	3	2.16	2.19	2.17	1	
NO. OF TIRES ON TRAILER (NO./UNIT)	16	16	16	2.17	2.17	2.17	0	
FIXED								
CAPITAL COST OF TRAILER ($/UNIT)	70000	100000	85000	2.13	2.21	2.17	4	
TRAILER SERVICE LIFE (YRS)	8	12	10	2.16	2.2	2.17	2	
TRAILER SALVAGE VALUE ($/UNIT)	13000	20000	17000	2.17	2.18	2.17	0	
TRAILER LICENSE COST ($/YR)	22	22	22	2.17	2.17	2.17	0	
OVERHEAD								
ANNUAL INSURANCE - TRACTOR & TRAILER ($/YR)	5000	7000	6000	2.16	2.19	2.17	1	
ADMIN. COSTS % OF TOTAL VEHICLE COSTS (%)	0	0	0	2.17	2.17	2.17	0	
PROFITS AS % TOTAL VEHICLE PLUS ADMIN. (%)	0	0	0	2.17	2.17	2.17	0	

(1) VALUES WOULD BE EXPECTED TO FALL OUTSIDE THE RANGE DEFINED BY THE MINIMUM AND MAXIMUM 2% OF THE TIME.
(2) IN SOME CASES DRIVERS ARE PAID A MILEAGE RATE FOR DRIVING AND AN HOURLY RATE FOR LOADING / UNLOADING / WAITING.

Figure 5.
Productivity vs. Vehicle Configuration Tankers

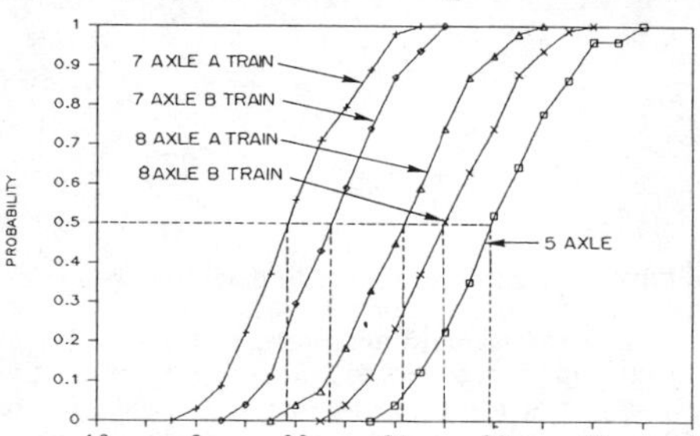

Figure 5 do not overlap very much—there is a high probability that there is a significant difference between the means). Put another way, the difference between the expected productivities for the two vehicles is large relative to the uncertainties associated with the expected values.

The previous example has illustrated how productivity might be enhanced within a particular jurisdiction by increasing allowable gross vehicle weights and thereby permitting the use of larger, differently configured vehicles. By moving from a permitted GVW of 37,500 kgs to 53,500 kgs, productivity could be increased by 15% (2.6 − 2.2 = .4; .4/2.6 = 15%).

It is of potential interest to also compare the productivity of particular fleets operating in different regulatory environments across the country. Such a comparison is illustrated in Table 3 and Figure 6 for tankers transporting petroleum products in various jurisdictions across the country. As summarized in Table 3, different vehicle configurations operating at different axle and gross vehicle weights dominate the haul in environment in which they must operate. Figure 6 illustrates the implications of these various regulatory environments on vehicle productivity in fuel transportation. It is interesting to note that of the possible configurations analyzed, the tractor—semi trailer operating in New Brunswick is the least productive while the same vehicle configuration operating at higher axle and gross vehicle weights in Quebec is among the most productive (i.e. compare curve 1 and curve 2 in Figure 6). It is also interesting to note that the productivity of this fleet varies about 20% due to the difference in regulatory environments across the country.

Figure 6.
Probability Distributions for Petroleum Haul Across Canada for Typical Vehicle Configuration

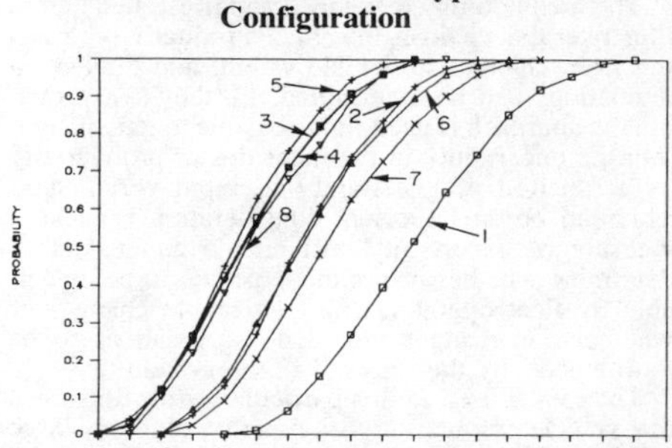

LEGEND: 1. NEW BRUNSWICK, 3-S3 at 50,000 kg.
2. QUEBEC, 3-S3 at 54,203 kg.
3. ONTARIO (1), 3-S3-S2 at 63,500 kg.
4. ONTARIO (2), 3-S2-S2 at 62,100 kg.
5. ONTARIO(3), 3-S4 at 61,230 kg.
6. MANITOBA, 3-S2-2 at 54,875 kg.
7. SASKATCHEWAN, 3-S2-2 at 53,500 kg.
8. BRITISH COLUMBIA, 3-S3-S2 at 63,500 kg.

TABLE 3
Typical Truck Configurations for Petroleum Products Hauling

DESCRIPTION	N.B.	QUEBEC	ONTARIO (1)	ONTARIO (2)	ONTARIO (3)	MANITOBA	SASK.	B. C.
CONFIGURATION	(TRACT-SEMI TRAILER) 3-S3	(TRACT-SEMI TRAILER) 3-S3	B-TRAIN 3-S3-S2	B-TRAIN 3-S2-S2	(TRACT-SEMI TRAILER) 3-S4	A-TRAIN 3-S2-2	A-TRAIN 3-S2-2	B-TRAIN 3-S3-S2
NUMBER OF AXLES	6	6	8	7	7	7	7	8
REGISTERED GVW (kg.)	50,000	54,203	63,500	62,100	61,230	54,875	53,500	63,500
TARE - TRACTOR (kg.)	9,000	8,618	8,810	8,810	9,000	9,000	9,000	9,072
- TRAILER (kg.)	9,000	8,437	11,100	10,150	9,600	6,833	6,833	11,453
- TOTAL (kg.)	18,000	17,055	19,910	18,960	18,600	15,833	15,833	20,525
PAYLOAD (kg.)	32,000	37,147	43,590	43,140	42,630	39,042	37,667	42,975
AXLE WEIGHTS (kg.)								
- 1 STEERING	6,000	6,258	7,005	7,500	6,500	4,675	4,500	5,500
- 2								
	17,000	18,960	17,055	18,500	18,310	16,000	15,750	17,000
- 3								
- 4	9,000				9,155			7,000
				19,100		16,000	15,750	
- 5		28,985	20,860		9,155			
	18,000							17,000
- 6						9,100	8,750	
				17,000	18,310			
- 7						9,100	8,750	
			18,580					17,000
- 8								

VIII. CONCLUSIONS

Increases in productivity and efficiency are the primary motivating forces behind changes in vehicle weight and dimension regulations which permit larger and/or heavier vehicles to operate on Canada's public highway system.

The methodology developed and illustrated herein illustrates that changes in vehicle productivity resulting from changes in vehicle weight and dimension regulations can be determined. Further if a probabilistic approach is used, it is possible to gain insight into the uncertainty in the measures of productivity as a function of uncertainty in input parameters. This can be an important consideration because a measure of uncertainty will enable an analyst to determine whether or not the expected to be detectable by fleet operators (i.e. noticeable change) or whether the changes would be so small as to be "swamped" by the "noise" in the system.

The model used in this particular effort illustrates that vehicle productivity is a function of a very large number of identifiable parameters (see Figure 1). Values for any one of these parameters can vary substantially (i.e. the capital cost of a tractor capable of pulling a semi trailer could vary from $120,000 for a new unit to $25,000 for an old one; the tare weight of a particular type of vehicle can vary by 15-25% depending on the design and the materials used). The high potential variability in values for various parameters necessitates the development of "typical" scenarios in order to determine and compare the productivity of vehicles.

Using the scenario approach, it was possible to identify the parameters to which vehicle productivity are most sensitive. Further it was possible to compare productivity of various vehicle configurations within particular jurisdictions as well as across jurisdictions.

In general terms, vehicle productivity for a given permitted gross weight is most sensitive to:

1) vehicle tare weight,
2) annual utilization in terms of annual distance travelled,
3) wages in terms of distance wage rate, and
4) fuel consumption.

Vehicle productivity was found to be highly sensitive to allowable gross vehicle weights and vehicle configuration (i.e. 5 axle semis versus 7 axle A-Trains etc.). Combination units (i.e. 7 and 8 axle A and B-Trains) were found to be typically 15-25% more productive than 5 axle semis operating at the same axle weights but higher gross weights. Given the fact that gross vehicle weights are typically a constraint in existing regulations, 7 axle units were found to be in some cases more productive than 8 axle units. This is attributable to their lower tare weights (and therefore higher payloads) and somewhat lower capital costs.

Vehicle productivity is very sensitive to permitted axle weights. Increasing axle weights translates directly to increased payloads and therefore increased productivity. Increasing gross vehicle weights on the other hand usually translates into different vehicle configurations (i.e. more trailers and more vehicles) and while increasing payload, there is also an associated increase in tare weight. This increase in tare weight in part offsets the increase in payload relative to the increase in allowable gross vehicle weight.

To illustrate, it was found that the 5 axle configuration in the fuel haul is potentially one of the most inefficient vehicles used (i.e. under New Brunswick's allowable axle weights) and potentially one of the most productive (i.e. under Quebec's allowable axle weights).

In conclusion, vehicle weight and dimension regulations do have an important impact upon vehicle productivity. These differences in productivity are associated primarily with allowable axle weights which directly affect payloads and allowable gross vehicle weights that affect vehicle configurations which in turn affect payloads. Within existing regulatory regimes in Canada, productivity differences of 15-25% are observed over various vehicle configurations (i.e. 5 axle versus 7 and 8 axle) and over various jurisdictions (i.e. for the same vehicle in different provinces).

ACKNOWLEDGEMENT

This paper draws upon material assembled as part of a project undertaken for the Transportation Development Centre, Transport Canada. The information and opinions presented herein are those of the authors and not those of Transport Canada.

ENDNOTES

* University of Saskatschewan.

1. This model draws heavily in a structural sense upon one prepared by Trimac Consulting Services Ltd. for Transport Canada. The report is entitled: "Operating Costs of Trucks in Canada".
2. This draws heavily upon a probabilistic motor carrier costing model developed at the University of Saskatchewan, Saskatoon.
3. "Study of Vehicle Weights and Dimension Regulations and Canada's Trucking Industry: Background Paper #5, Productivity Implications of Vehicle Weight and Dimension Regulations". Prepared for the Transportation Development Centre, Transport Canada, by Clayton, Sparks & Associates Ltd., November 1986.

Institutional Constraints on the Movement of Canadian Grain to Export by Alternative Routes

*by A. G. Wilson and P. Miller**

I. INTRODUCTION

Numerous routes exist for the movement of Canadian grain from the prairies to export destination. Currently all this grain moves by domestic routes to seaboard. These routes include movement by rail to Vancouver/Prince Rupert for transfer to ocean vessel, movement to Thunder Bay by rail for transfer to laker and ultimate transfer to ocean vessel at ports on the lower St. Lawrence, direct rail movement to eastern ports for transfer to ocean vesel, limited ocean vessel pickup at Thunder Bay and movement by rail to Churchill for transfer to ocean vessel. Transfer between modes is accomplished through use of terminal and transfer elevators.

Research by Fruin at the University of Minnesota and by Miller at the University of Mabitoba showing the economic potential for use of routes through the United States has stimulated interested in using such routes. This interest has been sharpened by the desire to circumvent the effects of interruptions on current routes such as those arising from strikes and lockouts of elevator employees or longshoremen in port areas, accidents on the St. Lawrence Seaway, and strikes or lockouts of sailors staffing lake vesels on the Seaway. Potential routes for the movement of prairie grain through the United States include: direct rail movement from the prairies to Duluth or ports on the eastern seaboard; direct rail movement to ports on the Gulf of Mexico and to ports in the Pacific Northwest; and a combination of rail and barge movement to ports on the Gulf.

Before routes through the United States can be utilized, several institutional constraints need to be overcome. These constraints take several forms and reflect the regulation put in place with producer support to overcome some perceived problem in the marketing of grain. As a result of this regulation movement through the United States is made more complex than would otherwise be the case.

The constraints affecting movement by U.S. routes include certain aspects of the grading system, the location of cleaning, segregation of grain enroute, inspection of grain at export position, and issuance of the Certificate Final. Other potential constraints include restriction on the use of Canadian government hopper cars, use of the U.S. handling and transportation system in periods of heavy domestic use and the need for the movement of Canadian grain in bond through the United States. In addition, the current definition of "export" under the *Western Grain Transportation Act* precludes eligibility for subsidy on movements to the United States. Opposition from Canadian owners of handling and transportation facilities to any diversion of grain can be expected, particularly when these facilities remain underutilized.

II. HISTORY OF GRAIN MOVEMENTS THROUGH THE UNITED STATES

Movement of Canadian grain through the United States would not be a new development. Canadian grain has moved through that country in significant volumes in the past. In each instance this movement occurred more in response to a limitation of the Canadian system than from the competitiveness of the U.S. system.

Substantial volumes of Canadian grain moved through U.S. ports in the early years of this century. This movement peaked in the early 1920's when up to 50% of Canadian grain moved through Portland, Maine and Buffalo, New York, Table I. U.S. facilities were used since they provided an economic outlet for large volumes of grain. Some grain moved to the U.S. for consumption in that country during the war years. During the period 1946/47 to 52/53 significant volumes of wheat moved through U.S. ports, Duluth in particular, Table II. Canadian terminal elevators at Thunder Bay had insufficient available storage capacity to expedite shipment. This situation was exacerbated in 1951/52 by a low grade wheat crop, 40 per cent of which was harvested tough or damp. During this period Canadian wheat moved in rail cars in bond to Duluth where it was cleaned and stored in sealed bins, storage facilities being leased so the grain could be stored in these binds. Canadian Grain Commission inspectors were stationed at Duluth, the grain being transshipped by lakers to ports on the lower St. Lawrence. The following year a significant volume, 3.7 million bushels, was re-exported through U.S. Atlantic ports. The competitiveness of this U.S. route remained such that the Federal Government deemed necessary the establishment of the At and East (of Buffalo) rates, rail rates from Georgian Bay ports and those on Lake Erie and Lake Ontario to ports on the St. Lawrence and also to Halifax and St. John, which were competitive with U.S. rail rates at and east of Buffalo.

U.S. grain continues to flow through Canadian facilities on the Lower St. Lawrence. U.S. grain is allowed to occupy 40 per cent of the space in the transfer elevators at these ports, this grain moving to these elevators by vessel from ports on the Great Lakes. In order to ensure the American grain is up to standard upon export an office of the U.S. Federal Grain Inspection Service is maintained in Montreal. The Inspection Service posts inspectors in these

TABLE I
Comparative Exports of Canadian Grain Via Domestic and United States Ports by Crop Year 1918 to 1937

Crop Year	Volume of Exports Domestic Ports	U.S. Ports	Proportion of Exports Domestic Ports	U.S. Ports
	thousand tonnes		per cent	
1918/19	960.5	519.9	64.9	35.1
1919/20	1315.7	354.4	78.8	21.2
1920/21	891.8	1474.9	37.7	62.3
1921/22	1158.6	2721.8	29.9	70.1
1922/23	2364.3	3534.5	40.1	59.9
1923/24	3450.6	3839.5	47.3	52.7
1924/25	1870.1	2043.1	47.8	52.2
1925/26	3345.3	3869.3	46.4	53.6
1926/27	2922.9	3707.9	44.1	55.9
1927/28	3924.6	3699.4	51.5	48.5
1928/29	5179.9	4192.1	55.3	44.7
1929/30	2147.4	1893.0	53.1	46.9
1930/31	3557.7	2441.9	59.3	40.7
1931/32	3528.2	1325.2	72.7	27.3
1932/33	5035.7	1491.4	77.1	22.9
1933/34	3411.0	1216.0	73.7	26.3
1934/35	2465.2	1053.7	70.1	29.9
1935/36	3524.3	1998.8	63.8	36.2
1936/37	3287.1	1067.0	75.5	24.5
1937/38	1713.5	325.2	84.0	16.0

SOURCE: Processed from Grain Trade of Canada, Crop Years 1918/19 to 1937/38 inclusive.

TABLE II
Overseas Clearance of Canadian Grains Via United States Ports by Crop Year, 1946 to 1956

Crop Year	Volume of Exports Wheat	Oats	Barley	Rye	Proportion of Exports Wheat	Oats	Barley	Rye
	thousand tonnes				per cent			
1946/47	325.1	19.6	61.9	18.4	7.3	5.8	41.2	13.8
1947/48	428.0	0.4	4.1	33.4	11.8	0.5	7.0	12.8
1948/49	266.5	-	20.0	-	5.3	-	4.2	-
1949/50	3.9	-	-	-	0.1	-	-	-
1950/51	43.1	-	-	77.3	0.8	-	-	32.5
1951/52	108.5	22.1	15.6	21.7	1.3	2.1	1.0	12.5
1952/53	102.0	-	-	2.4	1.1	-	-	1.1
1953/54	3.5	-	-	-	0.1	-	-	-
1954/55	1.0	-	-	29.5	1/	-	-	12.5
1955/56	6.4	-	-	-	0.1	-	-	-
1956/57	13.4	-	-	-	0.2	-	-	-

1/ less than 0.05

SOURCE: Processed from Canadian Grain Exports, Crop Years 1946/47 to 1955/56 inclusive.

Canadian transfer elevators for the purpose of overseeing weighing, grading and the issuance of a certificate final upon export.

Use of the handling facilities of each country has been supplementary to the movement of the host country's grain. In general such movements have been jointly advantageous. The experience gained from these precedents could well be beneficial to any future large volume movement of Canadian grain through U.S. ports.

III. INSTITUTIONAL RESTRAINTS AND THE MOVEMENT OF CANADIAN GRAIN

Before proceeding to an assessment of the potential for the various institutional constraints deterring the movement of Canadian grain by U.S. routes, reference to the means used to guarantee uniformity of this grain should be made. This will enable some of the constraints to be placed in perspective. A rigorous grading system has evolved for Canadian grain. Rigid grade specifications are in place for each class or type of grain. Varietal standards exist for most of the classes. Any new variety to be accepted must meet as a minimum the characteristics of the standard variety. The requirement for cleanliness in the grain usually far exceeds that of the grain as received from producers necessitating that the grain be cleaned, this usually being accomplished in the terminal elevators. The grain from many areas is melded when it flows into the terminal elevators. There it is officially graded and weighed by inspectors of the Canadian Grain Commission. Upon export a Certificate Final identifying the weight and grade of the grain is issued by the Commission. Upon export a Certificate Final identifying the weight and grade of the grain is issued by the Commission. The melding and cleaning of the grain results in a high degree of uniformity within grades, this being a characteristic sought after by buyers and one on which Canada capitalizes when merchandising. The Canadian Wheat Board is responsible for the marketing of wheat, oats and barley for export and for domestic human consumption. Other grains are marketed by private traders. The movement of grain from the Canadian prairies to Thunder Bay and Vancouver/Prince Rupert for "export" falls under the *Western Grain Transportation Act*. The Federal Government is committed under the *Act* to pay a significant share of the cost of the movement to the railways with the shipper paying the remainder. The Grain Transportation Agency, established by the *Act*, performs a general administrative role while exercising a degree of control over the allocation of railway rolling stock to the movement.

Each of the constraints identified is discussed below:

1. Grain Cleaning

Cleaning is one of the constraints affecting movement of Canadian grain through the United States. Cleaning, to be economic, must be done in volume, both from a physical standpoint and also to enable development of a market for the screenings. The majority of the cleaning capacity has therefore been installed at terminal elevators, the screenings having a ready demand in both the domestic and export markets. Cleaning becomes a major profit centre for the terminals, a cleaning charge being imposed at the primary elevator level and the screenings becoming the property of the grain company upon payment of the shippers freight rate to the terminal. Grain diverted to U.S. routes will not pass through these terminals but take a more direct route to seaboard. Terminal elevators in the United States, in general, have only limited and rudimentary cleaning capability. This reflects the approach of that country to dockage in grain. The cleaning requirement for Canadian grain moving through the U.S. must therefore be met within the local areas of production. On the prairies relatively few elevators, including the "interior terminals," have the ability to clean grain to the standard required upon export. Consequently before grain could move in volume within prairie cleaning facilities would have to be expanded and upgraded.

2. Segregation of Grain Enroute Through the United States

In order for Canadian grain to be segregated from U.S. grain enroute, shipments must be made in volume. This is particularly so in the case of combination rail/barge movement when one barge has the capacity of several railway cars. Essentially, the grain should move in trainload lots. Relatively few delivery points on the prairies can support unit train movements of grain. This follows from the current organization of the primary elevator collection system which reflects transportation costs related to distance rather than cost on other than a collective basis.

Currently, Canadian National has applied, along with several grain companies, for rates \$1.50 per tonne lower than those applicable on a system basis for shipments of a minimum of 18 cars from a single origin to a single destination. The impact of this, even though approved by the Canadian Transport Commission, is by no means certain. The tendency will no doubt be, however, towards further consolidation of the railway/elevators network thereby rendering more feasible volume shipments. This consolidation would result in the cleaning of grain on the prairies becoming more economic.

3. Movement in Bond

Canadian grain moving for export through the U.S. must travel in bond through that country. Such segregation is feasible in the case of sealed railway cars which travel directly to seaboard. A different situation exists where trans shipment occurs as would be the case for barge movements. Transfer facilities from rail to barge would also have to be segregated—sealed bins for the Canadian grain being one alternative. In all cases, the grain would have to be segregated at the seaboard terminal to maintain its identity and ensure that no diversion occurs within the United States.

4. Inspection of Grain at Export Position

In order to ensure the grain exported conforms to the grade indicated, the grain would have to be inspected. In the past, Canadian Grain Commission inspectors have been located at U.S. elevators for grading purposes. This procedure can be expensive in relation to the volume of grain shipped. A feasible alternative would be for the Canadian Grain Commission to reach a reciprocal agreement with the Federal Grain Inspection Service whereby Commission inspectors would grade U.S. grain according to that country's specifications when that grain moved through Canada while the Inspection Service inspectors would grade Canadian grain by Canadian standards as that grain passed through the United States. This would enable each country's grain to be exported under its own Certificate Final.

5. Use of Canadian Government Hopper Cars

Prairies grain moves to Thunder Bay and Vancouver/Prince Rupert largely in government-owned hopper cars. The degree of use of these cars is reflected in the rate charged shippers for the movement of grain. These cars have been dedicated to shipment to these ports. While this restriction has recently been relaxed to allow their use in the movement of prairie grain beyond Thunder Bay, close control is maintained over these cars. Further relaxation of the restriction would be required for movements through the United States. While such cars would be ideal for unit train movement, extreme care would have to be exercised to ensure these cars were returned promptly and not diverted into the U.S. system. The additional hauling distance to the Gulf could result in a longer average turnaround time with an attendant increase in the number of hopper cars required, these having to be supplied by the railways and thus impadcting further on rail rates.

6. Definition of Export

"Export" as defined under the *Western Grain Transportation Act* in respect to grain "means shipment by vessel within the meaning of the *Canadian Shipping Act* to any destination outside Canada and shipment by any other mode of transport to the United States for use of the grain in that country and not for shipment out of that country."

This definition essentially removes any possibilities for grain moving for export through the U.S. from qualifying for the rates under the *Act*. Attempts to have this restriction removed have no as yet met with success. For the rail link between Canada and the United States, the grain, therefore, would move at commercial rates. Discussions with the major railway companies who would be involved in the movement indicate their willingness to quote through unit train rates for the entire journey, any increase in the rates from those under the *Act* adversely affecting movement through the United States.

7. Access to the United States System

In previous periods the comparatively limited flow of Canadian grain through the U.S. system posed few problems. The latter system was not operating at maximum capacity. The same situation applies today since U.S. export shipments of grain have fallen from a peak of 5042.4 million bushels in 1981 to 3017.7 million bushels in 1986. Decline in exports by route in millions of bushels are as follows: Great Lakes 1978 to 1986, 555.0 to 193.8; Atlantic 1982 to 1986, 606.4 to 168.0; Gulf 1982 to 1986, 3119.3 to 2057.8; Pacific 1984 to 1986, 925.1 to 545.8, Table III. It is apparent that a substantial volume

TABLE III
United States Grain Exports by Port Area

Year	Great Lakes	Atlantic	Gulf	Pacific	Total[1]
			million bushels		
1978	555.0	507.8	2617.3	517.1	4197.2
1979	510.5	585.8	2779.7	689.1	4565.1
1980	476.9	528.8	2909.6	917.9	4951.7
1981	491.7	516.6	3034.4	898.2	5042.4[2]
1982	391.2	606.4	3119.3	628.3	4757.9[2]
1983	296.9	367.4	2928.6	760.3	4454.9[2]
1984	336.4	342.3	2887.8	925.1	4576.8[2]
1985	218.9	349.2	2249.6	658.4	3551.8[2]
1986	193.8	168.0	2057.8	545.8	3017.7[2]

1 Includes interior exports
2 Includes sunflower seeds

SOURCE: Processed from Grain Transportation Situation, March 9, 1987.

of capacity remains unused at the present time in all transportation modes. Canadian grain moving to export is welcome in the United States at this time, some terminal elevators being used for storge rather than throughput. Large numbers of barges on the Mississippi remain idle and the potential revenue from Canadian grain would be of assistance to the river traffic. There is no guarantee, however, that this situation will persist indefinitely. With parts of the U.S. system (the river development projects in particular) constructed at public expense for strategic purposes, domestic users can be expected to have priority in periods when the capacity of the system is under pressure. Any diversion of Canadian grain through the U.S. system, therefore, involves a degree of risk. While U.S. grain is guaranteed access to Canadian transfer elevators, no reciprocal arrangements are in place since movement of Canadian grain through the U.S. has not occurred in recent years. A regular movement would require access to the U.S. system to be guaranteed, either by ownership of facilities or by other arrangements which may be easy to obtain if the present bilateral free trade negotiations are successful.

8. Pricing of Grain for Export

At the present time Canadian grain is exported under Canadian standards, identity of the grain being maintained to port of import. The grain is priced basis of port of export from Canada. Exports through the United States would require grain to be priced at point of export from that country. For Canadian Wheat Board grain this would essentially require the grain to be forwarded as agency stocks when in the United States. Since the Canadian Wheat Board has the right to price grain on a C.I.F. (cost, insurance and freight) basis at the import destination, it would seem reasonable to expect that the Board would price grain at U.S. ports. This price would incorporate the cost of movement from the prairies. This would represent an extension of present procedures though in a different country. Certain complexities would be introduced in that the cost of movement through the U.S. system varies over time and according to the volume shipped.

9. Underutilization of the Canadian Transportation System for Export Grain

The Canadian grain handling and movement system for export grain has benefitted from significant infusions of government funds accomplished by a variety of means, whether for the construction of elevators, increasing the lake fleet or the development of the St. Lawrence Seaway. The system, therefore, represents a complex combination of government and private funding and initiative. The whole is bound together by various labour and other agreements. Any diversion of Canadian grain through the U.S. system will reduce volume through a presently underutilized system further raising unit costs. The political fall out from such a diversion could be such as to give rise to Canadian restrictions on movement through the United States.

IV. ACTION ON THE CONSTRAINTS

Each of the constraints identified can be overcome. For some constraints this can be achieved collectively while others must be handled on an individual basis. Problems arising from the grain cleaning and segregation constraints can be jointly dealt with. The volumes required for grain cleaning and for proper segregation can be accumulated by further centralization of the handling and transportation system, a process which is on-going. Fewer delivery points result in a greater volume for each point rendering in-house cleaning more economic and enabling shipments to be made in volume. Relatively few delivery points at present are capable of sustaining large volume movements but the number can be expected to incresae progressively in accordance with the economics of the handling and transportation system, such being made apparent by release of data on the cost of maintaining certain branch lines and the savings attainable by volume movements. In addition, increased sophistication of equipment renders cleaning to export standard at other than terminal elevators feasible. Volume movement assists in the segregation of grain. Unit trains segregate grain from origin to destination, which could be from the point of shipment to point of export from the United States. For movements involving transfers from train to barge, large volumes would also assist in segregation since a unit train of grain will fill about six barges.

Administration of grain movements in bond is made easier where many cars are moved at one time notwithstanding that each car will be sealed. Barges can also be sealed upon transfer, car seals being relinquished for the barge seals. Inspection of grain to ensure grade is up to standard at point of export by Canadian Grain Commission inspectors becomes economically feasible with large volume movements. A reciprocal arrangement between Commission inspectors and those of the Federal Grain Inspection Service (F.G.I.S.) appears nonetheless to hold many potential advantages.

While use of Canadian Government hopper cars should result in lower rail rates on movement through the U.S. this is not a necessary condition for the movement. Substantial numbers of presently underutilized railway hopper cars are available, the supply far exceeding prospective demand. Backhauls would be available for a number of these cars on their return journey, for example, clay, salt and phosphate from New Orleans. This would also avoid the Government cars being diverted into U.S. traffic.

Guaranteed access to U.S. handling and transportation capacity could perhaps be accomplished by reciprocal arrangements covering U.S. movements using Canadian facilities. There would also be merit in the lease of U.S. elevator space so that access would be guaranteed, such space being available at the present time. No problem is foreseen in pricing Canadian grain at U.S. ports. For C.W.B. grains, it would merely involve incorporating the handling and transportation costs beyond the prairie interior elevator while making an adjustment for the transportation cost to the appropriate export position in

Canada. The diversion of grain through the U.S. system can be expected to result in opposition from principals in the Canadian movement system, primarily concerning underutilization and, therefore, reduced earnings. This opposition, on the other hand, should be offset, at least in part, by any additional producer net returns arising from use of the U.S. system. Furthermore, in many cases a different arm of the same firm may be involved in the movement as in the case of the railways. Indeed, if ore and grain traffic again expands to test the capacity of the St. Lawrence Seaway, any diversion of grain will delay or negate any need for extensive investment of additional capital in the Seaway plant.

Each of the constraints identified on movement of Canadian grain through the U.S. can be overcome. This will only be accomplished, however, if there is the will and initiative to seek out and use the alternatives available and to introduce new procedures by which the marketing of Canadian grain can be expedited. The economics of movement are rendering changes essential to the welfare of grain producers. The marketing system has shown the ability to adapt in the past. Consequently the constraints identified can be expected to be overcome as movement through the U.S. is demonstrated to be a viable economic alternative.

REFERENCES

1. Board of grain Commissioners for Canada, *Canadian Grain Exports,* Crop Years 1946/47 to 1955/56.
2. Canadian Wheat Board, *Annual Report,* 1951/52 and 1953/54.
3. Parliament of Canada, *Western Grain Transportation Act,* 1983.
4. Statistics Canada, *Grain Trade of Canada,* Crop Years 1918/19 to 1937/38.
5. U.S. Department of Agriculture, *Grain Transportation Situation,* March 9, 1987.

ENDNOTE

*University of Manitoba Transport Institute.

The Economic Cost of Transporting Grains From Farm to Market

*by N. Meyer and G. Sparks**

I. ABSTRACT

The impact on producers of adjustments in the Canadian grain collection system depends heavily on how the system evolves. If the system maintains the numerous delivery points which presently exist, producer transport distances will be relatively short but the benefits of system rationalization will elude them. However fewer assembly points mean generally longer hauls, although not necessarily higher transport costs for producers. Economies of size and scale in grain delivery can actually reduce total and variable transport costs for producers assembling their grain. The net result could be higher net margins to producers for grains.

Modifying the grain assembly system can bring about a number of changes which are beneficial to producers. First, because they have adequate on farm storage, they can deliver their grains to market throughout the year. Using larger trucks, producers can actually reduce the per unit cost of moving grains to the assembly points even if the distance is up to four times farther. This is because more use of transport capacity reduces per unit fixed costs. In cases where custom or commercial trucking is used costs per unit decline even more because equipment has more annual use.

II. INTRODUCTION

Many of the most recent innovations in grain handling have not been incorporated into the Canadian grain handling system. Rationalization of the Canadian grain handling system would be very likely to incorporate many of these innovations. That action has important implications for several different groups in Canadian society. These changes are likely to bring some costs and some changes in rural lifestyles to different segments of society. Therefore, there are strong vested interests in any change of the system.

Producer's perspective: Producers often envision two different aspects from the rationalization of the grain handling system. The first thought in many producer's mind is the additional distance to haul grains if local elevators are closed. That means additional expenses and time to transport grain. It may mean upgrading and resizing a truck or it may mean constructing additional on-farm storage to be able to keep combines running during harvest. Additional traffic on rural roads may also mean higher maintenance costs to rural municipalities and thus higher taxes. All of these imply additional costs. Producers want to know specifically what are the benefits to the individual producer of the proposed changes in the system? If producers receive some of the grains from efficiency, they may support changes in the system, if not they will oppose changes.

Grain firms perspective: The grain firms also have potential gains and costs associated with changes in the system. Presently they have a system setup for collection, processing and storage of grains. To change the system renders some of the facilities obsolete with few alternative uses. In addition, what about labor contracts and the need for fewer managers of a smaller number of high thru-put facilities. These firms have the potential of greatly increasing efficiency but at a cost of reduced facility values and labor contract settlements. Both can have a significant impact on the firm's profitability in the short run.

Rural municipalities perspective: A major concern of rural municipalities in system rationalization is abandoning branch lines. The concern is for additional traffic on roads causing increased maintenance costs. Abandoning rail lines could reduce the property tax base in some rural municipalities while increasing the tax base in communities where the assembly facilities are maintained or expanded. Reducing the service levels at elevators on lines to be abandoned could result in the transfer of visits to town by rural residents to the community retaining the assembly point. Rural municipalities are concerned that revenue to support additional road maintenance demands will not be available from the tax base. The net result could be inadequate revenue to meet additional expenses.

Province's perspective: Research has shown that the majority of the additional traffic resulting from hauling grain additional distance will be on the provincial highways leading to the rail sites where rapid loading of grains is possible (Calkins and Meyer). Maintenance and improvement of these roads is the responsibility of the province. Their concern is: will they end up with additional costs and no additional revenues?

National perspective: Grains are an important earner of foreign exchange for Canada. With the depressed grain prices each extra dollar of cost to get grains to tidewater inhibits Canada's ability to compete in world markets without government subsidies. Extra costs in the system reduce the price which can be paid to producers, shifts production out of these crops and puts additional financial pressure on them. A related national question is: could the funds being put into the grain handling system be used in another manner and improve the overall economy of Canada?

III. LITERATURE REVIEW

Sparks, G. "Investment Options Related to Rail Line Rehabilitation", CANADIAN TRANSPORTATION RESEARCH FORUM, 1985 pp. 334-358. First, rehabilitation of branchlines may actually cost more than the value of the grain being moved. A second issue is raised concerning can a duplicate system be afforded in a time of scarce resources? The problem is presented as the provinces providing the roads, therefore being reluctant to encourage activities which increase traffic without some form of compensation. The rural municipalities are also reluctant because they expect additional costs without compensation. The costs for branchline upgrading are paid by the federal government. Branchline abandonment could mean additional costs to rural municipalities and provincial governments and no compensating revenues. That is why local and provincial governments tend to oppose it.

Irwin, N. A., L. S. Sims and E. E. Hopkin, "The Role of Trucking in The Grain Industry", CANADIAN TRANSPORTATION RESEARCH FORUM, 1985 pp. 359-378. This report is an overview of changes likely to occur in transporting Canadian grains by 1990/91. They assume elevator numbers will continue to decrease while throughput per facility continues to increase. Farm to elevator transport costs are estimated at $2.35/tonne (Authors' note: our analysis indicated that only with producers working for zero wages could grain be transported for $2.30 per tonne), elevation charges at $6.50/tonne and rail charges are estimated to be $22.52/tonne for a total estimated transport cost of $31.37/tonne to an export point 1200 km distant. Farm to elevator costs are estimated to be 8% of total transport costs; elevation charges are 21%; and rail charges are 72% of transport costs for 1990/91.

Chaudhary, G. N., "Impacts of Reforms in the Statuary Rates on Trucking", CANADIAN TRANSPORTATION RESEARCH FORUM, 1986, pp. 14-25. Farmers are replacing their 2 axle (6,000 kg) trucks with 3 axle (14,000 kg) in Alberta. The number of 3 axle increased from 1% to 5.4% of registered trucks from 1976/77 to 1983/84. During the same period 2 axle trucks decreased by 7.7%. Commercial truckers can generally move grain for lower rates because the annual capital costs are spread over more annual miles and tonnes. However because the Western Grain Transportation Act pays the railroads rather than the producers, there is no incentive for the changes necessary to gain the benefits to occur.

Sparks, G. A., PRODUCTIVITY/OPERATIONAL IMPLICATIONS OF VEHICLE WEIGHT AND DIMENSION REGULATIONS, Clayton Sparks & Associates Ltd. Saskatoon, Sask. July, 1986. Sparks research points out that the variables of capacity (weigh or volume), distance wage rate and annual use have the greatest effect on cost variation in transporting commodities. In the case of bulk commodities, fuel consumption was also an important variable affecting cost per unit for transportation. Sparks developed a probabilistic model for estimating costs. He then goes on to show the efficiency gains to be derived by service suppliers if axle and gross vehicle weight restrictions were removed. Vehicle configuration (5 axle vs. 7 axle vs. 8 axle) also had a strong influence on productivity. The type of truck (flat bed, bulker or tanker) did not particularly effect productivity. A and B trains showed efficiency gains of 15 to 25 percent over 6 axle semi's. Axle and gross vehicle weight restrictions are given in the appendices.

Calkins, B. L. and N. L. Meyer, ADOPTION OF NEW MARKETING METHODS BY IDAHO GRAIN PRODUCERS, Bulletin 636, Ag. Experiment Station, Univ. of Idaho, 1985. Based on survey data, the adaptations of Idaho producers to the introduction of 3-5 and 25-26 car rail rates and barge traffic to Lewiston Idaho are described. When multicar rates were introduced, facilities were upgraded and new facilities constructed to load trains within the required loading time. The benefits of lower rates were shared among those providing transportation, handlers, and producers. Producers adapted to the new environment by transporting grain greater distances to take advantage of the higher bid prices being offered for grain. In North Idaho, with ready access to the Snake River, the average transport distance from on farm storage to market was 38 miles. The average transport distance for those selling to points capable of loading barges or multicar trains was 104 miles which was 66 miles greater in distance. South Idaho producers, who are dependent on rail and truck transport, shipped grains an average of 34 miles to traditional markets while those marketing to locations capable of loading multicar trains transported grains an average of 85 miles one way, a distance 51 miles farther. In spite of the greater distances, by using larger trucks which traveled greater distances annually, the total and variable cost of getting grains to locations capable of loading rail cars decreased on a per unit basis as compared to 2 axle farm truck deliveries.

Karadininis, K. & G. G. Storey WESTERN CANADIAN GRAIN HANDLING AND TRANSPORTATION SYSTEM; THE IMPACTS OF VARIABLE FREIGHT RATES, Dept of Ag. Economics, Univ. of Sask. Saskatoon, Sask. July, 1986. A model is developed to simulate the introduction of 20, 40, and 100 car variable freight rates for grain. The discounts are respectively 1.50, 1.75, and 3.00 dollars per tonne. The effect of these rate breaks was to concentrate deliveries to fewer points. Alternative one resulted in an additional 31% of the delivery points being closed. Alternative two resulted in 20 more delivery points being closed. Efficiency gains come from reducing the number of miles of track to be maintained and from increases in volumes handled by existing elevators. Farmers greater distances from delivery points would be required to haul grains greater distances thus enduring additional transportation costs. The authors do not make adjustment for technical change in producer delivery of grains in this study. Lack of cost information from the railroads also makes the estimation of gains to the railroads somewhat speculative.

IV. METHODOLOGY

Objective: To estimate the producer's cost of delivering grains to assembly points under alternative assumptions. Thus permitting evaluation of the impacts resulting from rail line and assembly elevator rationalization. The following cases were used as a basis for estimation.

(1) base case—two axle farm truck,
(2) three axle truck,
(3) three axle tandem pulling a pup,
(4) five axle semi and
(5) A-train

Three systems of service provision were evaluated: a producer hauling his own grain, a producer hauling his own grain plus that of six other producers, and commercial service.

Theory: Costs of transporting grain from farm to assembly point can be classified into fixed costs and variable costs. Fixed costs refer to costs having a constant annual value unrelated to the level of utilization. Variable costs refer to those varying with the quantity of use. Total cost is the summation of fixed and variable costs as shown in equation 1.

(1) TC = FC + VC
TC = total cost
FC = fixed cost
VC = variable cost

(2) FC = D + In + L + I + T
D = annual depreciation
In = annual insurance
L = licenses
I = interest on average investment
T = annual cost for tarps

Fixed costs (2) are divided by the annual kilometers of truck use to determine the per kilometer cost. Grain transports part of the total cost are then estimated by the number of kilometers traveled to deliver the producer's annual grain sales.

(3) VC = L + Ti + RM + F
L = labor cost to transport grain
Ti = annual tire cost for grain transport
RM = repairs and maintenance costs for grain transport
F = fuel cost to transport grain

Variable costs (3) increase or decrease with the level of use. Labor is based on the time required to load, transport to the delivery point, unload, and return to the loading point. In the commercial operation an additional fee is paid to the driver for loaded tonne kilometers. Tire cost is based on the return kilometers covered in delivering grains to assembly point and a per kilometer cost. Repairs and maintenance are also based on a per unit rate and the distance covered hauling grain. Fuel is the efficiency (Km/1) times the return distance traveled transporting grain and the cost per liter of fuel.

Economies of scale theory indicates there are gains from specialization. Free trade theory is based on the assumption that the maximum of scale economies can be gained with each producer specializing in what he can do best. Economies of size refers to individual long run average cost curves. Each average cost curve has a minimum cost point for production. This analysis deals with economies of size for individual Prairie grain producers as well as economies of scale in permitting some to specialize in grain production and others in custom or commercial grain transport.

Scale economies dictate that a given service can be provided in alternative ways but there is usually a least cost method at each level of output. In the physical movement of grains, the most efficient tool depends on several factors. Distance, capital investment, operating costs and flexibility are important factors to evaluate. As capital investment increases, the utilization rate becomes more important. As operating costs increase, it becomes more important to control the increases. This is particularly true in the cases of fuel and labor costs for the transport of farm grains to market. Theory tells us that larger loads, more productive kilometers per year, and more rapid traveling speeds help to increase the efficiency of trucks transporting grains.

The movement of grain from on-farm storage to the assembly point for loading unto rail cars can be divided into three parts. The first part is the loading of grains from farm storage to trucks. The second part is the movement by truck from the farm storage location to the assembly point. The third part is the unloading of grains at the assembly point. A fourth part of domestic transportation is from the assembly point to export or domestic utilization location. This last part of the transportation may be entirely by rail, truck or a combination of rail, truck and water transportation. This paper deals with the assembly of grains, the first three parts.

Farm storage to delivery point shipping requires loading the truck at the farm. The time required for this operation depends on the equipment used. Larger trucks are assumed to require more time to load. Loading of grain is assumed to be on a units per time period basis. Loading includes setting augers, opening bins, loading and cleaning up complete. These values differ for the quantity of grain shipped, the number of different types handled and size of the storage facility. Total cost and variable cost per unit would be expected to change with the situation.

Transporting grains to the assembly point involves movement from the farm on a grid road to the provincial highway and then transport on the highway to the assembly elevator. This distance will vary for deliveries to local elevators and to more distant multicar loading sites (Calkins & Meyer). Travel distance from farm storage points to the assembly points will likely be greater to larger assembly points than to the local assembly locations. This travel distance will generally be covered at highway speeds. The additional distance on provincial highways can be covered much more rapidly and the additional cost is only the cost of moving down the road, no waiting, loading or unloading time.

Transporting grains involves fewer trips if the trucks are larger for transporting the same quantity of grain. These larger trucks are less likely to have overloaded axles because they have more axles under them. Load on roads per square inch of tire is likely to be less because larger trucks must pass over scales and are more likely to travel on main roads. Larger loads require fewer trips over the road, therefore fewer vehicles to meet. Costs to the producer may also be less depending on the haul distance and truck size.

The third portion of the cost is the unloading cost at the assembly point. Because many of the locations accepting greater quantities of grain have larger capacity facilities, the required unloading time will be less. These are the reasons: the larger trucks are of bottom dump design which unload faster; because of larger loads fewer trucks need to be unloaded; and larger capacity pits and legs permit faster handling of grains. These factors translate into potentially increased capacity and lower costs for grain handling.

V. ANALYSIS

To analyze the options which producers have for moving their grains under alternative arrangements five vehicles were evaluated under assumptions of delivery to assembly points 10 km. up to 160 km distance from the farm. These alternatives were evaluated with respect to different wage rates for the driver and with different levels of vehicle utilization. The impact of fuel efficiency on the cost of grain assembly was then evaluated. Combining all the options defines a decision tree which can help policy makers, as well as producers decide which alternatives are best for a given situation.

A decision tree is an outline of the different steps in making a decision (Howard). It notes the alternatives available at each decision point. The number of variables at any decision point as well as alternative outcomes entered into the decision model. Figure 1 outlines a decision tree for a producer deciding how to deliver his grain to an assembly point. First he has the options of doing it himself with his present equipment or hiring it done. If he decides to hire, then it is a matter of who is available to provide the service, what quality of service is likely to be offered and what the cost will be. If the producer decides to do it himself, then he must decide if it should be done with his current equipment or with different equipment. if the producer decides different equipment is needed, should it be new or used? What size should it be? What is the cost trade-off versus the probability of breakdown with new versus used equipment? Is there enough volume to warrant the addition of a pup to the main trucking unit to increase capacity? Should custom work for neighbors be done to increase utilization and reduce per unit costs? Each of these alternatives is a potential answer to the producers question of what is the best for my overall situation?

BASE CASE

A basic model was set up to evaluate the costs of delivering grain various distances in alternative transport units. The traditional two axle farm truck was used as a base. (The basic assumptions and data are available from the authors upon request). For comparison, producers were assumed to deliver their grain 10, 20, 40, 80, and 160 km. to the assembly elevators. Each farm was assumed to have 353 tonnes of grain to transport annually to market. Data from the University of Saskatchewan Top Management Program indicate each farm had an average of 1.6 trucks being used to deliver grain in 1985. Based on the assumptions in the model, if the average delivery distance is 10 km. (6.2 mi.), the producer's estimated variable and total delivery cost would be respectively $4.54 and $7.90 per tonne (Table 1). Converting to a used tandem would reduce a producer's total cost $2.45 per tonne to $5.45 for the same distance haul. Adding to pup behind the 3-axle truck would permit hauling over 20 kilometers for less than is presently being paid. A neighbor hauling with a used semi or a used A-train could transport the grain up to 40 kilometers for less than the present cost of transporting with a two axle truck. The most economical for the producer would be to use a commercial A-train or hire custom transport from a neighbor. Both of these could transport his grain up to 80 kilometers (50 mi.) for less than the present estimated total cost for transporting grain. A producer choosing not to be paid for his labor could transport grain the 10 kilometers for a total cost of $5.67/tonne and a variable cost of $2.30/tonne.

Figure 1

Decision tree for a producer concerning grain transport from farm to assembly point

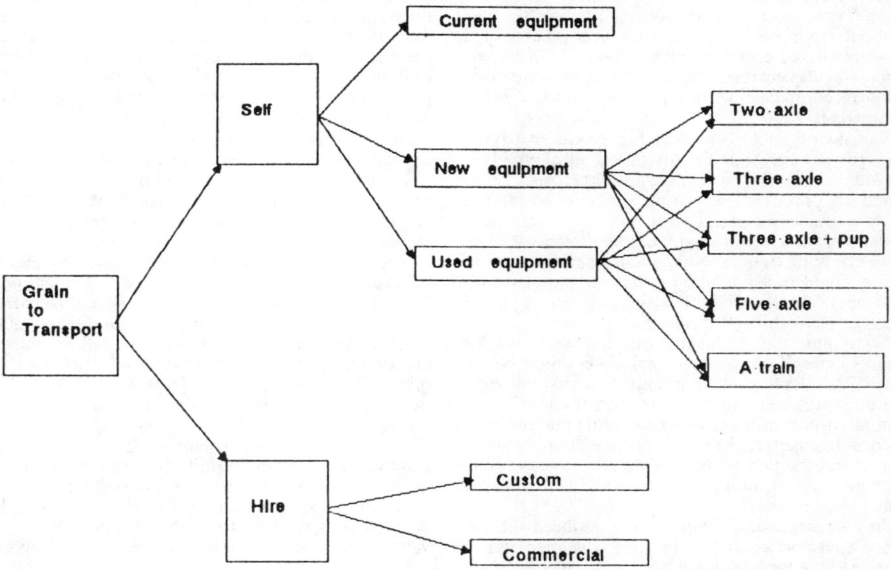

Table 1
Total and Variable cost per tonne for grain assembly

		Kilometers				
Equipment	Costs	10	20	40	80	160
2-axle	total	7.90	11.16	15.34	21.75	38.26
	variable	4.54	6.42	9.38	14.77	27.03
3-axle	total	5.45	8.70	13.21	18.83	27.86
	variable	2.84	4.32	6.61	9.95	16.12
3-axle	total	5.15	7.39	11.57	15.96	22.19
/pup	variable	3.45	4.30	5.91	7.96	11.32
5-axle	total	5.25	8.94	14.70	22.45	31.61
	variable	2.52	4.01	6.42	9.94	14.79
A-train	total	3.69	6.47	11.15	18.16	27.17
	variable	1.76	2.84	4.70	7.60	11.68
Custom						
3-axle	total	2.96	3.93	5.35	7.79	13.37
+pup	variable	1.68	2.14	2.98	4.76	8.47
5-axle	total	3.88	5.21	7.58	12.16	20.19
	variable	1.98	2.66	3.99	6.80	12.00
A-train	total	2.86	4.11	5.58	8.71	15.17
	variable	1.47	2.04	2.81	4.62	8.47
Commercial						
A-train	total	**	**	3.55	5.60	9.60
	variable	**	**	2.65	4.40	7.83

** A commercial operator could not generate adequate revenue to cover cost transporting these short distances.

Custom service could deliver it 20 km. in a 3 axle truck plus pup or a 5 axle semi for less than the cost of delivering it 10 km. in a 2 axle unit.

The total cost, variable cost and variable cost without labor cost show the decision points for a producer with a two axle truck (Figure 2). Decisions start with the current equipment and variable cost. That is compared with total costs for all other alternatives.

Considering total costs and a producer paying himself $10 per hour for his labor, all alternatives considered reduce per unit transport costs. Compared to producer per tonne variable costs, the custom and commercial total transport costs were lower for delivering grain equal distances. The greatest reduction is with a commercial A-train which could transport the producer's grain over 40 kilometers for less than he would expect to pay in variable costs doing it himself.

Examining the producer's total transport cost for grain (Table 2) shows that total costs could be reduced by adapting other trucking alternatives over the present predominant 2-axle truck system. Doubling transport distance to 20 km while purchasing a used 3-axle or hiring custom service could actually reduce transport costs. In the case of a custom 3-axle with pup or a custom A-train, it could be by almost half.

In all cases using a larger truck reduced the per tonne transport costs for the same distance and in certain cases for additional distances. Figures 3 and 4 show cost reductions were greatest when trucks are used for custom work or commercial service (high km annual usage example). In that case the estimated costs per tonne for delivery were lower even though distances were up to four times greater. In the cases of custom 3-axle plus pup and commercial A-train the distance up to eight times greater cost less per tonne to deliver to the assembly point. The differences tended to get larger as the delivery distance increased. Even in cases where the producer did not charge a labor cost for transporting his own grain, the commercial and custom services cost less than he presently spends to transport his grain to the assembly point 10 km from his present grain storage.

The majority of producers have at least one truck, therefore the decision criteria for an individual producer is the variable cost which must be covered. If the variable cost of delivering grain to the assembly point in the producers truck is less than the commercial or custom rate, it pays the producer to haul his grain with his present unit. Table 3 shows the additional cost of moving grains by alternative size trucks additional distances. Larger trucks reduce the transport costs for all distances. Use of custom or commercial services permits hauling grains up to four times greater distance for less cost. If a new or different truck is needed, then full costs must be used as a basis for decision making rather than variable cost because new unit costs must be paid.

Figure 2
Two axle farm truck grain per tonne delivery cost

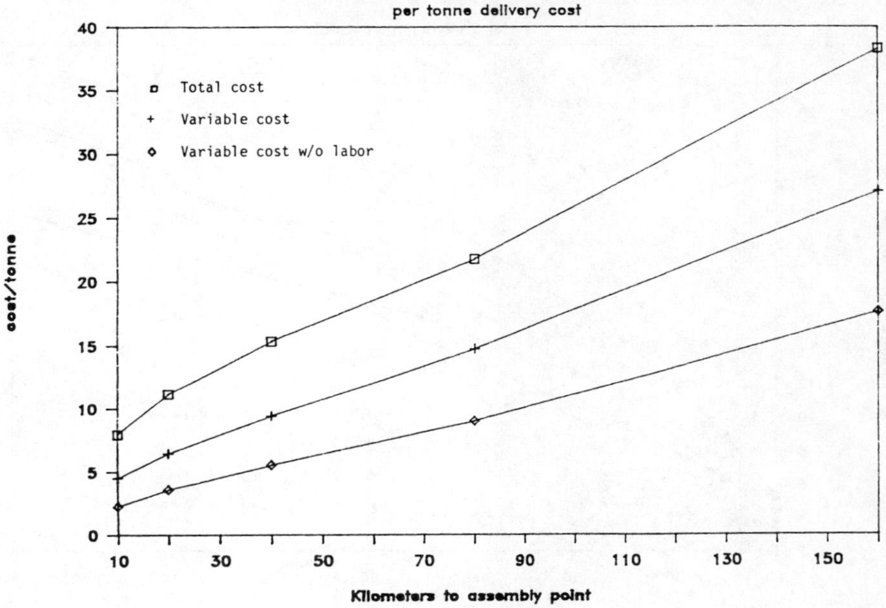

Table 2
Total cost to deliver 353 tonnes to assembly points

Equipment	Units	Kilometers				
		10	20	40	80	160
2-axle (5.6)	total cost cost/tonne	2789 7.90	3940 11.16	5415 15.34	7678 21.75	13507 38.26
3-axle (12)	total cost cost/tonne	1922 5.45	3070 8.70	4464 13.21	6646 18.83	9834 27.86
3-ax + pup (24)	total cost cost/tonne	1818 5.15	2608 7.39	4083 11.57	5632 15.96	7833 22.19
5-axle (22)	total cost cost/tonne	1854 5.25	3157 8.94	5188 14.70	7925 22.45	11156 31.61
A-train (35)	total cost cost/tonne	1303 3.69	2283 6.47	3937 11.15	6411 18.15	9592 27.17
Custom 3-ax + pup (24)	total cost cost/tonne	1045 2.96	1387 3.93	1889 5.35	2749 7.79	4721 13.37
5-axle (22)	total cost cost/tonne	1368 3.88	1859 5.27	2677 7.58	4292 12.16	7129 20.19
A-train (35)	total cost cost/tonne	1008 2.86	1452 4.11	1969 5.58	3076 8.71	5353 15.17
Commercial A-train (35)	total cost cost/tonne	400 1.13	605 1.72	1013 2.87	1816 5.15	3409 9.66

Figure 3
Two axle variable cost versus total costs for alternatives

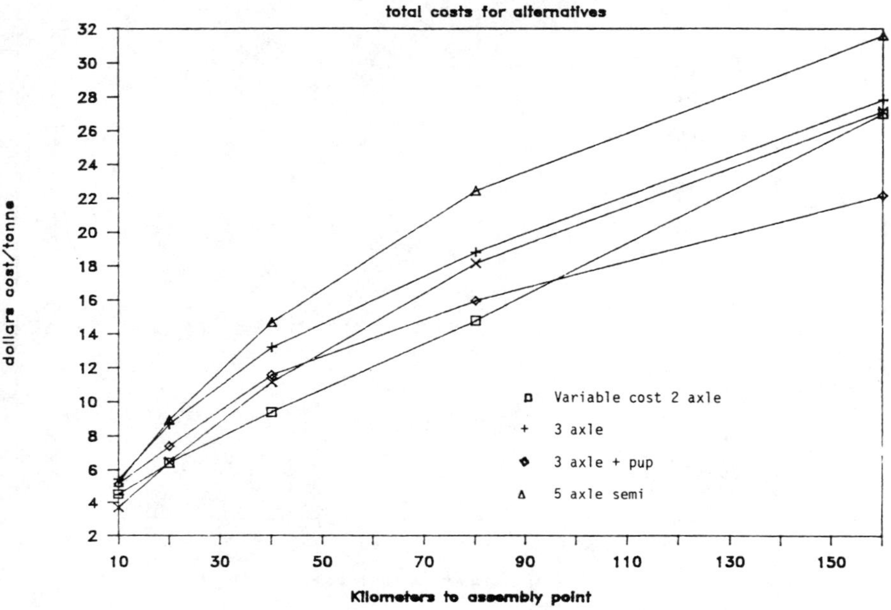

Figure 4
Two axle variable versus custom/commercial total costs alternatives

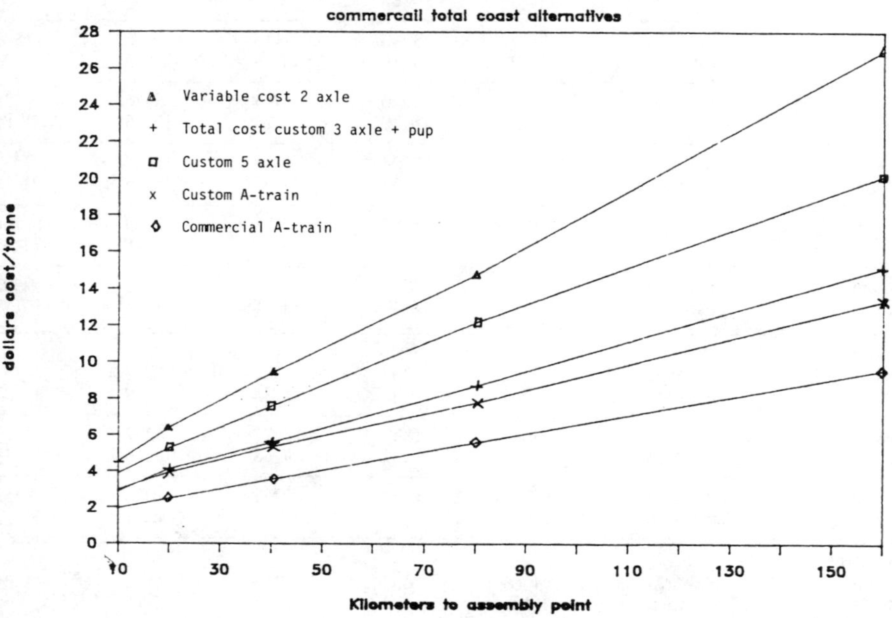

Table 3

Additional total cost of transporting 353 tonnes of grain additional distances over base cost of $2789.

Equipment	Base	Kilometers			
	10 km	20	40	80	160
	($)	($)	($)	($)	($)
2-axle	0	1151	2626	4889	10718
3-axle	−867	−282	1876	3857	7045
3-axle + pup	−971	−181	1294	2843	5044
5-axle	−935	368	2399	5137	8368
A-train	−1486	−506	1148	3622	6803
Custom					
3-axle + pup	−1744	−1402	−900	−39	1932
5-axle	−1421	−930	−112	1503	4340
A-train	−1781	−1337	−821	287	2565
Commercial					
A-train	N/A	N/A	−1776	−973	621

SENSITIVITY

Solution sensitivity to various parameters was tested to determine how changes in various factors effect the per tonne cost of transportation. Subjectively, to estimate sensitivity to changes in parameters, a 20 percent over and under variation in each variable was tested. The most important factor was truck capacity for 2-axle and 3-axle units hauling 40 kilometers. This resulted in a 42 percent cost per tonne kilometer variation for a 2-axle truck, and a 39 percent variation for a 3-axle truck, both hauling 40 kilometers (Tables 4 and 5). The next two most important variables for both units were the annual kilometers driven and the purchase price of the unit.

As can be expected commercial truck operations were most sensitive to annual kilometers, delivery distance, truck capacity and drivers wages. A plus or minus twenty percent variation in each of these resulted in a 28.6 percent variation in cost per tonne kilometer (Table 6).

Sensitivity to increasing or decreasing the fuel efficiency was small. It effected total delivery cost

Table 4

Deterministic sensitivity for operating costs of a two axle farm grain truck hauling 40 kilometer (24 miles)

Inputs	Typical value			Associated productivity			
				costs/tonne kilometer			
	−20%	most likely	+20%	min	expected	max	%
Tonnes haul annually	282	353	424	.38	.38	.38	0.0
Annual travel	5424	6780	8136	.44	.38	.35	23.6
Delivery distance	64	80	96	.41	.38	.37	10.5
Delivery time (min)	104	130	156	.36	.38	.43	18.4
Truck cap (tn)	4.5	5.6	6.7	.48	.38	.32	42.1
Plates & Ins	118	147	176	.38	.38	.39	2.6
Truck cost	13920	17400	20880	.34	.38	.43	23.7
Service life	8	10	12	.42	.38	.36	15.8
Salvage value	2088	2610	3132	.38	.38	.38	0.0
Tarp cost	800	1000	1200	.38	.38	.38	0.0
Tarp service	8	10	12	.38	.38	.38	0.0
Fuel cons (km/1)	1.7	2.1	2.5	.40	.38	.37	7.9
Fuel cost/1	.27	.34	.41	.37	.38	.40	7.9
Tire cost ($/)	202	252	302	.39	.38	.38	2.6
Tire life (km)	64000	80000	96000	.38	.38	.38	0.0
Interest rate	.09	.11	.13	.37	.38	.39	5.3
Drivers wages/hr.	8.00	10.00	12.00	.36	.38	.40	10.5

Table 5

Deterministic sensitivity for 3 axle farm grain truck delivering 40 km (24 mi).

Inputs	Typical value			Associated productivity			
	−20%	expect	+20%	min	expect	max	% var
Tonnes /year	282	353	424	.33	.33	.33	0.0
Km/ year	3792	4740	5688	.39	.33	.29	30.3
Del dis (ret)	64	80	96	.34	.33	.32	6.1
Del time (min)	112	140	168	.32	.33	.34	6.1
Truck cap tn	9.6	12	14.4	.41	.33	.28	39.4
License & ins.	120	151	181	.33	.33	.33	0.0
Price truck	24000	30000	36000	.28	.33	.38	30.3
Service life	8	10	12	.37	.33	.30	21.2
Salvage value +	3600	4500	5400	.33	.33	.33	0.0
Tarp cost	800	1000	1200	.33	.33	.33	0.0
Tarp ser-yrs	8	10	12	.33	.33	.33	0.0
Fuel con (km/1)	1.7	2.1	2.5	.34	.33	.33	3.0
Fuel cost/1	.27	.34	.41	.32	.33	.34	6.1
Tire cost	298	373	448	.33	.33	.33	0.0
Tire life	96000	120000	144000	.33	.33	.33	0.0
Interest rate	.09	.11	.13	.32	.33	.34	6.1
Driver wages	8.00	10.00	12.00	.32	.33	.34	6.1

per tonne 7.9 percent for the two axle, 3 percent for the 3-axle and insensitive for the commercial A-train.

One option simulated was to have a producer do commercial work for six additional producers. This was called a custom operator. This increased the volume over which the fixed costs were allocated and reduced per kilometer costs. Figure 3 shows the effect of having more units over which to allocate the fixed costs. Doing custom work for others reduces the per unit cost of transporting the producer's grain because it reduces per unit fixed cost. Certainly a question to consider is, does the individual have adequate labor to begin another business. If so, it could be a viable part time employment for a producer. Certainly in all cases the costs per unit transported were reduced by hauling greater volumes with the same unit.

TRIPS

For many producers the second thought concerning grain assembly point consolidation is that of road maintenance and expected higher taxes associated with longer hauls. One result of using larger trucks is fewer loads are required to transport the same quantity of grain. As shown in Table 7, it takes 10 A-trains to move the same quantity of grain as 63 2-axle farm trucks. Traffic on roads could be reduced by 84 percent by switching grain transport to larger trucks. However there would still be traffic concentrations near larger capacity assembly points. Road damage rates would be expected to decline rapidly a few miles distant from the assembly point.

VI. SUMMARY AND POLICY IMPLICATIONS

The above factors reduce the marginal per unit cost of moving grain the additional distance to higher capacity loading sites. The net result is that under certain circumstances the cost of delivering grain can decrease permitting a greater net return to the producer despite the greater transport distances. Trucks transporting the grain may be commercial or custom operations or producer owned. The key to greater efficiency and lower per unit cost is higher

Table 6
Deterministic sensitivity for high annual use A-train grain truck delivering 40 km (24 mi)

Input	min	expect	max	min	exp	max	var
Tonnes yearly	282	353	424	.07	.07	.07	0.0
Km/ year	80000	100000	120000	.08	.07	.07	28.6
Delivery dist	64	80	120	.08	.07	.06	28.6
Delivery time (min)	152	190	228	.07	.07	.07	14.3
Truck cap (tonne)	28	35	42	.08	.07	.06	28.6
License & ins	2758	3448	4138	.07	.07	.07	0.0
Truck price	104000	130000	156000	.07	.07	.08	14.3
Service life	8	10	12	.07	.07	.07	0.0
Salvage value	15600	19500	23400	.07	.07	.07	0.0
Tarp cost	2000	2500	3000	.07	.07	.07	0.0
Tarp life	8	10	12	.07	.07	.07	0.0
Fuel con (km/1)	1.7	2.1	2.5	.07	.07	.07	0.0
Fuel cost/1	.25	.31	.37	.07	.07	.07	0.0
Tire cost	385	481	577	.07	.07	.07	0.0
Tire ser	149000	187000	224000	.07	.07	.07	0.0
Pup tire cost	184	230	276	.07	.07	.07	0.0
Pup tire ser	104000	130000	156000	.07	.07	.07	0.0
Interest rate	.09	.11	.13	.07	.07	.07	0.0
Driver wages	15.07	18.84	22.61	.06	.07	.08	28.6

Table 7
Trips required to transport 353 tonnes of grain with different capacity trucks.

Truck axles	Net capacity tonnes	Trips	Axle trips
2	5.6	63	126
3	12	30	90
3 + pup	24	15	75
5	22	16	80
7	35	10	70

annual volumes to spread fixed costs over a larger number of units to be moved.

Hauling grains greater distances does not necessarily cost more. It depends totally on how the producer hauls the grain the additional distance. If the producer is expected to change the manner in which he is doing business and be exposed to new risks, he must receive compensation for his efforts. If he does not, he will not willingly change. Presently producers have little incentive to change. Will the efficiencies to be gained by elevator rationalization be shared with producers to an extent they are more than compensated for their extra transport expense? This analysis indicates that the additional cost to producers of transporting grain varies widely depending on the assumptions used for equipment and transport distance. Larger equipment with adequate volume reduces total transport cost for the producer. The cost reductions more than compensate for the additional transport distances. This analysis does not consider road maintenance costs but does point out that the larger trucks with more axles require fewer trips to move the same quantity of grain. That means fewer vehicles on the road.

VII. NEEDS FOR FURTHER RESEARCH

This paper reports on estimates of additional costs to producers of transporting grains greater distances under several assumptions. This research needs to be evaluated in conjunction with other parts of the grain handling system to access the impact system rationalization will have on producers and how the benefits need to be allocated among participants in the system for equitable treatment.

REFERENCES

1. Calkins, B. L., and N. L. Meyer, Adoption of New Marketing Methods by Idaho Grain Producers, Bulletin 636, University of Idaho Agricultural Experiment Station, Moscow, Idaho, 1985.
2. Chaudhary, G. N., "Impacts of Reforms in the Statuary Rates on Trucking", Canadian Transportation Research Forum, 1986, pp. 14-25.
3. Irwin, N. A., L. S. Sims and E. E. Hopkin. "The Role of Trucking in the Grain Industry", Canadian Transportation Research Forum, 1985, pp. 359-378.
4. Karadininis, K. and G. G. Storey. Western Canadian Grain Handling and Transportation System: The Impacts of Variable Freight Rates, Department of Agricultural Economics, Technical Bulletin TB 86-02, University of Saskatchewan, Saskatoon, Saskatchewan, July, 1986.
5. Sparks, G. A. Productivity/Operational Implications of Vehicle Weight and Dimension Regulations, Clayton, Sparks & Associates Ltd., Saskatoon, Saskatchewan, July, 1986.
6. Sparks, G. A., "Investment Options Related to Rail Line Rehabilitation", Canadian Transportation Research Forum, 1985, pp. 334-358.
7. Appendices (Methods of data development, copies of the data and the model used for this analysis are available from the authors.

ENDNOTE

* Department of Civil Engineering, University of Saskatchewan.

Transit Service Distribution—Evaluating Level of Service

by M. A. Gallagher and M. C. Ircha***

I. INTRODUCTION

Municipal public services compete for scarce government funding at both the provincial and municipal levels. Government subsidization is prevalent in the financing of urban public transit systems and has prompted the need for a systematic framework to analyze transit level of service by geographic area and urban population segmentation. The results of such an analysis addresses some of the distributional equity questions such as "who receives the transit services" and "who pays for urban transit". Equity oriented data will provide an information base for decision-making when funding levels are being set and budgets being established.

A procedure to evaluate distributional equity of urban public transit was developed for the regional transit system (Metro Transit) in Halifax, Nova Scotia in 1985. The procedure is applicable to other Canadian transit systems and indeed can be adapted to analyze distributional issues for most 'measurable' urban public services. The two main data requirements for this analysis are transit operation statistics and demographic (Census) characteristics. The locations and frequencies for Metro Transit's 39 bus routes formed the basis for assessing the supply of transit service available to area residents. Geographic and socioeconomic segmentation was carried out using a combination of data sources: 1981 Statistics Canada Census, a 1981 Regional Transportation Travel Demand and Household Survey (1) and 1985 demographic statistics.

Metro Transit serves the four municipal units of Halifax, Dartmouth, Bedford and the urban portion of Halifax County. For analysis purposes, the region has been subdivided into 129 traffic zones. These zones are also aggregated into 19 residential areas for ease of presentation ad comparing data results for this paper.

II. DEFINING THE POPULATION

Using various social and economical indicies, the urban population was defined both spatially and by socioeconomic status (SES). The Blishen Scale (2) was employed which assigns a numerical value to a population group based on reported occupation. The value represents each occupation's level of education, income and prestige. The 1981 Canada Census data set of 22 occupational groups by sex and by census tract is the information source used in the calculation of the Blishen Scale for the Metro Transit area. Figure 1 shows the data manipulation process used to derive an average SES for the traffic zones.

500 Blishen Scale Occupations (by sex)
↓
22 Major Occupational Groups as defined by Statistics Canada (by sex)
↓
Labour Force 15 years and over by Major Occupation Groups (by sex) for Metro Area
↓
Weighted Average SES (by sex)
↓
Weighted Average SES by Census Tract
↓
Average SES by Traffic Zone (using conversion factor)

Figure 1. Socioeconomic Status Calculation Procedure

The choropleth diagram (Figure 2) depicts the Metro region using the Blishen Scale. Note that these values have been updated to reflect 1985 demographic statistics. Similar visual displays of the Metro Transit area can be produced based on population and employment densities, family income levels, age and other characteristics.

III. DEFINING THE LEVEL OF TRANSIT SERVICE

For the purposes of this research, level of transit service was defined as transit capacity and accessibility for a given geographic area. Transit capacity or the number of available transit seats for a given time period and transit travel time for area residents are the yardsticks used to define level of service. ribbon or catchment band was used. This assumes that a transit route within 450 metres (as the crow flies) is considered accessible. Figure 3 shows those portions of the Metro region which are within the defined transit service catchment area.

Transit capacity was calculated using transit segments or links which were defined for the entire Metro Transit system. Using a computer program designed for this research, the number of available bus seats per hour was tallied for each transit segment within each traffic zone. A series of transit route frequency data sets were accessed depending on the time of day and day of week. These data files therefore consisted of the segment numbers within each traffic zone, the routes operating on each segment and the route capacity (frequency X bus seats per unit) for each time period.

Service coverage for a traffic zone was considered to be the percentage of the area's population within

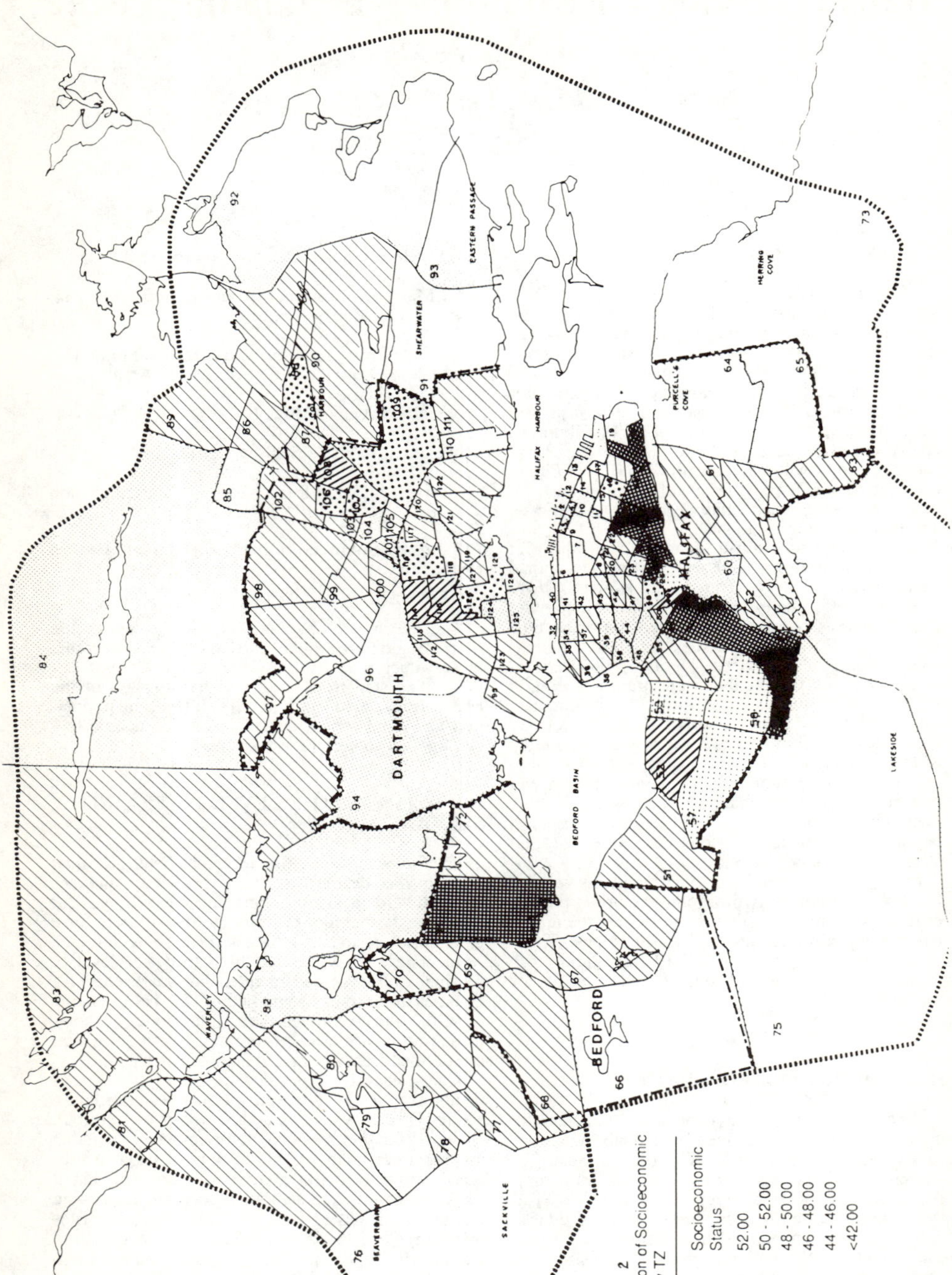

FIGURE 2
Distribution of Socioeconomic Status by TZ

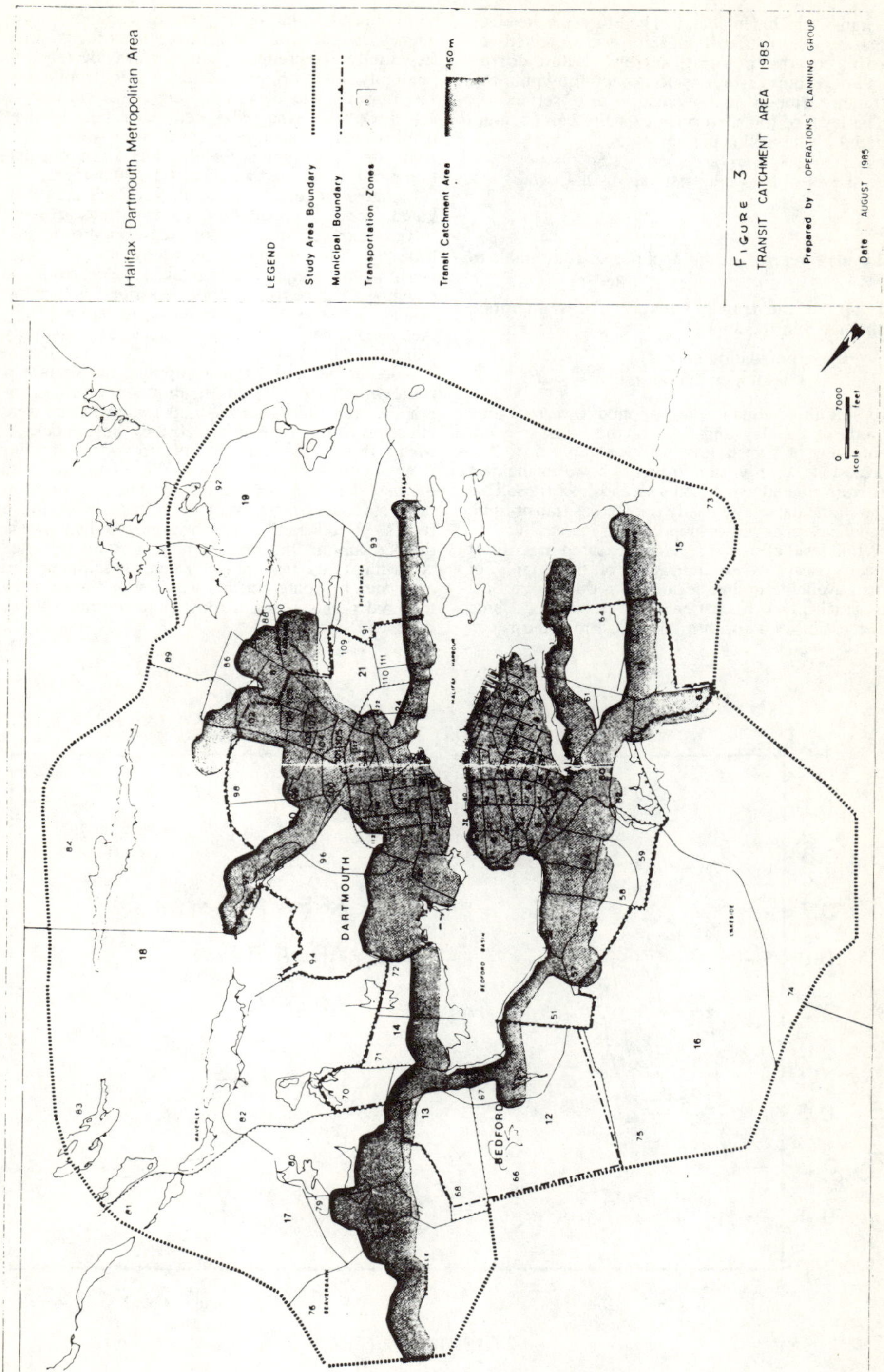
Halifax - Dartmouth Metropolitan Area
LEGEND
Study Area Boundary
Municipal Boundary
Transportation Zones
Transit Catchment Area
450m
FIGURE 3
TRANSIT CATCHMENT AREA 1985
Prepared by: OPERATIONS PLANNING GROUP
Date: AUGUST 1985
DARTMOUTH
BEDFORD
0 7000
scale feet

the transit catchment band. Therefore the level of transit service or transit capacity was weighted according to the population served and correspondingly, areas which were not totally within the 450 metre 'ribbon' had reduced level of service.

The level of transit service capacity can be summarized by the following equation:

$$\text{Level of Service}_N = (\text{Cap.}_j^i)_N \text{ X Cov.}_N$$

where:

Level of Service$_N$ is the # of bus seats available to area N

(Cap.$_j^i$ is the transit capacity for segments i through j in the area N

Cov.$_N$ is $\left\{\frac{\text{population served}}{\text{total population}}\right\}$ of area N

This simplified equation was applied to various time periods of the day and days of the week to yield summaries of transit level service capacities. The calculations were written in Basic, however the data sets were created using Lotus spreadsheet files. This allowed the data to be easily updated as transit route frequencies changed or populations increased.

While level of transit service measured in terms of capacity provides an indication of the supply of transit available to area residents, it does not reflect the actual travel trip patterns and trip times. Using an accessibility algorithm, existing trip patterns can be simulated resulting in plots of cumulative frequency diagrams of transit travel times for an area. Essentially, the computer program was designed to randomly select origin-destination pairs from a data set, however the data was weighted according to actual travel patterns in the Metro area. For instance, if 40 percent of all transit travel from Area i is to/from Area j, then the probability that origin-destination pair ij is selected would also be 40 percent. The cumulative frequency diagrams for each area provided a visual comparison of the trip times involved in transit travel for different areas. Figure 4 illustrates the results of comparing three areas in Halifax-Dartmouth which are a similar distance from the downtown business core. If one considers 60 percent of all transit travel as the criterion for comparison, then Area A has superior travel times or accessibility than Areas B and C.

Another measure of transit supply is the amount of route kilometres within the boundary of a geographic area. Odometer readings for each route were allocated to the areas in which they were incurred. Metro Transit maintains an inventory of route kilometres, however similar data could be collected with the use of a planimeter and a system map or computer storage of the x and y coordinates of digitized routes. "Deadhead" or non-revenue kilometres were divided among the areas which a route travelled according to the percentage of revenue kilometres in each area by route. In this way, no one area at the terminal points on each route bore the majority of deadhead kilometres.

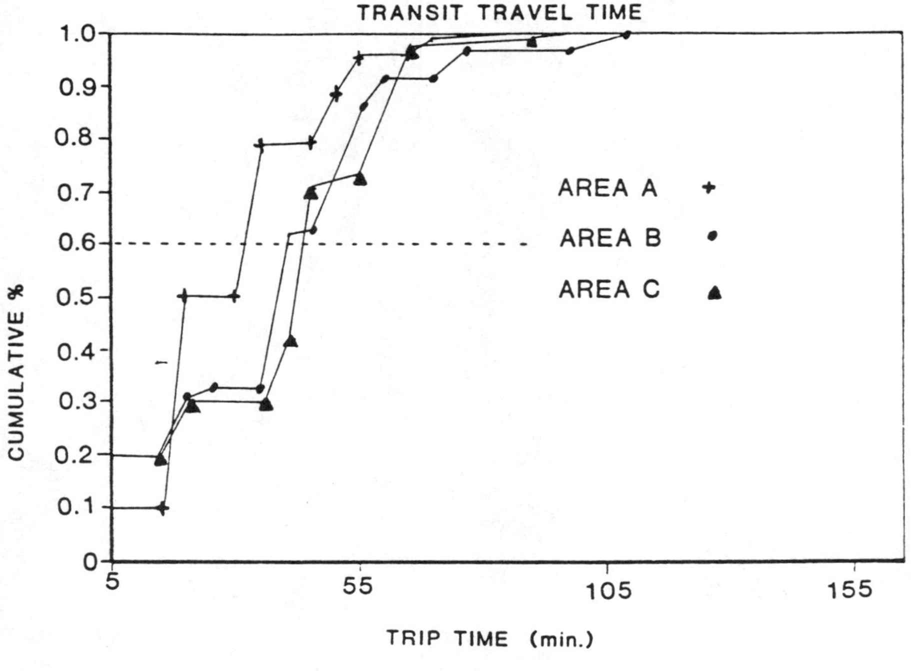

Figure 4. Accessiblity

IV. DEFINING TRANSIT USAGE

For all transit operations, the predominant productivity measure is system ridership. Either a system wide or route level, ridership statistics provide a meaningful indication of performance and transit usage. For an area by area analysis however, ridership must be further refined to indicate where passengers alight and board each route. Electronic farebox and passenger counting systems are available which can allocate passenger ridership to a specific bus stop location. For area analysis a less sophisticated method of data collection was used, i.e., the passenger loadcount surveys. Loadcount surveys or on/off counts are normally gathered on a route by route basis for various station points. From the loadcount data, annual passenger ridership (revenues plus non-revenue passengers) figures by route can be allocated to specific areas based on the percentage of route ridership boarding in each area. Figure 5 illustrates the accounting procedure used to divide passenger loads by area.

Figure 5 illustrates an ideal situation where loadcount stations are located on the area boundaries. In reality, as in the Metro Transit system, the data collection stations do not conform exactly to area boundaries and thus the information was massaged to reflect actual passenger loading in each of the 19 residential areas.

Passenger ridership was recorded for each area at the station of boarding, therefore a return trip from a suburban area to the downtown core would be allocated as a trip in *both* trip end location areas. The ridership summaries therefore indicated a high percentage of trips accredited to major trip attraction centres such as the CBD.

A more elaborate and expensive method of defining transit demand is through household surveys of transportation patterns. In 1981, such a Travel Demand Survey was carried out in the Halifax-Dartmouth area. From this survey, origin-destination tables were produced including transit travel tables. From the O-D table transit trips or transit usage can be tallied on an area basis. To update the O-D table through correlation with annual ridership data, the base year O-D table must include all transfer transit

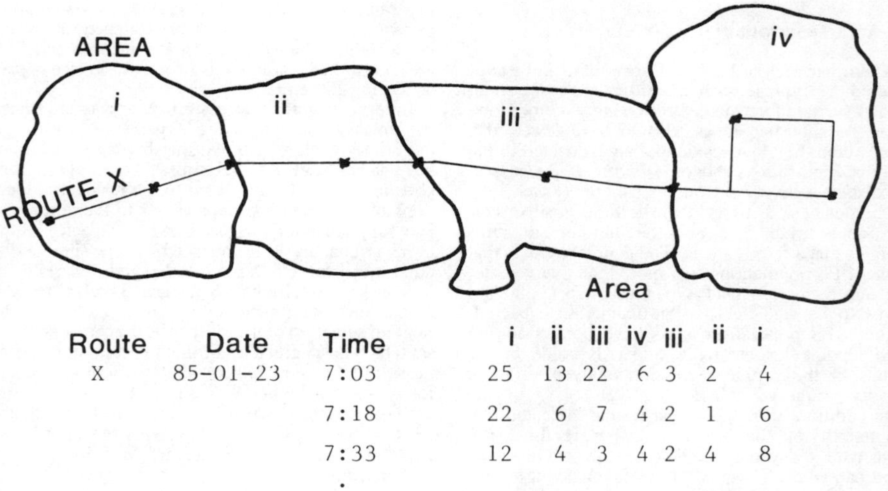

			Area						
Route	Date	Time	i	ii	iii	iv	iii	ii	i
X	85-01-23	7:03	25	13	22	6	3	2	4
		7:18	22	6	7	4	2	1	6
		7:33	12	4	3	4	2	4	8
		⋮							
		23:03	2	0	4	13	7	2	1
	85-07-12	7:03	22	8	6	2	8	5	9
	⋮	⋮							
		TOTAL T	t_i	t_{ii}	t_{iii}	t_{iv}	t_{iii}	t_{ii}	t_i
		%	$\frac{t_i}{T}$	$\frac{t_{ii}}{T}$	$\frac{t_{iii}}{T}$	$\frac{t_{iv}}{T}$	$\frac{t_{iii}}{T}$	$\frac{t_{ii}}{T}$	$\frac{t_i}{T}$

Figure 5. Allocation of Passenger Ridership by Area

trips, not just the travel trip ends. Therefore the O-D survey questionnaire should be structured to include origins and destinations as well as the transfers required to complete each transit trip.

The travel demand survey results also provided transit trip generation statistics such as trips per household, by sex and income group, travel patterns by time of day, and trip purpose. This type of comprehensive study clearly defines transit usage patterns and provides a good socioeconomic breakdown of the transit patrons.

V. ANALYSIS OF TRANSIT SERVICE DISTRIBUTION

Three factors have been described which form the basis for transit service distribution analysis:

1. area socioeconomic description
2. transit supply by area
3. transit demand/usage by area

A comparison of these elements yielded interesting results pertaining to the supply and demand of urban public transit in the Metro area.

A. Area Description

The urban area of Halifax-Dartmouth can be considered a 'typical' urban centre with suburban sprawl surrounding the central business core. However, the suburban areas tend to have lower SES values than the older residential areas located in the Halifax Peninsula as shown in Figure 2. Because of the distances involved from the suburbs (lower SES) to the business districts, there is little positive correlation between SES and the numbers of automobiles that a family owns. The general trend is that higher SES populations own more than one vehicle, but our regression analysis of this aspect did not prove to be statistically significant. As expected, higher SES population areas have corresponding high income levels. Table 1 details some of the results of the simple regression analysis of these socioeconomic variables.

In addition, crosstabulation results indicated that the majority of elderly persons live nearer the downtown core and youth populations are greater in the suburban areas. These two transit dependent groups are therefore situated at different geographic locations in the Metro Transit area.

B. Transit Level of Service

A comparison of transit supply and the socioeconomic status of area residents did not yield results which were not statistically significant, but did indicate trends. Thus on a global basis, transit in the Metro area was not distributed to favour particular SES population groups. The data trends indicated that with increasing SES, transit capacity also increased. This can be verified by the fact that as distance to the CBD lessens, SES increases and transit capacity increases. The level of service capacity distribution by time of day for weekday periods followed a similar trend. Sunday and Holiday service level capacities tended to serve the older residential neighbourhoods and correspondingly higher SES areas.

Graphical interpretation of supply by area and socioeconomic groups can also be accomplished by using a series of overlay choropleth drawings. By setting the shading grid to indicate similar transit dependent groups with similar shading intensities, overlain transparencies will highlight the most transit dependent areas. As shown in Figure 6 transit capacity illustrated by columnar bar graphs overlaying transit dependency areas provides a visual comparison of supply by area. A brief glance shows that those lightly shaded areas (lower income and SES areas) also had lower levels of transit service capacity during the a.m. peak hours.

In terms of transit accessibility, or average transit trip times, an analysis using 60 percent of all trips as the criterion, demonstrated that higher SES groups have lower transit travel times. Once again, this phenomena can be attributed to the residential locations of higher SES groups closer to the core areas and supplied with increased transit capacity. However, a subanalysis of areas which are similar distance from the CBD produced regression statistics indicating that higher SES groups also receive higher level of transit accessibility. Note that the random selection of transit trips is drawn from a data set which is weighted according to actual transit trips taken and thus may indicate that travel times are less for persons of higher SES due to the fact that they are less willing to travel on time-consuming transit trips. If the origin-destination pairs are chosen entirely at random with no regard to existing travel patterns, then those areas of similar distances from the CBD with higher SES still experience less travel times, but the trend is less significant.

TABLE 1

Socioeconomic Variables
Simple Linear Regression Results*

Variables	Correlation Coef. r	Level of Significance**
SES—Household Income	(+) 0.825	1%
SES—Low-Income Families	(−) 0.560	1%
SES—Population Density	(+) 0.129	REJECT
SES—Elderly Population	(+) 0.185	REJECT
SES—Distance to CBD	(−) 0.386	1%
SES—Auto Ownership	(+) 0.119	REJECT

* Further statistical tests have also been performed but are not presented here.

** Degrees of freedom will vary depending on the sample size. The data sets of the variables are analyzed on a census tract basis (r = 56) and on a traffic zone basis (r = 117).

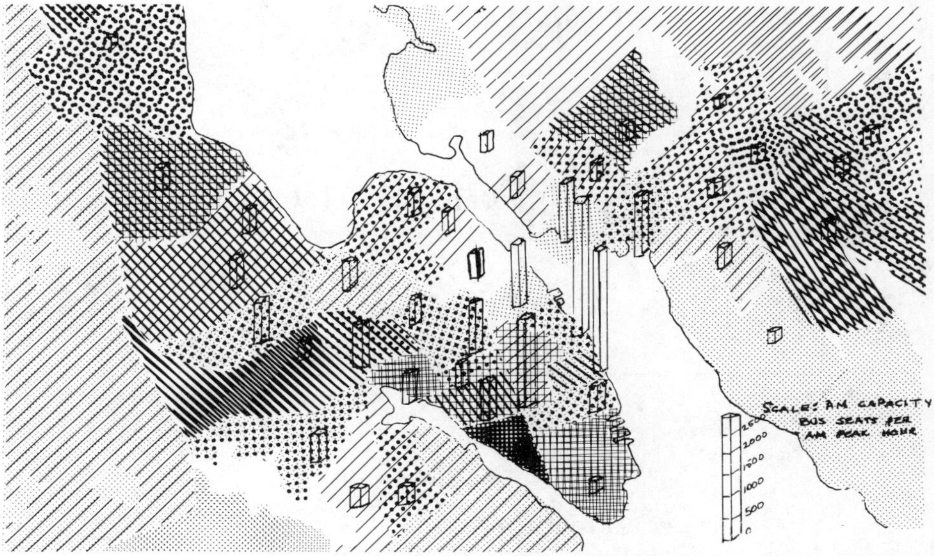

Figure 6. SES and A.M. Peak Capacity (Bus Sector)

C. Transit Demand/Usage

Using the 1981 Household Travel Demand Survey and ridership by area data sets, transit travel patterns can be reviewed. As one would expect, as the number of automobiles per household increases, the number of daily transit trips per household decreases. Also, a simple regression run of SES and a.m. peak hour transit trips/household indicates a reverse correlation whereby higher SES households take fewer transit trips. This trend is also evident in off-peak travel and Sunday and Holiday transit travel. These statistical patterns help to define the transit dependent population as those with lower SES values and a lower incidence of auto ownership.

To further examine transit use by household groups, those areas with similar levels of transit capacity available to them have been aggregated for analysis purposes. Low service levels (less than 150 bus seats per a.m. peak hour or a frequency of 30 minutes and less) for area residents showed that the lower SES groups were most apt to increase the number of transit trips as capacity levels increased.

SES and the number of transit trips per a.m. peak hour were negatively correlated, but the increase in transit use rose sharply with increased capacity areas (between 450 and 1200 bus seats per a.m. peak hour of frequencies of 10 and 4 minutes), the negative correlation between SES and transit trips remained, however any increase in service capacity had less of an affect on the travel patterns of lower SES groups ($\Delta_3 < \Delta_1$ in Figure 7B). Mathematical interpretation of the propensity to travel by transit for various for various income groups as a function of the available level of service can therefore be derived. Those areas with the highest transit capacity are omitted from this subanalysis because travel to these areas is a function of the trip attraction centres, not the residence based trips of the area. The Halifax core traffic zone has a capacity of 5400 bus seats per a.m. peak hour and a resident population of only 700 persons (1985). In this instance, employment populations would be a more realistic variable to assess transit trips.

The manipulation of crosstabulation of the variables by areas of similar capacities produced useful factors which can be used in transit trip behaviour modelling. For a given transit capacity, values for expected transit trip generation rates by population groups of different SES, income levels, distance from the CBD or auto ownership can be found. Rigorous statistical analysis is required to ensure that the regression results are in fact correlated and not attributable to causation only.

As previously outlined, ridership by residential area was tallied from passenger loadcount surveys. An average fare per rider was assigned to yield total passenger revenue by area. System costs by area were calculated from route kilometres in the area multiplied by an average cost per kilometre of service. A resulting value of total transit subsidy per area and per capita was then determined. It should be noted that in areas of major trip generation such as the CBD, the high levels of revenue recorded for trip beginnings tend to be offset by the costs of large number of route kilometres throughout the area.

A simple regression analysis comparing SES of the residential area with the subsidy per capita in the area indicated a slight trend that higher SES groups receive less transit subsidy per person, however the trend could not be proven beyond the r^2 coefficient of regression. The SES groups receiving the highest subsidy were the middle income residents living away from the CBD. The highest SES groups received the least subsidy and in fact, two such areas generate revenues in excess of the cost of providing

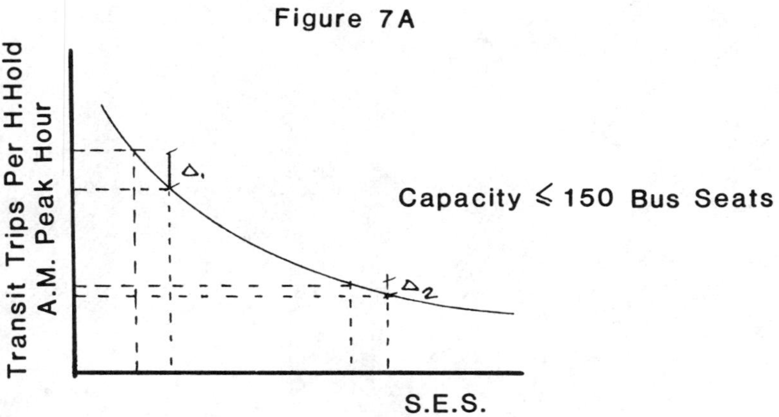

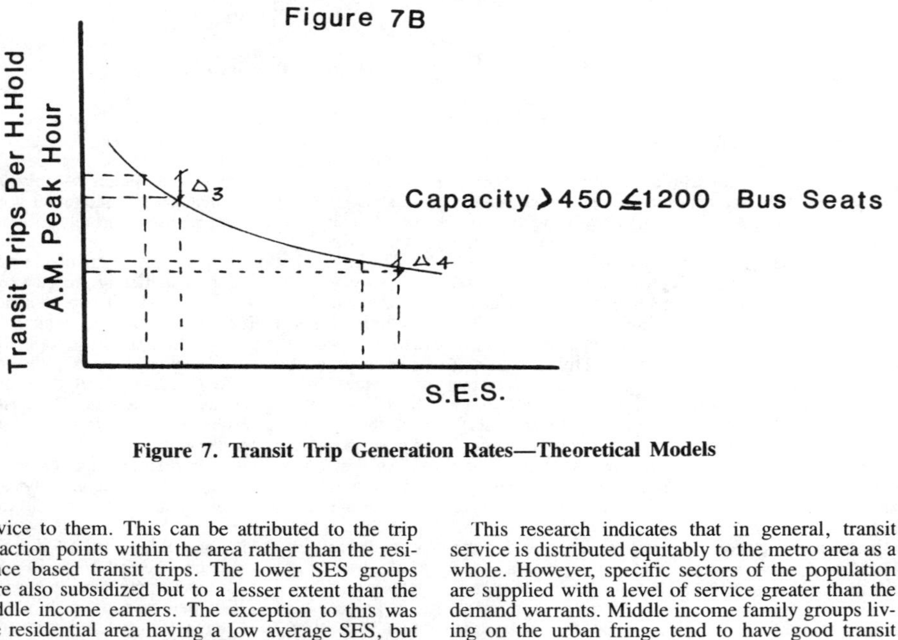

Figure 7. Transit Trip Generation Rates—Theoretical Models

service to them. This can be attributed to the trip attraction points within the area rather than the residence based transit trips. The lower SES groups were also subsidized but to a lesser extent than the middle income earners. The exception to this was one residential area having a low average SES, but the highest calculated subsidy per person of $169 annually. This is an indication of a purely "social" transit service which is not based on a capacity equivalent to the demand/use for the transit service.

VI. SUMMARY

A practical framework has been presented outlining supply and demand models of transit service distribution. The case study of Halifax-Dartmouth was used to illustrate the simplicity of the data requirements and the modelling approach.

This research indicates that in general, transit service is distributed equitably to the metro area as a whole. However, specific sectors of the population are supplied with a level of service greater than the demand warrants. Middle income family groups living on the urban fringe tend to have good transit access to routes which travel between the suburban areas and the downtown core and yet ridership revenue does not yield much of the service kilometre costs in the area, therefore subsidy per capita levels in these areas exceed the system average. Subsidy levels are less for higher SES groups and lower SES populations.

Most of the mathematical interpretation of the variables correlated indicates that for level of service indicies, as SES values climb, so does the level of service. As explained, this observation is attributed to the location of the higher SES groups living closer to the CBD. Transit trip times or accessibility echo this trend and the model developed illustrates that increasing SES groups have lower trip times by transit. Subarea analysis of those traffic zones of similar distance to the CBD indicates that higher SES groups have shorter transit travel times, but the level of significance is less than with the regional results.

Transit trip behaviour observations also became apparent from this research. The lower SES groups are more likely to increase their propensity to travel as the level of service capacity increases (low frequency of < 30 minutes) than higher SES populations. Factors such as trip generation rates by household for a given level of service can also be calculated from the data collected for the study. Other trip rates are easily tallied from the disaggregate analysis of the various models presented.

This paper outlines a simplified process consisting of data manipulation models for use as a decision-making tool in the distribution of transit service. It is hoped that future transit service will be distributed based on sound transit planning principles, however an analysis such as this helps to point out the inefficiencies and inequities of existing and proposed transit service. The models are easily set up on personal computers with BASIC language and spreadsheet software packages. LOTUS has been used to establish the data sets of socioeconomic and transit systems operations variables and accessed via BASIC programming. The editing and graphics capabilities of LOTUS added to the ease of data manipulation. Although this analysis is data intensive, the variables can be obtained from the 1986 Statistics Canada Census material by census tract and from the operating transit system. These two sources of data provide adequate information to carry out the analysis without the necessity of a comprehensive travel demand and household information survey.

The paper highlights the modelling involved and some of the statistics used to evaluate the equity of transit service distribution on Halifax-Dartmouth Metro Transit. This decision-making tool can be used for transit planning of service options when operating with scarce government funds.

ACKNOWLEDGEMENT

We wish to acknowledge the support of an operating grant provided by the Natural Science and Engineering Council (NSERC) for our continued studies of equity in the provision of urban services.

REFERENCES

[1] *Regional Transportation Travel Demand and Household Survey,* Metro Authority, Halifax, NS, 1981.

[2] Blishen, B.R., "Socioeconomic Status—500 Occupational Codes", unpublished, 1986; See also, Blishen, B.R. and Carroll, W.K., "Sex Differences in a Socioeconomic Index for Occupation in Canada", *Canadian Review of Sociology and Anthropology,* 15:3, 1978; and, Blishen, B.R. and McRoberts, H.A., "A Revised Socioeconomic Index for Occupations in Canada", *Canadian Review of Sociology and Anthropology,* 13:1, 1976.

ENDNOTES

* London Transit Commission.

** University of New Brunswick, Fredericton, NB, Canada.

Port State Control: Developing a Canadian Position

*by Mary R. Brooks and Wade Elliott**

I. INTRODUCTION

The harsh competitive environment created by tonnage overcapacity in all international shipping markets in recent years has created fundamental changes in the nature of vessel ownership, registry and management, in the seafaring labour market and in the *politics* of shipping.

In the early 1980s, surplus shipbuilding capacity, cheap yard credits and easy access to government subsidies and ship finance lured more players into the shipowning arena and by 1985 surplus capacity was estimated at 24.3 percent of the total merchant fleet.[1] More and more of the traditional maritime fleets are flagging out in an attempt to cut costs and survive in this era of severely depressed freight rates. The ensuing result has meant that links between ownership/management and the flag state are frequently less well defined as many of the open registries have neither the intention nor the resources to exert greater control.[2]

Concurrently, increasing competition amongst open registries, crew-supplying nations (predominately Third World), classification societies and underwriting syndicates have abetted a further deterioration of crew welfare and safety standards. Perhaps at the root of the problem is the inevitable rogue owner who, one step ahead of his creditors, runs his ships to the ground.

A direct result of these changes has been a considerable increase in the operation of substandard unsafe ships, and the consequent problems of ship-generated marine pollution, strandings, founderings, collisions, and human rights abuses at sea. Any list of defects, deficiencies or infractions which might be generated to identify such ships would be enormous, and include such factors as inoperable (or non-existent) fire fighting and life saving equipment, the refusal to pay back wages or to repatriate crew, and the absence of certificated personnel or oil pollution prevention certificates. The objective of Port State Control (PSC), which seeks to redress some of these problems, might best be summed up in the phrase "safe ships and clean seas".

Where once the establishment and maintenance of standards on merchant ships was largely the responsibility of the country of ship's registry, coastal states and host ports are increasing their efforts to prevent the problems caused by substandard ships. In July 1985 the Paris Memorandum of Understanding (MOU) was signed by 14 European nations, and articulates the principle of Port State Control (PSC) in which the host country accepts the responsibility for enforcing internationally-accepted standards of shipboard operation regardless of the flag of registry and the maritime practice of that registry. Under this principle, deficient ships can be detained until remedial action is taken. Japan, and more recently Korea, have begun implementing PSC based on the European model and other developed nations, like Australia, are showing increasing interest in the concept. In the tanker trades, the US has already unilaterally instituted a measure of port state control through its US Coast Guard Tanker Boarding Program and this move is supported by the American shipowning community.[3] The US has begun talks with the European MOU states and Japan about possible future implementation of full PSC. This paper will address the question: What should be the Canadian position on Port State Control?

II. THE DEVELOPMENT OF PORT STATE CONTROL

On March 17, 1978, the heavily laden VLCC *Amoco Cadiz* grounded on the Brittany coast of France, spilling 230,000 tonnes of crude, and focussing world attention on the issue of marine pollution. The ship was owned by Amoco International Oil Company, a subsidiary of the American multinational Standard Oil, and Amoco was subsequently found negligent in the maintenance and repair of the vessel, and in the failure to adequately train its crew.[4] It was this disaster which served as the final incentive for the adoption of PSC principles in January of 1982 through the adoption of the Memorandum of Understanding on Port State Control. Those countries initially party to the MOU are listed in Exhibit 1.

EXHIBIT 1

Parties to the MOU on Port State Control

Belgium	Denmark
Finland	France
Federal Republic of Germany	Greece
Ireland	Italy
The Netherlands	Norway
Portugal	Spain
Sweden	
United Kingdom of Great Britain and Northern Ireland	

While the *Amoco Cadiz* disaster played a predominant role in the accelerated acceptance of Port State Control, the concept itself is by no means a new one, having been firmly founded in maritime law during the time of the Roman Empire and its modern basis can be traced to the Digest accompanying the Justinian Code.[5] The concept of PSC has been furthered by the general adoption of the Plimsol mark and the 1930 Loadline Convention, the 1914 and the 1929 Safety Conferences arising from the loss of the *Titanic,* and the 1948 Conference which adopted the first modern Convention on Safety of Life and Sea. In more recent times, the 1969 Civil Liability Convention placed the onus on its contracting parties to apply its measures to tankers of non-contracting states. The move from the Grotian principle of freedom of the seas to one of national management was most evident in the inclusion of this concept in Articles 216, 218, 220 and 226 of the 1982 Law of the Sea Convention.[6]

The objective of the Memorandum is to assist in securing compliance of ships with international standards regarding the safety of life at sea, the prevention of marine pollution, and the working and living conditions on-board ship. The standards are those established by the conventions listed in Exhibit 2 and are referred to as the 'relevant instruments' of the MOU.

EXHIBIT 2

International Conventions and Amendments Enforced Through Port State Control

International Convention on Load Lines, 1966
International Convention for the Safety of Life at Sea, 1974 as amended by the Protocol of 1978 (SOLAS 1974/78)
International Convention for the Prevention of Pollution from Ships, 1973, as modified by the Protocol of 1978 (MARPOL 1973/78)*
International Convention on Standards of Training, Certification and Watchkeeping for Seafarers, 1978. (STCW 1978)**
Convention on the International Regulations for Preventing Collisions at Sea, 1972. (COLREG 1972)
Merchant Shipping (Minimum Standards) Convention, 1976 (ILO Convention No. 147)*

* Ratified by all MOU parties but Portugal and Ireland
** Ratified by all MOU parties but Italy, Netherlands and Portugal

A significant feature of the Memorandum lies in its concept of 'no more favorable treatment' whereby Port States can enforce even those relevant instruments to which the flag state may not be a party. However, a Port State can apply only those relevant instruments which are in force and to which it is a party. It is this which would prove to be the primary hurdle in gaining Canadian implementation of PSC as it exists in Europe.

III. IMPLEMENTATION AND EFFECTIVENESS IN EUROPE

The parties to the MOU practice PSC by means of the inspection process. Each country is committed to inspecting 25 percent annually of individual *foreign* flag merchant vessels which enter its ports. The inspector is qualified and authorized by the port state concerned, and the inspection consists of a preliminary examination of certificates and documents relevant to the conventions of importance to the MOU. A detailed inspection of the ship is only conducted in those cases where the documents are not in order, and there are clear grounds for believing that the ship does not comply with the standards.

In the United Kingdom, during the first two years of PSC enforcement, 62 substandard ships were detected and detained.[7] Overall, the MOU reports that 428 ships were detained or delayed due to serious deficiencies in 1984-85, down only slightly from the 436 detained in 1983-84.[8] PSC can only be considered effective if evidence of a declining detention rate exists, and the number of minor deficiencies uncovered drops.

As can be seen in Exhibit 4, the effectiveness to date has been questionable given these particular criteria. To quote the Third Annual Report on the implementation of the MOU: "The number of deficiencies and of substandard ships have not yet declined significantly . . . an unflagging effort will remain necessary."[10] However, to be fair, neither has the magnitude of the substandard shipping problem been determined nor has the target inspection rate been met.[11]

EXHIBIT 4

Deficiency Record

	Percentage of Total		
Major Deficiencies[9]	1982-83	1983-84	1984-85
Ship's Certificates	6.93	9.53	13.11
Life Saving Appliances	29.69	25.81	23.68
Fire-fighting Appliances	16.89	18.83	16.87
Safety in general	12.09	10.38	10.58
Load Lines	6.27	5.32	4.94
Navigation	11.39	13.51	12.13
Subtotal	83.26	83.38	81.31
Other	16.74	16.62	18.69
Total	100.0%	100.0%	100.0%
Total Number of Deficiencies	**8352**	**13436**	**14002**

Sources: Annual Reports, MOU, various years.

The principles of PSC must also be more universally adopted if they are to be effective in a global sense; we cannot rely solely on the MOU states to promote the principles of "safe ships and clean seas" for all, and certainly more time is needed to determine if PSC will be effective in achieving its intent.

IV. THE CANADIAN EXPERIENCE WITH SUBSTANDARD SHIPS

On January 27, 1985 the Colombian-registered general cargo vessel *Monica B* limped into Halifax harbour with a 15 to 20 degree list under the watchful eye of Canada Customs, which has suspected the vessel of carrying drugs. After pumps were put aboard to lighten the ship, it was soon apparent that the cargo the vessel was to load was non-existent. The Halifax Port Corporation held the ship but became frustrated in its efforts to trace the owner. The crew were owed back wages and were living in unsanitary conditions without heat or adequate food and water. Efforts to trace the owner(s) were unsuccessful and the vessel was eventually sold. In the meantime the crew of the *Monica B* were stranded in Halifax for over nine months.

In a similar case, the Panamanian registered *Devon III* ran aground 50 nautical miles south of Halifax on November 17, 1983 but was soon refloated and towed to a local shipyard. Its owner vanished after receiving more than $500,000 in insurance claims, and the ship was arrested for debts to the salvage and drydock companies, a Saint John shipping agent, and the motel which accommodated the crew after the rescue effort. The crew of Guyanese, Filipinos and Grenadians, some of whom had not been home for more than two years, were owed more than seven months of back wages and sought to be repatriated. After another year and a half sitting in dock accumulating more bills, the *Devon III* was sold at a sheriff's sale for $6,000 and six months later was towed out to sea and sunk by the Navy.

In a more recent incident, a Charlottetown pilot experienced difficulty in docking a Panamanian-registered bulk carrier because of problems with the ship's electrical power supply. The pilot reported the incident to Canadian Coast Guard's Ship Safety Branch (SSB). Subsequent Coast Guard inspections later revealed a host of defects: an inoperable generator and another with operating difficulties created an unreliable supply of power; defective wires and inoperable brakes on derrick winches made cargo handling dangerous; severe corrosion of decks, bulwarks, hatch openings and hatch covers as well as problems with hatch securing devices jeopardized the watertight integrity of the ship; and many more.

The vessel was declared unseaworthy and detained; Canada Customs clearance withheld until the defects were corrected. Both the Panamanian Consulate and the vessel's classification society were informed. Repairs were undertaken and the ship eventually permitted to sail with a cargo of pulpwood destined for Europe.

It is difficult to ascertain the extent and degree of foreign flag substandard shipping activity in Canadian waters. When incidences are identified, we are left to wonder how much activity goes undetected, the potential costs of which may be quite staggering. Besides local port, repair, salvage and other charges, elusive owners often avoid the costs of relief, repatriation and medical treatment of their abandoned crews. Furthermore, if substandard shipping activity escalates, the costs of pollution clean-up and search and rescue activity will likely increase.

At present, responsibility for monitoring foreign flag shipping activity primarily rests with the Ship Safety Branch of the Canadian Coast Guard (SSB). Its responsibilities are derived from the international conventions and domestic legislation in Exhibit 5.

EXHIBIT 5

The Basis for Existing Port State Control Powers in Canada

Regulation 19 of Chapter I of SOLAS 1974/78;
Article 21 of the 1966 Load Line Convention;
Non-Canadian Ships Compliance Certificate Regulations;
Non-Canadian Ships Safety Order;
Navigation Appliance Regulations;
Ships Deck Watch Regulations;
Safe Manning Regulations;
Charts and Publications Regulations;
Certain Provisions of COLREG, 1972;
Oil Pollution Prevention Regulations.

Involvement can be initiated in one of two ways: (1) as a result of a routine inspection by a Coast Guard surveyor to verify that there are valid certificates on board; or (2) as a result of a complaint or report which can come from any individual with an interest in the safety of the ship, its crew and passengers. For example, a crew member might bring his concerns about defective life-saving equipment on board his ship to a local representative of a seafarer's welfare/religious organization who in turn passes this information along to the SSB. Or a local port authority employee might inform SSB of a pool of oil surrounding a berthed ship. It is only when suspicion of a decline in standards exists that an inspection of the vessel can take place. Otherwise, certificates are accepted at face value.

However, because Canada has not ratified MARPOL 1973/78, STCW 1978 or ILO 147, the Coast Guard is frustrated in its efforts to enforce these conventions on foreign flag vessels visiting Canadian ports. In particular, not having adopted ILO 147, Canada is especially restrained in preventing abuses of basic human rights regarding working and living conditions. Furthermore, Canada does not have access to the data bank employed by MOU members and therefore has only very limited information on the trading activities of substandard ships. A key to successful implementation will be an information link with other PSC authorities.

V. THE ISSUES AFFECTING THE CANADIAN POSITION

1. The position on PSC taken by our trading and continental partners

As can be seen in Exhibit 6, although the total value of our exports moving by water has not grown, there has been a shift away from Europe towards greater trade with Asia, which has generally not adopted PSC measures. Import patterns show a strengthening of our trade with Asia as well, although not at the expense of our European partners. (See Exhibit 7.)

It is in our trade with Europe, for example, that the Greek flag has been prominent after OECD flags, and it is the Greeks who have been shown to be among the more notorious owners and operators of substandard vessels.[12] Fully one-third of vessels flying the Greek flag in 1980 were manned by foreign nationals from developing countries.[13] In 1982, flag of convenience vessels carried 35 percent of Canada's deep-sea trade.[14] (It must be said, however, that flags of convenience and substandard shipping are *not* synonymous.)

Therefore, Canadian concern should focus not so much on the vessels involved in regular liner service to Canadian ports as on those engaged in tramp activity, as this the least controlled of the shipping sectors and the core activity of many an un-

EXHIBIT 6

Waterborne Exports—Merchandise Trade
(millions of dollars)

	1985		1980	
Region	By Water	%	By Water	%
Western Europe	6272	24.8	9122	35.8
Eastern Europe	1872	7.4	2051	8.0
Middle East	889	3.5	625	2.5
Africa	933	3.7	949	3.7
Asia	8884	35.1	6676	26.2
Oceania	590	2.3	678	2.7
Central/South America	1716	6.8	2710	10.6
North America	4173	16.5	2678	10.5
Total	25329	100%	25490	100%

Note: Numbers may not add due to rounding error.
Source: Exports Merchandise Trade, Statistics Canada (65-202), 1980, 1985.

EXHIBIT 7

Imports—Merchandise Trade
(in millions of dollars)

Region	1985	(Share)	1980	(Share)
Western Europe	12554	(12.0)	7006	(10.1)
Eastern Europe	296	*	307	*
Middle East	371	*	3027	(4.4)
Africa	1035	(1.0)	538	*
Asia	11218	(10.7)	5030	(7.3)
Oceania	604	*	696	(1.0)
Central/South America	4455	(4.2)	4048	(8.4)
North America	74380	(70.9)	48475	(70.1)
Total	104914	(100%)	69128	(100%)

Note: Numbers may not add due to rounding error. * = less than one percent.
Source: Imports Merchandise Trade, Statistics Canada (65-203), 1980, 1985.

EXHIBIT 8

Exports—Top 15 Bulk Commodities 1978
(in millions of tons)

Flag to	Group of 77	OECD	COMECON	Other	Total
OECD	4.7	55.0	1.4	1.8	62.9
Greek share	2.5	3.5	0.9	0.8	7.7
Group of 77	4.3	19.3	0.5	1.5	25.6
Liberian share	2.2	15.6	0.3	0.6	18.7
COMECON	0.2	0.5	3.3	—	4.1
Other	0.4	0.8	0.2	2.4	3.7
Total	9.6	75.6	5.4	5.8	96.4

Source: Maritime Transport of Bulk Commodities in Canadian Trade, CTC Research (1982/OE7), p.17.

scrupulous owner. This is particularly true when insurance rates are low, sale and scrap markets are poor, and owners are awaiting better prices.

2. Canada's record on environmental issues, safety of life at sea, and human rights

In December 1981, Ted McDorman documented Canada's position on marine environmental issues in a very interesting paper, *Record of an Environmental Nation: Canada 1967-1982*[15] In this paper he indicates the Canadian interest in marine environmental affairs at the international level to be marginal at the beginning of the period, and that this interest abruptly grew after the 1969 voyage of the Manhattan and the resultant 1970 Arctic Waters Pollution Prevention Act (AWPPA). It was at this time that Canada emerged as a strong advocate of expanded coastal state control over vessel-source pollution. It appears that the hidden agenda was to gain support for the AWPPA and Canada's radicalism was seen by some as an effort focussed on jurisdictional gain, particularly as Canada has failed to ratify the IMO Conventions that were developed to address marine pollution during this period. It is the lack of Canadian support for these conventions which may hinder any formal Canadian adoption of port state control principles although Canada has clearly little difficulty with the concept itself.

Canada boasts a proactive role in the protection and promotion of human rights throughout the world and evidence shows that it has become a cornerstone of Canadian foreign policy.

A recent study has shown that abuses of human rights at sea are on the increase and are intimately connected with substandard shipping and flag of registry.[16] ILO Convention 147[17] focusses on the prevention of human rights abuses but has not received support from the Canadian quarter.

Insofar as existing Canadian procedures for control of foreign registered substandard ships (based on our obligations under SOLAS 1974/78 and the 1966 Load Line Convention) provide for the inspection of certificates to ensure compliance with minimum ship construction, safety equipment and emergency procedures, foreign seafarers do receive some protection but these focus on technical inadequacies. By ratifying ILO Convention 147, Canada would be able to provide more effective control over working and living conditions on foreign flag ships, major "human causes" of casualties and accidents at sea.

3. Administrative Issues

Should Canada decide to strengthen its commitment to the control of substandard shipping by implementing PSC procedures, certain organizational, administrative and legislative issues first need to be addressed.

Coast Guard's existing Ship Information Reporting System should be modified to reflect the data requirements of PSC procedures and a data link between the Coast Guard and the MOU for the exchange of data on substandard shipping activity and ship inspections activated. Canadian ship inspectors will also need to be trained in accordance with the syllabus for training of European inspectors.

In order to implement PSC, Canada should then ratify MARPOL 1973/78 and STCW 78, after the passage of Bill C-39 (amendments to the Canada Shipping Act). In addition, the ratification of ILO Convention 147 requires compliance with the various conventions listed in its appendix and which set other minimum standards of living and working conditions. Because much of Canadian law relating to the internal order and economy of vessels under the Canadian flag is at least substantially equivalent to, and often supersedes, the standards proposed by the appended conventions to ILO 147, some may only need to be cross-referenced. The guidelines for inspections laid down in IMO Resolutions 466, 481 and 542 must also be adopted.

VI. SHOULD CANADA ADOPT PORT STATE CONTROL?

There would obviously be costs involved in the implementation of full Port State Control. However, the Coast Guard's existing efforts towards curbing substandard shipping activity mean that much of the personnel, organizational infrastructure and computer requirements is already in place. The costs of prevention must be weighed against the costs of cure.

To be most effective, Canada should establish an integrated PSC network with the US and Europe. This will require consultations amongst the three to harmonize procedures and data formats and create uniformity of inspection standards.

This effectiveness would be diminished if Canada were not to ratify a number of the necessary conventions. If Canada is to reaffirm its moral leadership as a nation dedicated to the protection and promotion of human rights in the international political arena, ratification of ILO Convention 147 would be an important first step.

The names *Devon III* and *Monica B* are but a tiny sample of Canadian experience with substandard a shipping in recent years but there are fears that if we do not follow the lead of the Europeans (and especially if the Americans become members of the MOU), these types of vessels will become even more familiar sights; indeed, Canadian ports could become "ports of convenience" with cargoes and substandard ships diverted to them. But at what price?

ENDNOTES

* Associate Professor, School of Business Administration, and a Research Associate, respectively with the International Institute for Transport and Ocean Policy Studies, Dalhousie University, Halifax, Canada.

1. *Maritime Transport 1985*. Paris: OECD, p. ix
2. The United Nations Convention on Conditions for Registration of Ships produced in February 1986, makes an attempt at strengthening the "genuine link" between ships and their flag states in areas of ownership, management and crewing. However, many of these provisions are vague and flexible and a genuine link may be difficult to establish.
3. Philip J. Loree, Chairman of the Federation of American Controlled Shipping, remarks at Port State Control Conference, New York, April 4, 1986.
4. *Seatrade*, May 1984, p. 3.
5. Frank Wiswall, President and Chief Counsel of Liberian Services Inc., remarks at Port State

Control Conference, New York: April 2 - 4, 1986.
6. Tsuneo Akaha, *Japan in Global Ocean Politics,* University of Hawaii Press, 1985, pp. 8-9.
7. Captain R. L. Newbury, "Implementation of Port State Control", *Proceedings: The Legal Rights of Seafarers,* 5-6 June 1985, Cardiff, Wales, p 124.
8. Only 271 were delayed or detained in the first year of implementation, but it would only be fair to say that in the first year, the member countries were not yet "up to speed" in their inspection process.
9. A deficiency is considered to be a major one if it constitutes 5 percent or more of the total number of deficiencies reported in that year.
10. From page 2.
11. The target inspection rate of 25% has not been reached and has tended to stabilize at around 20% mainly because of a shortage of personnel in the national shipping inspection services of MOU members. This essentially means that 60% of all ships plying the region are inspected, rather than the 90% coverage which would occur if the 25% target were reached.
12. C.V.D. Smith *Trends and Patterns in Human Rights Violations at Sea,* M.Sc. thesis, Plymouth Polytechnic, 1986, p. 93.
13. *Task Force on Deep-Sea Shipping,* Ottawa: Supply and Services (T22-68/1985E), April 1985, p.37.
14. *Ibid*. p. 23.
15. Published by the Dalhousie Ocean Studies Programme, Dalhousie University, Halifax, N.S. Canada in December 1981.
16. C.V.D. Smith *Trends and Patterns in Human Rights Violations at Sea,* M.Sc. thesis, Plymouth Polytechnic, 1986
17. This convention sets standards for the placement of seafarers' articles of agreement, minimum age of employment, medical examinations, wages, hours of work and manning, health and welfare, food, catering and accommodation, sanitation, repatriation and social security.

Benefit Cost Analysis of a Toll Highway—British Columbia's Coquihalla

*by W.G. Waters II and Shane J. Meyers**

ABSTRACT

The Coquihalla is a major new toll highway bisecting existing routes to British Columbia's "Interior." This ex ante benefit cost study estimates the user benefits to traffic diverting to the new highway as well as reduced congestion benefits to those remaining on the old routes. Tolls reduce the number diverting to the new highway, therefore reduce the overall net benefits. The benefit cost analysis is conducted both from a global and provincial perspective. The project shows a modest net benefit or loss depending on the perspective adopted, the discount rate used, and other assumptions. Optimal timing is also considered; the highway would have a slightly higher net benefit if construction had been deferred for a few years.

I. INTRODUCTION

The Coquihalla Highway project involves the construction of 189 kilometres of four-land divided highway through the Cascade mountains in south-eastern British Columbia. The highway connects the town of Hope and the Lower Mainland area of B.C. with interior cities of Kamloops and Merritt. The new highway saves a considerable distance for travellers destined to Kamloops and points north, and those travelling on the Trans Canada highway from or to the Lower Mainland. Exhibit 1 is a map of the surrounding area.

The highway is being constructed in two phases. Phase I, which connects Hope to Merritt, was completed in early 1986. Phase II consists of the remaining link between Merritt and Kamloops; it will open

Exhibit 1
Map of Coquihalla Highway and Nearby Routes

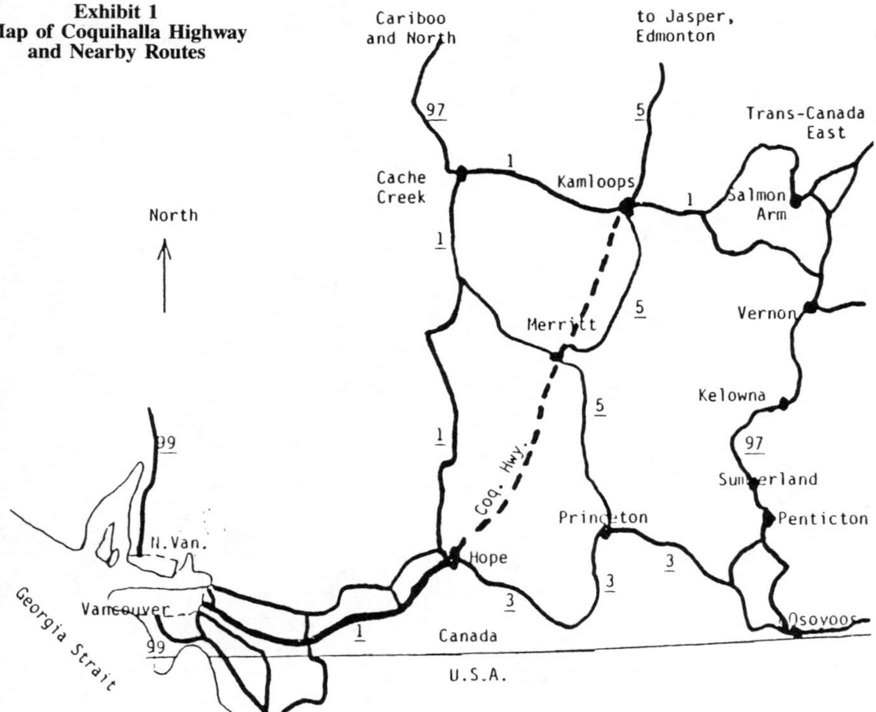

in 1987. (A third phase which would connect eastward toward Kelowna and the Okanagan area is still in the planning stage and is not considered in this paper).

This paper summarizes a social benefit cost analysis of the Coquihalla highway. Unlike some public projects, there are few externalities and social impacts to be included. Most of the benefits accrue directly to users of the project. There are some interesting features concerning the evaluation of this project. The new highway lies between two existing major highways; hence estimating project benefits require that interaction or network effects of the new highway be considered. Another interesting feature is that this is a toll highway. This case provides a useful illustration of the effect of pricing and finance policies on the economic benefits of a project. Although the project will generate a substantial cash flow from toll revenues, the presence of tolls causes the overall economic benefits of the project to be reduced.

Although the highway is partially open, this analysis is an ex ante rather than ex post study. It uses information and forecasts developed before the highway was built as this is more relevant to assessing the original decision to build the highway. A few issues are addressed which have arisen subsequent to the highway's construction.

The following sections outline the project costs, the benefit categories, the forecasts of traffic levels and diversions among routes, and the estimations of benefits by categories of users. The base case benefits are estimated without tolls. A subsequent section incorporates the impact of tolls on net benefits on the highway. Distinguishing between provincial and non-provincial users provides an alternate calculation of net benefits. From the Province's perspective, toll revenues from non-residents are a benefit, while other benefits which accrue to non-residents are excluded. Nonetheless, the results still show that the imposition of tolls reduces the overall net benefits of the project. The remaining sections of the paper explore the question of optimal timing of the project (was now the right time to build it?) and raise a few benefit and cost considerations which warrant further study.

II. PROJECT COSTS

There are three broad categories of costs: construction costs, ongoing maintenance and snow removal costs, and environmental costs associated with construction and ongoing use of the highway. In addition, the rerouting of traffic will entail diversion of road-related business (e.g. gasoline sales) from one route to another. As is customary, these transfers of economic activity are not included in the analysis (although, not surprisingly, the secondary effects of diverted spending by highway users have generated controversy in the affected regions).

The total undiscounted costs of the project (in 1984 dollars) are $375 million. In lieu of more precise budgeting figures, it was assumed that the construction outlays over the years were proportional to the published estimates of the number of direct jobs associated with the project:

1984: 2750 jobs—$105,225,000
1985: 4700 jobs—$179,850,000
1986: 2350 jobs—$89,925,000

The present value of the construction costs (at a 7.5 per cent real discount rate) is $338 million.

For this project there was no shadow pricing of construction costs. That is, it is assumed that the amounts paid to contractors reflect the actual opportunity costs of the resources employed on the highway project. This may be an overstatement. The B.C. economy was not healthy during this period, so some of the employment and equipment used may not have had comparable opportunities for employment elsewhere. Therefore, a case could be made to "shadow price" at least part of the inputs used in constructing the highway. (There is a counter argument that, during times of unemployment, the alternative to this public project would be some other public project; in order to accurately evaluate the relative merits of the different projects, it would be appropriate to value the resources used at a consistent price such as their usual market price).

The Coquihalla is a new addition to the road system and will incur costs of routine maintenance and snow removal. The latter is potentially important for this route. Examination of snow removal and maintenance costs for several mountain highway districts revealed annual costs between $2500 and $7500 per kilometre. A conservative figure of about $2600 per kilometre or $500,000 per year was used for this cost category.

Environmental costs are of several types. The construction is through mountainous terrain with many river crossings. The Coquihalla is a valuable river for steelhead, salmon and trout. The impacts of highway construction could have been severe. However, by careful timing of construction activity and special efforts to repair any encroachment on the river (including relocating and reconstructing river habitat for fish), most of these potential environmental impacts have been prevented. That is, the costs of mitigating environmental impacts have been incorporated into the construction cost estimates. There may be ongoing environmental impacts such as the impact of greater surface water runoff and sedimentation of the river, but these effects are unknown at this time.

The other environmental impact is the inevitable encroachment of civilization on this area which was previously largely a wilderness area. There will be impacts on wildlife through greater hunting and fishing pressure as well as animals killed by cars. However, this is the paradox of all such wilderness access. Clearly, there are costs of harming the wilderness, yet it is also clear that people value the opportunity to penetrate these areas. That is, there are both benefits and costs of access to regions with limited previous access. These environmental benefits and costs have not been included in this study.

III. BENEFIT CATEGORIES

The primary benefits of the project accrue to the users of the highway in the form of savings in operating costs and time. There are also significant safety benefits associated with the project, and there are benefits to road users who remain on other highways but who experience lower congestion levels once the Coquihalla has opened.

The Coquihalla cuts 72 kilometres off the previous Hope-Kamloops route. At a cost of $0.10 per vehicle kilometre, this is a saving of over $7.00 per trip for automobiles. For large trucks, the saving is even greater, about $33.50 per trip.[1]

The reduction in distance and increased average speed means significant time savings to traffic who divert.[2] The valuation of travel time is often contentious in transportation studies. Usually this is because only small time savings are involved. The time savings per vehicle are substantial in the case of the Coquihalla. The value of time for leisure travelers was set at 25 per cent of the gross wage; multiplied by the typical passengers per car on B.C. Highways results on a value of time of $6.68 per vehicle hour, or a time savings worth about $8.00 per trip on average for cars diverting to the Coquihalla. Higher values of time were used for automobile drivers assumed to be traveling on business ($12 per hour) and truck drivers ($14 per hour) (the former based on the average wage in the service sector as reported by the Statistics Canada, and the latter from Trimac Consulting Ltd., 1982).

Building the Coquihalla will relieve traffic volumes on the existing highways Number 1 and Number 3. The effect will be minor during periods of light to moderate traffic, but there will be significant benefits during congested periods. This is a difficult category of benefits to estimate. It requires knowledge of the number of hours of various degrees of congestion, and the impact on operating speeds (hence time and cost savings) of reducing the level of congestion. Only rough approximations of this benefit category were possible; they are described later.

The other major benefit category is safety. Use of the Coquihalla saves substantial distance (thus reduces the chance of accidents). In addition, the Coquihalla is a divided four lane highway with adequate shoulders. Such a highway has a significantly better accident record than regular two lane highways. One set of estimates are as follows:

Effect on Accident Rates of Changing to Divided Highway*
(Accidents per million vehicle kilometres)

	fatal accident rate	injury accident rate	property damage only	total accident rate
two-lane	.0316	.323	.472	.825
divided expressway	.0236	.273	.472	.769
freeway	.0155	.168	.304	.490

* For rural highways, partial access control; converted to metric.
Source: Pacquette and Wright, (1979) p. 73.

These figures show that fatal accidents can be reduced by 25 to 50 percent by changing from two-lane undivided to multi-lane divided highways. We use an estimate 33 percent for our calculations. For accidents involving injuries and property damage only accidents, we assume reductions of 33 and 15 percent, respectively.

The safety benefits arise for two reasons: (1) reduced distances travelled; and (2) travel on a safer highway.

In the base year of 1987, approximately 130 million vehicle-kilometres could be saved due to the shorter distances travelled. For a fatality rate of .0316 per million vehicle kilometres, this implies a saving of 4.1 "statistical" lives. (By way of comparison, the B.C. death rate per million vehicle kilometres is .027 from Government of British Columbia, 1984). Similarly, the reduced distances travelled imply a reduction of about 42 accidents involving injuries, and 61 fewer non-injury accidents.

The other safety benefits are due to the use of a safer highway. The base year calculation estimates 313 million kilometres diverted to the new highway. Using the estimated accident reductions noted above, this results in 3.3 additional lives saved, 3.4 fewer accidents involving injuries, and 22 fewer property damage only accidents.

Therefore, total safety benefits for the base year are 7.4 lives saved, 76 fewer accidents involving injuries, and 83 fewer p.d.o. accidents. These benefits can be expected to grow with increases in traffic volumes.

The valuation of safety benefits is a highly controversial subject, but at least some estimate needs to be included for the study. The most contentious figures are those involving loss of life. Arbitrarily using a value of $500,000 per statistical life[3] results in a base year benefit of $3.7 million.

The valuation of non-fatal accidents depends on accident and injury severity. Christofferson (1984), citing several sources including work by Lawson, used $10,780 for the average cost of an accident involving injuries. The more comprehensive study by Miller, Reinhert and Whiting calculated a range from about $2300 to $261,000 depending on accident severity. This study uses a single figure of $11,000 for an accident involving injuries. Property damage only accidents entail lower total costs. Christoferson (1984), citing several Canadian sources, settled on $2000 per such accident. This is the figure used here. The base year accident reduction benefits add about $1.0 million to the benefits of lives saved, for a total annual safety benefit of $4.7 million dollars.

A final category on the benefit side is the residual value of the project. The highway will last much longer than the 20 year life used for the analysis of benefits and costs. In principle, the residual value for such a project would be the capitalized value of benefit accruing after 20 years, less all reconstruction costs which would be necessary to keep the road in operation, all discountered back to the present. A simple procedure often done is to include all or part of the initial construction costs as a terminal value, on the assumption that much of the initial road investment will still be useful at the end of the assumed project life. The latter practice was used here. A residual value of 75 per cent of initial construction cost was included as a terminal value.

IV. FORECASTS OF TRAFFIC DIVISION

There are two forecasting tasks: first is forecasting the overall volume of traffic, and second, their assignment among alternate routes. B.C. Highway Officials had made such forecasts as part of the original environmental impact study of the project. The B.C. Highway's forecast were for a base year (1986) with and without the Coquihalla Highway. These forecasts are illustrated in Exhibit 2. In order to calculate a net present value, it was necessary to extend the

Exhibit 2
Traffic Assignment

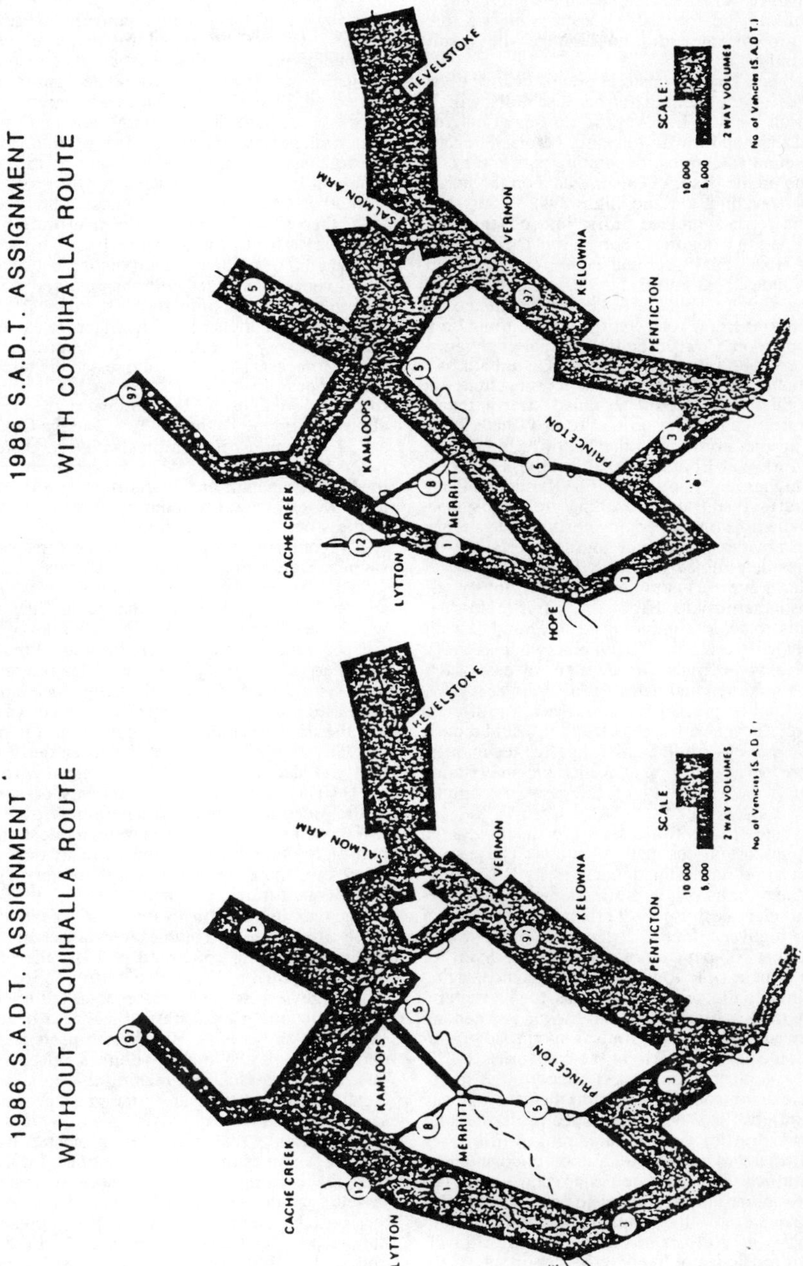

forecast 20 or more years into the future. Projection methods were used based on recent traffic growth but with a "EXPO bulge" for traffic in 1986. Sensitivity analysis was used to assess the significance of the results to the forecasts; variations in the forecasted growth rates did not change the results significantly.

It was necessary to construct price-sensitive diversion rates to the Coquihalla because tolls will be charged on the new highway. To do this effectively required that total traffic forecasts be disaggregated into different categories. Separating cars and trucks was done on the bases of existing data on the mix of vehicles travelling on the highways. It was also important to disaggregate traffic into different origin-destination categories. For example, Highway 1 north of Hope (the Fraser and Thompson River canyons) includes local traffic and traffic going on north at Cache Creek. This traffic is not candidate for diversion to the new road. Of the traffic going on to Kamloops, some traffic either originates or terminates here. The time saving of the Coquihalla is a more significant portion of total journey time and cost for this "Kamloops-determined" traffic than it is for traffic continuing on the Trans-Canada highway or turning north up to the Yellowhead highway (Jasper Park and Edmonton). Typically, one would expect higher diversion rates for Kamloops-determined traffic than traffic travelling greater distances because the time (and operating costs) savings represent a larger share of total journey costs for the Kamloops-determined traffic.

Similar arguments apply to traffic previously taking the southern route, Highway 3. Traffic destined along this route or continuing to the Southeast are not likely to divert to the Coquihalla, whereas traffic which rejoins the Trans-Canada highway east of Kamloops would be candidates for diversion.

Predicting the precise diversion rates for different toll charges is very difficult. Ideally it would require extensive surveys or other investigative techniques, and even there would be considerable uncertainty concerning the reliability of forecasts. A simpler approach was adopted.

Aggregate traffic forecasts were disaggregated into various origin-destination groups. These were based on an origin-destination survey by B.C. Highways. The various origin-destination pairs were reviewed to identify those most affected by opening of the new highway. In the absence of tolls, it was assumed that 100 per cent of traffic would divert for those O-D pairs able to use the full distance saving of the Coquihalla. This traffic was then disaggregated by truck, commercial auto travel, and general or leisure auto travel, and grouped into broad origin-destination categories. Two of the categories, traffic travelling on north and traffic destined to the southeast, were assumed not to divert to the new highway at all. Local traffic was also excluded from estimates of diverted traffic. For the remaining traffic, they must be separated into diverted, undiverted and generated traffic. The latter refers to new traffic attracted to the roads because of the improved travel conditions. Because the time and operating costs of travelling is only a part of the total costs of travel, generated traffic is not likely to be important except on the Coquihalla route itself, and this will primarily be for the "Kamloops-determined" traffic (ignoring tolls, the time and operating costs savings via the Coquihalla rather than via Highway 1 represent a 28 percent saving in total trip cost; this is sufficient to stimulate some response in the total amount of travel by this traffic category).

The proportion of traffic diverting to the new highway—and the potential amount of generated traffic—will be influenced by the level of tolls. Estimating price and time elasticities of demand is a difficult task. For this study some approximations were used. First of all, for each traffic group affected by the Coquihalla, the potential time and operating cost savings were calculated. It was assumed that if the toll charged was exactly equal to the average time and costs saved, that the traffic group would split 50/50 between the two routes. If there were no tolls, then a much higher percent of traffic divert, 100 percent for traffic which realize significant savings via the Coquihalla. Interpolating between these two diversion rates and toll charges (zero tolls and tolls equal to total time and cost savings) for each group gave predictions of diversion rates for each traffic category for different toll charges.

Our forecasts of traffic volumes on the three highways appear to be very close to B.C. Highways' forecasts. With tolls, our estimates of the traffic allocation among Highways 1, 3, and the Coquihalla are 34, 35, and 31 percent, respectively. B.C. Highways' figures are 32, 34, and 34 percent, respectively (Kasianchuk and Warburton, 1986). Without tolls, we estimate that about 50 percent of through-traffic would use the Coquihalla.

For generated traffic, the size of time and cost savings were compared to total costs and travel (adding estimates of food and lodging costs) to get an idea of the percentage change in "trip price" which faced travellers. For commercial vehicles, this percentage change in price would be multiplied by a derived elasticity of demand for freight transport, typically a fairly low elasticity (for example, if the elasticity of demand for final products was unity, and the transportation costs represent 10 percent of the delivered price of goods, then the derived elasticity of demand for freight transport would be −.1). The only significant amounts of generated traffic were for Kamloops-determined traffic, and levels of generated traffic were modest in comparison to other traffic volumes already on the road. Therefore, the overall results are not very sensitive to generated traffic estimates.

The remaining traffic estimates of potential importance are the total volumes of traffic on the old highways. During congested periods, the presence of the Coquihalla means that traffic congestion on other highways will be reduced, and all undiverted traffic (including local traffic) realize a benefit. To calculate this benefit to undiverted users requires a comparison of the traffic volumes relative to the rated road capacity, and relating this to speed-flow conditions to estimate any time and operating cost savings to undiverted traffic. Surveys of highway use by time of day had been projected by B.C. Highways. An estimate was made of the local annual time the existing highways would encounter congested conditions ("Level D" and "Level E" service, unstable vehicular flow at 55-70 kph and 55 kph, respectively, Government of B.C., 1978, p. 17) Traffic diverting to the Coquihalla will greatly reduce congestion levels for traffic which remain on the old routes. A 20 minute saving was estimated for through-vehicles on the old routes (traffic figures were based on volumes away from settled areas, this tends to exclude traffic which realize smaller benefits from reduced congestion). The benefits to undiverted traffic were proportional to the number of cars diverting to the Coquihalla, and the latter is a function of the tolls charged.

V. BASE CASE EVALUATION

The base case consists of the Coquihalla highway with no tolls. Exhibit 3 summarizes the benefit and cost categories using a 7.5 percent real discount rate and an arbitrary project life of 20 years.[4] A positive terminal value for the project was assumed at the end of the calculation period.

The "bottom line" shows a positive net benefit for the project although not a very large one. The internal rate of return (IRR) of the base case is about 11 percent.

Exhibit 3 also lists the benefits (and costs) with current toll charges of $40.00 maximum for trucks ($35. used as an average) and $8.00 for cars. The project costs are largely unaffected by tolls (there are the annual costs of toll collection), but the project benefits are smaller because fewer travellers divert when tolls are levied. There are also costs and time delays imposed while collecting tolls (this cost is not included). The toll revenue itself represents a transfer of benefits from users to the government or tax payers, therefore toll revenues are not included in calculations because it would be double counting (but see the discussion from a "Provincial perspective," below).

The overall project with tolls are positive but small at a 7.5 percent discount rate. Exhibit 4 plots the IRR's for the project with and without toll charges. This exhibit is also useful for illustrating the sensitivity of project benefits to the discount rate.

Traffic diversion rate for other toll levels were interpolated between the base case (zero toll) and the currently-imposed tolls. Therefore, the sensitivity of net benefits to different toll charges are estimated as a straight forward proportion of the difference between the two net benefit calculations and any proportionate change in the overall level of tolls charged. Different categories of users were assumed to have different elasticities of demand, for example, truck and commercial automobile travel are expected to be less sensitive to tolls than are leisure travellers. It would be difficult to discriminate among the two types of auto traffic, but a shift of tolls charged to increase those on commercial automobiles and decrease them on leisure travellers could generate the same toll revenue but with less reduction in project net benefits (up to a point). The same argument could apply to toll charges for commercial trucks relative to leisure cars, but additional information on their respective demand elasticities would be needed to settle this.

Exhibit 3
Summary of Project Benefits and Costs
With and Without Tolls

(7.5% discount rate)
(Million 1984 dollars)

Project Costs:	No Tolls		With Tolls	
Construction	338.1		338.1	
Maintenance	7.6		7.6	
Toll Collection	-		8.4	
Toll Booth Const.	-		.3	
	345.7		354.4	
Project Benefits:	**No Tolls**	**%**	**With Tolls**	**%**
Coquihalla Traffic	389.8	75.3	290.4	73.6
Residual Value of Coq.	53.3	10.3	53.3	13.5
Safety Benefits	52.1	10.1	36.5	9.3
Traffic decongestion (#1 & #3)	21.9	4.2	14.1	3.6
Generated Traffic	.8	.2	.3	.1
Total Benefits	517.9		394.5	
Net Benefits	172.2		40.1	

Exhibit 4
Net Benefits vs. Discount Rate
(with and without tolls)

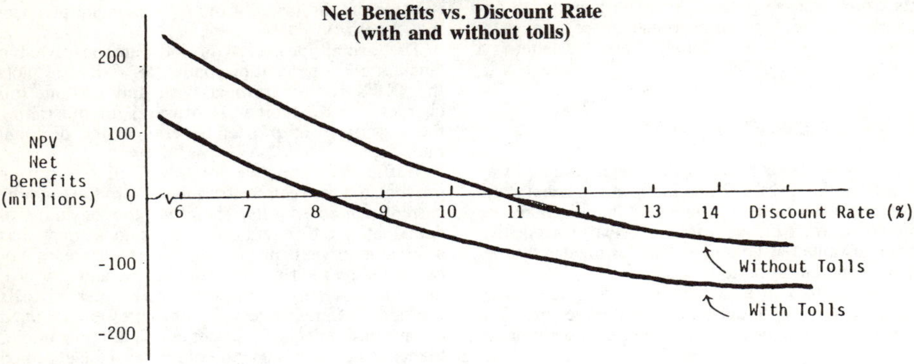

VI. PROJECT FINANCE AND BENEFITS FROM A PROVINCIAL PERSPECTIVE

The relationship between project benefits and project pricing is an illustration of basic economic principles which date back to Dupuit, 1844. The Coquihalla is a classic example of a public project with largely fixed costs and low variable costs with use. Such a project gives rise to a paradox because financial and economic criteria diverge. The optimal policy from an economic point of view is to keep the price low in order to encourage use of the project because the incremental costs of additional use are very low. But then the project runs a financial deficit. A better method of finance for such a project is to levy a fixed charge such as via general taxation rather than specific tax (toll) which discourages use of the facility.

However, there is still the necessity to finance the project. The usual illustration of the foregoing principle applies to "first best" world where prices equal marginal costs everywhere and there are few or no taxes. If taxation is discussed at all in textbook illustrations of the problem, taxes are limited to a head tax or some other non-distorting tax. But in practice, a variety of taxes are imposed, so the decision to add another public project such as the Coquihalla may not necessarily leave total taxation unaffected. Whatever form of taxation is used to finance the project will have some resource allocation effects. The estimate of the reductions in net benefits due to tolls is an over-estimate of the actual loss because it ignores any economic costs (resource allocation effects) of the ultimate forms of taxation used to finance the project. But the basic principles are still valid: levying a specific tax (toll) on a particular group (users) causes reductions in project benefits not compensated for by reductions in project costs.[5]

The other interesting facet of project benefits, costs and finance is to focus on the project from a purely provincial perspective. The project is financed by B.C. residents, yet many of the benefits accrue to non-residents. Without a toll, the residents of B.C. would be incurring costs for the benefit of non-residents. The tolls are a way of capturing some of the benefits for B.C. and non-B.C. residents. This is shown in Exhibit 5.

Surveys in the 1970's showed that out-of-province license plates constitute over 30 percent of traffic on Highways 1 and 3, especially the former. However, the proportion of out-of-province vehicles is less during winter months than it is during the rest of the year. We used an arbitrary figure of 25 percent of traffic on the Coquihalla as out-of-province vehicles. The provincial benefits are defined as the time savings and operating costs savings to provincial car users, and toll revenues from non provincial vehicles. From the purely provincial perspective, net benefits of the project without tolls are reduced compared to the base case calculations; this is because a provincial perspective does not count benefits to non-B.C. users. Counting the toll revenues from non-B.C. residents as a benefit still results in even lower overall net benefits. That is, the impact of tolls causes a sufficient number of B.C. drivers not to use Coquihalla thereby reducing the benefits to B.C. users, and the reduction in benefits to B.C. users because of tolls is not offset by the toll revenues collected from non-B.C. travellers. The Provincial net benefit without tolls are positive whereas the net benefits with tolls are slightly negative at the 7.5 percent discount rate. Sensitivity analysis showed that assuming a higher proportion of non-B.C. travellers made little change to this basic result.

Another possible approach would be to only levy tolls on non-B.C. users. Setting aside the probable negative reaction by travellers and the possible harm to B.C.'s tourist reputation, such an approach would probably yield a small net benefit, although it would be less than the base case net benefits without tolls (it is less because compared to the base case without tolls, the benefits to non-B.C. users are replaced by toll charges paid by this group and the tolls are less than the savings realized by this group).

VII. OPTIMAL TIMING OF THE PROJECT

Most project evaluations concentrate solely on the question of whether or not to build a project, that is, are the discounted net benefits positive? An often neglected question concerns the optimal timing of a project, viz, is it appropriate to undertake the project immediately or would the net benefits be even larger if the project were deferred for a few years? Setting

Exhibit 5
Benefits and Costs from a Provincial Perspective
(excludes benefits to non-residents but
includes their toll revenues paid)
7.5% discount rate
(Millions 1984 dollars)

Project Costs:	No Tolls		With Tolls	
Constructions	338.1		338.1	
Maintenance	7.6		7.6	
Toll Collection	-		8.4	
Toll booth Const.	-		.3	
	345.7		354.4	
Provincial Benefits:	**No Tolls**	**%**	**With Tolls**	**%**
Toll Revenues	-	-	37.0	10.7
Coquihalla Traffic	292.3	76.1	217.8	62.9
Residual Value	53.3	13.3	53.3	15.4
Safety Benefits	39.1	9.7	27.4	7.9
Benefits to Traffic (#1 & #3)	16.4	4.1	10.6	3.1
Generated Traffic	.6	.1	.2	.1
	401.7		346.3	
Net Provincial Benefits	56.0		(-8.1)	

aside the short term macroeconomic concerns of employment generation during an economic downturn (which were a consideration in the Coquihalla construction), the net benefit calculation can be carried out for different years' construction and all results discounted back to the same base year. The optimal year for construction is the year for which the NPV is largest.

Most transportation projects will realize a larger undiscounted benefit stream by postponing the project. Traffic is usually growing over time, therefore larger user benefits (and reduced congestion effects) will arise if a project is built later. The key point for determining the optimal timing of a project is to compare the change in benefits (discounted to a common year) by postponing a project, with the opportunity cost of capital. The costs of postponing the project are the foregone benefits which are primarily those of the first year; the benefits of postponement are the opportunity costs or use of the capital for the postponed period, and the present value of the increased benefits that would result from the higher traffic volumes over the life of the project.

Exhibit 7 plots the net benefits of the project for various years' opening. The optimal year appears to be about 1990, i.e., the economic analysis based on existing forecasts suggests that the highway was built a few years early. However, because the NPV curve in Exhibit 7 is fairly flat, this indicates that there is very little loss by premature construction. The question of optimal timing is sensitive to assumed growth rates, which are somewhat uncertain, therefore not too much confidence can be placed on precise dates for the optimal timing of this project. If one then considers the need for employment stimulation and the enthusiasm for facilitating EXPO 86, the earlier construction to open in 1986 may be understandable.

VIII. CAVEATS AND QUESTIONS FOR FURTHER ANALYSIS

This paper was prepared during construction of the Coquihalla, but following the opening of Phase I there are a few developments which warrant comment. First, there is a general caveat which must accompany nearly all benefit cost studies. Finding a positive NPV is not a guarantee that there might not be alternate projects with even higher social returns. This study has also been confined to examining the given scale of the project. The Coquihalla is a four lane divided highway for most of its length. This paper has not inquired into whether or not a phased construction might generate higher net benefits. For example, this study did not consider building a two lane highway and later expanding it as traffic grew. The Ministry of Highways reportedly did consider this but found that the cost savings of two construction periods were very small (Kasianchuk and Warburton, 1986).

Exhibit 6
Provincial Benefits vs. Discount Rate
(with and without tolls)

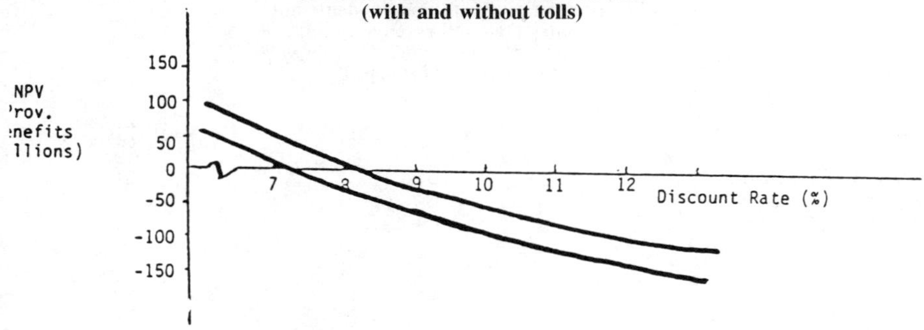

Since the opening of Phase I, there have been some comments about construction and vehicle operations which warrant mention. A high priority was placed on opening at least part of the highway (Phase I) in time to handle expected traffic for EXPO 86. The accelerated construction required greater use of overtime and weekend construction, which raised costs. Because bids for construction work on the highway were reportedly less than in previous years, the project has remained largely on budget. Nonetheless, the greater amounts of overtime and higher cost working hours suggests that the project could have been built at a lower total cost if construction was spread over a longer period.

A related problem are allegations that some paving was done without adequate time for the road base to stabilize. If so, this may require additional maintenance and/or premature reconstruction at some time in the future. However, it has not been possible to substantiate or estimate these possible costs.

In the use of the Coquihalla, there have been complaints, especially from truckers, concerning the very long grades on the highway (Vancouver Sun, June 14, 1986). The grades are no steeper than on other major mountain highways; however, allegedly the grades are so long that cars and trucks are more prone to overheating and similar mechanical problems. If so, the expected operating cost benefits may not be as high as forecast.

IX. CONCLUSION

The Coquihalla makes an interesting case study of the economic evaluation of highway projects. It is a largely scale project with significant effects on traffic flows on the highway network. Substantial time and cost savings accrue to users. While there is always some debate over the valuation of these savings, the amounts of time and costs saved by individual vehicles using the Coquihalla are much larger than the magnitude of these benefits on most other road projects. Therefore, there is more confidence in the economic significance of these benefits than comparable figures for many other road projects.

The project has a modest net present value at a 7.5 per cent discount rate without tolls. Tolls reduce the level of benefits (although the net benefits are still positive at a 7.5 per cent discount rate) because tolls reduce the amount of traffic diverting to the highway.

The highway construction was rushed for at least partial completion in time for EXPO 86. The hasty

Exhibit 7
Optimal Timing
of the Project

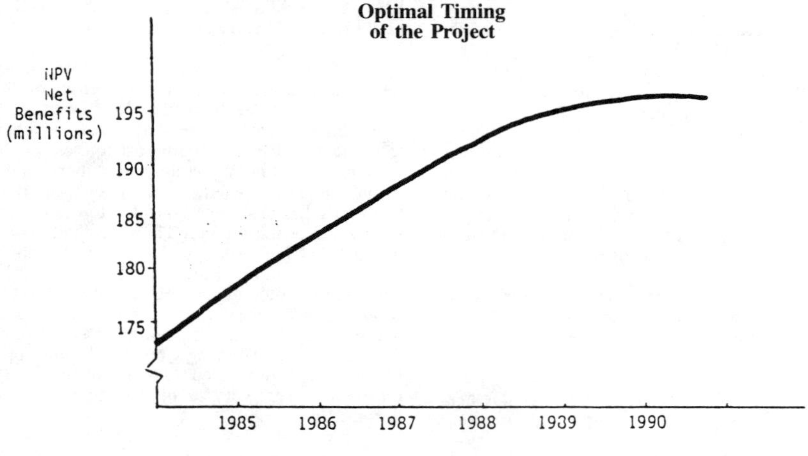

construction suggests that the project costs could have been less if construction had been spread over a long period. Based purely on the user benefits and construction costs, the highway may have been constructed a few years earlier than warranted by projected traffic growth. However, given the uncertainty of growth forecasts, the need for employment stimulation in the B.C. economy in the mid 1980's, and the concern for facilitating EXPO 86, the actual project timing is not surprising.

REFERENCES

Bailey, M.J. (1980) *Reducing Risks to Life: Measurement of the Benefits* (Washington, D.C.: American Enterprise Institute for Public Policy Research).

Burgess, D.F. (1981) "The Social Discount Rate for Canada: Theory and Evidence," *Canadian Public Policy* (Summer, 1981).

Christofferson, Kevin C. (1984) "A Benefit Cost Analysis of Reconstructing the Alaska Highway to R.A.U. 100 Status" unpublished M.Sc. (Bus. Admin.) thesis, University of British Columbia, Vancouver.

Depuit, J. (1844). "On the Measurement of the Utility of Public Works", *Annales des Ponts et Chaussees,* second series, Vol. 8; translated by R.H. Barback and reprinted in D. Munby, ed., *Transport* (Penguin) 1968), pp 19-57.

Government of British Columbia, (1978). Ministry of Highways, Design and Surveys Branch, *Coquihalla (Hope-Merritt) Highway: Preliminary Environmental Report*. November, 1978.

Government of British Columbia, Ministry of Transportation and Highway, (Annual). *Summer Traffic Volumes*.

Government of British Columbia, 1984, Ministry of Transportation and Highways: Motor Vehicle Branch, *1984 British Columbia Traffic Incident Statistics,* p. 13.

Government of Canada (1976), Treasury Board Secretariat, *A Guide to Benefit Cost Analysis*. (Ottawa: Queen's Printer).

Jenkins, Glen P. (1981). "The Public Sector Discount Rate for Canada: Further Implications", *Canadian Public Policy* (Summer, 1981).

Jones-Lee, M.W., M. Hammerton and P.R. Philips (1985) "The Value of Safety: Results of a National Sample Survey" *Econ. Journal* (March, 1985).

Kasianchuk, D.A. and William P. Warburton (1985). "Forecasting Highway Traffic: The Coquihalla Highway - A Case Study" in Canadian Transport Commission, Research Seminar Series, Spring, 1985, Volume 11, pp. 20-27.

Miller, Ted R., K.A. Reinhert, B.E. Whiting, (1984) *Alternative Approaches to Accident Cost Concepts: State of the Art* (Washington, D.C.: N.T.I.S. PB84 196648.

Pacquette, Radnor J., and Paul H. Wright (1979) *Highway Engineering* (New York: Wiley & Sons).

Rockerbie, R.A. (1980) *CounterAttack—Perspectives and Assessment* Policy Planning Division, Ministry of the Attorney General, Government of British Columbia, Victoria.

Sumner, M.T. (1980) "Benefit Cost Analysis in Canadian Practice", *Canadian Public Policy* (Spring, 1980), pp. 389-393.

Trimac Consulting Ltd., (1982) *Operating Costs of Trucks in Canada*

END NOTES

* The authors are, respectively, Associate Professor, Faculty of Commerce and Business Administration, the University of British Columbia, and Honours graduate in Economics, University of British Columbia, now with Pemberton, Houston and Willoughby, Vancouver.

The authors wish to thank British Columbia Ministry of Transportation and Highways Officials for answering various questions and offering comments. Thanks also to George Federoff for helpful research assistance in constructing the initial spreadsheets for evaluating the Coquihalla. The analysis and conclusions of this paper are the sole responsibility of the co-authors.

1. The vehicle operating cost figure was selected following perusal of figures estimated by the Canadian Automobile Association. The figures for truck operating costs were based on Trimac Consulting LTD. (1982) figures.
2. The time savings are based on design speeds of the new and old highways, taking into account traffic volumes. If people exceed the speed limit on the new highway, the time savings would be greater than calculated.
3. This figure is slightly below the figure (in current dollars) cited by the Insurance Corporation of British Columbia as the social costs of a fatal accident (Rockerbie, 1980).

   Bailey (1980) surveyed a number of studies estimating the valuation of life for decision making purposes. He found a range of figures, from about $170,000 to $584,000 per life saved (1978$). Studies based on direct interviews often produce higher estimates, e.g. see Jones-Lee, Hammerton and Phillips, 1985.
4. The standard discount rate recommended by the Treasury Board Guide (Government of Canada, 1976) is 10 per cent. However, it has been argued that the appropriate social discount rate for Canada is closer to 7.5 per cent (e.g. Burgess, 1981; Jenkins, 1981; Summer, 1980). The latter discount rate is used here; however, the sensitivity of results to the choice of discount rate can be seen in the plots of internal rates of return (Exhibits 4 and 6).
5. For projects with substantial external benefits such as most urban transit projects, where most of the benefits accrue to non users via reductions in congestion, the folly of trying to pay for projects by increasing user charges (therefore, decreasing use and losing the external benefits) is fairly clear. In the case of the Coquihalla, the vast majority of benefits are those which accrue directly to users' via time and operating costs savings. Therefore, the loss of benefits from toll charges is far less of a loss (reduced net benefits relative to total benefits without tolls of only 23 percent) than for many other public projects with greater non-user benefits. Therefore, given the practical necessity to finance large projects by a government facing financial constraints, the toll policy may be an understandable approach.

Forecasting for Subsidy Purposes: Idiosyncrasies in the Western Grain Transportation Act

*by E. W. Tyrchniewicz and E. R. Wansbutter**

I. INTRODUCTION

A long standing criticism of subsidy programs generally is that payments are made long after expenditures are incurred. The Western Grain Transportation Act attempts to get around this problem by incorporating forecasts in establishing annual subsidy levels. Specifically, the Canadian Transport Commission reviews the railways' projections of input price indices and makes its own forecast of these indices. Similarly, the Grain Transportation Agency is required to forecast the volume of grain that will be moved in the forthcoming crop year.

This volume forecast must be made in February for the period August to July (6 to 18 months into the future). This must be done before the crops are planted, let alone harvested! Prairie grain and oilseed production is extremely variable. Over the last nine crop years (1978-79 to 1986-87), production of the six major crops has ranged from 31.1 million tonnes to 52.5 million tonnes. Put another way, the rate of increase in Prairie crop production for the 25 year period 1960-85 has been two percent; the standard error around that trend line has been 4.92 million tonnes. Although year to year variations in plantings are a factor in this production variability, the major variable is climatic condition during the crop growing season, e.g., seasonal rainfall, heat, time of frost, etc.[1] Couple this with the vagaries of the international grain market, and the task of forecasting the volume of grain to be moved is indeed a difficult one.

This paper sets out some desirable criteria in developing forecasts for subsidy purposes, describes the volume forecasting procedures as established by the Western Grain Transportation Act, assesses how well these requirements have been met, and suggests improvements in the volume forecasting procedures.

II. DESIRABLE CRITERIA FOR DEVELOPING FORECASTS

A number of criteria should be considered in developing forecasts for subsidy purposes.[2] Some of these include:

1. *Understanding the purpose for which forecasts are being developed*. Clearly, forecasts are not an end in themselves. Forecasts can be and are used for many purposes: guides for production and investment, transmission of market signals, influencing policies, and even manipulating the outcome of policies. In the case of forecasts for subsidy purposes, the developers and users of the forecasts must clearly understand the purpose of the subsidy as laid out in legislation.

2. *Objectivity*. A credible forecast must be free of conflict of interest—real or apparent. Forecasters should not be in a position to either gain or lose from the results of their work. Also, forecasts should not over-react to short-run deviations from long-term trends.

3. *Developing and using a good data base*. A good data base should be accurate, relevant to the task, and timely for forecasting purposes.

4. *Maintaining consistency*. Short-run forecasts should be consistent with and lead into long-term forecasts.

5. *Recognizing and accommodating uncertainty*. No matter how sophisticated the analytical techniques used in forecasting, uncertainty cannot be eliminated. Nevertheless, forecasting procedures should take account of uncertainty and attempt to minimize the impact of uncertainty and its ramifications.

III. VOLUME FORECASTING PROCEDURES IN THE WGTA

The Western Grain Transportation Act which came into effect in January 1984 clearly established, among other things, that:

1. the railways would be compensated for their costs of moving grain, doing away with the increasing losses experienced under the Crow's Nest rates;
2. the federal government would commit itself to an annual subsidy payment of $658.6 million, defined as the Crow benefit, and its share of inflation; and,
3. the producer would be responsible for future cost increases.

The intent of the WGTA was that the federal government would limit its payments and that the producer would pay a proportionately larger and larger share of the rate over time. The railways would be compensated for the movement of grain not years after the fact as had been the experience with the branch line subsidy program, but immediately and for *all* costs incurred. The railways are, in general, compensated the month after the movement has taken place, on a monthly basis, based on annual rates established for the crop year by the CTC.

Space does not permit a thorough explanation of the forecasting and rate-making procedures specified by the Western Grain Transportation Act. Figures 1 and 2, taken from Kraft and Karman[3] do provide a schematic illustration of the procedures.

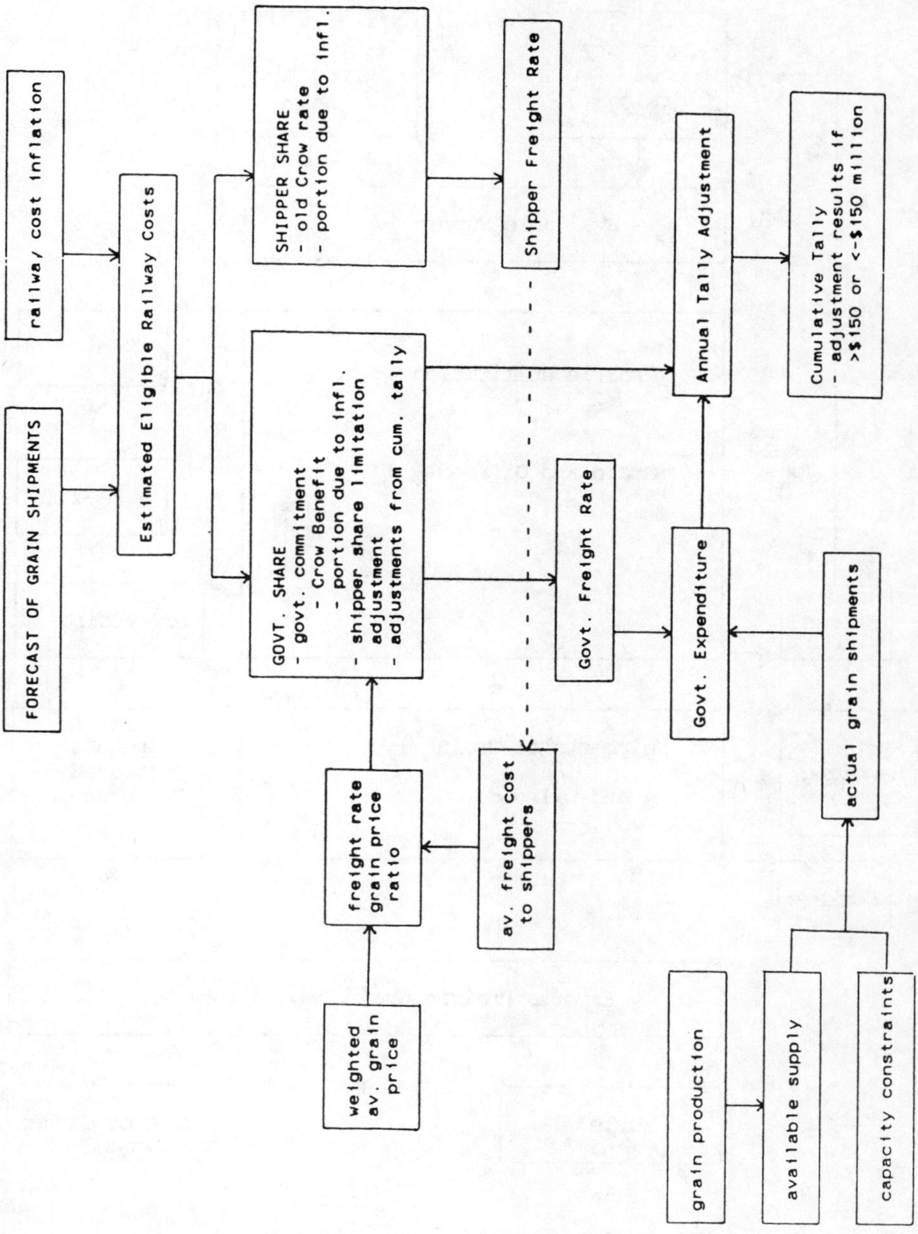

Figure 1. The Rate Setting Process Under the WGTA

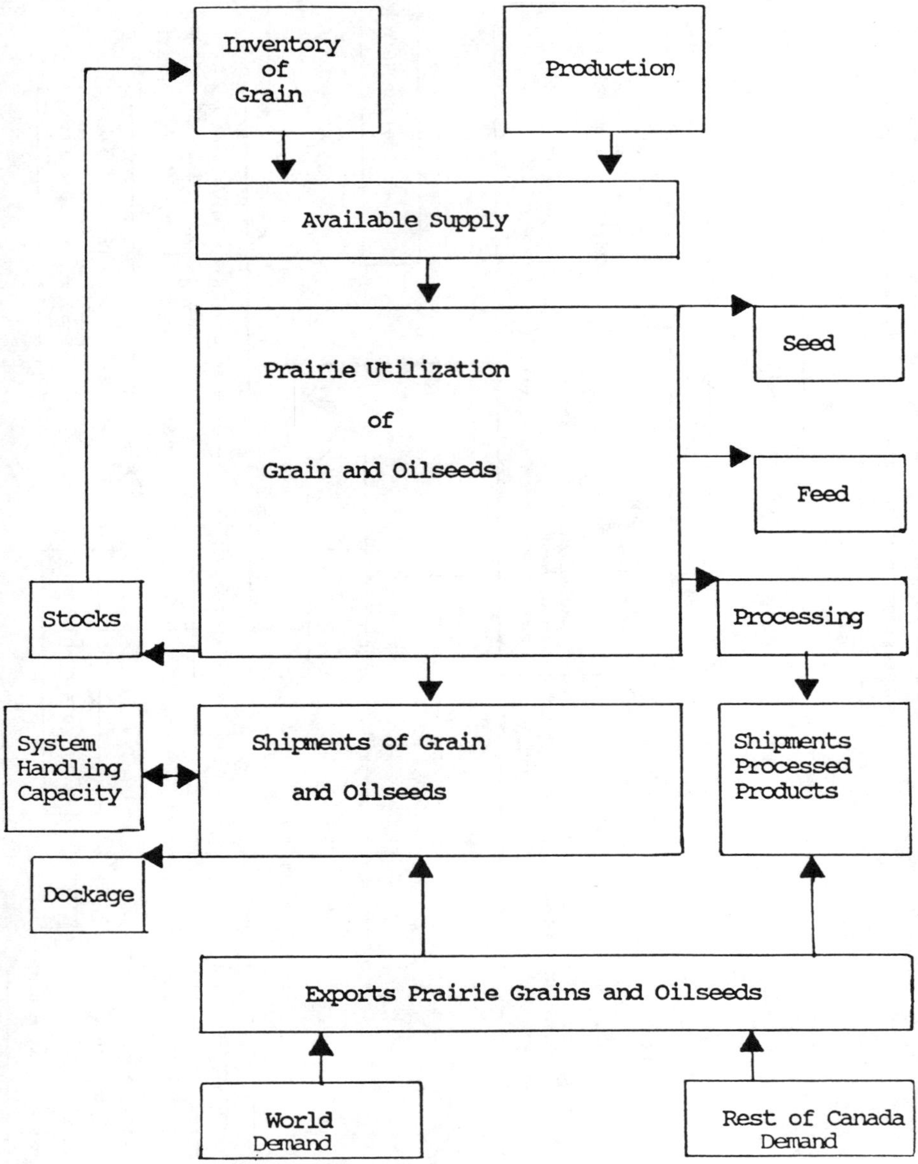

Figure 2. Components of Forecasting Grain Shipments

One of the key procedures in the establishment of the annual grain freight rate is the forecast of grain shipments. Section 18(1) of the WGTA states that the Administrator of the Grain Transportation Agency:

"(a) on or before February 28th of each calendar year provide the Commission (CTC) and the Committee (Senior Grain Transportation Committee) with a tonnage forecast for the movement of grain in the next crop year."

This tonnage forecast is critical because the WGTA specifies that the producer will be responsible for the full cost of transportation in excess of the government commitment. Accordingly, as the tonnage forecast rises so does the producer's absolute freight rate as well as his share of the total per tonne freight rate. At the same time, the government share of the freight rate on a per tonne basis would fall. To complicate matters a little further, there are two types of costs estimated by the CTC in the calculation of the rate—the volume related variable costs and the volume independent line related costs. Again, as the tonnage forecast increases, the portion of the rate tied to line related costs falls on a per tonne basis.

An example may help illustrate the issues mentioned above. (See Table 1). For the 1986-87 crop year the volume related costs were $23.06 per tonne and the total line related costs were approximately $104.0 million. The contribution to constant costs is calculated as 20 percent of volume related costs, another cost component which is affected by the volume forecast. As the volume forecast increases, so does the volume related variable cost and, in turn, the constant costs.

Based on a tonnage estimate of 28.5 million tonnes, the producers' average freight rate would be $5.97 per tonne with the federal government share at $24.97 per tonne for a total of $30.94 per tonne.

In the case where the tonnage forecast is 34.0 million tonnes, the total freight rate falls marginally to $30.43 per tonne, as the line related costs are spread over a greater tonnage. However, the producer/federal government share of the freight rate changes significantly. The producer now pays a freight rate of $9.17 per tonne (an increase from 19.3 percent to 30.1 percent of the total) while the government pays $21.26 per tonne (a decrease from 80.7 percent to 69.9 percent of the total).

These examples illustrate two problems related to forecasting volumes of grain movement: (1) what the prodcer and federal government have paid versus what they should have paid, and (2) what the railways receive versus what they should have received. Assuming that the movement for the 1986-87 crop year will equal 34.0 million tonnes (it actually appears that this movement will be attained or exceeded), the payout, basis a 28.5 million tonne forecast, and a 34.0 million tonne movement is shown below:

Government $24.97 × 34.0 =	$ 848.98 million
Producer 5.97 × 34.0 =	$ 202.98 "
Railways' Receive	$1,051.96 "

However, basis a 34.0 million tonne forecaast, the payouts and receipts would have been:

Government $21.26 × 34.0 =	$ 722.84 million
Producer 9.17 × 34.0 =	$ 311.78 "
Railways' Receive	$1,034.62 "

With the rate set on a forecast of 28.5 million tonnes but with an actual movement of 34.0 million tonnes, the government has overpaid by $126.14 million and the producers have underpaid by $108.8

TABLE 1

1986-87 Freight Rate Calculations

Volume Related Variable Cost	**$ 23.06/tonne**	**$ 23.06/tonne**
Volume Forecast (million tonnes)	**28.5**	**34.0**
Total Volume Related Costs	**$ 657.21 million**	**$ 784.04 million**
Contribution to Constant Costs (20%)	**131.44 "**	**156.81 "**
Line Related Costs	**104.00 "**	**104.00 "**
Total Eligible Costs[a] (Less CN adjustment of $10.4 million)	**$ 882.25 "**	**$1034.45 "**
Government Commitment (Less CN adjustment of $10.4 million)	**$ 710.30 "**	**$ 722.98 "**
TOTAL FREIGHT RATE[b]	**$ 30.94 / tonne**	**$ 30.43 / tonne**
Government Rate	**$ 24.97 / "**	**$ 21.26 / "**
Producer Rate	**$ 5.97 / "**	**$ 9.17 / "**
Government Share	**80.7%**	**69.9%**
Producer Share	**19.3%**	**30.1%**

[a]Costs are reconstructed and may not correspond precisely with all CTC procedures and calculations.

[b]Average freight rate as per CFAW Tariff 4310.

million with the railways receiving an overpayment of $17.34 million for line related costs.

In drafting the WGTA, the legislators clearly recognized these potential problems. In Section 54 of the WGTA provisions were made to account for over or under payment by the producer or federal government in the next year's freight rate. That is, the government commitment would be adjusted accordingly the following year to take into acocunt any over or under payment.

However, no adjustments were put into place to account for railways' over or under payment due to the apportionment of line related costs. The intent of the WGTA was that the over and under payments to the railways would balance themselves out over time. It must be remembered that the reverse happens as well. Every time the actual movement is less than the estimated or forecast movement, the railways will be underpaid.

IV. ASSESSMENT OF VOLUME FORECASTING PROCEDURES AND SUBSEQUENT ADJUSTMENTS

What type of success has been achieved? Average producer freight rates per tonne rose from $5.72 in 1983-84 to $7.70 in 1984-85 and fell to $6.00 in 1985-86. Naturally there was confusion as to how the freight rate to the producer was being set. As noted earlier, everyone expected the producers' freight rate to rise. Yet in three short years the freight rose to a high of $7.70 and then fell to $6.00. The problem, if that is the way to define it, is that given the differences between the actual movement and forecast movement, the interim adjustments could materially affect the following year's freight rate. Very simply, a $60 million overpayment in one year could lead to a $2.00 per tonne reduction in the freight rate the following year (basis a 30 million tonne forecast) as the government's commitment would be reduced by that amount. That, in fact, is what happened in the setting of the 1984-85 rate. For that crop year the federal share, after taking into account the CN adjustment, was reduced to $560.9 million, basis a 31.8 million tonne forecast. This meant that the federal share of the freight rate, on a per tonne basis, declined from $19.52 per tonne to $17.64—a difference the producer had to make up.

While the forecast and the subsequent interim and final adjustment achieved the objectives of compensating the railways in a timely fashion, these were difficulties that were hard to ignore. Concern was rapidly increasing about the fact that, in the short history of the WGTA, the producer freight rate exhibited some significant (some would say strange) gyrations.

Attention was turned to providing a more stable rate environment; that is, regular but gradual freight rate increases. In 1985, in consultation with the Senior Grain Transportation Committee, the Grain Transportation agency developed a "tally system" which became part of Bill C-44, an Act to amend the Western Grain Transportation Act, which was passed on June 27, 1985.

The concept was fairly straightforward. The Agency would continue to forecast grain movement as it had, but rather than try and "balance the books" in the following year through freight rate adjustments, a tally would be kept of over and under payments. The expectation was that, over time, any over and under payments due to differences in actual and forecast movement would be equal to zero. While the tally may be technically appealing, its "success" was even less than that of the original interim and final adjustments. The intended purpose of the tally didn't even survive a year.

For the 1985-86 crop year the Grain Transportation Agency forecast a movement of 31.8 million tonnes. However, 1985-86 turned out to be the second poor crop in a row. Ultimately, 27.5 million tonnes of grain, or 4.3 million tonnes less than the forecast 31.8 million tonnes was moved. This resulted in a build-up of $60 million of producer money in the tally. For 1986-87, the GTA's initial volume forecast of 30.64 million tonnes (less than the previous year due to inventory adjustments) would have resulted in a producer freight rate of $7.50 per tonne, up $1.50 from the then (1985-86) freight rate of $6.00 per tonne, a not unexpected or unrealistic freight rate increase.

However, production, prices and world markets for 1986-87 were expected to be poor and the GTA was criticized by most groups in the grain handling and transportation industry for its unrealistic forecast. Given this criticism and the fact that over $60 million of producers' money was in the tally, pressure was brought to bear on the federal government and ultimately the GTA was directed to lower its forecast. A volume of 28.5 million tonnes was selected which in turn led to a producer freight rate of $5.97 per tonne—a rate nearly identical to the previous year's $6.00.

What will the effect be on the tally? Using the example mentioned earlier the government will have overpaid by $126 million and the producers will have underpaid by $108 million with the railways having received an overpayment of $17 million for line related costs. However this will not happen. Bill C-44, which introduced the tally system, also has a number of clauses which will materially affect what is put into the tally and what the railways will receive.

Section 43.1 states that..."the portion of the rate payable by the shipper . . . (for) . . . the 1986-87 crop year shall not exceed the portion of the rate that was so filed and published by the railway company in respect of the 1984-85 crop year for such movement." Section 55.2 also states . . . "where the transitional tally adjustment in respect of a crop year is a positive amount, the Minister shall reduce any payments due to the railway companies under this Act by the amount by which the transitional tally exceeds . . . "

No one was concerned about these sections when the 1986-87 rate was being established as it appeared the movement would be below normal levels with the resultant producer rate below $7.70 per tonne. However, it is estimated that grain movement this year will be in excess of 34 million tonnes—well above average. Had the forecast been 34 million tonnes, these clauses would have come into effect. At 34 million tonnes the producers share would have been in the order of $9.17 per tonne and the federal portion of $21.26 per tonne, but Section 43.1 would have limited the producer portion to $7.70 per tonne. In addition, the federal share of the rate would have equalled $21.26 and not $24.97 per tonne. As a result, it is now estimated that the railways will have to pay back some $90 million.

These clauses were retroactive. In effect almost the entire railway contribution to constant costs will be eliminated for 1986-87.

Some have chosen to interpret the 1986-87 situation as one where the railways have simply been overpaid. It must be made clear that the 1986-87 situation is not an overpayment to the railways for their costs but an overpayment through a definition that tried to limit the increase in the producers rate. It is also a definition that will not apply beyond the 1986-87 rate setting exercises.

For 1987-88, the producer freight rate will be in the order of $8.50 per tonne—a dramatic increase from $5.97. However, had the intent and principles of the tally and forecasting mechanism been followed, the increase in the producer share of the freight rate would have been gradual—from $6.00 in 1985-86 to $7.50 in 1986-87 to $8.50 in 1987-88.

In essence, the tally has some potentially serious drawbacks. First, the cumulative tally may not always return to zero within a reasonable time horizon. There are five reasons why this is so:

1. The farmer never pays less than the old "Crow Rate;" the government underpayment in low volume years can be less than the overpayment in correspondingly high volume years; the tally is therefore biased.
2. Although there is an equal probability of overestimating or underestimating the volume, there may not be equal probabilities attached to the size of the error. Again, the tally could therefore be biased in whichever direction lies the chance of the larger error.
3. If the long term growth rate changes, there will be a period when volume estimates are either consistently overestimated or underestimated, and the tally will be driven one way or the other.
4. A particularly high or low volume year can result in an addition to the tally from which it is unlikely to recover.
5. Although there is no lower limit to volume shipped (in the event of a crop failure), there is an upper limit imposed by the capacity of the grain handling and transportation system.

Second, once the cumulative tally reaches its ceiling (now set at $150 million) the effectiveness is destroyed, and the "tally" adjustments have the same effect on rates as did the previous "interim" and "final" adjustments.

Third, the prospect of something in excess of $100 million being owed to the farmer is unpalatable to some grain industry participants as was evidenced during the establishment of the 1986-87 freight rates.

The current forecasting and freight rate setting mechanism compensates the railways almost immediately after the movement takes place, with the major exception of the 1986-87 crop year. However, the current procedures, even with the tally, are quite incompatible with the objective of trying to stabilize producer freight rates.

Returning to the desirable criteria for forecasts mentioned earlier, how does the WGTA volume forecasting procedure rate? With respect to understanding the purpose for which forecasts are being developed, it would appear that the GTA and most industry groups have a clear understanding of the role of volume forecasts as envisaged in the WGTA. However, pressure on the federal government to "adjust" the volume forecast in order to minimize producers' freight rate increases in 1986-87 led to producer freight rates that are quite inconsistent with the original intent of the WGTA.

The economic difficulties of Prairie grain producers are a matter of great concern. It is more appropriate to deal with these difficulties through other policy measures such as emergency income support programs, and not through changes to the intent and procedures in volume forecasting under the WGTA.

On the matter of maintaining consistency, the "adjustments" to volume forecasts have resulted in short-term freight rate adjustments which are not consistent with long-term trends. The tally system attempts to cope with this, but its degree of success is questionable.

Recognizing and accommodating uncertainty is perhaps the most difficult area facing the volume forecasters. No amount of improvement in analytical techniques will eliminate the variability in Prairie grain production. Clearly, what is necessary is a reduction in the need for volume forecasts in establishing subsidy levels. Some preliminary suggestions along these lines are contained in the next section.

Turning to objectivity and a good data base, there appear to be no "threats to objectivity" with respect to the procedures followed by the forecasters in volume forecasting. Also, the data base available in the Prairie grain production and marketing sector is well-developed and is being used effectively.

V. SOME RECOMMENDATIONS FOR CHANGE

Given the requirement of a fixed (in current dollars) federal government commitment for payment of the Crow benefit and the reality of fluctuating grain volumes, fluctuating freight rates to grain producers are inevitable. It is quite clear that the current tally system is not able to overcome these difficulties.

A recent report by Kraft and Karman[4] concluded that the shortcomings of the volume forecast did not rest with the methodology but rather with the task at hand, and that a procedure less reliant upon foretelling the future would be preferable. More specifically, they recommend establishing a rate based on a volume unlikely to be exceeded in any crop year and then allowing a rebate to producers at the end of each shipping season where shipments fail to equal this volume. However, this would likely lead to a regular underpayment to the railways for line related costs. Also, the practicality, and acceptability, of having the producer always paying a larger than expected freight rate is questionable.

Two other possible adjustments that would lessen some of the fluctuations in freight rates would be:

1. A separate payout to the railways for line related costs. This would avoid over or under-payment to the railways due to differences in the volume forecast and actual grain movements.
2. A holdback of a portion of the Crow benefit, with any residual to be paid the following year in the form of a final payment. This would be similar to the Canadian Wheat Board's concept of the initial and final payment for grains marketed through the CWB. The concept is that the producers' freight rate

would be based on a government commitment of, say, $500 million rather than the $658.6 million. This would result in the producer paying a higher initial freight rate. However, at the end of the year, once the movement was known, he could, potentially, receive a rebate. If actual movement was much higher than forecast, then the rebate would be small or non-existent—as the holdback would be used to cover government payments. If less than expected, he would receive a larger portion of the $158 million held back as it would not have been spent. Again, this proposition has a similar limitation to the Kraft and Karman suggestion, namely questionable producer acceptability.

Notwithstanding the political sensitivity of the "method of payment" issue, many of the problems inherent with the present legislation with respect to volume forecasting, rate setting, and producer/government sharing of freight costs could be minimized by amending the legislation to permit a "pay the producer" approach. For example, a payout of the federal government commitment, as recommended by the GTA[5] would result in the producer being responsible for the entire freight rate with the only changes being due to changes in costs. No volume forecasts would be necessary, and no division would have to be made between the federal and producer share.

Another option under the "pay the producer" umbrella would be to incorporate the federal government commitment into the Western grain Stabilization Program. This approach would explicitly take account of the instability associated with Prairie grain production and producers' ability to pay higher freight rates. This is explored in more detail in another paper at this Conference.[6]

Although the Western Grain Transportation Act has achieved, in general, one of its objectives, namely to compensate the railways for their costs of moving grain, it has been less than successful in bringing about a reasonable adjustment in freight rates that reflect cost changes. Part of that problem lies in the difficulty inherent in forecasting grain shipments. More of the problem lies with interference in the technical procedures of forecasting and rate making.

ENDNOTES

* University of Manitoba Transport Institute and Grain Transportation Agency respectively. The views expressed in this paper are those of the authors, and are not necessarily endorsed by UMTI and GTA.

1 For a detailed discussion of the role of climatic factors influencing grain production, see D.F. Kraft and J. Karman, *Analysis of Procedures used to Forecast Grains and Oilseeds Shipments,* University of Manitoba, November 1986.

2 For a more general discussion of forecasting for subsidy purposes, see L.A. McCoomb, "Criteria for Credible Forecasting," CTC Research Seminar Series, Spring 1985.

3 Kraft and Karman, op. cit.

4 Ibid.

5 Grain Transportation Agency, *Review of the Western Grain Transportation Act,* April 1986.

6 P.D. Earl and E.W. Tyrchniewicz, "An Alternative Mechanism for Paying the Crow Benefit: The Western Grain Stabilization Program."

Modelling the Effect of Passenger/Freight Interference on Railway Costs

by Vivian Wei[*1]

I. INTRODUCTION

The purpose of this paper is to investigate the potential application of railway capacity and cost models to the estimation of incremental costs incurred by interaction between passenger and freight trains. In addition to the use of existing models, this study also develops computer models to aid in computations of the interference cost. These models are designed to examine the route specific impact on the cost of freight service of different options for rail passenger service, including the removal of passenger trains and the incremental expansion of passenger train service.

Both the cost and the capacity implications of running freight and passenger trains on the same track have been subjects of much interest in various railway studies. However, they are usually dealt with separately, with costing studies examining passenger service costing methodologies and railway capacity models trying to determine the impact of passenger trains on the average train trip time and required improvement in existing plant. The present study attempts to combine the methodologies that have been developed in these two fields to examine delay-induced increases in costs of freight services at different levels of passenger traffic.

Although passenger/freight interference currently exists in railway operations, the traditional railway costing methodology does not specifically deal with capacity issues nor interference costs. Rather, these costs are hidden costs included in annual actual expenditures, and they are accounted for in the calculation of systems average unit costs.

The concept of interference cost associated with passenger trains is one of current importance. Since VIA acquired the railway passenger operations in Canada, freight and passenger services have become two separate services offered by different firms. While the traditional method of railway costing in Canada is considered as satisfactory in costing freight and passenger services on a system basis, it lacks flexibility in dealing with the problem of allocation of costs for individual passenger services, e.g. the corridor, regional and remote passenger services. This is particularly true in the allocation of indirect cost because the cost is prorated to each service based on output units and system average unit costs.

As the cost effect of passenger/freight interference is dependent on the railway line capacity, its significance could vary from one rail segment to the other. Thus, the interference cost analysis could provide useful information supplemental to the traditional costing method.

In this paper, freight train delays are calculated with the railway capacity model developed by the Canadian Institute of Guided Ground Transport (CIGGT). Two different approaches are used in the estimation of interference costs. One follows the concept of railway costing as normally applied in Canada where unit costs of delay affected railway service units are calculated and applied. The other adopts largely the method of the World Bank's Rail model in the interference cost analysis.

II. INTERFERENCE ANALYSIS

A. Railway Line Capacity Analysis

The CIGGT's LINE Model is used to investigate the impact of passenger trains on the capacity of the railway line to provide freight train services. LINE is an analytical model that calculates the expected transit time for each train type at given traffic levels on a specific line segment.[2] The expected transit time is estimated from the free running time plus the interference delays. The length of interference delay is influenced by the frequency of meets and overtakes, which is, in turn, affected by a number of factors in railway systems and operations. The following is a list of factors that affect interference delay; correspondingly, a list of input requirements to the model is also provided.

Factors Affecting Train Delays	Inputs to the LINE Model
1. Traffic level and mix	—Trains per day by train classes in each direction.
2. Track configuration	—Single or double trackage, mileages.
3. Train speed	—Free running speed for each type of train in each direction.
4. Signal System	—e.g. Automated Train Control (ATC), Centralized Train Control (CTC). Number of intermediate signals, etc.
5. Train priority	—Train priority matrix indicating the relative priority between a pair of trains.
6. Switch time	—Switch delays for each priority class.
7. Safe stopping distance	—Minimum headway between trains.
8. Changes in traffic level	—The class of trains for which traffic can be increased for capacity measurements.

At the start of the analysis, it is necessary to define line capacity. Depending on the constraints set by the analyst, capacity can be defined in various ways. For the purposes of this study, the capacity is considered reached if any of the following conditions hold:

1. Average total delay reached 50% of free running time.
2. Average total delay for any class of train reached 60% of that class' free running time.
3. Any train or class of trains had a transit time that exceeded 10 hours.[3]

A case study is established based on railway operations on the CN tracks between Kamloops and Hydro in British Columbia. This rail segment is 227 miles in length. The signal system is CTC. This route is chosen partly because traffic density on this rail segment is relatively high and partly because detailed data are available in a recent study of joint-track usage of this area.[4]

In order to measure delays due to interference between passenger and freight trains, the daily frequencies of freight trains are held constant while the number of passenger trains per day is varied.

The base case assumes 17.2 freight and 2 passenger trains per day. The freight traffic level represents the railway's 1984 operations; the daily run of passenger trains on this CN route did not start until the middle of 1985. Thus, the traffic level in the base case is hypothetical, but not unlike the actual operation in the middle of 1980's. The alternative cases include the removal of passenger trains and the incremental increase in the number of passenger trains, two at a time (one in each direction) until the capacity is reached.

The results of the rail line capacity analysis is summarized in Table 1. The table contains two parts. The first part shows transit times and speeds of freight trains when there is no passenger train. It is noted that even without passenger trains, there could be interference delays due to interaction between different freight trains. This is shown in the last two columns. In this paper, the freight train interference delay is held constant.

The second part of the table shows train transmit times when passenger trains are added. The first column shows the number of passenger trains per day in each alternative case. Based on the CIGGT model, the capacity as above defined will be reached at the level of 12 passenger trains (including both directions) in addition to the daily freight trains. The columns numbered 1 to 10 represents the five classes of trains in each direction as listed in the first part of the table.

B. Analysis of Train Delays

Interference delays are reflected in the changes in transit times of freight trains at different passenger service levels. The interference delays represent the first order delay resulting from more frequent meets and overtakes. A second order delay may also be generated due to increased track maintenance times at higher traffic levels. Work program slow orders may in turn affect first order delays. However, in this study, only the interference delay is estimated.

To obtain yearly delay due to passenger/freight interference, changes in average freight train trip times are multiplied by the total number of freight trains per day and by 365 days for a given traffic level as compared to the base case.

Delays in train hours can be directly converted to car-days and locomotive hours based on the average number of cars and locomotives per train respectively. Train equipment and traffic of the route studied are estimated based on CN 1984 operating statistics. In Table 2, the interference delays measured in train hours, locomotive hours and car days are presented.

TABLE 1

Effect of Passenger Trains on Freight Train Transit Times
Kamloops–Hydro Based On CIGGT Capacity Model

I. FREIGHT TRAIN TIME AND SPEED - WITHOUT PASSENGER TRAINS

Train Code	Dir.	Type of Trains	Trains* per day	Free Running Time: Time (hr)	Free Running Time: Speed (ml/hr)	With Interference: Time (hr)	With Interference: Speed (ml/hr)
1	EB	Slow Freight	3.4	7.28	31.2	8.92	25.5
2	EB	Solid/Unit	3.7	7.10	32.0	8.38	27.1
3	EB	Fast Freight	1.4	7.03	32.3	7.88	28.8
4	EB	Local	.2	6.85	33.2	8.43	26.9
5	EB	Passenger	.0	6.13	37.0	6.42	35.4
6	WB	Passenger	.0	6.03	37.7	6.25	36.4
7	WB	Local	.2	6.87	33.1	8.32	27.3
8	WB	Fast Freight	1.8	7.03	32.3	7.63	29.8
9	WB	Solid/Unit	3.8	7.38	30.8	8.70	26.1
10	WB	Slow Freight	3.1	7.55	30.1	8.95	25.4

II. TRANSIT TIME IN HOURS - WITH PASSENGER TRAINS

Passenger Trains/Day	Train Code 1	2	3	4	5	6	7	8	9	10
Free Time	7.28	7.10	7.03	6.85	6.13	6.03	6.87	7.03	7.38	7.55
2	9.28	8.69	8.13	8.77	6.52	6.30	8.67	7.83	9.07	9.34
4	9.67	9.05	8.38	9.14	6.62	6.36	9.01	8.01	9.44	9.69
6	10.11	9.40	8.60	9.51	6.71	6.44	9.38	8.24	9.84	10.16
8	10.59	9.83	8.93	9.91	6.83	6.54	9.82	8.42	10.31	10.61
10	11.13	10.25	9.26	10.37	6.89	6.66	10.26	8.67	10.78	11.15
12	11.67	10.71	9.56	10.91	7.02	6.76	10.80	8.90	11.33	11.69

* See Footnote 7.

TABLE 2

Measures of Interference Delay Impact of Passenger Trains on Freight Operators

KAMLOOPS - HYDRO

	Annual Increase in		Increase in Car Days by Train Type				
Psg. Trn Per Day	Freight Trn Hrs	Loco Hrs	SL Frt	Unit	Fast Frt	Local	Total
0	-2 103	-4 838	-3 727	-3 848	-844	-44	-8 464
2							
4	2 146	4 935	3 666	4 117	820	43	8 646
6	4 537	10 435	8 114	8 409	1 688	89	18 301
8	7 193	16 544	12 725	13 524	2 621	143	29 014
10	10 078	23 180	18 060	18 640	3 735	201	40 636
12	13 072	30 065	23 396	24 342	4 717	269	52 725

III. THE COST OF INTERFERENCE DELAY

Two approaches are used in the cost estimation. One is based largely on the traditional railway unit cost method; the other estimates the interference cost based on delay effect on equipment cycle time and productivity.

A. The Railway Unit Cost Approach

This approach involves three steps: firstly, the cost items affected by the train delay are determined; secondly, a set of unit costs related to train delay measurements is developed; and finally, interference costs are calculated by multiplying the quantities of delay time units by the appropriate unit costs.

Cost Items Affected by Train Delay. In Table 3, major cost items associated with delay time units are identified. This table is constructed based on the unit cost system adopted by Canadian railways. Train interferences and delays are considered to have a direct impact on costs associated with the following five groups: train control and communication, crew wage, freight car, locomotive and fuel.

Traditionally, crew wages are computed on a train-run (or distance) rather than on an hourly basis. However, in the long run, unions negotiate for income levels would probably press for an increase in mileage rates should hours per train-run rise. Ultimately, it is hourly income that matters. Therefore, in this analysis, hourly crew wages are applied to train delay and included in the determination of the total interference cost.

In the traditional railway unit cost development, most operations and maintenance expenditures are related to output units, by either direct analysis or regression analysis. In addition, many of the indirect expenses including administration, supervisory and overhead expenses are related to other sets of expenses. For example, a part of the maintenance of shops and engine houses is explained by freight car maintenance expenses which is, in turn, related to

TABLE 3

Cost Items Affected By Train Delay

Train Delay	Cost Directly Affected	Cost Indirectly Affected	Service Units* Associated with Cost
Train Hours	- Signals, Communication and Train Control, Maintenance, and Ownership	- Overhead	Train hour (CP), Train mile (CN)
	- Crew Wages		Train hour
Car Days	- Freight Car Maintenance and Ownership	- Shop Machine and Equipment Maintenance and Ownership - Overhead	Car days
Locomotive Hours	- Locomotive Maintenance and Ownership	- Shop Machine and Equipment - Overhead	Diesel Unit Miles
	- Fuel	- Fuel Station Maintenance and Ownership	Gallons

* The service or output units are for Directly Affected Cost Items; the unit costs of indirect cost items are normally associated with the expenditure of direct costs.

car days and car miles. This process is known as "cost pyramiding." Thus, one work unit may be directly and indirectly linked to a number of expense accounts in the railway system.[5]

Unit Cost Development. In estimating interference costs, some distance or other physical units have to be converted to time-related units. These include the conversion from train miles to train hours, from diesel unit miles to locomotive hours and relating gallons of fuel consumption to delay in locomotive hours.

An average speed of 12 miles per hour is used to convert distance to hours. This speed is derived based on an analysis of the train acceleration profile produced by the AAR's Train Energy Model. The speed represents the average train speed during the first 10 minutes after the starting of a train, on level track.

Accordingly, the unit costs of train miles and diesel unit miles calculated from costing models are adjusted to unit costs of train and locomotive hours respectively.

The fuel consumption per locomotive hour of delay is also determined based on the Energy Report of the AAR model which calculates fuel consumption of a locomotive unit at a given notch position or speed. As train delay in each meet or overtake involves train slow down, speed up and wait time, fuel consumption in interference delay is calculated assuming 25% acceleration time and 75% train idle time. The results show that fuel consumption per locomotive hour of delay is approximately 58% of fuel consumption (in gallons per hour) at normal speed. The unit cost of fuel per locomotive hour is adjusted accordingly for interference cost estimation.

Results of the Unit Cost Model. Results of the unit cost model are summarized in Table 4. The table shows the total interference cost and breakdowns by the five cost groups. The case with 0 passenger trains can be interpreted from two points of view. On the other hand, it represents the interference cost that can be saved by removing the passenger trains. On the other hand, it represents the hidden cost borne by the freight service because of the existence of the passenger train runs.

B. The Productivity Approach

Train delay results in longer car and locomotive trip times. Thus, it increases the rolling stock cycle (turnaround) time and reduces the annual mileages that can be generated per car and per locomotive. In other words, to carry the same traffic level of ton-miles, more rolling stock is required. The relationships, between average train speed and equipment cycle time, and between cycle times and rolling stock requirement, can be calculated with the World Bank's RAIL Model.

The RAIL model was designed in the late 1970's to investigate alternative investment policies for developing nations. The RAIL is a large scale and multi-function model by construction. It permits the user to define parameters of eight basic ELEMENT types in a railway system. The types of elements are Project, Locomotive, Wagon, Route, Service, Traffic by service/route, Line and Yard. Furthermore, RAIL provides a range of analysis types which represent varying levels of complexity.

The model calculates the locomotive and car turnaround times using the following two formulae.

LTURN = (TIME + WKDEL + LOCDEL)/24
CTURN = (TIME + WKDEL + DELSHN + NSHUNT)/24 + CDEL

TABLE 4
The Unit Cost Approach Estimated Interference Costs

KAMLOOPS - HYDRO

Yearly Estimates in 1983 Costs
Dollars

Passenger Trains/Day	Signal & Comm.	Loco	Freight Cars	Fuel	Crew Wages	Total
0	-67266	-159652	-50824	-253760	-102669	-634170
2						
4	68620	162865	51874	258867	104736	646962
6	145092	344367	109899	547357	221457	1368171
8	230027	545955	174183	867773	351094	2169031
10	322298	764956	244029	1215867	491930	3039081
12	418022	992150	316598	1576982	638035	3941787

Where LTURN = locomotive turnaround time in days
TIME = average road time over route
WKDEL = average working delay
LOCDEL = locomotive terminal delay per trip
CTURN = car turnaround time in days
DELSHN = average delay per shunt
NSHUNT = number of shunts en route
CDEL = car delays at rail terminals per trip in days

In the above formulae, train interference delay is included in the average road time (TIME). In order to measure the effect of train itneraction on locomotive and car turnaround times, the other delay components are held constant. Specifically, based on CN freight operating data and train schedules, average work program delay and locomotive and car terminal delay are estimated. These delay components are assumed unchanged in the interference analysis.

For the first formula, an average locomotive terminal delay of 40 hours and a work delay of 8 hours are assumed. Thus the total delay for these two components is assumed 2 days.

In the second formula, terminal delay is the most important component affecting car turnaround time. The terminal delay includes car loading and unloading. For the purpose of this study, a terminal delay of 4 days per trip is assumed. Car shunting delay is also a major factor that affects car cycle time; however, it is not affected by train congestion on the main line. A total of 1 day in car shunt and working delay per trip is assumed.

The requirements for rolling stock are calculated using:

LSERV (I) = (NTR/DAYS) * NLOC * TURN
LSTOCK (I) = (LSERV(I) * 100) /LAVAIL
CSERV (I) = (NTR/DAYS) * NTCAR * CTURN
CSTOCK (I) = (CSERV(I) * 100) /CAVAIL

where LSERV (I) = Number of locomotives required in service of Case I
LSTOCK (I) = number of locomotives required in stock
CSERV (I) = number of cars required in service
CSTOCK (I) = number of cars required in stock
NTR = number of train runs a year
NLOC = number of locomotives per train
LAVAIL = percent of time locomotive available for service
DAYS = number of days per year service operates
NTCAR = number of cars per train
CAVAIL = percent of time car available for service

Inputs to the World Bank Model are presented in Table 5. To demonstrate the methodology, the base

TABLE 5
Operating Statistics—Inputs

KAMLOOPS–HYDRO

DESCRIPTION	INPUTS*	PASSENGER TRAINS PER DAY 2	0	4	6
System Trackage	ML	227	227	227	227
Length of Haul	ML	227	227	227	227
Annual Traffic	TONS (000)	26591	26591	26591	26591
Average Car Load	TONS	78	78	78	78
Loaded Cars/Train	CARS	54	54	54	54
Empty/Loaded Ratio		0.784	0.784	0.784	0.784
Average Speed	ML/HR	25.65	26.58	24.70	23.74
Average Road Time	HRS	8.85	8.54	9.19	9.56
Locomotives/Train	LOCOS	2.31	2.31	2.31	2.31
Loco. Availability**	%	95	95	95	95
Locomotive Weight	TONS	180	180	180	180
Empty Car Weight	TONS	28	28	28	28
Car Availability**	%	90	90	90	90
	INTERMEDIATE OUTPUTS				
Car Cycle Time	DAYS	9.58	9.55	9.60	9.63
Loco. Cycle Time	DAYS	2.37	2.36	2.38	2.40

* Estimated based on CN 1984 operating data.

** The availability figures in the table apply to existing equipment only. For new equipment, a 100% availability is assumed.

case and three alternative cases are analyzed. The latter include the case with passenger trains removed, and the cases with 4 and 6 passenger trains per day respectively. it is noted that among the inputs, only two variables, average speed and average transit time, are different in each case.

The Rail Operations Report shown in Table 6 includes annual freight traffic statistics and rolling stock requirement estimated by the model.

Cost Estimation from the Productivity Approach Under this approach interference cost is analyzed from the point of view of costs incurred due to reduced productivity. Train delay increases the number of rolling stock units required in service, the cost implication is investigated here.

The World Bank Model has bult-in mathematical equations to calculate revenues and costs based on route specfici cost and traffic data. In this analysis, the World Bank Model method is used wherever possible. In addition, a method of calculating the incremental costs in train control and rail operation administration is proposed.

Four major cost groups are identified for the interference analysis; they are rail operation costs, maintenance costs, depreciation and capital charges. Data on cost components are presented in Table 7.

1. Rail Operation Costs

In the World Bank's LINE model, line haul costs are the variable cost of operating a train and include crew wages, fuel and oil. The crew labour cost is assumed to be distance rather than time related. Oil cost is estimated as a percentage of the fuel cost. The equation used to calculate fuel cost per train is of interest in estimating interference delay cost; it is specified as:

FUEL C (I) = FCON * FCOST * HORSEP * NLOC * ((.15 + .85 * AVSPD (I)/MXSPD) * FRTIME + .15 * (WAIT (I)/60-FRTIME + .25 * LOCDEL))

A detailed description of this equation and its variables can be found in reference.[6] Essentially, this equation specifies fuel cost per train as a function of the free running time (FRTIME), train speeds (AVSPD and MXSPD), road time (WAIT) which

TABLE 6
Freight Operation Report

KAMPLOOPS–HYDRO

DESCRIPTION	PASSENGER TRAINS PER DAY 2 (Base Case)	0	4	6
Number of Trains	6313	6313	6313	6313
Locomotive Miles (000)	3310	3310	3310	3310
Train Miles (000)	1433	1433	1433	1433
Net Ton Miles (000)	6036157	6036157	6036157	6036157
Gross Ton Miles (000)	10497651	10497651	10497651	10497651
Car Miles (000)	138058	138058	138058	138058
Equipment Required in Service				
Locomotives	94.6	94.1	95.2	95.8
Freight Cars	8945.7	8924.2	8969.3	8995.0
Equipment Required in Stock				
Locomotives	99.6	99.1	100.2	100.8
Freight Cars	9939.7	9918.2	9963.3	9989.0
Increase in Rolling Stock Requirement				
Locomotives		-0.52	0.57	1.18
Freight Cars		-21.52	23.61	49.29
Annual Miles per Locomotive	34978	35170	34770	34576
Annual Miles per Car	15433	15470	15392	15348
GTM/Locomotive	6296106	6330626	6258674	6218443
GTM/Car	1106873	1109542	1103959	1100807
Train Hours	55871	53914	58018	60354

TABLE 7
Cost Data

COST COMPONENTS	VAR OR FX	UNIT	COSTS $*
Crew	V	Trn-HR	48.81
Fuel	V	Trn-HR	208
Other Rail Operation	V	GTM	0.0015
Maintenance			
Locomotive	V	Loco-ML	1.3534
	F	Annual/Loco	79838
Freight Car	V	Car-ML	0.0750
	F	Annual/Car	807
Way and Structure	V	Trn-ML	10.2834
	F	Annual/ML	4123
Depreciation			
Locomotive	V	Loco-ML	0.7153
Freight Car	V	Car-ML	0.0219
Way and Structure	V	GTM	0.0007
Capital Charges**			
Loco	V	Locomotive	10%
Wagon	V	Car	10%
Way and Structure	V	ML	10%

* Derived based on CN system data from the 1983 Railway Financial and Traffic Statistics. Statistics Canada Report 52-215.

** The opportunity cost of capital is assumed at ten percent.

includes delays, and other variables related to locomotive characteristics. With some mathematical operations, the equation can be reduced to show that the fuel cost per unit of delay time can be calculated as 15% of the cost for free running time. This ratio is a low estimate compared with the factor of 58% developed in the unit cost approach. The fact that this formula gives lower estimates of fuel consumption in delay time than a previously used formula is considered unimportant, because the main purpose of this study is to investigate alternative methods for the estimation of interference costs.

In this paper, a set of formulae is developed to estimate the possible changes in the other rail operation cost, i.e. administration and train control. These formulae are described below.

First, it is assumed that total annual gross ton miles are not affected by interference delay, only the requirement for new rolling stock. Thus:

(1) $GTM = GTM_{oj} + GTM_{nj}$

where GTM is annual gross ton miles, GTM_{oj} represent the GTM carried by existing equipment (including cars and locomotives) and GTM_{nj} represents the GTM carried by added equipment at a given traffic level j. In the base case:

$GTM = GTM_{oj}$ and $GTM_{nj} = 0$.

the unit cost per gross ton-mile for a given traffic level is adjusted as:

$RCGTMj = RCGTM_o * (GTM/GTM_{oj})$

where $RCGTM_j$ and $RCGTM_o$ represent rail operation costs per GTM at traffic level j and the base case o respectively. The interference cost is calculated as:

$IRC = (RCGTM_j - RCGTM_o) * GTM = RCGTM_j * GTM_{nj}$

where IRC represents the incremental railway operating cost.

2. Maintenance Costs

In the World Bank Model, maintenance costs for rolling stock are specified as a combination of fixed annual cost per unit plus per mile travelled. The delay induced rolling stock maintenance costs are calculated based on number of new locomotives and cars required in each alternative case.

3. Depreciation Costs

In this analysis, depreciation is calculated on rolling stock usage and replacement value. Thus, the delay induced increases in equipment depreciation are calculated as a function of new rolling stock requirement, annual mileage per equipment, replacement cost and equipment life in distance.

4. Capital Charges

Capital charges are defined as the opportunity cost of the undepreciated capital invested in rolling stock.

The annual capital charge is the average book value (equal to half the replacement value) times the required rate of return. The interference costs are calculated based on new equipment requirement in each alternative case.

Results of the Productivity Approach. Under this approach, train delay times are built into equipment cycle times and the incremental costs associated with the resulted new equipment requirement are calculated. The results of interference costs based on the simulated operation on CM Kamloops segments are presented in Table 8.

It is noted that for two reasons, the general expense group e.g. overhead and other expenses is not included in the productivity approach: First, it is debatable whether this cost group should be included in the incremental cost analysis, and second, it would require further analysis to develop functional relationship between the overheads and the interference delay. Consequently, the total interference costs presented in Table 9 are lower than those estimated under the unit cost approach where system overheads are allocated to output units.

However, the lower cost estimates are also attributable to the different mathematical equations adopted in the two approaches. In particular, fuel costs estimated in the previous section are about three times higher than those presented here; this is due to the use of different fuel consumption formulae rather than the unit cost of fuel, i.e. cost per gallon. Furthermore, the crew cost is included in the unit cost approach based on a long term perspective, while in the World Bank Model, crew wage is assumed not changed in short-run marginal cost calculations.

IV. CONCLUSIONS

The preliminary modelling effort made in this study indicates the feasibility of estimating the interference cost for individual train services. However, further development effort would be necessary to make use of the methodology as alternative costing approach. The following is an evaluation of the two cost models illustrated here.

Firstly, the advantage of the unit cost approach is that the "pyramiding" unit cost method provides a link between train delays and the cost items directly and indirectly affected by the delays. The major difficulty of this approach is that most of the unit costs currently adopted by the railways were developed based on system total costs. These unit costs may lead to an overestimate of incremental or marginal cost. Thus, to develop more precise interference costs it would be necessary to review and reconstruct the unit costs in the railway costing system.

Secondly, the World Bank Model is found very useful in analyzing the effect of interference delay on rolling stock requirement and costs. Furthermore, the Model's ability to analyze the marginal cost on a route or service basis is particularly useful for the route-specific interference analysis. However, more

TABLE 8

The Productivity Approach Estimated Interference Costs

KAMLOOPS–HYDRO

Yearly Estimates in 1983 Costs

	Remove Passenger Trains	Add Two Passenger Trains	Add Four Passenger Trains
RAIL OPERATING COST			
Fuel	-61061	66970	139849
Other	-40517	44646	93476
MAINTENANCE			
Locomotive	-41203	45190	94367
Freight Car	-17369	19049	39780
DEPRECIATION			
Locomotive	-12983	14078	29209
Freight Car	-7307	7974	16604
CAPITAL CHARGES			
Locomotive	-25804	28301	59099
Freight Car	-32284	35408	73940
TOTAL	-238528	261616	546324

case studies are required to draw conclusions on its cost estimates.

Overall, the interference cost approach may provide useful information supplementary to the traditional railway costing method. It is an aspect of importance to a railway for its internal investment evaluation and traffic development purposes. Moreover, it may be of potential use in service contract negotiations between passenger and freight railways. However, it remains to emphasize that this is a research project which concentrates on methodologies. Further studies are required to deal with their practical applications in railway costing systems.

ENDNOTES

*Transport Canada.

1. The views expressed in this paper are those of the author and do not reflect the views of the Transport Canada.
2. English, G.W. and G. Schwier (1986) "User Analyst Guide CIGGT Linemodel V. 2.1." Canadian Institute of Guided Ground Transport, Queen's University, Kingston.
3. The ten hours limit on train-run is based on crew shift specified in industry labour agreements. See Spector, A.N. "An assessment of railway line capacity models available to the CTC," Canadian Transport Commission, 1986.
4. Delcan, Deleun Chather, Canada Ltd., "Joint Track Usage Study, Kamloops—Mission, B.C." Report No. 32-1727, November, 1985.
5. Hariton, G., "Railway Costing—A Review," Research Branch, Canadian Transport Commission, 1984, WP 60-84-04.
6. Taborga, P.M., "Railway Analysis Interactive Language, Rail User's Manual" The Work Bank, March, 1986.
7. Table 1, the trains per day are inputs to the Model. They are annual averages derived from train schedules. For local trains, the average number of daily trips was weighted by the proportion of the line that was traversed. The passenger train inputs were .001.

The Feasibility of Submarines for Transportation of Arctic Resources

*by D. Molyneux**

ABSTRACT

Over the years, there has been a growing interest in the transportation of mineral and hydrocarbon resources from mines and wells in the Arctic circle to markets in more temperate areas. So far, this has only been done by pipeline or by ship. Neither of these methods is completely satisfactory, but they have the advantage of demonstrating that it is technically possible to carry out such functions. Another potential solution to the problem is the cargo carrying submarine. This alternative has several attractive features, especially since it can sail beneath the ice. The speed losses caused by variations in ice thickness which inevitably affect the schedule of a surface ship, are therefore avoided.

However, since there are no cargo carrying submarines in existence, it is important to review the technical alternatives and evaluate their economics before proposing the submarine as a viable alternative to conventional methods. This paper reviews the various design proposals in the literature and considers the feasibility of the principal features of the transportation system such as cargo, hull structure, propulsion machinery and terminal operations. It also considers the various routes for which a submarine cargo vessel could be used.

The study concludes that a submarine cargo ship is technically feasible and economically comparable with surface ship systems under certain conditions. There remains, however, a tremendous amount of work to be done before such a system could become operational. The paper makes recommendations on features which are likely to make the submarine systems more competative.

I. INTRODUCTION

Since the discovery of oil and gas in Prudhoe Bay in the late sixties, and the opening of a mine at Nanisivik on Baffin Island, NWT, in 1974 there have been several attempts to demonstrate the feasibility of shipping resources from the Arctic to markets in temperate climates. In 1969, the ice strengthened tanker 'Manhattan' negotiated the North-West passage. This was soon followed by the construction of the Trans-Alaska pipeline system (TAPS). In 1978 the 'Arctic' shipped ore southwards during the summer season. All these systems were extensions of conventional technology, but had some drawbacks when developed for Arctic use. The TAPS was five times more expensive to build than its estimates. It also raised objections from native groups who were concerned over disruptions to the environment. In 1977 the Berger commission placed a 10 year moratorium on the development of new pipeline systems in the Canadian Arctic. The pipeline also limits the market potential of Canadian crude oil, which is just as likely to end up in Europe or Japan rather than the lower 48 states of the USA.

If the pipeline is undesirable from economic and environmental considerations, the alternative must be some form of vessel system. Ships have operated in Arctic waters for many years, but so far there has only been one (MV Arctic) operating as a conventional transportation system in Canadian waters. There are several disadvantages to a surface ship system. Variations in ice features can cause speed fluctuations which can result in considerable disruption to a ship's schedule. There are also the environmental considerations of the broken channel left by the ship, which can disrupt traditional migration routes for animals and so effect the lifestyles of humans in the area.

One possible method of avoiding the surface ice is by travelling underneath it. Military submarines have operated, without any apparent difficulties, under the Arctic ice cap for many years, but so far there have been no attempts to develop submarine cargo systems beyond the feasibility study stages. There are good reasons why the use of submarine cargo ships is not more widespread. The difficulty is a fundamental problem of hydrodynamics. The resistance of a ship may be assumed to be made up of two compounds, one due to the viscosity of the fluid in contact with the body, and the other due to the movement of the fluid around the body. For a surface ship, this component also includes the force due to wavemaking at the free surface. The potential advantage of a submarine is the elimination of the wavemaking component of the resistance.

The ratio of the two resistance components, as fractions of the total is not constant throughout the speed range, and only at high speeds does the wavemaking resistance become significant. These speeds are typically outside the range of those considered economical for bulk cargos. The low speed performance of a submarine will be worse than that of the equivalent surface ship, since its wetted area is higher. The submarine is uncompetitive if operation in open water alone is considered. However, if the submarine can operate in circumstances which the surface ship cannot, there is the possibility that it could be economically competitive. Such a situation could be the shipping of resources from the Arctic. The purpose of this paper is to review the technical and economic considerations of this option when compared to the more conventional systems. In addition we must anticipate some of the regulations which may apply to submarine cargo ships, since they may have an effect on the operating economics.

II. PRINCIPLES OF SUBMARINE DESIGN

The design of a submarine is more complex than that of a surface ship. The principle of Archimedes states that a submarine must displace its own volume and mass of fluid in order to remain in equilibrium. The external volume of the submarine is constant, but the mass changes during a voyage as fuel and stores are consumed. The problem is further complicated by the thermal expansion or contraction of the cargo. To accommodate all these changes, the submarine must be capable of taking on and discharging ballast during the voyage.

Submarines operate at increased hydrostatic pressure relative to surface ships, due to the increased depth of water. The submarine structure must be designed to accommodate these extra loads, or it will collapse. The option taken by military submarines is to build a complete pressure hull, capable of withstanding the variation in hydrostatic loading over the operating range. This construction technique becomes very expensive as submarines increase in size since very thick high grade plates must be used.

If the pressure inside the submarine can be maintained at the same level as the external hydrostatic pressure, then the outer skin of the submarine simply becomes a membrane to contain the cargo. It can be built for a much lower cost than the pressure hull, since lower grade materials can be used. The type of cargo most suited to this approach is a liquid, since its internal pressure is naturally the same as the hydrostatic pressure, provided that the liquid fills the container. This concept can be modified slightly to provide pressure hulls inside the submarine for machinery and accommodation.

The other area which needs to be addressed is the design of a system to allow for a variable weight of the submarine or variable volume of the cargo. It is extremely unlikely that the submarine can make a round trip full of cargo in each direction. Therefore it must carry something, to maintain its neutral buoyance. Most surface bulk carriers use sea water as ballast, since it is easily available, cheap and easy to dispose of at the end of the journey. This would also be the obvious choice for submarines. However, unless the cargo and the ballast had exactly the same densities and thermal expansion properties, it is necessary to provide empty spaces in the submarine to allow for these features. These spaces must be pressure hulls in order that they may operate partly full. The problem is greatly simplified for the designer if the cargo spaces can be used for carrying ballast, and the density of the cargo is close to that of sea water. The most effective cargo then becomes a liquid cargo, such as refined petroleum products, crude oil or methanol derived from liquified natural gas (LNG). Solid cargo would be extremely difficult to carry, since it would not withstand the hydrostatic pressure. Coal and ore could be carried in the form of a slurry, but it would require extra processing before loading. For a design to carry LNG there are additional problems, since ballast water cannot be carried in the cargo tanks. This means that the whole cargo and ballast system is built from pressure vessels, which would be expensive, or some other form of ballast must be used, which can be carried in the cargo tanks.

The most efficient hull structure for a submarine minimizes the use of expensive pressure vessel construction to those areas which are essential, and all other tanks are based on a membrane approach.

III. LITERATURE REVIEW

Work on the development of submarines began around the beginning of the 18th century. By the middle of the 19th century, the military submarine as we know it today had been born. Development of this type of ship has continued up to the present day and sizes have been greatly increased and propulsion units made more reliable. The largest submarines sailing belong to the USSR, and have a submerged displacement of approximately 25,000 tonnes. The role of the military submarine is well developed and the vessels are highly sophisticated in terms of control and navigation systems. They have also shown that nuclear power is a practical method of propelling submarines. However there has only been one submarine built with the intention of carrying cargo on a regular basis. The submarine, called 'Deuthschland', was built in 1916 for carrying strategic goods for the Kaiser's war effort. The deadweight of this ship was only 700 tonnes. There are accounts of submarines carrying fuel oil in ballast tanks during the second world war, but again this was for strategic reasons, not commercial ones.

The earliest reference to cargo carrying nuclear powered submarines was given by Russo, Turner and Wood [1]. This paper presented a parametric study covering ship deadweights between 20,000 and 40,000 tonnes at operating speeds between 20 and 40 knots. This design was not developed for Arctic operation, and as discussed above this is not likely to be competitive with surface ships. The concept was designed to carry refined petroleum products from gasoline to bunker. The general structural design was based on the concept of minimizing the amount of pressure vessel construction within the hull. The hull form was of a rectangular cross section with a beam approximately twice the depth. This was chosen, despite the degradation of resistance relative to a square section, to allow access to existing ports and terminals.

As the development of Arctic resources became a realistic concept, the potential benefits of submarine transportation systems were realized, and the basic concept described above was modified to carry crude oil from Alaska to the Eastern Seaboard of the United States. This was described by Jacobsen [2], and the study looked at two basic sizes of submarines. The smaller submarine studied had a deadweight of 170,999 tonnes, and was considered to be the largest which could be built at that time. The larger submarine had a deadweight of 250,000 tonnes, and was considered to be the largest vessel which could safely navigate the North-West Passage fully submerged. The total transportation concept included a submarine loading dock, so that the submarine would never have to surface through the ice.

The basic design was extended again in 1975 [3], to an increased deadweight of 280,000 tonnes. Since the deadweight had been increased, it was no longer possible to use the North-West Passage, and so the only practical route was under the polar ice cap. This study was used by several authors for comparisons with surface tankers and pipelines [4, 5, 6]. The basic concept of the design was still the same as the original proposal, with minimum pressure hull construction, a beam to depth ratio of two, and a submarine loading dock. However, since the deadweight had been increased, the idea of a trans-shipment terminal was introduced. The location proposed was one of the Norwegian fjords, which had sheltered deep water, relatively free of ice and was therefore ideal for the construction of such a terminal. This concept minimized the route of the expensive submarine to just the section covered by ice. The remainder of the journey could be completed by a conventional surface tanker.

The concept was modified by Court, Kumm and O'Callaghan [7] to carry methanol derived from Arctic natural gas. The methanol was processed on a floating barge system towed north and installed near the production site for the gas. The propulsion unit was changed from a nuclear powered steam turbine to a fuel cell. The ship size was reduced to a deadweight of 160,000 tonnes, but the route chose remained under the polar ice cap, rather than using the North-West Passage. This gave the potential for diversifying into the European market as well as the United States.

Another version was given by Jacobsen and Murphy [8] where the design was modified for carrying liquified natural gas, with a non-nuclear power option. The propulsion system was fueled by methane or distilate fuel oil, but the final drive system was still a stream turbine. This reference also presented a design for a 180,000 tonne deadweight crude oil tanker, with the same propulsion unit option. The design had to be modified for the transportation of LNG since ballast water could no longer be carried in the cargo spaces. This meant the hull had to be fitted with extra pressurized ballast tanks, and was therefore less efficient than the crude oil carrier.

All the above references were derived from the initial work developed by Russo, Turner and Wood. The hull design was chosen to minimize the amount of pressure vessel construction, and hence capital costs. A midship section for this type of construction is shown in Figure 1 and a typical general arrangement in Figure 2. All of the references propose a submarine docking system for loading and unloading cargo, which is shown in Figure 3.

An alternative approach was taken by MacPhail [9] in his proposal for an LNG carrier. He suggested building the complete hull as a pressure vessel. This design left very little room for ballast tanks, and so he proposed using liquid nitrogen as ballast for the north-bound trip, which would be discharged at the northern terminal. The design was powered by a gas turbine modified to burn the boil-off from the LNG. In the paper, MacPhail presents two hull options. One is a conventional submarine and the other is a barge for cargo transportation, pushed by a removable propulsion unit, which could be transferred between barges.

The original work performed by Russ et al [1], included some model experiments to establish the basic propulsive performance of the design. Although the models used were small, the results should be useful in preliminary design. A range of hull form proportions were considered, with breadth to depth ratio varying from 1.0 to 2.5. This study showed that there was little change in form drag once

Figure 1
Midship Section Through Membrane Type Submarine

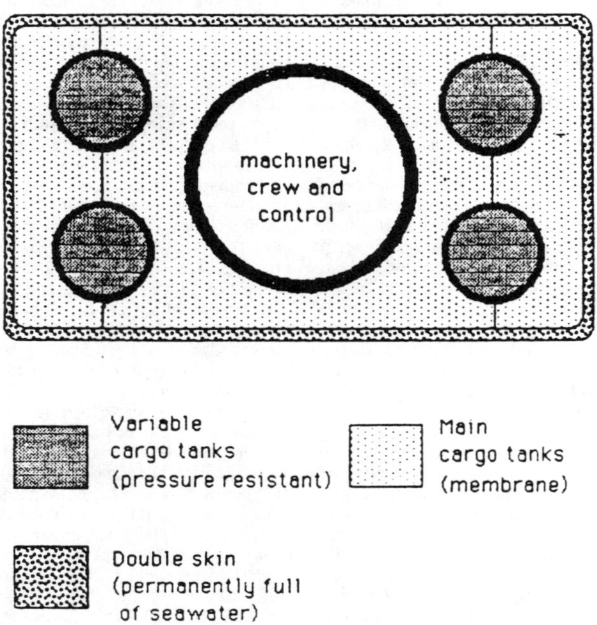

Figure 2
General Arrangement—Membrane Construction

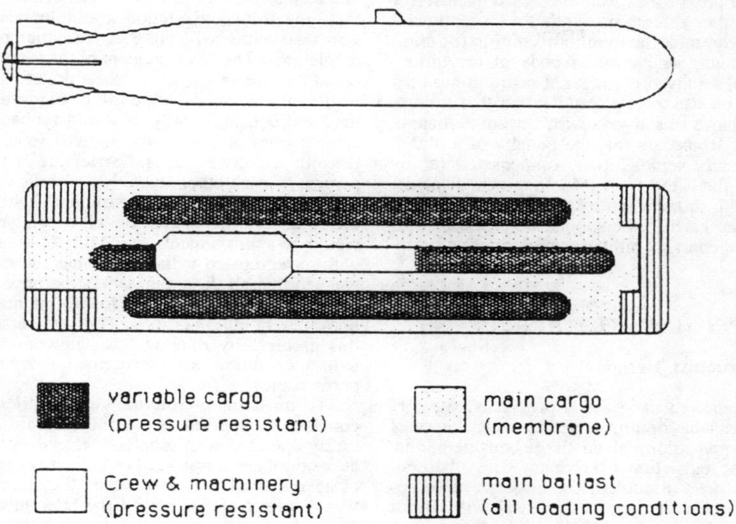

Figure 3
Submarine Docking Arrangement

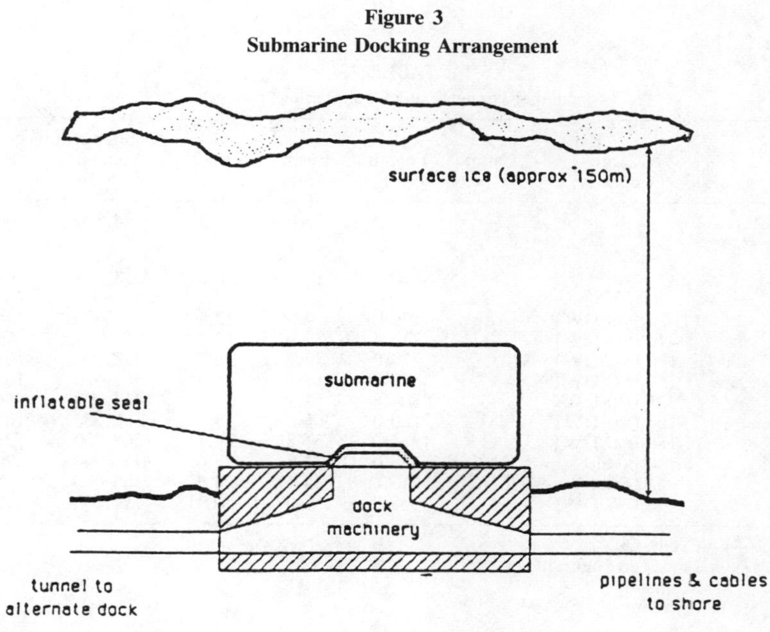

the breadth to depth ratio was greater than 1.5. However, the ratio was constrained to be 2.0, so that the draught was not too great to enter conventional ports. These proportions were then used exclusively by all the other references, with the exception of MacPhail, who took the optimum solution for minimum resistance, which was a body of revolution. Spencer [10] removed the draught restriction on the form given by Russo et al, and found that the optimum resistance was always with a beam to draught ratio of 1.0. It appears that the penalty of a higher form resistance is more than compensated for by minimizing the wetted area for a given displacement. Table 1 shows a comparison of the principal dimensions of each of the submarine proposals, together with the largest military submarines currently sailing.

IV. TECHNICAL EVALUATION

A. Hull Structure Design

Having reviewed the various proposals, we may now consider what options have the most chance of successful application, given these considerations, and potential legislation which may affect the concept. As we have discussed, the most probable application of a submarine cargo vessel will be for liquid cargos, such as crude oil or methanol. The hull structure is reasonably efficient provided that ballast water can be carried in the cargo tanks. This was the approach taken by Russo et al, and with one exception, has been generally adopted as the most feasible method. However, since that proposal was prepared, the IMO introduced legislation concerning the segregation of ballast from cargo tanks for all tankers with deadweights greater than 20,000 tonnes. It also requires the ships to have double skins, to minimize the risk of cargo spillage if the outer hull is damaged.

Although the IMO regulations do not apply specifically to submarines, it is reasonable to assume that submarine construction would not be allowed to go ahead without meeting at least the spirit of the regulations. The basic concept of the membrane hull could be extended to allow for a double skin, but it would of course have to be full of sea water for both load and ballast literally, it would probably kill the submarine concept before it started. Since the densities of, say, crude oil and water are within a few percent of each other, the volume of the cargo tanks and ballast tanks would be similar. This means that the volume of the submarine with segregated ballast would be approximately twice that of the submarine with unsegregated ballast. This has the extra disadvantage that all these tanks must be pressurized, so that they can be run partly full to maintain neutral buoyancy in all conditions. The surface ship avoids this problem by running at a reduced draft in the ballast condition with virtually no degradation in performance.

The most likely solution to this dilemma is to convince the legislators that the submarine could be safely operated with a ballast water treatment plant at the northern terminal. The ballast water would be cleaned and oil residue removed, before clean water was disposed of. Several of the later authors [8, 9, 10] make allowance for the double skin in the general arrangements, but only Spencer makes any allowance for the treatment of the ballast.

The other regulations which apply to ships operating in the Canadian Arctic are the Canadian Arctic Pollution Prevention Regulations (CASPPR). These regulations cover such aspects of ship design as installed power, shell plating strength, and navigation and safety systems. Again these regulations are not formulated for submarines, and in this case most

TABLE 1
Principal Dimensions of Submarine Proposals

Reference	Cargo Capacity	Cargo Type	Length (m)	Beam (m)	Depth (m)	Submerged Volume (m_3)	Installed Power (kw)
Typhoon Class (USSR)	—	—	170.0	23.0	11.5	24,500	60,000
Ohio Class (USA)	—	—	170.7	12.8	10.8	18,200	45,000
[1]	20,000 DWT	P	170.0	24.0	12.2	46,000	14,000
[1]	40,000 DWT	P	216.4	36.6	12.2	79,000	53,300
[2]	170,000 DWT	C	274.0	42.7	26.8	248,000	101,000
[2]	250,000 DWT	C	310.0	52.0	28.5	361,000	101,000
[3]	278,000 DWT	C	304.8	54.9	29.0	414,000	84,000
[7]	160,000 DWT	M	260.0	46.6	24.7	260,000	160,000
[8]	180,000 DWT	C	311.0	51.8	24.7	368,000	74,500
[8]	140,000m^3	LNG	448.0	74.3	28.6	721,000	74,500
[9]	52,000m^3	LNG	213.4	24.4	24.4	83,000	15,000
[10]	440.000 DWT	C	335.0	50.0	50.0	714,000	112,000

Cargo Type—C: Crude Oil
LNG: Liquefied Natural Gas
M: Methanol
P: Petroleum Products

of the problems encountered by surface ships are avoided. It is likely that the legislation would be updated to cover hull strengths for emergency surfacing in ice covered water or grounding situations where the submarine accidentally sits on the bottom. Other areas are likely to be crew evacuation and safety systems, especially if the submarine was nuclear powered.

B. Main Machinery Arrangement

It is unlikely that existing fossil fuel burning systems could be adapted for use in large submarines. There are two problems with such systems. The first is the requirement for oxygen to support the combustion of the fuel, and the second is the disposal of the products of combustion. The first is solved by carrying quantities of liquid oxygen in addition to the fuel. Although valuable cargo capacity is lost, it is technically feasible and relatively easy to implement. The second problem is potentially more difficult. Exhaust gases cannot be disposed of without compressing them to greater than the hydrostatic pressure. Since this pressure is very high the compression system is a large capital investment. One alternative to compression is cleaning systems, which would overcome combustion products and allow the exhaust gases to be recycled. Another option would be to dissolve the exhaust products in water which could then be discharged easily. Some work has been done on both these options, but again extra capital investment is required for the treatment equipment.

The one option of submarine propulsion which has demonstrated that it can operate completely submerged for extended periods is the nuclear powered steam turbine. This type of unit first came into operation in the mid-fifties, and soon demonstrated that it could be used for under ice operation. The first submarine to sail under the Arctic ice cap was the USS Nautilus in 1957. Such units could easily be supplied to provide the power requirements for large cargo carrying submarines, even allowing for the other demands on the system, such as accommodation, ballast handling pumps and other ship systems. Although most of the operational data for such ships is based on military applications, where there is considerable secrecy, there appears to be a reasonable safety record for such systems.

Although nuclear powered submarines have demonstrated that they can operate for long periods under water, there is a growing reaction against them from environmental considerations. The fuel cell is one system which shows potential for solving some of the problems of submarine propulsion without using nuclear power. The fuel cell is a device similar to the lead acid battery, but with the fundamental difference that it does not need recharging by electro-mechanical methods. Instead, the system is based on a chemical reaction. The reactants are hydrogen at the anode and oxygen at the cathode. The system is recharged by replacing the oxygen at the anode on a continuous basis, and the products of the reaction are DC electric power, water and heat. Phosphoric acid is used as an electrolyte and hydrogen is provided from a fuel source, either directly or from a hydrogen rich gas such as methane. The oxygen is provided directly. The dimensions of such a system are similar to those of the equivalent nuclear powered system, and therefore suitable for use in submarines. Although this system has only been developed on a small scale there are several potential advantages for submarine systems. The products of reaction are easy to dispose of, unlike fossil fuel burning systems. In addition the system lends itself to small scale experimentation more readily than nuclear power, and so would be more suitable for pilot projects. Possible disadvantages are the need to carry two potentially hazardous fuels, both of which are highly flammable. The development needed to realize such a system is described by Court et al.

V. TERMINAL DESIGN AND ROUTE SELECTION

There are two proven sources of oil in the Canadian Arctic. Firstly there is the Beaufort Sea, and secondly there is the Sverdrup Basin, where natural gas has been the main find to date, but significant oil reserves have been identified. In addition, seismic surveys of Lancaster Sound and Baffin Bay look promising for future development. The indications are that the three fields of the Beaufort Sea are the most likely for initial development. There are three major markets for Canadian oil. The domestic North American market would most likely be served with an east coast terminal in Canada or the United States. The European market could be serviced from Rotterdam. The other likely market is Japan, who, with no natural resources of its own, is looking for politically stable sources of oil and gas. The Beaufort Sea is approximately the geographic centre of these three markets, with distances of approximately 3700 nautical miles to each one.

The North American ports have relatively shallow water, approximately 12m, and this would be too shallow for submarines and large surface ships. One option to solve this problem would be to use a single point mooring in deep water offshore. The vessel could tie up to the point, either a gravity based tower or a floating buoy, to discharge cargo. The European and Japanese ports have very deep water, and so access directly to the port would be possible but only in surface operation. If the route of the expensive submarine or icebreaker was to be limited to the northern section, and the journey completed in conventional surface tanker, the requirement to include a trans-shipment terminal would have to be considered.

The selection of the appropriate terminal type is important to the overall operability of the transportation system. Since the development of Arctic resources is a very recent idea, there are no existing terminal facilities for surface ships or submarines. It would be necessary to construct the appropriate facility for either system. To maximize efficiency, as much construction as possible would probably be done in temperate regions and the prefabricated units would be towed to the final erection site. It is important to review the design alternatives for the various terminal options to see which would be the most feasible.

In order to avoid the complex control systems needed to maintain neutral buoyancy when loading or discharging cargo, it is preferable to design the submarine with positive or negative buoyancy for this condition. The options then become either a surface system or a submerged system. A submarine system for the Arctic terminal has the advantage of never having to surface the submarine through the ice sheet. In addition, it maintains the local climate, which would be effected by the presence of quantities of open water. It also means that the amount of reinforcing of the submarine hull is minimized, which is desirable when using the 'soft' hull option. However, there are some considerations which work against the submarine docking system. Not least is the one of manoeuvring such a large submarine into a specific location underwater. The inertia of such a vessel is considerable, and docking procedures would have to be initiated up to 35 km away from the site. A sophisticated network of sonar sensors linked to the ships control system would be needed to direct the submarine onto its dock. Since any failure of this system could be catastrophic, there would have to be fail-safe back-up systems with high levels of redundancy.

The cost of the submarine and surface terminal options may not be greatly different. As discussed above the submarine tanker would not be effective unless it could carry cargo in ballast tanks. It would therefore need some facilities at the terminal for cleaning and storing dirty ballast water to meet IMO regulations for ballast discharge. The surface ship option would however, need to be equipped with larger storage tanks than the submarine system, since the probability of the tanker not being able to get to the dock would be higher. Similar considerations would have to be made for the unloading or trans-shipment port at the southern end. Since this is likely to be in ice-free water some of the above considerations are irrelevant. If conventional terminals could be adapted to submarine cargo ships there could be some potential cost savings. However, since these ports are usually very busy, it would be necessary to surface a long way from the terminal to avoid collisions. In addition, if a submarine loading system was used at the northern end, it would mean the added complication of having the ship negatively buoyant at the northern end, and positively buoyant at the southern end.

There are several factors which must be considered when selecting the route for the submarine. The submarine must have adaquate water depth for safe navigation. It is important to consider both the clearance between the keel of the vessel and the ocean floor and the clearance between the top of the submarine and the bottom of the ice cover including pressure ridges. A reasonable clearance in each direction would be one hull depth, making the minimum hull water depth between the bottom of the ice and the ocean floor three hull depths. This may be too small to avoid the effects of the boundaries on the fluid, and so the preferred water depth would be up to seven hull depths, for the majority of the route.

If we take this depth of water, we find that two possible routes from the Arctic are not feasible for submarines. There is insufficient water depth in the North-West Passage and the Bering Strait for the safe operation of submarines. This leaves the only possible route to be under the polar ice cap. So the shipping of Arctic oil to Japan by submarine is not likely to be economical, simply because of the geography. As already discussed it would not be easy to provide submarine access to conventional ports, and so a dedicated deep water trans-shipment terminal would have to be considered. There are two possible locations for such a terminal. One is in Trinity Bay, Newfoundland, and the other is Bodo, Norway. Both ports have very deep water, are sheltered, and relatively ice free. The trans-shipment terminal could be surface or submerged depending on what option was chosen for the northern terminal.

Other considerations for the route can be related to social, environmental and economic factors. The route should create as little disturbance as possible to the delicate Arctic environment, and especially breeding grounds for animals and fish. Busy shipping lanes and areas of offshore activity should be avoided, and so should areas of high iceberg concentration. All of these factors indicate that a route under the polar ice cap to Norway, or around Greenland to Newfoundland are the most suitable for submarines.

The discussion above relates entirely to the technical aspects of the design of a submarine transportation system. It appears that the system is feasible, allowing for a reasonable extrapolation of existing technology. However, there is only one practical route for a submarine tanker, and that is directly under the polar ice cap, as shown in Figure 4.

VI. ECONOMIC EVALUATION

Having determined that a submarine transportation system is technically feasible, we must now review its operating economics in relation to the surface ship and pipeline options. The initial work by Russo et al [1] did not make any attempt at any economic evaluation. There was some discussion when the paper was presented, which indicated that the submarine option would be up to 4.5 times more expensive than the equivalent surface tanker. It is important to remember that this was not based on Arctic operation, and so the surface tankers were only to normal specification. In addition, it was before the IMO legislation was introduced, and so the tanker construction costs would be lower than the equivalent IMO tanker today.

Jacobsen [2] compares his two submarine proposals with an icebreaking surface tanker, of 250,000 tonnes deadweight. Although there were only two submarines and one icebreaker, the study presented the required freight rates for a range of acquisition cost and ship speed for each case. In general the acquisition cost of the submarine option was more than twice the cost of the surface icebreaker. However, if the surface tanker could not average a speed of more than six knots, the submarine option should have a lower transportation cost. If the surface tanker can maintain an average speed of between 6 and 10 knots the two systems are comparable. It is difficult to establish absolute values, since no details of the economic evaluation are given. Also, no allowance is made for the capital cost of the terminals at either end of the route.

The basic design prepared in [3] was used for three seperate comparisons with alternative transportation methods. Taylor and Montgomery [4] compare the submarine tanker, using either a direct route to the Eastern Seaboard or trans-shipment in Norway, with an icebreaking tanker and a pipeline. The icebreaking tanker uses either a direct route through the North-West Passage, or trans-shipping to conventional tankers in Greenland or Iceland. The pipeline options are either trans-Alaska with tankers to the eastern seaboard, or trans-Canada directly to

Figure 4
Routes for Submarine Tankers

the east coast. The results showed that even though the capital cost was approximately twice that of the surface ship, the required freight rate for the submarine system was comparable to the icebreaking tanker. If we use the icebreaker as the standard for both ships going direct, the submarine option was 17% higher than the surface ship. If trans-shipment was used, the submarine was 8% lower. The pipeline is considered to be up to 55% higher than the best surface ship option. The same information was presented by Montgomery and Jordan [5], but with the addition of a mobile trans-shipment terminal. This shortened the route of the submarine, and reduced the required freight rate of the submarine to 16% below the surface icebreaking tanker.

Several surface ship options were compared with the submarine system by German, Macpherarson, Meakin and Park [6], for carrying crude oil or methanol from the Beaufort Sea to the Eastern Seaboard. The surface ship options were divided into two basic categories. The first consists of Class 3 tankers loading in the Beaufort Sea, and travelling westward through the Bering Strait. These then had the option of trans-shipping at the Unimak Pass, lightering into vessels capable of going through the Panama Canal, or going around Cape Horn, to ports on the Eastern Seaboard. The second option was to use Class 8 tankers through the North-West Passage, directly to the Eastern Seaboard, or trans-shipping in Greenland. The submarine option was as described above in [4]. The surface ship sizes were varied from 60,000 to 400,000 tonnes deadweight for the crude oil option. The methanol option assumed that the ships would be the same size, but the deadweight range would be reduced to 54,000 to 360,999 tonnes. The results showed that the required freight rate for the submarine and the Class 3 icebreakers were comparable if the cargo was crude oil. The Class 8 tankers were slightly more expensive, by about 20%. If the cargo was methanol, the submarine was approximately 12% more expensive than the best Class 3 tanker but the submarine and the Class 8 tanker had similar required freight rates. Another option was considered for the surface tanker, and that was LNG carriers of the same classification on the same routes. This was not compared with the submarine since these authors did not consider it economical to ship LNG by submarine. The ship sizes for LNG varied between 75,000 and 175,000 cubic metres. The surface ship route options are shown in Figures 5 and 6.

The analysis of the methanol carrying tanker given by Court et al, using the polar route and trans-shipping in Norway, concludes that it is more economical to ship by submarine than by surface ship. It also considers it to be more economical to ship methanol from the Arctic by submarine than by conventional tanker from the Persian Gulf. The rider to the study is that it is only economical to ship the methanol by submarine if it is used as an automobile fuel or additive, based on the cost of gasoline at the time of the study. There are, however, substantial

Figure 5
Routes for Class 3 Tankers

Figure 6
Routes for Combination of Class 8 and Unstrengthened Tankers

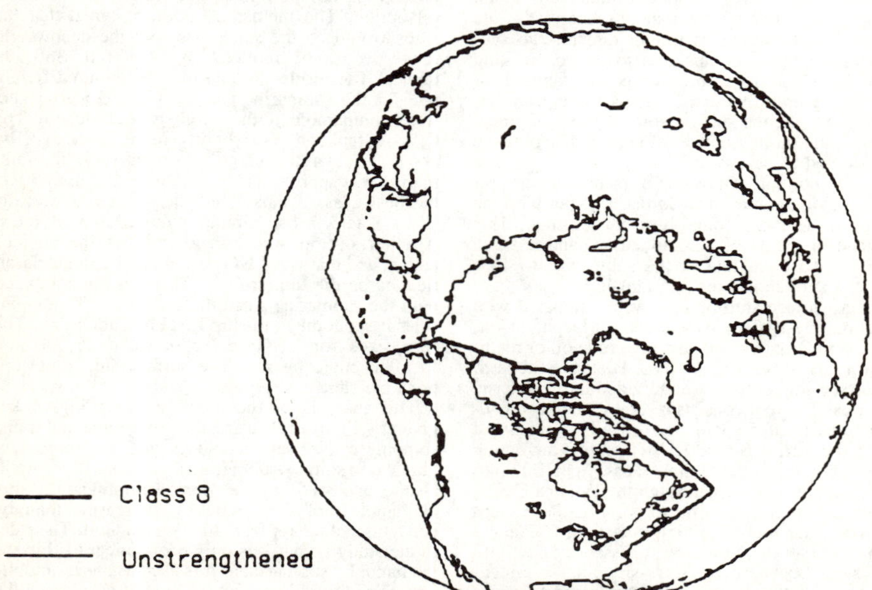

reserves of natural gas in southern Canada and the United States, which could also be processed for automobile fuel. Although not covered by the study, intuitively one feels it would be cheaper to produce methanol from these sources rather than from Arctic gas. In addition, it concludes that shipping LNG directly would not be cost effective.

The results of the economic analyses generally agree that shipping of Arctic crude oil or methanol by submarine is comparable with the icebreaking tanker, but it shows no significant advantage. It is also accepted that the acquisition cost of a submarine is much higher than a surface ship of equivalent deadweight. There is some contradictory evidence, for example the transportation of methanol, but even in that case neither system appeared to have a significant advantage over the other. Neither appeared to have a significant advantage over the other. Neither of the proposals for the shipping of LNG by submarine directly [8, 9] include an economic analysis, but other authors [6, 7] state that it would not be economically viable when compared to other technical options. It is difficult to make a full review of the economics, since very little data is given in the references, but it is hoped that the relative comparisons within a given reference are not biased in favour of one system or another. The one independent study available was given by Spencer, and this also indicated that the submarine system had a lower required freight rate for the transportation of crude oil than surface icebreaker.

VII. THE FUTURE OF SUBMARINES FOR ARCTIC TRANSPORTATION

It appears that the Arctic submarine is technically feasible given the current state of the art of the major ship systems, and there is evidence to indicate that it could be at least as economical a transportation system as the surface icebreaker. The final question to ask is, will it be accepted by the industry as a reasonable alternative to existing systems? The transportation of natural resources from the Arctic by submarine represents a new concept in transportation systems. It is also a complete departure from techniques which have evolved from the shipping industry, such as ice breaking bulk carriers or pipelines. These systems have demonstrated that they are able to operate in the Arctic environment. The economic evaluation of the concept suggests that to be competitive with surface ships, the submarine must have a submerged volume at least ten times larger than the largest submarine built to date. Since this step is so large, it would be more likely that an uneconomic pilot project would be built to develop the design, construction, and operating techniques on a smaller scale.

Even a pilot project could experience some difficulties. The reduced size may simplify hull construction, but the selection of propulsion unit may pose some problems. The only propulsion unit which has demonstrated that it can operate in a polar environment under water is the nuclear powered steam turbine. The disadvantage of such a unit is that it is not economical for small scale operation, since its capital cost per unit of power is much higher than the equivalent non-nuclear system. The fuel cell system may be preferred since the environmental risk is smaller, and it would be easier to produce a small scale unit, but more development work would have to be done in order to provide a functioning system of the powers required. An existing military system of nuclear power could be used, but this would not provide the required power output, unless the pilot project was for a very small design, or a substantial loss of speed was accepted.

The submarine is a very inflexible system. The requirement to maintain neutral buoyancy means that the hull structure, cargo type and ballast arrangement must be very carefully tuned. This means that it is difficult to change cargo type, even between different liquids such as crude oil and methanol. In addition there is only one practical route out of the Arctic ocean which can be guaranteed to have sufficient water depths for all ice conditions, and that is directly under the polar ice cap. An icebreaking bulk carrier however, if built to the appropriate ice class, could navigate any of several routes out of the Arctic ocean, and if necessary, it could also carry several cargos. The M.V. Arctic was recently converted from a dry bulk carrier to a combined oil and bulk cargo carrier. The surface icebreaker could carry a wide variety of equipment on deck which would resupply northern social and industrial requirements. This would be extremely difficult to incorporate into a submarine design.

The economic evaluation of the submarine concept is based entirely on estimates. None of the studies published are beyond the feasibility stage, and so detail design work still has to be done. The TAPS when finally completed cost approximately five times the original estimates, but now there is expertise and experience in that field. Similarly for the icebreaking tanker, there is a certain amount of experience, based on supply boats, government icebreakers and one bulk carrier. There is no actual data for the construction costs of a submarine cargo ship together with the terminal and trans-shipment ports. It is likely that there will be unforseen complications and costs, which could well make the submarine system less competitive than the surface ship.

Given all the above considerations, it is extremely unlikely that a submarine system would be used in the initial development of Arctic resources. The system could be more efficient than a surface system, but it is unlikely that any company would take a risk with so much unproven technology until the development of the Arctic is more established. Perhaps the best hope for the submarine lies with the legislators. The advantage of the submarine is that it can operate without disturbing the surface ice. Any disruptions to the surface ice conditions could permanently change factors such as breeding grounds for wildlife and hunting and transportation patterns for Inuit. If legislation was passed to prevent the disturbance of the Arctic environment, then the submarine would be the system with the least disruption to the delicate Arctic infra-structure.

REFERENCES

[1] Russo, V.L., Turner, H. and Wood F.W. "Submarine Tankers", Trans, Society of Naval Architects and Marine Engineers, Vol. 68, 1960, p. 263.

[2] Jacobsen, L.R. "Subsea Transport of Arctic Oil—A Technical and Economic Evaluation". Offshore Technology Conference, 1971.

[3] Arctic Submarine Transportation System—1975", U.S. Marine Administration. Marad Contract 4:37032, January 1975 (7 volumes).

[4] Taylor, P.K. and Montgomery, J.B. "Arctic Submarine Tanker System", Offshore Technology Conference, 1977.

[5] Montgomery, J.B. and Jordan, C.R. "Commercial Marine Transportation of Arctic Resources", Sixth Ship Technology and Research Symposium, Society of Naval Architects and Marine Engineers, 1981, Paper No. 4.

[6] German, J.G., Macpherson, M.D., Meakin, J. and Parker, C.W. "Marine Transportation of Oil and Gas in the Alaskan Arctic", Sixth Ship Technology and Research Symposium, Society of Naval Architects and Marine Engineers, 1981, Paper No. 3.

[7] Court, K.E., Kumm, W.H. and O'Callagan, J.E. "Fuel-cell Propelled Submarine Tanker System Study", Report DOE/FE/15086.1, June 1982.

[8] Jacobsen, L.R. and Murphy, J.J. "Submarine Transportation of Hydrocarbons from the Arctic", Cold Regions Science and Technology, Vol. 7, 1983.

[9] Macphail, D.C., "Some Considerations Concerning the Transportation by Gas Burning Submarine of Natural Gas from the Canadian Arctic Islands", National Research Council Canada, DME MISC. 43. 1979.

[10] Spencer, D. "Comparison of Submarine Tankers and Icebreaking Tankers for the Transportation of Canadian Arctic Oil", M.Sc. Thesis, University of Newcastle-Upon Tyne, 1983.

ENDNOTE

*National Research Council

APPENDICES

STATEMENT OF PURPOSE

The Transportation Research Forum is an independent organization of transportation professionals. Its purpose is to provide an impartial meeting ground for carriers, shippers, government officials, consultants, university researchers, suppliers and others seeking an exchange of information and ideas related to both passenger and freight transportation. The Forum provides pertinent and timely information to those who conduct research and those who use and benefit from research.

The exchange of information and ideas is accomplished through international, national, and local TRF meetings, and by publication of professional papers related to numerous transportation topics.

The TRF encompasses all modes of transport and the entire range of disciplines relevant to transportation, including:

Economics	Urban Transportation and Planning
Marketing and Pricing	Governmental Policy
Financial Controls and Analysis	Economic Development
Labor and Employee Relations	Equipment Supply
Carrier Management	Regulation
Organization and Planning	Safety
Technology and Engineering	Environment and Energy
Education	Physical Distribution
Transport Operations	Intermodal Transportation

HISTORY AND ORGANIZATION OF THE TRF

A small group of transportation researchers in New York started the New York Transportation Forum in March 1958. Monthly luncheon meetings were established at that time and still continue. The first organizing meeting of the American Transportation Research Forum was held in St. Louis, Missouri, in December 1960. The New York Transportation Research Forum sponsored the meeting and became the founding chapter of the ATRF. The Lake Erie, Washington, D.C. and Chicago chapters were organized soon after and have been joined by chapters in San Francisco, Los Angeles, Philadelphia, Boston, Roanoke, Seattle, Hampton Roads, Southwest Virginia and North Carolina. The TRF currently has over 700 members.

With the expansion of the organization into Canada, the name was shortened to the Transportation Research Forum. The Canadian Transportation Forum now has approximately 300 members.

TRF organizations have also been established in Australia and Israel. In addition, an International Chapter was organized for TRF members interested particularly in international transportation and transportation in countries other than the U.S. and Canada.

Interest in specific transportation-related areas has recently encouraged some members of TRF to form other special interest chapters, which do not have geographical boundaries—Agricultural and Rural Transportation, Personal Computer Users, High-Speed Ground Transportation, and Cost Analysis. TRF members may belong to as many geographical and special interest chapters as they wish.

A student membership category is provided for undergraduate and graduate students who are interested in the field of transportation. Student members receive the same publications and services as other TRF members.

TRF ANNUAL MEETINGS

In addition to monthly meetings of the local chapters, national meetings have been held every year since TRF's first meeting in 1960. Annual meetings generally last three days with 25 to 35 sessions. They are held in various locations in the United States and Canada, usually in October or early November. The Canadian TRF also holds an annual meeting, usually in the spring.

Each year at its annual meeting, the TRF presents an award for the best graduate and undergraduate student papers. Recognition is also given by TRF annually to an individual for Distinguished Transportation Research.

Annual TRF meetings generally include the following features:

- Members are addressed by prominent speakers from government, industry, and academia.
- The research papers are published and distributed before the meeting so those interested can read them in advance and be prepared to discuss and critique them.
- Speakers typically summarize (not read) their papers, then discuss the principal points with the members.
- Members are encouraged to participate actively in any session; sufficient time is allowed for discussion of each paper.
- Some sessions are organized as debates or panel discussions.

The *Journal of the Transportation Research Forum* is a hard-bound volume of approximately 500 pages. A copy of the current *Journal of the Transportation Research Forum* is mailed to each member before the annual meeting. The annual *Journal of the Transportation Research Forum* is widely recognized as an indispensable resource document in transportation. Copies of recent *TRF PROCEEDINGS* are available from the headquarters office.

TRANSPORTATION RESEARCH FORUM 1987 COUNCIL ROSTER

PRESIDENT
Frederick C. Dunbar
National Economic Research Associates, Inc.

EXECUTIVE VICE PRESIDENT
Joedy W. Cambridge
Leeper, Cambridge & Campbell, Inc.

PROGRAM VICE PRESIDENT
Thomas N. Harvey
US DOT/TSC

PROGRAM VICE PRESIDENT-ELECT
Frederick J. Beier
University of Minnesota

ACADEMIC VICE PRESIDENT
Anthony M. Pagano
University of Illinois-Chicago

CHAPTER RELATIONS VICE PRESIDENT
H. Chris Nuckols, Jr.
Norfolk Southern Corporation

PUBLIC RELATIONS VICE PRESIDENT
Josephine E. Olson
University of Pittsburgh

INTERNATIONAL VICE PRESIDENT
James R. Snitzler
Office of Transportation, USDA

MEMBERSHIP VICE PRESIDENT
Richard Beilock
University of Florida

GENERAL COUNSEL
Fritz R. Kahn, Esquire
Verner, Liipfert, Bernhard,
McPherson and Hand

SECRETARY
Joanne F. Casey
American Trucking Associations

TREASURER
Carol E. Lutz
Association of American Railroads

COUNCIL-AT-LARGE MEMBERS
W. Bruce Allen
The Wharton School

Joseph F. Folk
Conrail

Robert Gallamore
Sanford C. Bernstein & Company, Inc.

Robert Hannus
Port of Seattle

Larry L. Herndon
Union Tank Car Company

Brian C. Kullman
Citicorp Management Logistics

Craig E. Philip
Ingram Barge Company 0901

Jack S. Ventura
Interstate Commerce Commission

IMMEDIATE PAST PRESIDENT
Carl D. Martland
Massachusetts Institute of Technology

EXECUTIVE DIRECTOR
Joan Walsh Cassedy
Transportation Research Forum

TRF CHAPTER PRESIDENTS

New England
Carl A. Zellner
Carl A. Zellner & Associates, Inc.

New York
Robert E. Gallamore
Sanford C. Berstein & Company, Inc.

Northern California
Daniel Smith
Manalytics

Southern California
Frederic I. Enfield
Flying Tiger Line

Chicago
Larry L. Herndon
Union Tank Car Company

Lake Erie
Arthur W. Todd
Cleveland, OH

Philadelphia
Thomas J. Judge
Consolidated Rail Corporation

Washington, DC
Porter K. Wheeler

Pacific Northwest
Robert Hannus
Port of Seattle

Southwest Virginia
H. Chris Nuckols, Jr.
Norfolk Southern Corporation

Hampton Roads, VA
A. Don Bourquard
Norfolk Southern Corporation

North Carolina
Abdussalam Addus
North Carolina AT&T University

International
Tillman Neuner
The World Bank

Agricultural & Rural Transportation
Richard Beilock
University of Florida

Personal Computer Users
Jim Blaze
Conrail

High Speed Ground
Herbert Bixler (contact)

Cost Analysis
Joseph L. Barnett
Norfolk Southern Corporation

1987 TRF ANNUAL MEETING

PROGRAM COMMITTEE

Thomas N. Harvey, Chairman
U.S. Department of Transportation

Frederick Beier, Vice Chairman
University of Minnesota

J. Richard Jones, Local Coordinator
Trinity University

W. Bruce Allen
University of Pennsylvania

Richard Beilock
University of Florida

Lisa J. Botnick
Transportation Research Forum

Joedy W. Cambridge
Leeper, Cambridge and Campbell, Inc.

Joanne Casey
American Trucking Associations

Joan Walsh Cassedy
Transportation Research Forum

Frederick C. Dunbar
National Economic Research Associates, Inc.

Joseph F. Folk
Conrail

Robert E. Gallamore
Sanford C. Bernstein and Company, Inc.

Robert Hannus
Port of Seattle

Larry L. Herndon
Union Tank Car Company

William A. Jessiman
American Airlines

Brian G. Kullman
Citicorp Management Logistics

Carl D. Martland
Massachusetts Institute of Technology

Robert M. Montbriand
Burlington Northern Railroad

Anthony M. Pagano
University of Illinois-Chicago

Craig E. Phillip
Ingram Barge Company

Michael W. Pustay
Texas A&M University

James R.Snitzler
U.S. Department of Agriculture

Jack S. Ventura
Interstate Commerce Commission

PAST PRESIDENTS

TRANSPORTATION RESEARCH FORUM

1986 Carl D. Martland
1985 Allan D. Schuster
1984 Douglas McKelvey
1983 William B. Tye
1982 Michael S. Bronzini
1981 Jay A. Smith, Jr.
1980 Samual E. Eastman
1979 Lana R. Batts
1978 Carl J. Liba
1977 Gerald Kraft
1976 Edward Morlok
1975 Barry A. Brune
1974 Richard Shackson
1973 Harvey M. Romoff
1972 Arthur W. Todd
1971 Herbert E. Bixler
1970 Paul H. Banner
1969 George W. Wilson
1968 Donald P. MacKinnon
1967 David L. Glickman
1966 Edward Margolin
1965 Robert A. Bandeen
1964 Gayton Germane
1963 Herbert O. Whitten
1962 Herbert O. Whitten
1961 Herbert O. Whitten
1960 Herbert O. Whitten

1986-87 BOARD OF DIRECTORS OF THE CANADIAN TRANSPORTATION RESEARCH FORUM

PRESIDENT
Craig S. Dickson
Atlantic Provinces Transportation Commission

EXECUTIVE VICE PRESIDENT
Wm. G. Waters, II
University of British Columbia

SECRETARY
E. J. O'Brien

TREASURER
Ray Bassett
Alberta Economic Development

VICE PRESIDENT EXTERNAL AFFAIRS
Andrew Elliott
PCS Sales

VICE PRESIDENT MEETINGS
E. M. Ludwick
E. M. Ludwick & Associates, Inc.

VICE PRESIDENT ORGANIZATION & DEVELOPMENT
W. H. Sheffield
Abitibi-Price, Inc.